Bee Products – Chemical and Biological Properties

José M. Alvarez-Suarez
Editor

Bee Products – Chemical and Biological Properties

Second Edition

Springer

Editor
José M. Alvarez-Suarez
Laboratorio de Investigación en Ingeniería en Alimentos (LabInAli)
Departamento de Ingeniería en Alimentos, Colegio de Ciencias e Ingenierías
Universidad San Francisco de Quito (USFQ)
Quito, Ecuador

Laboratorio de Bioexploración
Colegio de Ciencias Biológicas y Ambientales
Universidad San Francisco de Quito (USFQ)
Quito, Ecuador

ISBN 978-3-031-89048-2 ISBN 978-3-031-89049-9 (eBook)
https://doi.org/10.1007/978-3-031-89049-9

This Springer imprint is published by the registered company Springer Nature Switzerland AG
The registered company address is: Gewerbestrasse 11, 6330 Cham, Switzerland

From the Egyptians who offered it to their gods to the Indigenous peoples of the Americas who venerated it as a divine gift, honey has united cultures and generations in a song to nature and life.

Introduction

Since ancient times, the honeybee (*Apis mellifera*) has been an invaluable resource for humanity. Evidence of the relationship between humans and bees can be found in cave paintings in Spain and prehistoric caves in Africa, where honey collection is depicted as a vital activity for early communities. Over centuries, apiculture, or beekeeping, evolved into a sophisticated practice embedded in the cultural and economic traditions of many civilizations. The beehive, often seen as a "living factory," has provided a wide range of substances with unique properties that have fascinated humanity for millennia. Among these, honey holds a central place. This golden liquid, crafted by bees from flower nectar, has been revered not only as food but also as medicine, a cultural symbol, and a sacred resource. Civilizations such as the Egyptians and Greeks attributed divine qualities to honey, using it in rituals, as a preservative, and as a treatment for various ailments. In African traditions, particularly in the Yoruba religion, honey is imbued with spiritual significance, often employed in ceremonies to attract prosperity, peace, and harmony.

The planet's botanical diversity is reflected in the sensory and chemical characteristics of honey, whose composition includes sugars, enzymes, vitamins, minerals, and phenolic compounds. These elements grant it antioxidant, antimicrobial, and anti-inflammatory properties that vary based on floral origin and climatic conditions. In addition to its role as a nutritious food, honey has historically been used to treat wounds, burns, and respiratory ailments. However, the treasures of the hive extend beyond honey. Propolis, a resin collected by bees from tree buds, has a long history of medicinal use. Ancient Egyptians utilized it in embalming, while the Greeks and Romans applied it to treat wounds and skin conditions. Its antimicrobial, antifungal, and antioxidant properties continue to attract scientific interest for their potential in managing infections and inflammatory diseases.

Royal jelly, the exclusive diet of queen bees, is another remarkable product of the hive. Its complex composition, including vitamins, minerals, proteins, and fatty acids, endows it with stimulating and regenerative properties, leading to its use in food supplements, cosmetics, and other industries. Similarly, bee pollen, often referred to as a superfood, is prized for its high content of proteins, essential amino acids, and micronutrients. Consumed by athletes and individuals seeking to enhance their health and physical performance, pollen is also linked to positive effects on the immune and metabolic systems. Beeswax, another vital product, has been utilized since antiquity in making candles, waterproofing materials, and creating sculptures. In modern times, its applications have expanded to include cosmetics, pharmaceuticals, and apiculture.

Finally, bee venom, though often feared for its painful sting, has gained attention in biomedical research for its therapeutic potential. Its anti-inflammatory and analgesic properties have shown promise in treating conditions like arthritis and chronic pain. Despite its risks, bee venom is now employed in apitherapy as a natural treatment option. Each of these hive products has been the subject of numerous scientific studies, revealing their diverse benefits for human health. In an era increasingly focused on natural alternatives for disease prevention and treatment, the substances derived from the hive present a promising field of exploration.

This book aims to provide a comprehensive and updated overview of bee products, delving into their chemical properties, therapeutic applications, and potential health benefits. By compiling the latest research and highlighting practical implications, we hope to deepen scientific understanding of these remarkable substances, inspire new lines of study, and promote the development of high-quality bee-derived products that contribute to human well-being in innovative and impactful ways.

Universidad San Francisco de Quito
Quito, Ecuador
e-mail: jalvarez@usfq.edu.ec

José M. Alvarez-Suarez

Contents

Part I
Honey

Chapter 1
Botanical Classification

Estefanía Sánchez-Reyes, José Sánchez-Sánchez, and Silvia Sánchez-Durán

Contents

Abstract The study of pollen content and fungal spores in honey (melissopalynology), plays a key role in identifying the botanical and geographical origins of this hive product. With this aim, it is essential to describe the features of pollen such as its morphology, including polarity and symmetry, size and color, shape, sculpturing, and aperture system. Quantitative and qualitative analyses are used to determine honey origins, aiding in quality control and fraud detection. Quantitative methods classify honey based on pollen concentration, while qualitative analysis identifies pollen types through microscopy, allowing for the categorization of honeys as monofloral, polyfloral, or honeydew, based on pollen content and botanical sources. However, where they exist, regulatory standards vary with different pollen percentage thresholds for the classification of monofloral honey. Furthermore, it is important to note that honeydew honey, unlike floral honey, originates from tree sap processed by insects, containing fungal spores and other honeydew elements instead of significant pollen. Accurate classification enhances market value and strengthens consumer confidence, ensuring that the quality and authenticity of the product they are purchasing can be trusted. As a result, pollen analysis proves to be a valuable tool in the honey industry.

E. Sánchez-Reyes (✉) · J. Sánchez-Sánchez · S. Sánchez-Durán
Department of Botany and Plant Physiology, University of Salamanca, Unamuno Campus, Salamanca, Spain
e-mail: fani_sanchez@usal.es

J. M. Alvarez-Suarez (ed.), *Bee Products – Chemical and Biological Properties*,
https://doi.org/10.1007/978-3-031-89049-9_1

Keywords Melissopalynology · Pollen morphology · Monofloral honey · Honeydew honey · Quality control

1.1 Melissopalynological Analysis

1.1.1 Palynology: A Young Science—Applications

In 1944, Hyde and Williams coined the term *palynology* to name the science responsible for the study of pollen grains and fungal spores. It was from the middle of the twentieth century onward that the study of these particles progressed considerably, thanks to the appearance of the electron microscope (Knoll and Ruska 1932), which improved the resolution and magnification reached with the optical microscope at the beginning of the seventeenth century.

Size or type, number or position of apertures, or even sporoderm ornamentation of pollen grains and fungal spores can be used as taxonomic characters for the identification, differentiation and classification of plant species or fungi they are coming from. For that reason, the study of pollen grains and fungal spores has its application in disciplines apparently as disparate as vegetal taxonomy, paleopalynology (study of palynomorphs present in different sediments for paleoclimatic and paleoecological reconstructions), aeropalynology (airborne pollen and fungal spore content), medicine (related to allergies) or even forensic palynology, agronomy (control of crop pollination or phytopathogen detection, among others), biodeterioration, and of course, the subject that concerns us in this book, melissopalynology (study of pollen content and fungal spores in honeys and other hive products).

In honey, pollen grains come mainly from the plant species foraged by honey bees (Ohe et al. 2004); so palynological studies provide a good fingerprint of the plant species where the honeys come from (botanical origin).

However, we could find pollen grains in the nectar as a consequence of secondary (pollen from the inside of the hive), tertiary (inclusions of pollen grains during the extraction process), and quaternary (aerial contamination) enrichment (Corvucci et al. 2015). As such, it appears exactly in Directive 2014/63/EU as "*additional pollen in honey can come from pollen on bees' hair, from pollen in the air inside the hive and from pollen that was packed in cells by bees and released as a result of the accidental opening of those cells during the extraction of honey by food business operators.*"

On the other hand, the entire pollen spectrum is consistent with the flora of a particular region, depending thus on the agricultural and forest conditions where a honey was produced, giving an idea of the geographical origin (Ohe et al. 2004). In that sense, markers of geographical origin are generally pollen grains found in lower percentage, as Conceição Silva and Ribeiro Dos Santos (2014) and Corvucci et al. (2015) reported.

Quantitative and qualitative analysis are carried out in order to determine the botanical and geographical origin of honeys, royal jelly, bee pollen, and propolis, which contribute to the quality control of hive products (together with sensory analyses). In addition, they allow the identification of possible fraudulent adulteration or contamination and provide important information on honey extraction and/or filtration.

1.1.2 Pollen Grain

Pollen grain or *microgametophyte* is the male germ element responsible for the fertilization in flowers and subsequent seed formation. It is the male gametophyte of seed plants (spermatophytes), and it is formed in the androecium of the flower, composed of *stamens* (*microsporophylls*) consisting of a *filament* and the *anther*. Anthers are usually two-*lobed*. Each anther lobe develops two *pollen sacs* or *microsporangia*.

Pollen sacs are composed of sporogenous tissue formed by a mass of cells called *pollen mother cells* or *arquesporium*, which undergo reduction division (meiosis). The four haploid cells generated form a tetrad within the mother cell until their maturation.

Inside each of these microspores, an uneven mitotic division occurs, generating two morphologically and functionally different nuclei: the *vegetative nucleus*, larger in size, with vacuoles and reserves, is responsible for the formation of the *pollen tube,* and it performs a nutritional function for the *generative nucleus* (which could be placed in an eccentric position), smaller and fusiform, which will generate (through mitosis) two *sperm nuclei* or *male gametes*.

Simultaneous to the formation of the generative and vegetative nuclei, they will make their own wall, so that the mother cell wall degenerates. Subsequently, the wall is thickened, and it is possible to differentiate between two layers: an external, or *exine,* and an internal, or *intine* (that is generated after the aforementioned exine), which give pollen grain a great resistance against the adverse atmospheric conditions to which it will be exposed, thus avoiding possible dehydration.

Once the pollen grains reach this stage of development, they will be disseminated, thanks to the dehiscence of the anther, which will occur in response to certain environmental conditions. In general, pollen grains are released in isolation (monad), although sometimes, in some species, they are released in groups of four (tetrad), as in the case of the family Ericaceae and even in groups of 16 grains (poliads) in the genus *Acacia* Mill. or in pollinia as occurs in Orquidaceae and Asclepiadaceae.

1.1.3 *Pollen Morphology*

A specific terminology, accepted internationally by palynologists, is used for the determination and description of the microscopic characteristics of pollen grains. Below is a summary of the main pollen characteristics that form the basis for the identification of pollen grains: polarity and symmetry, size and color, shape, sporoderm, ornamentation, and apertural system.

1.1.3.1 Polarity and Symmetry

As indicated above, pollen grains are three-dimensional structures (with volume), so they may have polarity and symmetry. If within a tetrad, we analyze a monad individually (Fig. 1.1), we can define the zone that is closest to the center of the tetrad as the *proximal pole* and the *distal pole* as the one furthest from the tetrad. The line connecting the two poles, passing through the center of the monad and the tetrad, coinciding with the axis of rotation of an imaginary ellipsoid, is called the

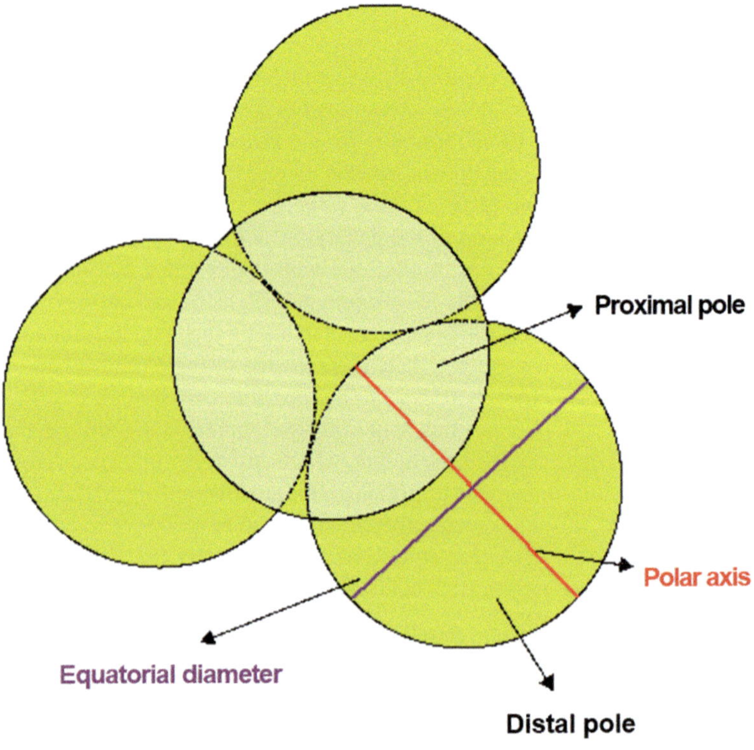

Fig. 1.1 Pollen tetrad. (Taken from Reyes and Sanchéz (2017). Modified from Punt et al. (2007))

polar axis. The *equatorial diameter* corresponds to the perpendicular axis to this polar axis that crosses the grain through its middle part.

We say a pollen grain is in polar view when the polar axis coincides with the optical axis of the observer. However, an equatorial view of a pollen grain is one in which the equatorial plane is directed toward the observer.

Erdtman (1952) defined as *apolar* those grains in which no pole is seen once liberated from the anther, whereas the *polar* ones present these zones more or less differentiated. Within the latter, we differentiate the *isopolar* grains, when the equatorial plane determines two polar zones more or less similar, and the *heteropolar* ones, in which the poles are clearly different either in shape, ornamentation, or apertural system.

With respect to symmetry (always defined in polar view), we can differentiate the *asymmetric* pollen grains (when they do not present any plane of symmetry) and the *symmetrical* ones (which do have planes of symmetry). These are divided into two types: those with *bilateral* symmetry (if they only have two planes of symmetry, but of different length) or with *radial* symmetry (if they have more than two planes of symmetry or only two but of equal length) (Erdtman 1969).

Therefore, taking into account the polarity and the symmetry, several combinations arise between these concepts (Fernández and Díez 1990), which can considerably complicate the description of pollen grains.

1.1.3.2 Size and Color

In order to establish the average size of a pollen type, it is necessary to perform multiple measurements and a subsequent statistical treatment of the obtained values. However, it is a very useful characteristic from the taxonomic point of view, since it is kept constant between species, which allows to separate those next to each other in some cases. Nevertheless, it is possible to appreciate important variations according to the different varieties, geographic areas, or the physiological state of the plant (cultivated plants that grow in optimal conditions usually develop larger grains). It can even be influenced by the pretreatment of the pollen grains, which are able to adapt to changes in volume since the openings also have a harmomegathic function (Wodehouse 1935), remaining practically constant in the fossilized and acetolyzed pollen grains.

It is usually defined by its diameter when the pollen grain is spherical and by the lengths of the polar and equatorial axes if it is ellipsoid, regardless of the size of the ornamentation it possesses (about 0.5 μm in length in general).

Erdtman (1952) defined the size of the pollen grains as a function of the mean length of their major axis (either polar or equatorial): *very small* (<10 μm), *small* (10–25), *medium* (25–50), *large* (50–100), *very large* (100–200), and *gigantic* (>200).

We can find grains from 5 μm as in the case of some forget-me-not (genus *Myosotis* L.) to some of more than 200 μm produced by the four o'clock flower

(*Mirabilis jalapa* L.), most of them being between 25 and 50 μm (Sánchez Sánchez and Baldi Coronel 2010).

This could be a very interesting parameter to consider in relation to filtration processes since the larger pollen grains would be removed depending on the filters used.

Color is of relative importance as it is imperceptible in microscopic studies. However, it can be seen when the grains are grouped in a mass. In fact, the bee pollen loads have very variable colors, which fundamentally depend on the chemical composition of the pollen grain and ultimately on its botanical origin. For this reason, it is possible to find bee pollen loads that are even black or bluish-violet, although commercial bee pollen usually has a brownish-yellow color hue.

1.1.3.3 Shape

Pollen grains are three-dimensional structures, so they can have different shapes depending on the view we have (polar or equatorial); shape is one of the most important characteristics for their identification. It is a variable property because it is subject to changes due to the physical discontinuity of the wall that covers them (which causes them to deform) or because of the volume variations that the mechanism of harmomegathy cannot neutralize. For this reason, the palynologists have based morphological descriptions mainly on acetolyzed grains.

In order to achieve a better description of pollen grains, Reitsma (1970) classified them according to the contour of both equatorial and polar views, which may be either non-angular (circular or elliptical) or angular, corresponding to some rhomboid, rectangular, triangular, quadrangular, pentagonal, or hexagonal polygons, with sharp or obtuse edges and with flat, concave, or convex sides.

Likewise, the shape of a pollen grain can be defined according to the relationship between the length of the polar axis and the equatorial diameter. In this way, both Erdtman (1943, 1952) and Reitsma (1970) separated nine morphological types, which they named using different terminology (Table 1.1).

Table 1.1 Pollen terminology regarding to morphology proposed by Erdtman (1943, 1952) and Reitsma (1970) based on polar axis/equatorial diameter ratio (P/E)

P/E	Erdtman	Reitsma
<0.50	Peroblate	Pertransverse
0.50–0.75	Oblate	Transverse
0.75–0.88	Suboblate	Semitransverse
0.88–1.00	Oblate-spheroidal	Subtransverse
1.00	Spheroidal	Adequate
1.00–1.14	Prolate-spheroidal	Suberect
1.14–1.33	Subprolate	Semi-erect
1.33–2.00	Prolate	Erect
<2.00	Perprolate	Pererect

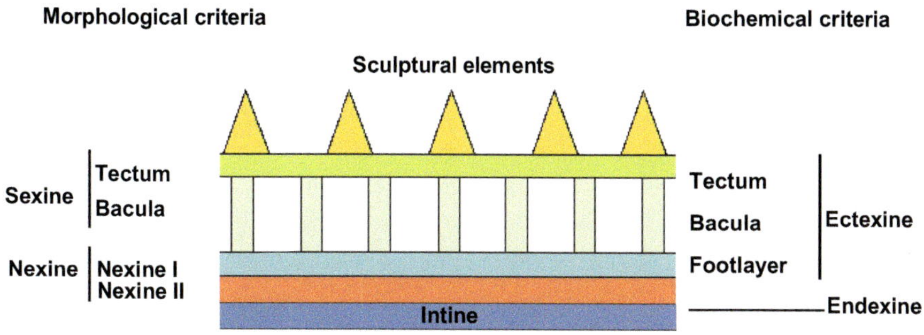

Fig. 1.2 Sporoderm structure according to morphological or biochemical criteria. (Taken from Reyes and Sanchéz (2017). Modified from Punt et al. (2007))

1.1.3.4 Sporoderm

This name is given to the layers surrounding the live content of the pollen grain that act as protective covers against drying, abrasion, and the mutagenic effect of ultraviolet rays. The *intine* corresponds to the innermost cellulosic layer, which is homogeneous, the thickness and position of which are the main characteristics to take into account. It is the only stratum that develops to originate the pollen tube. The *exine* is the most external stratum and the one that really protects the pollen grain because it is very resistant, being able to withstand the action of strong bases and acids, as well as the heating up to 300 °C. Its degradation occurs only after the action of some microorganisms or certain very strong oxidants.

According to biochemical criteria (Fægri and Iversen 1950; Fægri 1956), it is possible to differentiate the *ectexine*, with a strong coloration after staining with fuchsin, from the weakly stained *endexine* (Fig. 1.2). The ectexine is formed by a set of radial elements called *bacula*, *columellae,* or *infratectum*. Generally, the expansion and welding of its upper ends give rise to a structure that receives the name of *tectum*, whereas its inferior ends fuse, giving origin to a common support or *foot layer*. The *endexine* is located below the foot layer.

From a morphological point of view (Fig. 1.2), the exine is composed of two layers: the *sexine*, which is the outer structured layer, constituted by the *tectum* and the *bacula*, and the *nexine* or the unstructured layer, in which we differentiate the *nexine I* and *nexine II,* which correspond to the foot layer and the endexine, respectively (Erdtman 1952, 1966).

1.1.3.5 Sculpturing

As has been described above, tectum is the outermost layer of the pollen grain on the surface of which the sculptural elements, if any, are to be found, both being structures of vital importance in the identification of the different pollen types.

Tectum is not always present though, and pollen grains without tectum are named *intectates*, unlike those that present it, either partially (*semitectates*) or completely (*tectates*).

Likewise, pollen grains may have ornamentation without sculptural elements. When sculptural elements do exist, they are located on the tectum or on the foot layer when there is no tectum. They remain constant in the same species, and they are described in Punt et al. (2007) and Reyes and Sanchéz (2017):

Spine, a pointed element >3 µm in length with a height greater than the width (*equinate* or *spinate* pollen, e.g., *Helianthus annus* L.); *spinule*, small spines, <3 µm in length (*equinulate* or *spinulate* pollen, e.g., Campanulaceae); *baculum*, a non-pointed element >1 µm in length and less than this in diameter (*baculate* pollen, e.g., Nymphaea alba type—not written in italics since it is a pollen type not a species name); *pilum*, a non-pointed element >1 µm in length, with a height greater than the width, sharply swollen in the apical part (*pilate* pollen, e.g., *Jatropha* L.); *clava*, a club-shaped element that is higher than 1 µm in length, with diameter smaller than the height and thicker at the apical part than the base (*clavate* pollen, e.g., Ilex aquifolium type); *verruca*, a wart-like element >1 µm wide, that is broader than it is high, which is not constricted at the base (*verrucate* pollen, e.g., Plantago coronopus type); *gemma*, a non-pointed element, higher than 1 µm in length that has approximately the same width as its height, which is constricted at its base (*gemmate* pollen, e.g., Cupressaceae); and *granulum*, a rounded element that is less than 1 µm in all directions (*granulate* pollen, e.g., Alnus glutinosa type).

In contrast to the *psilate* pollen, which presents a smooth surface, we can find the sculptural elements distributed in a more or less elongated form. In that sense, they define the *striate* pollen when distributed parallelly and separated by grooves, and *rugulate* if they are arranged in an irregular pattern. Pollen *insulate* when the elements are grouped in circular or polygonal areas separated by grooves.

1.1.3.6 Apertures

Other chief criteria in the identification of pollen grains are the type, number, and position of the apertures they can present. Apertures are areas on the walls of a pollen grain, where the wall is thinner and/or softer, that function as sites of germination (Erdtman 1969). They also act as regulators of pollen grain volume changes, produced as a consequence of hygroscopic variations, thus presenting harmomegathy function (Wodehouse 1935).

It is possible to distinguish two main types according to their morphology: *colpus,* or *furrow*, when it is at least twice as long as wide, and *pore*, a circular or elliptic aperture with a length/breadth ratio less than 2, which define *colpate* and *porate* grains, respectively (Punt et al. 2007; Reyes and Sanchéz 2017). They are described as *simple* if they are present in only one wall layer of the exine or *compound* if they affect more than one layer of the wall. In compound apertures, the shape of apertures may differ between layers, giving rise to *colporate* grains when there is a colpus type of ectoapertura and *pororate* if it is pore type. However, the designation of simple aperture is maintained if both apertures coincide in shape and/or size.

Those pollen grains that do not have any apertures are designed as *inaperturate*; as their number increases, they are described using the following prefixes before the terms that define the type (Erdtman 1969): *mono-*, *di-*, ..., *poly-* (more than six apertures).

Considering the arrangement of the apertures on the surface of the grain, different particles are added between the terms that define number and type: *-cata-*, if they appear in the proximal pole; *-anacata-*, if they are located in the two poles; *-ana-*, if they appear in the distal pole; -zono-, if they are at the equator; -dizono- if they are in two more or less parallel zones of the equator; and -panto-, if the apertures occur throughout the surface.

1.1.4 Methodology

In order to determine botanical and geographical origin of honeys, both quantitative and qualitative analysis of pollen sediment are carried out and are of great importance for quality control processes.

1.1.4.1 Quantitative Analysis

Together with the technique we normally use in our laboratory (Sánchez Sánchez and Baldi Coronel 2010), there are many methodologies used to date (Louveaux et al. 1978; Ohe et al. 2004, among others), although they are all based on the number of pollen grains and honeydew elements per unit of weight, usually 10 g of honey. In this way, honeys are classified according to different classes established by Maurizio in 1939.

Class I (less than 20,000 elements/10 g honey) includes monofloral honeys with underrepresented pollen; *class II* (20,000–100,000) includes most of multifloral honeys, honeydew honeys, and a mixture of flower and honeydew honeys; *class III* (100,000–500,000) includes monofloral honeys with overrepresented pollen and honeydew honeys; *class IV* (500,000–1,000,000) includes monofloral honeys with strongly overrepresented pollen and some pressed honeys; and *class V* (more than 1,000,000) includes almost only pressed honey.

In those honeys where it is important to reflect the honeydew elements content, counts are customarily expressed by a formula such as 37/36/0.5-II, which means that 10 g of honey contains 37,000 pollen grains, 36,000 fungal spores, and 500 algae cells. The 73,500 plant constituents places the honey in class II (Louveaux et al. 1978).

Although the quantitative study is not decisive, we believe that a high amount of sediment could indicate a pressed honey, whereas a low amount of sediment may indicate honey that is naturally poor in pollen (e.g., *Citrus* spp.) or has undergone excessive filtering or falsification (e.g., by feeding sugar to the bees) as Louveaux et al. already pointed out in 1978.

The mean of the pollen absolute number for the 15 main European monofloral honeys can be found in the work of Persano Oddo and Piro (2004), with values concordant in any case with the results obtained after an extensive bibliographical review of previous European works (Piazza and Persano Oddo 2004), which would indicate a good characterization of these honeys at this level.

The results showed that citrus honey from *Citrus* spp., sunflower honey from *Helianthus annuus* L., rhododendron honey from *Rhododendron* spp., black locust honey from *Robinia pseudoacacia* L., rosemary honey from *Rosmarinus officinalis* L., and lime honey from *Tilia* spp. belong to class I; rapeseed honey from *Brassica* spp., calluna heather honey from *Calluna vulgaris* L. (Hull), dandelion honey from *Taraxacum officinale* group, and thyme honey from *Thymus* spp. belong to class II, whereas chestnut honey from *Castanea sativa* Miller, eucalyptus honey from *Eucalyptus* spp., and honeydew honeys belong to class III (except for metcalfa honeydew honey that belongs to class II).

In any case, ranges between classes are very high, and, in fact, the information obtained from their classification is not very relevant.

1.1.4.2 Qualitative Analysis

The acetolysis method was first introduced by Erdtman in 1960, but it is still followed throughout the world for qualitative analysis, as most pollen descriptions in palynological literature are based on acetolyzed material (Louveaux et al. 1978; Simeão et al. 2015). It is based on the treatment of pollen grains with the acetolysis mixture (consisting of nine parts of acetic anhydride and one of sulfuric acid), which destroys the living content but preserves the external part (exine) in which the apertures and the ornamentation are located, allowing to observe and identify the different pollen types. It also prevents the deformation of the grains as a result of the changes in humidity, allowing to standardize size values through biometric studies.

This methodology presents certain disadvantages though; for example, acetolysis destroys fungal spores, hyphae, algae, etc., the identification and quantification of which are associated with honeydew honeys (Louveaux et al. 1978), so these structures should be observed during the performance of the quantitative analysis.

Subsequent modifications were incorporated and accepted due to the complexity of the method, as proposed by Hideux (1972) or Louveaux et al. (1978), the bases of which are still used today (La Sena-Ramos et al. 2002; Soria et al. 2004; Rodríguez de la Cruz et al. 2013; Conceição Silva and Ribeiro Dos Santos 2014; Simeão et al. 2015; Atanassova et al. 2016).

Considering the proposals made by Louveaux et al. in 1978, the International Honey Commission (IHC) elaborated and proposed another method that is well-established nowadays in most European laboratories involved in routine honey analyses (Ohe et al. 2004), for the practical purpose of verifying if the pollen spectrum complies with the declared botanical (and even geographical) origin. Moreover, this method was used to produce the data for the European monofloral honey descriptions (Persano Oddo and Piro 2004) and has been used in many studies worldwide (Ponnuchamy et al. 2014; Belay et al. 2015; Corvucci et al. 2015; Juan-Borrás et al. 2015; Kadri et al. 2016). Pollen grains are extracted from a solution of honey and distilled water by centrifugation; however, the use of sulfuric acid is still contemplated even in the current legislation of some countries like Spain (Orden de 12 de junio de 1986).

At the European level, it is established that "*Member States shall, whenever possible, use internationally recognized validated methods of analysis such as those approved by the Codex Alimentarius to verify compliance with the provisions of this Directive*" (Directive 2014/63/EU). However, the revised Codex Alimentarius Standard for Honey (Codex Alimentarius Commission 2001) only specifies that "*samples should be prepared in accordance with AOAC 920.180*" in relation to the methods of sampling and analysis (7.1 point of the Codex), but there is no specification regarding the melissopalynological analysis itself.

On the other hand, a collection of reference slides is an indispensable supplement to the literature on melissopalynology, in order to identify different pollen types. Different keys and pollen atlases must also be used, both of native and foreign flora, always considering defined pollen types and taking into account that these can present different taxonomic ranges (at the species, genus, or even family level). Indeed, the critical point of melissopalynological analysis remains the correctness of pollen identification and the subsequent interpretation of the results (Ohe et al. 2004), depending greatly on the performance and the experience of the operator (Louveaux et al. 1978) being necessarily specialized personnel (Persano Oddo and Bogdanov 2004; Schievano et al. 2015).

We believe that a minimum of 1200 pollen grains per sample should be counted and identified for the certification of both the botanical and geographical origin of a honey (Louveaux et al. 1978; Orden de 12 de Junio de 1986; Sánchez Sánchez and Baldi Coronel 2010), which will make it necessary to perform and analyze more than one microscopic slide in the case of underrepresented honeys. In honeydew honeys, poor in pollen grains, counts of only 600 grains would be necessary as Louveaux et al. (1978) pointed out. However, there are frequent papers in which the analyzed number of pollen grains was lower (e.g., 500 pollen grains in Belay et al. (2015), Corvucci et al. (2015), Atanassova et al. (2016), and Kadri et al. (2016), among others).

The number of pollen grains to be counted might be more important if we are trying to certify the geographical origin of a honey since we should try not to ignore any pollen type that is a geographical marker.

1.1.4.3 Development of Result Report: Quality Control

The sample must be perfectly identified in the final result report, including the date of receipt of the sample, data included in the label, as well as any other additional information considered relevant.

According to the quantitative analysis, the total number of plant constituents in 10 g of honey should be expressed, thus indicating into which Maurizio's class it is placed.

Regarding the qualitative study, a list with the complete pollen spectrum must be included, which will determine the geographic origin of the sample. It should be accompanied by the relative frequency as the percentage with respect to the total number of pollen grains counted/identified.

However, for the determination of the botanical origin, it is necessary to recalculate the relative frequencies including only pollen types from nectariferous species since they actually contribute to the elaboration of a certain honey. For that reason, these are the percentages that have to be taken into account to determine the monoflorality of a honey. That is, pollen types from nectar-less (*Plantago* spp., Amaranthaceae, Cistaceae, etc.) and/or wind-pollinated plants (Poaceae, Cyperaceae, pollens of conifers, etc.) have to be removed from the final spectrum.

Abortive and misshapen pollen grains should be counted as far as they can be identified (Louveaux et al. 1978). Non-identifiable or non-identified grains should be noted (spoiled, folded, etc.) but never exceed a significant percentage of the sample.

Last, a final evaluation must be made taking into account all that is stated above.

It may be the case that the calculated percentages indicate the possible monoflorality of two honey types simultaneously, e.g., a sample that presents 20% of *Citrus* spp. and 50% of *Helianthus annus* L., which would indicate that complementary sensory analysis will be essential to confirm monoflorality in one direction or another.

1.2 Monofloral, Polyfloral, and Honeydew Honey

It is necessary to perform physicochemical, palynological, and sensory analyses for the classification and, therefore, the correct labeling of monofloral honeys (Persano Oddo et al. 1995), although the last two types of studies are the most relevant in order to decide on their monoflorality.

The palynological study of a honey, as already mentioned, can help in its botanical characterization since it will allow us to define a honey as monofloral, multifloral (when it is not possible to define the honey as monofloral because the sample does not come either mainly from a single nectar source or from honeydew), or honeydew (if it presents honeydew elements such as mold hyphae and fungal spores).

In general, a honey is considered as coming predominately from a given botanical origin (monofloral) if the relative frequency of the pollen type of that taxon exceeds 45% (Maurizio 1975), but each honey type has its own pollen spectrum, which has to be taken into account before issuing a conclusion about its botanical origin. In that sense, all the botanical species giving monofloral honeys in Europe are listed in the work of Persano Oddo et al. (2004).

There are countries like Argentina, where the legislation considers 45% as the minimum percentage of a certain pollen type that a honey must contain in order to be considered monofloral (Resolution 274/95 SAGPyA 1998). Some exceptions are contemplated though, such as *Citrus* spp. honey (10–20%) and *Eucalyptus* spp. (70%), among others.

Other countries, without enforced legislation, set the threshold values for their monofloral honeys according to recommendations made by specialists (such as the national associations of palynologists).

Such is the case of Spain, whose quality standard for honey states that "honey will have its normal content of pollen that cannot be removed in the filtering process" (Orden de 5 de agosto de 1983; Real Decreto 1049/2003), but it does not define what must be considered as "normal."

At this point, we consider, for example, that eucalyptus honey must contain more than 80% of pollen from these plants (Rodríguez de la Cruz et al. 2013) since it is an overrepresented pollen type (there are too many pollen grains per unit of nectar) being determined even greater for Italian eucalyptus honeys (Persano Oddo and Piro 2004).

Castanea sativa Miller is also likely to be overrepresented, and only honeys containing 90% or more of this pollen type can be regarded as *Castanea* honey in many European laboratories (Louveaux et al. 1978; Persano Oddo et al. 1995), as our expertise confirms. Conversely, other nectar sources can be underrepresented, which leads us to consider as monofloral those honeys with percentages around 15–20% and those with even lower percentages such as *Citrus* spp., Labiatae, and *Tilia* spp., among others (Piazza and Persano Oddo 2004).

Levels higher than 10% are acceptable in our laboratory for *Citrus* spp. honeys because of the increment in the cultivation of sterile hybrid varieties, which are characterized by their small amounts of pollen (Juan-Borrás et al. 2015).

We would like to highlight the case of sunflower honey since there are varieties that produce little pollen and nectar, so the threshold is set at 25% and even with levels above 15%, even if the morphology of flower and pollen grains would not justify any underrepresentation (Persano Oddo and Piro 2004).

At the European level (2001/110/EC, EU Council 2002), regarding pollen content of honeys, it is only mentioned that "*no pollen or constituent particular to honey may be removed except where this is unavoidable in the removal of foreign inorganic or organic matter*" (Annex II) or "*the product names may be supplemented by information referring to floral or vegetable origin, if the product comes wholly or mainly from the indicated source and possesses the organoleptic, physicochemical and microscopic characteristics of the source*" (article 2b), but these properties are not defined through the prescription of given limits or any legal criteria that guarantee an efficient control of the denominations (Persano Oddo and Bogdanov 2004).

In that sense, in 2004, Persano Oddo and Piro characterized the 15 most important monofloral European honey types (in terms of abundance of production or commercial interest), determining the mean percentage of specific pollen obtained in 21 countries of the European geographical area.

Values obtained were as follows: lavandula honey from *Lavandula* spp.: 8.2%, dandelion honey from *Taraxacum officinale* group: 17.2%, citrus honey from *Citrus* spp.: 18.6%, lime honey from *Tilia* spp.: 22.9%, black locust honey from *Robinia pseudoacacia* L.: 28.1%, rosemary honey from *Rosmarinus officinalis* L.: 28.7%, thyme honey from *Thymus* spp.: 36%, calluna heather honey from *Calluna vulgaris* L. (Hull): 37%, rhododendron honey from *Rhododendron* spp.: 38.6%, sunflower honey from *Helianthus annuus* L.: 56.7%, rapeseed honey from *Brassica* spp.: 82.8%, chestnut honey from *Castanea sativa* Miller: 94.5%, and eucalyptus honey from *Eucalyptus* spp.: 94.8%. Honeydew honeys obtained a honeydew elements/pollen (HDE/P) ratio of 1.5 except for metcalfa honeydew honey (4.8).

Moreover, the results obtained in IHC work were compared to other European studies, being fairly consistent with the literature available (Piazza and Persano Oddo 2004). However, the results have not been reflected in the European legislation to date, as it was mentioned before.

Different criteria are accepted at international and national levels (Atanassova et al. 2016), which may affect the trade of monofloral honeys (Persano Oddo and Bogdanov 2004).

1.2.1 Honeydew Honey

It is necessary to consider honeydew honeys apart from what has been discussed so far since they are clearly different from nectar honeys.

They are produced by a wide range of different sucking insects on a big number of trees, Coniferae (fir, spruce, and pine trees) and Latifoliae (produced in most of the European countries, mainly from different *Quercus* species) mainly (Persano Oddo et al. 2004).

On the other hand, according to our expertise in the laboratory, in many European countries, a mixture of nectar and honeydew honey is still marketed, labeled as *forest honey*, which may have abundant honeydew elements but without reaching the minimum value of electrical conductivity (0.8 mS/cm) established in the European legislation (2001/110/EC, EU Council 2002) for honeydew honeys.

They do not have a floral origin, so they practically lack pollen grains, although they do have those named honeydew elements (HDE) consisting of fungal spores, mold hyphae, microscopic algae, tricomas, etc. and pollen from nectar-less plants including anemophilous pollen (by air pollution).

Honeydew index is determined according to the HDE/P ratio (Louveaux et al. 1978), where HDE is the total number of honeydew elements counted and P is the total number of pollen grains counted. It is considered to be predominately honeydew if the ratio exceeds 3 (Louveaux et al. 1978), although Persano Oddo and Piro (2004) often found a HDE/P ratio below that limit for European monofloral honeys.

As a conclusion, it could be said that the botanical and geographical classification of honeys substantially determines their value in the market, the processes of quality control and certification being of utmost importance for both consumers and the honey industry.

References

Atanassova J, Pavlova D, Lazarova M, Yurukova L (2016) Characteristics of honey from Serpentine area in the Eastern Rhodopes Mt., Bulgaria. Biol Trace Elem Res 173:247–258

Belay A, Solomon WK, Bultossa G, Adgaba N, Melaku S (2015) Botanical origin, color, granulation, and sensory properties of the Harenna forest honey, Bale, Ethiopia. Food Chem 167:213–219

Codex Alimentarius Commission (2001) Revised Codex Standard for Honey, Codex STAN 12–1981, Rev. 1 (1987), Rev. 2 (2001)

Conceição Silva AP, Ribeiro Dos Santos FA (2014) Pollen diversity in honey from Sergipe, Brazil. Grana 53(2):159–170

Corvucci F, Nobili L, Melucci D, Grillenzoni FV (2015) The discrimination of honey using melissopalynology and Raman spectroscopy techniques coupled with multivariate analysis. Food Chem 169:297–304

Directive 2014/63/EU of the European Parliament and of the Council of 15 May 2014 amending Council Directive 2001/110/EC relating to honey. OJEU L164:1–5

Erdtman G (1943) An introduction to pollen analysis. The Ronald Press Co, New York, pp 1–239

Erdtman G (1952) Pollen morphology and plant taxonomy. Angiosperms. Almqvist and Wiksell, Stockholm, pp 1–539

Erdtman G (1960) The acetolysis method. Sven Bot Tidskr 54:561–564

Erdtman G (1966) A propos de la stratification de l'exine. Pollen Spores 8:5–7

Erdtman G (1969) Handbook of palynology—an introduction to the study of pollen grains and spores. Hafner Publishing Co, Copenhague, pp 1–486

EU Council (2002) 2001/110/EC of 20 December 2001 relating to honey. OJEC L10:47–52

Fægri K (1956) Recent trends in palynology. Bot Rev 22:639–664

Fægri K, Iversen J (1950) Textbook of modern pollen analysis. Munksgaard, Copenhagen, pp 1–169

Fernández I, Díez M (1990) Algunas consideraciones sobre terminología palonológica. I, Polaridad y simetría. Lagascalia 16(1):51–60

Hideux M (1972) Techniques d'étude du pollen an MED: effects comparés des différents traitements physicochimiques. Micron 3:1–31

Hyde HA, Williams DA (1944) The right word. In: Sears PB (ed) Pollen analysis circular no. 8. Department of Botany, Oberlin College, Oberlin, Ohio

Juan-Borrás M, Periche A, Domenech E, Escriche I (2015) Correlation between methyl antranilate level and percentage of pollen in Spanish citrus honey. Int J Food Sci Technol 50:1690–1696

Kadri SM, Zaluski R, Pereira Lima GP, Mazzafera P, de Oliveira OR (2016) Characterization of *Coffea arabica* monofloral honey from Espírito Santo, Brazil. Food Chem 203:252–257

Knoll M, Ruska E (1932) Das Elektronenmikroskop. Z Phys 78:318

La Sena-Ramos I, Méndez Pérez B, Gómez Ferreras C (2002) Pollen spectra of different unifloral honeys from La Palma (Canary Islands, Spain). Grana 41:48–57

Louveaux J, Maurizio A, Vorwhol G (1978) Methods of melissopalynology. Bee World 59(4):139–157

Maurizio A (1939) Untersuchungen zur quantitativen pollenlyse des honing. Mitt Lebensmittelunters Hyg 30:27–69

Maurizio A (1975) Microscopy of honey. In: Crane E (ed) Honey: a comprehensive survey. Heinemann in cooperation with the International Bee Research Association, London, pp 240–257

Ohe WVD, Persano Oddo L, Piana ML, Morlot M, Martin P (2004) Harmonized methods of melissopalynology. Apidologie 35:S18–S25

Orden de 12 de junio de (1986) por la que se aprueban los métodos oficiales de análisis para la miel. BOE 145:22195–22202

Orden de 5 de agosto de (1983) por la que se aprueba la norma de calidad para la miel destinada al mercado interior. BOE 193:22384–22386

Persano Oddo L, Bogdanov S (2004) Determination of honey botanical origin: problems and issues. Apidologie 35:S2–S3

Persano Oddo L, Piro R (2004) Main European unifloral honeys: descriptive sheets. Apidologie 35(Suppl. 1):S38–S81

Persano Oddo L, Piazza MG, Sabatini AG, Accorti M (1995) Characterization of unifloral honeys. Apidologie 26:453–465

Persano Oddo L, Piana L, Bogdanov S, Bentabol A, Gotsiou P, Kerkvliet J, Martin P, Morlot M, Ortiz Valbuena A, Ruoff K, Ohe KVD (2004) Botanical species giving unifloral honey in Europe. Apidologie 35:S82–S93

Piazza MG, Persano Oddo L (2004) Bibliographical review of the main European unifloral honeys. Apidologie 35:S94–S111

Ponnuchamy R, Bonhomme V, Prasad S, Das L, Patel P, Gaucherel C, Pragasam A, Anupama K (2014) Honey pollen: using melissopalynology to understand foraging preferences of bees in Tropical South India. PLoS One 9(7):e101618

Punt W, Hoen PP, Blackmore S, Nilsson S, Le Thomas A (2007) Glossary of pollen and spore terminology. Rev Palaeobot Palynol 143:1–81

Real Decreto 1049/2003, de 1 de agosto, por el que se aprueba la Norma de calidad relativa a la miel. BOE 186:30181–30183

Reitsma TJ (1970) Suggestions towards unification of descriptive terminology of Angiosperm pollen grains. Rev Palaeobot Palynol 10:39–60

Reyes ES, Sanchéz JS (2017) Botanical classification. In: Alvarez-Suarez J (ed) Bee products – chemical and biological properties. Springer, Cham

Rodríguez de la Cruz D, Sánchez Reyes E, Sánchez Durán S, Sánchez Sánchez J (2013) Análisis palinológico de mieles comerciales monoflorales. Bot Complut 37:171–180

SAGPyA (Secretaría de Agricultura, Ganadería, Pesca y Alimentación) (1998) Miel, Buenas prácticas de manufactura. Guía de aplicación Normas y legislación vigentes. Resolución 274(95):72–73

Sánchez Sánchez J, Baldi Coronel B (2010) Características microscópicas y origen botánico. In: Baldi B (ed) La miel. Una mirada científica. EDUNER. Universidad Nacional de Entre Ríos, Córdoba, pp 1–221

Schievano E, Finotello C, Mammi S, Belci AI, Colomban S, Navarini L (2015) Preliminary characterization of monofloral *Coffea* spp. honey: correlation between potential biomarkers and pollen content. J Agric Food Chem 63:5858–5863

Simeão CMG, Silveira FA, Sampaio IBM, Bastos EMAF (2015) Pollen analysis of honey and pollen collected by *Apis mellifera linnaeus*, 1758 (Hymenoptera, Apidae), in a mixed environment of *Eucalyptus* plantation and native cerrado in Southeastern Brazil. Braz J Biol 75(4):821–829

Soria AC, González M, de Lorenzo C, Martínez-Castro I, Sanz J (2004) Characterization of artisanal honeys from Madrid (Central Spain) on the basis of their melissopalynological, physicochemical and volatile composition data. Food Chem 85:121–130

Wodehouse RP (1935) Pollen grains. Their structure, identification and significance in science and medicine. McGraw-Hill, New York, pp 1–57

Chapter 2
Sensory Studies

Silvia Sánchez-Durán, José Sánchez-Sánchez, and Estefanía Sánchez-Reyes

Contents

Abstract Honey is a natural food with significant nutritional and medicinal benefits, whose properties depend on the flora and environment where it is produced. Sensory analysis is crucial to assess its quality, considering aspects such as color, aroma, taste, and texture. In addition to physicochemical and pollen analyses, organoleptic analysis helps verify the authenticity of honey, especially monofloral honeys. This product has a positive impact on the Sustainable Development Goals,

---

S. Sánchez-Durán (✉) · J. Sánchez-Sánchez · E. Sánchez-Reyes
Department of Botany and Plant Physiology, University of Salamanca, Miguel de Unamuno Campus, Salamanca, Spain
e-mail: sylvias8@usal.es; fani_sanchez@usal.es

J. M. Alvarez-Suarez (ed.), *Bee Products – Chemical and Biological Properties*,
https://doi.org/10.1007/978-3-031-89049-9_2

promoting beekeeping and protecting pollinators. The sensory analysis process of honey is divided into different phases to evaluate its organoleptic attributes, including both qualitative and quantitative methods, using hedonic or intensity scales to measure sensory attributes. This process requires several stages in order: preparation, visual evaluation, olfactory evaluation, taste evaluation, tactile evaluation, and finally, the recording and processing of the results obtained. These sensory analyses are fundamental in the food industry as they allow us to understand how the senses influence the consumer experience. In the case of honey, educational tastings are an excellent way to explore the wide variety of honey types and promote the appreciation of its quality. These activities also provide networking opportunities and contribute to social responsibility, while raising awareness about the importance of pollinators.

Keywords Sensorial analysis · Honey · Tasters · Tasting room · Sensorial scale

2.1 Introduction

Honey, the sweet food produced by different species of bees, is a natural product and highly prized for its high nutritional and medicinal value. It is a complex delicacy that reflects the flora and the environment in which it originated, which have given it unique characteristics; it is not a simple sweetener.

It contains mineral elements, mainly potassium (especially in chestnut (*Castanea sativa* Mill.) and eucalyptus (*Eucalyptus* sp. L'Hér) honeys (Di Rosa et al. 2019)), which are related to the color, flavor, acidity, and botanical and geographical origin of honey; vitamins, which are found in small amounts and come from pollen grains; aromatic substances, which serve for the characterization of honeys; and phenolic compounds that can help as markers of botanical origin and are beneficial to humans because they are considered as anticarcinogenic, antioxidant, antiseptic, or anti-inflammatory (Eteraf-Oskouei and Najafi 2022; Moore et al. 2024; Ranneh et al. 2021). Therefore, the composition and physical characteristics of honey are closely related to its botanical origin (Tixe 2024) and the geographical area of collection (Araujo et al. 2020; Machado et al. 2020).

In short, sensory analysis is a fundamental tool for the evaluation of the quality of honeys, as it makes it possible to identify and characterize the organoleptic properties that differentiate them from each other.

In honey, as in other products, producers must consider the whole process, from the time it is formed by the bees in the honeycombs until it reaches our table as consumers. Therefore, the traceability or path followed by the honey must be controlled, analyzing the dangers and critical points that can be found throughout its life cycle, as honey can undergo modifications due to various causes such as poor processing, prolonged storage, or incorrect heating.

Honey is produced in the world of excellent quality, but it must be controlled. In order to evaluate it, physicochemical, pollen, and organoleptic analyses are carried out. Physicochemical analyses (moisture, sucrose, pH, or conductivity, among others) are easy to interpret and many of them are simple to perform (such as moisture), although others are less so (such as diastatic activity). These analyses can define the botanical origin of honeys, such as electrical conductivity or the proportion of glucose and fructose (Geană et al. 2020). Pollen analyses are also necessary; pollen grains are mainly (not exclusively) incorporated during nectar pecoration by worker bees and will therefore depend on the plants they have visited (Sánchez Reyes and Sánchez Sánchez 2017). All honeys should contain naturally occurring pollen grains of different species and in different amounts, although some honeys are derived from honeydew or extrafloral nectaries. In some countries, total filtering of honey is allowed for (mainly) sanitary reasons.

Quality also depends on consumer appreciation, which is largely influenced by the types of products we are used to consuming. We have found that a type of honey is considered to be of better quality if it is consumed by people living in the areas where it is produced; for example, dark honeys such as those of holm oak (*Quercus ilex* L.) or oak (*Quercus pyrenaica* Willd.) are more appreciated in the Sierra de Salamanca (Spain) where they are produced. In other areas, these dark honeys could be considered as low quality.

Sensory analysis can identify the botanical and geographical origin of the honey due to the interaction of its chemical and organoleptic components. It is therefore used to certify whether what appears on the product label is true, always together with pollen and physicochemical analyses. The correct labeling of the product is very important as it is the first thing that the consumer sees before purchasing it, so it has to be totally in accordance with what is inside the packaging.

In the case of geographical origin, it is very important that the label contains accurate information and complies with the regulations on the indication of the country of origin. In the European Union (EU), if the honey comes from only one country, this must be indicated on the label, and if the honey comes from more than one country of origin, the indications "blend of EU honeys," "blend of non-EU honeys," and "blend of EU and non-EU honeys" can be used.

In terms of botanical origin, in recent years, there has been an increasing consumer demand for monofloral honey (honey that comes predominantly from one nectar source). This honey can achieve a distinctly higher market value than multifloral honey due to its pharmacological properties (Geană and Ciucure 2020; Tsagkaris et al. 2021) and because it possesses unique organoleptic characteristics, such as color, smell, aromas, and flavors of its own, and can therefore be considered a premium product (Machado et al. 2020; Na Li et al. 2024; Soares et al. 2017).

For example, manuka honey (*Leptospermum scoparium* (J.R.Forst. & G.Forst)) has gained great popularity for its significant benefits and nutritional value, but its price is high, and its production is very limited. This implies that many high value monofloral honeys are adulterated (Tsagkaris et al. 2021).

Verifying authenticity is the goal of the global honey industry. However, current quality standards and analytical methods are still ineffective in determining and assessing the veracity of honey labeling, creating challenges for the market. Identifying the authenticity of honey requires well-trained and experienced sensory panelists and statistical evaluation of observations; this sensory expertise is a key factor in ensuring the reliability of the results; but sensory analysis is only suitable for preliminary judgments and should not be used as a definitive assessment of honey authenticity as there are no globally authenticated expert sensory analysis panels.

This would be important as honey is a very complex product that requires different analyses; for example, the authenticity of avocado (*Persea americana* Mill.) honey can be determined based on its composition, mineral content, and sensory attributes, having a very dark brown or dark amber color and strong, fruity odors and flavors (Rodríguez et al. 2019).

Apart from knowing the quality and authenticity of the honey and the strictness of the labeling, another objective of carrying out sensory analysis of honeys is to teach people involved in this product at any level: producers, buyers, packers, sellers, consumers, etc., the precise techniques so that they can quickly judge the botanical origin of a honey and issue an opinion on its quality.

We need to raise awareness of this product, which has high nutritional and medicinal values, such as antioxidant, antibacterial, hepatoprotective, gastroprotective, anti-inflammatory, and anticarcinogenic properties (Ranneh et al. 2021).

During honey storage and processing, color, aroma, or physicochemical parameters can be modified independently of honey type, as well as microbial potential (Da Silva et al. 2016a; Combarros-Fuertes et al. 2019, 2020) (there may even be a relationship between honey color and microbial activity (Marić et al. 2021; Russo et al. 2023)), which helps in the search for alternative agents capable of controlling infections, in view of the recent prevalence of antibiotic resistance (Bouzo et al. 2020). Honey has the ability to kill or inhibit the growth of a broad spectrum of microorganisms, including multidrug-resistant pathogens (Cilia et al. 2020), and its antioxidant, antibacterial and anti-inflammatory properties have been used for therapeutic purposes (Zhang et al. 2021), mainly for skin conditions.

Honey is produced and marketed all over the world because of all these properties but with great differences in nutritional value and biological importance from place to place. The way of proceeding with sensory analysis is the same regardless of its botanical and geographical origin, the way it is obtained, its treatment, and its physical state (Sánchez Durán and Sánchez Reyes 2024). Our research group has spent many years teaching people interested in differentiating honeys and raising awareness of the importance of promoting beekeeping and protecting all pollinators.

The sensory analysis of honeys is regulated by several international regulations that establish a standardized framework for the sensory evaluation of foods, including the analysis of honeys (some of them are listed in this chapter). In some countries, specific regulations are applied detailing quality, purity, and labeling requirements for honey. Food information provided to consumers must be regulated and honey must be classified according to its floral origin. But if we want to make a descriptive analysis of honey, it is important to select appropriate descriptors and apply a standardized vocabulary.

Honey is such an important product that it is related to some of the Sustainable Development Goals such as 1 (end of poverty), 2 (zero hunger), or 8 (decent work and economic growth) according to the Food and Agriculture Organization (FAO) of the United Nations. The promotion of this product in developing countries such as Rwanda or Vietnam contributes to achieving these objectives. (Mugwiza 2024), and it also promotes an important ecosystem service such as pollination.

This chapter discusses the sensory analysis of honeys in depth, going into the regulations, the senses involved, the particularities of the tasting room, the recruitment of tasters, the phases of sensory analysis and its importance, and the organoleptic characteristics of different honeys from around the world.

2.2 The Senses

Our senses: sight, hearing, taste, smell, and touch, are fundamental tools for perceiving the world around us, capturing information from the environment, processing it, and generating appropriate responses, as each of them is associated with specialized organs that convert specific stimuli into neural signals that are interpreted by the brain.

They also play an essential role in the execution of sensory analyses, as the interpretation of the results obtained in these studies is intrinsically influenced by the interaction of the senses.

This interaction is multisensory; the sensory experience is not limited to a single sense, but the result is the sum of multiple stimuli. For example, the visual presentation of a honey can impact the perception of taste.

On the other hand, the senses evoke memories and emotions that can affect our evaluation of a product, in this case, honey. A familiar aroma can induce nostalgia, thus altering perceptions in the rest of the tasting phases. In addition, as people are repeatedly exposed to a stimulus, their response may diminish, and this is important in honey tasting sessions, where sensory fatigue can affect the results.

The sensory characteristics of honey mainly include color, odor, aroma, taste, and texture (Da Silva et al. 2016a; Ng et al. 2023), so the senses we employ in organoleptic analyses of honeys, which were already described in the previous edition of this chapter (Sánchez Durán and Sánchez Sánchez 2017), are sight, smell, taste, and touch when honey is introduced in the mouth.

Furthermore, the appearance of honey and its aromatic properties influence consumer preferences at the beginning, while taste, which is appreciated later, is one of the most important sensory parameters that determines its acceptability at the end (Na Li et al. 2024); therefore, sensory analysis of honeys involves the integration of all these senses.

2.2.1 Sight

Visual perception influences the evaluation of honey, allowing the appreciation of transparency, color and general appearance. The presentation of the product is the first thing the consumer sees and makes him decide which honey to choose; it is the first indicator of quality. The color of honey is one of the characteristics used to preliminarily determine its vegetable origin, as different monofloral honeys can have a unique color (Castiglioni et al. 2017; Sogut and Seydim 2020).

Not only variations in color can be identified but also its clarity or turbidity and viscosity, which can also be related to the floral origin and age of the honey, providing information on its quality and typicity.

The visual phase includes mainly color and physical condition, cleanliness, homogeneity, deterioration, and even defects. The "honey color" is always associated with amber color, but in the spectrum of colors, depending on the floral variety from which it comes, we can find a great variety of colors from very soft, almost transparent yellows, intense amber to brown, copper, and coal black; these last shades seem to have a higher content of polyphenols than the clear ones (Da Silva et al. 2016b; Karabagias et al. 2020).

2.2.2 Smell

Olfactory perception allows the detection of volatile chemicals in the air that reach us through the nose (odors) or via the retronasal route (aromas).

This sense is fundamental in sensory analysis, as smells and aromas can trigger memories and emotions, as well as contributing greatly to taste perception, as the aroma of honey reaches us at the moment we taste the sample.

The olfactory properties of honey are directly related to the type and content of volatile compounds from the nectar. Several studies have combined sensory and chemical analyses to investigate the volatile and aromatic compounds active in honey (Seisonen et al. 2015; Siegmund et al. 2017; Kortesniemi et al. 2018), being able to characterize different volatile compounds that were correlated with the geographical origin of the honeys (Patrignani et al. 2018) or discriminate honey on the basis of aromatic compounds (Tahir et al. 2016).

2.2.3 Taste

Of all the sensory attributes of honey, smell and especially taste are often the main ones used to identify the authenticity of honey (Berriel and Perdomo 2019; Debela and Belay 2021; Zhang et al. 2023).

In many honeys, the unique flavor of the plants from which they originate can be appreciated. For example, citrus honey has a strong citrus flavor, with a sweet and slightly sour taste, while the taste of rapeseed honey (*Brassica napus* L. var. oleifera) is reminiscent of the flowers that the bees forage for this honey. Those most appreciated by the consumer have a sweet taste and floral aroma; however, less popular honeys have intense acidity, astringency, and/or bitterness, probably attributed to high levels of organic acids and flavonoids (Kang et al. 2023).

Taste perception is the ability to detect flavors through the taste buds on the tongue. The identification of primary flavors, especially sweet, sour, and bitter, is done continuously during the taste phase of honey tasting, but we also need to know what these flavors remind us of floral, smoky, etc.. Although salty is also another primary taste, it is not very important as few honeys have salty notes.

Sweetness: Sweetness levels in honeys are determined by the concentration of simple sugars. Honeys of floral origin such as orange blossom honey (*Citrus* sp. L.) or acacia honey (*Robinia pseudoacacia* L.) tend to have a more pronounced sweetness, while honeys from resinous woods, honeydew honeys, may have a less intense and more complex sweetness, such as holm oak honey.

Acidity: Acidity in honey is mainly due to the presence of organic acids, such as gluconic acid and acetic acid. A slight acidity can balance sweetness and enhance other taste impressions, providing a sensory complexity that is desirable in certain varieties of honey.

Bitterness: Although bitterness is not a predominant flavor in all honeys, some varieties, particularly those with darker hues or honeydew, may have bitter notes that complement the sweetness.

Flavor profiles can vary significantly between monofloral and multifloral honeys and between different monofloral honeys.

2.2.4 Touch

Tactile perception involves the appreciation of textures, temperatures, and pressures through the skin and, in this case, in the mouth. In sensory evaluation, the texture of a food can significantly affect its acceptability. The mouthfeel, which includes the texture (stickiness, viscosity, and crystallization) of honey, affects the taste experience. More viscous honeys offer a completely different sensory experience than more liquid honeys.

One of the most remarkable alterations, but which in this case is a totally natural process of all honeys, is crystallization. The speed of crystallization depends on several factors such as temperature, viscosity, water content (with low humidity they tend to crystallize), and sugar composition since the higher the amount of fructose, the lesser the crystallization (as is the case with acacia honey).

Small wax fragments, dust, or air bubbles can act as crystallization nuclei causing it not to form homogeneously.

Advertising can sometimes influence the preferences of consumers who may consider a honey that does not crystallize (even with an anti-drip system) as being of better quality than others that are crystallized, when this is not the case.

On the other hand, more and more advances are being made to try and eliminate the inevitable subjective part of sensory analysis, for example, by using artificial senses (e-eye, e-nose, and e-tongue) that can confirm the impact of climatic factors and genetic characteristics on the composition of honey (Russo et al. 2023) and whose use can be compared to the analysis of its organoleptic characteristics, such as electronic nose and aromas (Tian et al. 2018).

We must bear in mind that our senses must necessarily be trained; even so, they must be sharpened from the outset so that the evaluation can be reliable.

2.3 Tasting Panel: Tasters

In order to form a tasting panel, we need people willing to train their senses for the correct quality analysis of the products. These tasters should have adequate training and high organoleptic sensitivity, but they should also have experience, and it would be advisable for them to have theoretical knowledge of honey palynology and honey biochemistry in order to understand the principles of beekeeping and the variables that affect its quality. They should also have previous training in the identification of specific honey descriptors. The selection of tasters can include different experiences, from beekeepers to food specialists.

Regular practice and ongoing training are essential to refine sensory skills and maintain a high level of tasting accuracy.

Initially, more people will be selected than are needed to form the final tasting panel, and they will be trained and selected, reducing their number. The purpose of these candidate selection tests is to determine the perception, recognition, and differential thresholds (value of the minimum perceptible change in the intensity of a stimulus).

The tasters who finally make up the tasting panel must describe the sensations perceived in an objective manner, so it is very important that there is no distraction and that their senses are not affected, being careful, for example, with the ingestion of too much coffee (the threshold of bitterness would be too high), drinking alcohol, or smoking since, in the latter case, the senses of smell and taste are atrophied. Hygiene products that may leave residual odors that interfere with the results (such as colognes or creams and soaps with odor) should not be used either.

Tasters will learn and train on the different phases of organoleptic analysis discussed below by means of standardized scales, the different evaluation points of each phase and in the development and use of descriptors, in order to enhance their sensory skills. Standardized protocols are used to train the senses. These scales include the identification of color, odors, primary and secondary flavors, textures, and defect recognition. We recommend the use of the standardized scales (González and De Lorenzo 2002a, b) with the modifications proposed in this chapter, which also appear in the previous edition.

After training with the different scales and descriptors, tasters must perform blind tests (examinations) of different flavors and textures to give the correct evaluation (specific point of the different scales of primary flavors and adhesiveness, viscosity, and crystallization). If it is not correctly assessed, they will proceed to recall again the specific points of the scale with which they made the mistake.

At the end of this training, the tasters start tasting honeys, first, without analyzing each of the phases and without giving a complete evaluation but by performing duo-trio tests (ISO 5495 2005) or the triangular test (ISO 4120 2004). In this case, it is only necessary to determine whether the given honeys are the same or different from each other.

After these tests, with a positive result, the complete organoleptic analysis of both millifloral and monofloral honeys will be carried out, where the expert tasters will provide a reasoned judgment.

It is very important that these experts carry out continuous practices, participating in guided tasting sessions that allow the comparison between different types of honey and the exchange of observations between experienced tasters.

There are several challenges in the training tasters process as the huge diversity of honeys on the international market, all of which have unique characteristics that must be correctly identified and valuated (Fig. 2.1); or the subjectivity in the evaluation since it is a sensory perception, so a standardized approach is required to minimize individual biases and continuous training due to the constant evolution of the market and production techniques.

Even so, the training of expert tasters in honeys is crucial for the preservation of the quality and authenticity of this product. Through good theoretical training, the development of sensory skills, and continuous practice, it is possible to create a network of experts capable of maintaining quality standards in beekeeping.

Fig. 2.1 Honey samples in glass cups prepared to be analyzed

2.4 The Tasting Room

The circumstances in which sensory analysis is carried out, such as the physical and social environment, can modify the experiences and share different types of responses. Lighting, temperature, and noise are variables to be considered during sensory tests, so there is a specific standard on the specific characteristics that a tasting room should have at an international level (ISO 8589 1988). Some of them are:

- The ambient temperature should be pleasant and kept in a constant range, between 20 and 22 °C, to maximize the fluidity and organoleptic properties of the honey, releasing aromas, but without altering them.
- The illumination should be soft and diffuse, as well as uniform, to avoid interference in the perception of color.
- The sample should be at room temperature so that its organoleptic qualities are not altered.
- The room must be free of external odors; there cannot be secondary odors that can interfere in the evaluation (cooked food, smoke, paint, etc.), making it necessary that there is a prudential time between a session and the following one so that these possible parasitic odors do not interfere.
- It is essential that there is good ventilation to renew the air.
- Each taster should have an individual booth to avoid distractions and noise.
- The room should be spacious enough to accommodate a group of tasters and allow adequate circulation, with large tables and enough space between them to facilitate interaction without interference.
- It is advisable to have a room nearby in which to give instructions and prepare the sensory analysis sessions.

Ultimately, the design and organization of a tasting room for sensory analysis of honeys are determining factors in the quality of the evaluation. Every element, from the location and design of the space to the equipment and environment, must be carefully considered to create a conducive environment that allows tasters to make accurate and reliable evaluations, contributing to a better understanding of the quality and characteristics of honeys, fundamental for both the beekeeping industry and scientific research. Tasting room design is not only a logistical aspect but an essential part of the sensory analysis process.

2.5 The Material

Among the necessary elements for the good achievement of a honey tasting, it is important, first of all, to select the honey samples to be analyzed; these must be representative and must be labeled in an adequate way but always avoiding that the information influences the sensory analysis. It is necessary to decide the order in which the honeys will be analyzed since, in the first place, it is necessary to start with the honeys with softer textures and flavors and go progressively to the strongest honeys.

These honeys will be introduced in "balloon"-type glass cups to better perceive the volatile particles when the nose is introduced. As mentioned above, the honey (approximately 35 g in each glass) will be at room temperature. These cups will be covered with lids or aluminum foil to avoid the loss of volatile molecules or the influence of parasitic odors. The glasses will have a code whose meaning will only be known by the panel leader so as not to influence the tasters during the sensory analysis.

As for the rest of the material, chemical reagents and standard foods are needed to prepare all the scales with which the tasters are trained, such as citric acid, caffeine, or sucrose to prepare the acidic, bitter, or sweet scales, and foods such as dulce de leche, cheese, or butter to prepare the viscosity or adhesiveness scales.

For the preparation of these scales, the following will also be necessary: glass laboratory equipment (beakers, pipettes, rods, etc.), a precision balance, and vials or opaque containers (one for each point of the different scales).

Each taster will have at his or her disposal tasting utensils such as disposable odorless spoons or spatulas to taste both the scales and the honeys, glass beakers for water, plates, paper napkins, or knives. For preparation and palate cleansing between samples, there will always be low-mineral mineral water, low-salt bread, and Granny Smith apples.

Finally, it will also be important to provide tasters with a pen and documentation and answer sheets to be filled in during the tasting process (Fig. 2.2).

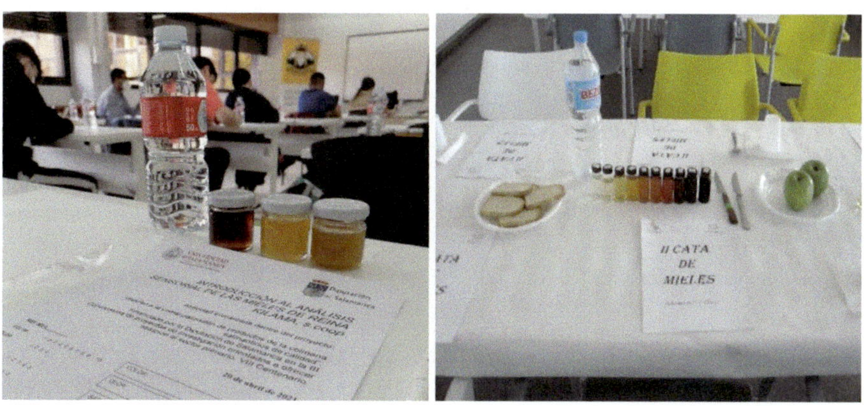

Fig. 2.2 Documentation for different sensory analysis sessions: (**a**) consumer appreciation analysis of three honeys from a specific beekeeping cooperative, (**b**) informative sensory analysis

2.6 The Honey Tasting Process

The evaluation of a honey involves the application of sensory methods that can be qualitative (a detailed study of each sample is not carried out) or quantitative (a complete description of the honeys is made), where each sample will be scored.

Hedonic or intensity scales are used to measure the perception of sensory attributes. The descriptive methodology used in studies of honeys is based on the identification of these specific attributes, which also allow determining consumer preferences. Discrimination tests are also part of this process, assessing whether evaluators can distinguish between different types of honey.

The process of sensory analysis of honeys is complex as it involves multiple phases that we will see in detail, all of them being crucial to guarantee the quality and authenticity of the honey. These phases must be carried out in order: prior preparation by the head of the panel (selection of samples and preparation of the tasting conditions), giving an initial talk (to provide information about the honeys and how to proceed with the tasting, in the case of tasters who are not experts), visual evaluation (analysis of the color, intensity, and homogeneity of the honey), olfactory evaluation (identification of primary and secondary odors and nuances), gustatory evaluation (analysis of the flavor, including sweetness, acidity, and bitterness, as well as aromas), tactile evaluation (observation of adhesiveness, viscosity, and crystallization), data recording (annotations and results of the final evaluation), and interpretation of the results (comparison of the evaluations with norms and standards).

In addition, once the data from the sensory analysis day has been collected, a statistical analysis of the results will be performed, allowing the identification of significant patterns and preferences. This interpretation of the results not only includes the comparison of different samples but also the contextualization of the findings in relation to the botanical and geographical characteristics of the plants that the bees visit and the preferences of the taster. Generating a clear and accessible data analysis that summarizes the conclusions of the tasting is essential for the dissemination of the results.

Finally, feedback between tasters and validation of the methods used are crucial steps in order to make improvements in the methodology and adaptations based on the experience of the sensory panel.

Below, the tasting phases are detailed, evaluating each one of them.

2.6.1 Visual Phase

Color is one of the attributes of honey by which many buyers are guided when choosing one or another in the market. This color varies significantly among the different varieties, but it can also have slight nuances within the same type of honey.

Fig. 2.3 Fermentation on the surface of a honey sample

During this phase, the tonality, color intensity, or brightness are analyzed but also other assessments such as whether it is homogeneous, if it has remains or impurities, if it is well crystallized or not, or even if it has started a fermentation process (Fig. 2.3), among other things.

To compare the color results, we can use devices such as colorimeters, which are used to identify the colors and their shades, achieving objective and accurate measurements. Within quality control, their applications are very varied, and they can even guarantee that the analyzed sample is within the established color tolerances. In the case of honey, it is recommended to use a color classification system, such as the Pfund color scale, which measures the color of honey in degrees (between 0 and 140 mm), already discussed in the previous edition of this book.

The appearance of honey, its color and clarity, if analyzed by expert tasters, is a valuable initial indicator that serves to evaluate honey quality and determine possible adulteration (Brčina et al. 2021).

The color of honey is not only related to the flora of origin but also to the content of anthocyanins and carotenes, compounds that influence the visual perception of the product.

Apart from color, it is also interesting to analyze the clarity (which is an indicator of its purity) and transparency, since a cloudy honey could indicate the presence of impurities.

Crystallization can also be appreciated during the visual evaluation, as the presence of crystals can be a significant factor. Whether liquid or crystallized, the honey

Fig. 2.4 (**a**) Heterogeneous crystallization in honey sample and (**b**) honey samples separated into two phases (liquid and crystallized)

Fig. 2.5 Correlation between two samples of crystallized honey and the corresponding color scale points

should be homogeneous, and the size and distribution of crystals should be evaluated. Uniform crystallization may be indicative of quality, while irregular or large crystals, as well as if honey is separated into two phases, may suggest improper storage or handling processes (Fig. 2.4). It is important to know that when a honey crystallizes, its color becomes lighter.

The visual phase of sensory analysis of honeys is crucial to establish a first impression of the quality and organoleptic characteristics of the product. Tasters can obtain valuable information that will serve as a basis for the following phases. A rigorous approach during this visual evaluation not only contributes to the quality of the sensory analysis but also promotes the appreciation and understanding of the diversity of the honeys.

In this phase, color will be evaluated according to the scales with which the tasters have been trained, in our case, one for liquid honeys and another for crystallized honeys (Fig. 2.5, Tables 2.1 and 2.2), although crystallization itself will be evaluated in the tactile analysis phase.

Table 2.1 Color scale for liquid honey

Liquid honey		
N° scale	N° dissolution	Dissolutions[a]
10	1	Liquid caramel (Royal)
9	2	250 ml water: 35 ml liquid caramel
8	3	Cocoa liqueur (Bardinet)
7	4	Cocoa liqueur: water (1:1)
6	5	Dissolution 4: water (1:1)
5	6	Almond liqueur: cocoa liqueur (1:16)
4	7	Almond liqueur: banana liqueur (1:1)
3	8	330 ml banana + 33 ml cocoa + 637 ml water
2	9	Dissolution 8: water (1:1)
1	10	Dissolution 9: water (1:1)

[a]*Trademarks:* liquid caramel (Royal), cocoa liqueur (Bardinet), almond liqueur (Teichenné, S.A.), and banana liqueur (Frutaysol)

Table 2.2 Color scale for crystallized honey

Crystallized honey		
N° scale	N° dissolution	Dissolutions[a]
10	1	5 g soluble coffee + 10 ml almond liqueur
9	2	6 g soluble coffee + 20 ml distilled water
8	3	Dissolution 2: distilled water (1:1)
7	4	Dissolution 3: distilled water (1:1)
6	5	Dissolution 4: distilled water (65:35)
5	6	Dissolution 4: distilled water (1:1)
4	7	Dissolution 5: distilled water (1:1)
3	8	Dissolution 6: distilled water (1:1)
2	9	Dissolution 7: distilled water (1:1)
1	10	Dissolution 8: distilled water (1:2)

[a]*Trademarks:* almond liqueur (Teichenné, S.A.) and soluble coffee (Hacendado)

2.6.2 Olfative Phase

The evaluation of volatile compounds in honey allows us to know its quality and its relationship with consumer preferences.

As already mentioned, the volatile molecules can reach us directly through the olfactory pathways so that we appreciate the smell or through the retronasal pathway where we value the aromas of the honey at the moment we have it in our mouth (taste phase).

This complex mixture of volatile compounds will depend both on the flowers that have been picked and on the honey transformation process. In the first case, the composition of these volatile particles is influenced by factors such as climate and soil type, which makes it possible to establish a connection between the environment and the organoleptic characteristics of the honey. In the second case, these particles are influenced by changes that occur during honey maturation and storage. Compounds generated through these stages can modify or complement floral aromas, resulting in a richer and more complex sensory experience.

Therefore, odor is of crucial importance. Evaluating olfactory nuances can help identify floral variety and determine freshness; these odors are key components that contribute to product acceptance and quality.

To carry out an effective olfactory evaluation phase, it is recommended to follow a systematic protocol: honey should be at room temperature, as too cold or too hot temperatures can affect the volatility of the compounds. It should be carried out in a controlled environment, free from external odors that could interfere with olfactory perception.

Tasters should take short inhalations followed by a slow, deep inhalation to capture the full range of odors. After this, it is essential to score the honey according to odor intensity (Table 2.3) and persistence scales and also to perform a descriptive analysis, noting down their perceptions in a systematic way, identifying and quantifying specific olfactory attributes (floral, fruity, aromatic, animal, etc.), using standardized vocabulary to be able to quantify and compare the results. It is important to avoid sensory saturation, taking breaks between evaluations and drinking water or eating bread between samples (Table 2.3).

This olfactory phase is a process that allows the quality and characteristics of this product to be evaluated. Understanding the relationship between the honey manufacturing process and the odors given off in this phase helps promote more sustainable and efficient practices for the production of high-quality honey.

Some honeys have more intense and easily recognizable odors, such as orange blossom honey, which is reminiscent of the smell of the flowering season in citrus groves. Other good examples of markedly aromatic honeys would be thyme (*Thymus* sp. L.) or eucalyptus, the latter being of great interest due to its aroma and medicinal properties and being marketed worldwide (Bobis et al. 2020).

At the other extreme, we have the varieties with a very faint odor, such as acacia honey.

In addition, the identification of off-flavor (or unpleasant) odors can indicate problems in the honey extraction or storage process, highlighting the importance of olfactory analysis to ensure the quality of the final product. However, there are some honeys that have naturally unpleasant odors.

Table 2.3 Olfactory intensity scale

Smell intensity		
N° scale	Sensory descriptor	Dissolutions[a]
1	Not intense	Distilled water
2	Slightly intense	Banana liqueur: water (1:6)
3	Moderately intense	Banana liqueur: water (1:1)
4	Extremely intense	Banana liqueur

[a]*Trademark:* banana liqueur (Frutaysol)

To reduce possible subjectivity and establish a universal form of identification, there is the Aroma Wheel (International Honey Commission of Apimondia (IHC), 2001, with several subsequent updates) that unifies the criteria that describe the smells and aromas perceived.

2.6.3 Taste Phase

Honeys can vary in taste from sweet to very bitter, with a complexity that can include fruity, floral, and spicy notes. Taste evaluation is essential for a complete organoleptic analysis and is one of the most important phases.

Again, the honey should be at room temperature, and great care should be taken to avoid parasitic odors and flavors that may interfere with taste perception, as mentioned above. It is therefore important to use tasting spoons made of neutral material (e.g., plastic or glass) to avoid contamination that could alter the taste and take a small amount of honey for evaluation.

As in the previous phase, weakly mineralized water and low-salt bread should be taken to clean the mouth between samples to avoid sensory fatigue.

We should place the honey on the tongue and allow it to melt to stimulate the taste buds, noting the first impression according to the scale of primary tastes (especially sweetness, acidity, and bitterness, see Tables 2.4, 2.5, and 2.6) in a systematic way. In a more detailed way, we will then identify the flavor nuances that emerge during tasting and their intensity and persistence, as well as the aromas that reach our nose retronasally. These aromas can be similar to the odors perceived directly in the nose in the previous phase or have very different nuances.

It is interesting to consider texture and viscosity, which can influence flavor perception: more viscous honeys tend to have a stronger and longer-lasting flavor profile.

Table 2.4 Primary taste range: sweet

Sweetness		
N° scale	Sensory descriptor	Dissolutions
1	Not sweet	Mineral water (Bezoya)
2	Slightly sweet	25 g saccharose/liter
3	Moderately sweet	50 g saccharose/liter
4	Very sweet	100 g saccharose/liter
5	Extremely sweet	200 g saccharose/liter

Table 2.5 Primary taste range: acidity

Acidity		
N° scale	Sensory descriptor	Dissolutions
1	Not acidic	Mineral water (Bezoya)
2	Slightly acidic	0.500 g tartaric acid/liter
3	Moderately acidic	0.750 g tartaric acid/liter
4	Extremely acidic	1 g tartaric acid/liter

Table 2.6 Primary taste range: bitter

Bitterness		
N° scale	Sensory descriptor	Dissolutions
1	Not bitter	Mineral water (Bezoya)
2	Moderately bitter	0.200 g caffeine/liter
3	Extremely bitter	0.400 g caffeine/liter

The data extracted will be recorded on the evaluation sheet, including specific descriptors that may be useful for future categorizations, such as "floral," "fruity," "spicy," and "woody," among others.

This taste phase is crucial to understand the particularities of each variety, with primary flavors playing a fundamental role in the taste experience. A rigorous and structured evaluation of these flavors will allow the classification of honeys.

Everyone can educate and develop their palate over time, so taste is something that can grow.

2.6.4 Touch Phase

The tactile phase refers to the perception obtained through the sense of touch and is fundamental in the evaluation of textures and consistencies of food substances. In the case of honey, these properties, perceived in the mouth, are keys for an adequate characterization, since they can influence consumer preference, but also in the identification of honeys and in the verification of their authenticity.

Texture influences the sensory perception of honey at the moment it is introduced in the mouth. A creamy honey will have a different sensation than a liquid honey.

It is relevant to consider aspects such as stickiness, which can vary from a very creamy, highly adhesive honey to a nonadhesive or liquid honey. Creamy honeys tend to have a smoother sensation and can soften more intense flavors, balancing the perception of sweetness; viscosity, since honeys with a high viscosity generally offer a richer and more persistent tactile experience, which favors a more intense perception of flavors; and crystallization, where the presence or not of crystals and how they are also greatly influence the final perception of the honey.

These properties of the honeys are capable of modifying the way in which the primary flavors are perceived and, therefore, the overall experience of the taster; therefore, this tactile phase complements the primary and secondary flavors previously evaluated. For example, a honey that exhibits crystallization may accentuate the perception of the bitter taste, while a liquid honey may allow the sweetness to prevail.

The temperature of the honey during the tasting of this phase can significantly alter the tactile perception, as higher temperatures can reduce viscosity and modify the intensity of the flavors.

The tactile phase is determined not only by the physical structure of the honey but also by its chemical composition, which, in turn, is influenced by the nectar sources. Honeys from different flower species exhibit tactile variations that can be correlated with sugar, mineral, and amino acid content. For example, acacia honey takes longer to crystallize or does not crystallize at all; however, honeys such as rosemary (*Rosmarinus officinalis* L.) honey tend to crystallize.

Viscosity (Table 2.7) is a fundamental rheological property that influences the sensory perception of honey. Viscosity not only affects the texture and fluidity of honey (Fig. 2.6) but also impacts the overall experience of the taster and the preservation of its organoleptic properties.

During sensory analysis, it is essential to consider that certain components affect viscosity such as chemical composition: sugars, water, organic acids, enzymes, and phenolic compounds; for example, the relationship between water content and sugar

Table 2.7 Touch scale: viscosity

Viscosity		
N° scale	Sensory descriptor	Reference sample (trademark)
1	Not viscous	Mineral water (Bezoya)
2	Very slightly viscous	Extra-virgin olive oil (Koipe)
3	Slightly viscous	Hot chocolate (Ram)
4	Moderately viscous	Chocolate syrup (Royal)
5	Viscous	Liquid caramel (Royal)
6	Very viscous	Dulce de leche (La Lechera)
7	Extremely viscous	Dulce de leche (Mardel)

Fig. 2.6 Tasting panel leader showing the fluidity of a honey in an informative sensory analysis session

concentration is crucial (a higher concentration of sugars generally results in higher viscosity).

Product temperature also plays a role; viscosity decreases significantly with increasing temperature, which could alter the way the honey's flavors and aromas are perceived.

It is advisable to integrate viscosity into the sensory panel by means of scoring scales that evaluate texture and mouthfeel. This phase is evaluated by chewing the honey and appreciating the effort required to separate the teeth.

One of the honeys with high viscosity is calluna honey (*Calluna vulgaris* (L.) Hull), which also shares a characteristic that not all honeys possess, namely thixotropy (liquefying when agitated).

Adhesivity The evaluation of stickiness in honeys is a fundamental aspect of sensory analysis that, although often underestimated, plays a crucial role in quality perception and consumer experience.

Adhesiveness, defined as the ability of a material to adhere to surfaces, is a property that influences the texture and palatability of honey, affecting the sensory perception of the taster. The taster must appreciate the force required to take all the honey in the spoon (Fig. 2.7) (Table 2.8).

Adhesiveness analysis can also be performed using a texture analyzer, which measures the force required to separate the honey from a surface, providing quantitative data.

Fig. 2.7 Training scale on adhesiveness in closed cans to be transported

Table 2.8 Touch scale: adhesiveness

Adhesiveness		
N° scale	Sensory descriptor	Reference sample (trademark)
1	Not adhesive	Mineral water (Bezoya)
2	Very slightly adhesive	Butter (Flora)
3	Slightly adhesive	Cheese (El Caserío)
4	Moderately adhesive	Nocilla (Nutrexpa)
5	Very adhesive	Peanut butter (Capitán Maní)
6	Extremely adhesive	Caramel (Sugus)

Proper assessment of stickiness will not only enrich the consumer experience but will also provide a valuable tool for beekeepers in improving their production processes.

Crystallization This process, which manifests itself in the transformation of honey from a liquid to a solid state, has significant implications on the sensory perception of the final product.

Crystallization in honeys is mainly due to the concentration of sugars, mainly glucose and fructose, and the presence of suspended particles, such as pollen. Glucose tends to crystallize more easily than fructose, so, depending on the ratio between these sugars, honey can present different crystallization patterns. In addition, temperature, humidity, and duration of storage are also factors that influence this process.

Within the sensory analysis, the crystallization of a honey can influence the other organoleptic characteristics such as color, odor, aroma, flavor, and texture. For example, crystallized honey has a lighter color and a granular or creamy texture. It is essential that panelists identify these differences and evaluate the uniformity and consistency of crystallized honey (Fig. 2.8) (Table 2.9).

An organoleptic characterization can allow researchers to classify honeys not only according to their floral origin but also according to their crystallization potential. This, in turn, can influence marketing decisions and the formulation of honey-based products.

2.6.5 Final Phase Evaluation Sheets

After evaluating the sample in all the phases mentioned above, the final evaluation is carried out by filling out a sheet containing all the information of each phase (the numerical score of the standardized scales, the appreciation of each phase, etc.) and a final evaluation of the honey as a whole.

We proceed somewhat differently to describe unfamiliar honeys, to decide the quality within a type already well-known sensorially by the tasting panel or to evaluate the best honey of a given type among several, for example, by holding

Fig. 2.8 Training scale on crystallization

Table 2.9 Touch scale: crystallization

Crystallization		
N° scale	Sensory descriptor	Reference sample (trademark)
1	No crystallization	Mineral water (Bezoya)
2	Very fine crystallization	Icing sugar (Azucarera española)
3	Fine crystallization	Brown sugar <700 μm (Az. esp.)
4	Moderate crystallization	Brown sugar 700–900 μm
5	Slightly coarse crystallization	Brown sugar 900–1.12 μm
6	Coarse crystallization	Brown sugar >1.12 mm

competitions of monofloral honeys. In the latter case, the tasters must know this type of honey very well, and proceeding in the aforementioned manner, they will make their diagnosis as to which is the best within this monofloral type.

2.7 Utility of a Honey Sensory Analysis

In gastronomy and product quality, the senses play a crucial role. Understanding how the senses interact and how they influence consumer experiences can substantially improve product evaluation and development practices. The multisensory approach in these analyses not only enriches the consumer experience but also provides relevant data that can guide innovation and ensure quality in various sectors.

A honey tasting, in an informative way, can be done anywhere; you can even move the material to the facilities of companies, associations, or cooperatives that may be interested. This way, you can get a higher economic return, if that is your purpose. There are many companies that do so and even sell packs with several honeys so that you can do a tasting yourself at home; this helps more and more people value the importance and quality of honeys and know that there are many different varieties. What is not known is not valued.

A honey tasting is a unique experience that stands out in the business world. It offers an opportunity for networking, strengthens relationships, and aligns your company with corporate social responsibility. It is also an educational activity that raises environmental awareness.

Other ideas to achieve a higher economic return could be to enhance the experience by expanding the offerings, for example, by conducting mixed tastings of honey and wine, honey and chocolate, etc. or by including a guided tour to see the hives or the honey extraction plant.

These analyses are also very important for locating possible adulterations, as it is possible to organoleptically identify a high-priced monofloral honey mixed with large amounts of syrup and/or low-quality monofloral or multifloral honey. However, it can be complicated to identify adulterated honey that closely resembles pure honey (Price et al. 2019) due to the large number of adulterations that exist, leading to locating reliable markers for different monofloral honeys, such as by analyzing the volatile compounds characteristic of a monofloral honey regardless of its origin (Machado et al. 2020).

Encouraging consumers to taste different honeys also helps prevent the extinction of pollinators and will change the way this product is perceived and enjoyed. We will become more aware of the importance of bees, not only because they provide us with this precious delicacy but also due to their vital role, along with other pollinators, in an ecosystem service as important as pollination.

In this regard, it is crucial that beekeepers comply with the laws and keep their hives well-separated so that the ecosystem is not affected. This is particularly important in the case of transhumant hives, which can change the habitat they arrive in by suddenly increasing the number of pollinators.

Sensory analyses are also essential in various industries, including food, cosmetics, pharmaceuticals, and textiles (Severiano-Pérez 2019). Their goal is to understand how consumers perceive and value different properties. They are used to evaluate product quality (detecting defects and consistency in production and identifying consumer preferences are aspects addressed through this methodology), develop new products or packaging (in the innovation process, sensory tests are conducted to gather consumer feedback and adjust flavors, textures, aromas, and presentations before launching a product to the market), and help in market research (consumer preferences and expectations are evaluated through sensory studies, which can include preference tests, acceptance, and sensory profiling, providing valuable data for strategic decision-making).

These analyses are essential for standardizing the production of honey products, identifying botanical varieties, and this information is crucial not only for properly positioning products in the market but also for fostering sustainable beekeeping that respects local ecosystems and responsible agricultural practices.

The integration of this methodology into modern beekeeping not only promotes the quality and diversity of honey but also builds bridges between science, industry, and environmental conservation, highlighting the interdependence between ecosystem health and the production of quality food.

2.8 Attributes of Some Honeys

In this last section, different honeys are described organoleptically, discussing their sensory characteristics. This section is divided by color (Fig. 2.9), with additional information available in the same chapter of the previous edition (Sánchez Durán and Sánchez Sánchez 2017).

2.8.1 *Light Honeys*

These honeys, with a light hue, typically have a mild, floral aroma and are among the sweetest, without bitterness and with a slight acidic taste.

Some examples include acacia honey (highly valued for its slow crystallization); orange blossom honey (it is generally considered a citrus honey, without specifying if it is from orange or lemon trees); clover honey (*Trifolium* sp. L.), particularly in North America, ideal for sweetening without altering the taste of foods; rosemary honey (often found in cream form); and ulmo honey (*Eucryphia cordifolia* Cav.) (a whitish Chilean honey with grayish tones, which is delicious).

Fig. 2.9 Examples of light, medium, and dark honeys

2.8.2 Medium Honeys

These honeys have a more intense smell and taste than the previous ones; they are not as sweet, are still without bitterness, and feel less acidic. These are the ones from which the term "honey color" was coined.

This group includes multifloral honeys (which typically have this tone due to their mix); thyme honey (characteristic of the Mediterranean region, with an intense aroma and a peculiar herbal flavor, with antibacterial properties); honeys from various lavender species, even cultivated (*Lavandula* sp. L.) (which can vary in tone depending on the accompanying species); eucalyptus honey (slightly reddish, with a scent of wet dog or animal sweat, possibly with menthol notes); and manuka honey (from New Zealand, with a strong aroma and intense earthy flavor, which is one of the most expensive honeys in the world, attributed with medicinal and antibacterial properties).

2.8.3 Dark Honeys

These honeys, from dark brown to black, are less sweet than the previous groups, with certain bitter tones, a much stronger taste, and some balsamic notes.

Dark honeys include heather honey (*Erica* sp. L.) (found in northern Spain, with an amber or sometimes reddish color), oak or holm oak honey (slightly more bitter than heather honey), chestnut honey (even more bitter), and if you want to try a super bitter honey, strawberry tree honey (*Arbutus unedo* L.) (in our sensory tastings, we always bring strawberry tree honey for those who say they do not like honey because it is too sweet).

These varieties exemplify the richness of flavors and aromas found in honeys around the world, emphasizing the relationship between local flora and the sensory profile of the product.

The sensory analysis of honeys is a complex discipline that combines the science and art of tasting. With a meticulous approach and in-depth knowledge of organoleptic characteristics, we can better appreciate the diversity of honeys worldwide.

The combination of regulations, taster training, and consideration of the senses allows us to explore not only the sweet taste of honey but also its history and connection with nature. Honeys are highly complex from a sensory perspective, as they vary in color, aroma, flavor, and texture. Now all that is left is to try many different honeys to decide which one is the highest quality for you.

We must promote honey consumption for its properties and quality to help the bees, which are vital for the future of humanity and the rest of the species in both flora and fauna, considering that many of the ecosystems where honey is produced are damaged and unprotected.

References

Araujo D, Ruiz Pérez-Cacho P, Serrano S, Dios-Palomares R, Galán-Soldevilla H (2020) Sensory profile and physco – chemical properties of artisanal honey from Zulia, Venezuela. Foods 9(3):339. https://doi.org/10.3390/foods9030339

Berriel V, Perdomo CH (2019) Differentiating pasture honey from eucalyptus honey based on carbon isotopic data in Uruguay. Heliyon 5:e01228. https://doi.org/10.1016/j.heliyon.2019.e01228

Bobis O, Moise AR, Ballesteros I, Sánchez Reyes E, Sánchez Durán S, Sánchez Sánchez J, Cruz-Quintana S, Giampieri F, Battino M, Álvarez-Suárez JM (2020) Eucalyptus honey: quality parameters, chemical composition and health promoting properties. In: Food Chemistry, vol 325. Elsevier, Reino Unido, pp 1–14

Bouzo D, Cokcetin NN, Li L, Ballerin G, Bottomley AL, Lazenby J, Whitchurch CB, Paulsen IT, Hassan KA, Harry EJ (2020) Characterizing the mechanism of action of an ancient antimicrobial, manuka honey, against *Pseudomonas aeruginosa* using modern transcriptomics. mSystems 5:106–120. https://doi.org/10.1128/msystems.00106-20

Brčina T, Seferović S, Husejnagić D, Cvrk R, Bojanović L (2021) Health safety and sensory properties of honey from two different areas in Tuzla Canton. Int J Res Appl Sci Biotechnol 8(6):20. https://doi.org/10.31033/ijrasb.8.6.5

Castiglioni S, Stefano M, Astolfi P, Carloni P (2017) Chemometric approach to the analysis of antioxidant properties and colour of typical Italian monofloral honeys. Int J Food Sci Technol 52(5):1138–1146. https://doi.org/10.1111/ijfs.13397

Cilia G, Fratini F, Marchi M, Sagona S, Turchi B, Damchuk L, Felicioli A, Kacániová M (2020) Antibacterial activity of honey samples from Ukraine. Vet Sci 7:181. https://doi.org/10.3390/vetsci7040181

Combarros-Fuertes P, Valencia-Barrera RM, Estevinho LM, Dias LG, Castro JM, Tornadijo ME, Fresno JM (2019) Spanish honeys with quality brand: a multivariate approach to physico-chemical parameters, microbiological quality and floral origin. J Apic Res 58:92–103. https://doi.org/10.1080/00218839.2018.1494918

Combarros-Fuertes P, Fresno JM, Estevinho MM, Sousa-Pimenta M, Tornadijo ME, Estevinho LM (2020) Honey: another alternative in the fight against antibiotic-resistant bacteria? Antibiotics 9:774

Da Silva PM, Gauche C, Gonzaga LV, Costa ACO, Fett R (2016a) Honey: chemical composition, stability and authenticity. Food Chem 196:309–323. https://doi.org/10.1016/j.foodchem.2015.09.051

Da Silva CI, Aazza S, Faleiro ML, Miguel MG, Neto L (2016b) The antibacterial, anti-biofilm, anti-inflammatory and virulence inhibition properties of Portuguese honeys. J Apic Res 55(4):292–304. https://doi.org/10.1080/00218839.2016.1243294

Debela H, Belay A (2021) Caffeine, invertase enzyme and triangle test sensory panel used to differentiate Coffea arabica and *Vernonia amygdalina* honey. Food Control 123:107857. https://doi.org/10.1016/j.foodcont.2020.107857

Di Rosa AR, Leone F, Cheli F, Chiofalo V (2019) Novel approach for the characterisation of Sicilian honeys based on the correlation of physico-chemical parameters and artificial senses. Ital J Anim Sci 18:389–397. https://doi.org/10.1080/1828051X.2018.1530962

Eteraf-Oskouei T, Najafi M (2022) Uses of natural honey in cancer: an updated review. Adv Pharm Bull 12(2):248. https://doi.org/10.34172/apb.2022.026

Geană EI, Ciucure CT (2020) Establishing authenticity of honey via comprehensive Romanian honey analysis. Food Chem 306:125595. https://doi.org/10.1016/j.foodchem.2019.125595

Geană EI, Ciucure CT, Costinel D, Ionete RE (2020) Evaluation of honey in terms of quality and authenticity based on the general physicochemical pattern, major sugar composition and $\delta^{13}C$ signature. Food Control 109:106919. https://doi.org/10.1016/j.foodcont.2019.106919

González M, De Lorenzo C (2002a) Calidad sensorial de las mieles de Madrid (I). Configuración de un grupo de cata y obtención de escalas normalizadas. Alimentaria 97:97–102

González M, De Lorenzo C (2002b) Calidad sensorial de las mieles de Madrid (II). Correlación con el análisis instrumental. Alimentaria 103:103–109

IHC, International Honey Commission of Apimondia. Minutes of the IHC meetings (Louvain-la-Neuve) (2001)

ISO 4120 (2004) Sensory analysis: methodology. Triangle test. International Organization for Standardization, Geneva. http://www.iso.org

ISO 5495 (2005) Sensory analysis. Methodology. Paired comparison test. International Organization for Standardization, Geneva. http://www.iso.org

ISO 8589 (1988) Sensory analysis. General guidance for the design of test rooms. International Organization for Standardization, Geneva. http://www.iso.org

Kang MJ, Kim KR, Kim K, Morrill AG, Jung C, Sun S, Lee DH, Suh JH, Sung J (2023) Metabolomic analysis reveals linkage between chemical composition and sensory quality of different floral honey samples. Food Res Int 173:113454. https://doi.org/10.1016/j.foodres.2023.113454

Karabagias IK, Badeka A, Kontominas MG (2020) A decisive strategy for monofloral honey authentication using analysis of volatile compounds and pattern recognition techniques. Microchem J 152:104263. https://doi.org/10.1016/j.microc.2019.104263

Kortesniemi M, Rosenvald S, Laaksonen O, Vanag A, Ollikka T, Vene K, Yang B (2018) Sensory and chemical profiles of Finnish honeys of different botanical origins and consumer preferences. Food Chem 246:351–359. https://doi.org/10.1016/j.foodchem.2017.10.069

Li N, Song M, Li H, Liu Z, Jiang A, Lang Y, Chen L (2024) Advancement of foodomics techniques for honey botanical origins authentication: past decade (2013–2023) and future perspectives. Trends Food Sci Technol 147:104458. https://doi.org/10.1016/j.tifs.2024.104458

Machado AM, Miguel MG, Vilas-Boas M, Figueiredo AC (2020) Honey volatiles as a fingerprint for botanical origin-a review on their occurrence on monofloral honeys. Molecules 25(2):374. https://doi.org/10.3390/molecules25020374

Marić A, Jovanov P, Sakač M, Novaković A, Hadnađev M, Pezo L, Mandić A, Milićević N, Đurović A, Gadžurić S (2021) A comprehensive study of parameters correlated with honey health benefits. RSC Adv 11:12434–12441. https://doi.org/10.1039/d0ra10887a

Moore G, Brooks P, Pappalardo L, Boufridi A (2024) Phenolic profiles of Australian monofloral Eucalyptus, Corymbia, Macadamia and Lophostemon honeys via HPLC-DAD analysis. Food Chem 462:140900. https://doi.org/10.1016/j.foodchem.2024.140900

Mugwiza O (2024) Rwanda, Vietnam, Benin y Chile participan en la iniciativa de la FAO Un país, un producto prioritario con la miel designada como producto elegido. © FAO. https://www.fao.org/newsroom/story/making-honey-a-priority/es, last accessed 2024/12/28

Ng WJ, Ee KY, Hew PS, Chin HY, Chin R (2023) Sensory, physicochemical and antioxidant properties coupled with chemometric analysis of ginger (*Zingiber officinale Roscoe* var. Bentong) enriched *Apis cerana* honey. Food Chem Adv 3:100459. https://doi.org/10.1016/j.focha.2023.100459

Patrignani M, Fagúndez GA, Tananaki C, Thrasyvoulou A, Lupano CE (2018) Volatile compounds of Argentinean honeys: correlation with floral and geographical origin. Food Chem 246:32–40. https://doi.org/10.1016/j.foodchem.2017.11.010

Price EJ, Tang R, El Kadri H, Gkatzionis K (2019) Sensory analysis of honey using Flash profile: a cross-cultural comparison of Greek and Chinese panels. J Sens Stud 34(3). https://doi.org/10.1111/joss.12494

Ranneh Y, Akim AM, Hamid HA, Khazaai H, Fadel A, Zakaria ZA, Albujja M, Bakar MFA (2021) Honey and its nutritional and anti-inflammatory value. BMC Complement Med Ther 21:1–17. https://doi.org/10.1186/s12906-020-03170-5

Rodríguez I, Cámara-Martos F, Flores JM, Serrano S (2019) Spanish avocado (*Persea americana* Mill.) honey: authentication based on its composition criteria, mineral content and sensory attributes. LWT Food Sci Technol 111:561–572. https://doi.org/10.1016/j.lwt.2019.05.068

Russo N, Di Rosa AR, Pino A, Mazzeo G, Liotta L, Caggia C, Randazzo CL (2023) Assessment of sensory properties and *in vitro* antimicrobial activity of monofloral Sicilian honey. Food Biosci 52:102449. https://doi.org/10.1016/j.fbio.2023.102449

Sánchez Durán S, Sánchez Reyes E (2024) La miel de Salamanca, más que un endulzante. Revista de Estudios, 68. La Universidad de Salamanca y su territorio más próximo, pp 217–234

Sánchez Durán S, Sánchez Sánchez J (2017) Sensory studies. In: Alvarez-Suarez JM (ed) Bee products-chemical and biological properties. Springer, pp 21–41. https://doi.org/10.1007/978-3-319-59689-1

Sánchez Reyes E, Sánchez Sánchez J (2017) Botanical classification. In: Alvarez-Suarez JM (ed) Bee products-chemical and biological properties. Springer, pp 3–19. https://doi.org/10.1007/978-3-319-59689-1

Seisonen S, Kivima E, Vene K (2015) Characterisation of the aroma profiles of different honeys and corresponding flowers using solid-phase microextraction and gas chromatography–mass spectrometry/olfactometry. Food Chem 169:34–40. https://doi.org/10.1016/j.foodchem.2014.07.125

Severiano-Pérez P (2019) ≐Qué es y cómo se utiliza la evaluación sensorial? Interdisciplina 7(19):47–68. https://doi.org/10.22201/ceiich.24485705e.2019.19.70287

Siegmund B, Urdl K, Jurek A, Leitner E (2017) "More than Honey": investigation on volatiles from monovarietal honeys using new analytical and sensory approaches. J Agric Food Chem 66(10):2432–2442. https://doi.org/10.1021/acs.jafc.6b05009

Soares S, Amaral JS, Oliveira MBPP, Mafra I (2017) A comprehensive review on the main honey authentication issues: production and origin. Compr Rev Food Sci Food Saf 16:1072–1100. https://doi.org/10.1111/1541-4337.12278

Sogut E, Seydim AC (2020) Classification of honeys collected from different regions of Anatolia by chemometric methods. J Food Process Preserv 44. https://doi.org/10.1111/jfpp.14960

Tahir HE, Xiaobo Z, Xiaowei H, Jiyong S, Mariod AA (2016) Discrimination of honeys using colorimetric sensor arrays, sensory analysis and gas chromatography techniques. Food Chem 206:37–43. https://doi.org/10.1016/j.foodchem.2016.03.032

Tian H, Shen Y, Yu H, Chen C (2018) Aroma features of honey measured by sensory evaluation, gas chromatography-mass spectrometry, and electronic nose. Int J Food Prop 21(1):1755–1768. https://doi.org/10.1080/10942912.2016.1213744

Tixe SE (2024) Evaluación *in vitro* del efecto antimicrobiano de la miel de eucalipto sobre *Mammaliicoccus sciuri*. Documento final del proyecto de investigación como requisito para obtener el grado de médico veterinario. Universidad Técnica de Ambato Facultad de Ciencias Agropecuarias Ceballos, Ecuador. 71pp

Tsagkaris AS, Koulis GA, Danezis GP, Martakos I, Dasenaki M, Georgiou CA, Thomaidis NS (2021) Honey authenticity: analytical techniques, state of the art and challenges. RSC Adv 11:11273–11294. https://doi.org/10.1039/D1RA00069A

Zhang YZ, Si JJ, Li SS, Zhang GZ, Wang S, Zheng HQ, Hu FL (2021) Chemical analyses and antimicrobial activity of nine kinds of unifloral Chinese honeys compared to Manuka Honey (12+ and 20+). Molecules 26(9). https://doi.org/10.3390/molecules26092778

Zhang XH, Gu HW, Liu RJ, Qing XD, Ni JF (2023) A comprehensive review of the current trends and recent advancements on the authenticity of honey. Food Chem X 19:100850. https://doi.org/10.1016/j.fochx.2023.100850

Chapter 3
Chemical Composition of Honey

Celestino Santos-Buelga and Ana M. González-Paramás

Contents

Abstract Honey is one of the most popular natural sweet substances. From a chemical point of view, it could be defined as a natural food mainly composed of sugars and water (W), along with other minor constituents such as minerals, vitamins, amino acids, organic acids, flavonoids, and other phenolic compounds and aromatic substances. Its composition is particularly variable, depending on its botanical and geographical origins. In addition to beneficial components, certain honeys may contain natural toxic compounds derived from specific botanical

C. Santos-Buelga · A. M. González-Paramás (✉)
Grupo de Investigación en Polifenoles (GIP-USAL), Facultad de Farmacia, Universidad de Salamanca, Campus Miguel de Unamuno, Salamanca, Spain
e-mail: paramas@usal.es

J. M. Alvarez-Suarez (ed.), *Bee Products – Chemical and Biological Properties*,
https://doi.org/10.1007/978-3-031-89049-9_3

sources, which can pose health risks if consumed in high quantities. The stability of these bioactive and potentially toxic compounds, particularly in relation to storage conditions and the effects of processing, is crucial for ensuring honey's nutritional value and safety. Analytical methods such as high-performance liquid chromatography (HPLC), gas chromatography–mass spectrometry (GC-MS), inductively coupled plasma mass spectrometry (ICP-MS), and nuclear magnetic resonance (NMR) spectroscopy are commonly employed for both identifying honey's major compounds and detecting contaminants. The aim of this chapter is to describe the principal families of compounds present in honey, review the stability of these compounds, and discuss the analytical methods employed for their analysis, as well as their utility in discriminating honeys according to their origin and quality.

Keywords Honey · Chemical composition · Analysis · Geographical or botanical discrimination

3.1 Introduction

Honey is the best-known primary product of beekeeping, and the history of its use is parallel to the history of mankind; in every culture, evidence can be found of the use of honey as a food source but also as a symbol employed in religious, magic, and therapeutic ceremonies (Krell 1996).

Nowadays, honey is the most important primary product of beekeeping, both from a quantitative and an economic point of view. Recent statistics estimate annual global honey production at over 1.8 million metric tonnes, with a market value exceeding $8 billion. China continues to lead as the world's top producer and exporter, contributing more than 470,000 tonnes annually, valued at approximately $400 million. Other major exporters include Turkey, Iran, India, and Argentina, all of which hold prominent positions in the international market. Notably, in 2024, Ukraine emerged as the primary supplier to the EU, surpassing China (FAOstat 2023).

According to Codex Alimentarius (2001), "honey is the natural sweet substance, produced by honeybees from the nectar of plants or from secretions of living parts of plants, or excretions of plant-sucking insects on the living parts of plants, which the bees collect, transform by combining with specific substances of their own, deposit, dehydrate, store and leave in honeycombs to ripen and mature." Although there are other honeybee species that make honey, and other bees and even wasps that store different kinds of honeys as their food reserves, in the EU, according to the legal definition, honey can only be referred to as such when it is produced by *Apis mellifera* honeybees (EU Council 2002).

Generally, honey is classified according to its provenance in blossom and honeydew honeys. Blossom honey is obtained predominantly from the nectar of flowers, while honeydew honey is produced by bees after collecting the secretions of aphids or other plant sap-sucking insects, which pierce plant cells, ingest plant sap, and then, secrete it again (Bradbear 2009). Among floral honeys, the most commercially

Table 3.1 Chemical composition of honey

	Mean	SD	Range
Majority constituents (%)			
Water	17.90	3.16	13.21–26.50
Fructose	39.44	2.11	37.07–42.65
Glucose	28.15	5.74	18.20–32.10
Sucrose	3.19	3.81	0.36–16.57
Other sugars	8.5	–	0.1–16.0
Minority constituents (%)			
Minerals	0.36	0.18	0.11–0.72
Total protein	1.13	1.22	0.22–2.93
Acids (such as gluconic acid)			0.17–1.17
Vitamins, enzymes, aromas	<0.1		
Phenolic compounds	0.1		0.02–0.2

Adapted from Solayman et al. (2016) and Bradbear (2009)

available ones are multifloral honeys, meaning that they come from several botanical sources, none of which is predominant. However, it is also possible to find monofloral honeys made primarily from the nectar of one type of flower, and they are named according to that plant.

From a chemical point of view, honey could be defined as a natural food mainly composed of sugars and water (W), together with other minor constituents such as minerals, vitamins, amino acids, organic acids, flavonoids, and other phenolic compounds and aromatic substances (Table 3.1). This composition, and especially minority components, is related to the botanical and geographical origins and can undergo significant changes depending on the storage time and conditions.

Although because of its unique composition and chemical properties, honey is suitable for long-term storage, some compositional variations usually occur due to different chemical and biochemical processes, such as fermentation, oxidation, or dehydration of sugars, leading to changes in acidic content and the formation of compounds like 5-hydroxymethylfurfural (5-HMF), which give rise to modification in sensorial properties and reduce the quality of honey (see 3.10 for details).

3.2 Carbohydrates

Honey is mostly a concentrated aqueous solution composed of a mixture of fructose (F) and glucose (G), but it also contains some amounts of nearly 30 other complex sugars (Table 3.2). Due to its high sugar content, more than 95% of the solids of honey are carbohydrates, and honey is an excellent energy source (304 kcal/100 g). The production of this supersaturated sugar solution is only possible at hive temperature (30 °C) and by the addition of the enzyme invertase, secreted from hypopharyngeal glands of bees, which inverts sucrose, from the nectar or honeydew, into

Table 3.2 Main sugars described in honeys

Monosaccharides		
Glucose		
Fructose		
(Galactose)[a]		
Disaccharides		
Majority	Minority	Traces
Isomaltose	Cellobiose	Isomaltulose
Kojibiose	Gentiobiose	Laminaribiose
Maltose	Maltulose	Leucrose
Sucrose	Nigerose	Melibiose
Turanose	Palatinose	Trehalose
Trisaccharides		
Majority	Minority	Traces
Erlose	Isomaltotriose	Centose
Theanderose	Isopanose	1-kestose
Panose	Melezitose	Laminaritriose
Maltotriose	Raffinose	Planteose
		α-3′-glucosyl-isomaltose
Higher oligosaccharides		
		Traces
		Isomaltotetraose
		Maltotetraose
		Isomaltopentaose
		Nystose

Adapted from Alves-Moreira and Bastos-de Maria (2001), Alvarez-Suarez et al. (2010b), and Ruiz-Matute et al. (2010).

[a]Occasionally cited in trace levels Goldschmidt and Burkert (1955) and Val et al. (1998)

glucose and fructose (Crane 1990). Monosaccharides represent about 75% of the sugars found in honey, along with 10–15% disaccharides and small amounts of other sugars (da Silva et al. 2016).

Different analytical approaches, especially using chromatographic techniques, have been employed to characterize the sugar profile of honeys. However, the structural similarity of sugars present, derived mainly from the combination between glucose and fructose linked through different glycosidic bonds, makes the chromatographic resolution of many of the compounds difficult (see Sect. 3.11.2 for details).

Fructose and glucose are the predominant monosaccharides in honey, accounting for approximately 65–85% of its total soluble solids (Tedesco et al. 2020). Fructose is generally the dominant sugar in most honeys (from 42% of the sugar fraction in some unifloral honeys such as Robinia or *Thymus* to around 32% in honeydew honey), but in some particular unifloral honeys, such as rapeseed (*Brassica napus*) or dandelion (*Taraxacum officinale*), the glucose contribution is greater than that of fructose (Crane 1990; Persano-Oddo and Piro 2004).

Honey disaccharides are mainly constituted by regioisomers of α-glucosyl-glucose and α-glucosyl-fructose; disaccharides with β-glycosidic linkages are present in minor amounts, while fructosyl-fructoses are very scarce. Among disaccharides, maltose, isomaltose, kojibiose, and turanose are the most abundant ones in nectar honeys and represent between 8% and 10% of total sugars. In blossom honeys, sucrose is also important, although due to the invertase enzyme action, its content represents less than 3% of total sugars. The most abundant trisaccharides are derivatives of sucrose (de la Fuente et al. 2011). Relatively high amounts of the trisaccharides melezitose, erlose, and raffinose have been reported in honeydew honey (Doner 1977; Bogdanov et al. 2008).

Sugar composition can be influenced by the botanical/geographical origin of honeys. Thus, the concentrations of fructose (F) and glucose (G), in addition to the ratio between them, have been proposed as indicators to distinguish honeydew honey from nectar honeys (Terrab et al. 2001; Sanz et al. 2005; Escuredo et al. 2014). In general, honeydew honeys have the lowest mean concentration of both glucose and fructose but present a high F/G ratio. However, there would be many exceptions, as some nectar honeys also present relatively high F/G ratios, like heather honey (*Calluna vulgaris* L.) (Waś et al. 2011) or *Robinia* and *Thymus* honeys (Persano-Oddo and Piro 2004). Multivariate analysis on physicochemical parameters and sugar composition has also been applied to differentiate blossom and honeydew honeys (Bentabol-Manzanares et al. 2011), as well as to try to establish correlations between the carbohydrate's profiles and the nectar sources (Sanz et al. 2004b; Nozal-Nalda et al. 2005a, b; Senyuva et al. 2009; Ouchemoukh et al. 2010). Using linear Spearman rank correlation, Escuredo et al. (2014) found positive correlations between the contents of maltose with *Rubus* honey and of melezitose with both *Castanea sativa* and *Rubus* honeys. Different carbohydrate combinations have been suggested to be characteristic of some important honey types in Spain (i.e., *Citrus*, heather, *Eucalyptus*, rosemary, *Echium,* and Rosaceae), although they did not always allow an unambiguous classification of the main unifloral sources, possibly as some honey samples assigned as unifloral by pollen analysis might include contributions from other floral sources (de la Fuente et al. 2011). According to Kaškonienė and Venskutonis (2010), sugar composition would only be a reliable indicator for honey classification in the case of unifloral honeys with a high amount of a dominating sugar(s). Other authors conclude that sugar composition alone is not enough to identify the botanical or geographical origin of honeys (Bogdanov and Baumann 1988; Földházi 1994; Kaškonienė et al. 2010). Indeed, references related to geographical discrimination of honey using sugar composition are scarce and not conclusive (Sancho et al. 1991; Sanz et al. 1995; Gómez Bárez et al. 2000; Consonni et al. 2013).

During storage, sugar composition of honey continues changing due to the action of enzymes and temperature. When the honey is stored at temperatures below those in the hive, some of the sugars, especially glucose and fructose, may crystallize. This phenomenon is largely influenced by the relative proportions of the main sugars in honey. Honeys with a high F/G ratio, greater than 1.33, do not crystallize for long periods, while honeys with an F/G ratio less than 1.11 crystallize quickly

(Smanalieva and Senge 2009). In addition, the ratio between glucose and water contents (G/W) was also proposed in 1975 to predict the granulating behavior of honey (White 1975), and it has largely been used by other groups all around the world with different satisfaction (Zamora and Chirife 2006). In general, slow crystallization of honey should occur when the G/W ratio is less than 1.7, whereas, when the ratio is greater than 2.0, the phenomenon would be fast and complete (Dobre et al. 2012). More recently, through the analysis of 136 honeys (including bramble, chestnut, eucalyptus, heather, acacia, lime, rape, sunflower, and honeydew), Escuredo et al. (2014) have confirmed that fructose, glucose, and moisture contents and F/G and G/W ratios are the main value indicators to predict the crystallization phenomenon in honey. The size of sugar crystals significantly influences the texture of crystallized honey, with smaller crystals creating a creamy consistency favored by most consumers, while larger crystals result in a coarse mouthfeel. Recently, a simple and efficient method for measuring sugar crystal size was developed (Weber et al. 2022).

3.3 Proteins, Enzymes, and Amino Acids

The number of studies on *proteins* in honey samples is relatively low, mainly because of their presence in small amounts, the difficulty in extracting them from the sugar-rich environment, and the complexity of their characterization by conventional approaches (Chua et al. 2013). Most of the studies determine protein contents by general methods such as Bradford's or Kjeldhal's, although only 40–80% of the total nitrogen is in the protein fraction, and the rest might be attributed to free amino acids (Azeredo et al. 2003; Escuredo et al. 2013).

Honey proteins can originate from nectar and pollen of flowers and, occasionally, also from the sap of plants, but more often they are derived from secretions of cephalic glands of honeybees that are responsible for the enzymatic breakdown of pollen and nectar. The mean protein content of honey is around 0.2–0.7% with molecular weights ranging from 22 to 75 kDa, although this content is quite variable depending upon different factors, such as the honeybee species, the honeydew-producing insects, and the vegetal source of nectar and pollen. Protein concentration ranges from 0.2% to 1.6% in the honey from *Apis mellifera* and from 0.1% to 3.3% in that from *Apis cerana* (Lee et al. 1998). Regarding the vegetal origin, European honeydew and chestnut honeys have been shown to possess higher protein content (around 1 g of proteins per 100 g of honey) than others, like eucalyptus (0,6%), blackberry (0,73%), or polyfloral honeys (0,7%) (Escuredo et al. 2013). Some types of honey, such as manuka and heather honeys, may present unusually high contents of protein, near to 1.5%, which cause an elevated viscosity that makes difficult their extraction by centrifugation (Crane 1990). Quantitative and qualitative changes in protein composition may occur during the storage of honey by the formation of protein–polyphenolic complexes (Brudzynski et al. 2013).

The main protein identified in honey is the so-called major royal jelly protein 1 (MRJP1), which is similar to apalbumin-1. The average amount of MRJP1 measured by ELISA relating to the total protein content was estimated to be 23.4% (Bilikova and Simuth 2010). This protein shows distinct molecular weights and surface structures depending on the honeybee species, due to differences in the sugar chain length and glycosylation pattern, and this feature can be applied for the discrimination of honey according to the honeybee species (Won et al. 2009). Moreover, up to nine major royal jelly proteins have been detected in honey samples from different botanical and geographical origins (Simuth et al. 2004; Won et al. 2009; Bilikova and Simuth 2010; Rossano et al. 2012).

Trace proteins have been used to discriminate the botanical or geographical origin of honey (Cimpoiu et al. 2013). Baroni et al. (2002) applied immunoblot assays for the analysis of proteins from sunflower and eucalyptus honeys, concluding that the pollen proteins could be used as a marker for the authentication of floral origin. A recent study examined the influence of geographical and floral origins on the antioxidant activity of Saudi honey, highlighting the role of short peptides derived from honeybee proteins. Honey from Jazan had the highest short peptide content, which correlated with increased antioxidant activity and suggested potential health benefits such as antimicrobial and antidiabetic properties (Alarjani and Mohammed 2024). Similarly, a study identified 48 proteins in Lithuanian monofloral and polyfloral honeys, with variations in protein content between red clover and rapeseed honeys. Proteins from aphids and lactic acid bacteria were found in all samples, showing strong correlations between certain bacteria, pollen types, and protein quantities. The findings suggest a connection between honey protein composition, microbiota diversity, and plant pollination, emphasizing the need for further research into these relationships (Čeksterytė et al. 2024).

Honey protein that originated from bees mostly consists of *enzymes* secreted from the salivary and hypopharyngeal glands of worker bees (Table 3.3). The presence of various enzymes in honey has been known for years (White 1978). The most prominent one is α-glucosidase (invertase or saccharase), present in the hypopharyngeal gland of the forager bee, reaching about 50% of the total protein of the gland, whereas amylase and glucose oxidase were each estimated to account for 2–3% (Ohashi et al. 1999).

Enzyme activities are related to the intensity of the nectar flow, i.e., the amount of nectar that flows from the flowers and the concentration and composition of the nectar. Honeys from rich nectar sources, such as acacia, often contain low natural enzyme activities (Wehling et al. 2006). Enzyme content can be considerably diminished by honey processing, heating, and prolonged storage so that the measurement of their activities can provide information about the intensity of the heat treatment carried out and the aging degree of the product (White 1978; Sánchez et al. 2001; Serrano et al. 2007).

α-glucosidase is an exo-carbohydrase that hydrolyzes the nonreducing terminal α-glucosidic bond and releases α-glucose from the substrate. In the presence of high concentrations of substrate, the enzyme also catalyzes the transferring reaction to form oligosaccharides (Chiba 1998). Up to 18 isoenzymes have been described for

Table 3.3 Main honey enzymes and their activities. (Belitz et al. 2009)

Enzymes	Activity	
α-glucosidase (invertase)	Converts sucrose to glucose and fructose (inverts sugar)	7.5–10 g saccharose hydrolyzed by 100 g honey per hour at 40 °C
α- and β-amylase (diastase)	Transform starch to other carbohydrates (dextrins and oligo-, di-, and monosaccharides)	16–24 g starch degraded by 100 g honey per hour at 40 °C
Glucose oxidase	Converts glucose to gluconolactone, which, in turn, yields gluconic acid and hydrogen peroxide	80.8–210 μg H_2O_2 formed per g honey/h
Catalase	Converts hydrogen peroxide to water and oxygen	0–86.8 catalytic activity/g honey
Acid phosphatase	Removes phosphate from organic phosphates	5.07–13.4 mg P/100 g honey released in 24 h
Proteases	Hydrolyze proteins and polypeptides to yield peptides of lower molecular weight	
Esterases	Break down ester bonds	

α-glucosidase; most of them (13 out of 18) have a constant ratio of hydrolysis activity upon disaccharides such as sucrose and maltose (White and Kushnir 1967). α-glucosidases from both Japanese honeybee (*Apis cerana japonica*) and Western honeybees (*Apis mellifera*) have been purified and characterized, showing that diverse isoenzymes are different in molecular size, pH- and temperature-stability, substrate specificity, and organ localization (Wongchawalit et al. 2006). α-glucosidase activity is believed to be responsible for most of the chemical changes taking place during the conversion of nectar to honey by providing the required supersaturated sugar solution (Babacan and Rand 2005).

Some of the best-known enzymes in honey are *amylases* or diastase (mixture of amylases). Although their presence in honey has been sometimes associated with pollen or nectar, based on the observation of amylase activity in honey produced by sugar-fed bees, the most accepted theory assumes that honey diastase derives from the salivary secretions of bees (Stadelmeier and Bergner 1986; Persano-Oddo et al. 1990). Despite its bee origin, the activity of diastase may vary depending on the honey type. Some unifloral honey (e.g., black locust and citrus honeys) and honeys from tropical regions have a naturally low diastase activity (Flanjak et al. 2016). As for invertase, diastase activity is usually employed as an indicator of freshness and storage duration, and the minimum values for fresh, unprocessed honey are regulated in international and national standards (Codex Alimentarius 2001; Council Directive 2002).

The estimation of the molecular weights of amylases has been a matter of discrepancies, possibly due to the ionic interaction between the enzyme and plant proteins present in honey that leads to the formation of multiple forms (Ohashi et al. 1999). A molecular weight of 57,000 Da was estimated by Babacan and Rand (2005) for the major α-amylase in honey following its isolation and purification using a combination of ultrafiltration, ultracentrifugation, and ion-exchange chromatography.

Honey adulteration undermines its quality and authenticity, often involving the addition of foreign amylases to mask reduced diastase activity. A study using label-free nano-liquid chromatography mass spectrometry (LC-MS)/MS proteomics analyzed 45 honey samples and identified foreign amylases from *Aspergillus niger*, *Bacillus amyloliquefaciens*, and *Bacillus licheniformis* in four samples, confirming their presence with high specificity. Additionally, plant-related proteins, aphid proteins, and catalase-related proteins were detected, providing new insights into non-honey bee protein contributions. The proteomics approach demonstrated here not only detects adulteration but also advances honey characterization by identifying natural markers and creating curated protein sequence databases for future research (Erban et al. 2021).

Glucose oxidase is another important enzyme incorporated into honey by bees that was already partially purified and characterized in the 1960s (Schepartz and Subers 1964; White and Kushnir 1967). Like invertase and diastase, it plays a role in the formation of honey in the hive through the oxidation of glucose to gluconolactone, which is further transformed to gluconic acid and hydrogen peroxide (H_2O_2), thus contributing to the acidity and the antibiotic activity of honey (White et al. 1963). Honey also contains *catalase* that decomposes hydrogen peroxide to water and oxygen and therefore has an opposite role of glucose oxidase. Unlike this latter, catalase is originated from flower pollen (Dustman 1971); therefore, the level of hydrogen peroxide in a honey will depend on how much pollen is collected by the bees, the floral source of the pollen, and the catalase activity of that pollen (Weston 2000).

The presence of *acid phosphatase* in honey was first described by Giri (1938). This enzyme is a hydrolase that catalyzes the formation of inorganic phosphates from organic phosphates, whose activity is related to the fermentation processes of honey. Acid phosphatase originates mainly from nectar and pollen and can be used as a parameter for honey characterization (Alonso-Torre et al. 2006).

Although the enzymes in honey mainly originate from the bee, some relationships have been established between the enzyme activity observed in different honey types and their botanical origin. Most likely the enzyme content is a result of several factors including floral source, abundance of nectar flow, nectar sugar content, temperature, diet, physiological stage of the bee, and the strength of the colony (Persano-Oddo et al. 1999). Enzyme activities have sometimes been used to indicate honey botanical origin, but this is only possible for fresh honeys since the activity decreases after processing and storing honey (Kenjerić et al. 2006; Primorac et al. 2009). Recently, four of the most abundant honey types produced in Croatia (black locust (*Robinia pseudoaccacia* L.), sage (*Salvia officinalis* L.), chestnut (*Castanea sativa* Mill.), and honeydew honeys) have been characterized according to the protein and proline content and enzyme activities. The results showed higher proline content and diastase, invertase, and glucose oxidase activities in dark honey types (honeydew and chestnut honeys) than in lighter ones. Although black locust honey showed naturally low enzyme activity, it exhibited the highest acid

Table 3.4 Average contents of free amino acids in honey. (Belitz et al. 2009)

Amino acid	mg/100 g honey (dry weight basis)	Amino acid	mg/100 g honey (dry weight basis)
Aspartic acid (Asp)	3.44	Tyrosine (Tyr)	2.58
Asparagine + glutamine (Asn + Gln)	11.64	Phenylalanine (Phe)	14.75
Glutamic acid (Glu)	2.94	β-alanine (β-Ala)	1.06
Proline (Pro)	59.65	γ-aminobutyric acid (γ-Abu)	2.15
Glycine (Gly)	0.68	Lysine (Lys)	0.99
α-alanine (α-Ala)	2.07	Ornithine (Orn)	0.26
Cysteine (Cys)	0.47	Histidine (His)	3.84
Valine (Val)	2.00	Tryptophan (Trp)	3.84
Methionine (Met)	0.33	Arginine (Arg)	1.72
Isoleucine (Ile)	1.12	Others	24.53
Leucine (Leu)	1.03	Total	118.77

phosphatase activity among the analyzed honey types, while honeydew honey, otherwise known to possess high proline content and enzyme activity, had a low total protein content comparable to black locust honey (Flanjak et al. 2016). Nevertheless, honey's proteome, including honey proteins and enzymes, is still very poorly researched, mostly because of the low amount of protein present in honey.

Honey also contains free *amino acids* at a level of around 1% (W/W) of their components (Table 3.4). Three possible sources have been considered that may contribute to the amino acid composition in honey: nectar (Baker and Baker 1986; Wunnachit et al. 1992), pollen (Marshall and Williams 1987; Sing and Sing 1996), and bees themselves, especially for proline (Petrov 1974; Ball 2007; Truzzi et al. 2014).

A total of 26 free amino acids have been described in honey. In addition to those listed in Table 3.4, serine (Ser), threonine (Thr), homoserine (Hser), taurine (Tau), and α-aminoadipic acid have also been reported in some honey types (González-Paramás et al. 2006). Proline is the prevalent one, representing between 50% and 85% of this fraction, followed by phenylalanine (Belitz et al. 2009). Proline content has been used as a criterion for the evaluation of the maturation of honey and, in some cases, of the adulteration with sugar. A minimum value of 180 mg/Kg is accepted as the limit value for genuine honey (Bogdanov et al. 1999). Recently, the presence of D-amino acids in honey has been suggested as a way to distinguish adulterations (del Campo et al. 2016).

During storage, some changes in the free amino acid contents could occur. Iglesias et al. (2006) indicated that the contents of most of the free amino acids decrease with storage time, with the greatest reduction observed in the first 9 months. In contrast, the contents of the amino acids aspartic acid, α-alanine, and proline increased in the first few months of storage, reaching maximum values at 6 months, possibly due to the activity of proteolytic enzymes. A reason for the amino acid decrease could be their involvement in Maillard reactions; actually, the presence of

Amadori's compounds derived from the amino acids lysine, proline, γ-aminobutyric acid, and arginine has been described in honey (Sanz et al. 2003). Manuka honey, rich in antibacterial methylglyoxal (MGO), forms protein-bound Maillard reaction products (MRPs) like Nε-carboxyethyllysine (CEL) and MG-H1 during storage, as shown by high-performance liquid chromatography mass spectrometry (HPLC-MS)/MS analysis. The formation of these MRPs, influenced by the MGO content, progresses over time and offers insights into the honey's biochemical properties, potentially aiding its authentication (Thierig et al. 2023).

Since pollen is one of the most important sources of amino acids in honey, different approaches based on the use of free amino acid profiles have been suggested for the discrimination of honeys according to their botanical and/or geographical origin. Early contributions in this respect were made by Davies (1975, 1976) that, following analysis of the amino acid profile in 98 honeys from different botanical sources coming from six geographical origins, proposed the ratios of aspartic acid/proline and amides/phenylalanine to distinguish samples from different countries. Conte et al. (1998) analyzed the amino acid composition of 92 honey samples from 17 botanical and four different geographical origins; although some trends could be observed for thyme (high levels of Ser, Tyr, and Lys) and rosemary honeys (Tyr was the most abundant), they concluded that the application of linear discriminate analysis including other parameters than amino acids was required for distinguishing honeys. In contrast, the application of statistical methods to the contents of 13 amino acids determined in 56 honey samples from three different regions in Argentina allowed Cometto et al. (2003) to cluster the samples into nine groups related to the collecting regions and more strictly to the flora around apiaries. They concluded that between 5 and 7 amino acids, selected by principal component analysis (PCA), could be sufficient to verify both the botanical or geographic origin of honey. Canonical variate analysis (CVA) was also used to differentiate samples from different countries (Argentina, Australia, UK, and Canada) based on their amino acid profile (Gilbert et al. 1981). Senyuva et al. (2009) suggested that phenylalanine and tyrosine contents might contribute to distinguishing honeys, with valine, leucine, and isoleucine being other important amino acids for recognizing botanical and geographic origins.

Regarding the botanical origin, in a study carried out by Pirini et al. (1992) on different types of honeys (acacia, citrus, chestnut, rhododendron, rosemary, and lime), the presence of some amino acids like arginine, tryptophan, or cystine was found to be characteristic for some honeys, although a single amino acid or a group of amino acids could not be selected as suitable for the recognition of a particular kind of honey. Hermosín et al. (2003) described that lavender honey from Spain contains a high concentration of tyrosine, while Cotte et al. (2004a), analyzing samples from France, pointed to high amounts of threonine and phenylalanine as characteristics of sunflower and lavender honeys. Amino acid profiles, especially glutamic acid and tryptophan, have been suggested as useful for the discrimination of honeydew honey and floral honey (Iglesias et al. 2004). In another study, 40 monovarietal honeydew (ilex and oak) or nectar (heather and chestnut) samples were analyzed for their free amino acid profiles. Proline was always the principal amino acid, followed by γ-aminobutyric acid in chestnut-tree (22.4%) and oak (7.2%), asparagine in ilex, and phenylalanine in heather honeys. The discrimination between ilex and oak honeys is of particular interest since neither the melissopalinological analysis—absence of pollen—nor the physicochemical one enables their differentiation (González-Paramás et al. 2006). More recently, lysine, arginine, and histidine were reported as the majority amino acids in honeydew honey, in contrast with rhododendron honeys where the main detected amino acid was aspartic acid (Silici and Karaman 2014). The free amino acid profiles of 192 samples of seven different floral types of Serbian honey (acacia, linden, sunflower, rape, basil, giant goldenrod, and buckwheat) from six different regions were analyzed in order to distinguish honeys by their botanical origin. Principal component analysis revealed that basil honey samples form a well-defined cluster imposed with phenylalanine content. Furthermore, the model obtained by linear discriminant analysis could be used to distinguish basil honey from the rest of the samples and also showed moderate predictive power to separate genuine acacia, linden, sunflower, and rape honeys (Kešckeš et al. 2013). In honeys from Poland, high concentrations of aspartic acid and asparagine were suggested as indicators of raspberry and buckwheat honeys (Janiszewska et al. 2012). The application of combined nuclear magnetic resonance (NMR) and chemometric data analysis approach allowed the characterization of wildflower honeys based on their higher concentrations of phenylalanine and tyrosine (Boffo et al. 2012).

In summary, amino acid profiles might be, in some occasions, useful for the determination of the botanical and/or geographical origin and more suitable than protein composition. However, the methods must be employed in conjunction with other techniques, namely statistical analysis, in order to obtain a reliable characterization (Anklan 1998; Kaškonienė and Venskutonis 2010).

3.4 Organic Acids

Organic acids are present in honey at levels around 0.5–0.6% and contribute to its organoleptic and physicochemical properties (acidity, pH, and electrical conductivity) (Mato et al. 2006). These acids are obtained directly from nectar or derived from sugars by the activity of enzymes secreted by honeybees when transforming nectar into honey. Many authors have reported the increase of acidity over time as well as during fermentation as a result of the transformation of honey sugars and alcohols into acids by the action of honey yeasts (Cavia et al. 2007).

Up to 32 organic acids have been identified in honey including acetic, butyric, citric, formic, fumaric, glyoxylic, propionic, lactic, maleic, malic, oxalic, and succinic acids (Wilkins et al. 1995). Although formic and citric acids were originally described as predominant, it is now clear that gluconic acid (produced by the action of glucose oxidase on glucose) is the majority acid in honey, with some exceptions like fir honey, where galacturonic acid has been indicated as the main acid (Daniele et al. 2012).

The interest of organic acids in honey is diverse and includes their antibacterial and antioxidant activities, their utility as possible indicators of incipient fermentation, and the potential use for treatment of Varroa infestation. Organic acids have been deployed to discriminate honeys according to their botanical and/or geographical origins. Daniele et al. (2012) proposed a qualitative and quantitative method, based on ion chromatography with electrochemical detection followed by isotope ratio mass spectrometry (IRMS) determination of $^{13}C/^{12}C$, to analyze organic acids in honeys. The principal component analysis of the results obtained regarding the content of 14 organic acids enabled good discrimination among monofloral honeys from seven botanical origins (acacia, chestnut, rapeseed, lavender, fir, linden, and sunflower). Even though galacturonic acid has been indicated to be the main acid in fir honeys, through PCA analysis, these honeys could be easily distinguished from the other varieties by their higher contents of pyruvic and citric acids, whereas pyroglutamic acid allowed distinguishing chestnut honey (Daniele et al. 2012). The authors also suggested that the organic acids could be used as internal standards instead of proteins in order to detect honey adulteration, especially when ultrafiltration is employed in honey processing.

^{1}H-NMR analysis of Polish multifloral and monofloral honeys (heather, buckwheat, lime, rape, and acacia) allowed their differentiation according to the botanical origin with the only exception of acacia's honey. Based on the results, the authors proposed phenylacetic acid and dehydrovomifoliol as markers for heather honey, formic acid, and tyrosine for buckwheat honey and 4-(hidroxy-1-methylethyl) cyclohexane-1,3-dienecarboxylic acid for lime honey (Zieliński et al. 2014).

3.5 Mineral and Trace Elements

Minerals originate in the soil, are transported into plants through the roots, and get into honey via the nectar or the honeydew (Anklan 1998), although they may also come from anthropogenic sources, such as environmental pollution, or be influenced by beekeeping practices and honey processing (Bogdanov et al. 2007; Pohl 2009; Silici et al. 2016). Table 3.5 collects the main mineral elements described in honey; most of them are metals, some of which are present only in minute amounts (i.e., trace elements).

The total weight of the mineral elements in honey (total ash) varies within a wide range of values (0.02–1%), the largest of any honey component (Silva et al. 2009), and has been systematically investigated in honey with the aim to certify their botanical or geographical origin. Up to 54 minerals have been reported in honey, with the research presented by Batista et al. (2012) possibly being the most complete one, describing the presence of 42 chemical elements in Brazilian honeys. Almost all macrominerals were commonly found in honeys from all countries, with the exception of Cl, which was only detected in honey samples from Spain (Solayman et al. 2016). Among them, K was the predominant one representing one-half to three-quarters of the total mineral content (Abu-Tarboush et al. 1993; Terrab

Table 3.5 Principal mineral elements described in honey

Majority minerals Average contents in light to dark honey (ppm)		Trace elements
Potassium (K)	40–1350	Copper (Cu)
Chlorine (Cl)	52–427	Chromium (Cr)
Sulfur (S)	15–100	Lithium (Li)
Sodium (Na)	3–237	Nickel (Ni)
Calcium (Ca)	5–218	Lead (Pb)
Phosphorus (P)	29–119	Tin (Sn)
Magnesium (Mg)	2–564	Osmium (Os)
Silicon (as SiO_2)	9–41	Beryllium (Be)
Iron (Fe)	0.4–224	Vanadium (V)
Zinc (Zn)	0.2–74	Zirconium (Zr)
Manganese (Mn)	0.3–4	Silver (Ag)
		Barium (Ba)
		Gallium (Ga)
		Bismuth (Bi)
		Gold (Au)
		Germanium (Ge)
		Strontium (Sr)

Based on Crane (1990) and Solayman et al. (2016)

et al. 2004; Kadri et al. 2017), followed by Ca, Na, or Mg depending on the honey types (González-Paramás et al. 2000; Atanassova et al. 2012; Mondragón-Cortez et al. 2013). Other minerals such as Fe, Si, and S were present in intermediate amounts (around 0.05 mg/g). Trace and ultratrace elements were present at levels below 1 μg/g, and apparently, when some of them (e.g., Cd, Cu, Co, Ni, or Pb) are not detected in honey, it is an indication that no relevant contamination has happened during processing, shipping, and storage due to its contact with steel or galvanized containers (Baroni et al. 2009). Updated tables with bibliographic data of mineral composition of honeys from different countries have been compiled by Solayman et al. (2016).

Due to differences in soils, floral source, and nectar and pollen composition, some minerals cannot be detected in some types of honey collected from certain areas, whereas they are common in others. Therefore, the mineral composition of honey can actually be used to identify its geographical origin. One of the first works in this sense was done by Morse and Lisk (1980), who studied the elemental composition of honeys from different countries (USA, Mexico, El Salvador, and China), finding significant differences among them. Feller-Demalsy et al. (1989) undertook an ambitious study to differentiate honeys from the ten Canadian provinces by the application of principal component analysis to the data of 20 mineral elements, achieving 92% correct classification; they found that these differences were independent of the floral origin of the honeys and seemed to be very complex, probably since they were related to maritime influence coupled with the extent of rainfall rather than being dependent upon soil composition. The provenance of honey samples from northern and southern Argentina was also successfully evaluated by Baroni et al. (2009) by the application of stepwise discriminant analysis (SDA) and κ-nearest-neighbor (κ-NN) to their mineral profiles and chemical properties. Similarly, multielement analysis was applied to the classification of Brazilian honey samples according to the region based on their trace element composition, using chemometric tools (Batista et al. 2012). The mineral composition has also been used to typify honeys from different regions of Spain. The contents of Na, K, Ca, Mg Cu, Fe, Mn, P, Cl, Si, and S were determined in samples of raw honeys from Galicia, concluding that they showed, in general, higher mineral contents than those reported in the literature for other Spanish regions (Rodriguez-Otero et al. 1994). In another study, 60 honey samples from six different production zones in western Spain (provinces of Salamanca, Zamora, and Caceres) were analyzed for 13 common legal physicochemical parameters and 17 mineral elements (13 cations and four anions); the application of linear stepwise discriminant analysis allowed discrimination of honeys from all six zones with a result of 91.38% of correctly classified samples (González-Paramás et al. 2000). Similar approaches were used to characterize honeys from the Canary Islands (Hernández et al. 2005) and Andalusia under the protected designation of origin "Miel de Granada" (de Alda-Garcilope et al. 2012). The levels of 14 minerals were investigated in 23 varieties of honey from Saudi Arabia and six other countries, although, in this case, no correlation between geographical origin and composition was tried (Alqarni et al. 2014).

Although the contents of macro- and microelements would reflect rather the geographical provenance and the environmental contamination in the respective locations than the botanical origin (Atanassova et al. 2012), some attempts have also been made to use minerals to classify honeys according to the plant origin. Thus, Nozal-Nalda et al. (2005a, b) achieved a global success rate close to 90% in the classification of ling (*Calluna vulgaris*), heather (*Erica* sp.), rosemary (*Rosmarinus officinalis*), thyme (*Thymus vulgaris*), lavender (*Lavandula* sp.), and oak (*Quercus* sp.) honeydew honeys by applying discriminant analysis to their metal content data (15 minerals) and other common physicochemical parameters. Similarly, the concentration of Zn, Mn, Mg, and Na were reported to be strongly dependent on the botanical origin in eucalyptus, heather, orange blossom, and rosemary honeys from Spain (Fernandez-Torres et al. 2005). Based on their trace element profile, honeydew, buckwheat, and rape honeys from Poland were correctly classified as honeydew or nectar samples (Chudzinska and Baralkiewicz 2011). Twelve mineral elements were determined by inductively coupled plasma mass spectrometry (ICP-MS) in 163 Chinese honey samples from different plant sources (linden, vitex, rape, and acacia), allowing their classification according to the botanical origin by the application of principal component analysis, which reduced the variables to four principal components and could explain 93.06% of the total variance (Chen et al. 2014). Recently, a study combined elemental fingerprinting with machine-learning techniques to differentiate the geographical origins of unifloral and multifloral honeys from neighboring regions, Sardinia (Italy) and Spain. Using ICP-MS, the elemental compositions of 247 honey samples were analyzed. Among the tested models, random forest (RF) outperformed linear discriminant analysis (LDA) in stability and classification accuracy. The most effective classification based on geographical origin achieved 90% accuracy, utilizing key elements such as Na, Mg, Mn, Sr, Zn, Ce, Nd, Eu, and Tb as predictors (Mara et al. 2024). Other authors, however, could not find statistically significant differences among honey samples of different botanical origins (tualang, gelam, acacia, and forest honeys) based on two-factor ANOVA and cluster analysis, even though the minerals and trace element composition could vary depending on the type of honey (Chua et al. 2012).

Mineral content is frequently correlated with the color of honey. Dark and amber honeys contain higher amounts of certain major, minor, and trace elements when compared to pale-colored honeys (González-Miret et al. 2005; Pisani et al. 2008). Positive correlation between the mineral content of honey and its electrical conductivity has also been demonstrated (Bogdanov et al. 2007; Silva et al. 2009; Kropf et al. 2008; Vanhanen et al. 2011), as well as their contribution to the intracellular redox balance (Alvarez-Suarez et al. 2012). Interestingly, Perna et al. (2012) found a statistically significant correlation ($P < 0.05$) between metal content and both phenolic content and antioxidant activity.

In summary, the analysis of trace element profiles in combination with modern statistical data evaluation techniques could be suitable for the characterization of geographical and/or botanical origin of honey (Anklam 1998). However, some risks should be carefully considered when using microelements for the authentication of honey. Thus, whereas most minerals are of floral and soil origin, some metals, particularly Pb and Cd, may be present in honey as a result of the environmental contamination, and therefore, their concentration may be more dependent on contamination than on the floral or geographical origin of the honey (Kaškonienė and Venskutonis 2010).

3.6 Vitamins

Honey contains small but detectable quantities of vitamins, mostly water-soluble, owing to the aqueous nature of the matrix, although it should not be considered a good source of vitamins at all, because their concentrations are in the parts per million (ppm) range. Most of the vitamins are from the pollen grains present in suspension, so commercial filtration of honeys dramatically reduces their contents due to the almost complete removal of pollen (Ciulu et al. 2011).

Ascorbic acid is the predominant vitamin, and it is usually found in almost all types of honey; mean contents have been indicated to be around 2 mg/100 g (Álvarez-Suarez et al. 2010a). Vitamin C analysis is one of the indicators often used to evaluate the nutritional quality of a food since it is very vulnerable to chemical and enzymatic oxidation, further accelerated by factors such as light, oxygen, or heat (León-Ruiz et al. 2013). Different B complex vitamins are also present in honey, including thiamine (B1), riboflavin (B2), nicotinic acid (B3), pantothenic acid (B5), biotin (B8 or H), and folic acid (B9).

Viñas et al. (2004) determined vitamins B6 and B2 in different types of honeys. Vitamin B6, such as pyridoxine, was found in one eucalyptus honey and in a multifloral honey at levels lower than 0.021 mg/100 g, whereas it was not detected in rosemary or orange blossom honeys. Riboflavin vitamers were detected in chestnut-tree, eucalyptus, cane, and multifloral honeys at levels up to 0.074 mg/100 g. Ciulu et al. (2011) optimized a gradient reversed-phase mode (RP)-HPLC method to analyze vitamins C, B2, B3, B5, and B9 in 28 honey samples (mainly from Sardinia, Italy). The presence of vitamin B3 was detected in all the studied samples, although in some of them, the concentration was under the limit of quantification. Vitamins B2 and B9 were quantified in about half of the samples, whereas vitamin B5 was much less common than the others. Neither vitamin B1 nor B6, which often appear in honey, were determined by the proposed method. Interestingly, although the concentration of water-soluble vitamins in honey was very low, there might be a potential correlation with the origin of honeys, in particular, vitamin B3 that seemed to be strongly dependent on the botanical origin. An isocratic RP-HPLC method was developed by León-Ruiz et al. (2013) for the quantification of seven water-soluble vitamins (B1, B2, B3H, B3N, B5, B6, and C) that was applied to their analysis in different monofloral honeys (rosemary, thyme, lavender, chestnut, and *Echium*) and a honeydew honey. The results confirmed the presence of vitamin C in all samples, highlighting the thyme honeys with a content ten times higher than the other ones. Vitamin B1 was also found in all honey samples with the exception of rosemary. In samples belonging to the Protected Denomination of Origin "La Alcarria Honey" (Spain), only vitamin C was detected in rosemary's, and B1, B3H, and C in lavender's honeys (León-Ruiz et al. 2011).

3.7 Aroma Compounds

Volatile and semi-volatile compounds are present in honeys and are responsible for aroma and flavor qualities. The honey flavor is produced by complex mixtures of volatile compounds, which may differ depending on the nectar or honeydew; conversion by honeybees of plant constituents into other compounds with volatile properties; processing conditions, especially heating; and storage (da Silva et al. 2016). The changes in the volatile fraction occurring in honey during storage have been attributed to two principal causes: compounds that are labile and may be destroyed and volatile compounds produced by nonenzymatic browning (Maillard reaction) (Wootton et al. 1978; Castro-Vázquez et al. 2008; Moreira et al. 2010).

More than 600 volatile compounds, which originated from various biosynthetic pathways, have been described in honey. Their chemical families include: hydrocarbons, aldehydes, superior alcohols, ketones, fatty and other carboxylic acids, esters, benzene and its derivatives (furan and pyran), norisoprenoids, sesquiterpenes and their derivatives, and sulfur and cyclic compounds (Manyi-Loh et al. 2011a, b). The length of the carboxylic acid carbon chains provides different flavors that can range from spicy to rancid. The short-chain carboxylic acids, such as acetic acid, have a spicy aroma and flavor, while butanoic and hexanoic acids provide a rancid aroma (Barra et al. 2010). Benzene derivatives are found at high concentrations in all honeys; these include benzaldehyde, benzyl alcohol, 2-phenylethanol, and phenylacetaldehyde (Radovic et al. 2001). Compounds such as acetic acid (de la Fuente et al. 2005) and acetone (Pérez et al. 2002) are also found at important levels in multiple types of honeys. There are, however, volatile compounds that could be used to discriminate monofloral honeys from different floral origins, and the diversity of volatile compounds that may occur in honey also gives rise to distinct profiles that might represent fingerprints for particular types of honey (Manyi-Loh et al. 2011a, b). For instance, the presence of important levels of benzaldehyde and furfural has been reported in clover, blueberry, wildflower (Overton and Manura 1994), and thistle honeys (Bianchi et al. 2011). Methyl anthranilate, sinensal isomers, lilac alcohol, lilac aldehyde, limonene diol, and 1-*p*-menthen-9-al have been proposed as characteristics of citrus honey (Piasenzotto et al. 2003; Alissandrakis et al. 2005; Castro-Vázquez et al. 2009). In chestnut honey, Bonaga et al. (1986) identified 3-aminoacetophenone as the main component of the volatile fraction and suggested it as a specific marker of this type of honey. Radovic et al. (2001) proposed 4-methyl acetophenone as a candidate marker for lime tree honeys, which was later confirmed by Piasenzotto et al. (2003) and Wardencki et al. (2009). 3,9-epoxy-1-*p*-mentadieno, t-8-*p*-menthan-oxide-1,2-diol, and *cis*-rose have been indicated as markers for lemon honey (Castro-Vázquez et al. 2007). Diketones, sulfur compounds, nonanal, and nonanoic acid are typical of eucalyptus honey. Hexanal and heptanal are the main compounds in the aroma of lavender honeys (Radovic et al. 2001), although nerolidol oxide was proposed to typify lavender honey from Spain (Castro-Vázquez et al. 2009). Vomifoliol and coumarin were indicated to be characteristics of *Prunus mahaleb* L. honey, while methyl-4-hydroxy-3,5-dimethoxybenzoate could be a potential marker for asphodel honey (Jerković et al. 2011a; Jerković et al. 2011b). More than 300 compounds were identified in honeydew honey from Slovakia, four of which (2,3-butanediol, 3-hydroxy-2-butanone, acetic acid, and methyl ester of 2-hydroxibenzoic acid) were established as markers of that type of honey (Jànoškovà et al. 2014). Previously, methyl butanal and lilac aldehyde had been proposed as marker compounds in Polish honeydew honey (Wardencki et al. 2009), and n-decane, nonanal, α-α-dimethylphenyl acetate, nonanol, and 2-methyl heptanoic acid in Turkish honeydew honey (Senyuva et al. 2009). High acetic acid concentrations have also been reported to be characteristics of honeydew honeys, although its formation by microbial metabolism cannot be ruled out, so it would not be a good marker compound (Campos et al. 2000). Other examples of volatile compounds proposed as characteristics of some unifloral honeys are given in Table 3.6.

Despite all those observations, there is disagreement about which compounds could actually serve as floral markers for a given honey type, owing to the differences among plant varieties, geographical origins, and beekeeping practices. According to the published data, only a few types of honeys might contain a characteristic compound, while the majority of honey samples would be characterized

Table 3.6 Specific volatile compounds that are proposed to typify unifloral honeys

Honey source	Geographical origin	Compound	Reference
Acacia	Poland	Hexanal	Wardencki et al. (2009)
	European countries	*Cis*-linalool oxide and heptanal	Radovic et al. (2001)
Eucalyptus	Spain	3-caren-2-ol, *p*-cymene, and its derivative alcohol	Castro-Vázquez et al. (2009)
		Acetoin, diacetyl, 2,3-pentanedione, and dimethyldisulfide	de la Fuente et al. (2005)
	Turkey	Nonanal, ethylphenyl acetate, and phenethyl alcohol	Senyuva et al. (2009)
	European countries	1-octene or 2,3-pentanedione	Radovic et al. (2001)
	Italy	Nonanoic acid and acetoin	Piasenzotto et al. (2003)
Thyme	Italy	Ethenyl phenyl acetate	Piasenzotto et al. (2003)
	Turkey	3,4,5-trimethoxybenzaldehyde	Mannas and Altug (2007)
	Greece	1-phenyl-2,3-butanedione, 3-hydroxy-4-phenyl-2-butanone, 3-hydroxy-1-phenyl-2-butanone, phenylacetonitrile, and carvacrol	Alissandrakis et al. (2007)
Pine	Turkey	3-carene and another unidentified compound	Tananaki et al. (2007)
		Nonanal, benzene, 4-hexen-3-ol, alpha-pinene, and 2-heptanone	Silici (2011)
Sunflower	European countries	α-pinene or 3-methyl-2-butanol	Radovic et al. (2001)
	Slovakia	*Trans*-furanoid linalool	Pazitna et al. (2012)
Buckwheat	Poland	Pentanal, furfural, and 2-ethylhexanol	Wardencki et al. (2009)
	USA and England	3-methylbutanal, 3-hydroxy-4,5-dimethyl-2(5H)-furanone (sotolon), and β-damascenone	Zhou et al. (2002)
Fir honeydew	Hungary	Acetonitrile, methyl-2-buten-1-ol, n-hexanol, 3-hexanol, 1-propyne, 2-furanmethanol, 5-methyl-2(5H)-furanone, 4-methylphenol, hexadecanoic acid, and methylheptanoate	Lušić et al. (2007)
Oak honeydew	Spain	*Trans*-oak lactone, aminoacetophenone, and propylanisol	Castro-Vázquez et al. (2006)

by several compounds (Kaškonienė and Venskutonis 2010). Clearly, the focus to choose markers for the discrimination of honey's floral origins must be put on plant-derived compounds and their metabolites, such as benzene, terpenes, and their derivatives and norisoprenoids (Manyi-Loh et al. 2011a, b). Possibly, further developments in the analytical techniques could enable us to identify and quantify the very minority volatile constituents in honey that might be specific and provide more accurate information on the floral source. The reliability of the results may also be improved by increasing the number of samples and comparing with honeys from other floral origins. Nevertheless, it should also be taken into account that volatile compounds present in very low concentrations may be the result of the contribution of co-occurring minor floral nectars and could therefore be misinterpreted (Kaškonienė and Venskutonis 2010). The volatile profiles of 51 Portuguese monofloral honey samples from 12 types were analyzed, revealing significant variation by botanical origin. Hydrodistillation (HD) extracted more volatiles than solid-phase microextraction (SPME). Major compounds included alkanes (e.g., n-nonadecane and n-heneicosane), fatty acids (e.g., palmitic and linoleic), oxygenated monoterpenes (e.g., hotrienol), and aromatic amino acid derivatives (e.g., benzene acetaldehyde). Using 12 key volatiles, classification tree analysis discriminated 92% of the honey types, demonstrating the effectiveness of volatile profiling for botanical identification (Machado et al. 2021).

A systematic review identified 187 distinct volatile compounds across 33 honey varieties, including terpenoids, aldehydes, alcohols, esters, acids, ketones, and phenols. These compounds were linked to 63 odor attributes, with dominant notes such as sweet, floral, honey-like, fruity, green, and herbal. Key aroma-defining compounds like phenylacetic acid, γ-butyrolactone, ethyl phenylacetate, and furaneol contribute significantly to honey's characteristic scent. However, the complexity of honey's aroma profile, influenced by its botanical and geographical origins, highlights the need for continued research to fully elucidate the specific aroma-active compounds responsible for these variations (Mahmoud et al. 2024).

Volatile compounds have also been proposed for the geographical discrimination of honey samples. Radovic et al. (2001) analyzed honey samples from England, Denmark, Spain, and Portugal and suggested the existence of some typical patterns, such as the presence of 1-penten-3-ol in English honeys or the absence of 3-methylbutanal in Danish ones, which could be used as a geographical marker. Nevertheless, the limited number of analyzed samples makes those results to be considered just tentative. The influence of the geographical origin in Spanish chestnut honeys was investigated by Castro-Vázquez et al. (2010) through the comparison of their sensory characteristic and the volatile composition; the application of principal component analysis to the whole data allowed successful differentiation among samples from north-east, north-west, and south-east areas of Spain.

3.8 Phenolic Compounds

Phenolic compounds are one of the largest groups of plant secondary metabolites, biosynthesized mainly for protection against biotic and abiotic stress and oxidative damage and transferred via the nectar to honey. The phenolic compounds of honey can be classified into two main families: phenolic acids and flavonoids.

Honey may contain various aromatic and aryl-aliphatic carboxylic acids, mainly hydroxyl and methoxy-derivatives of benzoic and cinnamic acid (Fig. 3.1) that contribute to their sensory properties (Anklam 1998). Among others, the presence of benzoic, hydroxybenzoic, salicylic, gallic, vanillic, syringic, protocatechuic, caffeic, coumaric, ferulic, chlorogenic, phenylacetic, and rosmarinic acids has been described in honey (Dimitrova et al. 2007; Mato Trautvetter et al. 2009).

The flavonoid family comprises several thousand compounds sharing a common C6-C3-C6 phenylchromane skeleton (Fig. 3.2). Based on the oxidation level of the ring C, different flavonoid classes are distinguished: flavones, flavonols, flavanones, flavan-3-ols, anthocyanins, dihydroflavonols, and isoflavones, although in honey, mainly flavanones, flavones, flavonols, and dihydroflavonols have been identified (Fig. 3.3), such as pinocembrin, chrysin, galangin, pinobanksin, apigenin, genistein, quercetin, kaempferol, and myricetin (Tomás-Barberán et al. 2001). Although for many years, it was assumed that only flavonoid aglycones were present in honey, today it is well established that *C*- and *O*-glycosyl derivatives are also present (Truchado et al. 2011).

The total phenolic content in honey, calculated using unspecific methods like Folin–Ciocalteu, has been estimated to range from 20 to 193 mg gallic acid equivalents (GAE)/100 g of honey and the flavonoid content from 1.1 to 7.5 mg quercetin equivalents (QE)/100 g (Pontis et al. 2014; Ciappini and Stoppani 2014; Alvarez-Suarez et al. 2010a).

Phenolic compounds, as well as other organic compounds, are degraded in honey, depending on the environmental conditions to which they are subjected. For instance, some of the flavonoid glycosides are not stable under slight alkaline conditions and are sensitive to oxidation in the presence of oxidizing agents like hydrogen peroxide present in honey (Truchado et al. 2008). However, in general terms, the impact of industrial processing, including thermal treatments and storage, on the polyphenolic profile is less important than the logical variability due to the year of harvesting (as a consequence of environmental factors) and botanical origin (Escriche et al. 2014).

The qualitative and quantitative dissimilarities in the phenolic profile of honeys are direct consequences of the natural variability of these compounds in the plant from which they originate. The botanical origin may affect the profile of flavonoids and phenolic compounds sufficiently to permit discrimination based on the predominance of some individual components or a group of compounds in honey. One of the most significant examples of specificity of a single phenolic compound as chemical marker of the floral origin is the case of homogentisic acid for the strawberry tree (*Arbutus unedo* L.) honey, proposed by Cabras et al. (1999), and that has

Phenolic acids

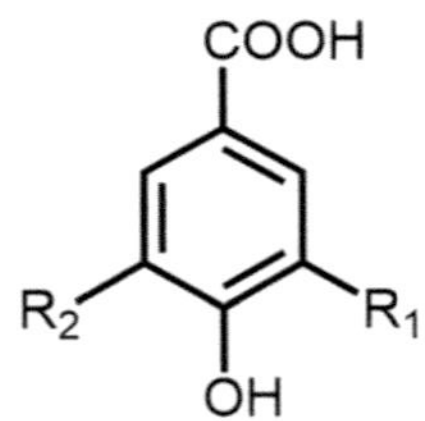

Benzoic acids

Acid	R_1	R_2
4-hydroxybenzoic	H	H
Protocatechuic	H	OH
Gallic	OH	OH
Vanillic	H	OCH_3
Syringic	OCH_3	OCH_3

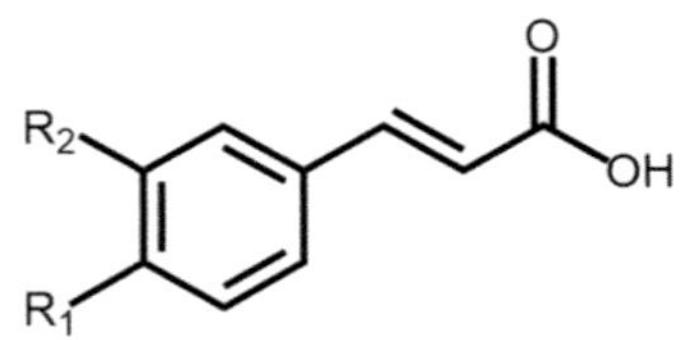

Cinnamic acids

Acid	R_1	R_2
Cinnamic	H	H
p-Coumaric	OH	H
Caffeic	OH	OH
Ferulic	OH	OCH_3

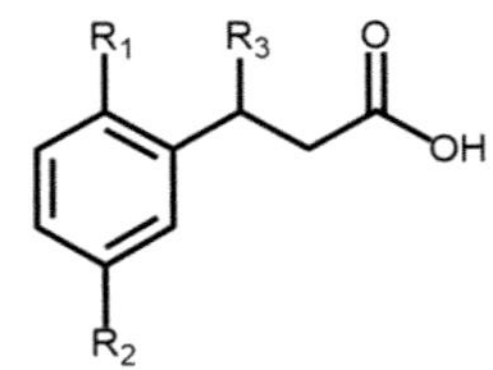

Other acids

Acid	R_1	R_2	R_3
Phenylacetic	H	H	H
Mandelic	H	H	OH
Homogentisic	OH	OH	H

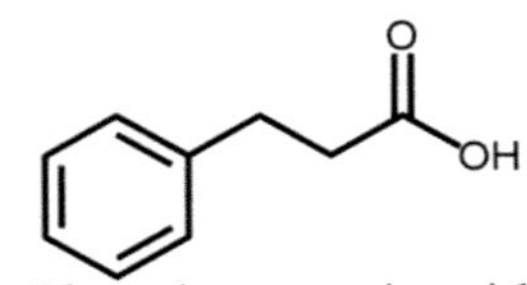

Phenylpropanoic acid

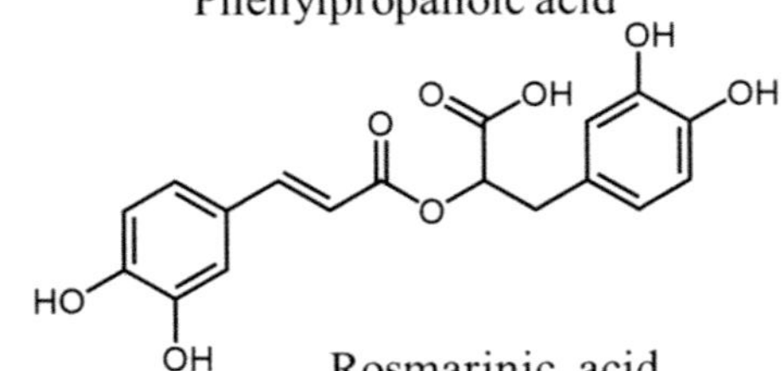

Rosmarinic acid

Fig. 3.1 Structures of the principal phenolic acids described in honey

Fig. 3.2 Phenylchromane skeleton

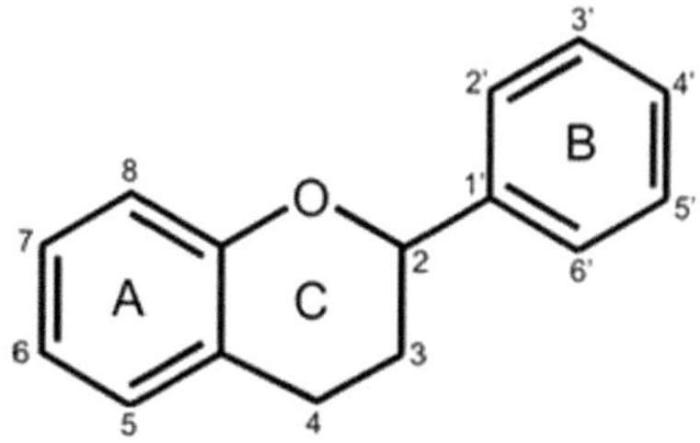

never been found in any other unifloral honey. More recently, the same research group has also proposed the presence of *cis, trans*- and *trans, trans*-abscisic acids, together with the volatile compound unedone, as a complementary marker for strawberry tree honey (Tuberoso et al. 2010). The occurrence of those isomers of abscisic acid was described in acacia honeys (*Robinia pseudoacacia*) (Tomás-Barberán et al. 2001), heather honey (*Erica* spp.) from Portugal (Ferreres et al. 1996), and sage honeys (*Salvia officinalis* L) (Kenjeric et al. 2008) but not in such high amounts and with a constant ratio of about 1:1 as in arbutus honey.

Although many other compounds have been proposed as chemical markers, most of them have been later found also in honeys of other botanical origins, thus losing any aspect of strict specificity toward a single unifloral honey (Ciulu et al. 2016). This is, for example, the case of hesperetin, described as a typical marker for citrus honey by Ferreres et al. (1993) and was later detected in other families of honey such as *Asteraceae*, *Fabaceae*, *Lamiaceae,* and *Rosaceae* (Kurtagić et al. 2015).

Nowadays, almost all the studies are focused on the quantification of one or more compounds, rather than the presence/absence of a specific chemical marker, and the subsequent analysis of the results using chemometrics tools. For example, the phenolic profiles of four Hungarian unifloral honeys (acacia, milkweed, linden, and goldenrod) have been analyzed to identify marker compounds and distinguish honey types. Twenty-five phenolic substances were examined, with pinobanksin, chrysin, and galangin identified as the most abundant. Specific markers included quercetin and p-syringaldehyde for acacia honey, taxifolin for milkweed honey, and syringic acid for goldenrod honey. Principal component analysis demonstrated the utility of phenolic profiles for floral origin determination, though results suggested geographical factors also significantly influence honey composition (Farkas et al. 2023). A minimum level of 122.6 mg/kg for methyl syringate (MSYR) can be considered as a chemical marker in asphodel honeys (*Asphodelus microcarpus* Salzm. et Viv.) (Tuberoso et al. 2009); although this compound is present in several honeys like robinia, rape, chestnut, clover, linden blossom, dandelion, sunflower, thyme, and fir, in all of them, the described levels were < 5 mg/kg (Tuberoso et al. 2009). In manuka honeys (*Leptospermum scoparium*), methyl syringate is also the major compound, representing approximately 44% of the total phenolic content; however, its content is around 20 mg/kg (Alvarez-Suarez et al. 2016), far enough from the level proposed as marker in asphodel honeys.

Kaempferol and 8-methoxykaempferol were proposed as markers for rosemary honeys (Tomás-Barberán et al. 2001), although in a recent paper, chrysin,

Flavonoids

Flavanones

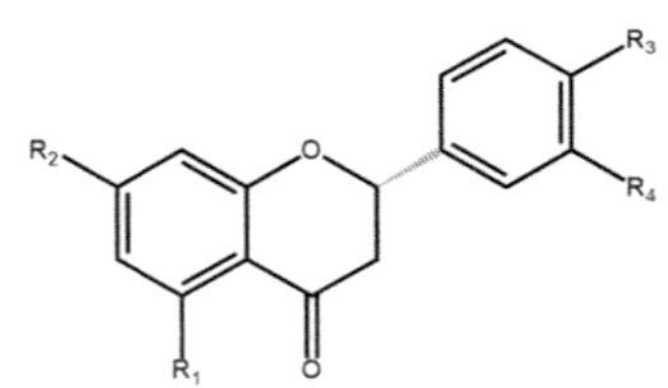

	R_1	R_2	R_3	R_4
Hesperetin	H	H	OCH_3	OH
Pinocembrin	OH	OH	H	H
Naringenin	OH	OH	OH	H

Flavonols

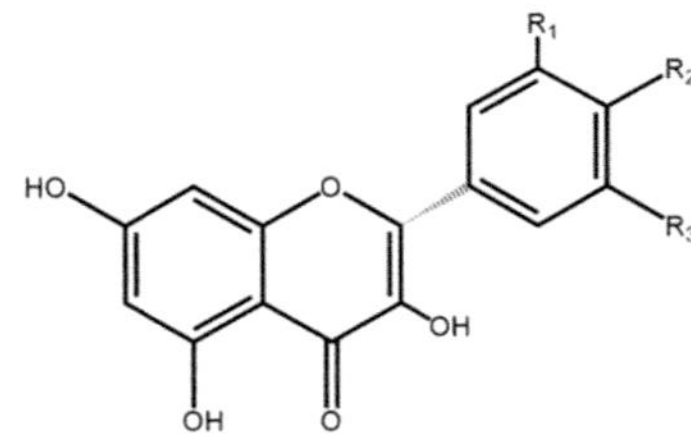

	R_1	R_2	R_3
Galangin	H	H	H
Kaempferol	H	OH	H
Quercetin	OH	OH	H
Isorhamnetin	OCH_3	OH	H
Myricetin	OH	OH	OH

Flavones

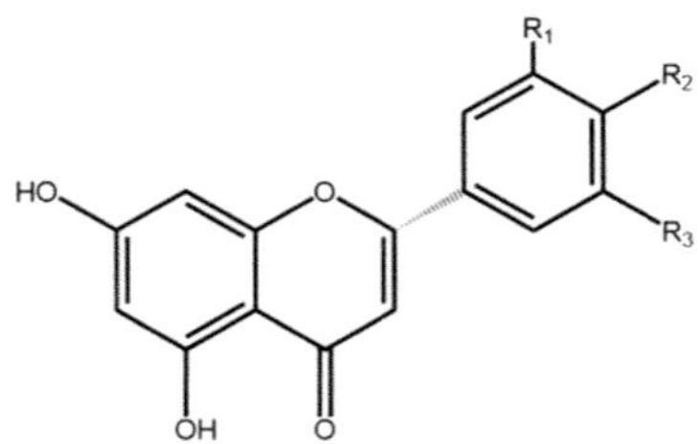

	R_1	R_2	R_3
Chrysin	H	H	H
Apigenin	H	OH	H
Luteolin	OH	OH	H
Tricetin	OH	OH	OH

Dihydroflavonols

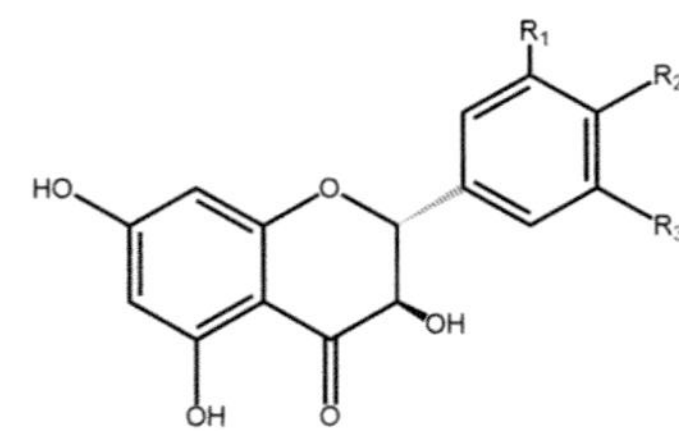

	R_1	R_2	R_3
Pinobanksin	H	H	H

Fig. 3.3 Structures of the principal flavonoids described in honey

pinocembrin, caffeic acid, and naringenin have been indicated as complementary markers (Escriche et al. 2014). Quercetin was suggested as a floral marker for sunflower (*Helianthus annuus*) honey by Tomás-Barberán et al. (2001) and later supported by other authors (Hogenboom et al. 2009). More recently, eriodictyol, together with quercetin, has been suggested as a floral marker for Serbian sunflower (Kečkeš et al. 2013). Relatively larger amounts of gallic acid (maximum value 1.45 mg/kg) have also been reported in sunflower honeys compared to acacia, linden (*Tilia cordata*), basil (*Ocimum basilicum*), buckwheat (*Fagopyrum esculentum*), oilseed rape (*Brassica napus*), and goldenrod (*Solidago virgaurea*) honeys (Kečkeš et al. 2013). The same authors described the presence of a relevant amount of apigenin (0.97 mg/kg) as a characteristic of buckwheat honey. The significant occurrence of *p*-coumaric acid and the flavonoids myricetin, quercetin, and galangin in honeydew honey has been pointed out by Escriche et al. (2014), whereas the abundance of hydroxycinnamates (caffeic, *p*-coumaric, and ferulic acids) was indicated as typical for chestnut honey (Cherchi et al. 1994).

In acacia honeys, there is no consensus among the scientific community, and different markers have been suggested to typify acacia honeys from different geographic locations, such as *cis, trans*- and *trans, trans*-abscisic acids together with ellagic acid (Tomás-Barberán et al. 2001); cinnamic acid derivatives (Dimitrova et al. 2007); or kaempferol rhamnosides (Truchado et al. 2008). A similar situation occurs with heather honeys, for which different markers have been proposed, e.g., tricetin and different myricetin derivatives (Ferreres et al. 1994b); ellagic, *p*-hydroxybenzoic, syringic, and *o*-coumaric acids (Andrade et al. 1997); or high amounts of phenyllactic (800 mg/kg) and phenylacetic (170 mg/kg) acids (Dimitrova et al. 2007).

There are also examples in the literature regarding the use of the phenolic profile for discrimination of honeys that are indistinguishable by means of a melissopalynological analysis. Stephens et al. (2010) observed that 2-methoxybenzoic and trimethoxybenzoic acids were characteristic of manuka honey, while 4-methoxyphenyllactic acid was distinctive of the phenolic composition of the close relative kanuka honey (*Kunzea ericoides*). The analysis of flavonoids (naringenin, hesperetin, chrysin, galangin, kaempferol, luteolin, pinocembrin, and quercetin) and phenolic acids (caffeic acid and *p*-coumaric acid) and the subsequent application of principal component analysis was shown useful for the botanical authentication of lemon blossom honey (*Citrus limon*) and orange blossom honey (*Citrus* spp.) (Escriche et al. 2011).

Approaches based on the use of two- and three-dimensional fluorescence spectroscopy coupled to chemometric data treatment methods have also been applied to the differentiation and classification of unifloral honeys according to their botanical origin based on the analysis of characteristic fluorescence spectra, as determined by the presence of different phenolic phluorophores (Sergiel et al. 2014; Lenhardt et al. 2015).

Attempts of geographical discrimination of honey according to differences in the phenolic profile have also been described in the literature. Thirty-five samples of thyme honey from four different locations in Greece were correctly classified based on the phenolic composition and conventional physicochemical parameters by means of multivariate analysis of the variance (Karabagias et al. 2014). The concentrations of gallic, chlorogenic, caffeic, *p*-coumaric, and ferulic acids were related as a function of the production site of Italian sulla (*Hedysarum* spp.) honeys (Gambacorta et al. 2014). Furthermore, because sunlight exposure promotes the production of phenolic compounds in plants, a significant difference in the content of phenolic compounds of honeys produced in non-arid and arid regions has been found (Habib et al. 2014).

In conclusion, the revised literature suggests different markers based on the comparison of the phenolic profiles of a limited number of samples and honey types. Consequently, although the definition of the qualitative and quantitative profiles of phenolic compounds in unifloral honeys could contribute to the characterization of their botanical and/or geographical origins, more datasets should be compared at the same time in order to establish adequate phenolic markers able to unequivocally distinguish a unifloral honey type from the rest.

3.9 Natural Toxins

Natural toxins in food represent a significant yet often overlooked health risk, challenging the widespread perception that natural products are inherently safe. Honey, a globally consumed natural sweetener distinguished for its nutritional value, can also carry such risks. In addition to globally distributed toxins such as pyrrolizidine alkaloids (PAs) and tropane alkaloids (TAs), honey can contain toxins specific to certain regions. When bees gather nectar from plants with toxic properties—such as *Rhododendron* species, *Tripterygium wilfordii Hook* F, and *Coriaria arborea* —the honey produced may contain plant-derived toxins like grayanotoxins (GTXs) or diterpenoid alkaloids (Yan et al. 2022). These naturally occurring compounds pose potential risks to human health, especially in higher concentrations.

Numerous studies have detected *pyrrolizidine alkaloids* (PAs) in honey, especially in Italy and Germany, with higher contamination levels observed in Southern Europe due to the prevalence of PA-producing plants such as *Echium* (Lucatello et al. 2021). PAs are typically found in concentrations below 20 μg/kg, though some samples have reported levels exceeding 1 mg/kg (Brugnerotto et al. 2021). The European Food Safety Authority (EFSA) has expressed concern that honey consumption could contribute to chronic PA exposure, particularly among children (EFSA 2017). In addition, *tropane alkaloids* (TAs), including hyoscyamine and scopolamine, are toxic compounds found in plants like *Atropa belladonna* and *Datura* species, which can also be transferred to honey. Although acute toxicity is rare, recent studies have found low concentrations of atropine and scopolamine in honey, with one Spanish sample containing 27 μg/kg of scopolamine (Romera-Torres et al. 2020).

Additionally, certain toxins are more region-specific, often linked to the local flora, which presents unique challenges for honey safety. One such example is grayanotoxins found in *Rhododendron* species, particularly prevalent in "mad honey" from Turkey and Nepal. Although the genus *Rhododendron* (Ericaceae) includes over 800 species distributed from Japan to Europe and the USA, only a subset—such as *R. ponticum*, *R. luteum*, and *R. flavum*—produces nectar with toxic properties. These neurotoxic compounds, specifically grayanotoxins I, II, and III, are absorbed into honey when bees collect nectar from these toxic plants. The consumption of honey contaminated with grayanotoxins can cause symptoms like dizziness, nausea, hypotension, and in severe cases, unconsciousness, though death is rare. A recent EFSA opinion (EFSA 2023) highlighted the risks of grayanotoxins I and III in honey, establishing a protective threshold of 0.05 mg of GTX I and III per kg of honey, though this does not account for genotoxicity or other related compounds. Notably, despite the global distribution of *Rhododendron* species capable of producing these toxins, the majority of reported poisoning cases from "mad honey" have occurred in Turkey. This suggests that factors such as local beekeeping practices, higher concentrations of grayanotoxins in Turkish Rhododendron species, or even consumption habits of local populations may contribute to these higher poisoning rates (Gunduz et al. 2023; Aygun et al. 2017).

In China, *Tripterygium honey*, also known as thunder god vine honey, has been associated with several poisoning incidents, particularly during the late summer months when *Tripterygium* species become the primary nectar source for honeybees. The toxic compound *triptolide*, found in *Tripterygium wilfordii* and *Tripterygium hypoglaucum*, is recognized for its potent medicinal properties, including its use in treating autoimmune diseases and rheumatoid arthritis. However, in high doses, triptolide poses serious health risks, including acute renal failure and liver toxicity. Recent studies show that triptolide levels in honey decrease over time, with significant reductions in toxicity observed after three months of storage, possibly due to enzymatic activity. While honey contaminated with triptolide is relatively safe when consumed in small amounts, honey from regions abundant in *Tripterygium* plants carries a higher potential for toxicity (Zhang et al. 2017). Additionally, *celastrol*, another toxic compound found in *Tripterygium* honey, exacerbates safety concerns. Known for its potential to cause severe liver and heart damage, celastrol can interact with triptolide, increasing the toxicity of the honey. Interestingly, the transfer of these toxins to honey is not always directly linked to the proximity of beehives to *Tripterygium* plants, highlighting the complex role of environmental and geographical factors in toxin transfer. This underscores the critical need for monitoring beekeeping practices and educating beekeepers on how to identify and remove *Tripterygium* pollen from honey to mitigate health risks (Xiao et al. 2024; Végh et al. 2024).

On the other hand, *matrine-type alkaloids*, found in *Sophora* species, are key components of traditional Chinese medicine, recognized for their anticancer, anti-inflammatory, and neuroprotective properties. However, they are also linked to toxic effects such as hepatotoxicity and neurotoxicity. Alkaloids like *matrine* and *oxymatrine* are used as natural pesticides and can enter the food chain through honey derived from *Sophora* plants. A study of 567 Chinese honey samples detected these compounds in 61% of the cases, particularly in acacia honey from *Sophora viciifolia*, with concentrations reaching up to 2774 µg/kg (Wang et al. 2023).

In New Zealand, honey derived from *Coriaria arborea* (tutu), a plant known for producing the neurotoxin *tutin*, has been linked to symptoms such as dizziness, nausea, and, in extreme cases, seizures and coma. The toxicity of tutu honey is mainly influenced by the ratio of tutin to its less toxic derivative, *hyenanchin*, with tutin being up to 1200 times more toxic than hyenanchin. To prevent toxicity, New Zealand has set a regulatory limit of 0.7 µg/g for tutin in honey. Recent studies have confirmed that both tutin and hyenanchin are plant-derived toxins, not insect-derived as previously believed. This highlights the importance of monitoring honey from regions where *Coriaria* species are present to protect consumers from unsafe toxin levels (Watkins et al. 2018).

Moreover, *gelsemium alkaloids*, found in *Gelsemium elegans* and *G. sempervirens*, contribute to honey toxicity. These compounds, such as *gelsemine* and *humantenidine*, are present in the plant's roots, stems, and flowers and can severely impair the nervous, digestive, and respiratory systems. While studies in Southern China have detected varying levels of gelsemine and other gelsemium alkaloids in honey, no research has been conducted on honey derived from American *G. sempervirens* plants. Symptoms of poisoning from these alkaloids include dizziness, vomiting, and abdominal pain, with severe cases leading to convulsions or even death (Long et al. 2023; Wang et al. 2024).

Overall, the global distribution of various toxic compounds in honey, along with region-specific toxins linked to local vegetation, highlights the need for more comprehensive sampling, regulation, and monitoring. In particular, there is a critical need to educate beekeepers on identifying and managing the presence of toxic plants, as local practices can significantly influence honey safety. Several recent review papers provide more detailed information on these topics and offer further insights into improving honey safety standards (Végh et al. 2024; Yan et al. 2022).

3.10 Stability of the Compounds Present in Honey

The stability of compounds in honey is a critical aspect of its quality and safety, as it directly influences the preservation of both nutritional and toxic components. Honey is known for its natural preservative properties due to its low moisture content, high acidity, and the presence of antimicrobial compounds such as hydrogen peroxide and phenolic acids. However, the stability of specific compounds in honey, including bioactive components like flavonoids, phenolic acids, and toxic substances like grayanotoxins, can be affected by storage conditions, time, and the composition of the honey.

One of the key factors affecting honey's nutritional and sensory qualities is *temperature*. The browning that occurs during heating or storage results from a chemical reaction between reducing sugars and amino groups, which produces compounds such as 5-hydroxymethylfurfural (5-HMF) and furfural. These reactions, part of the

Maillard process, can either develop slowly over time or accelerate under high temperatures. 5-HMF, formed during sugar dehydration, is a major intermediate in this process and can lead to changes in both color and nutritional quality. In honey, this reaction causes darkening during storage, particularly when heat treatments are used to prevent crystallization or microbial growth. Factors such as temperature, composition, and pH influence this browning, with acidic conditions favoring the formation of 5-HMF. The reaction also generates Amadori products, key intermediates in the browning process, which undergo further rearrangement to produce additional reactive compounds (Alaerjani et al. 2022). The Maillard reaction diminishes the nutritional value of food by breaking down essential amino acids like lysine, which reacts with sugars, contributing to further nutrient loss (Yan et al. 2023). The contents of most of the free amino acids were found to decrease with storage time, with the greatest reduction observed in the first 9 months (Iglesias et al. 2006).

Temperature also affects the stability of honey's bioactive compounds. Thermal treatments, such as moisture reduction at 55 °C, significantly reduce the content of phenolic compounds in honey, with phenolic acids being particularly sensitive to heat compared to flavonoids, which exhibit greater thermal stability (Wu et al. 2022). The degradation of phenolic compounds and antioxidant activity in food follows first-order kinetics, with the rate of degradation influenced by temperature, soluble solids, and pH (Zapata et al. 2022). Antioxidant activity declines in parallel with the loss of phenolic compounds, underscoring the role of these compounds as key contributors to honey's antioxidant capacity. These findings highlight the need to minimize heat exposure during honey processing to preserve its bioactive properties and nutritional value. The half-life of phenolic content and antioxidant activity varies significantly under different storage conditions, emphasizing the importance of proper storage practices to maintain honey's health benefits (Zapata et al. 2022). Thermal treatments of *Tetragonisca angustula* honey provide an interesting perspective on managing honey quality under heat exposure. In contrast to the commonly observed degradation of bioactive compounds in honey, studies show that thermal processing at temperatures between 52 and 71 °C minimally alters its physicochemical profile. Notably, the phenolic profile undergoes changes, including an increase in certain compounds and the emergence of new ones like isoquercetin and rutin, likely due to the conversion of glycosides to aglycones. These changes positively influence the honey's antioxidant capacity. Moreover, 5-hydroxymethylfurfural (HMF), a marker of heat damage, was undetectable after heating, suggesting that careful application of thermal processing can reduce fermentation risks in high-moisture stingless-bee honey without compromising its quality (Braghini et al. 2020).

Similarly, high temperatures can also impact the stability of enzymes, such as diastase and invertase, which are crucial for honey's quality and its ability to remain free from microbial contamination (Kamboj et al. 2024). Both HMF and diastase activity are modified over time, even when honey is stored at room temperature for 18 months. During this period, HMF content increases significantly, while diastase activity is notably reduced. Additionally, the pH of honey slightly decreases, and its water content remains relatively stable throughout the storage time (Živkov Baloš et al. 2023).

In addition to temperature, the *pH* of honey and the presence of *moisture* play crucial roles in the stability of its compounds. Honey's natural acidity helps maintain the stability of many bioactive compounds, but fluctuations in pH during storage or processing can destabilize sensitive compounds, including certain amino acids and vitamins. For instance, studies have shown that prolonged storage can lead to a gradual decrease in pH, which may result in the increased production of organic acids, such as gluconic acid, potentially affecting honey's quality and microbial stability (Živkov Baloš et al. 2023). Excessive moisture, on the other hand, can promote the growth of yeast and bacteria, leading to fermentation and the degradation of both nutrients and antimicrobial properties. Research indicates that when honey's moisture content exceeds 20%, it accelerates fermentation and can cause a decrease in quality (Vijayakumar et al. 2021). Therefore, managing both pH and moisture content during storage is essential to preserve honey's bioactive properties and prevent microbial contamination.

Hydrocarbons in honey generally show a decrease in concentration over time, primarily due to volatilization. As their molecular weight increases, the rate of decrease is more pronounced. Oxidative processes during storage can break down hydrocarbons into smaller molecules, such as alcohols, which are more volatile. The increased presence of alcohols may result from lipid oxidative degradation or reduction processes catalyzed by aldehyde reductase from honeybees or microorganisms (da Silva et al. 2016).

A study on unprocessed *Apis mellifera* honey from Brazil evaluated the stability of volatile compounds over 18 months of storage at 20 ± 4 °C. Cis- and trans-linalool oxide and hotrienol increased in concentration during this period, while other compounds, such as ethyl acetate and various acids, were detected in some samples after 12 months. These findings suggest that cis- and trans-linalool oxide and hotrienol could serve as indicators of honey degradation due to their consistent presence across all samples and time points (da Silva et al. 2020). Similar observations have been reported following heat treatments, which can generate compounds like hotrienol and its precursors, along with other molecules such as terpendiol I, 8-hydroxylinalool, and lilac aldehydes, formed through oxidation processes (Kaškonienė et al. 2008).

In conclusion, while honey is inherently stable due to its natural properties, the stability of its bioactive components and nutrients can be influenced by storage conditions, including temperature, pH, and moisture content. Understanding these factors is crucial for maintaining honey's health benefits and ensuring its safety for consumption.

3.11 Analysis of Honey Constituents

Honey is a complex natural product composed of a wide range of compounds, including sugars, proteins, amino acids, phenolic compounds, minerals, and volatile substances, all of which contribute to its nutritional and sensory properties. Each component plays a significant role in defining the quality and functionality of honey, making their accurate analysis essential for assessing authenticity, freshness, and safety. Over the years, advancements in analytical technologies have enabled precise identification and quantification of these components, employing both conventional and cutting-edge techniques. These methods not only help ensure compliance with regulatory standards but also deepen our understanding of honey's composition. In the following sections, we will explore the main analytical techniques applied to evaluate the different components of honey, highlighting their principles, applications, and relevance in quality control and research.

3.11.1 Sample Preparation

Classical methodologies for honey sample preparation typically involve steps such as dilution, homogenization, and filtration, particularly to address honey's high viscosity and complex sugar mixtures before further analysis. Initial sample treatments aim to dissolve honey in water or specific solvents, ensuring that analytes of interest are accessible for detection. Sample preparation plays a critical role in ensuring the accuracy, reproducibility, and precision of analytical results, significantly contributing to experimental success while minimizing errors.

In the analysis of honey bioactive components, these preparatory steps are essential for maximizing extraction yields while preserving the integrity of the honey matrix. Several techniques are employed to extract desired compounds, particularly phenolic compounds, including ultrasound-assisted extraction (UAE), microwave-assisted extraction (MAE), pressurized liquid extraction (PLE), and QuEChERS (Quick, Easy, Cheap, Effective, Rugged, and Safe), each with its unique advantages.

UAE, for instance, is an environmentally friendly technique that uses ultrasonic waves to enhance solvent penetration, efficiently extracting bioactive compounds—especially polyphenols—making it a cost-effective and rapid alternative to traditional methods (Pedisić et al. 2023). Similarly, MAE reduces extraction time and solvent use, offering high selectivity, particularly for thermolabile compounds (Biesaga and Pyrzyńska 2013). PLE employs pressurized fluids to shorten extraction times and minimize solvent volumes (Zhang et al. 2023a, b), while QuEChERS, initially developed for pesticide residue analysis, has gained popularity in honey analysis due to its simplicity and safety (Lewiński et al. 2023). Collectively, these techniques ensure high-quality, reproducible honey extracts suitable for subsequent analysis, optimizing both environmental and economic impacts.

3.11.2 Analysis of Sugars

Sugars are the predominant components of honey, accounting for approximately 95% of its dry weight. They play a crucial role in defining its nutritional value, authenticity, and quality. Sugar analysis serves various purposes, including detecting adulteration with external sweeteners, determining botanical and geographical origins, and ensuring compliance with international standards such as those set by the Codex Alimentarius. The fructose-to-glucose ratio, for instance, can help differentiate floral sources, while unexpected deviations in sugar profiles may indicate tampering.

The analysis of sugars in honey follows two primary strategies, depending on the desired outcome. When the goal is to determine the total sugar content, simpler methods such as titration or colorimetric assays are commonly employed. These techniques provide an aggregate measure of sugar levels and are particularly useful for routine quality control. Titration methods estimate reducing sugar content (Aljohar et al. 2018), while colorimetric assays, using agents like dinitrosalicylic acid, quantify total sugars based on their chemical reactivity (Cavaco et al. 2021). These approaches are straightforward, cost-effective, and widely used but lack specificity for individual sugar components.

On the other hand, when the focus is on accurately quantifying major sugars such as glucose, fructose, and sucrose, as well as detecting minor sugars and derivatives that contribute to honey's unique profile, more sophisticated techniques are required.

Gas chromatography (GC) has long been used in this field, specifically gas chromatography with flame-ionization detection (GC-FID), a method that provides high sensitivity after the derivatization of sugars into trimethylsilyl (TMS) derivatives (Sweeley et al. 1963). Over the years, improvements such as hydroxylamine reactions to create oximes for reducing sugars have simplified chromatographic profiles, allowing for better resolution (Mason and Slaver 1971). GC-FID remains a powerful tool for separating and quantifying oligosaccharides, although it requires derivatization, which complicates the process and increases analysis time (Battaglini and Bossi 1972; Low and Sporns 1988; Serra and Ventura 1995; Gómez-Bárez et al. 1999; Cotte et al. 2004b). Gas chromatography–mass spectrometry (GC-MS) has also been used to analyze tri- and tetrasaccharides in honey (Sanz et al. 2004a), though its use is less common compared to other methods like HPLC.

High-performance liquid chromatography (HPLC), coupled with refractive index detection (RID), has become the most widely used method due to its reliability and precision in separating individual sugars like glucose, fructose, and sucrose (Valverde et al. 2022). HPLC offers excellent sensitivity but may struggle with sugars that are structurally similar. In contrast, high-performance anion-exchange chromatography with pulsed amperometric detection (HPAEC-PAD) is particularly effective for detecting sugars at lower concentrations (i.e., melezitose, maltose, isomaltose, turanose, erlose, raffinose, and trehalose) in honeys (Ouchemoukh et al. 2010; Escuredo et al. 2014). This method offers superior resolution and sensitivity and has been successfully used to classify honeys by their botanical and geographical origins with high accuracy, such as distinguishing acacia honeys from Europe and China with a 100% correlation index when combined with chemometric analysis (Carabetta et al. 2020). In addition to these techniques, mass spectrometry (MS) and matrix-assisted laser desorption/ionization mass spectrometry (MALDI-MS) have been explored for sugar analysis in honey. These methods allow for the identification of sugar isomers and can be valuable tools for detecting sugar adulteration or determining the botanical origins of honey. However, their use is still limited due to high sugar concentrations in honey and challenges with ionization (Liangliang et al. 2019; Zhang et al. 2023a, b).

Other analytical techniques such as nuclear magnetic resonance (NMR) or Raman and infrared spectroscopy have also been applied for characterization of honey sugars. Mazzoni et al. (1997) published the first ^{13}C NMR study for the analysis of carbohydrates in honey, quantifying ten different saccharides. Cossonni et al. (2012) were able to identify up to 19 saccharides in Italian honey samples by ^{1}H and ^{1}H-^{13}C NMR spectroscopy, on the basis of their anomeric proton. Automating NMR analysis using machine learning models, such as logistic regression, deep learning, and gradient boosting, has been successfully demonstrated for detecting sugar adulterants like brown rice syrup, corn syrup, and jaggery syrup, improving efficiency for food chemistry researchers and the food industry (Rachineni et al. 2022). Near-infrared spectroscopy (NIRS) is being explored for rapid, on-site analysis to accurately determine the contents of fructose, glucose, sucrose, and maltose in commercial honeys and has also proven useful for detecting adulterants (Qiu et al. 1999). Additional information on the application of NIRS in honey analysis can be found in the recent review by Biswas and Chaudhari (2024).

Capillary electrophoresis (CE) is another viable option, delivering high resolution and rapid analysis with minimal sample preparation. For example, CE coupled with light-emitting diode–induced fluorescence detection (LEDIF) has been used to efficiently separate and quantify sugars in honey, allowing for rapid analysis with excellent sensitivity. This method has successfully differentiated fructose/glucose ratios across various honey types, emphasizing its potential for honey analysis (Andrasi et al. 2023).

In conclusion, sugar analysis in honey requires a tailored approach depending on the desired goal, whether it be for routine quality control, detecting adulteration, or identifying specific sugar profiles linked to honey's origin. While significant advancements have been made, challenges remain, such as honey's viscosity, the potential co-occurrence of oligosaccharides and sugar alcohols, and the need to account for variations due to season, botanical origin, and geographical location.

3.11.3 Analysis of Proteins, Enzymes, and Amino Acids

Several methods have been employed to analyze proteins and amino acids in honey, depending on the specific compound being studied. For protein quantification, common methods include the Kjeldahl, Bradford, and Association of Official Analytical

Chemists (AOAC) methods. These methods are well-established, with the Kjeldahl method being the most widely used for nitrogen determination in food matrices. However, due to honey's relatively low protein content compared to other bee products like pollen, these methods are rarely applied in honey analysis. The AOAC method, using isotope ratio mass spectrometry (IRMS), has been primarily applied to detect C4 sugar adulteration in honey but has also been used to assess the authenticity of honey through protein analysis (Dong et al. 2018). Instead, more advanced chromatographic techniques are preferred for amino acid analysis.

Gel electrophoresis, particularly Sodium Dodecyl Sulfate-Polyacrylamide Gel Electrophoresis (SDS-PAGE), has been widely used to estimate the molecular size of proteins in honey. Early studies, such as by Marshall and Williams (1987), identified protein bands, while Rodríguez-Otero et al. (1990) observed proteins like glucosidases, amylase, and glucose oxidase in Spanish honeys. Recent advances include combining SDS-PAGE with mass spectrometry (MS) for more detailed identification. Chua et al. (2015) found that membrane dialysis offered the best protein extraction from honey. Proteins predominantly from *Apis mellifera* were identified in various honeys, while plant-derived proteins were mostly degraded by bee enzymes (Di Girolamo et al. 2012; Rossano et al. 2012). The use of bidimensional zymography allowed the detection of enzyme activities specific to honey types, aiding in honey classification and authenticity verification (Rossano et al. 2012). More recently, Valachová et al. (2016) developed an ELISA method for quantifying defensin-1, which may further help verify honey authenticity.

For amino acid analysis, high-performance liquid chromatography (HPLC) and ultrahigh-performance liquid chromatography (UHPLC) are commonly employed, often coupled with fluorescence detection (FLD), diode-array detection (DAD), or mass spectrometry (MS). These techniques typically involve derivatization, with reagents like 9-fluorenylmethyl chloroformate (FMOC) or o-phthaldialdehyde (OPA) used to enhance the detection of amino acids. DANS-Cl, PTIC, and FMOC react with primary and secondary amines, but each has limitations. DANS-Cl requires high temperatures and long reaction times, and its derivatives are unstable, complicating chromatograms. PTIC produces stable derivatives with low detection limits but requires time-consuming drying steps, affecting reproducibility. OPA and FMOC create highly fluorescent derivatives for high sensitivity, though FMOC's production of fluorenilmethylchloroformate is a drawback. OPA does not react with secondary amino groups like proline when 2-mercaptoethanol is present, requiring careful control of reaction times and injections (González-Paramás et al. 2006). Mass spectrometry, including MS and MS/MS, offers high sensitivity and selectivity, making it a suitable method for analyzing complex mixtures of amino acids, though it is more expensive and requires derivatization (Biluca et al. 2019). A recent application demonstrated a simple and efficient dilute-and-shoot strategy using hydrophilic interaction liquid chromatography tandem mass spectrometry (HILIC-MS/MS) for determining 20 free amino acids (FAAs) in honey. The method requires minimal sample preparation, relying on direct dilution with water to mitigate matrix effects and enhance FAA extraction. With a rapid 5-minute analysis time, high sensitivity, and wide linear range, the method was validated and applied to honey samples of various botanical origins, successfully detecting 19 FAAs. This approach highlights its potential for fast and reliable FAA analysis in honey and other complex matrices (Ma et al. 2022).

Capillary electrophoresis (CE) is another method used for amino acid analysis, providing rapid separation and high resolution with minimal sample preparation. CE coupled with DAD has been used to analyze honey samples from various regions, successfully detecting proline concentrations higher than 200 mg/kg in all samples analyzed (Dominguez et al. 2016). In addition to these methods, gas chromatography–mass spectrometry (GC-MS) has been employed to analyze free amino acids in honey, with some studies demonstrating the technique's ability to separate over 30 amino acids in less than seven minutes (Brugnerotto et al. 2024).

Recent advances in nuclear magnetic resonance (NMR) spectroscopy have also provided a noninvasive, rapid method for analyzing amino acids in honey, with studies showing the quantification of phenylalanine and tyrosine as well as the relationship between amino acid content and storage time (Boffo et al. 2012; Mannina et al. 2015)). NMR offers the advantage of requiring minimal sample preparation but can be more expensive compared to other methods (Laghi et al. 2014).

The selection of analytical techniques for amino acids in honey largely depends on the analytical objectives, target compounds, and desired sensitivity. HPLC is widely favored for its precision in quantifying individual amino acids, while CE offers a faster and more economical option with excellent resolution. Advanced methods such as NMR and MS provide heightened sensitivity and are particularly well-suited for handling complex analyses, albeit requiring greater resources and expertise. Each method's applicability varies, emphasizing the importance of aligning the technique with the specific requirements of the analysis.

3.11.4 Analysis of Volatile Components

The extraction of volatile compounds from honey is complex due to their low concentration and diverse chemical structures within a sugar-dominated matrix (Castro-Várquez et al. 2003). Traditional techniques like simultaneous distillation–extraction (SDE) system developed by Likens and Nickerson (1964) and its modified version (Godefroot et al. 1981) have been widely used; however, they require heat, which may alter honey's aroma by forming compounds such as furfural and hydroxymethylfurfural. Modifications to SDE, including vacuum-based protocols, have addressed these drawbacks by preserving fresher honey notes (Maignial et al. 1992; Bouseta and Collin 1995).

Ultrasonic solvent extraction (USE) has been applied to enhance efficiency and shorten extraction times but often employs toxic solvents, which can lead to contamination of the analytical equipment (Cuevas-Glory et al. 2007). Further advancements include refining the volatile isolation process using solvent systems like chloroform and n-hexane, followed by silica gel purification to improve the extract quality (Manyi-Loh et al. 2010).

To overcome the limitations of solvent-based methods, techniques such as solid-phase microextraction (SPME) and headspace solid-phase microextraction (HS-SPME) have gained popularity. These methods require minimal preparation, are environmentally friendly, and are easily integrated with GC or MS systems for analysis (Pontes et al. 2007). HS-SPME, in particular, protects the fiber from sugar interference, making it highly suitable for honey matrices. Optimal fiber coatings, such as divinylbenzene/carboxen/polydimethylsiloxane (DVB/CAR/PDMS), have been identified for volatile compound extraction (Plutowska et al. 2011).

Innovative techniques like selected ion flow tube mass spectrometry (SIFT-MS) enable real-time analysis of volatile mixtures, assisting in differentiating honeys by floral origin using chemometric tools. This technique performs real-time analysis of complex volatile mixtures by utilizing three positive precursor ions H_3O^+, NO^+, and O_2^+ to ionize volatile compounds (Agila and Barringer 2012). Electronic noses (e-noses) offer rapid and straightforward aroma profiling, making them promising tools for honey quality control despite their inability to identify individual volatile compounds. The simplicity, speed, and accuracy of e-nose technology highlight its potential for routine applications in monitoring honey authenticity and assessing quality parameters (Gonçalves et al. 2023).

More recently, gas chromatography coupled with quadrupole time-of-flight mass spectrometry (GC-QTOF-MS) has facilitated the identification of numerous volatile and semi-volatile compounds in honey, enhancing the resolution of compounds with similar nominal masses but different molecular formulas (Castell et al. 2023). These advancements underscore the continuous innovation in methodologies to refine volatile compound analysis in honey.

3.11.5 Analysis of Minerals and Trace Elements

The analysis of minerals and trace elements in honey requires specialized sample preparation distinct from the approaches described for other honey components, such as sugars or amino acids (Sect. 3.11.1). Due to the low concentration of minerals and their potential interactions within honey's complex matrix, specific procedures are necessary to ensure accurate measurements and minimize interferences.

Sample preparation typically begins with wet acid digestion, a critical step to break down the organic matrix and release bound mineral elements. This process involves treating honey samples with concentrated nitric acid, often combined with perchloric acid or hydrogen peroxide, under controlled conditions (González-Paramás et al. 2000). The choice of acid combination and digestion parameters depends on the analytical technique and the minerals of interest. This digestion step eliminates sugars and other organic substances that could interfere with the detection and quantification of minerals, especially when using techniques like atomic absorption spectroscopy (AAS) or inductively coupled plasma (ICP) spectrometry (Valverde et al. 2022). Following digestion, the sample is usually diluted with ultrapure water to achieve the appropriate concentration range for the chosen analytical method. This dilution step ensures compatibility with the instrument's calibration range and reduces potential matrix effects that could compromise the accuracy of the results (Ploegaerts et al. 2023). Proper handling and storage of prepared samples are essential to maintain their stability until analysis (Solayman et al. 2016).

After sample preparation, different techniques can be employed to analyze minerals, metals, and nonmetals in honey samples, each offering unique advantages depending on the analytical goals. The most commonly used methods include atomic absorption spectroscopy (AAS) and inductively coupled plasma (ICP) spectrometry, which are frequently chosen for their sensitivity and reliability.

AAS is often the preferred choice for routine analyses of major minerals and some trace elements due to its relatively low operational costs and straightforward methodology. Several atomization techniques are available within AAS, including flame atomization, which is simple and fast; graphite furnace atomization, offering higher sensitivity and reduced interference; and cold vapor AAS, particularly effective for mercury analysis due to its ability to measure the element in its vapor state. However, AAS is limited to analyzing one element at a time, which can extend the overall analysis duration for multielement determinations (Mititelu et al. 2023).

ICP techniques, such as ICP-optical emission spectrometry (ICP-OES) and ICP-mass spectrometry (ICP-MS), have gained prominence for multielement analysis. ICP-OES provides robust detection capabilities for a broad range of elements and is particularly suitable for higher concentrations (Karabagias et al. 2017). On the other hand, ICP-MS is widely recognized for its exceptional capability in trace element analysis, providing extremely low detection limits and a broad linear dynamic range (Ploegaerts et al. 2023). While ICP-MS is costlier and resource-intensive, its capacity to simultaneously analyze multiple elements makes it invaluable for comprehensive mineral profiling. Additionally, it has been widely used to investigate the mineral content in honey of different origins and environmental contexts, including the quantification of heavy metals such as cadmium and lead (Frazzoli et al. 2007).

Emerging alternatives include total reflection X-ray fluorescence (TXRF), which allows the simultaneous detection of several elements with minimal sample preparation. TXRF has been successfully applied to analyze up to 12 elements in honey, providing reliable results in a cost-effective manner (Ribeiro et al. 2014). Capillary electrophoresis (CE) has also been explored for determining specific cations such as K^+, Na^+, Ca^{2+}, Mg^{2+}, and Mn^{2+}. CE offers the advantages of requiring minimal solvents, rapid analysis, and high resolution, although it is less commonly employed compared to spectrometric techniques (Rizelio et al. 2012).

Each technique presents distinct strengths and limitations, and the choice of method often depends on the target elements, the required sensitivity, and the available resources.

3.11.6 Phenolic Compound Analysis in Honey

The analysis of phenolic compounds in honey is complex and involves several methodological steps tailored to specific research goals. Broadly, these approaches include isolation from the matrix, separation, identification, and quantification of phenolic acids and flavonoids. These processes aim to ensure accuracy and sensitivity while minimizing interference from sugars and other polar substances prevalent in honey.

Extraction methods are critical for isolating phenolic compounds from honey's complex matrix. Traditional liquid–liquid extraction using solvents like ethyl acetate or ethanol is common for aglycone analysis, while more recent methods like dispersive liquid–liquid microextraction (DLLME) provide higher recovery rates, often exceeding 70% (Campone et al. 2014; Campillo et al. 2015). Solid-phase extraction (SPE), particularly with resins like Amberlite XAD-2, offers an effective cleanup step, removing sugars and concentrating phenolic compounds (Ferreres et al. 1994a). Modern sorbents, such as multi-walled carbon nanotubes and carbamate-embedded silica, have also shown high efficiency in recent studies (Liu et al. 2016; Badjah Hadj Ahmed et al. 2014).

After extraction and purification, various techniques can be employed for the identification and quantification of phenolic compounds. Two main strategies dominate phenolic compound analysis in honey: spectrophotometric methods for total phenolic content (TPC) and total flavonoid content (TFC) and HPLC/UHPLC combined with MS/MS for individual compound profiling. While spectrophotometric assays are accessible and cost-effective, chromatographic methods provide more detailed and specific insights into honey's phenolic composition, which is vital for quality, authenticity, and nutritional assessments. Emerging green analytical techniques further demonstrate the ongoing evolution in this field, emphasizing efficiency and environmental sustainability.

Spectrophotometric assays remain widely used for rapid and less precise measurements of the total content of phenolic compounds or specific families of these compounds. These methods are advantageous for their simplicity and cost-effectiveness, making them suitable for routine analysis. Among the most common assays, the Folin–Ciocalteu *method* (FCM) is used for determining total phenolic content (TPC). It relies on the reaction between phenolic compounds and the Folin–Ciocalteu reagent at basic pH, resulting in a blue coloration that is measured spectrophotometrically (Pérez et al. 2023). However, this method lacks specificity as it reacts with any reducing substances, not exclusively phenolics (Ares et al. 2018). Similarly, the $AlCl_3$ colorimetric assay is employed to measure total flavonoid content (TFC). This method takes advantage of the ability of flavonoids to form stable complexes with aluminum ions in methanol, reducing interference from other phenolic compounds, particularly phenolic acids (Ares et al. 2021). Another less common approach involves the Glorieś method, a spectrophotometric technique using ethanol and HCl for flavonoid quantification, with quercetin as the standard (Can et al. 2015). These techniques are widely favored due to their straightforward sample preparation, which typically involves dilution with water or an acidic solution. However, they are less suitable for detailed profiling or the identification of individual phenolic compounds.

When a more detailed analysis of the phenolic compound profile is required, rather than just their total concentration, more specific and advanced techniques such as high-performance liquid chromatography (HPLC) or HPLC coupled with mass spectrometry (HPLC-MS) are employed. HPLC, particularly in reversed-phase mode (RP-HPLC), is the most prevalent technique for analyzing individual phenolic compounds in honey (Alvarez-Suarez et al. 2009). In these conditions, the usual elution order of phenolic compounds is benzoic acids, cinnamic acids, flavanone glycosides followed by flavonol and flavone glycosides, and then the free aglycones in the same order (Pyrzynska and Biesaga 2009). Commonly used detectors include diode-array detector (DAD), electrochemical detector (ECD), and mass spectrometry (MS/MS) (Ciulu et al. 2016; Pyrzynska and Biesaga 2009). Advances in HPLC, like ultrahigh-performance liquid chromatography (UHPLC), have significantly reduced analysis times while improving sensitivity and selectivity (Wabaidur et al. 2015). MS/MS detectors further allow structural elucidation of compounds through fragmentation analysis, essential for differentiating structurally similar compounds like glycosides (Truchado et al. 2015; Truchado et al. 2009). Some recent applications can be found in the works of Nyarko et al. (2023) and Wang et al. (2023).

Capillary electrophoresis (CE) has also been employed for phenolic compound analysis, offering high resolution with minimal solvent use (Arráez-Román et al. 2006). Moreover, newer techniques such as sugaring-out assisted liquid–liquid extraction (SULLE) and miniaturized vortex-ultrasound-assisted extraction methods are gaining attention for their green chemistry principles and efficiency in reducing solvent use while maintaining high analytical performance (Vazquez et al. 2021; Zhu et al. 2019; Xu et al. 2016). QuEChERS, initially developed for pesticide analysis, has been adapted for phenolic compound analysis in honey, combining simplicity with high extraction efficiency (Silva et al. 2019).

Although less common, GC-MS/MS has been used for analyzing phenolic compounds in powdered honey. This method provides excellent sensitivity and has been employed in innovative product development, such as spray-dried honey powders (Kozłowicz et al. 2020).

3.11.7 Analysis of Other Compounds

The determination of *vitamins* in honey involves a combination of analytical techniques, predominantly spectrophotometric and HPLC-based methods, depending on the properties of the target compounds. Sample preparation is essential to ensure accurate results and typically involves simple dilution with aqueous solutions or acids for water-soluble vitamins like vitamin C (Alvarez-Suarez et al. 2018). For more complex cases, solvent extraction (SE) or solid-phase extraction (SPE) may be employed to enhance selectivity and reduce matrix effects, particularly for lipid-soluble vitamins like E and K2 (Kim and Brudzyski 2018; Sawicki et al. 2020). Advanced protocols sometimes integrate size exclusion chromatography (SEC) with SPE for further purification (Kim and Brudzyski 2018).

Spectrophotometric methods, such as colorimetric assays, remain widely used for their simplicity and cost-effectiveness. For example, vitamin C is frequently analyzed by its reaction with hydrazine derivatives or 2,6-dichloroindophenol, leading to color changes detectable via UV-Vis spectrophotometry. These approaches are particularly suited for routine assessments but may lack the specificity required for structurally similar compounds (Wu et al. 2023).

HPLC with various detection systems offers greater precision and versatility. DAD and FLD are commonly used for routine quantification, while HPLC-MS/MS provides high sensitivity and structural confirmation, especially useful for compounds at low concentrations or with overlapping spectral characteristics (Sawicki et al. 2020). For instance, vitamin E and its isoforms, as well as vitamin B2 and its metabolite lumichrome, have been analyzed using these techniques in different studies (Tuberoso et al. 2012).

Despite significant advancements, the simultaneous determination of a broad range of vitamins remains rare, primarily due to their diverse solubility and stability profiles. Methods are therefore often tailored to the specific vitamins of interest. While simpler extraction procedures suffice for routine analyses, more sophisticated techniques like HPLC-MS/MS are indispensable for detailed characterization and trace-level quantification.

On the other hand, the analysis of *organic acids* in honey is fundamental due to their role as bioactive compounds with antimicrobial and antioxidant properties, as well as markers for authenticity and indicators of adulteration (Seraglio et al. 2021). Historically, enzymatic methods were used for their sensitivity and specificity; however, they allow the quantification of only one acid at a time. Advances in analytical techniques have favored chromatographic methods, including gas chromatography (GC) and liquid chromatography (LC), which enable the profiling of multiple organic acids simultaneously. These methods, however, require significant sample preparation, such as derivatization for GC to handle nonvolatile acids and solid-phase extraction (SPE) or specific column configurations to reduce sugar interferences in LC (Cherchi et al. 1994; Suárez-Luque et al. 2002).

Recent methods have incorporated modern technologies like capillary electrophoresis (CE), ion chromatography (IC), and nuclear magnetic resonance (NMR). Capillary electrophoresis (CE) has gained attention for its high resolution, rapid analysis, minimal sample preparation, and reduced solvent consumption, effectively separating low-molecular-weight organic acids while offering lower operational costs (Seraglio et al. 2021). IC with conductimetric detection is particularly useful for quantifying formic acid due to its capability to analyze organic acids in the presence of inorganic anions (Matysiak et al. 2018). NMR has been applied to investigate both the concentration and structural characteristics of organic acids, demonstrating variability in acid profiles depending on the honey type (del Campo et al. 2016).

High-performance liquid chromatography (HPLC) remains one of the most widely used techniques, often coupled with detectors like DAD or MS/MS for improved sensitivity and specificity (An et al. 2020; Villacrés-Granda et al. 2021). MS/MS, especially when combined with UHPLC or two-dimensional HPLC, has been instrumental in detailed profiling, identifying gluconic acid as the predominant organic acid alongside others such as malic and citric acids (Suto et al. 2020).

In conclusion, the choice of analytical technique for organic acids in honey depends on study objectives, analyte concentrations, and resources. HPLC-DAD is ideal for routine analysis due to its simplicity, while MS/MS with UHPLC or advanced chromatography offers exceptional precision for complex profiling. CE and IC are cost-effective alternatives, and NMR is suited for detailed structural studies despite its complexity and cost.

The various analytical techniques utilized for honey characterization, along with their specific applications, advantages, and limitations, are summarized in Table 3.7, providing a comprehensive overview to guide the selection of appropriate methods based on the study's objectives and resource availability.

A comprehensive evaluation of analytical methods for honey underscores the importance of tailoring techniques to the specific compounds of interest and study objectives. Spectrophotometric assays, while rapid and economical, are generally limited to estimating total component content without detailed profiling. For more precise analyses, chromatographic methods such as HPLC and UHPLC, especially when coupled with advanced detectors like MS or MS/MS, stand out for their versatility and sensitivity, particularly in the study of phenolics, amino acids, and organic acids.

Techniques like ICP-MS and AAS are indispensable for trace element analysis, providing robust accuracy for mineral profiling. Simultaneously, emerging technologies such as QuEChERS, two-dimensional chromatography, CE, and IC have introduced efficiency, reduced environmental impact, and improved analytical precision. Noninvasive approaches, including NIR spectroscopy and electronic noses, offer innovative possibilities for on-site honey quality assessments.

By integrating traditional and modern methodologies, researchers can achieve a balance between simplicity, cost, and analytical depth, highlighting the evolution of honey analysis toward greater accuracy, reproducibility, and environmental sustainability.

Table 3.7 Overview of analytical techniques for honey characterization: applications, advantages, and limitations

Technique	Applications	Advantages	Disadvantages
Spectrophotometry (UV-Vis)	Total phenolic content (TPC) via Folin–Ciocalteu assay	Simple and cost-effective	Limited specificity
	Enzyme activities (e.g., diastase, invertase) Antioxidant activity (e.g., DPPH, ABTS, FRAP assays)	Rapid bulk measurements	Not suitable for complex profiling
High-performance liquid chromatography (HPLC)	Sugars: using refractive index or evaporative light-scattering detectors	Versatile and widely applicable	Requires careful sample preparation
	Amino acids: quantification after derivatization- Phenolic compounds: often paired with diode-array or mass spectrometry detection Vitamins: water-soluble types like vitamin C analyzed via UV detection	High precision and sensitivity	Relatively expensive setup
Gas chromatography (GC)	Volatile compounds: identifies aldehydes, ketones, and esters with GC-MS	Excellent for volatile/semi-volatile components	Nonvolatile analytes require derivatization
	Sugars: profiles less common carbohydrates after derivatization Contaminants: detects residues like pesticides	High resolution in compound separation	Complex sample preparation
Mass spectrometry (MS)	Phenolic compounds: identifies specific compounds with LC-MS	High sensitivity and selectivity	Expensive equipment
	Amino acids: profiles free amino acids/peptides Contaminants: detects trace residues and adulterants	Enables structural analysis of complex compounds	Requires skilled operators
Fluorescence spectroscopy	Amino acids: detects compounds like proline linked to authenticity Vitamins: measures riboflavin and	High sensitivity and selectivity	Limited to fluorescent compounds
	others due to fluorescence properties	Simple methodology	Requires expertise for interpretation
Capillary electrophoresis (CE)	Sugars: separates simple sugars and oligosaccharides	Rapid and cost-effective	Limited by detection systems
	Phenolic compounds: analyzes flavonoids and phenolic acids Amino acids: high-resolution profiling	Minimal sample preparation and solvent consumption	Requires optimization for some sample types

(continued)

Table 3.7 (continued)

Technique	Applications	Advantages	Disadvantages
Solid-phase microextraction (SPME)	Aroma profile: detects esters, alcohols, and aldehydes	Requires no solvents	Fiber coatings limit analyte selection
	Degradation indicators: identifies markers like linalool oxide for honey aging	Easily coupled with GC-MS	Suitable primarily for volatile compounds
Inductively coupled plasma (ICP)	ICP-OES: quantifies major minerals like potassium and magnesium	Multielement analysis	High operational costs
	ICP-MS: detects trace metals like lead or cadmium	Extremely low detection limits (ICP-MS)	Sample mineralization often required
Atomic absorption spectroscopy (AAS)	Minerals: quantifies elements like iron, calcium, and zinc	Cost-effective for single-element analysis	Sequential analysis only (one element per run)
		Fast and robust	Interferences from matrix composition
Fourier-transform infrared spectroscopy (FTIR)	Adulteration detection: identifies added sugars or syrups	Nondestructive	Lower sensitivity compared to chromatography
	Moisture content: quantifies water content through absorption bands	Rapid analysis with minimal sample preparation	Limited to specific applications
Stable isotope ratio analysis (SIRA)	Adulteration detection: Differentiates natural honey from syrups via isotopic ratios (e.g., C13/C12, H2/H1)	High specificity	High costs
		Powerful tool for detecting adulteration	Requires advanced instrumentation and expertise

References

Abu-Tarboush HM, Al-Kahtani HA, El-Sarrage M (1993) Floral-type identification and quality evaluation of some honey types. Food Chem 46:13–17

Agila A, Barringer S (2012) Application of selected ion flow tube mass spectrometry coupled with chemometrics to study the effect of location and botanical origin on volatile profile of unifloral American honeys. J Food Sci 77:C1103–C1108

Alaerjani WMA, Abu-Melha S, Alshareef RMH, Al-Farhan BS, Ghramh HA, Al-Shehri BMA, Bajaber MA, Khan KA, Alrooqi MM, Modawe GA, Mohammed MEA (2022) Biochemical reactions and their biological contributions in honey. Molecules 27(15):4719

Alarjani WMA, Mohammed MEA (2024) Antioxidant activities of Saudi honey samples related to their content of short peptides. Sci Rep 14:24318

Alissandrakis E, Tarantilis PA, Harizanis PC, Polissiou M (2005) Evaluation of four isolation techniques for honey aroma compounds. J Sci Food Agric 85:91–97

Alissandrakis E, Tarantilis PA, Harizanis PC, Polissiou M (2007) Comparison of the volatile composition in thyme honeys from several origins in Greece. J Agric Food Chem 55:8152–8157

Aljohar HI, Maher HM, Albaqami J, Al-Mehaizie M, Orfali R, Orfali R, Alrubia S (2018) Physical and chemical screening of honey samples available in the Saudi market: an important aspect in the authentication process and quality assessment. Saudi Pharm J 26:32–942

Alonso-Torre SR, Cavia MM, Fernández-Muiño MA, Moreno G, Huidobro JF, Sancho MT (2006) Evolution of acid phosphatase activity of honeys from different climates Food Chemistry 97:750–755

Alqarni AS, Owayss AA, Mahmoud AA, Hannan MH (2014) Mineral content and physical properties of local and imported honeys in Saudi Arabia. J Saudi Chem Soc 18:618–625

Alvarez-Suarez JM, Tulipani S, Romandini S, Vidal A, Battino M (2009) Methodological aspects about determination of phenolic compounds and in vitro evaluation of antioxidant capacity in the honey: a review. Curr Anal Chem 5:293–302

Alvarez-Suarez JM, González-Paramás AM, Santos-Buelga C, Battino M (2010a) Antioxidant characterization of native monofloral Cuban honeys. J Agric Food Chem 58:9817–9824

Alvarez-Suarez JM, Tulipani S, Romandini S, Bertoli E, Battino M (2010b) Contribution of honey in nutrition and human health: a review. Mediterr J Nutr Metab 3:15–23

Alvarez-Suarez JM, Giampieri F, Damiani E, Astolfi P, Fattorini D, Regoli F, Quiles JL, Battino M (2012) Radical-scavenging activity, protective effect against lipid peroxidation and mineral contents of monofloral Cuban honeys. Plant Foods Hum Nutr 67:31–38

Alvarez-Suarez JM, Giampieri F, Cordero M, Gasparrini M, Forbes-Hernández TY, Mazzoni L, Afrin S, Beltrán-Ayala P, González-Paramás AM, Santos-Buelga C, Varela-Lopez A, Quiles JL, Battino M (2016) Activation of AMPK/Nrf2 signalling by Manuka honey protects human dermal fibroblasts against oxidative damage by improving antioxidant response and mitochondrial function promoting wound healing. J Funct Foods 25:38–49

Alvarez-Suarez JM, Giampieri F, Brenciani A, Mazzoni L, Gasparrini M, González-Paramás AM, Santos-Buelga C, Morroni G, Simoni S, Forbes-Hernández TY, Afrin S, Giovanetti E, Battino M (2018) *Apis mellifera* vs *Melipona beecheii* Cuban polifloral honeys: a comparison based on their physicochemical parameters, chemical composition and biological properties. LWT Food Sci Technol 87:272–279

Alves-Moreira RF, Bastos-de Maria CA (2001) Glicídios no mel. Quim Nova 24:516–525

An N, Cai WJ, Zhu QF, Wang W, Hussain D, Feng YQ (2020) Metabolic profiling of organic acids in honey by stable isotope labeling assisted liquid chromatography-mass spectrometry. J Food Compos Anal 87:103423

Andrade P, Ferreres F, Amaral MT (1997) Analysis of honey phenolic acids by HPLC, its application to honey botanical characterization. J Liq Chromatogr Relat Technol 20:2281–2288

Andrasi M, Gyemant G, Sajtos Z, Nagy C (2023) Analysis of sugars in honey samples by capillary zone electrophoresis using fluorescence detection. Separations 10:150

Anklam E (1998) A review of the analytical methods to determine the geographical and botanical origin of honey. Food Chem 63:549–562

Ares AM, Bernal JL, Nozal MJ, Bernal J (2021) Aroamatic herbs in food analysis of herbal bioactives IN: Aromatic herbs in food. Galanakis Cm (Ed). Academic Press pp 201–232

Ares AM, Valverde S, Bernal JL, Nozal MJ, Bernal J (2018) Extraction and determination of bioactive compounds from bee pollen. J Pharm Biomed Anal 147:110–124

Arráez-Román D, Gómez-Caravaca AM, Gómez-Romero M, Segura-Carretero A, Fernández-Gutiérrez A (2006) Identification of phenolic compounds in rosemary honey using solid-phase extraction by capillary electrophoresis-electrospray ionization-mass spectrometry. J Pharm Biomed Anal 41:1648–1656

Atanassova J, Yurukova L, Lazarova M (2012) Pollen and inorganic characteristics of Bulgarian unifloral honeys. Czech J Food Sci 30:520–526

Aygun A, Sahin A, Karaca Y, Turkmen S, Turedi S, Ahn SY, Kim S, Gunduz A (2017) Grayanotoxin levels in blood, urine and honey and their association with clinical status in patients with mad honey intoxication. Turk J Emerg Med 18:29–33

Azeredo LC, Azeredo MAA, de Souza SR, Dutra VML (2003) Protein contents and physicochemical properties in honey samples of *Apis mellifera* of different floral origins. Food Chem 80:249–254

Babacan S, Rand AG (2005) Purification of amylase from honey. J Food Sci 70:C413–C418

Badjah Hadj Ahmed AY, Obbed MS, Wabaidur SM, Al Othman ZA, Al-Shaalan NH (2014) High-performance liquid chromatography analysis of phenolic acid, flavonoid, and phenol contents in various natural Yemeni honeys using multi-walled carbon nanotubes as a solid-phase extraction adsorbent. J Agric Food Chem 62:5443–5450

Baker HG, Baker I (1986) The occurrence and significance of amino acids in floral nectar. Plant Syst Evol 151:175–186

Ball DW (2007) The chemical composition of honey. J Chem Educ 84:1643–1646

Baroni MV, Chiabrando GA, Costa C, Wunderlin DA (2002) Assessment of the floral origin of honey by SDS-page immunoblot techniques. J Agric Food Chem 50:1362–1367

Baroni MV, Arrua C, Nores ML, Faye P, Diaz MP, Chiabrando GA, Wunderlin DA (2009) Composition of honey from Cordoba (Argentina). Assessment of North/South provenance by chemometrics. Food Chem 114:727–733

Barra MPG, Ponce-Díaz MC, Venegas-Gallegos C (2010) Volatile compounds in honey produced in the central valley of Ñuble province, Chile. Chilean J Agric Res 70:75–84

Batista B, Da Silva L, Rocha B, Rodrigues J, Berretta-Silva A, Bonates T, Gomes V, Barbosa R, Barbosa F (2012) Multi-element determination in Brazilian honey samples by inductively coupled plasma mass spectrometry and estimation of geographic origin with data mining techniques. Food Res Int 49:209–215

Battaglini MB, Bossi G (1972) Determination of glucides by GLC and its possibilities for honey qualification. Apiacta 1

Belitz HD, Grosch W, Schieberle P (2009) Sugars, sugar alcohols and honey. In: Food chemistry, 4th edn. Springer-Verlag, Berlin, pp 883–890

Bentabol-Manzanares A, Hernández-García Z, Rodríguez-Galdón B, Rodríguez-Rodríguez E, Díaz-Romero C (2011) Differentiation of blossom and honeydew honeys using multivariate analysis on the physicochemical parameters and sugar composition. Food Chem 126:664–672

Bianchi F, Mangia A, Mattarozzi M, Musci M (2011) Characterization of the volatile profile of thistle honey using headspace solid-phase microextraction and gas chromatography–mass spectrometry. J Food Chem 129:1030–1036

Biesaga M, Pyrzyńska K (2013) Stability of bioactive polyphenols from honey during different extraction methods. Food Chem 136:46–54

Bilikova EK, Simuth J (2010) New criterion for evaluation of honey: quantification of royal jelly protein Apalbumin 1 in honey. J Agric Food Chem 58:8776–8781

Biluca FC, Bernal J, Valverde S, Ares AM, Gonzaga LV, Costa ACO, Fett R (2019) Determination of free amino acids in stingless bee (Meliponinae) honey. Food Anal Methods 12:902–907

Biswas A, Chaudhari SR (2024) Exploring the role of NIR spectroscopy in quantifying and verifying honey authenticity: a review. Food Chem 445:138712

Boffo EF, Tavares LA, Tobias ACT, Ferreira MMC, Ferreira AG (2012) Identification of components of Brazilian honey by ^{1}H NMR and classification of its botanical origin by chemometric methods. LWT Food Sci Technol 49:55e63

Bogdanov S, Baumann E (1988) Determination of sugar composition of honeys by HPLC. Mitt Geb Lebensm Hyg 79:198–206

Bogdanov S, Lüllmann C, Martin P, Von der Ohe W, Russmann H, Vorwohl G et al (1999) Honey quality and international regulatory standards: review by the International Honey Commission. Bee World 80:61e69

Bogdanov S, Haldimann M, Luginbuhl W, Gallmann P (2007) Minerals in honey: environmental, geographical and botanical aspects. J Apic Res 46:269–275

Bogdanov S, Jurendic T, Sieber R, Gallmann P (2008) Honey for nutrition and health: a review. J Am Coll Nutr 27:677–689

Bonaga G, Giumanini AG, Gliozzi G (1986) Chemical composition of chestnut honey: analysis of the hydrocarbon fraction. J Agric Food Chem 34:319–326

Bouseta A, Collin S (1995) Optimized Likens-Nickerson methodology for quantifying honey flavors. J Agric Food Chem 43:1890–1897

Bradbear N (2009) Definition and uses of honey. In: Bees and their role in forest livelihoods. FAO, Rome, pp 81–88. http://www.fao.org/3/a-i0842e.pdf. Accessed Oct 2016

Braghini F, Biluca FC, Gonzaga LV, Vitali L, Costa ACO, Fett R (2020) Effect thermal processing in the honey of Tetragonisca angustula: profile physicochemical, individual phenolic compounds and antioxidant capacity. J Apic Res 60:290–296

Brudzynski K, Sjaarda C, Maldonado-Alvarez L (2013) A new look on protein–polyphenol complexation during honey storage: is this a random or organized event with the help of dirigent-like proteins? PLoS One 8:1–9

Brugnerotto P, Seraglio SKT, Schulz M, Gonzaga LV, Fett R, Costa ACO (2021) Pyrrolizidine alkaloids and beehive products: a review. Food Chem 342:128384

Brugnerotto P, Fuente-Ballesteros A, Martín-Gómez B, María Ares A, Valdemiro Gonzaga L, Fett R, Carolina Oliveira Costa A, Bernal J (2024) Free amino acid profile in *Mimosa scabrella* honeydew honey from Brazil and chemometric analysis for geographical discrimination. Food Res Int 177:113856

Cabras P, Angioni A, Tuberoso C, Floris I, Reniero F, Guillou C, Ghelli S (1999) Homogentisic acid: a phenolic acid as a marker of strawberry-tree (*Arbutus unedo*) honey. J Agric Food Chem 47:4064–4067

Campillo N, Viñas P, Férez-Melgarejo G, Hernández-Córdoba M (2015) Dispersive liquid-liquid microextraction for the determination of flavonoid aglycone compounds in honey using liquid chromatography with diode array detection and time-of-flight mass spectrometry. Talanta 131:185–191

Campone L, Piccinelli AL, Pagano I, Carabetta S, Di Sanzo R, Russo M, Rastrelli L (2014) Determination of phenolic compounds in honey using dispersive liquid-liquid microextraction. J Chrom A 1334:9–15

Campos G, Nappi U, Raslan D, Augusti R (2000) Substâncias voláteis em mel floral e mel de melato. Cienc Tecnol Aliment 20:18–22

Can Z, Yildiz O, Sahin H, Turumtay EA, Silici S, Kolayli S (2015) An investigation of Turkish honeys: their physico-chemical properties antioxidant capacities and phenolic profiles. Food Chem 180:133–141

Carabetta S, Di Sanzo R, Campone L, Fuda S, Rastrelli L, Russo M (2020) High-performance anion exchange chromatography with pulsed Amperometric detection (HPAEC–PAD) and chemometrics for geographical and floral authentication of honeys from southern Italy (Calabria region). Food Secur 9:1625

Castell A, Arroyo-Manzanares N, Guerrero-Núñez Y, Campillo N, Viñas P (2023) Headspace with gas chromatography-mass spectrometry for the use of volatile organic compound profile in botanical origin authentication of honey. Molecules 28:4297

Castro-Várquez L, Pérez-Coello MS, Cabezudo MD (2003) Analysis of volatile compounds of rosemary honey. Comparison of different extraction techniques. Chromatographia 57:227–233

Castro-Vázquez L, Díaz-Maroto MC, Pérez-Coello MS (2006) Volatile composition and contribution to the aroma of Spanish honeydew honeys. Identification of a new chemical marker. J Agric Food Chem 54:4809–4813

Castro-Vázquez L, Díaz-Maroto MC, Pérez-Coello MS (2007) Aroma composition and new chemical markers of Spanish citrus honeys. Food Chem 103:601–606

Castro-Vázquez L, Díaz-Maroto MC, González-Viñas MA, de la Fuente E, Pérez-Coello MS (2008) Influence of storage conditions on chemical composition and sensory properties of citrus honey. J Agric Food Chem 56:1999–2006

Castro-Vázquez L, Díaz-Maroto MC, González-Viñas MA, Pérez-Coello MS (2009) Differentiation of monofloral citrus, rosemary, eucalyptus, lavender, thyme and heather honeys based on volatile composition and sensory descriptive analysis. Food Chem 112:1022–1030

Castro-Vázquez L, MC DÍ-M, de Torres C, Pérez-Coello MS (2010) Effect of geographical origin on the chemical and sensory characteristics of chestnut honeys. Food Res Int 43:2335–2340

Cavaco T, Figueira AC, González-Domínguez R, Sayago A, Fernández-Recamales Á (2021) Evolution of physicochemical parameters during the thermal-based production of Água-mel, a traditional Portuguese honey-related food product. Molecules 27:57

Cavia MM, Fernández-Muino MA, Alonso-Torre SR, Huidobro JF, Sancho MT (2007) Evolution of acidity of honeys from continental climates: influence of induced granulation. Food Chem 100:1728–1733
Čeksterytė V, Kaupinis A, Aleliūnas A, Navakauskienė R, Jaškūnė K (2024) Composition of proteins associated with red clover (Trifolium pratense) and the microbiota identified in honey. Life (Basel) 14:862
Chen H, Fan C, Chang Q, Pang G, Hu X, Lu M (2014) Chemometric determination of the botanical origin for Chinese honeys on the basis of mineral elements determined by ICP-MS. J Agric Food Chem 62:2443–2448
Cherchi A, Spanedda L, Tuberoso C, Cabras P (1994) Solid-phase extraction and high-performance liquid chromatographic determination of organic acids in honey. J Chromatogr A 669:59–64
Chiba S (1998) α-Glucosidase. In: The Amylase Research society of Japan (ed) Handbook of amylases and related enzymes. Pergamon Press, Oxford, pp 104–116
Chua LS, Abdul-Rahaman NL, Sarmidi MR, Aziz R (2012) Multi-elemental composition and physical properties of honey samples from Malaysia. Food Chem 135:880–887
Chua LS, Lee JY, Chan GF (2013) Honey protein extraction and determination by mass spectrometry. Anal Bioanal Chem 405:3063–3074
Chua LS, Lee JY, Chan GF (2015) Characterization of the proteins in honey. Anal Lett 48:697–709
Chudzinska M, Baralkiewicz D (2011) Application of ICP-MS method of determination of 15 in honey with chemometric approach for the verification of their authenticity. Food Chem Toxicol 49:2741–2749
Ciappini MC, Stoppani FS (2014) Determination of antioxidant capacity, flavonoids, and total phenolic content in eucalyptus and clover honeys. J Apic Sci 58:103–111
Cimpoiu C, Hosu A, Miclaus V, Puscas A (2013) Determination of the floral origin of some Romanian honeys on the basis of physical and biochemical properties. Spectrochim Acta A Mol Biomol Spectrosc 100:149–154
Ciulu M, Solinas S, Floris I, Panzanellia A, Pilo MI, Piu PC, Spanoa N, Sanna G (2011) RPHPLC determination of water-soluble vitamins in honey. Talanta 83:924–929
Ciulu M, Spano N, Pilo MI, Sanna G (2016) Recent advances in the analysis of phenolic compounds in unifloral honeys. Molecules 21:451–482
Codex Alimentarius (2001) Revised codex standard for honey. Codex Stan 12-1981, Rev.2
Cometto PM, Faye PF, di Paola RD, Rubio MA, Aldao MAJ (2003) Comparison of free amino acids profile in honey from three Argentinian regions. J Agric Food Chem 51:5079–5087
Consonni R, Cagliani LR, Cogliati C (2012) NMR characterization of saccharides in Italian honeys of different floral sources. J Agric Food Chem 60:4526–4534
Consonni R, Cagliani LR, Cogliati C (2013) Geographical discrimination of honeys by saccharides analysis. Food Control 32:543e548
Conte LS, Miorini M, Giomo A, Bertacco G, Zironi R (1998) Evaluation of some fixed components for unifloral honey characterization. J Agric Food Chem 46:1844–1849
Cotte JF, Casabianca H, Giroud B, Albert M, Lheritier J, Grenier-Loustalot MF (2004a) Characterization of honey amino acid profiles using high-pressure liquid chromatography to control authenticity. Anal Bioanal Chem 378:1342–1350
Cotte JF, Casabianca H, Chardon S, Lheritier J, Grenier-Loustalot MF (2004b) Chromatographic analysis of sugars applied to the characterisation of monofloral honey. Anal Bioanal Chem 380:698–705
Council Directive 2001/110/EC of 20 December 2001 relating to honey. (2002) Official Journal of European Community L, 10: 47–52
Crane E (1990) The traditional hive products: honey and beeswax. In: Crane E (ed) Bees and beekeeping: science, practice and world resources. Heinemann Newnes, Oxford, pp 388–413
Cuevas-Glory LF, Pino JA, Santiago LS, Sauri-Duch LF (2007) A review of volatile analytical methods for determining the botanical origin of honey. Food Chem 103:1032–1043
da Silva PM, Gauche C, Gonzaga LV, Costa ACO, Fett R (2016) Honey: chemical composition, stability and authenticity. Food Chem 196:309–323

da Silva PM, Gonzaga LV, de Azevedo MS, Biluca FC, Schulz M, Costa ACO, Fett R (2020) Stability of volatile compounds of honey during prolonged storage. J Food Sci Technol 57:1167–1182

Daniele G, Maitre D, Casabianca H (2012) Identification, quantification and carbon stable isotopes determinations of organics acids in monofloral honeys. A powerful tool for botanical and authenticity control. Rapid Commun Mass Spectrom 26:1993–1998

Davies AMC (1975) Amino acid analysis of honeys from eleven countries. J Apic Res 14:29–39

Davies AMC (1976) The application of amino acid analysis to the determination of the geographical origin of honey. J Food Technol 11:515–523

de Alda-Garcilope C, Gallego-Picó A, Bravo-Yagüe J, Garcinuño-Martínez R, Fernández-Hernando P (2012) Characterization of Spanish honeys with protected designation of origin "Miel de Granada" according to their mineral content. Food Chem 135:1785–1788

de la Fuente E, Martínez-Castro I, Sanz J (2005) Characterization of Spanish unifloral honeys by solid phase microextraction and gas chromatography-mass spectrometry. J Sep Sci 28:1093–1100

de la Fuente E, Ruiz-Matute AI, Valencia-Barrera RM, Sanz J, Martínez-Castro I (2011) Carbohydrate composition of Spanish unifloral honeys. Food Chem 129:1483–1489

del Campo G, Zuriarrain J, Zuriarrain A, Berregi I (2016) Quantitative determination of carboxylic acids, amino acids, carbohydrates, ethanol and hydroxymethylfurfural in honey by 1H NMR. Food Chem 196:1031–1039

Di Girolamo F, D'Amato A, Righetti PG (2012) Assessment of the floral origin of honey via proteomic tools. J Proteome 75:3688–3693

Dimitrova B, Gevrenova R, Anklam E (2007) Analysis of phenolic acids in honeys of different floral origin by solid-phase extraction and high-performance liquid chromatography. Phytochem Anal 18:24–32

Dobre I, Georgescu LA, Alexe P, Escuredo O, Seijo MC (2012) Rheological behavior of different honey types from Romania. Food Res Int 49:126–132

Doner LW (1977) The sugars of honey – a review. J Sci Food Agric 28:443–456

Dong H, Xiao K, Xian Y, Wu Y (2018) Authenticity determination of honeys with non-extractable proteins by means of elemental analyzer (EA) and liquid chromatography (LC) coupled to isotope ratio mass spectroscopy (IRMS). Food Chem 240:717–724

Dominguez MA, Jacksen J, Emmer A, Centurion ME (2016) Capillary electrophoresis method for the simultaneous determination of carbohydrates and proline in honey samples. Microchemical Journal 129:1–4

Dustman JH (1971) Über die katalaseaktivität in bienenhonig ausder tracht der heidekrautgewächse (Ericaceae). Z Lebensm Unters Forsch 145:292–295

EFSA CONTAM Panel (2023) Scientific opinion onthe risks for human health related to the presence of grayanotoxins in certain honey. EFSA J 21:7866, 120 pp

Erban T, Shcherbachenko E, Talacko P, Harant K (2021) A single honey proteome dataset for identifying adulteration by foreign amylases and mining various protein markers natural to honey. J Proteome 239:104157

Escriche I, Kadar M, Juan-Borrás M, Domenech E (2011) Using flavonoids, phenolic compounds and headspace volatile profile for botanical authentication of lemon and orange honeys. Food Res Int 44:1504–1513

Escriche I, Kadar M, Juan-Borrás M, Domenech E (2014) Suitability of antioxidant capacity, flavonoids and phenolic acids for floral authentication of honey. Impact of industrial thermal treatment. Food Chem 142:135–143

Escuredo O, Míguez M, Fernández-González M, Seijo MC (2013) Nutritional value and antioxidant activity of honeys produced in a European Atlantic area. Food Chem 138:851–856

Escuredo O, Dobre I, Fernández-González M, Seijo MC (2014) Contribution of botanical origin and sugar composition of honeys on the crystallization phenomenon. Food Chem 149:84–90

EU Council (2002) Council Directive 2001/11 O/EC of 20 December 2001 relating to honey. Off J Eur Communities L10:47–52

European Food Safety Authority (2017) Risks for human health related to the presence of pyrrolizidine alkaloids in honey, tea, herbal infusions and food supplements. EFSA J 15:4908

Farkas A, Horváth G, Kuzma M, Mayer M, Kocsis M (2023) Phenolic compounds in Hungarian acacia linden milkweed and goldenrod honeys Current Research in Food Science 6100526

Food and Agriculture Organization of the United Nations. FAOSTAT. 2023. https://www.fao.org/faostat. Accessed Oct 2024

Feller-Demalsy MJ, Vincent B, Beaulieu F (1989) Teneur en minéraux et origine géographique des miels du Canada. Apidology 20:77–91

Fernandez-Torres R, Perez-Bernal JL, Bello-Lopez MA, Callejon-Mochon M, Jimenez-Sanchez JC, Guiraum-Perez A (2005) Mineral content and botanical origin of Spanish honeys. Talanta 65:686–691

Ferreres F, García-Viguera C, Tomás-Lorente F, Tomás-Barberán FA (1993) Hesperetin: a marker of the floral origin of citrus honey. J Sci Food Agric 61:121–123

Ferreres F, Tomás-Barberán FA, Soler C, García-Viguera C, Ortiz A, Tomás-Lorente F (1994a) A simple extractive technique for honey flavonoid HPLC analysis. Apidologie 25:21–30

Ferreres F, Andrade P, Tomás-Barberán FA (1994b) Flavonoids from Portuguese heather honey. Z Lebensm Unters Forsch 199:32–37

Ferreres F, Andrade P, Tomás-Barberán FA (1996) Natural occurrence of abscisic acid in heather honey and floral nectar. J Agric Food Chem 44:2053–2056

Flanjak I, Strelec I, Kenjerić D, Primorac L (2016) Croatian produced unifloral honeys characterised according to the protein and proline content and enzyme activities. J Apic Sci 60:39–48

Földházi G (1994) Analysis and quantitation of sugars in honey of different botanical origin using high performance liquid chromatography. Acta Alim 23:229–311

Frazzoli C, D'Ilio B, Bocca B (2007) Determination of Cd and Pb in honey by SF–ICP-MS: validation figures and uncertainty of results. Anal Lett 40:1992–2004

Gambacorta E, Simonetti A, Garrisi N, Intaglietta I, Perna A (2014) Antioxidant properties and phenolic content of sulla (*Hedysarum* spp.) honeys from Southern Italy. Int J Food Sci Technol 49:2260–2268

Gilbert J, Shephard MJ, Wallwork MA, Harris RG (1981) Determination of the geographical origin of honeys by multivariate analysis of gas chromatographic data on their free amino acid content. J Apic Res 20:125–135

Giri KV (1938) The chemical composition and enzyme content of indian honey. Madras Agri J 26:68–72

Godefroot M, Sandra P, Verzele M (1981) New method for quantitative essential oil analysis. J Chromatogr 203:325–335

Goldschmidt S, Burkert H (1955) Observation of some heretofore unknown sugars in honey. Z Physiol Chem 300:188–200

Gómez Bárez JA, Garcia-Villanova RJ, Elvira Garcia S, Rivas Palá T, González Paramás AM, Sánchez Sánchez J (2000) Geographical discrimination of honeys through the employment of sugar patterns and common chemical quality parameters. Eur Food Res Technol 210:437–444

Gómez-Bárez JA, García-Villanova RJ, Elvira-García S, González-Paramás AM (1999) Optimization of the capillary gas chromatographic analysis of mono- and oligosaccharides in honeys. Chromatographia 50:461–469

Gonçalves WB, Teixeira WSR, Cervantes EP, Mioni MSR, Sampaio ANCE, Martins OA, Gruber J, Pereira JG (2023) Application of an electronic nose as a new technology for rapid detection of adulteration in honey. Appl Sci 13(8):4881

González-Miret ML, Terrab A, Hernanz D, Fernández-Recamales MA, Heredia FJ (2005) Multivariate correlation between color and mineral composition of honeys and by their botanical origin. J Agric Food Chem 53:2574–2580

González-Paramás AM, Gómez-Bárez JA, García-Villanova RJ, Rivas Palá T, Ardanuy Albajar R, Sánchez J (2000) Geographical discrimination of honeys by using mineral composition and common chemical quality parameters. J Sci Food Agric 80:157–165

González-Paramás AM, Gómez-Bárez JA, Cordón-Marcos C, García-Villanova RJ, Sánchez-Sánchez J (2006) HPLC-fluorimetric method for analysis of amino acids in products of the hive (honey and bee-pollen). Food Chem 95:148–156

Gunduz A, Şimşek P, Ayaz FA (2023) Worldwide distribution and clinical characteristics of mad honey poisoning cases. Cent Eur J Public Health 31:69–73

Habib HM, Al Meqbali FT, Kamal H, Souka UD, Ibrahim WH (2014) Bioactive components, antioxidant and DNA damage inhibitory activities of honeys from arid regions. Food Chem 153:28–34

Hermosín I, Chicón RM, Cabezudo MD (2003) Free amino acid composition and botanical origin of honey. Food Chem 83:263–268

Hernández OM, Fraga JMG, Jiménez AI, Jiménez F, Arias JJ (2005) Characterization of honey from the Canary Islands: determination of the mineral content by atomic absorption spectrophotometry. Food Chem 93:449–458

Hogenboom AC, van Leerdam JA, de Voogt P (2009) Accurate mass screening and identification of emerging contaminants in environmental samples byliquid chromatography–hybrid linear ion trap Orbitrap mass spectrometry. J Chromatogr A 1216:510–519

Iglesias MT, de Lorenzo C, Polo MC, Martín-Álvarez PJ, Pueyo E (2004) Usefulness of amino acid composition to discrimínate between honeydew and floral honeys. Application to honeys from a small geographic area. J Agric Food Chem 52:84–89

Iglesias MT, Martín-álvarez PJ, Polo MC, de Lorenzo C, González M, Pueyo E (2006) Changes in the free amino acid contents of honeys during storage at ambient temperature. J Agric Food Chem 54:9099–9104

Janiszewska K, Aniołowska M, Nowakowski P (2012) Free amino acids content of honeys from Poland. Pol J Food Nutr Sci 62:85–89

Jànošková N, Vyviurska O, Špànik I (2014) Identification of volatile organic compounds in honeydew honeys using comprehensive gas chromatography. J Food Nutr Res 53:353–362

Jerković I, Marijanović Z, Staver MM (2011a) Screening of natural organic volatiles from *Prunus mahaleb* L. honey: coumarin and vomifoliol as nonspecific biomarkers. Molecules 16:2507–2518

Jerković I, Tuberoso CIG, Kasum A, Marijanović Z (2011b) Volatile compounds of *Asphodelus microcarpus* Salzm. et Viv. honey obtained by HS-SPME and USE analyzed by GC/MS. Chem Biodivers 8:587–598

Kadri SM, Zaluski R, de Oliveira OR (2017) Nutritional and mineral contents of honey extracted by centrifugation and pressed processes. Food Chem 218:237–241

Kamboj R, Singh RS, Nanda V (2024) Effect of heating and pH on hydroxymethylfurfural content, diastase and invertase activity of Dalbergia honey using response surface methodology. Food Chem Adv 5:100766

Karabagias IK, Vavoura MV, Badeka A, Kontakos S, Kontominas MG (2014) Differentiation of greek thyme honeys according to geographical origin based on the combination of phenolic compounds and conventional quality parameters using chemometrics. Food Anal Methods 7:2113–2121

Karabagias IK, Louppis AP, Karabournioti S, Kontakos S, Papastephanou C, Kontominas MG (2017) Characterization and geographical discrimination of commercial citrus spp. honeys produced in different Mediterranean countries based on minerals, volatile compounds and physicochemical parameters, using chemometrics. Food Chem 217:445–455

Kaškonienė V, Venskutonis PR, Čeksteytė V (2008) Composition of volatile compounds of honey of various floral origin and beebread collected in Lithuania. Food Chemistry 111(4):988–997

Kaškonienė V, Venskutonis PR (2010) Floral markers in honey of various botanical and geographic origins: a review. Compr Rev Food Sci Food Saf 9:620–634

Kaškonienė V, Venskutonis PR, Čeksteytė V (2010) Carbohydrate composition and electrical conductivity of different origin honeys from Lithuania. LWT Food Sci Technol 43:801–807

Kečkeš S, Gašić U, Veličković TĆ, Milojković-Opsenica D, Natić M, Tešic Z (2013) The determination of phenolic profiles of Serbian unifloral honeys using ultra-high-performance liquid chromatography/high resolution accurate mass spectrometry. Food Chem 138:32–40

Kenjerić D, Primorac L, Mandić ML, Bubalo D, Perl PA, Flanjak I (2006) Dalmatian sage (*Salvia officinalis* L.) honey characterisation. Deutcshe Lebensmittel-Rundschau 102:479–484

Kenjeric D, Mandic ML, Primorac L, Cacic F (2008) Flavonoid pattern of sage (*Salvia officinalis* L.) unifloral honey. Food Chem 110:187–192

Kešckeš J, Trifković J, Andrić F, Jovetić M, Tešić Ž, Milojković-Opsenica D (2013) Amino acids profile of Serbian unifloral honeys. J Sci Food Agric 93:3368–3376

Kim L, Brudzyski K (2018) Identification of menaquinones (vitamin K2 homologues) as novel constituents of honey. Food Chem 249:184–192

Kozłowicz K, Rózyło R, Gładyszewska B, Matwijczuk A, Gładyszewski G, Chocyk D, Samborska K, Piekut J, Smolewska M (2020) Identification of sugars and phenolic compounds in honey powders with the use of GC–MS FTIR spectroscopy, and X-ray diffraction. Sci Rep 10:16269

Krell R (1996) Value-added products from beekeeping. In: FAO agricultural services bulletin no. 124. http://www.fao.org/3/a-w0076e/index.html. Accessed Sept 2016

Kropf U, Jamnik M, Bertoncelj J, Golob T (2008) Linear regression model of the ash mass fraction and electrical conductivity for Slovenian honey. Food Technol Biotechnol 46:335–340

Kurtagić H, Barudanović S, Durmić V (2015) Determination of rutin, quercetin, naringenin and hesperetin in the honey from Bosnia and Herzegovina (B & H) in relation to the composition of pollen. J Environ Sci Eng A 4:615–622

Laghi L, Picone G, Capozzi F (2014) Nuclear magnetic resonance for foodomics beyond food analysis. TrAC Trends Anal Chem 59:93–102

Lee DC, Lee SY, Cha SH, Choi YS, Rhee HI (1998) Discrimination of native bee-honey and foreign bee-honey by SDS–PAGE. Korean J Food Sci 30:1–5

Lenhardt L, Bro R, Zeković I, Dramićanin T, Dramićanin MD (2015) Fluorescence spectroscopy coupled with PARAFAC and PLS DA for characterization and classification of honey. Food Chem 75:284–291

León-Ruiz V, Vera S, González-Porto AV, San Andrés MP (2011) Vitamin C and sugar levels as simple markers for discriminating Spanish honey sources. J Food Sci 76:C356–C361

León-Ruiz V, Vera S, González-Porto AV, Andrés MPS (2013) Analysis of water-soluble vitamins in honey by isocratic RP-HPLC. Food Anal Methods 6:488–496

Lewiński R, Hernik A, Liszewska M, Buckley B, Czaja K, Korcz W, Słomczyńska A, Struciński P (2023) Validation of a modified QuEChERS method for the determination of selected organochlorine compounds in honey. Molecules 28:842

Liangliang Q, Jiang Y, Huang X, Cui M, Ning F, Liu T, Gao Y, Dong W, Nie Z, Luo L (2019) J Agric Food Chem 67:11256–11261

Likens ST, Nickerson GB (1964) Detection of certain hop oil constituents in brewing products. Am Soc Brew Chem Proc, pp 5–13

Liu H, Zhang M, Guo Y, Qiu H (2016) Solid-phase extraction of flavonoids in honey samples using carbamate-embedded triacontyl-modified silica sorbent. Food Chem 204:56–61

Long J, Tang M, Zuo M, Xu W, Meng S, Liu Z (2023) The antianxiety effects of koumine and gelsemine two main active components in the traditional Chinese herbal medicine Gelsemium: A comprehensive review. Abstract Brain-X 1(4). https://doi.org/10.1002/brx2.v1.4; https://doi.org/10.1002/brx2.46

Low NH, Sporns P (1988) Analysis and quantitation of minor di- and trisaccharides in honey, using capillary gas chromatography. J Food Sci 53:558–561

Lucatello L, Merlanti R, Inacio LDJ, Bisutti V, Montanucci L, Capolongo F (2021) Pyrrolizidine alkaloid concentrations in local Italian and retail honeys of different origin: a scenario of human exposure. J Food Compos Anal 104:104182

Lušić D, Koprivnjak O, Curić D, Sabatini AG, Conte LS (2007) Volatile profile of Croatian lime tree (*Tilia sp.*), fir honeydew (*Abies alba*) and sage (*Salvia officinalis*) honey. Food Technol Biotechnol 45:156–165

Ma W, Yang B, Li J, Li X (2022) Development of a simple, underivatized method for rapid determination of free amino acids in honey using dilute-and-shoot strategy and liquid chromatography-tandem mass spectrometry. Molecules 27:1056

Machado AM, Antunes M, Miguel MG, Vilas-Boas M, Figueiredo AC (2021) Volatile profile of Portuguese Monofloral honeys: significance in botanical origin determination. Molecules 26:4970

Mahmoud MAA, Kılıç-Büyükkurt O, Aboul Fotouh MM, Selli S (2024) Aroma active compounds of honey: Analysis with GC-MS, GC-O, and molecular sensory techniques. Journal of Food Composition and Analysis 134 (106545)

Mahmoud MAA, Kılıç-Büyükkurt Ö, Aboul Fotouh MM, Selli S (2024) Aroma active compounds of honey: analysis with GC-MS, GC-O, and molecular sensory techniques. J Food Compos Anal 134:106545

Maignial L, Pibrot P, Bonetti G, Chaintrau A, Marion JP (1992) Simultaneous distillation-extraction under static vacuum: isolation of volatile compounds at room temperature. J Chromatogr 606:87–94

Mannas D, Altug T (2007) SPME/GC/MS and sensory flavour profile analysis for estimation of authenticity of thyme honey. Int J Food Sci Technol 42:133–138

Mannina L, Sobolev AP, di Lorenzo A, Vista S, Tenore GC, Daglia M (2015) Chemical composition of different botanical origin honeys produced by Sicilian black honeybees (*Apis mellifera* ssp. *sicula*). J Agric Food Chem 63:5864–5874

Manyi-Loh CE, Clarke AM, Ndip RN (2011a) Identification of volatile compounds in solvent extracts of honeys produced in South Africa. Afr J Agric Res 6:4327–4334

Manyi-Loh C, Ndip RN, Clarke AM (2011b) Volatile compounds in honey: a review on their involvement in aroma, botanical origin determination and potential biomedical activities. Int J Mol Sci 12:9514–9532

Mara A, Migliorini M, Ciulu M, Chignola R, Egido C, Núñez O, Sentellas S, Saurina J, Caredda M, Deroma MA, Deidda S, Langasco I, Pilo MI, Spano N, Sanna G (2024) Elemental fingerprinting combined with machine learning techniques as a powerful tool for geographical discrimination of honeys from nearby regions. Food Secur 13:243

Manyi-Loh CE, Clarke AM, Munzhelele T, Green E, Mkwetshana NF, Ndip RN (2010) Selected south african honeys and their extracts possess in vitro anti-Helicobacter pylori activity. Arch Med Res 41(5):324–331

Marshall T, Williams KM (1987) Electrophoresis of honey: characterization of trace proteins from a complex biological matrix by silver staining. Anal Biochem 167:301–303

Mason BS, Slaver HT (1971) A gas chromatographic method for the determination of sugars in food. J Agr Food Chem 19:551–554

Mato I, Huidobro JF, Simal-Lozano J, Sancho MT (2006) Analytical methods for the determination of organic acids in honey. Critical Reviews in Analytical Chemistry 36:3–11

Mato Trautvetter S, Koelling-Speer I, Speer K (2009) Confirmation of phenolic acids and flavonoids in honeys by UPLC–MS. Apidologie 40:140–150

Matysiak I, Balcerzak M, Michalski R (2018) Ion chromatography with conductometric detection for quantitation of formic acid in polish bee honey. J Food Compos Anal 73:55–59

Mazzoni V, Bradesi P, Tomi F, Casanova J (1997) Direct qualitative and quantitative analysis of carbohydrate mixtures using ^{13}C NMR spectroscopy: application to honey. Magn Reson Chem 35:S81–S90

Mititelu M, Udeanu DI, Docea AO, Tsatsakis A, Calina D, Arsene AL, Nedelescu M, Neacsu SM, Velescu BŞ, Ghica M (2023) New method for risk assessment in environmental health: the paradigm of heavy metals in honey. Environ Res 236(Pt 1):115194

Mondragón-Cortez P, Ulloa J, Rosas-Ulloa P, Rodríguez-Rodríguez R, Resendiz Vázquez J (2013) Physicochemical characterization of honey from the West region of México. CyTA J Food 11:7–13

Moreira RFA, Maria CAB, Pietroluongo M, Trugo LC (2010) Chemical changes in the volatile fractions of Brazilian honeys during storage under tropical conditions. Food Chem 121:697–704

Morse RA, Lisk DJ (1980) Elemental analysis of honeys from several nations. Am Bee J 120:522–523

Nozal-Nalda MJ, Bernal JL, Toribio L, Alamo M, Diego JC, Tapia J (2005a) The use of carbohydrate profiles and chemometrics in the characterization of natural honeys of identical geographical origin. J Agric Food Chem 53:3095–3100

Nozal-Nalda MJ, Bernal-Yagüe JL, Diego-Calva JCD, Martín-Gómez MT (2005b) Classifying honeys from the Soria Province of Spain via multivariate analysis. Anal Bioanal Chem 382:311–319

Nyarko K, Boozer K, Greenlief CM (2023) Profiling of the polyphenol content of honey from different geographical origins in the United States. Molecules 28(13):5011

Ohashi K, Natori S, Kubo T (1999) Expression of amylase and glucose oxidase in the hypopharyngeal gland with an age-dependent role change of the worker honeybee (*Apis mellifera* L.). Eur J Biochem 265:127–133

Ouchemoukh S, Schweitzer P, Bey MB, Djoudad-Kadji H, Louaileche H (2010) HPLC sugar profiles of Algerian honeys. Food Chem 121:561–568

Overton SV, Manura JJ (1994) Flavor and aroma commercial bee honey: a burge-and-trap thermal desorption technique for the identification and quantification of volatiles and semivolatiles in honey. Am Lab 26:45–53

Pazitna A, Janácová A, Špánik (2012) The enantiomer distribution of major chiral volatile organic compounds in Slovakian monofloral honeys. Journal of Food and Nutrition Research 51: 236–241

Pedisić S, Čulina P, Pavlešić T, Vahčić N, Elez Garofulić I, Zorić Z, Dragović-Uzelac V, Repajić M (2023) Efficiency of microwave and ultrasound-assisted extraction as a green tool for polyphenolic isolation from Monofloral honeys. PRO 11:3141

Pérez RA, Sánchez-Brunete C, Calvo RM, Tadeo JL (2002) Analysis of volatiles from Spanish honeys by solid-phase microextraction and gas chromatography-mass spectrometry. J Agric Food Chem 50:2633–2637

Pérez M, Dominguez-López I, Lamuela-Raventós RM (2023) The chemistry behind the Folin-Ciocalteu method for the estimation of (poly)phenol content in food: total phenolic intake in a Mediterranean dietary pattern. J Agric Food Chem 71:17543–17553

Perna A, Simonetti A, Intaglietta I, Sofo A, Gambacorta E (2012) Metal content of southern Italy honey of different botanical origins and its correlation with polyphenol content and antioxidant activity. Int J Food Sci Technol 47:1909–1917

Persano-Oddo L, Piro R (2004) Main European unifloral honeys: descriptive sheets. Apidologie 35:S38–S81

Persano-Oddo L, Baldi E, Accorti M (1990) Diastatic activity in some unifloral honeys. Apidologie 21:17–24

Persano-Oddo L, Piazza MG, Pulcini P (1999) Invertase activity in honey. Apidologie 30:57–65

Petrov V (1974) Quantitative determination of amino acids in some Australian honeys. J Apic Res 13:61–66

Piasenzotto L, Gracco L, Conte L (2003) Solid phase micro extraction (SPME) applied to honey quality control. J Sci Food Agric 83:1037–1044

Pilz-Güther D, Speer K (2004) Development of a GC method for the simultaneous determination of organic acids in honey. Dtsch Lebensmitt Rundsch 100:84–87

Pirini A, Conte LS, Francioso O, Lercker G (1992) Capillary gas chromatographic determination of free amino acids in honey as a means of discrimination between different botanical sources. J High Resolut Chromatogr 15:165–170

Pisani A, Protano G, Riccobono F (2008) Minor and trace elements in different honey types produced in Siena County (Italy). Food Chem 107:1553–1560

Ploegaerts G, Desmecht D, Ernould S, Dubois V (2023) Determination of metals and metalloids in traces in honey: evaluation of different sample preparation methods and assay techniques. J Trace Elem Miner 4:100070

Plutowska B, Chmiel T, Dymerski T, Wardencki WA (2011) Headspace solid-phase microextraction method development and its application in the determination of volatiles in honeys by gas chromatography. Food Chem 126:1288–1298

Pohl P (2009) Determination of metal content in honey by atomic absorption and emission spectrometries. Trends Anal Chem 28:117–128

Pontes M, Marques JC, Cámara JS (2007) Screening of volatile composition from Portuguese multifloral honeys using headspace solid-phase micro-extraction-gas chromatography-quadrupole mass spectrometry. Talanta 74:91–103

Pontis JA, Alves Da Costa LAM, Reis Da Silva SJ, Flach A (2014) Color, phenolic and flavonoid content, and antioxidant activity of honey from Roraima, Brazil. Food Sci Technol Campinas 34:69–73

Primorac L, Angelkov B, Mandić M, Kenjerić D, Nedeljko M, Flanjak I, Perl Pirički A, Arapceska M (2009) Comparison of the Croatian and Macedonian honeydew honey. J Cent Eur Agric 10:263–270

Pyrzynska K, Biesaga M (2009) Analysis of phenolic acids and flavonoids in honey. Trends Anal Chem 28:893–902

Qiu PY, Ding HB, Tang YK, Xu RJ (1999) Determination of chemical composition of commercial honey by near-infrared spectroscopy. J Agric Food Chem 47:2760–2765

Rachineni K, Rao Kakita VM, Awasthi NP, Shirke VS, Hosur RV, Chandra SS (2022) Identifying type of sugar adulterants in honey: combined application of NMR spectroscopy and supervised machine learning classification. Curr Res Food Sci 5:272–277

Radovic BS, Careri M, Mangia A, Musci M, Gerboles M, Anklam E (2001) Contribution of dynamic headspace GC-MS analysis of aroma compounds to authenticity testing of honey. Food Chem 72:511–520

Ribeiro ROR, Mársico ET, de Jesús EFO, Carneiro CS, Junior CAC, de Almeida E, Filho VFN (2014) Determination of trace elements in honey from different regions in Rio de Janeiro state (Brazil) by total reflection X-ray fluorescence. J Food Sci 79:T738–T742

Rizelio VM, Gonzaga LV, da Silva GCB, Maltez HF, Oliveira-Costa AC, Fett R (2012) Fast determination of cations in honey by capillary electrophoresis: a possible method for geographic origin discrimination. Talanta 99:450–456

Rodriguez-Otero JL, Paseiro P, Simal J (1990) Intento de caracterización de las mieles naturales de Galicia mediante las fracciones proteicas separadas por electroforesis. Anal Bromatol 42:83–98

Rodriguez-Otero J, Paseiro P, Simal J, Cepeda A (1994) Mineral content of the honeys produced in Galicia (north-west Spain). Food Chem 49:169–171

Romera-Torres A, Romero-González R, Vidal JLM, Frenich AG (2020) Comprehensive tropane alkaloids analysis and retrospective screening of contaminants in honey samples using liquid chromatography-high resolution mass spectrometry (Orbitrap). Food Res Int 133:109130

Rossano R, Larocca M, Polito T, Perna AM, Padula MC, Martelli G, Riccio P (2012) What are the proteolytic enzymes of honey and what they do tell us? A fingerprint analysis by 2-d zymography of unifloral honeys. PLoS One 7:e49164

Ruiz-Matute AI, Brokl M, Soria AC, Sanz ML, Martínez-Castro I (2010) Gas chromatographic–mass spectrometric characterization of tri- and tetrasaccharides in honey. Food Chem 120:637–642

Sánchez MP, Huidobro JF, Mato I, Muniategui S, Sancho MT (2001) Evolution of invertase activity in honey over two years. J Agric Food Chem 49:416–422

Sancho MT, Muniategui S, Huidobro JF, Simal J (1991) Provincial classification of Basque Country (nothern Spain) honeys by their chemical composition. J Apic Res 30:168–172

Sanz S, Pérez C, Herrera A, Sanz M, Juan T (1995) Application of a statistical approach to the classification of honey by geographic origin. J Sci Food Agric 69:135–140

Sanz ML, del Castillo MD, Corzo N, Olano A (2003) 2-furoylmethyl amino acids and hydroxymethylfurfural as indicators of honey quality. J Agric Food Chem 51:4278–4283

Sanz ML, Sanz J, Martinez-Castro I (2004a) Gas chromatographic–mass spectrometric method for the qualitative and quantitative determination of disaccharides and trisaccharides in honey. J Chromatogr A 1059:143–148

Sanz ML, Gonzalez M, de Lorenzo C, Sanz J, Martinez-Castro I (2004b) Carbohydrate composition and physico chemical properties of artisanal honeys from Madrid (Spain): occurrence of *Echium* sp honey. J Sci Food Agric 84:1577–1584

Sanz ML, Gonzalez M, de Lorenzo C, Sanz J, Martinez-Castro I (2005) A contribution to the differentiation between nectar honey and honeydew honey. Food Chem 91:313–317

Sawicki T, Bączek N, Starowicz M (2020) Characterisation of the total phenolic, vitamins C and E content and antioxidant properties of the beebread and honey from the same batch. Czech J Food Sci 38:158–163

Schepartz AI, Subers MH (1964) The glucose oxidase of honey. Biochim Biophys Acta 85:228–237

Senyuva HZ, Gilbert J, Silici S, Charlton A, Dal C, Gurel N, Cimen D (2009) Profiling Turkish honeys to determine authenticity using physical and chemical characteristics. J Agric Food Chem 57:3911–3919

Seraglio SKT, Bergamo G, Brugnerotto P, Gonzaga LV, Fett R, Costa ACO (2021) Aliphatic organic acids as promising authenticity markers of bracatinga honeydew honey. Food Chem 343:128449

Sergiel I, Pohl P, Biesaga M, Mironczyk A (2014) Suitability of three-dimensional synchronous fluorescence spectroscopy for fingerprint analysis of honey samples with reference to their phenolic profiles. Food Chem 145:319–326

Serra J, Ventura E (1995) Characterization of citrus honey (*Citrus Spp.*) produced in Spain. J Agric Food Chem 43:2053–2057

Serrano S, Espejo R, Villarejo M, Jodral ML (2007) Diastase and invertase activities in Andalusian honeys. Int J Food Sci Technol 42:76–79

Silici S (2011) Determination of volatile compounds of pine honeys. Turk J Biol 35:641–645

Silici S, Karaman K (2014) Chemometric approaches for the characterization of Turkish rhododendron and honeydew honeys depending on amino acid composition. J Liq Chromatogr Relat Technol 37:864–877

Silici S, Uluozlu OD, Tuzen M, Soylak M (2016) Honeybee and honey as monitors for heavy metal contamination near the thermal power plants in Mugla, Turkey. Toxicol Ind Health 32:507–516

Silva LR, Videira R, Monteiro AP, Valentaõ P, Andrade PB (2009) Honey from Luso region (Portugal): physicochemical characteristics and mineral contents. Microchem J 93:73–77

Silva B, Gonzaga LV, Fett R, Oliveira Costa AC (2019) Improved strategy based on QuEChERS method followed by HPLC/DAD for the quantification of phenolic compounds from Mimosa scabrella Bentham honeydew honeys LWT 116108471

Simuth J, Bilikova K, Kovacova E, Kuzmova Z, Schroder W (2004) Immunochemical approach to detection of adulteration in honey: physiologically active royal jelly protein stimulating TNF-alpha release is a regular component of honey. J Agric Food Chem 52:2154–2158

Sing RP, Sing PN (1996) Amino acid and lipid spectra of larvae of honey bee (*Apis cerana* Fabr) feeding on mustard pollen. Apidologie 27:21–28

Smanalieva J, Senge B (2009) Analytical and rheological investigations into selected unifloral German honey. Eur Food Res Technol 229:107–113

Solayman M, Islam MA, Paul S, Ali Y, Khalil MI, Alam N, Gan SH (2016) Physicochemical properties, minerals, trace elements, and heavy metals in honey of different origins: a comprehensive review. Compr Rev Food Sci Food Saf 15:219–233

Stadelmeier M, Bergner KG (1986) Proteine des bienenhonigs. VII. Eigenschaften und herkunft der honigamylase. Z Lebensm Unters Forsch 182:196–199

Stephens JM, Schlothauer RC, Morris BD, Yang D, Fearnley L, Greenwood DR, Loomes KM (2010) Phenolic compounds and methylglyoxal in some New Zealand manuka and kanuka honeys. Food Chem 120:78–86

Suárez-Luque S, Mato I, Huidobro JF, Simal-Lozano J, Sancho MT (2002) Rapid determination of minoritary organic acids in honey by high performance liquid chromatography. J Chromatogr A 955:207–214

Suto M, Kawashima H, Nakamura Y (2020) Determination of organic acids in honey by liquid chromatography with tandem mass spectrometry. Food Anal Methods 13:2249–2257

Sweeley CC, Bentley R, Makita M, Wells WW (1963) Gas-liquid chromatography of trimethylsilyl derivatives of sugars and related substances. J Am Chem Soc 85(16):2497–2507

Tananaki C, Thrasyvoulou A, Giraudel JL, Montury M (2007) Determination of volatile characteristics of Greek and Turkish pine honey samples and their classification by using Kohonen self organising maps. Food Chem 101:1687–1693

Tedesco R, Barbaro E, Zangrando R, Rizzoli A, Malagnini V, Gambaro A, Fontana P, Capodaglio G (2020) Carbohydrate determination in honey samples by ion chromatography–mass spectrometry (HPAEC-MS). Anal Bioanal Chem 412:5217–5227

Terrab A, Vega-Pérez JM, Díez MJ, Heredia FJ (2001) Characterisation of northwest Moroccan honyes by gas chromatographic-mass spectrometric analysis of their sugar components. J Sci Food Agric 82:179–185

Terrab A, Hernanz D, Heredia FJ (2004) Inductively coupled plasma optical emission spectrometric determination of minerals in thyme honeys and their contribution to geographical discrimination. J Agric Food Chem 52:3441–3445

Thierig M, Raupbach J, Wolf D, Mascher T, Subramanian, K, Henle T (2023) 3-Phenyllactic acid and polyphenols are substances enhancing the antibacterial effect of methylglyoxal in manuka honey. Foods 12(5)

Tomás-Barberán FA, Martos I, Ferreres F, Radovic BS, Anklam E (2001) HPLC flavonoid profiles as markers for the botanical origin of European unifloral honeys. J Sci Food Agric 81:485–496

Truchado P, Ferreres F, Bortolotti L, Sabatini AG, Tomás-Barberán FA (2008) Nectar flavonol rhamnosides are floral markers of acacia (*Robinia pseudoacacia*) honey. J Agric Food Chem 56:8815–8824

Truchado P, Ferreres F, Tomas-Barberan FA (2009) Liquid chromatography-tandem mass spectrometry reveals the widespread occurrence of flavonoid glycosides in honey, and their potential as floral origin markers. J Chrom A 1216:7241–7248

Truchado P, Vit P, Ferreres F, Tomas-Barberan F (2011) Liquid chromatography–tandem mass spectrometry analysis allows the simultaneous characterization of C-glycosyl and O-glycosyl flavonoids in stingless bee honeys. J Chromatogr A 1218:7601–7607

Truchado P, Vit P, Heard TA, Tomàs-Barberàn FA, Ferreres F (2015) Determination of interglycosidic linkages in O-glycosyl flavones by high-performance liquid chromatography/photodiode-array detection coupled to electrospray ionization ion trap mass spectrometry. Its application to *Tetragonula carbonaria* honey from Australia. Rapid Commun Mass Spectrom 29:948–954

Truzzi C, Annibaldi A, Illuminati S, Finale C, Scarponi G (2014) Determination of proline in honey: comparison between official methods, optimization and validation of the analytical methodology. Food Chem 150:477–481

Tuberoso CIG, Bifulco E, Jerkovic I, Caboni P, Cabras P, Floris I (2009) Methyl syringate: a chemical marker of asphodel (*Asphodelus microcarpus* salzm. et viv.) monofloral honey. J Agric Food Chem 57:3895–3900

Tuberoso CIG, Bifulco E, Caboni P, Cottiglia F, Cabras P, Floris I (2010) Floral markers of strawberry tree (Arbutus unedo L.) honey

Tuberoso CI, Jerković I, Bifulco E, Marijanovic Z, Congiu F, Bubalo D (2012) Riboflavin and lumichrome in Dalmatian sage honey and other unifloral honeys determined by LC-DAD technique. Food Chem 135(3):1985–1990

Val A, Huidobro JF, Sánchez MP, Muniategui S, Fernández-Muiño MA, Sancho MT (1998) Enzymatic determination of galactose and lactose in honey. J Agric Food Chem 46:1381–1385

Valachová I, Bučeková M, Majtán J (2016) Quantification of bee-derived peptide defensin-1 in honey by competitive enzyme-linked immunosorbent assay, a new approach in honey quality control. Czech J Food Sci 34:233–243

Valverde S, Ares AM, Stephen Elmore J, Bernal J (2022) Recent trends in the analysis of honey constituents. Food Chem 387:132920

Vanhanen LP, Emmertz A, Savage GP (2011) Mineral analysis of mono-floral New Zealand honey. Food Chem 128:236–240

Vazquez L, Armada D, Celeiro M, Dagnac T, Llompart M (2021) Evaluating the presence and contents of phytochemicals in honey samples: phenolic compounds as indicators to identify their botanical origin. Food Secur 10:2616

Végh R, Csóka M, Sörös C, Sipos L (2024) Underexplored food safety hazards of beekeeping products: key knowledge gaps and suggestions for future research. Compr Rev Food Sci Food Saf 23(5):e13404

Vijayakumar KT, Bhat NS, Neethu T, Nayimabanu T, Nithin Kumar HL (2021) Periodical changes in quality parameters of honey during storage and processing. Int J Chem Stud 9:19–24

Villacrés-Granda I, Coello D, Proaño A, Ballesteros I, Roubik DW, Jijón G, Granda-Albuja G, Granda-Albuja S, Abreu-Naranjo R, Maza F, Tejera E, González-Paramás AM, Bullón P, Alvarez-Suarez JM (2021) Honey quality parameters, chemical composition and antimicrobial activity in twelve Ecuadorian stingless bees (Apidae: Apinae: Meliponini) tested against multiresistant human pathogens. LWT 140:110737

Viñas P, Balsalobre N, López-Erroz C, Hernández-Córdoba M (2004) Liquid chromatographic analysis of riboflavin vitamers in foods using fluorescence detection. J Agric Food Chem 52:1789–1794

Wabaidur SM, Ahmed YBH, Alothman ZA, Obbed MS, Al-Harbi NM, Al-Turki TM (2015) Ultra high-performance liquid-chromatography with mass spectrometry method for the simultaneous determination of phenolic constituents in honey from various floral sources using multiwalled carbon nanotubes as extraction sorbents. J Separ Sci 38:2597–2606

Wang G, Ren Y, Su Y, Zhang H, Li J, Zhao H, Zhang H, Han J (2024) Identification of toxic Gelsemium elegans in processed food and honey based on real-time PCR analysis. Food Res Int 182:114188

Wang Z, Zu T, Huang X, Jiang X, Jia G, Xu J, Cui Z, Zhu F, Zhang J, Li J, Cao Y (2023) Comprehensive investigation of the content and the origin of matrine-type alkaloids in Chinese honeys. Food Chem 402:134254

Wardencki W, Chmiel T, Dymerski T, Biernacka P, Plutowska B (2009) Application of gas chromatography, mass spectrometry and olfactometry for quality assessment of selected food products. Ecol Chem Eng 16:287–300

Waś E, Rybak-Chmielewska H, Szczęsna T, Kachaniuk K, Teper D (2011) Characteristics of polish unifloral honeys. III. Heather honey (*Calluna vulgaris* L.). J Apic Sci 55:129–137

Watkins OC, Joyce NI, Gould N, Perry NB (2018) Glycosides of the neurotoxin Tutin in toxic honeys are from *Coriaria arborea* phloem sap, not insect metabolism. J Nat Prod 81:1116–1120

Weber M, Meixner M, Dasbach R, Rozhon W, Dasbach M (2022) Analysis of sugar crystal size in honey. MethodsX 9:101823

Wehling M, von der Ohe K, von der Ohe W (2006) Problemhonig robinie: Zu wenig invertaseaktivitä. Deutsch Bienen J 13:18–19

Weston RJ (2000) The contribution of catalase and other natural products to the antibacterial activity of honey: a review. Food Chem 71:235–239

White JW (1975) Physical characteristics of honey. In: Crane E (ed) From honey: a comprehensive survey. Heinemmann, London, pp 207–239

White JW Jr (1978) Honey. In: Advances in food research. Academic, New York, pp 287–364

White JW, Kushnir I (1967) The enzymes of honey: examination by ion-exchange chromatography, gel filtration, and starch-gel electrophoresis. J Apicult Res 6:69–89

White JW, Schepartz AI, Subers MH (1963) Identification of inhibine, antibacterial factor in honey, as hydrogen peroxide and its origin in a honey glucose-oxidase system. Biochim Biophys Acta 73:57–70

Wilkins AL, Lu Y, Tan S-T (1995) Extractives from New Zealand honeys. 5. Aliphatic dicarboxylic acids in New Zealand rewerewa (*Knightea excelsa*) honey. J Agric Food Chem 43:3021–3025

Won SA, Li C, Kim J, Rhee H (2009) Immunological characterization of honey major protein and its application. Food Chem 113:1334–1338

Wongchawalit J, Yamamoto T, Nakai H, Kim Y, Sato N, Nishimoto M, Okuyama M, Mori H, Saji O, Chanchao C, Wongsiri S, Surarit R, Svasti J, Chiba S, Kimura A (2006) Purification and characterization of α-glucosidase I from Japanese honeybee (*Apis cerana japonica*) and molecular cloning of its cDNA. Biosci Biotechnol Biochemist 70:2889–2898

Wootton M, Edwards RA, Faraji-Haremi R (1978) Effect of accelerated storage conditions on the chemical composition and properties of Australian honeys. 3. Changes in volatile components. J Apic Res 17:167–172

Wu F, Zhao H, Zhan Y, Sun J, Ji P, Liu C, Yang E, Cao W (2022) Effect of processing steps on phenolic profile of rape honey (Brassica napus) using HPLC-ECD. LWT 172:114183

Wu F, Xu F, Liu W, Chen S, Luo H, Cheng N, Zhao H, Cao W (2023) A high-performance liquid chromatography with electrochemical detection method developed for the sensitive determination of ascorbic acid: validation, application, and comparison with titration, spectrophotometric, and high-performance liquid chromatography with diode-Array detection methods. Food Secur 12:3100

Wunnachit W, Jenner F, Sedgley M (1992) Floral and extrafloral nectar production in *Anacardium occidentale* l. (anacardiaceae): an andromonoecious species. Int J Plant Sci 153:413–420

Xiao T, Yang L, Yang F, Nie G, Jin X, Peng X, Zhong X, Wang J, Lu Y, Zheng Y (2024) Traceability of chemicals from Tripterygium Wilfordii Hook. f. in raw honey and the potential synergistic effects of honey on acute toxicity induced by celastrol and triptolide. Food Chem 451:139407

Xu J-J, An M, Yang R, Cao J, Ye L-H, Peng L-Q (2016) Trace amounts of poly-cyclodextrin wrapped carbon nanotubes for the microextraction of flavonoids in honey samples by capillary electrophoresis with light-emitting diode induced fluorescence detection. Electrophoresis 37:1891–1901

Yan S, Wang K, Al Naggar Y, Vander Heyden Y, Zhao L, Wu L, Xue X (2022) Natural plant toxins in honey: an ignored threat to human health. J Hazard Mater 424(Pt D):127682

Yan S, Zhang M, Yuan Y, Mu G, Xu H, Zhao T, Wang Y, Xue X (2023) Chaste honey in long term-storage: occurrence and accumulation of Maillard reaction products, and safety assessment. Food Chem 424:136457

Zamora MC, Chirife J (2006) Determination of water activity change due to crystallization in honeys from Argentina. Food Control 17:59–64

Zapata JE, Sepúlveda CT, Álvarez AC (2022) Kinetics of the thermal degradation of phenolic compounds from achiote leaves (*Bixa orellana* L.) and its effect on the antioxidant activity. Food Sci Technol 42:e30920

Zhang Q, Chen X, Chen S, Ye Y, Luo J, Li J, Yu S, Liu H, Liu Z (2017) Fatal honey poisoning in Southwest China: a case series of 31 cases. Southeast Asian J Trop Med Public Health 48(1):189–196

Zhang XH, Wang MJ, Liu RJ, Qing XD, Nie JF (2023a) Green sample preparation techniques and their use in the extraction and separation analysis of phenolic compounds in honey. Sep Purif Rev 53:372–389

Zhang XH, Gu HW, Liu RJ, Qing XD, Nie JF (2023b) A comprehensive review of the current trends and recent advancements on the authenticity of honey. Food Chem X 19:100850

Zhou Q, Wintersteen CL, Cadwallader KR (2002) Identification and quantification of aroma-active components that contribute to the distinct malty flavor of buckwheat honey. J Agric Food Chem 50:2016–2021

Zhu Z, Zhang Y, Wang J, Li X, Wang W, Huang Z (2019) Sugaring-out assisted liquid-liquid extraction coupled with high performance liquid chromatographyelectrochemical detection for the determination of 17 phenolic compounds in honey. J Chromatogr A 1601:104–114

Zieliński L, Deja S, Jasicka-Misiak I, Kafarski P (2014) Chemometrics as a tool of origin determination of polish monofloral and multifloral honeys. J Agric Food Chem 62:2973–2981

Živkov Baloš M, Popov N, Jakšić S, Mihaljev Ž, Pelić M, Ratajac R, Ljubojević Pelić D (2023) Sunflower honey-evaluation of quality and stability during storage. Food Secur 12:2585

Chapter 4
Health Benefits and Uses of Honey in Medicine

Juraj Majtan, António Machado, Francesca Giampieri, and José M. Álvarez-Suárez

Contents

Abstract Honey has been used to treat a broad spectrum of injuries, including chronic wounds and burns, since ancient times. It is a widely available folk remedy with pronounced antibacterial, antibiofilm, anti-inflammatory, and reepithelialization properties. Characterization of its wound-healing properties and description of mechanisms of action allow its use in clinical practice. Indeed, medical-grade honey

J. Majtan (✉)
Laboratory of Apidology and Apitherapy, Department of Microbial Genetics, Institute of Molecular Biology, Slovak Academy of Sciences, Bratislava, Slovakia

Department of Microbiology, Faculty of Medicine, Slovak Medical University, Bratislava, Slovakia
e-mail: juraj.majtan@savba.sk

A. Machado
Centro de Biotecnologia dos Açores (CBA), Departamento de Biologia, Faculdade de Ciências e Tecnologia, Universidade dos Açores, Ponta Delgada, Portugal

Laboratorio de Bacteriología, Instituto de Microbiología, Colegio de Ciencias Biológicas y Ambientales (COCIBA), Universidad San Francisco de Quito (USFQ), Quito, Ecuador

J. M. Alvarez-Suarez (ed.), *Bee Products – Chemical and Biological Properties*,
https://doi.org/10.1007/978-3-031-89049-9_4

and its related products are used in wound care to prove clinical efficacy and safety. To increase its further clinical use, it is necessary to elucidate the honey-mediated wound-healing process in detail. This chapter aims to characterize recent advancements in the in vitro antimicrobial, anti-inflammatory, and wound-healing properties of honey and in clinical testing of medical-grade honey. Although an increasing number of clinical trials using honey as a wound remedy have been conducted, there is inconclusive evidence that honey is superior to standard care and/or other wound care products. Nevertheless, honey was shown to speed up the healing process and is the ideal wound dressing, positively affecting the entire process of wound healing. However, further research and clinical testing are needed to determine the overall efficacy of honey in wound care management.

Keywords Honey · Wound healing · Antibacterial · Anti-inflammatory · Clinical trials

4.1 Introduction

The use of natural products is becoming a more popular approach in both medical treatment and the preservation of foods. According to the World Health Organization (WHO) statistics, up to 80% of the population in some developed countries have used natural products in their primary health care. Natural products can be utilized in the discovery of new antimicrobial drugs and the treatment of infectious diseases. Scientists have found that natural materials are generally more acceptable to consumers. If alternative approaches are effective, they may reduce humanity's reliance

F. Giampieri
Department of Clinical Sciences, Polytechnic University of Marche, Ancona, Italy

Research Group on Food, Nutritional Biochemistry and Health, Universidad Europea del Atlántico, Santander, Spain

International Research Center for Food Nutrition and Safety, Jiangsu University, Zhenjiang, China

Joint Laboratory on Food Science, Nutrition, and Intelligent Processing of Foods, Polytechnic University of Marche, Ancona, Italy

Universidad Europea del Atlántico, Santander, Spain

Jiangsu University, Zhenjiang, China

J. M. Álvarez-Suárez (✉)
Laboratorio de Investigación en Ingeniería en Alimentos (LabInAli), Departamento de Ingeniería en Alimentos, Colegio de Ciencias e Ingenierías, Universidad San Francisco de Quito (USFQ), Quito, Ecuador

Laboratorio de Bioexploración, Colegio de Ciencias Biológicas y Ambientales, Universidad San Francisco de Quito (USFQ), Quito, Ecuador
e-mail: jalvarez@usfq.edu.ec

on synthetic substances. Moreover, in some cases, natural products/compounds can be used in combination with antibiotics to enhance their activity.

Behind the biological effect of natural products is usually a complex of bioactive substances rather than one active molecule. In fact, there is often a complex of active substances that act synergistically, guaranteeing a unique biological effect. This complexity may cause problems either in the identification of individual active substances or in the characterization of complex interactions with one another. The increase in the resistance of pathogenic bacteria to antibiotics is also an increasingly important factor behind the growing interest in the use of these natural compounds. Nowadays, one of the most attractive natural available products is honey. Honey has been used in many countries as a treatment for disease for several centuries, even before knowledge existed surrounding the causes of infection. In addition to its powerful antioxidant properties, which help neutralize harmful free radicals in the body, it is an effective immunomodulator that activates the cells and mechanisms of the immune system and skin cells (Majtan 2014). The use of honey as a topical drug with pronounced antimicrobial and antioxidant properties has been suppressed and replaced by comfortable antibiotic therapy.

The beneficial effects of honey on human health have long been recognized. Recently, the characterization of honey as a natural product with antioxidant and antibacterial properties has been studied in order to elucidate direct clinical usage in wound healing and its mode of action. When entering the words "honey" and "wound" into the MEDLINE database, we obtained 1515 current scientific outcomes, 58% of them published in the past 10 years (2014–2024). These findings suggest that the scientific and clinical community's interest in honey as a wound-healing product is still growing.

In recent years, the main antioxidant compounds in honey have been increasingly studied. Several flavonoids and phenolic acids have been identified as the main contributors to the antioxidant activity of honey. Several studies highlight the presence of several flavonoids (e.g., quercetin, kaempferol, luteolin, apigenin, chrysin, and galangin) and phenolic acids (e.g., gallic acid), which are known for their ability to scavenge free radicals and protect cells from oxidative stress (Alvarez-Suarez et al. 2010c). Additionally, it has been confirmed that the antioxidant capacity of honey is highly dependent on the floral source, with darker types of honey showing significantly higher levels of phenolic compounds compared to lighter varieties (Alvarez-Suarez et al. 2010b). These antioxidants, in addition to others in lower concentration, act synergistically and help reduce oxidative damage, further contributing to the therapeutic potential of honey. Ongoing research continues to uncover and better understand the various factors that influence this antioxidant activity and its full potential to promote health and fight disease.

Antimicrobial activity is the most studied biological property of honey, and recently, it has been proposed as a suitable qualitative parameter of honey (Majtan et al. 2021). The antibacterial activity of honey is determined by several distinct factors and molecules, such as low pH, a high sugar content (osmolarity), hydrogen peroxide (H_2O_2), methylglyoxal, and the bee antibacterial peptide defensin-1 (Def-1). Furthermore, certain types of phytochemicals, especially polyphenolic compounds, may elevate the antibacterial effect of honey (Bucekova et al. 2018; Thierig et al. 2023; Brudzynski 2023). A detailed characterization of the antibacterial effect of honey allows the effective use of topical honey-based formulations and medical-grade honey in wound care.

The major antibacterial compounds in honey have been identified in the past 10 years. Kwakman et al. identified the antibacterial honeybee peptide defensin-1 (Def-1) in honey and described its antibacterial potential against Gram-positive bacteria (Kwakman et al. 2010). A few years ago, in 2008, Thomas Henle and his group at Dresden University isolated and characterized the major antibacterial compound of manuka honey, namely methylglyoxal (MGO) (Mavric et al. 2008). MGO is a highly reactive molecule with antibacterial properties. The concentration of MGO in manuka honey varies significantly, but it is up to 100-fold higher than in different blossom honeys (Kwakman et al. 2010, 2011).

It has been known that honey is very effective against topical infection, and it promotes the healing of burn injuries and chronic wounds. As a result, many studies have analyzed the composition of honey and have studied the physicochemical and biological properties that may give rise to its ability to work against various microorganisms. Many different types of honey can be found around the world, and as the flora varies between regions, this will affect the production and activity of different types of honey (Alvarez-Suarez et al. 2010a; García-Tenesaca et al. 2017; Bucekova et al. 2023). It is possible to differentiate honey into two main types: floral honey made from the nectar of blossoms (blossom honey) and honeydew honey made from the secretions of living parts of plants or the excretions of plant-sucking insects. It has been suggested that some types of honey (e.g., honeydew and manuka) are more effective wound-healing agents than others. However, there is still no consensus on testing methods for comparing the wound-healing properties of different types of honey.

This chapter presents an update on several biological properties reported in honey, such as its antioxidant and antimicrobial properties. It also delves into current knowledge concerning the anti-inflammatory and wound-healing properties of topically applied honey based on in vitro and clinical studies carried out, preferably those published in the past 5 years. In addition, the chapter analyzes the effect of technological processing (including heat treatment and sterilization) on the wound-healing properties of honey.

4.2 Antioxidant Activity

Oxidative stress is defined as an imbalance between the production of free radicals (FRs) and the cellular capacity to counteract or scavenge them. The complex interplay of endogenous and exogenous factors contributes significantly to the generation of reactive oxygen species (ROS) and other free radicals (Chaudhary et al. 2023). While these molecules play essential roles in several physiological processes, their overproduction can disrupt the delicate balance of the antioxidant system, leading to oxidative stress. This state of imbalance has been linked to a wide variety of pathological conditions, including cancer, autoimmune disorders, aging, cataracts, rheumatoid arthritis, cardiovascular diseases, and neurodegenerative diseases (Chaudhary et al. 2023). To counteract the deleterious effects of oxidative stress, the human body employs a complex network of endogenous antioxidants and relies on the exogenous supply of these protective molecules through diet and supplementation (Sharifi-Rad et al. 2020). In fact, there is increasing scientific evidence pointing to the crucial role of antioxidants in mitigating the detrimental effects of oxidative stress (Muscolo et al. 2024). Numerous studies have explored the impact of dietary intake and supplementation with antioxidants on various aspects of health, demonstrating their potential to combat oxidative damage and promote overall well-being. Research indicates that antioxidant-rich foods, such as honey, may protect against the damaging effects of FRs and ROS and thus exhibit beneficial effects in preventing oxidative cellular damage (Muscolo et al. 2024). This ability is attributed to the presence of a complex mixture of bioactive compounds, as in honey, which act together through different action mechanisms to eliminate or block FRs.

The main antioxidant effects reported in honey include (*i*) the neutralization of free radicals through compounds such as flavonoids, phenolic acids, enzymes, ascorbic acid (vitamin C), and carotenoids (Gheldof et al. 2002; Gheldof and Engeseth 2002; Alvarez-Suarez et al. 2009); (*ii*) protection against lipid peroxidation (Gheldof and Engeseth 2002; Alvarez-Suarez et al. 2012); (*iii*) chelation of metal ions (Alvarez-Suarez et al. 2009); and (*iv*) increased activity of antioxidant enzymes and activation of genes related to antioxidant response (Alvarez-Suarez et al. 2016).

Phenolic acids and flavonoids, derived mainly from flower nectar and pollen, constitute the main class of antioxidants in honey (Wilczyńska and Żak 2024). The profile of these compounds varies significantly depending on the botanical origin of the honey, contributing to the unique antioxidant properties of different honey varieties. The main antioxidant phenolic compounds in honey can be divided into two main groups: phenolic acids and flavonoids. Among the most important phenolic acids are the hydroxybenzoic acids (gallic, ellagic, syringic, and benzoic acid) and hydroxycinnamic acids (caffeic, ferulic, p-coumaric, cinnamic, and chlorogenic acid). The main flavonoids present in honey include those within the family of flavanones (hesperetin and pinocembrin), flavones (apigenin, chrysin, and galangin), and flavonols (kaempferol and quercetin) (Alvarez-Suarez et al. 2009; Wilczyńska and Żak 2024). All these compounds contribute to honey's ability to neutralize free radicals and protect cells from oxidative damage. It is important to note that the phenolic composition of honey can vary significantly. While some compounds, such as gallic acid and quercetin, are found in most honeys, others, such as hesperetin and naringenin, are less common and are only present in certain varieties (Erejuwa et al. 2012). The amount and profile of phenolic antioxidants in honey are heavily influenced by its floral source. Generally, darker honeys exhibit a higher total phenolic content (TPC) and, consequently, a greater antioxidant capacity than lighter varieties (Alvarez-Suarez et al. 2010b; Otmani et al. 2021; Majewska et al. 2024). This is because darker honeys are often derived from nectar sources richer in these compounds, such as buckwheat or manuka flowers (Majewska et al. 2024). Furthermore, research has shown that dark honeys contain higher concentrations of other important antioxidants, including flavonoids like quercetin and apigenin, ascorbic acid (vitamin C), and β-carotene (Alvarez-Suarez et al. 2010b; Amessis-Ouchemoukh et al. 2021; Majewska et al. 2024). This variation in antioxidant content highlights the importance of considering the floral source and color of honey when seeking specific health benefits. It is also worth noting that factors beyond floral source, such as environmental conditions, processing, and storage, can also impact the antioxidant profile of honey.

In addition to polyphenols, there are other compounds in honey that, although to a lesser extent, contribute to its antioxidant activity. Some examples of these are enzymatic antioxidants, such as glucose oxidase and catalase, which are produced by the bees themselves and are incorporated into honey during their production. Glucose oxidase catalyzes the oxidation of glucose to gluconic acid and hydrogen peroxide, which, in low concentrations, have antimicrobial and antioxidant properties, while catalase breaks down hydrogen peroxide into water and oxygen, preventing its accumulation and possible cell damage (González-Ceballos et al. 2023). Ascorbic acid, or vitamin C, is another antioxidant found in honey (Alvarez-Suarez et al. 2010b). This vitamin functions as a powerful free radical scavenger. Carotenoids, derived primarily from pollen and propolis collected by bees, are fat-soluble pigments that contribute to the vibrant colors of honey (Alvarez-Suarez et al. 2010b). These compounds exhibit antioxidant activity by quenching free radicals and protecting against lipid peroxidation, particularly in lipid-rich environments such as cell membranes (Alvarez-Suarez et al. 2012). Amino acids and peptides, derived from pollen and bee metabolism, also contribute to honey's antioxidant capacity (Bryś and Strachecka 2024). Certain amino acids, such as cysteine and methionine, possess sulfur-containing groups that can effectively scavenge free radicals (Kim et al. 2020). Melanoidins, formed during the Maillard reaction, are responsible for the characteristic color and flavor of honey and have also been shown to influence its antioxidant capacity (Brudzynski and Miotto 2011). Melanoidins in honey are formed through complex interactions between proteins, polyphenols, and carbohydrates. It has been reported that heat treatment can accelerate the formation of melanoidins in honey by increasing the integration of phenols into the protein structure and promoting the creation of polyphenol–protein complexes. These results suggest a strong correlation between changes in the total phenolic content of melanoidins in heated versus unheated honeys and their respective antioxidant activity. This suggests a direct link between polyphenols and melanoidins, which influence the functional properties of melanoidins (Brudzynski and Miotto 2011).

The antioxidant capacity of honey is commonly assessed by several in vitro assays commonly used for this purpose, such as the 1,1-diphenyl-2-picrylhydrazyl (DPPH), 2,2-azino-bis (3-ethylbenzothiazoline-6-sulfonic acid) (ABTS), ferric reducing antioxidant power (FRAP), oxygen radical absorbance capacity (ORAC), cupric ion reducing antioxidant capacity (CUPRAC), reducing power method (RP), and ferrous ion-chelating (FIC) activity assay (Table 4.1). These methods measure

Table 4.1 Traditional methods used to evaluate antioxidant activity in honey

Principle	Assay	Reaction mechanisms	Quantification mode	References
Electron transfer	1,1-diphenyl-2-picrylhydrazyl (DPPH) free radical scavenging assay	The antioxidant donates a hydrogen atom to the DPPH radical, reducing it and decreasing its absorbance.	Spectrophotometry (measurement of the decrease in absorbance) (abs. 515 or 517 nm)	Beretta et al. (2005), Moniruzzaman et al. (2011), and Erejuwa et al. (2012)
	2,2-azino-bis (3-ethylbenzothiazoline-6-sulfonic acid) (ABTS) radical cation decolorization assay/ Trolox equivalent antioxidant capacity (TEAC) method	The antioxidant donates an electron to the $ABTS^{\bullet+}$ blue-green chromophore, and it is reduced to ABTS and discolored.	Spectrophotometry (measurement of the decrease in absorbance) (abs. 734 or 750 nm)	Moniruzzaman et al. (2011)
	Ferric reducing antioxidant power (FRAP) assay	Reduction of a ferric 2,4,6-tripyridyl-s-triazine complex (Fe^{3+}-TPTZ) to its ferrous, violet-blue form (Fe^{2+}-TPTZ) in the presence of antioxidants.	Spectrophotometry (measurement of the formation of the colored compound) (abs. 593 or 700 nm)	Beretta et al. (2005), Moniruzzaman et al. (2011), and Erejuwa et al. (2012)
	Cupric ion reducing antioxidant capacity (CUPRAC) assay	Bis(neocuproine) copper (II) chloride [Cu (II)-Nc] reacts with polyphenols, where the reactive Ar-OH groups of polyphenols are oxidized to the corresponding quinones, and Cu (II)-Nc is reduced to the highly colored Cu (I)-Nc.	Spectrophotometry (measurement of the formation of the colored compound) (abs. 450 nm)	Ulusoy et al. (2010)
Hydrogen atom transfer reaction	Oxygen radical absorbance capacity (ORAC) method	The antioxidant protects a fluorescent molecule (fluorescein) from oxidation by peroxyl radicals, maintaining its fluorescence.	Fluorimetry (measurement of fluorescence) (excit. 485 ± 20 nm emiss. 528 ± 20 nm)	Beretta et al. (2005), Moniruzzaman et al. (2011), and Erejuwa et al. (2012)

(continued)

Table 4.1 (continued)

Principle	Assay	Reaction mechanisms	Quantification mode	References
Electron transfer/ hydrogen atom transfer reaction	Reducing power (RP) method	Antioxidants, which have reduction potential, react with potassium ferricyanide (Fe^{3+}) to form potassium ferrocyanide (Fe^{2+}), which then reacts with ferric chloride to form ferric ferrous complexes.	Spectrophotometry (measurement of the formation of the colored compound) (abs. 700 nm)	Bellik and Selles (2017)
Metal-chelating activity	Ferrous ion-chelating (FIC) activity	Ferrozine can form a complex with a red color by forming chelates with Fe^{2+}. The reaction is restricted in the presence of other chelating agents and results in a decrease in the red color of the ferrozine–Fe^{2+} complexes.	Spectrophotometry (measurement of the formation of the colored compound) (abs. 562 nm)	Bellik and Selles (2017)

different aspects of antioxidant activity, such as the ability of antioxidants to scavenge free radicals, donate electrons and/or reduce iron, donate electrons and/or protect against oxidative damage to molecules, and sequester or bind ferrous ions. Other methods have also been proposed as indicators of the antioxidant capacity of honey, such as the total phenolic (phenols and polyphenols) content (TPC) and total flavonoid content (TFC) assays (Martinello and Mutinelli 2021). However, while these methods provide valuable information about individual antioxidant compounds in honey, they fail to capture the full antioxidant potential of this complex matrix. The protective effect of honey is derived not only from phenolic compounds but also from their synergistic interactions with other components, including enzymes, carotenoids, vitamins, and Maillard reaction products. Therefore, relying solely on TPC or TFC measurements may underestimate the true antioxidant capacity of honey.

In addition to conventional methods, several specific assays have emerged as crucial tools for a more comprehensive assessment of antioxidant activity in honey (Table 4.2). For example, the hydroxyl radical scavenging activity (OH•) assay has

Table 4.2 Other methods used in the study of the antioxidant activity of honey

Principle	Assay	Reaction mechanisms	Quantification mode	References
Hydroxyl radical (OH•) scavenging potential	Hydroxyl radical scavenging activity assay	By assessing the sample's ability to compete with deoxyribose for hydroxyl radicals generated from the Fe^{3+}-EDTA, ascorbic acid, and H_2O_2 reaction mixture, a reduction in the production of malondialdehyde-like compounds can be observed. This results in a decreased formation of the thiobarbituric acid chromophore.	Spectrophotometry (measurement of the decrease in absorbance) (abs. 520 or 532 nm)	Alvarez-Suarez et al. (2012)
Superoxide radical (O_2•-) scavenging potential	Superoxide radical scavenging activity assay	The antioxidant competes with deoxyribose for hydroxyl radicals generated from the Fe^{3+}-EDTA, ascorbic acid, and H_2O_2 reaction mixture, resulting in a reduction in the production of malondialdehyde-like compounds. This, in turn, leads to a lower formation of the thiobarbituric acid chromophore.	Spectrophotometry (measurement of the decrease in absorbance) (abs. 560 nm)	Henriques (2006) and Alvarez-Suarez et al. (2012)
Hydrogen peroxide (H_2O_2) scavenging potential	Direct current (DC) polarographic assay	The ability of the antioxidants present in honey to decrease the anodic current of the hydrogen peroxide complex, formed in an alkaline solution, at the potential of mercury dissolution.	DC polarography (DCP)	Gorjanović et al. (2013)
Hydrogen transfer reaction	β-carotene-linoleic acid bleaching assay (BCB)	Linoleic acid undergoes oxidation due to reactive oxygen species (ROS) generated by hydrogen peroxide. The resulting oxidation products trigger the degradation of β-carotene, causing the molecule to lose its double bonds and, consequently, its distinctive orange color.	Spectrophotometry (measurement of the decrease in absorbance) (abs. 434 nm)	Bouali et al. (2022)

been used to assess the ability of honey to neutralize hydroxyl radicals (Alvarez-Suarez et al. 2012). This assay provides insight into the potential protective effects of honey against oxidative damage in biological systems. Similarly, the superoxide radical scavenging activity ($O_2^{\bullet -}$) assay has been used to determine the ability of honey to neutralize superoxide anions (reactive oxygen species), reactive species closely related to cellular damage and oxidative stress (Henriques 2006; Alvarez-Suarez et al. 2012). Another technique used in the study of the antioxidant properties of honey is determining the capacity of honey to inhibit lipid peroxidation (thiobarbituric acid reactive substances and lipid hydroperoxides). This method allows one to evaluate honey's capacity to inhibit oxidative damage to lipids, which is essential in order to understand its antioxidant properties and protective effects against lipid peroxidation to membranes, a key factor in the development of various diseases and in food spoilage (Alvarez-Suarez et al. 2012). The hydrogen peroxide scavenging activity assay allows one to quantify honey's capacity to scavenge hydrogen peroxide, an ROS associated with oxidative damage and inflammation (Gorjanović et al. 2013). Additionally, the β-carotene-linoleic acid bleaching assay (BCB) evaluates honey's protective effect on β-carotene, which can oxidize in the presence of linoleic acid, thereby reflecting honey's antioxidant potential (Bouali et al. 2022). Incorporating these additional assays is essential for gaining a nuanced understanding of honey's bioactive properties as each method offers unique insights into different aspects of its ability to counteract oxidative stress, contributing to a more thorough assessment of its health benefits.

4.3 Antibacterial Activity

The increasing ineffectiveness of current antibiotics and the emergence of more virulent and resistant microorganisms have compelled the scientific community to explore therapeutic alternatives from natural sources. Consequently, plant and animal-derived products have become a focal point in the study of natural antimicrobials. Among these, honey and its derivatives stand out (Alvarez-Suarez 2017; Machado et al. 2023).

The antibacterial activity of honey has been extensively studied against both Gram-positive and Gram-negative pathogens related to wound, skin, and mucosal infections, including *Staphylococcus aureus, Pseudomonas aeruginosa, Escherichia coli*, and *Salmonella enterica* serovar Typhimurium (Diekema et al. 2019; Mejia et al. 2021; Montero et al. 2021; Abdel-Shafi et al. 2022). Efforts have been made to identify which compounds or properties in honey contribute to its antimicrobial activity (Fig. 4.1). Although the antimicrobial strength of *Apis mellifera* honey is well known (Alvarez-Suarez et al. 2018; Cabezas-Mera et al. 2023, 2024), the exact mechanisms of this honey remain partially unclear.

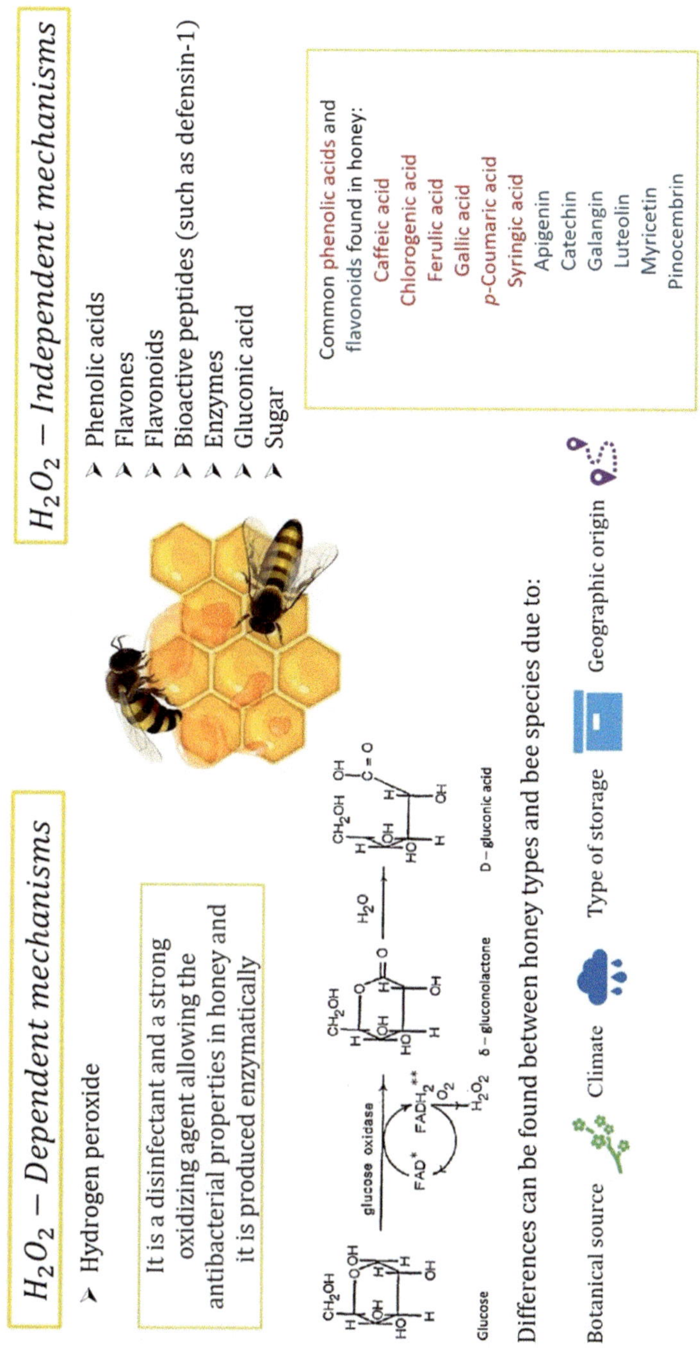

Fig. 4.1 Mechanisms and environmental factors influencing the antibacterial properties of honey and its key compounds

Apis mellifera honey and its compounds can control the growth of opportunistic pathogens and provide symbiotic benefits to the consumer (Chan-Rodríguez et al. 2012; Ewnetu et al. 2013; Alvarez-Suarez et al. 2018). Regarding Gram-negative bacteria, *Apis mellifera* honey exhibits significant antibacterial activity against both susceptible and antibiotic-resistant strains of bacteria, such as *Escherichia coli*, *Pseudomonas aeruginosa*, and *Klebsiella pneumoniae*. The honey's effectiveness is measured through inhibition zones, with higher concentrations generally showing stronger activity. Minimum inhibitory and bactericidal concentration (MIC and MBC) tests confirm its ability to inhibit bacterial growth and kill pathogens at various concentrations (Ewnetu et al. 2013). The antimicrobial activity of *Apis mellifera* honey against Gram-positive bacteria has been extensively studied, showing effectiveness against several pathogens, including *Staphylococcus aureus*, *Bacillus subtilis*, and *Streptococcus pyogenes*. For many Gram-positive bacteria, the MIC values for honey are typically between 2.5% and 20% (v/v), depending on the type of honey and the bacterial strain. For instance, *Staphylococcus aureus* has shown sensitivity to honey with MIC values of around 5–10%, while MBC values may range from 10% to 20% (v/v) (Tanuwidjaja et al. 2021; Kim et al. 2022).

Several studies have agreed that these damages result primarily from disruptive mechanisms of the cell membrane rather than other mechanisms such as inhibition of bacterial protein synthesis or enzymatic activity (Brudzynski and Sjaarda 2015; Cabezas-Mera et al. 2024). *Apis mellifera* honey also contains antimicrobial peptides (AMPs), which are known to interact with bacterial membranes. These peptides bind to membrane components, such as phospholipids, and insert themselves into the lipid bilayer. This results in pore formation and subsequent membrane depolarization, causing cytoplasmic leakage and the loss of vital cellular functions. AMPs in honey are particularly effective against Gram-positive bacteria like *Staphylococcus aureus* due to their thicker peptidoglycan layer (Hancock and Sahl 2006; Ghosh and Haldar 2015; Łyskowski et al. 2023). Moreover, defensin-1, a well-known AMP, plays a crucial role in its ability to fight bacterial infections. Defensin-1 is specifically known for its membrane-targeting action, where it binds to the bacterial cell wall, forming pores that lead to cell lysis (Sojka et al. 2016; Proaño et al. 2021).

4.4 Antifungal Activity

The antimicrobial potential of honey against phytopathogenic fungi is also known (Rozman et al. 2022), but reports involving yeasts and molds isolated from clinical infections are still scarce. Fungi can counteract honey's hyperosmolarity by consuming sugars (such as fructose, glucose, sucrose, maltose, and arabinose), adapting their metabolism with the help of different sugar-sensing systems (de Sousa et al. 2016; Ng et al. 2016; Pemmaraju et al. 2016; Weerasekera et al. 2017; Van Ende et al. 2019). However, the hyperosmolarity and low water activity limit fungal proliferation and spore germination. The antifungal activity of *Apis mellifera* honey has been explored in various studies, showing its effectiveness against a range of pathogenic fungi, including *Candida*, *Aspergillus*, and *Cryptococcus* species (Tanuwidjaja et al. 2021). Besides hydrogen peroxide, the *Apis mellifera* honey's antifungal properties are attributed to its complex mixture of bioactive compounds, such as phenolic acids and flavonoids. In addition, the high sugar concentration in honey creates a hyperosmotic environment, which inhibits fungal growth by dehydrating their cells. This low water activity limits fungal proliferation and spore germination (Kunat-Budzyńska et al. 2023).

One of the most studied clinically relevant pathogenic fungi is the *Candida* genus, mainly *Candida albicans*, due to its ability to produce nosocomial opportunistic infections through a wide repertoire of virulence factors, such as biofilm formation (Atiencia-Carrera et al. 2022a, b; Cangui-Panchi et al. 2022, 2023). Phenolic acids like caffeic acid and flavonoids such as quercetin and chrysin exhibit strong antifungal properties. These compounds interfere with fungal cell membranes and mitochondrial function, leading to disrupted energy production and ultimately cell death. For example, certain honeys have shown inhibitory effects against *Candida albicans* with minimum inhibitory concentrations (MICs) ranging from 5% to 15% (v/v) (Al Aboody and Mickymaray 2020; Tanuwidjaja et al. 2021; Łyskowski et al. 2023). Although the exact mechanisms are not completely understood at the molecular level, it is hypothesized that the presence of certain plant flavonoids (such as quercetin, luteolin, chrysin, and galangin) can produce morphological damage and affect the hyphal transition by modifying the mitochondrial and vacuolization activities (Candiracci et al. 2011; Canonico et al. 2014). Other flavonoids such as catechin and epigallocatechin gallate in co-treatment with fluconazole might be able to promote early apoptosis in *Candida* cells through the externalization of phosphatidylserine at the plasma membrane, mitochondrial depolarization, intracellular accumulation of ROS, cell condensation, and DNA fragmentation (Da Silva et al. 2014). Kim and Jin discovered that a peptide member of the major royal jelly prorein (MRJP) family called AcMRJP4–15 contained a high content of asparagine and positively charged (hydrophilic) amino acids showing a significant inhibition against *C. albicans* KCTC727 and an increment in membrane permeability with similar action as royalisin and jelleines (Kim and Jin 2019). This peptide was first observed in *Apis cerana* honey; however, homolog peptides have recently been reported in different types of honey (Ramón-Sierra et al. 2021). Some honey types produce hydrogen peroxide, which contributes to their antifungal action. Additionally, non-peroxide factors, such as methylglyoxal in manuka honey, exhibit potent antifungal activity by inducing oxidative stress and damaging the fungal cell wall and membrane (Ewnetu et al. 2013). Finally, *Apis mellifera* honey can inhibit the growth of *Candida* species with MIC values ranging from 3% to 10% depending on the fungal strain and honey type. In other cases, honeys with higher phenolic content have shown superior antifungal activity against *Aspergillus* species (Ewnetu et al. 2013; Cabezas-Mera et al. 2024).

4.5 Antibiofilm Activity

Biofilm formation by several pathogens on wounds, burns, or intravascular devices is a major concern worldwide due to the robust protection provided by biofilm structure against empirical standard treatments and host immune responses, thus making it a persistent source of microbial dissemination (Rajpaul 2015; Percival et al. 2015; Thomas and Thomas 2021; Cangui-Panchi et al. 2022, 2023). The antibiofilm activity of *Apis mellifera* honey is an important aspect of its antimicrobial properties, particularly against biofilm-forming microorganisms that are often resistant to conventional treatments. Biofilms are structured communities of monospecies or multispecies microorganisms encased in a self-produced extracellular matrix, making them difficult to eradicate. While few studies have been conducted on the possible antibiofilm effects of different honey samples, *Apis mellifera* honey has been shown to disrupt the formation of these biofilms and even break down established biofilms (Kim et al. 2022). For instance, *Staphylococcus aureus* and *Pseudomonas aeruginosa*, which are notorious biofilm formers, have shown reduced biofilm development in the presence of *Apis mellifera* honey, most likely due to the high osmotic pressure, along with the presence of phenolic compounds and hydrogen peroxide, which prevents bacteria from adhering to surfaces, a crucial initial step in biofilm formation (Rodríguez-Arias et al. 2022). *Apis mellifera* honey has been proven to be effective not only in preventing biofilm formation but also in disrupting mature biofilms. Literature shows that honey can penetrate the biofilm matrix and weaken its structural integrity, leading to the detachment of bacterial cells from the biofilm and increased susceptibility to antibacterial agents. For example, studies have reported that honey at concentrations of 10–20% (v/v) can significantly reduce the biofilm mass of *Pseudomonas aeruginosa* and *Staphylococcus epidermidis* (Ewnetu et al. 2013; Łyskowski et al. 2023). Honey has also been shown to enhance the effectiveness of antibiotics against biofilms. When combined with antibiotics like vancomycin or gentamicin, honey can improve the antibiotic's ability to penetrate biofilms and kill embedded bacteria. This synergistic effect has been observed in biofilms formed by *Staphylococcus aureus*, *Streptococcus mutans*, and *Escherichia coli*, indicating that honey could be used as adjuvant therapy in treating biofilm-associated infections (Ewnetu et al. 2013; Hasnamudhia et al. 2017; El-Allaky et al. 2020; Cabezas-Mera et al. 2024). Many studies underscore the potential of *Apis mellifera* honey as a natural agent in managing biofilm-related infections, offering an alternative or complementary strategy to conventional antimicrobial therapies (García-Tenesaca et al. 2018; Morroni et al. 2018; Proaño et al. 2021; Machado et al. 2023).

4.6 Antiviral Activity

Research on the antiviral properties of honey is limited, but existing studies suggest promising results. A crossover trial highlighted that topical treatment with honey was effective in managing pain and other symptoms of recurrent genital and labial herpes lesions, showing better results than acyclovir treatment with no side effects (Al-Waili 2004). Studies involving medicinal honey in addition to systemic acyclovir in zoster patients with impaired immune systems showed promising results in preventing secondary bacterial infections and speeding up the healing of herpetic lesions (Biluca et al. 2016; Ranneh et al. 2019; Mustafa et al. 2020; Ooi et al. 2021; Agussalim et al. 2022). Honey's antiviral activity can be attributed to several factors. One significant factor is the presence of methylglyoxal, which has been shown to exhibit antiviral properties. Additionally, honey contains various polyphenols and flavonoids, such as quercetin and kaempferol, which have been reported to inhibit the replication of viruses by interfering with their ability to infect host cells (Ahmed et al. 2018; Al-Hatamleh et al. 2020).

Numerous studies have demonstrated the antiviral properties of honey against a diverse array of viruses, including herpes simplex virus (HSV), human immunodeficiency virus (HIV), respiratory syncytial virus (RSV), varicella-zoster virus (VZV), adenovirus, and influenza viruses (Shahzad and Cohrs 2012; Hashemipour et al. 2014; Watanabe et al. 2014; Al-Hatamleh et al. 2020; Wu et al. 2021). These studies suggested that the antiviral effect was due to the honey's ability to enhance the immune response and directly interfere with viral replication. Watanabe et al. found that manuka honey efficiently inhibited influenza virus replication (which is related to its virucidal effects) and exhibited synergistic effects of honey with known anti-influenza-virus drugs such as zanamivir or oseltamivir, suggesting potential future pharmaceutical applications in managing virus-related infections (Watanabe et al. 2014).

Recent studies have also explored the potential of honey in the context of COVID-19. Honey's anti-inflammatory and immunomodulatory properties may help reduce the severity of symptoms in COVID-19 patients. Mustafa et al. highlighted that honey could modulate the cytokine response, potentially preventing the cytokine storm associated with severe COVID-19 cases (Mustafa et al. 2020). Furthermore, honey's antioxidant properties could mitigate oxidative stress, which is known to exacerbate viral infections (Ranneh et al. 2019). Despite these promising findings, more detailed, double-blind, randomized, controlled studies are warranted to investigate honey's antiviral effects comprehensively. The current literature lacks large-scale clinical trials that can provide conclusive evidence of honey's efficacy against viral infections. Additionally, factors such as the species of bee, botanical origin, geographical location, processing, and storage conditions of honey should be considered in ongoing research (Ahmed et al. 2018), as shown in Fig. 4.1.

4.7 Wound-Healing and Anti-inflammatory Effects: From In Vitro Evidence to Clinical Studies

4.7.1 Definition and Characterization of Medical-Grade Honey

Natural honey is considered an unsuitable remedy for a direct clinical application due to its content of harmful bacterial species or spores, which may negatively affect the wound-healing process. Medical-grade honey, which, in most cases, is 100% honey undergoing gamma-sterilization and laboratory/clinical testing, must fulfill the following criteria (Hermanns et al. 2020):

- Be organic and free of contaminants and toxic substances (e.g., pesticides and antibiotics).
- Be gamma-sterilized under standardized conditions and free of dangerous microorganisms.
- Be safe to implement in medical therapies.
- Follow strict production and storage standards and legal and safety regulations.
- Comply with the physicochemical criteria for the use of honey as a wound care product.

In addition, medical-grade honey should exhibit a certain level of biological properties, including antibacterial, antibiofilm, and anti-inflammatory activities. A very recent study revealed that the vast majority of clinical studies using honey as an oral remedy tested neither the basic quality criteria nor the biological potential of honey before its clinical testing (Majtan 2024). In light of massive fraudulent practices in the honey market (Gustafson et al. 2024), it is crucial to use honey that has retained its biological and health-promoting properties.

Medical-grade honey is not a food supplement but a medical device, and it must be manufactured under standardized conditions, including its handling, production, and storage. As stated above, medical-grade honey is either 100% honey or a mixture of honey and carriers to improve application. Like natural honey, medical-grade honey is sensitive to heat or light and improper storage, which can result in a decrease or even a loss of biological properties (e.g., antibacterial and wound-healing properties). Therefore, regularly testing honey at each stage of manufacture is necessary to maintain all of its quality parameters and biological activities. Figure 4.2 illustrates the schematic manufacturing process of medical-grade honey and the importance of regular testing before clinical use.

Although every type of raw honey (honey without any processing) displays a certain level of antibacterial, antioxidant, and anti-inflammatory activities, not every type is suitable for clinical use. It is very well-known that some types of honey are more effective than others in terms of eradicating bacterial wound pathogens and/or inhibiting wound inflammation due to the presence of active phytochemicals. Furthermore, not all medical-grade honeys work in the same way, and their mechanism of antibacterial action might be different. As an example, manuka honey, derived from the nectar of *Leptospermum scoparium*, exhibits strong antibacterial

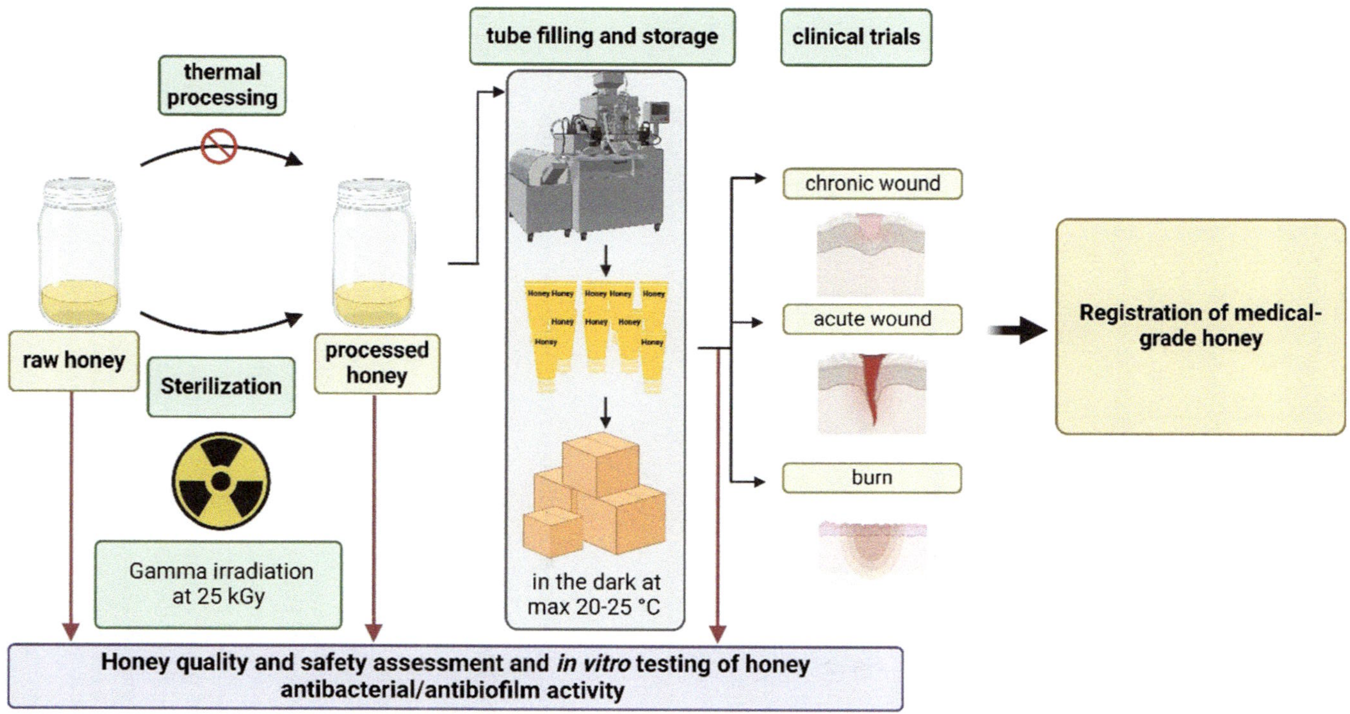

Fig. 4.2 Schematic production of medical-grade honey and recommended procedures before its registration

activity mediated through the action of MGO (Mavric et al. 2008; Adams et al. 2009; Atrott and Henle 2009). MGO, a highly reactive molecule, modifies defensin- 1, an antibacterial bee-derived peptide, the dominant protein in honey (MRJP1) (Majtan et al. 2012), and also the glucose oxidase (GOx) enzyme (Majtan et al. 2014) in manuka honey. Due to these MGO-induced structural modifications, defensin-1 and GOx lose antibacterial and enzymatic activities, respectively.

Phytochemicals, particularly plant-derived polyphenols (e.g., flavonoids and phenolic acids), are often reported in honey samples of different geographical and botanical origins; however, their concentrations vary between honeys (Becerril-Sánchez et al. 2021). It is proposed that phenolic compounds, according to various results reported, have great relevance to the biological and functional activities of honey. Polyphenols additionally enhance honey's antibacterial and antifungal potentials and are of primary importance for its other beneficial properties for human health, including antioxidant, anti-inflammatory, and antimutagenic activities. Moreover, these components are responsible for the sensory properties of honey, namely color, taste, and aroma.

There are currently several pharmaceutical products that are medical-grade honeys (Revamil®, Medihoney®, Melectis®, Actilitte™, and Vivamel®) based on sterile 100% honey of different botanical and geographical origins. A slightly different product based on honey has been recently developed and validated for wound care–bioengineered honey (SurgihoneyRO™). SurgihoneyRO™ was also developed into a novel synthetic product (Synthetic RO+), which mimics the honey's carbohydrate composition and contains GOx from *Aspergillus niger* as an alternative option to honey-based medical products (Matoke Holdings Ltd.).

4.7.1.1 Sterilization of Medical-Grade Honey–Based Products

The sterilization process of all medical-grade honeys is based on the action of ionizing radiation (gamma radiation), mostly at a dose of 20–30 kGy. The effect of ionizing radiation on the antibacterial properties of honey was first described in 1996, whereby the authors did not observe a reduction in the antibacterial activity of honey at doses of 25 and 50 kGy (Molan and Allen 1996). Other studies confirmed this observation for other types of honey and honey-incorporated wound dressings (Horniackova et al. 2017; Baldos et al. 2021; Wahyono et al. 2024). In addition, a recent meta-analysis showed that, except for moisture and vitamin C, gamma irradiation affected neither the biochemical characteristics of honey nor the viscosity as physical parameters (Ujilestari et al. 2023). Moreover, this type of sterilization significantly increased the content of flavonoids and phenolic compounds (Ujilestari et al. 2023).

Due to the viscous properties and heat-sensitive components of honey, ionizing radiation is the only way to effectively eliminate the vegetative and spore forms of microorganisms. Recent studies suggest that radiation does not negatively affect the physicochemical properties of honey (pH, color, acidity, or composition of sugars and minerals) but even increases the antioxidant potential of honey and the overall content of polyphenolic substances in honey (Hussein et al. 2011, 2014; Khalil et al. 2012). In addition, Horniackova et al. showed that gamma radiation at a dose of 10 kGy is effective in eliminating bacterial contamination without any negative effects on the antibacterial and antibiofilm activities of Slovak honeydew honey (Horniackova et al. 2017). Furthermore, their study showed that a dose of 10 kGy did not affect the content of defensin-1 in honeydew honey, whereas the higher doses of gamma radiation (above 10 kGy) significantly reduced its content.

4.7.2 Anti-inflammatory Effects of Topically Applied Honey

Inflammation is a key factor in the development of various chronic diseases, including cardiovascular diseases and diabetes, but also plays a key role in the chronicity of wounds. Uncontrolled inflammation is a common factor that contributes to extracellular matrix destruction, cellular senescence, and nonhealing cutaneous wounds. Several key factors/molecules responsible for inflammatory dysregulation, such as pro-inflammatory cytokines, growth factors, and matrix metalloproteinases (MMPs), have been identified (Liu et al. 2011). One of the major consequences of the persistent inflammatory response in the wound is unbalanced proteolytic activity, which overwhelms the protective mechanisms of local tissue (Eming et al. 2007; Widgerow 2012). Indeed, higher levels of proteolytic activity were found in patients with chronic wounds than in those with acute wounds (Tarnuzzer and Schultz 1996; Trengove et al. 1999). The expression and activity of various MMPs, particularly MMP-9, are highly upregulated in chronic wounds (Serra et al. 2013). Moreover, pro-inflammatory cytokines, the potent inducers of MMP expression in chronic wounds, have been shown to downregulate the expression of tissue inhibitors of MMPs (Gill and Parks 2008). On the other hand, MMPs are crucial in the wound-healing process if they are present in low concentrations where they are involved in each stage of the recovery process. Elevated concentrations of MMPs have been documented in diabetic foot ulcers (Fu et al. 2022).

Honey is considered an anti-inflammatory agent able to suppress the production of mediators of inflammation (e.g., pro-inflammatory cytokines) and support hemostasis (Majtan 2014). Various in vitro models of inflammation have been used to characterize the anti-inflammatory effects of honey. These effects can vary between honey samples of different botanical and geographical origins. A very recent study by Iosageanu et al. compared the anti-inflammatory and wound-healing properties of seven different honey varieties using various in vitro models developed in macrophage-like, fibroblast, and keratinocyte cell lines (Iosageanu et al. 2024). Interestingly, honey samples exhibited pro- and anti-inflammatory effects. Observed activities—such as fibroblast proliferation, keratinocyte migration, and collagen synthesis—were positively correlated with the content of total polyphenols in honey samples. In addition, the authors indicated that the obtained results may affect the reepithelialization and remodeling phases of wound healing. Our previous discoveries showed that acacia honey significantly enhanced the expression of MMP-9 mRNA in primary human keratinocytes and decreased the relative amount of collagen type IV in the basement membrane (Majtan et al. 2013). Honey was also found to stimulate the release of pro-inflammatory cytokines (TNF-α, IL-1, and IL-6) in isolated human peripheral blood monocytes (Tonks 2003). These findings contradict other results, where honey inhibited TNF-α-induced MMP-9 expression (Majtan et al. 2013; Lin et al. 2023). Recently, 3,6,7-trimethyllumazine, a newly discovered pteridine derivative unique to manuka honey, inhibited the gelatinase activities of both pro- and active MMP-9 (Lin et al. 2023). Honey exhibits both pro-inflammatory and anti-inflammatory properties (Fig. 4.3). It seems that honey acts according to the wound composition and local environments. When the level of pro-inflammatory mediators is low, honey stimulates the production of inflammatory cytokines and MMP-9 from keratinocytes. In contrast, when inflammation is in

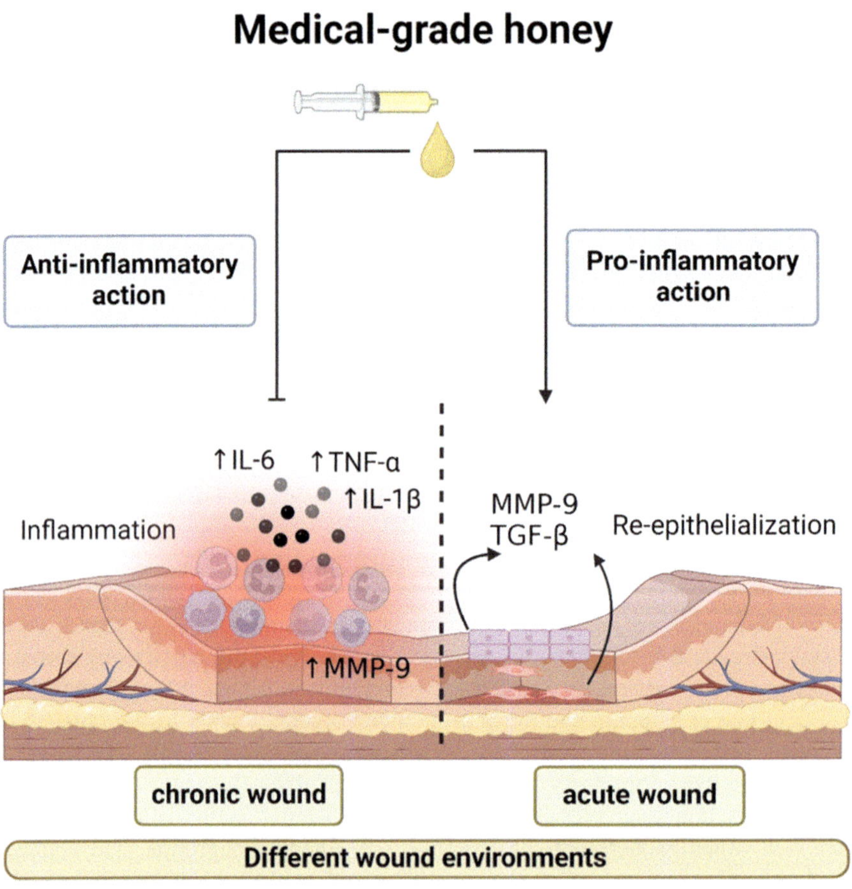

Fig. 4.3 Pro-inflammatory and anti-inflammatory actions of honey in wound-healing processes under different wound environment conditions. *IL* interleukin, *TNF* tumor necrosis factor, *MMP-9* matrix metalloproteinase 9

progress, honey suppresses the production of inflammatory cytokines and MMP-9 (Majtan 2014).

Inflammation reduction was observed following the application of honey, and these observations were supported by histological evidence of reduced numbers of inflammatory cells present in the wound tissue (Molan 2002). The anti-inflammatory effects of honey may be explained by several action mechanisms: *(i)* inhibition of the classical complement pathway (van den Berg et al. 2008), *(ii)* inhibition of ROS formation (van den Berg et al. 2008), *(iii)* inhibition of leukocyte infiltration (Leong et al. 2012), *(iv)* inhibition of cyclo-oxygenase-2 and inducible NO synthase expression (Hussein et al. 2012), *(v)* inhibition of MMP-9 and cell growth-promoting agents in chronic wound fluids in human keratinocytes (Majtan et al. 2013), *(vi)* reduction of inflammatory pain involving autonomic receptors (Owoyele et al. 2014), and *(vii)* modulation of the expression of nuclear factors, such as *NF-κB* and *p65* (Silva et al. 2021).

We performed a search for in vitro studies characterizing honey's anti-inflammatory effects, where honey was topically applied, and honey's active ingredients. The relevant studies were published in scientific databases (PubMed, Scopus, and Web of Science) between 2020 and 2024. Studies were excluded if honey was combined with other therapeutics or was part of natural product mixtures.

Fifteen studies were identified where the anti-inflammatory effect of honey and/or its active ingredients were examined. Individual studies with obtained outcomes are listed in Table 4.3. Broadly speaking, the phenolic compounds and proteins in honey are major groups of biologically active ingredients that suppress inflammation through different mechanisms. The most often used modes for evaluating the anti-inflammatory effects of honey were monocytes/macrophages.

Only two relevant in vivo studies published between 2020 and 2024 using honey as a topical anti-inflammatory agent have been identified (Table 4.4). Since honey is successfully used in wound care, further preclinical studies are not required, and human clinical studies are preferred.

4.7.3 Wound-Healing Effects of Topically Applied Honey

Honey has been a remedy and wound-healing agent since ancient times. It has a long tradition of use not only in Western medicine but also in traditional Chinese medicine and Ayurveda (Kuropatnicki et al. 2018). Nowadays, rediscovering honey with antimicrobial activity to combat bacterial resistance to current antibiotics is an attractive alternative that is gaining the attention of the scientific community and wound care practitioners. It has been postulated that bacteria cannot be resistant to honey due to its multifactorial action. However, a wider application of honey, particularly manuka honey in clinical medicine, might partially develop bacterial resistance to honey due to mutating genes involved in detoxifying MGO, which contributes to the antibacterial activity of manuka honey (Bischofberger et al. 2021).

Table 4.3 Summary of data collection from in vitro and in vivo studies using honey as a topically applied anti-inflammatory agent, published between 2020 and 2024

Honey type(s)	Experimental model/assay(s)	Major outcome(s)	Active ingredient(s)	References
Different varieties	Human THP-1 monocytes	Decreased TNF-α and IL-8 production in LPS-stimulated cells	Phenolic compounds	Iosageanu et al. (2024)
Centauri®	RAW 264.7 macrophages A549 lung cancer cells Normal colon cells	Reduction of LPS-induced cytokines from the model cell lines. NO-production inhibition	Naringenin Hydroxymethylglutaric acid	Santos Filipe et al. (2024)
Manuka	Gelatine zymography assay	Inhibition of MMP-9 gelatinase activity	Lepteridine™	Lin et al. (2023)
Coriander	Corneal limbal stem cells Stromal keratocytes	Downregulation of the expression of the IL-1β gene	Not available	Sanie-Jahromi et al. (2024)
Citrus	BSA denaturation assay	Protection of BSA against denaturation	Not available	Fratianni et al. (2023)
Citrus	Colorimetric COX inhibitor screening assay	Inhibition of the activity of COX-1 and COX-2	Not available	Tel-Çayan et al. (2023)
Chestnut	Colorimetric COX inhibitor screening assay	Inhibition of the activity of COX-1 and COX-2	Not available	Taş-Küçükaydın et al. (2023)
Astragalus	Colorimetric COX inhibitor screening assay	Inhibition of the activity of COX-1 and COX-2	Not available	Küçükaydın et al. (2023)
Chestnut and eucalyptus	Murine RAW 264.7 macrophages	Downregulation of the expression of LPS-induced Nfr2, iNOS, NF-κB, TNF-α, and IL-10 genes. Reduction of intracellular ROS and nitrite levels and apoptosis in LPS-stimulated cells	Not available	Cianciosi et al. (2021)
Ajwain	Gelatine zymography assay	Inhibition of MMP-2 and MMP-9 gelatinase activities	Not available	Kulkarni et al. (2021)
Stingless bee	Murine RAW 264.7 macrophages	Inhibition of NO production in LPS-stimulated cells	Not available	Chong et al. (2021)
Different varieties	Polymorphonuclear leukocytes ROS inhibition assay	Inhibition of ROS production by protein fractions	Unspecified honey proteins	Gohar et al. (2021)

(continued)

Table 4.3 (continued)

Honey type(s)	Experimental model/assay(s)	Major outcome(s)	Active ingredient(s)	References
Honeydew	RAW 264.7 macrophages	Decrease in the secretion of NOx, TNF-α, IL-6, MCP-1, IL-12p70, INF-γ, and IL-10 from LPS-stimulated cells	Phenolic compounds	Silva et al. (2020)
Manuka	HL-60 cells Primary human neutrophils	Increasing both anti- and pro-inflammatory signal release, depending on honey concentration	Not available	Minden-Birkenmaier et al. (2020)
Ziziphus	- Mouse macrophage cell line (J774.2) - Human whole blood phagocytes	Suppression of phagocyte oxidative burst and inflammatory markers and inhibition of NO	Unspecified honey proteins	Naqvi et al. (2023)

BSA bovine serum albumin, *COX* cyclooxygenase, *LPS* lipopolysaccharide, *MMP* matrix metalloproteinase, *NO* nitric oxide

Table 4.4 Summary of data collection from clinical studies using honey as a topically applied anti-inflammatory agent

Honey type source	Experimental model	Major outcome(s)	Active ingredients	References
Melipona beecheii	CD1 mouse	No significant difference was observed in terms of inflammation inhibition between honey and the control group.	Not available	Miranda et al. (2024)
Rhododendron	Wistar rats	Downregulation of the expression of TNF-α and MMP-9 genes. Enhancement of IL-10 gene expression.	Not available	Malkoç et al. (2020)

The wound-healing properties of honey involve antibacterial, antibiofilm, and anti-inflammatory activities, as well as immunomodulation and reepithelialization. The composition and physical properties of honey make it an ideal moist wound dressing that can positively affect each stage of the wound-healing process: coagulation and hemostasis, inflammation, proliferation, and wound remodeling. Honey is being clinically tested in the treatment of various acute and chronic wounds of different etiologies and burns; therefore, it is a versatile wound remedy. Besides medical-grade honeys and different types of natural honey have been used in case reports and clinical studies worldwide. However, some biological properties, at least antibacterial properties, should be tested in vitro prior to clinical use. In addition, the quality of conducted trials is disputable with a high rate of heterogeneity and bias. Clinical studies with consistent inclusion criteria and clearly defined efficacy parameters, including mean time to achieve wound healing, rate of complete wound healing, assessment of pain incidence, percentage of wound healing, duration of microorganism clearance in the wound, length of hospital stay, and treatment costs, allow for the assessment of the overall efficacy of honey in wound healing.

Since medical-grade honeys and their related wound care products are well-established and widely used in wound and burn care, in vitro and in vivo studies with honey as a wound-healing agent have mainly focused on *(i)* characterizing the antibacterial/antibiofilm potential of different honey types with manuka honey as a comparator, *(ii)* identifying honey's active ingredients and describing potential action mechanisms using different experimental models, and *(iii)* evaluating the wound-healing efficacy of honey mixtures with other substances, such as vitamins C and E and herbal extracts. According to a search of the Scopus database with the keywords "honey" and "wound healing," over 100 articles are published yearly. Therefore, this chapter focuses on the characterization of meta-analyses published in the past 5 years where the clinical efficacy of honey in wound care was analyzed.

4.7.3.1 The Clinical Efficacy and Safety of the Use of Honey in Wound Care

A detailed characterization of the biologically active compounds in each honey type is necessary to better understand the outcomes of clinical studies. According to the MEDLINE database, after the keywords "honey" and "wound healing" were entered and only clinical trials were selected, it was found that a total of 61 clinical studies have been conducted to date. Overall, a total of 26 meta-analyses have been performed where different types of wounds have been evaluated (e.g., chronic wounds, obstetric wounds, and diabetic foot ulcers). Considering the results from searching all databases (MEDLINE, Scopus, and Web of Science), a total of nine meta-analyses were identified for the period 2020–2024 (Table 4.5).

Honey as a safe wound remedy has been used in different wound types and compared with other wound agents and/or standard care. Chronic wounds, particularly diabetic foot ulcers, are the most common wound types to be treated with honey. A meta-analysis of diabetic foot treatment with honey was conducted in three out of nine studies. One study assessed the efficacy and safety of honey in the treatment of herpes wounds and compared the obtained results with those of standard care using acyclovir (Rocha et al. 2022).

Although topically applied honey has beneficial and positive effects on the healing of different wound types, there is inconclusive evidence to determine the effectiveness of honey on complete wound healing compared with standard care (e.g., gauze and saline) or other wound dressings such as alginate, hydrogel, and silver-impregnated dressings. Unfortunately, most clinical studies analyzed in meta-analyses were small (up to 100 participants), and all methods used are likely to

Table 4.5 Summary of data collection from meta-analyses analyzing honey as a topically applied agent in wound care

Wound type	Number of clinical studies analyzed (number of participants)[a]	Major outcomes	Year	References
Chronic	8 (906)	Increase in the percentage of wound healing Shortening of wound-healing time Decrease in VAS score Increase in painful discomfort during treatment No difference in healing rate, bacterial clearance, and hospital stay duration	2024	Tang et al. (2024)
Obstetric	5 (353)	No difference in healing rate Decrease in pain levels in the middle of the treatment Increase in personal satisfaction with honey treatment Decrease in the use of pain medication intervention	2024	Ferraz Barbosa et al. (2024)
Diabetic foot ulcer	15 (1008)	Increase in the ulcer healing rate Shortening of wound-healing time	2024	Monami et al. (2024)
Diabetic foot ulcer	16 (Not available)	Reduction of pain severity Increase in the rate of granulation development Reduction in the duration of wound recovery	2023	Yildiz Karadeniz and Kaplan Serin (2023)
Diabetic foot ulcer	9 (882)	Higher wound-healing rate Shortening of wound-healing time	2023	Li et al. (2023)
Surgical	11 (2561)	Reduction of hospitalization time Nonsignificant reduction of surgical site infection	2023	Dias et al. (2023)
Venous leg ulcer	10 (272)	No differences in the incidence of wound infection and complete wound healing compared to hydrogel	2022	Ribeiro et al. (2022)
Herpes	3 (808)	Increase in the healing rate compared to acyclovir Reduction of pain compared to using acyclovir	2022	Rocha et al. (2022)
Various etiologies	12 (1236)	Reduction of wound-healing time, length of hospital stays, and VAS score compared to povidone-iodine	2021	Zhang et al. (2021)

VAS: visual analog scale

[a] The number of clinical studies and participants may also include other wound agents—comparators

introduce errors in their results. The duration of follow-up in individual clinical studies was rather short (mostly ranging from 2 to 12 weeks), and studies were not designed to assess the time taken for the wound to completely heal. In addition, there is a very low quality of evidence, and therefore, further quantitative and qualitative clinical trials are required to obtain robust evidence, with consistent inclusion criteria, efficacy parameters such as duration of full recovery, microbiological clearance, the percentage decline in wound size, and adequate follow-up time to full recovery. Discrepancies among clinical studies may be caused, at least in part, by the use of different honey types. Some types could exhibit stronger therapeutic effects than others, and the action mechanisms could also be diverse.

One of the limitations of honey treatment in wound care is moderate transitioning pain that lasts up to 15 min after honey application (Dunford et al. 2000; Ahmed et al. 2003). A very recent retrospective study (Kenney et al. 2024) evaluated the pain intensity (daily pain score) reported by patients using the following burn wound dressing: manuka honey, silver nylon, silver sulfadiazine, negative pressure wound therapy, and 5% Sulfamylon solution. Manuka honey showed the highest daily average pain score; however, its use resulted in the lowest opioid consumption by patients during their hospital stay. On the other hand, other studies showed a significant decrease in pain scores using honey to treat a diverse spectrum of wounds (Biglari et al. 2013). Furthermore, Gulati et al. showed that patients with venous ulcers receiving honey-based wound dressings reported lower levels of pain and discomfort when compared with a povidone-iodine-based group (Gulati et al. 2012). In fact, honey can reduce pain by reducing both sensitization following inflammation-mediated prostaglandin synthesis and pressure on the tissue resulting from an edema (Al-Waili and Boni 2003; Kassim et al. 2010).

Though honey has also been used successfully in pediatric wound care, to the best of our knowledge, there are very few clinical studies describing the use of honey to treat wounds and/or burns in pediatric patients (Vardi et al. 1998; Ongom et al. 2004; Simon et al. 2006; Smaropoulos and Dimitriadou 2011; Sankar et al. 2021). According to a recent randomized controlled clinical study (Sankar et al. 2021), children's wounds treated with medical-grade manuka honey were about 1.9-fold more likely to heal if they were pressure injuries compared to the standard (routine) wound care group. Thus, honey decreased the wound-healing time in critically ill children with pressure injuries. In addition, neither allergic reactions nor other adverse events were observed.

It is shown here that honey has been intensively investigated as an anti-inflammatory agent and wound-healing remedy under in vitro and clinical conditions. Honey is an ideal wound remedy that benefits the wound-healing process at each step. Despite its beneficial effects on wound healing, meta-analyses provided limited evidence that honey is more effective than wound dressings. However, numerous case reports and noncontrolled clinical studies support honey as an effective and safe wound dressing for various nonhealing wounds, acute wounds, burns, and infected surgical wounds.

Acknowledgments The figures were generated with BioRender.com. This work was supported by the Scientific Grant Agency of the Ministry of Education of the Slovak Republic, the Slovak Academy of Sciences VEGA 2/0022/22, and the Slovak Research and Development Agency under Contract No. APVV-21-0262.

References

Abdel-Shafi S, El-Serwy H, El-Zawahry Y et al (2022) The association between icaA and icaB genes, antibiotic resistance and biofilm formation in clinical isolates of Staphylococci spp. Antibiotics 11:389. https://doi.org/10.3390/antibiotics11030389

Adams CJ, Manley-Harris M, Molan PC (2009) The origin of methylglyoxal in New Zealand manuka (Leptospermum scoparium) honey. Carbohydr Res 344:1050–1053. https://doi.org/10.1016/j.carres.2009.03.020

Agussalim, Umami N, Nurliyani, Agus A (2022) Stingless bee honey (Tetragonula laeviceps): chemical composition and their potential roles as an immunomodulator in malnourished rats. Saudi J Biol Sci 29:103404. https://doi.org/10.1016/J.SJBS.2022.103404

Ahmed AKJ, Hoekstra MJ, Hage JJ, Karim RB (2003) Honey-medicated dressing: transformation of an ancient remedy into modern therapy. Ann Plast Surg 50:143–147. https://doi.org/10.1097/01.SAP.0000032306.44107.C1

Ahmed S, Sulaiman SA, Baig AA et al (2018) Honey as a potential natural antioxidant medicine: an insight into its molecular mechanisms of action. Oxidative Med Cell Longev 2018:1–19. https://doi.org/10.1155/2018/8367846

Al Aboody MS, Mickymaray S (2020) Anti-fungal efficacy and mechanisms of flavonoids. Antibiotics 9:45. https://doi.org/10.3390/antibiotics9020045

Al-Hatamleh MAI, Hatmal MM, Sattar K et al (2020) Antiviral and immunomodulatory effects of phytochemicals from honey against COVID-19: potential mechanisms of action and future directions. Molecules 25:5017. https://doi.org/10.3390/molecules25215017

Alvarez-Suarez JM (ed) (2017) Bee products – chemical and biological properties, 1st edn. Springer International Publishing, Cham

Alvarez-Suarez JM, Giampieri F, Brenciani A et al (2018) Apis mellifera vs Melipona beecheii Cuban polifloral honeys: a comparison based on their physicochemical parameters, chemical composition and biological properties. LWT Food Sci Technol 87:272–279. https://doi.org/10.1016/j.lwt.2017.08.079

Alvarez-Suarez JM, Giampieri F, Cordero M et al (2016) Activation of AMPK/Nrf2 signalling by Manuka honey protects human dermal fibroblasts against oxidative damage by improving antioxidant response and mitochondrial function promoting wound healing. J Funct Foods 25:38–49. https://doi.org/10.1016/j.jff.2016.05.008

Alvarez-Suarez JM, Giampieri F, Damiani E et al (2012) Radical-scavenging activity, protective effect against lipid peroxidation and mineral contents of monofloral cuban honeys. Plant Foods Hum Nutr 67:31–38. https://doi.org/10.1007/s11130-011-0268-7

Alvarez-Suarez JM, González- Paramás AM, Santos-Buelga C, Battino M (2010a) Antioxidant characterization of native monofloral cuban honeys. J Agric Food Chem 58:9817–9824. https://doi.org/10.1021/jf1018164

Alvarez-Suarez JM, Tulipani S, Díaz D et al (2010b) Antioxidant and antimicrobial capacity of several monofloral cuban honeys and their correlation with color, polyphenol content and other chemical compounds. Food Chem Toxicol 48:2490–2499. https://doi.org/10.1016/j.fct.2010.06.021

Alvarez-Suarez JM, Tulipani S, Romandini S et al (2010c) Contribution of honey in nutrition and human health: a review. Med J Nutr Metab 3:15–23

Alvarez-Suarez JM, Tulipani S, Romandini S et al (2009) Methodological aspects about determination of phenolic compounds and in vitro evaluation of antioxidant capacity in the honey: a review. Curr Anal Chem 5:293–302. https://doi.org/10.2174/157341109789077768

Al-Waili NS (2004) Topical honey application vs. acyclovir for the treatment of recurrent herpes simplex lesions. Med Sci Monit 10:MT94–MT98

Al-Waili NS, Boni NS (2003) Natural honey lowers plasma prostaglandin concentrations in normal individuals. J Med Food 6:129–133. https://doi.org/10.1089/109662003322233530

Amessis-Ouchemoukh N, Maouche N, Otmani A et al (2021) Evaluation of Algerian's honey in terms of quality and authenticity based on the melissopalynology and physicochemical analysis and their antioxidant powers. Med J Nutr Metab 14:305–324. https://doi.org/10.3233/MNM-210561

Atiencia-Carrera MB, Cabezas-Mera FS, Tejera E, Machado A (2022a) Prevalence of biofilms in Candida spp. bloodstream infections: a meta-analysis. PLoS One 17:e0263522. https://doi.org/10.1371/journal.pone.0263522

Atiencia-Carrera MB, Cabezas-Mera FS, Vizuete K et al (2022b) Evaluation of the biofilm life cycle between Candida albicans and Candida tropicalis. Front Cell Infect Microbiol 12:953168. https://doi.org/10.3389/fcimb.2022.953168

Atrott J, Henle T (2009) Methylglyoxal in manuka honey – correlation with antibacterial properties. Czech J Food Sci 27:S163–S165. https://doi.org/10.17221/911-CJFS

Baldos D, Puno J, Tolentino L et al (2021) Radiation sterilization of Honey and Honey-Alginate Wound Dressing from Stingless Bee (Tetragonula biroi) Collected from Sta. Maria, Laguna, Philippines. J Trop Life Sci 11:217–223. https://doi.org/10.11594/jtls.11.02.11

Becerril-Sánchez AL, Quintero-Salazar B, Dublán-García O, Escalona-Buendía HB (2021) Phenolic compounds in honey and their relationship with antioxidant activity, botanical origin, and color. Antioxidants 10:1700. https://doi.org/10.3390/antiox10111700

Bellik Y, Selles SMA (2017) In vitro synergistic antioxidant activity of honey-Mentha spicata combination. J Food Meas Charact 11:111–118. https://doi.org/10.1007/s11694-016-9377-1

Beretta G, Granata P, Ferrero M et al (2005) Standardization of antioxidant properties of honey by a combination of spectrophotometric/fluorimetric assays and chemometrics. Anal Chim Acta 533:185–191. https://doi.org/10.1016/j.aca.2004.11.010

Biglari B, Moghaddam A, Santos K et al (2013) Multicentre prospective observational study on professional wound care using honey (Medihoney™). Int Wound J 10:252–259. https://doi.org/10.1111/J.1742-481X.2012.00970.X

Biluca FC, Braghini F, Gonzaga LV et al (2016) Physicochemical profiles, minerals and bioactive compounds of stingless bee honey (Meliponinae). J Food Compos Anal 50:61–69. https://doi.org/10.1016/J.JFCA.2016.05.007

Bischofberger AM, Pfrunder Cardozo KR, Baumgartner M, Hall AR (2021) Evolution of honey resistance in experimental populations of bacteria depends on the type of honey and has no major side effects for antibiotic susceptibility. Evol Appl 14:1314–1327. https://doi.org/10.1111/EVA.13200

Bouali N, Hamadou WS, Badraoui R et al (2022) Phytochemical composition, antioxidant, and anticancer activities of sidr honey: in vitro and in silico computational investigation. Life 13:35. https://doi.org/10.3390/life13010035

Brudzynski K (2023) Unexpected value of honey color for prediction of a non-enzymatic H_2O_2 production and honey antibacterial activity: a perspective. Meta 13:526. https://doi.org/10.3390/metabo13040526

Brudzynski K, Miotto D (2011) Honey melanoidins: analysis of the compositions of the high molecular weight melanoidins exhibiting radical-scavenging activity. Food Chem 127:1023–1030. https://doi.org/10.1016/j.foodchem.2011.01.075

Brudzynski K, Sjaarda C (2015) Honey glycoproteins containing antimicrobial peptides, Jelleins of the Major Royal Jelly Protein 1, are responsible for the cell wall lytic and bactericidal activities of honey. PLoS One 10:e0120238. https://doi.org/10.1371/JOURNAL.PONE.0120238

Bryś MS, Strachecka A (2024) The key role of amino acids in pollen quality and honey bee physiology—a review. Molecules 29:2605. https://doi.org/10.3390/molecules29112605

Bucekova M, Buriova M, Pekarik L et al (2018) Phytochemicals-mediated production of hydrogen peroxide is crucial for high antibacterial activity of honeydew honey. Sci Rep 8:9061. https://doi.org/10.1038/s41598-018-27449-3

Bucekova M, Godocikova J, Gueyte R et al (2023) Characterisation of physicochemical parameters and antibacterial properties of New Caledonian honeys. PLoS One 18:e0293730. https://doi.org/10.1371/journal.pone.0293730

Cabezas-Mera F, Cedeño-Pinargote AC, Tejera E et al (2024) Antimicrobial activity of stingless bee honey (Tribe: *Meliponini*) on clinical and foodborne pathogens: a systematic review and meta-analysis. Food Front 5:964. https://doi.org/10.1002/fft2.386

Cabezas-Mera FS, Atiencia-Carrera MB, Villacrés-Granda I et al (2023) Evaluation of the polyphenolic profile of native ecuadorian stingless bee honeys (Tribe: Meliponini) and their antibiofilm activity on susceptible and multidrug-resistant pathogens: an exploratory analysis. Curr Res Food Sci 7:100543. https://doi.org/10.1016/j.crfs.2023.100543

Candiracci M, Citterio B, Diamantini G et al (2011) Honey flavonoids, natural antifungal agents against Candida albicans. Int J Food Prop 14:799–808. https://doi.org/10.1080/10942910903453355

Cangui-Panchi SP, Ñacato-Toapanta AL, Enríquez-Martínez LJ et al (2022) Biofilm-forming microorganisms causing hospital-acquired infections from intravenous catheter: a systematic review. Curr Res Microb Sci 3:100175. https://doi.org/10.1016/j.crmicr.2022.100175

Cangui-Panchi SP, Ñacato-Toapanta AL, Enríquez-Martínez LJ et al (2023) Battle royale: immune response on biofilms – host-pathogen interactions. Curr Res Immunol 4:100057. https://doi.org/10.1016/j.crimmu.2023.100057

Canonico B, Candiracci M, Citterio B et al (2014) Honey flavonoids inhibit Candida albicans morphogenesis by affecting DNA behavior and mitochondrial function. Future Microbiol 9:445–456. https://doi.org/10.2217/FMB.14.17

Chan-Rodríguez D, Ramón-Sierra J, Lope-Ayora J et al (2012) Antibacterial properties of honey produced by Melipona beecheii and Apis mellifera against foodborn microorganisms. Food Sci Biotechnol 21(3):905–909. https://doi.org/10.1007/S10068-012-0118-X

Chaudhary P, Janmeda P, Docea AO et al (2023) Oxidative stress, free radicals and antioxidants: potential crosstalk in the pathophysiology of human diseases. Front Chem 11. https://doi.org/10.3389/fchem.2023.1158198

Chong KY, Chin NL, Yusof YA, Fakurazi S (2021) Anti-inflammatory activity of raw and processed stingless bee honey. Food Res 5:201–208

Cianciosi D, Forbes-Hernandez TY, Alvarez-Suarez JM et al (2021) Anti-inflammatory activities of Italian Chestnut and Eucalyptus honeys on murine RAW 264.7 macrophages. J Funct Foods 87:104752. https://doi.org/10.1016/J.JFF.2021.104752

Da Silva CR, De Andrade Neto JB, De Sousa CR et al (2014) Synergistic effect of the flavonoid catechin, quercetin, or epigallocatechin gallate with fluconazole induces apoptosis in Candida tropicalis resistant to fluconazole. Antimicrob Agents Chemother 58:1468–1478. https://doi.org/10.1128/AAC.00651-13/ASSET/E716FADE-D8D8-4553-827C-91CF3B0E1FCC/ASSETS/GRAPHIC/ZAC0031426350008.JPEG

de Sousa JMB, de Souza EL, Marques G et al (2016) Sugar profile, physicochemical and sensory aspects of monofloral honeys produced by different stingless bee species in Brazilian semi-arid region. LWT Food Sci Technol 65:645–651. https://doi.org/10.1016/J.LWT.2015.08.058

Dias TA, Fernandes DR, dos Santos BN et al (2023) Dressing to prevent surgical site infection in adult patients with cancer: a systematic review with meta-analysis. Support Care Cancer 31:1–13. https://doi.org/10.1007/S00520-022-07467-8/FIGURES/3

Diekema DJ, Hsueh PR, Mendes RE et al (2019) The microbiology of bloodstream infection: 20-year trends from the SENTRY antimicrobial surveillance program. Antimicrob Agents Chemother 63:1–10. https://doi.org/10.1128/AAC.00355-19

Dunford C, Cooper R, Molan P, White R (2000) The use of honey in wound management. Nurs Stand 15:63–68. https://doi.org/10.7748/NS2000.11.15.11.63.C2952

El-Allaky HS, Wahba NA, Talaat DM, Zakaria AS (2020) Antimicrobial effect of propolis administered through two different vehicles in high caries risk children: a randomized clinical trial. J Clin Pediatr Dent 44:289–295. https://doi.org/10.17796/1053-4625-44.5.1

Eming SA, Krieg T, Davidson JM (2007) Inflammation in wound repair: molecular and cellular mechanisms. J Invest Dermatol 127:514–525. https://doi.org/10.1038/SJ.JID.5700701

Erejuwa OO, Sulaiman SA, Ab Wahab MS (2012) Honey: a novel antioxidant. Molecules 17:4400–4423. https://doi.org/10.3390/molecules17044400

Ewnetu Y, Lemma W, Birhane N (2013) Antibacterial effects of Apis mellifera and stingless bees honeys on susceptible and resistant strains of Escherichia coli, Staphylococcus aureus and Klebsiella pneumoniae in Gondar, Northwest Ethiopia. BMC Complement Altern Med 13:269. https://doi.org/10.1186/1472-6882-13-269

Ferraz Barbosa B, de Moraes FCA, Alves A, da Silva B et al (2024) The use of honey for cicatrization and pain control of obstetric wounds: a systematic review and meta-analysis of randomized controlled trials. Nutrients 16. https://doi.org/10.3390/NU16020185

Fratianni F, Amato G, d'Acierno A et al (2023) In vitro prospective healthy and nutritional benefits of different Citrus monofloral honeys. Sci Rep 13(1):1–10. https://doi.org/10.1038/s41598-023-27802-1

Fu K, Zheng X, Chen Y et al (2022) Role of matrix metalloproteinases in diabetic foot ulcers: potential therapeutic targets. Front Pharmacol 13:1050630. https://doi.org/10.3389/FPHAR.2022.1050630

García-Tenesaca M, Navarrete E, Iturralde G et al (2017) Influence of botanical origin and chemical composition on the protective effect against oxidative damage and the capacity to reduce in vitro bacterial biofilms of monofloral honeys from the andean region of Ecuador. Int J Mol Sci 19:45. https://doi.org/10.3390/ijms19010045

García-Tenesaca M, Navarrete ES, Iturralde GA et al (2018) Influence of botanical origin and chemical composition on the protective effect against oxidative damage and the capacity to reduce in vitro bacterial biofilms of monofloral honeys from the andean region of Ecuador. Int J Mol Sci 19. https://doi.org/10.3390/ijms19010045

Gheldof N, Engeseth NJ (2002) Antioxidant capacity of honeys from various floral sources based on the determination of oxygen radical absorbance capacity and inhibition of in vitro lipoprotein oxidation in human serum samples. J Agric Food Chem 50:3050–3055. https://doi.org/10.1021/jf0114637

Gheldof N, Wang XH, Engeseth NJ (2002) Identification and quantification of antioxidant components of honeys from various floral sources. J Agric Food Chem 50:5870–5877. https://doi.org/10.1021/jf0256135

Ghosh C, Haldar J (2015) Membrane-active small molecules: designs inspired by antimicrobial peptides. ChemMedChem 10:1606–1624. https://doi.org/10.1002/cmdc.201500299

Gill SE, Parks WC (2008) Metalloproteinases and their inhibitors: regulators of wound healing. Int J Biochem Cell Biol 40:1334–1347. https://doi.org/10.1016/J.BIOCEL.2007.10.024

Gohar A, Dastagir N, Jabeen A, Azim MK (2021) Characterization of immunomodulatory activity of proteins of natural honeys. J Food Meas Charact 15:4475–4481. https://doi.org/10.1007/S11694-021-01033-2/TABLES/2

González-Ceballos L, Carlos Guirado-Moreno J, Utzeri G et al (2023) Straightforward purification method for the determination of the activity of glucose oxidase and catalase in honey by extracting polyphenols with a film-shaped polymer. Food Chem 405:134789. https://doi.org/10.1016/j.foodchem.2022.134789

Gorjanović SŽ, Alvarez-Suarez JM, Novaković MM et al (2013) Comparative analysis of antioxidant activity of honey of different floral sources using recently developed polarographic and various spectrophotometric assays. J Food Compos Anal 30:13–18. https://doi.org/10.1016/j.jfca.2012.12.004

Gulati S, Qureshi A, Srivastava A et al (2012) A prospective randomized study to compare the effectiveness of honey dressing vs. povidone iodine dressing in chronic wound healing. Indian J Surg 76:193. https://doi.org/10.1007/S12262-012-0682-6

Gustafson CR, Champetier A, Tuyizere O, Gitungwa H (2024) The impact of honey fraud information on the valuation of honey origin: evidence from an incentivized economic experiment. Food Control 155:110070. https://doi.org/10.1016/j.foodcont.2023.110070

Hancock REW, Sahl H-G (2006) Antimicrobial and host-defense peptides as new anti-infective therapeutic strategies. Nat Biotechnol 24:1551–1557. https://doi.org/10.1038/nbt1267

Hashemipour MA, Tavakolineghad Z, Arabzadeh SAM et al (2014) Antiviral activities of honey, Royal Jelly, and Acyclovir against HSV-1. Wounds 26:47–54

Hasnamudhia F, Bachtiar EW, Sahlan M, Soekanto SA (2017) The effect of CPP-ACP-propolis chewing gum on calcium and phosphate ion release on caries-active subjects' saliva and the formation of Streptococcus mutans biofilm. J Phys Conf Ser 884:012137. https://doi.org/10.1088/1742-6596/884/1/012137

Henriques A (2006) Free radical production and quenching in honeys with wound healing potential. J Antimicrob Chemother 58:773–777. https://doi.org/10.1093/jac/dkl336

Hermanns R, Mateescu C, Thrasyvoulou A et al (2020) Defining the standards for medical grade honey. J Apic Res 59:125–135. https://doi.org/10.1080/00218839.2019.1693713

Horniackova M, Bucekova M, Valachova I, Majtan J (2017) Effect of gamma radiation on the antibacterial and antibiofilm activity of honeydew honey. Eur Food Res Technol 243:81–88. https://doi.org/10.1007/S00217-016-2725-X/FIGURES/4

Hussein SZ, Mohd Yusoff K, Makpol S, Mohd Yusof YA (2012) Gelam honey inhibits the production of proinflammatory, mediators NO, PGE(2), TNF-α, and IL-6 in carrageenan-induced acute paw edema in rats. Evid Based Complement Alternat Med 2012:1. https://doi.org/10.1155/2012/109636

Hussein SZ, Yusoff KM, Makpol S, Yusof YAM (2011) Antioxidant capacities and total phenolic contents increase with gamma irradiation in two types of Malaysian honey. Molecules 16:6378–6395. https://doi.org/10.3390/MOLECULES16066378

Hussein SZ, Yusoff KM, Makpol S, Yusof YAM (2014) Does gamma irradiation affect physicochemical properties of honey? Clin Ter 165:e125. https://doi.org/10.7471/CT.2014.1695

Iosageanu A, Stefan LM, Craciunescu O, Cimpean A (2024) Anti-inflammatory and wound healing properties of different honey varieties from Romania and correlations to their composition. Life 14:1187. https://doi.org/10.3390/LIFE14091187/S1

Kassim M, Achoui M, Mansor M, Yusoff KM (2010) The inhibitory effects of Gelam honey and its extracts on nitric oxide and prostaglandin E(2) in inflammatory tissues. Fitoterapia 81:1196–1201. https://doi.org/10.1016/J.FITOTE.2010.07.024

Kenney CL, Thorpe CR, Schauer SG et al (2024) Evaluation of pain associated with the application of burn dressings. Int J Burns Trauma 14:84–89. https://doi.org/10.62347/RBJC6006

Khalil MI, Sulaiman SA, Alam N et al (2012) Gamma irradiation increases the antioxidant properties of tualang honey stored under different conditions. Molecules 17:674–687. https://doi.org/10.3390/MOLECULES17010674

Kim BY, Jin BR (2019) Antimicrobial activity of the C-terminal of the major royal jelly protein 4 in a honeybee (Apis cerana). J Asia Pac Entomol 22:561–564. https://doi.org/10.1016/J.ASPEN.2019.04.004

Kim B-Y, Kim Y-H, Choi Y-S et al (2022) Antimicrobial activity of Apidermin 2 from the Honeybee Apis mellifera. Insects 13:958. https://doi.org/10.3390/insects13100958

Kim J-H, Jang H-J, Cho W-Y et al (2020) In vitro antioxidant actions of sulfur-containing amino acids. Arab J Chem 13:1678–1684. https://doi.org/10.1016/j.arabjc.2017.12.036

Küçükaydın S, Tel-Çayan G, Çayan F et al (2023) Characterization of Turkish Astragalus honeys according to their phenolic profiles and biological activities with a chemometric approach. Food Biosci 53:102507. https://doi.org/10.1016/J.FBIO.2023.102507

Kulkarni SS, Mishra S, Patil SB et al (2021) Unifloral ajwain honey ameliorates differential inhibition of matrix metalloproteinases 2 and 9 protein, cytotoxicity, and antioxidant potential. J Ayurveda Integr Med 12:633–639. https://doi.org/10.1016/J.JAIM.2021.05.012

Kunat-Budzyńska M, Rysiak A, Wiater A et al (2023) Chemical composition and antimicrobial activity of new honey varietals. Int J Environ Res Public Health 20:2458. https://doi.org/10.3390/ijerph20032458

Kuropatnicki AK, Kłósek M, Kucharzewski M (2018) Honey as medicine: historical perspectives. J Apic Res 57:113–118. https://doi.org/10.1080/00218839.2017.1411182

Kwakman PHS, te Velde AA, de Boer L et al (2011) Two major medicinal honeys have different mechanisms of bactericidal activity. PLoS One 6:e17709. https://doi.org/10.1371/journal.pone.0017709

Kwakman PHS, Velde AA te, Boer L, et al (2010) How honey kills bacteria. FASEB J 24:2576–2582. https://doi.org/10.1096/fj.09-150789

Leong AG, Herst PM, Harper JL (2012) Indigenous New Zealand honeys exhibit multiple anti-inflammatory activities. Innate Immun 18:459–466. https://doi.org/10.1177/1753425911422263

Lin B, Nair S, Fellner DMJ et al (2023) The Leptospermum scoparium (Mānuka)-specific nectar and honey compound 3,6,7-Trimethyllumazine (LepteridineTM) that inhibits matrix Metalloproteinase 9 (MMP-9) activity. Food Secur 12. https://doi.org/10.3390/FOODS12224072

Li S, Xiao T, Ye N et al (2023) Effect of honey dressing in the management of diabetic foot ulcers: a meta-analysis. Int Wound J 20:2626–2633. https://doi.org/10.1111/IWJ.14135

Liu YC, Margolis DJ, Isseroff RR (2011) Does inflammation have a role in the pathogenesis of venous ulcers?: a critical review of the evidence. J Invest Dermatol 131:818–827. https://doi.org/10.1038/JID.2010.428

Łyskowski A, Miłek M, Dżugan M (2023) Assessing the antimicrobial properties of honey protein components through in silico comparative peptide composition and distribution analysis. Antibiotics 12:830. https://doi.org/10.3390/antibiotics12050830

Machado A, Zamora-Mendoza L, Alexis F, Álvarez-Suarez JM (2023) Use of plant extracts, bee-derived products, and probiotic-related applications to fight multidrug-resistant pathogens in the post-antibiotic era. Future Pharmacol 3:535–567. https://doi.org/10.3390/futurepharmacol3030034

Majewska E, Drużyńska B, Derewiaka D et al (2024) Comparison of antioxidant properties and color of selected polish honeys and manuka honey. Food Secur 13:2666. https://doi.org/10.3390/foods13172666

Majtan J (2014) Honey: an immunomodulator in wound healing. Wound Repair Regen 22:187–192. https://doi.org/10.1111/wrr.12117

Majtan J (2024) In vitro testing of honey quality and biological functionality: underestimated elements in the clinical testing of honey. Front Nutr 11. https://doi.org/10.3389/fnut.2024.1433786

Majtan J, Bohova J, Garcia-Villalba R et al (2013) Fir honeydew honey flavonoids inhibit TNF-α-induced MMP-9 expression in human keratinocytes: a new action of honey in wound healing. Arch Dermatol Res 305:619–627. https://doi.org/10.1007/S00403-013-1385-Y

Majtan J, Bohova J, Prochazka E, Klaudiny J (2014) Methylglyoxal may affect hydrogen peroxide accumulation in manuka honey through the inhibition of glucose oxidase. J Med Food 17:290–293. https://doi.org/10.1089/jmf.2012.0201

Majtan J, Bucekova M, Kafantaris I et al (2021) Honey antibacterial activity: a neglected aspect of honey quality assurance as functional food. Trends Food Sci Technol 118:870–886. https://doi.org/10.1016/j.tifs.2021.11.012

Majtan J, Klaudiny J, Bohova J et al (2012) Methylglyoxal-induced modifications of significant honeybee proteinous components in manuka honey: Possible therapeutic implications. Fitoterapia 83:671–677. https://doi.org/10.1016/j.fitote.2012.02.002

Malkoç M, Yaman SÖ, Imamoğlu Y et al (2020) Anti-inflammatory, antioxidant and wound-healing effects of mad honey in streptozotocin-induced diabetic rats. J Apic Res 59:426–436. https://doi.org/10.1080/00218839.2019.1689036

Martinello M, Mutinelli F (2021) Antioxidant activity in bee products: a review. Antioxidants 10:71. https://doi.org/10.3390/antiox10010071

Matoke Holdings Ltd GB201716984D0 – Antimicrobial compositions – Google Patents. https://patents.google.com/patent/GB201716984D0/en. Accessed 24 Oct 2024

Mavric E, Wittmann S, Barth G, Henle T (2008) Identification and quantification of methylglyoxal as the dominant antibacterial constituent of Manuka (*Leptospermum scoparium*) honeys from New Zealand. Mol Nutr Food Res 52:483–489. https://doi.org/10.1002/mnfr.200700282

Mejia L, Vela G, Zapata S (2021) High occurrence of multiresistant Salmonella infantis in retail meat in Ecuador. Foodborne Pathog Dis 18:41–48. https://doi.org/10.1089/fpd.2020.2808

Minden-Birkenmaier BA, Meadows MB, Cherukuri K et al (2020) Manuka honey modulates the release profile of a dHL-60 neutrophil model under anti-inflammatory stimulation. J Tissue Viability 29:91–99. https://doi.org/10.1016/J.JTV.2020.03.005

Miranda IR, Acevedo-Fernandez J, Negrete-Leon E et al (2024) In vivo wound-healing and anti-inflammatory activities of honey produced by Melipona beecheii Bees. Jundishapur J Nat Pharm Prod 19(2):19. https://doi.org/10.5812/JJNPP-143682

Molan PC (2002) Re-introducing honey in the management of wounds and ulcers – theory and practice. Ostomy Wound Manage 48:28–40

Molan PC, Allen KL (1996) The effect of gamma-irradiation on the antibacterial activity of honey. J Pharm Pharmacol 48:1206–1209. https://doi.org/10.1111/J.2042-7158.1996.TB03922.X

Monami M, Ragghianti B, Scatena A et al (2024) Effectiveness of different advanced wound dressings versus standard of care for the management of diabetic foot ulcers: a meta-analysis of randomized controlled trials for the development of the Italian guidelines for the treatment of diabetic foot syndrome. Acta Diabetol 61:1517. https://doi.org/10.1007/S00592-024-02320-2

Moniruzzaman M, Khalil M, Sulaiman S, Gan S (2011) Advances in the analytical methods for determining the antioxidant properties of honey: a review. Afr J Tradit Complement Altern Med 9:36. https://doi.org/10.4314/ajtcam.v9i1.5

Montero L, Irazabal J, Cardenas P et al (2021) Extended-spectrum beta-lactamase producing-Escherichia coli isolated from irrigation waters and produce in Ecuador. Front Microbiol 12:1–13. https://doi.org/10.3389/fmicb.2021.709418

Morroni G, Alvarez-Suarez JM, Brenciani A et al (2018) Comparison of the antimicrobial activities of four honeys from three countries (New Zealand, Cuba, and Kenya). Front Microbiol 9:1378. https://doi.org/10.3389/FMICB.2018.01378/BIBTEX

Muscolo A, Mariateresa O, Giulio T, Mariateresa R (2024) Oxidative stress: the role of antioxidant phytochemicals in the prevention and treatment of diseases. Int J Mol Sci 25:3264. https://doi.org/10.3390/ijms25063264

Mustafa MZ, Shamsuddin SH, Sulaiman SA, Abdullah JM (2020) Anti-inflammatory properties of stingless bee honey may reduce the severity of pulmonary manifestations in COVID-19 infections. Malays J Med Sci 27:165–169. https://doi.org/10.21315/MJMS2020.27.2.17

Naqvi F, Dastagir N, Jabeen A (2023) Honey proteins regulate oxidative stress, inflammation and ameliorates hyperglycemia in streptozotocin induced diabetic rats. BMC Complement Med Ther 23:14. https://doi.org/10.1186/S12906-023-03837-9

Ng TS, Desa MNM, Sandai D et al (2016) Growth, biofilm formation, antifungal susceptibility and oxidative stress resistance of Candida glabrata are affected by different glucose concentrations. Infect Genet Evol 40:331–338. https://doi.org/10.1016/J.MEEGID.2015.09.004

Ongom P, Kijjambu S, Mutumba S, Sebbaale A (2004) Comparison of honey-ghee dressing with collagen dressing in the management of superficial burn wounds in children. East Cent Afr J Surg 9:67–71

Ooi TC, Yaacob M, Rajab NF et al (2021) The stingless bee honey protects against hydrogen peroxide-induced oxidative damage and lipopolysaccharide-induced inflammation in vitro. Saudi J Biol Sci 28:2987. https://doi.org/10.1016/J.SJBS.2021.02.039

Otmani A, Amessis-Ouchemoukh N, Birinci C et al (2021) Phenolic compounds and antioxidant and antibacterial activities of Algerian honeys. Food Biosci 42:101070. https://doi.org/10.1016/j.fbio.2021.101070

Owoyele BV, Oladejo RO, Ajomale K et al (2014) Analgesic and anti-inflammatory effects of honey: the involvement of autonomic receptors. Metab Brain Dis 29:167–173. https://doi.org/10.1007/S11011-013-9458-3

Pemmaraju SC, Pruthi PA, Prasad R, Pruthi V (2016) Modulation of Candida albicans biofilm by different carbon sources. Mycopathologia 181:341–352. https://doi.org/10.1007/S11046-016-9992-8

Percival SL, McCarty SM, Lipsky B (2015) Biofilms and wounds: an overview of the evidence. Adv Wound Care (New Rochelle) 4:373. https://doi.org/10.1089/WOUND.2014.0557

Proaño A, Coello D, Villacrés-Granda I et al (2021) The osmotic action of sugar combined with hydrogen peroxide and bee-derived antibacterial peptide Defensin-1 is crucial for the antibiofilm activity of eucalyptus honey. LWT Food Sci Technol 136:110379. https://doi.org/10.1016/j.lwt.2020.110379

Rajpaul K (2015) Biofilm in wound care. Br J Community Nurs Suppl Wound Care:S6–S11. https://doi.org/10.12968/BJCN.2015.20.SUP3.S6

Ramón-Sierra JM, Villanueva MA, Rodríguez-Mendiola M et al (2021) Characterization of a non-glycosylated fraction from honey proteins of Melipona beecheii with antimicrobial activity against Escherichia coli O157:H7. J Appl Microbiol 130:1913–1924. https://doi.org/10.1111/JAM.14921

Ranneh Y, Akim AM, Hamid HA et al (2019) Stingless bee honey protects against lipopolysaccharide induced-chronic subclinical systemic inflammation and oxidative stress by modulating Nrf2, NF-κB and p38 MAPK. Nutr Metab (Lond) 16:1–17. https://doi.org/10.1186/S12986-019-0341-Z/FIGURES/12

Ribeiro CTD, Dias FAL, Fregonezi GAF (2022) Hydrogel dressings for venous leg ulcers. Cochrane Database Syst Rev 8:CD010738. https://doi.org/10.1002/14651858.CD010738.PUB2

Rocha MP, Amorim JM, Lima WG et al (2022) Effect of honey and propolis, compared to acyclovir, against Herpes Simplex Virus (HSV)-induced lesions: a systematic review and meta-analysis. J Ethnopharmacol 287:114939. https://doi.org/10.1016/J.JEP.2021.114939

Rodríguez-Arias RJ, Guachi-Álvarez BO, Montalvo-Vivero DE, Machado A (2022) Lactobacilli displacement and Candida albicans inhibition on initial adhesion assays: a probiotic analysis. BMC Res Notes 15:1–7. https://doi.org/10.1186/s13104-022-06114-z

Rozman AS, Hashim N, Maringgal B, Abdan K (2022) A comprehensive review of stingless bee products: phytochemical composition and beneficial properties of honey, propolis, and pollen. Appl Sci 12:6370. https://doi.org/10.3390/app12136370

Sanie-Jahromi F, Khaki M, Heydari M et al (2024) Effect of low dose honey on the apoptosis and inflammation gene expression in corneal limbal stem cells and keratocytes and its efficacy as an ophthalmic formulation in the treatment of dry eye: in-vitro and clinical study. Front Med (Lausanne) 11. https://doi.org/10.3389/FMED.2024.1359463

Sankar J, Lalitha AV, Rameshkumar R et al (2021) Use of honey versus standard care for hospital-acquired pressure injury in critically ill children: a multicenter randomized controlled trial. Pediatr Crit Care Med 22:E349–E362. https://doi.org/10.1097/PCC.0000000000002611

Santos Filipe M, Kowalczyk T, Kukula-Koch W et al (2024) Evaluating the quality, physicochemical properties, and biological activities of Centauri® honey from Turkey. Food Biosci 62:105028. https://doi.org/10.1016/J.FBIO.2024.105028

Serra R, Buffone G, Falcone D et al (2013) Chronic venous leg ulcers are associated with high levels of metalloproteinases-9 and neutrophil gelatinase-associated lipocalin. Wound Repair Regen 21:395–401. https://doi.org/10.1111/WRR.12035

Shahzad A, Cohrs RJ (2012) In vitro antiviral activity of honey against varicella zoster virus (VZV): a translational medicine study for potential remedy for shingles. Transl Biomed 3:2

Sharifi-Rad M, Anil Kumar NV, Zucca P et al (2020) Lifestyle, oxidative stress, and antioxidants: back and forth in the pathophysiology of chronic diseases. Front Physiol 11. https://doi.org/10.3389/fphys.2020.00694

Silva B, Biluca FC, Gonzaga LV et al (2021) In vitro anti-inflammatory properties of honey flavonoids: a review. Food Res Int 141:110086. https://doi.org/10.1016/J.FOODRES.2020.110086

Silva B, Biluca FC, Mohr ETB et al (2020) Effect of Mimosa scabrella Bentham honeydew honey on inflammatory mediators. J Funct Foods 72:104034. https://doi.org/10.1016/J.JFF.2020.104034

Simon A, Sofka K, Wiszniewsky G et al (2006) Wound care with antibacterial honey (Medihoney) in pediatric hematology-oncology. Support Care Cancer 14:91–97. https://doi.org/10.1007/S00520-005-0874-8

Smaropoulos ESR, Dimitriadou C (2011) Honey-based therapy for paediatric burns and dermal trauma compared to standard hospital protocol, vol 7. Wounds UK, London, pp 33–40

Sojka M, Valachova I, Bucekova M, Majtan J (2016) Antibiofilm efficacy of honey and bee-derived defensin-1 on multispecies wound biofilm. J Med Microbiol 65:337–344. https://doi.org/10.1099/JMM.0.000227

Tang Y, Chen L, Ran X (2024) Efficacy and safety of honey dressings in the management of chronic wounds: an updated systematic review and meta-analysis. Nutrients 16. https://doi.org/10.3390/NU16152455

Tanuwidjaja I, Svečnjak L, Gugić D et al (2021) Chemical profiling and antimicrobial properties of honey bee (Apis mellifera L.). Venom Molecules 26:3049. https://doi.org/10.3390/molecules26103049

Tarnuzzer RW, Schultz GS (1996) Biochemical analysis of acute and chronic wound environments. Wound Repair Regen 4:321–325. https://doi.org/10.1046/J.1524-475X.1996.40307.X

Taş-Küçükaydın M, Tel-Çayan G, Çayan F et al (2023) Chemometric classification of chestnut honeys from different regions in Turkey based on their phenolic compositions and biological activities. Food Chem 415:135727. https://doi.org/10.1016/J.FOODCHEM.2023.135727

Tel-Çayan G, Çiftçi BH, Taş-Küçükaydın M et al (2023) Citrus honeys from three different regions of Turkey: HPLC-DAD profiling and in vitro enzyme inhibition, antioxidant, anti-inflammatory and antimicrobial properties with chemometric study. Chem Biodivers 20:e202300990. https://doi.org/10.1002/CBDV.202300990

Thierig M, Raupbach J, Wolf D et al (2023) 3-Phenyllactic acid and polyphenols are substances enhancing the antibacterial effect of methylglyoxal in manuka honey. Food Secur 12:1098. https://doi.org/10.3390/foods12051098

Thomas RE, Thomas BC (2021) Reducing biofilm infections in burn patients' wounds and biofilms on surfaces in hospitals, medical facilities and medical equipment to improve burn care: a systematic review. Int J Environ Res Public Health 18:13195. https://doi.org/10.3390/IJERPH182413195

Tonks A (2003) Honey stimulates inflammatory cytokine production from monocytes. Cytokine 21:242–247. https://doi.org/10.1016/S1043-4666(03)00092-9

Trengove NJ, Stacey MC, Macauley S et al (1999) Analysis of the acute and chronic wound environments: the role of proteases and their inhibitors. Wound Repair Regen 7:442–452. https://doi.org/10.1046/J.1524-475X.1999.00442.X

Ujilestari T, Damayanti E, Sholikin MM et al (2023) A meta-analysis of the effects of gamma irradiation on biochemical and physical characteristics of honey. IOP Conf Ser Earth Environ Sci 1230:012053. https://doi.org/10.1088/1755-1315/1230/1/012053

Ulusoy E, Kolayli S, Osman SA (2010) Antioxidant and antimicrobial activity of different floral origin honeys from Turkiye. J Food Biochem 34:321–335. https://doi.org/10.1111/j.1745-4514.2009.00332.x

van den Berg AJ, van den Worm E, van Ufford HCQ et al (2008) An in vitro examination of the antioxidant and anti-inflammatory properties of buckwheat honey. J Wound Care 17:10.12968/JOWC.2008.17.4.28839

Van Ende M, Wijnants S, Van Dijck P (2019) Sugar sensing and signaling in Candida albicans and Candida glabrata. Front Microbiol 10:99. https://doi.org/10.3389/fmicb.2019.00099

Vardi A, Barzilay Z, Linder N et al (1998) Local application of honey for treatment of neonatal postoperative wound infection. Acta Paediatr 87:429–432. https://doi.org/10.1080/08035259850157048

Wahyono T, Benita AM, Pratama IM et al (2024) Effect of gamma irradiation and evaporation on physicochemical, antibacterial, and antioxidant activity of selected high-quality herbal honey. Radiat Phys Chem 214:111263. https://doi.org/10.1016/J.RADPHYSCHEM.2023.111263

Watanabe K, Rahmasari R, Matsunaga A et al (2014) Anti-influenza viral effects of honey in vitro: potent high activity of manuka honey. Arch Med Res 45:359–365. https://doi.org/10.1016/j.arcmed.2014.05.006

Weerasekera MM, Jayarathna TA, Wijesinghe GK et al (2017) The effect of nutritive and non-nutritive sweeteners on the growth, adhesion, and biofilm formation of Candida albicans and Candida tropicalis. Med Princ Pract 26:554–560. https://doi.org/10.1159/000484718

Widgerow AD (2012) Cellular resolution of inflammation–catabasis. Wound Repair Regen 20:2–7. https://doi.org/10.1111/J.1524-475X.2011.00754.X

Wilczyńska A, Żak N (2024) Polyphenols as the main compounds influencing the antioxidant effect of honey—a review. Int J Mol Sci 25:10606. https://doi.org/10.3390/ijms251910606

Wu Y, Yuan X, Li J, Kadowaki T (2021) DWV infection in vitro using honey bee pupal tissue. Front Microbiol 12. https://doi.org/10.3389/fmicb.2021.631889

Yildiz Karadeniz E, Kaplan Serin E (2023) Use of honey in diabetic foot ulcer: systematic review and meta-analysis. J Tissue Viability 32:270–278. https://doi.org/10.1016/J.JTV.2023.03.002

Zhang F, Chen Z, Su F, Zhang T (2021) Comparison of topical honey and povidone iodine-based dressings for wound healing: a systematic review and meta-analysis. J Wound Care 30:S28–S36. https://doi.org/10.12968/JOWC.2021.30.SUP4.S28

Part II
Propolis

Chapter 5
Phenolic Composition of Propolis

Celestino Santos-Buelga and Ana M. González-Paramás

Contents

Abstract Propolis is a resinous substance collected by honeybees from various plant sources, which largely define its chemical composition, particularly its phenolic constituents. Different types of propolis have been identified based on their phenolic profiles, such as poplar-type, rich in flavonoids like chrysin and pinocembrin; Brazilian green propolis, characterized by artepillin C and cinnamic acid derivatives from *Baccharis dracunculifolia*; Brazilian red propolis, distinguished by isoflavonoids and neoflavonoids from *Dalbergia* species; and Pacific-type, featuring phenolics from *Macaranga tanarius*. This chapter reviews the phenolic composition of these and other types of propolis—including flavonoids, phenolic acids, and other phenolic compounds—using data from advanced analytical methods like high-performance liquid chromatography-diode array detector-mass spectrometry (HPLC-DAD-MS) and gas chromatography-mass spectrometry (GC-MS). Furthermore, it examines the significance of phenolic markers in determining the botanical and geographic origins of propolis.

C. Santos-Buelga · A. M. González-Paramás (✉)
Grupo de Investigación en Polifenoles (GIP-USAL), Facultad de Farmacia, Universidad de Salamanca, Campus Miguel de Unamuno, Salamanca, Spain
e-mail: paramas@usal.es

J. M. Alvarez-Suarez (ed.), *Bee Products – Chemical and Biological Properties*,
https://doi.org/10.1007/978-3-031-89049-9_5

Keywords Propolis · Chemical composition · Flavonoids · Phenolic acids

5.1 Introduction

Propolis (bee glue) is a dark-colored resinous substance collected by the honeybees from living plants, which is mixed with wax and used to seal cracks, smoothen walls, and maintain stable moisture and temperature in the hive. Raw propolis is typically composed of 50% plant resins, 30% waxes, 10% essential and aromatic oils, 5% pollens, and 5% other organic substances (Huang et al. 2014). Its color varies from green to brown and reddish, depending on its botanical source (Kuropatnicki et al. 2013). Although it is generally accepted that in temperate zones, the bud exudates of *Populus* species and their hybrids are the main source of propolis, there is a long list of plants also used by bees as a source of propolis in different parts of the world. In addition to poplars, this list also includes resins of conifers, birch, alder, willow, palm, *Dalbergia ecastaphyllum*, etc. (Crane 1990; Bankova et al. 2000).

Propolis cannot be commercialized as a raw material; it must be purified by extraction with solvents. This process should remove the inert material and preserve the phenolic fraction. Traditionally, the purification of raw propolis is carried out by soaking, shaking, reflux, or Soxhlet extraction, using different solvents to obtain dewaxed propolis extracts rich in polyphenolic components (Pietta et al. 2002). Absolute ethanol is the solvent most commonly employed to prepare propolis extracts, although extraction with aqueous ethanol (70–95%) can result in wax-free tinctures containing higher amounts of phenolic substances (Park and Ikegaki 1998). Commercial propolis can also be obtained by turbo-extraction, which reduces the extraction time to less than 1 hour, without using heat (Cottica et al. 2011). Applying this technique, higher yields of total phenolic compounds can be obtained using lower percentages of ethanol (around 60%) and higher concentrations of crude propolis (near 30%, m/v); nevertheless, the highest concentrations of flavonoids in the extracts are obtained with higher ethanol percentages in the solvent (maximum 96%) (Cottica et al. 2011). Propolis extraction with pure water (Bankova et al. 1992), hexane and acetone (Pereira et al. 2000), methanol (Pietta et al. 2002; Cao et al. 2004) or chloroform (Negri et al. 2003) has also been used.

More than 500 compounds have been identified in propolis including phenolic components, terpenes, lipid-wax substances, beeswax, and other substances such as vitamins, proteins, amino acids, and sugars (Walker and Crane 1987; Kurek-Górecka et al. 2014). Among them, plant phenolics constitute the most abundant group of chemical components, including flavonoids, phenolic acids and aldehydes, simple phenols and their esters, coumarins, stilbenes, and lignans. The total flavonoid content has been proposed as a quality index for raw propolis; in this respect, Gardana et al. (2007) suggested that, except for some Brazilian propolis, whose flavonoid content is very low, crude propolis with a content less than 11% (w/w) should be considered to be of low quality, whereas propolis with contents of 11–14%, 14–17%, or >17% could be classified as of acceptable, good, and high quality, respectively. Phenolic contents between 6.4% and 15.2% (w/w) have been reported for ethanolic extracts of green Brazilian propolis (Woiski and Salatino 1998). Brazilian legislation establishes minimum flavonoid contents of 0.5% (w/w) and minimum phenolic contents of 5% (w/w) in crude propolis (Sawaya et al. 2011).

Different plant sources have an influence in the chemical composition of bee glue, especially in the phenolic profile. In the temperate zone (Europe, Asia, and North America), propolis (mostly poplar-type) is characterized by similar phenolic patterns, consisting of a high level of flavanones and flavones and a low level of phenolic acids and their esters (Marcucci 1995). In propolis from tropical regions, more complex profiles of phenolic compounds have been described, including prenylated p-coumaric acids, prenylated flavonoids, caffeoylquinic acid derivatives, and lignans (Basnet et al. 1996; Boudourova-Krasteva et al. 1997; Banskota et al. 1998).

5.2 Phenolic Compounds

Phenolic compounds are among the biggest and most widely distributed groups of secondary metabolites in plants. They differ from other plant metabolites that may also possess a phenolic structure (e.g., terpenes) in their biosynthetic origin. Biogenetically, phenolic compounds derive from two metabolic pathways: the shikimic acid pathway, where mainly phenylpropanoids are formed, and the acetic acid pathway, in which the main products are simple phenols. The combination of both pathways leads to the formation of flavonoids, the most abundant group of phenolic compounds in nature (Reis Giada 2013).

As above indicated, there are differences in the phenolic composition of propolis according to the plant source. The most characteristic compounds of temperate propolis, i.e., poplar-type propolis, are flavonoids without B-ring substituents, such as pinocembrin, pinobanksin, galangin, and chrysin (Walker and Crane 1987). Some recent publications pointed to the existence of three main types among poplar propolis according to their thin layer chromatography (TLC) behavior on silica gel 60 or NH_2 plates. Such a grouping is based on the color of the bands observed after derivatization with polyethylene glycol and 2-aminoethyldiphenylborinate and illumination under UV light at 366 nm (Ristivojević et al. 2015a). All kinds of poplar propolis show the presence of orange, blue, and green chromatographic bands after TLC separation; however, in the propolis called "orange type" (O-type), strong orange bands typical for flavonoids like quercetin are predominant, while in the "blue type" (B-type), light blue bands (characteristic of hydroxycinnamoyl derivatives) are dominant (Bertrams et al. 2013; Ristivojević et al. 2015b; Sârbu and Moţ 2011). The "green type" (G-type), identified in German propolis, is rare and exhibits bands similar to those of the O-type but of lower intensity, in addition to a weak green and one or two stronger green bands, which correspond to apigenin, galangin, and their methyl-ether derivatives (Bertrams et al. 2013).

In tropical propolis, flavonoids are less abundant than other phenolic compounds, such as prenylated derivatives of phenylpropanoids (Brazilian propolis) or benzophenones (characteristic of propolis from Venezuela, Cuba, and northeast regions of Brazil) and caffeoylquinic acids (Salatino et al. 2011). Geranyl flavanones have been reported as characteristic of the so-called Pacific-type propolis, derived from fruit exudates of the species of *Macaranga* (Bankova 2005); although, initially, those flavonoids were described for propolis samples from the Pacific region (Taiwan and Okinawa), they have also been found in propolis from African regions (Egypt and Kenya) where such *Macaranga* species are also distributed (Petrova et al. 2010). Propolis from some eastern Mediterranean regions may contain relevant levels of anthraquinones (e.g., chrysophanol) whose plant origins are still unknown (Salatino et al. 2011).

5.2.1 Flavonoids

The presence of flavonoids in propolis was first suggested by Jaubert (1926), who identified the flavone chrysin in bee glue and showed that it could contribute to the color of beeswax. The relation of the flavonoids present in propolis with the chemical composition of plant exudates was evidenced by Popravko et al. (1970), who identified two flavanones and another six flavonoid pigments in propolis. In parallel, Lavie's research group in France reported the isolation and identification of galangin, pinocembrin, tectochrysin, and izalpinin from propolis (Villanueva et al. 1964, 1970). Further advances in the chromatographic techniques allowed the separation of more components from propolis, and by the end of the 1980s, nearly 40 flavonoids had been described (Walker and Crane 1987); nowadays, more than 150 flavonoids are identified in propolis (Huang et al. 2014).

The composition of raw propolis has been more frequently studied than that of the commercial presentations. The usual preparation of crude propolis for analysis includes extraction with a solvent (the most usual proportion is 1:10, w/v), solvent evaporation under reduced pressure, and dissolution of the residue prior to compound determination. Methods described in the literature for the analysis of flavonoids in propolis include both non-separation, e.g., UV-Vis spectrophotometry or NMR, and separation techniques, such as gas chromatography (GC), high-performance thin layer chromatography (HPTLC), high-performance liquid chromatography (HPLC), or high-performance capillary electrophoresis (HPCE) (Gómez-Caravaca et al. 2006; Cuesta-Rubio et al. 2007) (see 5.3 for details).

Flavonoids belonging to the group's flavanones, flavones, flavonols, flavanonols, chalcones, dihydrochalcones, isoflavones, isodihydroflavones, flavans, isoflavans, and neoflavonoids have been reported in propolis. In general, the identified compounds are aglycones because while bees are collecting propolis, they secrete β-glucosidase that removes the sugar residues of the glycosylated flavonoids present in plants (Sahinler and Kaftanoglu 2005). Nevertheless, this enzyme could be inefficient to hydrolyze some types of glycosides, such as β-diglycosides (Zhang et al.

2011) or C-glycosides (Righi et al. 2011). Thus, the presence of rutin (quercetin 3-O-rutinoside) and isorhamnetin-3-O-rutinoside has been reported in propolis samples from different countries (Bonvehi and Call 1994; Popova et al. 2009), as well as flavone C-glycosides (Righi et al. 2011) and luteolin and naringenin glucosides (Cui-ping et al. 2014) in propolis samples from Brazil and China, respectively. A propolis with an uncommon phenolic profile rich in flavonoid glycosides was described in Portugal, showing more than ten glycoside derivatives, among which quercetin and kaempferol rutinosides were predominant, together with some glucuronides and glucosides (Falcão et al. 2013).

Flavanones are among the most significant groups of flavonoids present in propolis, with more than 40 different compounds of this class being described in this product. These compounds are common in poplar-type propolis from the temperate zone, both in the northern and southern hemispheres. Pinocembrin, pinostrobin, naringenin, sakuranetin, isosakuranetin, and liquiritigenin (Fig. 5.1) have been reported as the prevalent flavanones in European, Chinese, and North American propolis and propolis from nontropical regions of Asia (Villanueva et al. 1970; Greenaway et al. 1991; Marcucci 1995; Bankova et al. 2000; Shi et al. 2012; de Groot et al. 2014; Ristivojević et al. 2015a). Pinobanksin is frequently found in the form of methylated or esterified derivatives (acetates, propionates, butyrates, pentanoates, methylbutyrates, etc.) (Gardana et al. 2007; Falcão et al. 2010). Other hydroxy- and methoxy-flavanones that have been identified are (*2R, 3R*)-3,6,7-trihydroxyflavanone in propolis from Nepal (Shrestha et al. 2007a), (*2R, 3R*)-3,7-dihydroxy-6-methoxyflavanone and (*2R, 3R*)-3,5,7-trihydroxy-6-methoxyflavanone (alnustinol) in Brazilian propolis (Li et al. 2008), and 5-methoxy-3-hydroxyflavanone in propolis from Portugal (Falcão et al. 2010) and Mexico (Li et al. 2010). In addition, the corresponding chalcones of the principal flavanones have been detected in poplar-type propolis (Greenaway et al. 1991), Canadian samples from *Tacamahaca* poplars (Christov et al. 2006), and Brazilian red propolis from *Dalbergia ecastophyllum* (Li et al. 2008). In Portuguese propolis, Falcão et al. (2010) detected the presence of a dimethylated derivative of hesperetin and also identified six peculiar pinocembrin/pinobanksin derivatives containing a phenylpropanoic acid

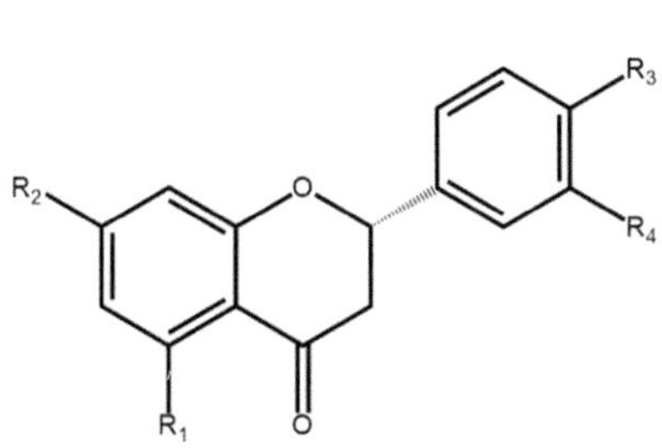

	R_1	R_2	R_3	R_4
Pinocembrin	OH	OH	H	H
Pinostrobin	OCH_3	OH	H	H
Naringenin	OH	OH	OH	H
Sakuranetin	OCH_3	OH	OH	H
Isosakuranetin	OH	OH	OCH_3	H
Liquiritigenin	H	OH	OH	H

Fig. 5.1 Structures of the principal flavanones described in propolis

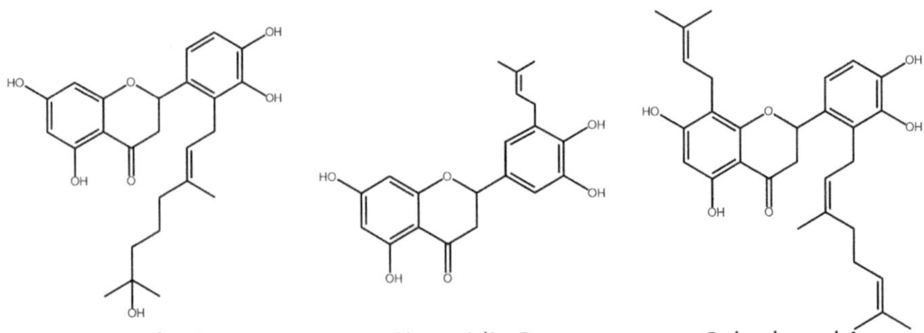

Fig. 5.2 Typical prenylated flavanones described in "Pacific-type" propolis

substituent. Two phenylallyl-pinobanksin derivatives with cytotoxic properties were also described by Li et al. (2011) in propolis from Mexico. Prenylated flavanones were shown to be common and abundant constituents in "Pacific-type" propolis from Taiwan, Japan, or Solomon Island. Prenylated flavanones such as propolin A (Fig. 5.2) and propolin B were first isolated and characterized by Chen et al. (2003) in Taiwanese propolis, where propolin E and sigmoidin B (Fig. 5.2) were also present. In propolis from Japan, geranylflavanone derivatives are abundant (Kumazawa et al. 2004), while solophenol A (Fig. 5.2) and sophoraflavanone A are typical in propolis from Solomon Island (Inui et al. 2012). In propolis from tropical zones, in particular those from the southeastern region of Brazil, where flavonoids are not the main phenolic constituents, some flavanones such as isosakuranetin have been described in propolis from *Baccharis dracunculifolia* (Park et al. 2002).

Although in minor extension than flavanones, *flavones and flavonols* (Fig. 5.3) are also commonly found in propolis. Some compounds of these families classically described are acacetin, apigenin, chrysin, tectochrysin, galangin, izalpinin, quercetin, and kaempferol and, in minor extension, others such as fisetin or pectolinarigenin (Marcucci 1995; Stevens et al. 1995). Also, the presence of kaempferide has been found in Brazilian green propolis (Simões et al. 2004) and luteolin in Chinese propolis (Cao et al. 2004). A new flavonol named macarangin (Fig. 5.4) was identified in Kenyan samples from *Macaranga* genus (Petrova et al. 2010). As commented above, propolis with an unusual flavonoid composition showing quercetin and kaempferol glycosides was described in Portugal (Falcão et al. 2013).

The presence of *isoflavones* and *isodihydroflavones* (Fig. 5.5) in European propolis is rare; however, compounds of this family (e.g., daidzein, formononetin, xenognosin B, vestitone, and calycosin, among others) were found in red Brazilian and red Cuban propolis, originated probably from leguminous plants, especially *Dalbergia ecastophyllum* in the Brazilian one (Piccinelli et al. 2005; Alencar et al. 2007; Campo Fernández et al. 2008; Li et al. 2008). *Isoflavans* like vestitol, mucronulatol, and neovestitol (Fig. 5.6) and *pterocarpins* (i.e., neoflavonoids; Fig. 5.7) were also found in those two types of red propolis (Piccinelli et al. 2005; Campo

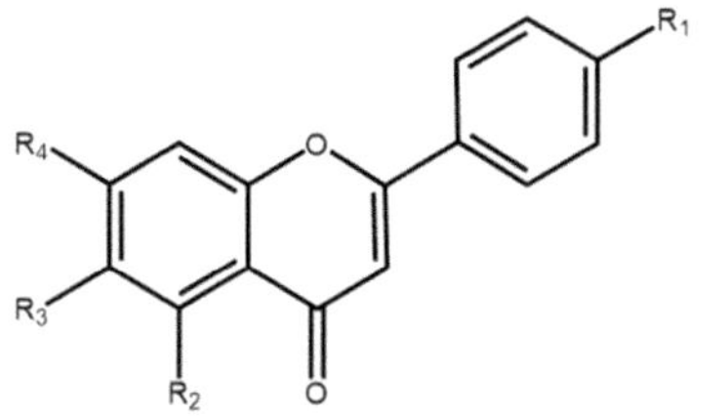

Flavones

	R_1	R_2	R_3	R_4
Chrysin	H	OH	H	OH
Tectochrysin	H	OH	H	OCH_3
Apigenin	OH	OH	H	OH
Acacetin	OCH_3	OH	H	OH
Pectolinarigenin	OCH_3	OH	OCH_3	OH

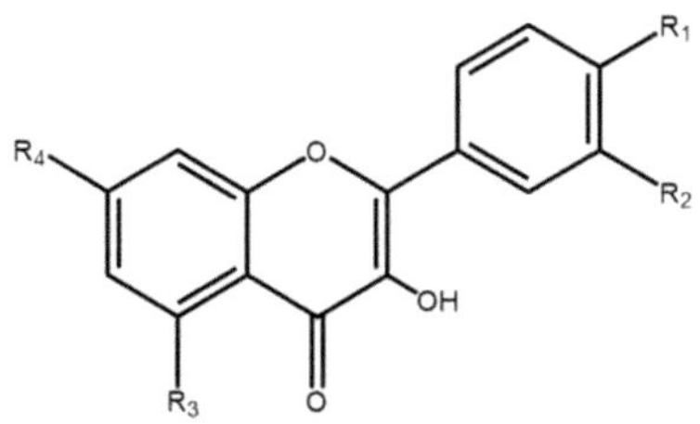

Flavonols

	R_1	R_2	R_3	R_4
Galangin	H	H	OH	OH
Kaempferol	OH	H	OH	OH
Quercetin	OH	OH	OH	OH
Fisetin	OH	OH	H	OH
Izalpinin	H	H	OH	OCH_3

Fig. 5.3 Main flavones and flavonols described in propolis

Fig. 5.4 Structure of macarangin

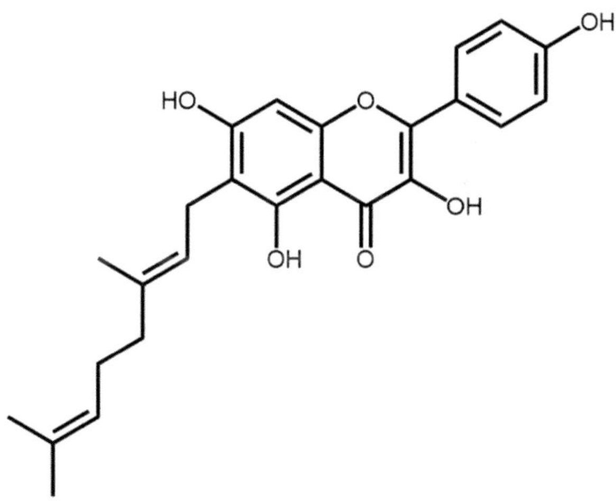

Fernández et al. 2008; Li et al. 2008). Finally, in samples from Nepal, up to fourteen unique open-chain neoflavonoids (Fig. 5.8) were identified and proposed as chemical markers for this type of propolis (Awale et al. 2005; Shrestha et al. 2007b).

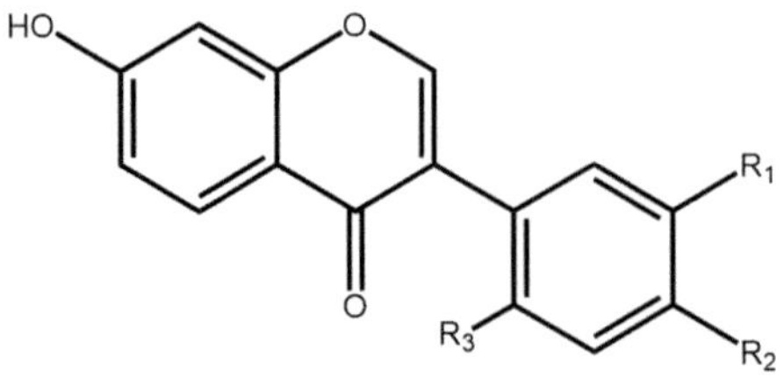

	R_1	R_2	R_3
Daidzein	H	OH	H
Formonetin	H	OCH_3	H
Calycosin	OH	OCH_3	H
Xenognosin B	H	OCH_3	OH

Fig. 5.5 Isoflavones described in propolis

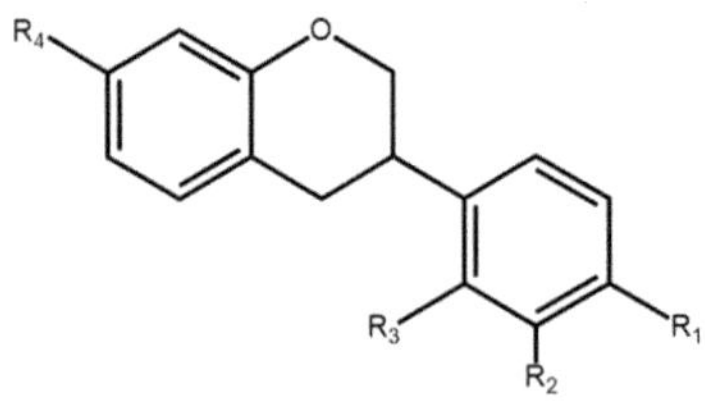

	R_1	R_2	R_3	R_4
Vestitol	OCH_3	H	OH	OH
Neovestitol	OH	H	OH	OCH_3
Mucronulatol	OCH_3	OH	OCH_3	OH

Fig. 5.6 Main isoflavans cited in propolis

Fig. 5.7 Basic structure of pterocarpins

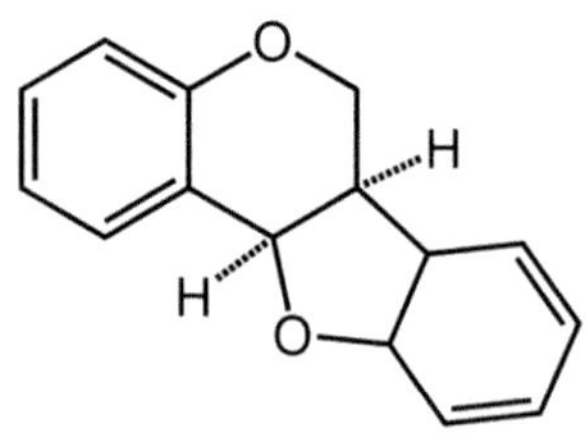

Fig. 5.8 Basic structure of open-chain neoflavonoids

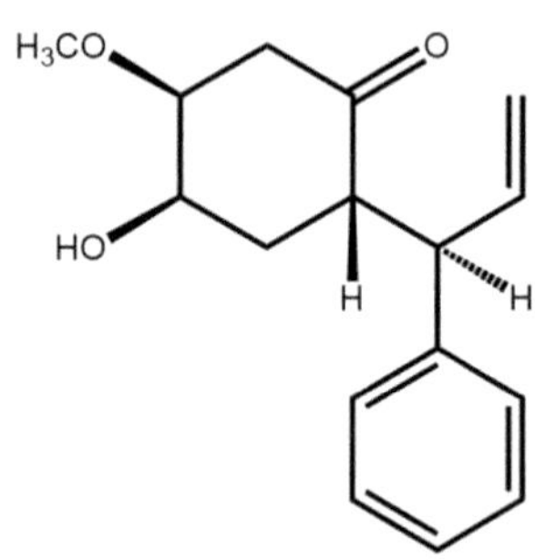

5.2.2 *Phenolic Acid Derivatives*

Phenolic acids in propolis are represented mainly by hydroxybenzoic acids (gallic, gentisic, protocatechuic, salicylic, and vanillic acids) and hydroxycinnamic acids (p-coumaric, caffeic, and ferulic acids). In addition to their free forms, phenolic acids are often present as benzyl-, methylbutenyl-, phenylethyl- and cinnamyl-esters, caffeic acid phenethyl ester (CAPE) (Fig. 5.9) being a major constituent in

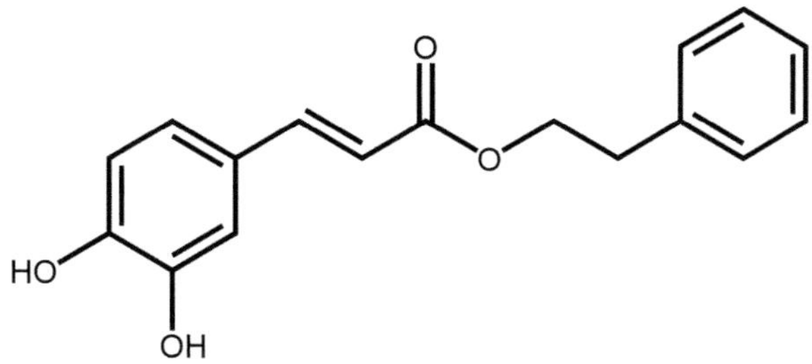

Fig. 5.9 Structure of caffeic acid phenethyl ester (CAPE)

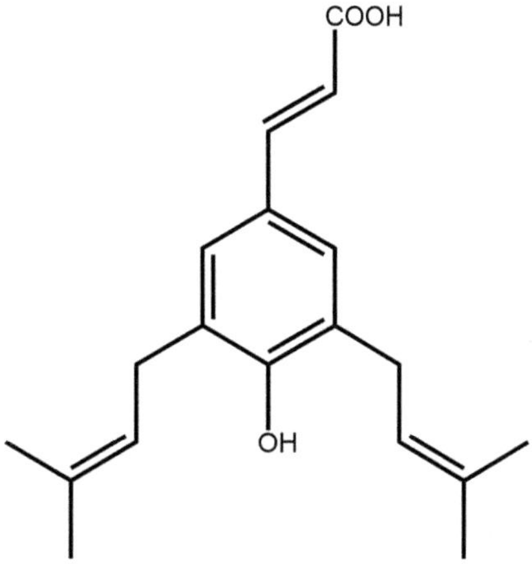

Fig. 5.10 Structure of artepillin C

temperate propolis (Marcucci 1995; Huang et al. 2014; Ristivojević et al. 2015a, b). Hydroxycinnamic acid derivatives were also found in Brazilian (Marcucci et al. 2001; Salatino et al. 2005) and Australian propolis (Abu-Mellal et al. 2012). In Brazilian propolis from *Citrus* spp. chlorogenic acids have been indicated to be particularly abundant (dos Santos et al. 2003). Furthermore, prenylated phenylpropanoids were shown to be very common in Brazilian green propolis (Salatino et al. 2005). Among them, two different classes of prenylated cinnamic acid–derived compounds have been described: one is characterized by the presence of one or two prenyl groups not involved in ring formation (e.g., artepillin C, Fig. 5.10) and the another one by the prenyl group forming a heterocycle ring giving rise to chromenes (Salatino et al. 2005). These prenylated cinnamic acids may also be present as esters, such as 3-prenyl-cinnamic acid allyl ester (Negri et al. 2003).

5.2.3 Other Phenolic Compounds

Although *stilbenes* are not very common in plants, there are few reports identifying this class of compounds in propolis from different geographical origins. Two geranyl stilbenes (scheweinfurthin A and B) and 5-farnesyl-3′-hydroxyresveratrol were isolated from propolis produced from *Macaranga* spp. in Kenya and Solomon Island, respectively (Petrova et al. 2010; Inui et al. 2012). In propolis from Australian Kangaroo Island, different stilbenes, especially prenylated, were identified (Abu-Mellal et al. 2012), suggesting that stilbene sources are not limited to only a few plants.

Lignans have been identified in a few propolis samples, especially from tropical zones, such as Canary Island (Bankova et al. 1998), Kenya (Petrova et al. 2010), and Brazil (Li et al. 2008), as well as in propolis from Chile (Valcic et al. 1998). In particular, in Brazilian red propolis, the presence of syringaresinol, pinoresinol, and their dimethyl ether derivative has been reported (Li et al. 2008).

Some *coumarins*, such as coumarin and daphnetin and the prenylated coumarin suberosin 1 were also cited in some propolis samples (de Castro 2001; Trusheva et al. 2010).

5.3 Methods for the Analysis of Phenolic Compounds in Propolis

Propolis has been used in traditional medicine since ancient times, valued for its diverse therapeutic properties. However, despite this long-standing usage, standardized methods for analyzing its bioactive components, especially phenolic compounds, remain scarce. The lack of standardized analytical protocols makes it challenging to assess and compare the phenolic composition of propolis samples across different studies, particularly given the variability in its composition due to geographic and botanical sources. This limitation highlights the necessity for robust and reproducible analytical methods capable of accurately reflecting the complex phenolic profile of propolis.

The analysis of phenolic compounds in propolis involves a series of critical stages, each designed to maximize the recovery, purification, and identification of these bioactive molecules. Initially, the extraction phase focuses on isolating phenolic compounds from the complex matrix of propolis using various solvents and techniques. Following extraction, a purification step is sometimes necessary to concentrate phenolics while removing unwanted substances, ensuring a cleaner profile for subsequent analysis. Finally, diverse analytical methods are employed to detect, quantify, and characterize the individual phenolic compounds present. These range from generic colorimetric assays to highly sensitive chromatographic techniques, such as high-performance liquid chromatography-diode array detector (HPLC-DAD) and high-performance liquid chromatography-mass spectrometry (HPLC-MS), which offer detailed insights into the phenolic composition of propolis.

5.3.1 Extraction

Propolis extraction differs from that of plants, as its resinous matrix partially dissolves during extraction, accelerating the release of bioactive compounds. This process not only targets plant-derived phenolics but also aims to remove wax, which constitutes up to 20% of propolis. The inherent low solubility of propolis in water and its affinity for organic solvents underscore the importance of optimizing solvent selection and extraction parameters for each specific propolis type, as its chemical composition varies with geographic and climatic factors (Bankova et al. 2021).

Polyphenol extraction from propolis requires selecting appropriate solvents based on polarity and solubility. Commonly used solvents include ethanol, methanol, acetone, and water in varying proportions, with ethanol being preferred due to its cost-effectiveness and scalability. Ethanol concentrations between 40% and 95% are typically used, with higher concentrations enhancing flavonoid yields, particularly for compounds such as galangin, pinocembrin, and chrysin. For instance, 70% ethanol has been found to extract higher phenolic content than 96%, yet both concentrations exhibit similar antioxidant activity (Woźniak et al. 2019). Other solvents, such as ethyl acetate and trichloromethane, may be used to target specific compounds, with their extraction capacities ranked as acetone > methanol > ethanol > diethyl ether > n-butanol > ethyl acetate > trichloromethane (Cui et al. 2022).

In addition to classical solvents, natural deep eutectic solvents (NADES) have recently been proposed for extracting bioactive compounds. NADES are biocompatible, biodegradable, and liquid at room temperature, offering advantages such as ease of preparation, adjustable properties, and wide combinability (Palos-Hernández et al. 2022). For poplar-type propolis, NADES like choline chloride–L(+)-tartaric acid and citric acid–propylene glycol have proven effective (Trusheva et al. 2024), while choline chloride–propylene glycol or choline chloride–lactic acid were suitable for Brazilian green propolis (Funari et al. 2019).

Once the appropriate solvent has been selected, it is essential to choose the most suitable extraction technique to maximize the yield and quality of bioactive compounds from propolis. Conventional methods, such as maceration and Soxhlet extraction, are widely used due to their simplicity and efficiency. However, advanced techniques like high hydrostatic pressure extraction (HHPE), ultrasound-assisted extraction (UAE), and microwave-assisted extraction (MAE) are gaining popularity as they offer significant advantages, including reduced extraction time, lower solvent consumption, and improved extraction efficiency. Also, a standardized extraction process has been developed to obtain phenolic extracts from poplar-type propolis. This patented method involves multiple steps, starting with an initial aqueous extraction from dewaxed raw propolis. Subsequent extractions are performed on the remaining residue using ethanol–water mixtures with progressively higher ethanol concentrations for each step (Zaccaria et al. 2019). According to Escriche and Juán-Borrás (Escriche and Juan-Borrás 2018), the type of propolis has a greater influence on the yield, total phenol content, and individual phenolic compounds than the extraction method itself. Their study highlights that ultrasonication, especially when used with double extraction, proves more efficient than maceration, significantly reducing analysis time while maintaining comparable results. However, it is important to report the results based on yield rather than raw propolis, as the latter may contain impurities. Bankova et al. (2021) recently conducted a review detailing the advantages and limitations of each extraction method, highlighting how the choice depends on factors such as the target compounds, extraction scale, and cost-efficiency.

Recently, several authors have used subcritical water extraction (SWE), an efficient method for extracting bioactive compounds from propolis by employing water at temperatures just below its critical point. This technique enhances the solubility of polar compounds, such as phenolics, and offers advantages such as reduced extraction time, minimal solvent use, and selectivity. Although eco-friendly, SWE requires careful control of temperature and pressure to prevent the degradation of sensitive compounds (Shin et al. 2023).

5.3.2 *Purification*

The purification stage is crucial for removing impurities and concentrating phenolic compounds to facilitate more precise analysis. Various techniques can be employed depending on the objectives and scale of the process.

Solid-phase extraction (SPE) is widely used for isolating active constituents from propolis extracts. SPE employs cartridges or resins with specific properties to selectively retain phenolic compounds while washing away impurities. For instance, flavonoids from Chinese propolis were extracted using microporous or microporous absorptive resins from ethanol extracts or water-soluble fractions, achieving purities exceeding 90% (Li et al. 2016). However, SPE is primarily suited for analytical purposes rather than large-scale preparative applications.

Liquid–liquid extraction (LLE) separates compounds based on their solubility in immiscible solvents but may lead to some compound loss. Homogeneous liquid–liquid extraction (HLLE) has gained attention for the multi-analyte analysis of phenolic compounds in propolis, using various acetonitrile–water-based methods. Techniques like salting-out, sugaring-out, hydrophobic-solvent, and subzero-temperature-assisted LLE have been shown to effectively partition phenolics, with more polar compounds benefiting from higher acetonitrile concentrations. Methods like NaCl-based Salt-Assisted Liquid-Liquid Extraction (SALLE) and Heat-Assisted Salt-Liquid-Liquid Extraction (HSLLE) offer selective extraction of low-polarity compounds, while others are suitable for broader extraction of multiple phenolic compounds (Chen et al. 2019). While effective, this method can result in some compound loss and is often less precise than SPE.

Membrane filtration is another option for purifying extracts, as it removes larger molecules and contaminants. However, careful handling is required to prevent degradation of sensitive compounds. Recently, a membrane-based ultrafiltration process was developed using Brazilian green propolis (Liang et al. 2024). After optimizing experimental conditions and parameters, a mathematical model was created to fit the process. The study concluded that the optimal solvent for retaining flavonoids and phenolic acids was 60% ethanol at pH 8–9, and the ideal membrane molecular weight cutoff (MWCO) was 30 KDa. This process provides an innovative alternative for extracting therapeutic compounds from propolis.

Each purification method has its own strengths and limitations, and the choice of technique depends on factors such as the specific extraction goals, the desired purity, and the characteristics of the propolis extract. Therefore, selecting the most appropriate method requires careful consideration of these variables to achieve the best results.

5.3.3 Analysis Methods

Once the extraction and purification stages are complete, analyzing propolis extracts becomes crucial to identifying and quantifying their bioactive phenolic compounds. Analytical methods can be categorized into general and specific approaches, each tailored to different objectives and levels of precision.

General methods are primarily used to obtain an overall profile of the extract, making them ideal for preliminary screenings or broad compositional assessments. Among these, the Folin–Ciocalteu assay is a widely used colorimetric technique for estimating total phenolic content (Ahn et al. 2004; Popova et al. 2004). However, this method lacks specificity as it can be influenced by other reducing substances present in the extract. Recent advancements have optimized this method specifically for propolis, incorporating multivariate techniques to enhance precision, reduce reagent use, and improve efficiency. This optimized approach, which also integrates X-ray and nephelometric analyses, is better suited for accurately quantifying phenolic content in propolis (Paganotti and Barbeira 2024). Two colorimetric methods have been used for the quantitative analysis of flavonoids in propolis based on the reaction with aluminum chloride for the determination of flavones and flavonols (Grassi et al. 2023) and with 2,4-dinitrophenylhydrazine (DNP) for the determination of flavanones and dihydroflavonols (Gómez-Caravaca et al. 2006). The sum of flavonoid contents determined by the above two methods closely represents the real content of total flavonoids (Chang et al. 2002). Similarly, the vanillin assay is used to determine condensed tannins through a reaction with vanillin that produces a measurable colored complex. Although tannins are less frequently reported in propolis, their content has been observed to range between 0.6% and 4.1% (Mayworm et al. 2014; Kabous and Ouhssine 2023). Thin layer chromatography (TLC) is another useful general method, serving as a rapid screening tool to separate phenolic compounds prior to detailed analyses (Milojković-Opsenica et al. 2022). UV-Vis spectrophotometry also provides a quick and straightforward method for general quantification but lacks the specificity necessary for analyzing complex mixtures such as propolis.

Specific methods are designed to precisely identify and quantify individual phenolic compounds, offering greater sensitivity and resolution. High-performance liquid chromatography (HPLC) is one of the most reliable techniques for this purpose, allowing the separation and accurate quantification of phenolics. When coupled with a diode array detector (HPLC-DAD), HPLC achieves high-resolution profiling of propolis compounds (Gardana et al. 2007; Pellati et al. 2011; Aldana-Mejía et al. 2021). Mass spectrometry coupled with HPLC (HPLC-MS) further enhances sensitivity, enabling both the identification and quantification of individual phenolics (Midorikawa et al. 2001). Liquid chromatography-tandem mass spectrometry (LC-MS/MS) takes this a step further, offering unparalleled precision and the ability to analyze molecular weights and fragmentation patterns, making it an ideal choice for detailed phenolic profiling (Gómez-Caravaca et al. 2006; Zhang et al. 2022). Recent innovations, such as nano-electrospray Ionization Mass Spectrometry (nanoESI-MS), have been applied to classify propolis samples by generating characteristic fingerprint spectra and integrating machine learning to achieve high predictive accuracies for differentiating samples based on climate zones and colors (Wang et al. 2024). Similarly, untargeted ultrahigh-performance liquid chromatography coupled with hybrid quadrupole-Orbitrap mass spectrometry (UHPLC-HRMS) has been used to discriminate between Greek and Chinese propolis, revealing significant variability in their quality. This method, combined with Partial Least Squares Discriminant Analysis (PLS-DA) and Orthogonal Partial Least Squares Discriminant Analysis (OPLS-DA) models, identified key features for sample differentiation (Stavropoulou et al. 2021).

Complementary techniques also play a role in propolis analysis. Gas chromatography-mass spectrometry (GC-MS) is effective for volatile compounds, although its use is limited in propolis analysis due to the nonvolatile nature of many phenolics. Capillary electrophoresis (CE) is considered an alternative to HPLC for flavonoid analysis, with the advantages of rapidity, high separation efficiency, and low cost (Jiang et al. 2008), although it is less versatile and the repeatability seems inferior to that of HPLC. There are some reports about the determination of flavonoids in propolis using CE (Volpi 2004; Fu et al. 2005; Jiang et al. 2008; Galeotti et al. 2021). Fourier transform infrared spectroscopy (FTIR), while suitable for rapid screening and functional group identification (Calegari et al. 2020), lacks the specificity required for quantitative profiling of individual phenolic compounds.

In conclusion, the choice of analytical method depends on the objectives of the study, the complexity of the target compounds, and the required level of detail. General methods are valuable for initial screenings, while advanced techniques like LC-MS/MS and UHPLC-HRMS are indispensable for comprehensive profiling of the phenolic composition in propolis.

5.4 Phenolic Compounds as Markers of Botanical Origin

Over 800 compounds have been identified in propolis, with flavonoids and terpenes being particularly abundant. These chemical profiles are shaped by both geographic and botanical origins and can be traced using phenolic markers. The geographic classification of propolis reflects its botanical sources, as the availability of regional flora determines the resins bees collect, linking plant species to the unique chemical composition of propolis.

Globally, propolis is categorized into eight types based on botanical origins and chemical constituents: *poplar-type* propolis from temperate regions, *aspen* propolis from northern Europe, *Brazilian green* propolis from southeastern Brazil, *South American* red propolis from Cuba and Brazil, *Mediterranean* propolis from southern Europe, *Pacific* propolis from Pacific islands, *Mangifera indica* propolis from Indonesia, and *hybrid* types such as aspen-poplar and Pacific-*Mangifera indica* propolis (Qiao et al. 2023).

Poplar-type propolis is one of the most widespread, dominating temperate regions across Europe, North America, and parts of Asia. The *Populus* genus, a major source of this propolis, is divided into five sections: *Tacamahaca* Spach, *Leuce* Duby, *Aigeiros* Duby, *Turanga* Bunge, and *Leucoides* Spach. Among these, *Populus nigra* L., part of the *Aigeiros* Duby section, has been identified as the primary botanical source in many European countries, including France, Great Britain, Spain, and Bulgaria, and North America. Despite its consistent association with *Populus nigra*, variations in chemical composition are observed both within and between countries. These differences are likely attributed to contributions from additional plant sources, such as *Pinus* spp., *Prunus* spp., *Acacia* spp., *Betula pendula*, *Aesculus hippocastanum*, and *Salix alba* (Dezmirean et al. 2021).

Even with such variability, European poplar-type propolis maintains a characteristic chemical profile dominated by poplar bud phenolics, including flavonoid aglycones, phenolic acids, and their esters, while phenolic glycosides and glycerides are rarely present (Bankova et al. 2002). Within the flavonol family, quercetin, kaempferol, and isorhamnetin derivatives are most common. The flavanonol family is represented primarily by pinobanksin and its derivatives, widely found in propolis from countries such as China, Serbia, Italy, Slovenia, and Germany. Regarding flavones, predominant compounds include luteolin, apigenin, chrysin, and methoxy-chrysin, while in the flavanone family, the main compounds are naringenin, liquiritigenin, pinostrobin, and pinocembrin (Ristivojević et al. 2015a). Nonetheless, unique regional variations exist. As commented before, Portuguese propolis, for instance, includes distinctive compounds such as p-coumaric ester derivatives, methylated and hydroxylated flavonoids, and uncommon pinobanksin/pinocembrin derivatives. Similarly, propolis from the Canary Islands contains furofuran lignans, which are rare in other European or Mediterranean propolis. Furthermore, temperate region propolis from Australia and Canada often contains chalcones, which are generally uncommon in poplar-type propolis. These variations underscore the profound influence of local flora on the chemical composition of propolis, reflecting its botanical and geographic diversity.

In China, where poplar species are abundant, propolis displays diverse origins within the *Populus* genus. Research identifies *Populus canadensis Moench* (a hybrid of *Populus nigra L.* × *Populus deltoides*), *Populus* × *euramericana cv. 'Neva'*, and *Populus simonii* × *P. nigra* as primary sources. These three types of poplar-type propolis are distinguished by their flavonoid profiles. *P. canadensis* is characterized by caffeate derivatives (benzyl, phenethyl, and cinnamyl), pinobanksin and its 5-methoxy and 3-acetate derivatives, cinnamyl p-cinnamate, chrysin, pinocembrin, and galangin (Wang et al. 2017). In contrast, the other two types contain fewer caffeoyl derivatives and are predominantly characterized by five flavonoids (5-methoxy pinobanksin, pinobanksin, pinocembrin, pinobanksin-3-acetate, and chrysin) or four (the same five flavonoids, excluding 5-methoxy pinobanksin) (Qiao et al. 2023).

In contrast, *Eurasian propolis* samples (mostly from Russia and Kazakhstan, Kyrgyzstan, Poland, Ukraine, and Slovakia) of propolis diverge significantly from typical European poplar-type propolis due to contributions from *Betula verrucose* (birch propolis) and *P. tremula* (aspen propolis). This results, in the case of *birch propolis*, in the presence of phenylpropanoids, cinnamic acid esters, phenylpropanoid glycerides, sesquiterpenols, flavonoids, and chalcones (Kasote et al. 2022). *Aspen propolis*, primarily derived from *Populus tremula* in the Eurasian region, is characterized by cinnamic acid glycerides, particularly 2-acetyl-1,3-di-p-coumaroyl glycerol (lasiocarpin A), previously identified in *P. lasiocarpa* buds (Asakawa et al. 1977). Unlike other poplar types, aspen buds have low flavonoid content and lack typical markers such as chrysin, pinocembrin, 3-O-acetyl-pinobanksin, and galangin, resembling the B-type German propolis. Phenolic acid monoesters, like p-coumaric and caffeic acid benzyl esters, are found in both poplars and aspens, while compounds such as CAPE and prenylated caffeic acid esters are specific to poplars (Okinczyc et al. 2018; Woźniak et al. 2022).

In the *Mediterranean* area, propolis rich in diterpenes, such as isocupressic acid, totarol, and 13-epi-manool, among others, is attributed to *Cupressus sempervirens*. Additionally, propolis from *Cistus ladanifer* is characterized by kaempferol-3,7-dimethyl ether as a marker (Falcão et al. 2013).

As indicated, propolis from different regions exhibits unique chemical profiles shaped by their botanical sources. *Brazilian Baccharis-type* (green Brazilian) propolis contains artepillin C (3,5-diprenyl-4-hydroxycinnamic acid), derived from the leaves of *Baccharis dracunculifolia*. In addition to artepillin C, the propolis also contains lower concentrations of other phenolic compounds, such as quercetin, kaempferol, p-coumaric acid, 3, 4-dicaffeoylquinic acid, 3, 5-dicaffeoylquinic acid, drupanin, cinnamic acid, and esters, which contribute to its bioactivity (Gomes et al. 2024). Higher artepillin C amounts have been detected in samples collected in the summer and autumn, while the main compounds of p-coumaric acid and quercetin were available in spring- and winter-sampled propolis, respectively (Tomazzoli et al. 2020). In contrast, *red propolis* from Brazil, Cuba, and Mexico is associated with *Dalbergia* species, which contribute distinctive phenolics, including flavonoids like formononetin, biochanin A, pinocembrin, medicarpin, isoliquiritigenin, neovestitol, and vestitol (Vieira de Morais et al. 2021), while *tropical Clusia-type* propolis, typical from secretions of flowers of *Clusia major* and *Clusia minor* in Venezuela and *Clusia nemorosa* or *Clusia rosea* in Cuba and Venezuela, is distinguished by benzophenones such as guttiferone E. Nemorosone is the predominant polyisoprenylated benzophenone found in the floral resins of *Clusia rosea*, a tree commonly found throughout Cuba. In contrast, scrobiculatones A and B have been identified exclusively in the floral resin of *Clusia scrobiculata* and in Venezuelan propolis from areas where honeybees utilized this plant as a resin source (Cuesta-Rubio et al. 2007).

Pacific propolis is predominantly derived from *Macaranga* species, leading to its classification as *Macaranga*-type propolis. Recent studies have also reported *Macaranga*-type propolis from African regions, originating from various *Macaranga* species. Despite geographic differences, these propolis types exhibit a high degree of chemical similarity, primarily due to the presence of prenylated flavonoids such as nymphaeol A (NA), nymphaeol B (NB), nymphaeol C (NC), and isonymphaeol B (INB). Other compounds, including geranylated flavonoids like mangiferolic acid and isomangiferolic acid, have also been identified. Specific markers, such as NA,

Table 5.1 Major types of propolis: plant sources and key phenolic constituents

Propolis type	Plant source	Phenolic markers	Geographic origin
Poplar propolis	*Populus* spp. (most often *P. nigra* L.)	Pinocembrin, pinobanksin, chrysin, galangin, phenolic acids, and their esters	Europe, North America, nontropical regions of Asia
Birch propolis	*Betula verucosa* Ehrh.	Cinnamic acids, phenylpropanoid, and sesquiterpenols	Russia
Green propolis	*Baccharis* ssp. (most often *B. dracunculifolia* DC.)	Prenylated p-coumaric acids and acetophenone	Brazil
Redpropolis	*Clusia* flower*Dalbergia ecastaphyllum*	Phenylpropene derivative, benzophenones, polyprenylated benzophenones, medicarpin, and biochanin A	Cuba, Mexico, Brazil, Venezuela, Amazonia
Mediterraneanpropolis	*Cupressaceae*/*Juniperus*/*Pin us*family), *Conifer* spp., *Ferula communis*, *Castanea sativa*, *Cistus* spp., *Quercus ilex* L., *Fraxinus ornus* L., and *Olea europaea*	Cupressic and isocupressic acids, totarol, diterpenes, pinobanksin esters, anthraquinones, and esters of caffeic and ferulic acids	Greece, Malta, Crete, Cyprus, Turkey, Algeria, Southern Italy
"Pacific" propolis	*Macaranga* plants,*Macaranga tanarius*	Prenylflavonoids	Okinawa, Taiwan, Japan

Adapted from Oršolić and Jazvinšćak Jembrek (2022)

NB, NC, and GN (30-geranyl-naringenin), are abundant in Pacific propolis, with *Macaranga tanarius* identified as a probable botanical source (Shahinozzaman et al. 2021).

A summary table (Table 5.1) has been included to highlight the predominant botanical species associated with each propolis type and its characteristic phenolic compounds. This overview emphasizes the role of specific plant sources, such as *Populus* species for European poplar-type propolis, *Baccharis dracunculifolia* for Brazilian green propolis, and *Macaranga tanarius* for Pacific propolis, in shaping the unique chemical profiles of these resins.

As a conclusion, the strong relationship between flora and the chemical composition of propolis underscores its botanical specificity, which is a critical factor in defining its bioactive properties. This connection not only reflects the influence of regional vegetation but also provides valuable insights into the potential pharmacological applications of different types of propolis based on their chemical makeup.

References

Abu-Mellal A, Koolaji N, Duke RK, Tran VH, Duke CC (2012) Prenylated cinnamate and stilbenes from Kangaroo Island propolis and their antioxidant activity. Phytochemistry 77:251–259

Ahn MR, Kumazawa S, Hamasaka T, Bang KS, Nakayama T (2004) Antioxidant activity and constituents of propolis collected in various areas of Korea. J Agric Food Chem 52:7286–7292

Aldana-Mejía J, Ccana-Ccapatinta G, Ribeiro V, Arruda C, Veneziani R, Ambrosio S (2021) A validated HPLC-UV method for the analysis of phenolic compounds in Brazilian red propolis and *Dalbergia ecastaphyllum*. J Pharm Biomed Anal 198:114029

Alencar SM, Oldoni TLC, Castro ML, Cabral ISR, Costa-Neto CM, Cury JA, Rosalen PL, Ikegaki M (2007) Chemical composition and biological activity of a new type of Brazilian propolis: Red propolis. J Ethnopharmacol 113:278–283

Asakawa Y, Takemoto T, Wollenweber E, Aratani T (1977) Lasiocarpin A, B and C, three novel phenolic triglycerides from *Populus lasiocarpa*. Phytochemistry 16:1791–1795

Awale S, Shrestha SP, Tezuka Y, Ueda JY, Matsushige K, Kadota S (2005) Neoflavonoids and related constituents from Nepalese propolis and their nitric oxide production inhibitory activity. J Nat Prod 68:858–864

Bankova V, Popova M, Bogdanov S, Sabatini AG (2002) Chemical Composition of European Propolis: Expected and Unexpected Results Zeitschrift für Naturforschung C 57:530–533

Bankova V (2005) Chemical diversity of propolis and the problem of standardization. J Ethnopharmacol 100:114–117

Bankova V, Christov R, Stoev G, Popov S (1992) Determination of phenolics from propolis by capillary gas chromatography. J Chromatogr 607:150–153

Bankova VS, Christov RS, Tejera AD (1998) Lignans and other constituents of propolis from the Canary Islands. Phytochemistry 49:1411–1415

Bankova VS, de Castro SL, Marcucci MC (2000) Propolis: recent advances in chemistry and plant origin. Apidologie 31:3–15

Bankova V, Trusheva B, Popova M (2021) Propolis extraction methods: a review. J Apic Res 60(5):734–743

Banskota AH, Tezuka Y, Prasain JK, Matsushige K, Saiki I, Kadota S (1998) Chemical constituents of Brazilian propolis and their cytotoxic activities. J Nat Prod 61:896–900

Basnet P, Matsushige K, Hase K, Kadota S (1996) Potent antihepatotoxic activity of dicaffeoyl quinic acids from propolis. Biol Pharm Bull 19:1479–1484

Bertrams J, Muller MB, Kunz N, Krammerer DR, Stintzing FC (2013) Phenolic compounds as marker compounds for botanical origin determination of German propolis samples based on TLC and TLC-MS. J Appl Bot Food Qual 86:143–153

Bonvehi JS, Call FV (1994) Phenolic composition of propolis from China and from South America. Z Naturforsch 49c:712–718

Boudourova-Krasteva G, Bankova V, Sforcin JM, Nikolova N, Popov S (1997) Phenolics from Brazilian propolis. Z Naturforsch 52c:676–679

Calegari MA, Ayres BB, dos Santos Tonial LM, de Alencar SM, Oldoni TLC (2020) Fourier transform near infrared spectroscopy as a tool for predicting antioxidant activity of propolis. J King Saud Univ Sci 32(1):784–790

Campo Fernández M, Cuesta-Rubio O, Rosado Perez A, De Oca M, Porto R, Márquez Hernández I, Piccinelli AL, Rastrelli L (2008) GC–MS determination of isoflavonoids in seven red Cuban propolis samples. J Agric Food Chem 56:9927–9932

Cao YH, Wang Y, Yuan Q (2004) Analysis of flavonoids and phenolic acid in propolis by capillary electrophoresis. Chromatographia 59:135–140

de Castro SL (2001) Propolis: biological and pharmacological activities. Annu Rev Biomed Sci 3:49–83

Chang CC, Yang MH, Wen HM, Chern JC (2002) Estimation of total flavonoid content in propolis by two complementary colorimetric methods. J Food Drug Anal 10:178–182

Chen CN, Wu CL, Shy HS, Lin JK (2003) Cytotoxic prenylflavanones from Taiwanese propolis. J Nat Prod 66:503–506

Chen W, Tu X, Wu D, Gao Z, Wu S, Huang S (2019) Comparison of the partition efficiencies of multiple phenolic compounds contained in propolis in different modes of acetonitrile water-based homogenous liquid-liquid extraction. Molecules 24(3):442

Christov R, Trusheva B, Popova M, Bankova V, Bertrand M (2006) Chemical composition of propolis from Canada, its antiradical activity and plant origin. Nat Prod Res 20:531–536

Cottica SM, Sawaya ACHF, Eberlin MN, Franco S, Zeoulae LM, Visentainera JV (2011) Antioxidant activity and composition of propolis obtained by different methods of extraction. J Braz Chem Soc 22:927–933

Crane E (1990) The traditional hive products: honey and beeswax. In: Crane E (ed) Bees and beekeeping: science, practice and world resources. Heinemann Newnes, Oxford, pp 388–413

Cuesta-Rubio O, Piccinelli AL, Campo Fernandez M, Márquez Hernández I, Rosado A, Rastrelli L (2007) Chemical characterization of Cuban propolis by HPLC-PDA, HPLC-MS, and NMR: the brown, red, and yellow Cuban varieties of propolis. J Agric Food Chem 55:7502–7509

Cui-ping Z, Shuai H, Wen-ting W, Shun P, Xiao-ge S, Ya-jing L, Fu-liang H (2014) Development of High-Performance Liquid Chromatographic for Quality and Authenticity Control of Chinese Propolis Abstract Journal of Food Science 79(7)

Cui J, Duan X, Ke L, Pan X, Liu J, Song X, Ma W, Zhang W, Liu Y, Fan Y (2022) Extraction, purification, structural character and biological properties of propolis flavonoids: a review. Fitoterapia 157:105106

Dezmirean DS, Paşca C, Moise AR, Bobiş O (2021) Plant sources responsible for the chemical composition and main bioactive properties of Poplar-type propolis. Plan Theory 10:22

Escriche I, Juan-Borrás M (2018) Standardizing the analysis of phenolic profile in propolis. Food Res Int 106:834–841

Falcão S, Vilas-Boas M, Estevinho LM, Barros C, Domingues MRM, Cardoso SM (2010) Phenolic characterization of northeast Portuguese propolis: usual and unusual compounds. Anal Bioanal Chem 396:887–897

Falcão SI, Vale N, Gomes P, Domingues MRM, Freire C, Cardoso SM, Vilas-Boas M (2013) Phenolic profiling of portuguese propolis by LC–MS spectrometry: uncommon propolis rich in flavonoid glycosides. Phytochem Anal 24:309–318

Fu SH, Yang MH, Wen HM, Chern JC (2005) Analysis of flavonoids in propolis by capillary electrophoresis. J Food Drug Anal 13:43–50

Funari CS, Sutton AT, Carneiro RL, Fraige K, Cavalheiro AJ, da Silva BV, Hilder EF, Arrua RD (2019) Natural deep eutectic solvents and aqueous solutions as an alternative extraction media for propolis. Food Res Int 125:108559

Galeotti F, Capitani F, Maccari F, Mantovani V, Volpi N (2021) Capillary electrophoresis separation of Artepillin C: determination in Brazilian green propolis. J Chromatogr Sci 59(10):994–1003

Gardana C, Scaglianti M, Pietta P, Simonetti P (2007) Analysis of the polyphenolic fraction of propolis from different sources by liquid chromatography-tandem mass spectrometry. J Pharm Biomed Anal 45:390–399

Gomes KO, Messias da Silva LCF, Dos Santos RD, Prado BA, da Silva MP, Silva Rodrigues LF, de Araújo MO, Bilac CA, Freire DO, Gris EF, Rodrigues da Silva IC, de Sá Barreto LCL, Orsi DC (2024) Chemical characterization and antibacterial activities of Brazilian propolis extracts from *Apis mellifera* bees and stingless bees (Meliponini). PLoS One 219(7):e0307289

Gómez-Caravaca AM, Gómez-Romero M, Arráez-Román D, Segura-Carretero A, Fernández-Gutiérrez A (2006) Advances in the analysis of phenolic compounds in products derived from bees. J Pharm Biomed Anal 41:1220–1234

Grassi G, Capasso G, Gambacorta E, Perna AM (2023) Chemical and functional characterization of propolis collected from different areas of south Italy. Food Secur 12(18):3481

Greenaway W, May J, Scaysbrook T, Whatley FR (1991) Identification by gas chromatography-mass spectrometry of 150 compounds in propolis. Z Naturforsch 46C:111–121

de Groot AC, Popova MP, Bankova VS (2014) An update on the constituents of poplar-type propolis. acdegroot publishing, Wapserveen. 11 pages. ISBN/EAN: 978-90-813233-0-7

Huang S, Zhang C-P, Wang K, Li GQ, Hu F-L (2014) Recent advances in the chemical composition of propolis. Molecules 19:19610–19632

Inui S, Shimamura Y, Masuda S, Shirafuji K, Moli RT, Kumazawa S (2012) A new prenylflavonoid isolated from propolis collected in the Solomon Islands. Biosci Biotechnol Biochem 76:1038–1040

Jaubert GF (1926) Origin of the colour of beeswax and the composition of propolis. Comptes Rendus Hebdomadaires des Seances de l'Académie des. Sciences 184:1134–1136. (French)

Jiang LQ, Fang GZ, Zhang Y, Cao GJ, Wang S (2008) Analysis of flavonoids in propolis and *Ginkgo biloba* by micellar electrokinetic capillary chromatography. J Agric Food Chem 56:11571–11577

Kabous KE, Ouhssine M (2023) Moroccan propolis: unraveling the phytochemical diversity, chemical composition, and antioxidant potential across two seasons. Trop J Nat Prod Res 7(9):3974–3980

Kasote D, Bankova V, Viljoen AM (2022) Propolis: chemical diversity and challenges in quality control. Phytochem Rev 21:1887–1911

Kumazawa S, Goto H, Hamasaka T, Fukumoto S, Fujimoto T, Nakayama T (2004) A new prenylated flavonoid from propolis collected in Okinawa, Japan. Biosci Biotechnol Biochem 68:260–262

Kurek-Górecka A, Rzepecka-Stojko A, Górecki M, Stojko J, Sosada M, Świerczek-Zięba G (2014) Structure and antioxidant activity of polyphenols derived from propolis. Molecules 19:78–101

Kuropatnicki AK, Szliszka E, Krol W (2013) Historical aspects of propolis research in modern times. J Evid-Based Complement Altern Med 2013:964149, 11 pages

Liang X-l, Wu Y-l, Chen Y-J, Zhang J-m, He J, Yuan M, Pan T-l, Pineda MA, Li K-p (2024) Membrane-Based preparation process and antioxidant and anti-ages activities of a novel propolis ultrafiltrate. Chem. Biodiversity 21, e202301333

Li F, Awale S, Tezuka Y, Kadota S (2008) Cytotoxic constituents from Brazilian red propolis and their structure-activity relationship. Bioorg Med Chem 16:5434–5440

Li F, Awale S, Tezuka Y, Esumi H, Kadota S (2010) Study on the constituents of Mexican propolis and their cytotoxic activity against PANC-1 human pancreatic cancer cells. J Nat Prod 73:623–627

Li F, He YM, Awale S, Kadota S, Tezuka Y (2011) Two new cytotoxic phenylallylflavanones from Mexican propolis. Chem Pharm Bull 59:1194–1196

Li A, Xuan H, Sun A, Liu R, Cui J (2016) Preparative separation of polyphenols from water-soluble fraction of Chinese propolis using macroporous absorptive resin coupled with preparative high performance liquid chromatography. J Chromatogr B Analyt Technol Biomed Life Sci 1012-1013:42–49

Marcucci MC (1995) Propolis: chemical composition, biological properties and therapeutical activity. Apidologie 26:83–99

Marcucci M, Ferreres F, García-Viguera C, Bankova V, De Castro S, Dantas A, Valente P, Paulino N (2001) Phenolic compounds from Brazilian propolis with pharmacological activities. J Ethnopharmacol 274:105–112

Mayworm M, Lima C, Tomba A, Fernandes-Silva C, Salatino M, Salatino A (2014) Does propolis contain tannins? Evid-Based Complement Alternat Med 2014:613647

Midorikawa K, Banskota AH, Tezuka Y, Nagaoka T, Matsushige K, Message D, Huertas AA, Kadota S (2001) Liquid chromatography-mass spectrometry analysis of propolis. Phytochem Anal 12(6):366–373

Milojković-Opsenica DM, Trifković JĐ, Ristivojević PM, Andrić FL (2022) Thin-layer chromatography in the authenticity testing of bee-products. J Chromatogr B Analyt Technol Biomed Life Sci 1188:123068

Negri G, Salatino MLF, Salatino A (2003) "Green propolis": unreported constituents and a novel compound from chloroform extracts. J Apic Res 42:53–56

Okinczyc P, Szumny A, Szperlik J, Kulma A, Franiczek R, Zbikowska B, Krzyzanowska B, Sroka Z (2018) Profile of polyphenolic and essential oil composition of polish propolis, black poplar and aspens buds. Molecules 23:1262

Oršolić N, Jazvinšćak Jembrek M (2022) Molecular and cellular mechanisms of propolis and its polyphenolic compounds against cancer. Int J Mol Sci 23(18):10479

Paganotti R, Barbeira P (2024) Accuracy enhancement of the Folin-Ciocalteu method for propolis. Am J Anal Chem 15:253–267

Palos-Hernández A, Gutiérrez Fernández MY, Escuadra Burrieza J, Pérez-Iglesias JL, González-Paramás AM (2022) Obtaining green extracts rich in phenolic compounds from underexploited food by-products using natural deep eutectic solvents. Opportunities and challenges. Sustain Chem Pharm 29:100773

Park YK, Ikegaki M (1998) Preparation of water and ethanolic extracts of propolis and evaluation of the preparations. Biosci Biotechnol Biochem 62:2230–2232

Park YK, Alencar SM, Aguiar CL (2002) Botanical origin and chemical composition of Brazilian propolis. J Agric Food Chem 50:2502–2506

Pellati F, Orlandini G, Pinetti D, Benvenuti S (2011) HPLC-DAD and HPLC-ESI-MS/MS methods for metabolite profiling of propolis extracts. J Pharm Biomed Anal 55:934–948

Pereira AD, Norsell M, Cardoso JN, Neto FRD (2000) Rapid screening of polar compounds in Brazilian propolis by high-temperature high-resolution gas chromatography–mass spectrometry. J Agric Food Chem 48:5226–5230

Petrova A, Popova M, Kuzmanova C, Tsvetkova I, Naydenski H, Muli E, Bankova V (2010) New biologically active compounds from Kenyan propolis. Fitoterapia 81:509–514

Piccinelli AL, Campo Fernandez M, Cuesta-Rubio O, Márquez Hernández I, de Simone F, Rastrelli L (2005) Isoflavonoids isolated from Cuban propolis. J Agric Food Chem 53:9010–9016

Pietta PG, Gordana C, Pietta AM (2002) Analytical methods for quality control of propolis. Fitoterapia 73:S7–S20

Popova M, Bankova V, Butovska D, Petkov V, Nikolova-Damyanova B, Sabatini AG, Marcazzan GL, Bogdanov S (2004) Validated methods for the quantification of biologically active constituents of poplar-type propolis. Phytochem Anal 15:235–240

Popova MP, Chinou IB, Marekov IN, Bankova VS (2009) Terpenes with antimicrobial activity from Cretan propolis. Phytochemistry 70:1262–1271

Popravko SA, Gurevich AI, Kolosov MN (1970) Isolation and identification of the main components of propolis In: Proceedings of the 22nd International Beekeeping Congress: Summaries. pp 163–164

Qiao J, Wang Y, Zhang Y, Kong L, Zhang H (2023) Botanical origins and antioxidant activities of two types of flavonoid-rich poplar-type propolis. Food Secur 12(12):2304

Reis Giada ML (2013) Food phenolic compounds: main classes, sources and their antioxidant power. In: Morales-González JA (ed) Oxidative stress and chronic degenerative diseases - a role for antioxidants. InTech. https://doi.org/10.5772/45722

Righi AA, Alves TR, Negri G, Marques LM, Breyer H, Salatino A (2011) Brazilian red propolis: unreported substances, antioxidant and antimicrobial activities. J Sci Food Agric 91:2363–2370

Ristivojević P, Trifković J, Andrić F, Milojković-Opsenica D (2015a) Poplar-type propolis: chemical composition, botanical origin and biological activity. Nat Prod Commun 10:1869–1876

Ristivojević P, Trifković J, Gašić U, Andrić F, Nedić N, Tešić Ž, Milojković-Opsenica D (2015b) Ultrahigh-performance liquid chromatography and mass spectrometry (UHPLC–LTQ/Orbitrap/MS/MS) study of phenolic profile of Serbian poplar type propolis. Phytochem Anal 26:127–136

Sahinler N, Kaftanoglu O (2005) Natural product propolis: chemical composition. Nat Prod Res 19:183–188

Salatino A, Weinstein Teixeira E, Negri G, Message D (2005) Origin and chemical variation of brazilian propolis. J Evid-Based Complement Altern Med 2:33–38

Salatino A, Fernandes-Silva CC, Righi AA, Salatino MLF (2011) Propolis research and the chemistry of plant products. Nat Prod Rep 28:925–936

dos Santos PA, de Miranda Pereira AF, Trugo LC, de Aquino Neto FR (2003) Distribution of quinic acid derivatives and other phenolic compounds in Brazilian Propolis. Z Naturforsch C 58:590–593

Sârbu C, Moţ AC (2011) Ecosystem discrimination and fingerprinting of Romanian propolis by hierarchical fuzzy clustering and image analysis of TLC patterns. Talanta 85:1112–1117

Sawaya ACHF, Cunha IBS, Marcucci MC (2011) Analytical methods applied to diverse types of Brazilian propolis. Chem Cent J 5:27

Shahinozzaman M, Obanda DN, Tawata S (2021) Chemical composition and pharmacological properties of Macaranga-type Pacific propolis: a review. Phytother Res 35(1):207–222

Shi H, Yang H, Zhang X, Yu LL (2012) Identification and quantification of phytochemical composition and anti-inflammatory and radical scavenging properties of methanolic extracts of Chinese propolis. J Agric Food Chem 60:12403–12410

Shin SB, Lee JK, Ko MJ (2023) Enhanced extraction of bioactive compounds from propolis (*Apis mellifera* L.) using subcritical water. Sci Rep 13:15038

Shrestha SP, Narukawa Y, Takeda T (2007a) Chemical constituents of Nepalese propolis (II). Chem Pharm Bull 55:926–929

Shrestha SP, Narukawa Y, Takeda T (2007b) Chemical constituents of Nepalese propolis: isolation of new dalbergiones and related compounds. J Nat Med 61:73–76

Simões LM, Gregório LE, Da Silva Filho AA, de Souza ML, Azzolini AE, Bastos JK, Lucisano-Valim YM (2004) Effect of Brazilian green propolis on the production of reactive oxygen species by stimulated neutrophils. J Ethnopharmacol 94:59–65

Stavropoulou MI, Termentzi A, Kasiotis KM, Cheilari A, Stathopoulou K, Machera K, Aligiannis N (2021) Untargeted Ultrahigh-Performance Liquid Chromatography-Hybrid Quadrupole-Orbitrap Mass Spectrometry (UHPLC-HRMS) metabolomics reveals propolis markers of Greek and Chinese origin. Molecules 26(2):456

Stevens JF, Hart HT, Wollenweber E (1995) The systematic and evolutionary significance of exudates flavonoids in *Aeonium*. Phytochemistry 39:805–813

Tomazzoli MM, Somensi Zeggio AR, Dal Pai NR, Specht L, Costa C, Rocha M, Yunes RA, Maraschin M (2020) Botanical source investigation and evaluation of the effect of seasonality on Brazilian Propolis from *Apis Mellifera* L. Sci Agric 77:e20180258

Trusheva B, Todorov I, Ninova M, Najdenski H, Daneshmand A, Bankova V (2010) Antibacterial mono-and sesquiterpene esters of benzoic acids from Iranian propolis. Chem Cent J 4:8

Trusheva B, Petkov H, Chimshirova R, Popova M, Dimitrova L, Zaharieva MM, Ilieva Y, Vasileva B, Tsvetkova I, Najdenski H, Miloshev G, Georgieva M, Bankova V (2024) Insight into the influence of natural deep eutectic solvents on the extraction of phenolic compounds from poplar type propolis: composition and in vitro biological activity. Heliyon 10(7):e28621

Valcic S, Montenegro G, Timmermann BN (1998) Lignans from Chilean propolis. J Nat Prod 61:771–775

Vieira de Morais D, Rosalen PL, Ikegaki M, de Souza Silva AP, Massarioli AP, de Alencar SM (2021) Active antioxidant phenolics from Brazilian red propolis: an optimization study for their recovery and identification by LC-ESI-QTOF-MS/MS. Antioxidants 10(2):297

Villanueva VR, Bogdanovsky D, Barbier M, Gonnet M, Lavie P (1964) Sur l'identification de la 3, 5, 7-trihydroxy flavones (galangine) a partir de la propolis. Ann Inst Pasteur 106:292–302

Villanueva VR, Barbier M, Gonnet M, Lavie P (1970) Les flavonoides de la propolis. Isolement d'une nouvelle substance bacteriostatique: la pinocembrine. Ann Inst Pasteur 118:84–87

Volpi N (2004) Separation of flavonoids and phenolic acids from propolis by capillary zone electrophoresis. Electrophoresis 25:1872–1878

Walker P, Crane E (1987) Constituents of propolis. Apidologie 18:327–334

Wang X, Hu H, Luo Z, Liu Y, Zhang H (2017) A plant origin of Chinese propolis: *Populus canadensis* Moench. J Apic Res 57:228–245

Wang R, Qu L, Wang Y, Qu Y, Xie Q, Liu H, Nie Z (2024) Rapid analysis and authentication of Chinese propolis using nanoelectrospray ionization mass spectrometry combined with machine learning. Food Chem 447:138928

Woiski RG, Salatino A (1998) Analysis of propolis: some parameters and procedures for chemical quality control. J Apicult Res 37:99–105

Woźniak M, Mrówczyńska L, Waśkiewicz A, Rogoziński T, Ratajczak I (2019) Phenolic profile and antioxidant activity of propolis extracts from Poland. Nat Prod Commun 14(5)

Woźniak M, Sip A, Mrówczyńska L, Broniarczyk J, Waśkiewicz A, Ratajczak I (2022) Biological activity and chemical composition of propolis from various regions of Poland. Molecules 28(1):141

Zaccaria V, Garzarella EU, Di Giovanni C, Galeotti F, Gisone L, Campoccia D, Volpi N, Arciola CR, Daglia M (2019) Multi dynamic extraction: an innovative method to obtain a standardized chemically and biologically reproducible polyphenol extract from poplar-type propolis to be used for its anti-Infective properties. Materials 12(22):3746

Zhang C-P, Zheng H-Q, Hu F-L (2011) Extraction, partial characterization, and storage stability of β-glucosidase from propolis. J Food Sci 76:C75–C79

Zhang Y, Cao C, Yang Z, Jia G, Liu X, Li X, Zongyan C, Li A (2022) Simultaneous determination of 20 phenolic compounds in propolis by HPLC-UV and HPLC-MS/MS. J Food Compos Anal 115:104877

Chapter 6
Updated Review on Propolis and Geopropolis Volatiles

Maria Graça Miguel and Ana Cristina Figueiredo

Contents

Abstract Propolis and geopropolis are bee products derived from honeybees and stingless bees activity, respectively. Both propolis types (propolis in general) are gaining more and more attention due to their alleged biological properties. Because composition and activity are interconnected, several works have addressed propolis

M. G. Miguel
MED—Mediterranean Institute for Agriculture, Environment and Development, CHANGE—Global Change and Sustainability Institute, Faculdade de Ciências e Tecnologia, Universidade do Algarve, Faro, Portugal
e-mail: mgmiguel@ualg.pt

A. C. Figueiredo (✉)
Centro de Ecologia, Evolução e Alterações Ambientais (CE3C) & CHANGE – Global Change and Sustainability Institute, Faculdade de Ciências da Universidade de Lisboa (FCUL), DBV, Biotecnologia Vegetal, Lisbon, Portugal
e-mail: acsf@fc.ul.pt

J. M. Alvarez-Suarez (ed.), *Bee Products – Chemical and Biological Properties*,
https://doi.org/10.1007/978-3-031-89049-9_6

chemical variability, but comparatively much less is known about their volatile chemical profile. The volatile fraction of propolis can be impacted in a variety of ways, from the available plants or other types of materials used by the various bee species and the type of bees, honeybees or stingless bees, to their ultimate isolation and analysis. This work reviews the existing literature on the volatile constituents of propolis and geopropolis, their origin, bee-producing species, plant sources, their isolation as well as chemical and sensory analysis, covering propolis and geopropolis volatile samples from 35 countries.

Keywords Propolis · Geopropolis · Volatile compounds · Chemical variability

6.1 Introduction

Volatile organic compounds (VOCs) are complex mixtures of molecules derived from both primary and secondary metabolism. These molecules play significant roles in biotic and abiotic interactions, and many of them also constitute important human resources. The study of volatile compounds emitted from diverse matrices has thus gained a huge interest because of their possible academic, social and economic impacts. Nevertheless, some matrices, such as propolis, are highly complex, and identification and interpretation of their volatile composition is a challenge.

Propolis, also known as bee glue, is a resinous and sticky bee product that bees use both as a building and sheltering material. Of varying odour, texture and colour (from light to dark), propolis is a complex mixture of plant buds and resins, as well as other plant secretions and volatiles, that honeybees (*Apis* spp.) blend with pollen and beeswax (the mixture being called cerum or cerumen). Occasionally, other constituents are also found in propolis, particularly when plant resources are low. Another type of propolis, geopropolis, is collected by stingless bee species (*Melipona* spp.) native to tropical countries. Geopropolis is composed by combining soil elements with plant exudates and beeswax (Bankova et al. 2000, 2014; Salatino et al. 2005, 2011; Toreti et al. 2013; Huang et al. 2014).

Although they account for only a minor portion of propolis constituents, volatiles impair particular aromas to propolis from different origins and have been considered as a possible quality criterion for propolis freshness (Woisky and Salatino 1998).

Increasing interest in the medicinal properties of propolis has inspired research on numerous topics related to this product, and several reviews have addressed the chemical composition, botanical origin and biological properties of propolis (Bankova et al. 2000, 2014; Salatino et al. 2011; Toreti et al. 2013; Huang et al. 2014; Siheri et al. 2017; El-Guendouz et al. 2019a). Despite the increasing interest in the past years, the literature addressing the volatile profile of propolis or geopropolis is still scarce compared to that of other chemical components, and only some reviews have focused on this subject (Bankova et al. 2014; Miguel and Figueiredo 2017; Cruz et al. 2020). The present work updates the review of Miguel and Figueiredo (2017) on the volatile constituents of propolis and geopropolis, their isolation as well as chemical and sensory analyses.

6.2 Propolis Volatile Fraction

As a bee product, the chemical composition of propolis in general, and of its volatiles in particular, is dependent on several factors, such as geographical origin; bee type; local flora, if sampling is direct from producer hives or of commercial origin, and the methodology used for the isolation and analysis of the volatiles (Miguel and Figueiredo 2017), as discussed below. Despite the use of different sampling and analytical techniques, they all allow for the characterization of the propolis volatile fraction in a more or less specific way, so they will all be considered here.

6.2.1 Propolis Origin

In total, propolis and geopropolis volatiles from 35 countries were evaluated, as shown in Table 6.1, accounting to 597 propolis samples. Propolis from different geographical origins showed different chemical profiles, as shown in Table 6.1, which usually correlate with the natural flora available at the hive location and the preferences of the bee type.

Most often, propolis types have been divided according to their continent of origin (Bankova et al. 2014; Miguel and Figueiredo 2017). Nevertheless, considering the diversity and specificity of the flora in different countries, for instance, those in northern and southern Europe, some additional subgroups may have to be considered.

6.2.2 Bee Type

Studies on Africanized bees (*Apis mellifera*) and European bees (*Apis mellifera ligustica*) showed more quantitative than qualitative variations in propolis volatiles. This may indicate that both types of bees collected propolis from the same plants, but, depending on when the propolis was collected, some plants may have been more dominant than others (Bankova et al. 1998a). According to Haile et al. (2012), the variability in the volatiles of propolis obtained from two locations in Ethiopia could in part be due to the dominance of *Apis mellifera monticola* in the Assela region compared to the Haramaya region, where *Apis mellifera jementica* was the dominant honeybee.

As with honeybees, stingless bees also collect plant exudates but produce their particular form of propolis, geopropolis, using different plant sources from those used by honeybees (Table 6.1). Foraging workers of three species of *Frieseomelitta* stingless bee have been shown to carry complex mixtures of plant-derived mono-, sesqui-, di- and tri-terpenes in the posterior tibia (Patricio et al. 2002). Although studies on volatiles from stingless bee propolis are limited, they have confirmed a different chemical profile from propolis obtained from honeybees of the same

Table 6.1 Studies on propolis and geopropolis volatiles, including the country of origin; type of bees, if mentioned; number of samples studied; method of volatile isolation and analysis and main volatile components

Propolis or geopropolis origin/ hive location	Hives' bee type[a]	[b]S	Volatile isolation/ type of extract	Volatile analysis	Propolis main volatile compounds (%, ≥ 3%)[c]	References
Albania (Tirana)		1	Steam distillation for 4 h followed by extraction with diethyl ether/pentane 1: 1	GC-MS	Cadinene (isomer) 11, methoxyacetophenone 9, heptacosane 7, tricosane 5, nonacosane 5, heneicosane 4, pentacosane 4, hentriacontane 4	Bankova et al. (1994)
Algeria (El-malha, Benibelaîd, Kaous)		3	Hydrodistillation for 4 h	GC-MS	2-Hexenal 1–11, myristic acid 2–6, cyclotetradecane nd-6, thymol nd-5, linoleic acid 2–5, isooctane nd-4, α-cedrol nd-3	Segueni et al. (2010)
Algeria (Oum El Bouaghi, Batna)	Honeybees: *Apis mellifera*	2	Hydrodistillation for 3 h	GC-MS	α-Pinene 4–56, cedrol nd-17, β-eudesmol nd-8, α-eudesmol nd-7, *cis*-verbenol 3–6, dimethylvinylcarbinol nd-5, γ-gurjunene nd-5, *trans*-caryophyllene nd-5, cyclohexene, 3-acetoxy-4-(1-hydroxy-1-methylethyl)-1-methyl 1–4, δ-cadinene 1–4, guaiol nd-4, 15,15′-bi-1,4,7,10,13-pentaoxacyclohexadecane nd-4, *trans*-pinocarveol 2–3, bulnesol nd-3, finasteride nd-3	Boulechfar et al. (2022)
Argentina (Andes)		1	HS-SPME for 30 min at 42 °C	HS-SPME-GC-MS	Similarity among the volatile profile of *Larrea nitida* and the corresponding studied propolis, with 17 constituents detected in both samples	Agüero et al. (2011)
Argentina		2	SPME for 30 min at 60 °C	HS-SPME-GC-MS	(2E,6Z)-Farnesol nd-33, heptadecane nd-27, γ-cadinene 11–14, γ-eudesmol 10–11, (2Z,6Z)-farnesol 7–7, (2Z,6E)-farnesol 5–6, T-muurolene 4–6, benzyl acetate 3–4, phenylethyl acetate nd-4, 6-camphenone 2–3, α-copaen-11-ol nd-3	Mohtar et al. (2018)

Argentina			DHS with charcoal-filtered air and Porapak Q column	HS-SPME-GC-MS	γ-Cadinene 12–19, benzyl acetate 10–13, 3-methyldecane 4–11, benzyl alcohol nd-10, α-curcumene nd-8, 2-phenylethyl acetate 6–7, L-calamenene 4–6, 3-methyl dodecane 3–6, camphenone nd-6, T-muurolene nd-6, 1-dodecene 4–5, β-chamigrene nd-5, β-eudesmol nd-5, prenyl acetate 4–4, 3-methyltridecane 2–3, α-cubebene nd-3, γ-eudesmol nd-3, champacol (α-guaiol) nd-3, α-cadinol nd-3	Mohtar et al. (2018)
Argentina (several areas in Amaicha del Valle, Tucuman)	Honeybees: *Apis mellifera*	11	HS-SPME 15 min at 50 °C from ethanolic extract	HS-SPME-GC-MS	Palmitic acid 37, tetracosane 32, eicosanol 4	González et al. (2019)
Argentina (several areas in Amaicha del Valle, Tucuman)	Honeybees: *Apis mellifera*		HS-SPME (powder) for 20 min at 75 °C	HS-SPME-GC-MS	Benzyl cinnamate 37, hexadecanol 30, tetracosane 5, nonadecanol 3	González et al. (2019)
Argentina (Paulo Neto-Nogueira)	Stingless bees: *Scaptotrigona* aff. *Postica*	1	HS-SPME for 40 min at 40 °C	HS-SPME-GC-MS	2-Ethylhexan-1-ol 27, 2-ethylhexyl acetate 13, acetic acid 8, ethyl propionate 7, propyl propionate 5, ethyl acetate 4, 3-carene 3, α-gurjunene 3, ethanol 3	Isidorov et al. (2022)
Argentina (Paulo Neto-Nogueira)	Stingless bees: *Tetragona clavipes*	1	HS-SPME for 40 min at 40 °C	HS-SPME-GC-MS	α-Pinene 12, β-pinene 9, *p*-cymene 7, ethyl acetate 5, camphene 3, terpinolene 3, β-caryophyllene 3, acetic acid 3, ethyl propionate 3, pyridine 3	Isidorov et al. (2022)
Argentina (Alto Uruguay)	Stingless bees: *Melipona quadrifasciata quadrifasciata*	1	HS-SPME for 40 min at 40 °C	HS-SPME-GC-MS	α-Pinene 21, β-pinene 10, limonene 7, styrene 7, *p*-cymene 6, myrcene 4, 3-carene 4, terpinolene 4, 2-ethylhexan-1-ol 4, camphene 3	Isidorov et al. (2022)
Argentina (Paulo Neto-Nogueira)	Stingless bees: *Tetragonisca fiebrigi*	1	HS-SPME for 40 min at 40 °C	HS-SPME-GC-MS	2-Ethylhexan-1-ol 17, acetic acid 12, ethanol 8, ethyl acetate 7, α-pinene 5, 3-carene 5, α-gurjunene 5, β-caryophyllene 3, ethyl propionate 3, ethyl butanoate 3	Isidorov et al. (2022)

(continued)

Table 6.1 (continued)

Propolis or geopropolis origin/hive location	Hives' bee type[a]	[b]S	Volatile isolation/type of extract	Volatile analysis	Propolis main volatile compounds (%, ≥ 3%)[c]	References
Brazil (Ceará, Paraná, São Paulo)		4	Steam distillation for 4 h followed by extraction with ethyl ether/*n*-hexane 1:1	GC-MS	2-*cis*,6-*trans*-Farnesol nd-17, diprenyl acetophenone nd-11, prenyl acetophenone nd-8, ledol 0.1–6, ethylphenol nd-5, decanoic acid nd-5, tricosane nd-5, pentacosane nd-4, 3-phenylpropanol nd-4, heneicosane nd-4, δ-cadinene 1–3, acetophenone nd-3, octadecane nd-3, nonadecane nd-3, heptacosane nd-3	Bankova et al. (1995)
Brazil (Botucatu)	Honeybees: *Apis mellifera* (Africanized honeybees)	1	SDE for 4 h	GC and GC-MS	Spathulenol 9–14, (2-*cis*,6-*trans*)-farnesol 8–12, benzoic acid 2–12, dihydrocinnamic acid 0.4–11, diprenyl acetophenone 0.2–9, prenyl acetophenone 7–8, benzyl benzoate 0.3–8, δ-cadinene 2–4, β-caryophyllene 1–3, ledol 2–3, (*trans*,*trans*)-8,10-dodecadienol 1–3, heptacosane nd-3	Bankova et al. (1998)
Brazil (Botucatu)	Honeybees: *Apis mellifera ligustica* (European honeybees)	1	SDE for 4 h	GC and GC-MS	Benzyl benzoate 1–18, spathulenol 3–12, (2-*cis*,6-*trans*)-farnesol 2–15, benzoic acid 0.2–5, diprenyl acetophenone 1–8, *p*-methoxy acetophenone 1–7, prenyl acetophenone 0.2–6, dihydrocinnamic acid nd-6, δ-cadinene 0.2–5, ledol 1–3, (*trans*,*trans*)-8,10-dodecadienol 1–3, pentacosane 1–3, β-caryophyllene nd-3	Bankova et al. (1998)
Brazil (Piaui and Paraná)	Stingless bees: *Melipona compressites*, *Tetragona clavipes*, *Melipona quadrifasciata anthidioides*	1	Essential oil	GC-MS	Aliphatic hydrocarbons C8–C25 11–28, spathulenol 1–10, nerolidol nd-12, p-cymen-8-ol nd-11, ethylphenol nd-10, verbenone nd-7, cinnamic acid 1–4, alkylbenzenes and alkylnaphtalenes nd-5, T-muurolol nd-4, ledol nd-3, verbenol nd-3, palmitic acid nd-3, dihydrocinnamic acid nd-3	Bankova et al. (1999)

Brazil (Minas Gerais)		1	Steam distillation for 5 h	NMR of the fractions obtained by column chromatography	2,2-Dimethyl-8-prenyl-6- vinylchromene, 2,6-diprenyl-4-vinylphenol, 2,2-dimethyl-6-vinylchromene, acetophenone, 2-prenyl-4-vinylphenol, 3,4-dimethoxy-styrene, 3,4-dimethoxy-allylbenzene, 4-hydroxy-3,5-diprenylbenzaldehyde, (−)-spathulenol	Kusumoto et al. (2001)
Brazil (Minas Gerais)	Honeybees: *Apis mellifera*	1	Chloroform extract for 24 h at room temperature	GC-MS	Derivatives of benzoic acid, benzaldehyde, prenylated and non-prenylated cinnamic acid, chromane, chromene	Negri et al. (2003a)
Brazil (São Paulo)	Honeybees: *Apis mellifera*	1	Chloroform extract for 24 h at 50 °C	GC-MS	α-Amorphene, β-cadinene, δ-cadinene, *trans*-caryophyllene, α-humulene, ledene, *trans*-nerolidol, spathulenol	Negri et al. (2003b)
Brazil	Honeybees: *Apis mellifera*	1	Distillation-extraction for 2 h	GC-MS	Propene, acetophenone, β-linalool, 2-methyl-1-hepten-6-one, caryophyllene	Atungulu et al. (2007)
Brazil (Minas Gerais)	Honeybees: *Apis mellifera*	1	Hydrodistillation for 3 h	GC and GC-MS	*trans*-Nerolidol 17, β-caryophyllene 13, selina-3,7(11)-diene 10, aromadendrene 8, germacrene A 8, δ-amorphene 5, spathulenol 5, *cis*-calamenene 4, α-copaene 4, viridiflorene 4, γ-muurolene 4, α-humulene 3	de Albuquerque et al. (2008)
Brazil (Southeastern)	Honeybees: *Apis mellifera*	1	Hydrodistillation for 1 h and methylene chloride extraction of decoction water	GC-MS	Nerolidol 7, 1-phenyl-ethanone 5, linalool 5, spathulenol 4, *trans*-caryophyllene 4, aromadendrene 3, δ-cadinene 3	Maróstica Junior et al. (2008)
Brazil (Piaui)		1	Micro-hydrodistillation for 3 h followed by extraction with dichloromethane	GC-MS	α-Pinene 0.3–34, 1,8-cineole 2–24, *trans*-caryophyllene 3–17, α-copaene 1–8, α-gurjunene 1–7, hexadecanoic acid 0.4–8, β-caryophyllene oxide 0.2–8, β-selinene 0.3–7, β-pinene nd-8, terpinen-4-ol nd-8, fenchone nd-6, *p*-menta-1,5-dien-8-ol nd-5, δ-cadinene 1–5, ethyl phenol nd-5, *trans*-pinocarveol nd-4, α-humulene nd-4, linalool nd-3, spathulenol 1–3, hexanoic acid nd-3, benzyl alcohol nd-3, nonanal nd-3, phenylethanol nd-3	Torres et al. (2008)
Brazil (São Paulo)	Stingless bees: from South America	1	Hydrodistillation for 4 h	GC and GC-MS	α-Pinene 18, β-pinene 7, δ-cadinene 7, germacrene D 6, α-copaene 4, *trans*-nerolidol 4, *trans*-pinocarveol 3, viridiflorol 3, isopimara-9(11)-15-diene 3	Ioshida et al. (2010)

(continued)

Table 6.1 (continued)

Propolis or geopropolis origin/hive location	Hives' bee type[a]	[b]S	Volatile isolation/type of extract	Volatile analysis	Propolis main volatile compounds (%, ≥ 3%)[c]	References
Brazil (Rio de Janeiro)	Honeybees: *Apis mellifera*	1	Steam distillation for 6 h	GC-MS	β-Caryophyllene 13, acetophenone 12, linalool 6, γ-elemene 6, γ-cadinene 6, γ-muurolene 4, α-copaene 3, spathulenol 3	Oliveira et al. (2010)
Brazil [green propolis]		1	Microwave-assisted extraction for 15 min at 35 °C	GC-MS	Longipinene 25, α-eudesmol 7, β-eudesmol 6, β-caryophyllene 5, guaiol 5, benzenecarboxylic acid 4, β-maaliene 4, 3,4-dihydro-2-naphthoic acid 4	Li et al. (2012)
Brazil (Minas Gerais)		8	HS 15 min at 100 °C	HS-GC-MS	Alkenes, alcohols, esters, terpenes, aldehydes and ketones	Nunes and Guerreiro (2012)
Brazil (Rio Grande do Sul)		3	Hydrodistillation for 4 h	GC and GC-MS and CPGC	α-Pinene 57–63, β-pinene 13–31, limonene nd-11	Simionatto et al. (2012)
Brazil		1	HS (static) 10 min at 50 °C	HS-GC-MS	α-Pinene 53, β-pinene 21, β-methyl crotonaldehyde 10, hexanal 3	Kaškoniene et al. (2014)
Brazil		1	Maceration with ethanol at room temperature followed by extraction with hexane	GC-MS	Sesquiterpenes, fatty acids and other compounds	Bittencourt et al. (2015)
Brazil (Mato Grosso)	Honeybees: *Apis mellifera*	1	Hydrodistillation	GC-MS	β-Caryophyllene 8, δ-cadinene 8, spathulenol 7, viridiflorene 5, α-copaene 4, aromadendrene 4, α-*trans*-bergamotene 4, *trans*-nerolidol 4, germacrene D 3, γ-muurolene 3	Fernandes et al. (2015)
Brazil (Minas Gerais)		1	Steam distillation	GC-MS	3-Prenylcinnamic acid allyl ester 26, spathulenol 23, 7-phenyl-5-oxo-heptanol 13, benzenepropanoic acid methyl ester 9, benzenepropanoic acid ethyl ester 6, δ-cadinene 4, β-caryophyllene 3, β-copaene 3	Fernandes-Silva et al. (2015)

Brazil (Bahia)	Honeybees: *Apis mellifera*	2	HS-SPME for 2 h at 40 °C	GC-MS	α-Bisabolol nd-37, acetic acid nd-13, limonene 7–7, nonanal 1–6, hexanoic acid 1–6, α-bergamotene nd-6, α-pinene nd-6, α-curcumene nd-5	Xavier et al. (2017)
Brazil (Paraná)		5	Hydrodistillation for 8 h	GC-MS	α-Pinene 86, β-pinene 12	Mayworm et al. (2017)
Brazil (commercial) [red propolis]		3	Hydrodistillation in a Likens–Nickerson-type apparatus for 4 h, followed by extraction with diethyl ether/n-pentane (1:1)	GC-MS	Elemicin 28–26, methyl eugenol 16–24, *trans*-methyl isoeugenol 9–12, isoelemicin 6–7, *trans*-anethole 4–7, methoxyeugenol 3–4, β-caryophyllene 1–3	Trusheva et al. (2017)
Brazil (Belém)	Stingless bees: *Frieseomelitta longipes*	1	Hydrodistillation for 3 h	GC-MS	β-Caryophyllene 20–35, γ-cadinene 7–8, (*Z*)-β-ocimene 1–7, germacrene D 3–7, α-copaene 4–6, α-humulene 4–6, δ-cadinene 3–6, β-*trans*-bergamotene 3–4, γ-muurolene 2–4, β-bisabolene 2–4, β-chamigrene 1–5, germacrene B 1–4, α-cubebene 1–3, valencene 1–3, α-*cis*-bergamotene 1–3, pseudolimonene 0.01–4, β-phellandrene 0.1–3	de Souza et al. (2018)
Brazil (Belém)	Honeybees: *Apis mellifera*	1	Hydrodistillation for 3 h	GC-MS	β-Caryophyllene 31, α-copaene 9, δ-cadinene 9, α-humulene 7, cyperene 6, γ-gurjunene 6, β-bisabolene 6, *E*-β-farnesene 4, γ-amorphene 4, β-*trans*-bergamotene 3, β-selinene 3, γ-cadinene 3	de Souza et al. (2018)
Brazil		1	SPME for 30 min at 60 °C	HS-SPME-GC-MS	Nerolidol 29, *β*-caryophyllene 17	Mohtar et al. (2018)
Brazil			DHS with charcoal-filtered air and Porapak Q column	HS-SPME-GC-MS	Acetophenone 32, D-germacrene 17, nerolidol 11, L-calamenene 9	Mohtar et al. (2018)

(continued)

Table 6.1 (continued)

Propolis or geopropolis origin/ hive location	Hives' bee type[a]	[b]S	Volatile isolation/ type of extract	Volatile analysis	Propolis main volatile compounds (%, ≥ 3%)[c]	References
Brazil (Sergipe) [red propolis]		1	Hydrodistillation for 1 h	GC-MS	Methyl eugenol 13, (*E*)-β-farnesene 3	Sena-Lopes et al. (2018)
Brazil (Maringá)	Stingless bees: *Scaptotrigona bipunctata*	1	Maceration in ethanol solution 70% (v/v) for 48 h/ ultrasonic bath at 40 kHz for 1 h, followed by filtration and concentration	GC-MS	Coumaran 7, isobutyl phthalate 6, 3-carene 6, benzyl alcohol 5, (+)-α-pinene 5, 4-acetylphenol 5, terpinen-4-ol 3, sobrerol 3, 2-acetoxy-1,8-cineole 3, 4,10-aromadendranediol 3, oplopanone 3, 3-ethylphenol 3, dihydroconiferyl alcohol 3, 2-phenylethanol 3	Dallabona et al. (2020)
Brazil (Minas Gerais) [brown propolis]	Stingless bees: *Scaptotrigona* aff. *Postica*	1	Hydrodistillation for 3 h	GC and GC-MS	Acetophenone 15, nerolidol 13, spathulenol 12, β-caryophyllene 6, α-copaene 5, aromadendrene 5, γ-cadinene 5, α-humulene 3, *allo*-aromadendrene 3, ledene 3, cadina-1,4-diene 3, α-muurolol 3	de Lima et al. (2019)
Brazil (Bahia, Minas Gerais, Paraná and Sergipe) [brown propolis]		4	SPME 20 min at 70 °C	HS-GC-MS	‡ β-Caryophyllene 13–29, α-copaene 0–9, limonene 8, humulene 3–6, δ-cadinene 4–5, β-bourbonene 4, α-cubebene 4, valencene 3, α-muurolene 3, γ-cadinene 3, α-bergamotene 3, β-cadinene 3, aromandendrene 3, *trans*-nerolidol 3, valencene 3, elemol 3	Olegário et al. (2019)
Brazil (Minas Gerais) [brown propolis]	Honeybees: *Apis mellifera*	4	Hydrodistillation for 4 h	GC-MS and NMR	Of one sample: nerolidol 10, spathulenol 7, 1,8-cineole 7, terpinen-4-ol 6, β-caryophyllene 6, γ-palmitolactone 6, hydrocinnamyl alcohol 5, α-copaene 5, γ-muurolene 5, acetophenone 4, γ-gurjunene 3	Ribeiro et al. (2021a)
Brazil (Minas Gerais) [brown propolis]	Honeybees: *Apis mellifera*	1	Hydrodistillation for 4 h, partition of hydrolate and oil with ethyl acetate	GC-MS	Nerolidol 13, spathulenol 9, cineole 7, β-caryophyllene 6, acetophenone 5, terpineol 5, γ-bisabolene 5, α-copaene 5, β-bourbonene 5, guaiacol 4, α-chamigrene 4, γ-muurolene 3, germacrene D 3,	Ribeiro et al. (2021b)

Brazil (Angatuba, São Paulo) [brown propolis]	Honeybees: *Apis mellifera*	1	Hydrodistillation	GC-MS	α-Pinene 18, β-pinene 10, 13-*epi*-manoyl oxide 9, caproic acid 6, β-caryophyllene 6, δ-cadinene 5, manool 5, spathulenol 4, eucalyptol 3, α-terpinyl acetate 3	Ribeiro et al. (2021c)
Brazil (Paraná)		7	SHS 180 °C for 15 min	SHS-GC-MS	α-Pinene 10–35, β-pinene 3–20, α-bisabolol 0.4–10, β-eudesmol nd-8, γ-eudesmol nd-6, myrcene 1–5, α-thujene 0.3–5, limonene 2–4, camphene 1–4, thuja-2.4(10)-diene 1–4, sabinene nd-4, γ-terpinene 1–3, *trans*-pinocarveol nd-3, α-cubebene nd-3, β-bourbonene nd-3, α-humulene 0.3–3, γ-curcumene nd-3, δ-cadinene 0.4–3, coumaran nd-3	Sartori et al. (2021)
Brazil (Paraná)		3	Hydrodistillation for 4 h	HS-GC-MS	α-Pinene 16–66, β-pinene 5–18, ethyl benzoate nd-8, thuja-2,4(10)-diene 2–5, zonarene nd-4, limonene 2–3, camphene nd-3, β-caryophyllene nd-3	Ikeda et al. (2021)
Brazil (Paraná)	Honeybees: *Apis mellifera*	7	SDE for 3 h	GC and GC-MS	α-Copaene 8–22, β-caryophyllene 6–22, (*E*)-α-bergamotene 1–22, δ-cadinene 4–13, cyperene nd-13, β-bisabolene 2–10, β-selinene 1–10, α-selinene 1–10, (*E*)-α-farnesene nd-8, γ-muurolene nd-7, naphthalene 1–5, α-cubebene 1–5, (*Z*)-α-bergamotene nd-5, α-humulene nd-4, 2-*epi*-β-funebrene nd-3, (*Z*)-muurola-4(14),5-diene nd-3	de Oliveira et al. (2021)
Brazil (Sobral, Ceará) [Caatinga green propolis]	Honeybees: *Apis mellifera*	4	HS-SPME	HS-SPME-GC-MS	δ-Cadinene 7–28, germacrene D 2–19, 7-*epi*-α-selinene 12–16, β-caryophyllene 12–15, β-selinene 2–9, α-*trans*-bergamotene nd-9, α-muurolene 2–8, eucalyptol nd-8, cyperene nd-5, α-copaene 2–4, α- caryophyllene 1–4, *trans*-cadina-1,4-diene nd-4, nonadecane 0.3–4, δ-selinene 3–3, D-limonene nd-3, γ-cadinene nd-3	Mendonça et al. (2021)
Brazil (Santa Catarina) [geopropolis]	Stingless bees: *Melipona quadrifasciata quadrifasciata*	8	Hydrodistillation for 4 h	GC and GC-MS	α-Pinene 32–51, β-pinene 7–27, limonene nd-16, *p*-cymene 1–8, α-copaene nd-8, α-terpineol 2–7, guaiol nd-7, terpinen-4-ol 1–6, camphene 2–5, verbenone nd-5, *trans*-pinocarveol nd-4, borneol nd-4, elemol nd-5, 5-*epi*-7epi-α-eudesmol nd-4, δ-cadinene/cis-calamene nd-3	Valcanaia et al. (2022)

(continued)

Table 6.1 (continued)

Propolis or geopropolis origin/ hive location	Hives' bee type[a]	[b]S	Volatile isolation/ type of extract	Volatile analysis	Propolis main volatile compounds (%, ≥ 3%)[c]	References
Brazil (Blumenau) [propolis]	Stingless bees: *Tetragonisca angustula*	1	Hydrodistillation for 4 h	GC and GC-MS	α-Pinene 89, β-pinene 8	Valcanaia et al. (2022)
Brazil (Bahia, Minas Gerais, Rio Grande do Sul, São Paulo) [brown, green and red propolis]	Honeybees: *Apis mellifera*	14	Direct thermal desorption	DTD–GC-MS	1-Methyl-octahydroanthracene nd-59, nonacosane nd-19, conipheryl alcohol nd-18, pentacosane 1–12, benzylcinnamate nd-10, tremetone nd-9, farnesyl bromide nd-8, lioxin nd-8, 2,5-dimethyl-γ-oxo-benzenebutanoic acid nd-8, hexatriacontane 0.3–7, α-pinene nd-7, α-copaene nd-7, anethole nd-6, lignocerane 1–6, *trans*-squalene nd-6, tetracontane nd-6, β-amyrin acetate nd-6, 2,3-dihydrobenzofuran nd-5, *o*-methoxy cinnamaldehyde nd-5, resibufogenin nd-5, 3-*O*-acetyl-6-methoxy-cycloartenol nd-5, methyl(triphenylphosphoranyl idene)acetate nd-5, 2′-hydroxy-4,7′-dimethoxyflavan nd-5, erucylamide nd-5, β-asarone nd-4, 9-methoxy-*S*-octahydrophenanthracene nd-4, 3-hydroxy-4-methoxycinnamic acid nd-4, 4,6,6-trimethyl-1-methylene-4-vinyloctahydro-1H-indene nd-4, myricyl alcohol nd-4, ursolic acid nd-4, methyl eugenol nd-3, δ-cadinene nd-3, elemicin nd-3, nerolidol nd-3, α-cadinol nd-3, isospathulenol nd-3, 4-hydroxy-2-methoxycinnamaldehyde nd-3, 8-amino-5,6-dimethoxy-4-methylquinoline nd-3, manoyl oxide nd-3, α-gurjunene nd-3, cembrene nd-3, 18-oxokauran-17-yl acetate nd-3, sclareol nd-3, cycloeucalenyl acetate nd-3, verticiol nd-3, isovestitol nd-3, heptacosanol nd-3, methyl commate A nd-3, lupenone nd-3, lupeol nd-3	Ribeiro et al. (2022)

Brazil (Paraná)	Honeybees: *Apis mellifera*	1	HS-SPME	HS-SPME-GC-MS	Benzenepropanoic acid ethyl ester 25, anthracene 16, benzoic acid ethyl ester 13, 2-propenoic acid 3-phenylethyl ester 4, decanoic acid ethyl ester 3, myrtenol 3	Bertotto et al. (2022)
Brazil (Jaguariúna, S. Paulo)	*Scaptotrigona depilis*	1	Hydrodistillation for 3 h	GC-MS	*cis*-β-Farnesene 16, α-guaiene 15, α-selinene 14, linalool isovalerate 9, α-bulnesene 8, *trans*-cadina-1,4-diene 6, α-cadinene 4, *cis*-β-guaiene 4, α-duprezianene 3	Souza et al. (2022)
Brazil (Blumenau) [geopropolis]	Stingless bees: *Melipona seminigra merrillae*	1	Hydrodistillation for 4 h followed by fractionation using vacuum liquid chromatography	GC-MS	δ-Cadinene 20, β-oplopenone 13, α-copaene 12, α-terpineol 9, γ-muurolene 8, α-muurolene 7, β-caryophyllene 7	Portal et al. (2023)
Brazil (Blumenau) [geopropolis]	Stingless bees: *Melipona quadrifasciata*	4	Hydrodistillation for 4 h followed by fractionation using vacuum liquid chromatography	GC-MS	α-Pinene 25–58, β-pinene 7–26, guaiol nd-20, limonene nd-10, α-copaene nd-9, sylvestrene nd-5, α-terpineol 2–4, *o*-cymene 1–4, β-caryophyllene 1–4, elemol nd-4, α-phellandrene nd-4, terpinen-4-ol nd-3, *p*-mint-2,4(8)-diene nd-3, β-nd-3, bulnesol nd-3	Portal et al. (2023)
Brazil (Blumenau) [geopropolis]	Stingless bees: *Melipona bicolor schencki*	4	Hydrodistillation for 4 h followed by fractionation using vacuum liquid chromatography	GC-MS	α-Pinene 14–41, β-pinene 5–28, δ-cadinene 3–21, α-copaene 1–12, limonene 5–11, terpinen-4-ol-nd-9, α-terpineol 2–8, *p*-mint-2,4(8)-diene nd-8, α-humulene nd-8, myrcene nd-3, *o*-cymene nd-6, β-caryophyllene 1–6, α-cubebene nd-4, germacrene D nd-4, guaiol nd-4, γ-muurolene nd-3	Portal et al. (2023)
Brazil (Caatinga) [geopropolis]	Honeybees: *Apis mellifera*	1	Hydrodistillation for 2.5 h	GC-MS	Lupenone 10, α-cedrene 6, α-eicosene 6, thunbergol 6, α-muurolene 5, α-retinene 5, clerodol 5, (*Z*)-longipinane 4, hexadecanoic acid methyl ester 4, drimenol 4, lupenyl acetate 4, γ-cadinene 3, caryophyllene oxide 3	Son et al. (2023)
Brazil [green propolis]		1	Hydrodistillation for 30, 60, 120, 240 and 360 min Hydrolate partitioned dichloromethane	GC and GC-MS	Benzyl propanoate 10–20, *trans*-nerolidol 11–16, estragole 7–17, *trans*-cadinene ether nd-14, caryophyllene oxide 1–7, spathulenol nd-7, β-caryophyllene 3–5, δ-cadinene 3–4, methyl-2-butenal nd-4, aromadendrene 2–3, linalool nd-3	Baptista Pereira et al. (2024)

(continued)

Table 6.1 (continued)

Propolis or geopropolis origin/hive location	Hives' bee type[a]	[b]S	Volatile isolation/type of extract	Volatile analysis	Propolis main volatile compounds (%, ≥ 3%)[c]	References
Bulgaria (Russe)		1	Steam distillation for 4 h followed by extraction with diethyl ether/pentane 1: 2	GC-MS	β-Eudesmol 9, δ-cadinene 5, γ-muurolene 5, tricosane 5, pentacosane 4, methoxyacetophenone 3, guaiol 3, cadinene (isomer) 3, heptacosane 3	Bankova et al. (1994)
Canada		1	Distillation-extraction for 2 h using dichloromethane	GC-MS	Benzoic acid 47, α-eudesmol 7, benzyl benzoate 6, γ-eudesmol 5, β-eudesmol 5	Okińczyc et al. (2018)
Canary Islands (Gran Canaria)		1	Distillation for 3 h using a Likens–Nickerson apparatus	GC-MS	Nerolidol 3–11, ethyl oleate 3–7, spathulenol 3–8, ledol 2–4, cinnamic acid 1–4, ethyl palmitate 1–4, δ-cadinene 1–3, aromadendrene 0.3–3	Bankova et al. (1998a, 1998b)
Canary (Tenerife)	Honeybees: *Apis mellifera*	3	HS 2 h at room temperature	HS-SPME-GC-MS	α-Pinene 6–19, limonene 2–13, nonanal 2–6, α-terpineol 0.1–8, sabinene nd-6, acetic acid 2–4, *p*-cymene 0.5–4, *n*-hexane nd-4, acetone 2–3, chloroform 1–4, β-caryophyllene 1–3, toluene 1–3, ethanol 0.03–3, α-thujene nd-3, bornyl acetate nd-3, formic acid 0.01–3	Isidorov et al. (2024)
China (23 different regions)		23	Trapped on Tenax	DHS and GC-MS and GC-O	(Concentrations in μg/L) 2-hydroxy-benzoic acid formate nd-96, acetic acid nd-45, 3-methyl-3-buten-1-ol nd-32, nonanal nd-31, benzyl alcohol 0.4–22, 2-methyl-2-buten-1-ol nd-20, phenylethyl alcohol 1–17, limonene 1–14, hexanal nd-14, guaiol nd-11, propanoic acid nd-11, benzaldehyde 2–10, eucalyptol nd-10, α-pinene nd-9, 2-methyl-3-buten-2-ol t-6, ethyl acetate nd-6, pentanal nd-6, toluene nd-3	Yang et al. (2010)

China (four different regions)		4	Thermal desorption and cryofocusing of volatile compounds. Incubation at 50 °C for 10 min	DHS and GC-MS and GC-MS-O	Acetic acid 9–63, benzyl alcohol 1–14, α-cedrene nd-17, 1-(1,5-dimethyl-4-hexenyl)-4-methyl-benzen 0.1–11, 3-methyl-3-buten-1-ol t-7, 3-methyl-2-buten-1-ol t-7, acetic acid methyl ester nd-7, 3-methyl-2-buten-1-ol acetate nd-5, acetic acid phenylmethyl ester 0.3–4, phenylethyl alcohol 0.3–4, 2-methyl-2-buten-1-ol nd-4, toluene nd-4, benzaldehyde 1–3, 2-methyl-2-butenoic acid nd-3, 3-methyl-3-buten-1-ol acetate nd-3, eucalyptol nd-3	Cheng et al. (2013)
China		1	HS (static) 10 min at 50 °C	HS-GC-MS	3-Methyl-3-buten-1-ol 40, 3-methyl-2-buten-1-ol 12, 4-penten-1-yl acetate 9, α-longipinene 9, 3-methyl-2-buten-1-ol acetate 4	Kaškoniene et al. (2014)
China (Taiwan) [green propolis]		1	Distillation for 4 h using a Likens–Nickerson apparatus, followed by extraction with diethyl ether/n-pentane (1:1)	GC-MS	β-Eudesmol 14, 6-methyl-3,5-heptadien-2-one 12, geranial 4, γ-eudesmol 4, 6-methyl-5-hepten-2-one 4, tricosane 4	Trusheva et al. (2017)
China (Shandong)	Honeybees: *Apis cerana*, *Apis mellifera*, *Ligustica spinola*	4	HS-SPME 10 min at 60 °C	HS-SPME-GC-MS	Kuromatsuene nd-13, benzyl alcohol nd-12, phenylethanol 7–8, benzoic acid nd-8, guaiol 1–4, benzaldehyde 1–3, bisabolol nd-4, caryophyllene nd-4, α-curcumene nd-4, citric acid nd-3, β-camphorol nd-3	Ding et al. (2021)
Colombia (Risaralda, Valle del Cauca, Boyacá, Huila)		4	Extraction with dichloromethane/ ethanol/water (1/1/1, v/v/v) for 15 min followed by three cycles of ultrasound at 60 kW for 15 min	GC, GC-MS	α-Bisabolol nd-36, β-caryophyllene nd-28, copaene nd-22, α-pinene nd-21, γ-cadinene nd-14, β-acoradiene nd-15, α-zingiberene nd-15, D-germacrene nd-12, dihydroeudesmol nd-11, β-elemene nd-10, β-bisabolene nd-9, α-cadinol nd-8, *ar*-curcumene nd-7, α-*trans*-bergamotene nd-5, linalool 0.4–5	Márquez et al. (2023)

(continued)

Table 6.1 (continued)

Propolis or geopropolis origin/hive location	Hives' bee type[a]	[b]S	Volatile isolation/type of extract	Volatile analysis	Propolis main volatile compounds (%, ≥ 3%)[c]	References
Croatia (Dalmatia and Slavonia)		2	Hydrodistillation for 4 h	GC-MS	Benzoic acid 1–27, benzyl alcohol 3–18, naphthalene nd-18, limonene 6–11, γ-terpinene 1–6, benzyl benzoate 4	Borčić et al. (1996)
Croatia (eight different areas)			HS-SPME of 40 min from sample heated for 1 h at 60 °C	GC and GC-MS	Thymol nd-76, α-bisabolol nd-23, γ-cadinene nd-21, vanillin nd-20, α-eudesmol nd-20, β-eudesmol nd-13, *cis*-calamenene nd-11, 2-phenylethanol nd-10, bornyl acetate nd-9, α-copaene nd-9, nonanal nd-7, guaiol nd-7, α-cadinol nd-7, γ-eudesmol nd-7, T-cadinol nd-7, *ar*-curcumene nd-6, benzyl alcohol nd-5, α-pinene nd-4, decanal nd-4, α-muurolene nd-4, β-bourbonene nd-3, α-guaiene nd-3, *allo*-aromadendrene nd-3, α-amorphene nd-3	Jercović et al. (2016)
Croatia (eight different areas)		8	Distillation-extraction using a Likens–Nickerson apparatus for 3 h	GC and GC-MS	Thymol nd-44, 4-vinyl-2-methoxyphenol nd-18, γ-cadinene nd-17, α-eudesmol nd-16, α-bisabolol nd-15, β-eudesmol nd-12, benzyl cinnamate nd-12, γ-eudesmol nd-11, T-cadinol nd-10, decanal nd-8, docosane nd-7, benzyl alcohol nd-5, nonanal nd-5, guaiol nd-5, tricosane nd-5, benzoic acid nd-4, bornyl acetate nd-4, *cis*-calamenene nd-4, α-muurolene nd-3, heneicosane nd-3	Jercović et al. (2016)

Croatia (Omišalj, Pinezići, Draga Baščanska, Rab, Pag, Biševo, Korčula)	Honeybees: *Apis mellifera carnica*	7	HS-SPME	HS-SPME-GC-MS	α-Pinene nd-53, benzoic acid nd-40, thymol nd-40, acetic acid nd-32, guaiol nd-29, δ-cadinene nd-22, benzaldehyde nd-18, bulnesol nd-17, nonanal nd-17, limonene nd-12, 2-phenylethanol nd-12, vanillin nd-11, 2-methoxy-*p*-cymene nd-9, γ-cadinene nd-9, 3-methylbut-2-enal nd-8, decanal nd-8, cedrol nd-8, acetone nd-7, benzyl alcohol nd-7, 6-methylhepta-3,5-dien-2-one nd-7, β-eudesmol nd-7, 2-methylbut-2-enoic acid nd-6, longicyclene nd-6, α-muurolene nd-6, α-cadinol nd-5, α-eudesmol nd-5, *trans*-verbenol nd-5, octanal nd-4, 3-methylbut-2-en-1-ol nd-4, octanal nd-4, α-amorphene nd-4, caryophyllene oxide nd-4, γ-eudesmol nd-4, pentanal nd-3, 2-methylbut-2-enal nd-3, styrene nd-3, prenyl acetate nd-3, 6-methylhept-5-en-2-one nd-3, linalool nd-3, benzyl acetate nd-3, verbenone nd-3, phenethyl acetate nd-3, bornyl acetate nd-3, α-longipinene nd-3, α-copaene nd-3, *trans*-β-caryophyllene nd-3, α-cadinene nd-3	Svečnjak et al. (2020)
Croatia (Omišalj, Pinezići, Draga Baščanska, Rab, Pag, Biševo, Korčula)	Honeybees: *Apis mellifera carnica*		Hydrodistillation for 2 h with recovery in pentane: diethyl ether (1:2)	GC-MS	(*Z*)-tricos-9-ene nd-3, tricosane 5–35, docosane 3–26, tetracosane nd-25, bulnesol nd-16, benzyl cinnamate nd-15, benzyl benzoate nd-14, guaiol nd-14, α-pinene nd-11, α-cadinol nd-10, β-eudesmol nd-10, α-eudesmol nd-9, manoyl oxide nd-9, caryophyllene oxide nd-6, γ-eudesmol nd-6, manool nd-6, hexadecanal nd-4, dehydroabietan nd-4, heneicosane nd-4, eicosane 0.1–3, *trans*-*p*-menth-2-ene-1,8-diol nd-3, benzyl salicylate nd-3, octadecan-1-ol nd-3, abietadiene nd-3	Svečnjak et al. (2020)

(continued)

Table 6.1 (continued)

Propolis or geopropolis origin/hive location	Hives' bee type[a]	[b]S	Volatile isolation/type of extract	Volatile analysis	Propolis main volatile compounds (%, ≥ 3%)[c]	References
Croatia		2	HS-SPME as in Jercović et al. (2016)	HS-SPME-GC-MS	Ethyl dodecanoate 8–19, benzoic acid nd-13, α-thujone nd-7, *cis*-calamenene nd-7, ethyl benzoate 4–5, benzyl alcohol 2–5, ethyl benzenepropanoic acid 2–3, 2-phenylethanol 0–5, 1,8-cineole nd-5, β-thujone nd-5, δ-cadinene nd-5, β-eudesmol nd-5, α-eudesmol nd-5, benzyl benzoate 0–4, ethyl tetradecanoate 0–4, menthone nd-4, menthol nd-4, methyl acetate nd-4, γ-cadinene 2–3, camphor nd-3, α-humulene nd-3, ethyl acetate nd-3, vanillin nd-3	Miljanović et al. (2022)
Cuba (three locations in Pinar del Río)	Honeybees: *Apis mellifera*	3	Distillation-extraction using a Likens–Nickerson apparatus for 1.5 h	GC-MS	(Data in ppm) 1,4-naphtoquinone 142–427, anethole 51–85, methyleugenol 30–64, dehydro-α-lapachole 17–33, eicosane 18–36	Bracho (2000)
Cyprus		7	SPME	HS-SPME-GC-MS	α-Pinene 34–69, sabinene nd-23, limonene nd-18, hexanal nd-7, carvacrol methyl ether nd-7, β-pinene nd-5, *p*-cymene nd-5, octanal nd-5, nonanal nd-7, cedrol 0.1–4, 2-propenoic acid, 6-methylhepthyl ester nd-4, decanal nd-4	Nalbantsoy et al. (2022)
England (Oxfordshire)		1	Trapped on Tenax GC, desorbed by flash heating	GC-MS	Benzoic acid 22, ester of 2-methylpropanoic acid 17, 3-methyl-3-butenyl acetate 15, 3-methyl-3-buten-1-ol 5, 3-methyl-2-butenyl acetate 5, acetic acid 5, benzyl alcohol 4, 3-methylbutyl acetate 3, 3-methyl-2-buten-1-ol 3, 2-methylbutanoic acid 3	Greenaway et al. (1991)
Estonia		1	HS (static) for 10 min at 50 °C	HS-GC-MS	Eucalyptol 26, α-pinene 21, benzaldehyde 11, β-pinene 9, heptane 5, benzyl alcohol 3	Kaškoniene et al. (2014)

Ethiopia (Assela)	Honeybees: *Apis mellifera monticola*	1	Hydrodistillation for 3 h	GC-MS	5,6,7,8-Tetramethylbicyclo [4,1,0]hept-4-en-3-one 15, acoradien 14, epicedrol 7, (6*E*,6*E*)-3,7,11,15-tetramethyl-1,6,10,14-hexadecatetraen-3-ol 6, α-cadinol 4, hexadecane 4, calamenene 4, 3-Isopropyl-6-methyl-2-oxo-1-(3-oxobutyl)-cyclohexanecarbaldehyde 5, *cis*-verbenol 3	Haile et al. (2012)
Ethiopia (Haramaya)	Honeybees: *Apis mellifera jemenitica*	1	Hydrodistillation for 3 h	GC-MS	Calamenene 14, 4-terpineol 9, *epi*-bicyclosesquiphellandrene 8, 4-(2-acetyl-5,5-dimethylcyclopent-2-enylidene)butan-2-one 8, 3-Isopropyl-6-methyl-2-oxo-1-(3-oxobutyl)-cyclohexanecarbaldehyde 6, α-himachalene 5, *cis*-verbenol 4, *trans*-pinocarveol 3, camphor 3, α-phellandren-8-ol 3, *p*-menth-1-en-8-ol 3, cedrol 3, 3-thujen-2-one 3, *p*-cymen-8-ol 3	Haile et al. (2012)
France (Berry and Lyonnais)		4	Hydrodistillation	GC and GC-MS	β-Eudesmol, guaiol, β-caryophyllene, β-selinene, α-guaiene, naphtalene, benzyl benzoate, benzyl cinnamate, benzyl acetate, trimethylbenzene, methylethylbenzene	Clair and Peyron (1981)
Germany (commercial sample)		1	Distillation-extraction for 2 h using dichloromethane	GC-MS	*ar*-Curcumene 11, γ-curcumene 7, α-eudesmol 7, 2-phenylethanol 6, γ-eudesmol 6, β-eudesmol 5, bulnesol nd-4	Okińczyc et al. (2018)
Greece (Chalkidiki, Agrinio, Arta, Preveza and Andros)		5	Hydrodistillation for 3 h	GC and GC-MS	α-Pinene 8–46, *n*-decanal 3–10, junipene nd-12, α-eudesmol nd-12, δ-cadinene 0.3–8, *trans*-β-terpineol 2–7, manoyl oxide 0.1–7, β-eudesmol 1–5, cedrol nd-6, *n*-nonanal nd-5, α-muurolene nd-5, guaiol nd-5, manool nd-5, hexadecen-1 nd-4, γ-eudesmol nd-4, α-cadinol nd-4, α-longipinene nd-3, γ-terpinene 1–3, *trans*-totarol 0.1–3, acetovanillone nd-3	Melliou et al. (2007)
Hungary (20 samples)		20	Steam distillation	GC and GC-MS	β-Eudesmol + sesquiterpene alcohol 10–60, benzyl benzoate 1–40, cyclohexyl benzoate 1–8, vanilin 1–6, benzyl alcohol 1–4	Petri et al. (1988)

(continued)

Table 6.1 (continued)

Propolis or geopropolis origin/hive location	Hives' bee type[a]	[b]S	Volatile isolation/type of extract	Volatile analysis	Propolis main volatile compounds (%, ≥ 3%)[c]	References
India (Maharashtra)	Honeybees: *Apis mellifera*	1	Hydrodistillation	GC and GC-MS	Tricosane 14, hexacosane 12, heptacosane 8, palmitic acid 8, linalool 7, geraniol 6, methyleugenol 6, palmitoleic acid 6, heneicosane 5, *cis*-ethyl cinnamate 5, catechol 3	Naik et al. (2013)
Italy (Pise and Livourne)		2	Steam distillation for 1 h	GC and GC-MS	Anethole, α-pinene and β-pinene for the Pise sample. Vanillin, eugenol and borneol for the Livourne sample	Corsi (1981)
Italy (one region, South)	Honeybees: *Apis mellifera*		Hydrodistillation	GC-MS	Similar volatile profile. HS-SPME-GC-MS technique allowed to obtain higher % relative peak area values for δ-cadinene, T-cadinol and α-cadinol, but lower for 3-methyl-3-buten-1-ol and 2-methyl-2-buten-1-ol (quantitative data not provided)	Pellati et al. (2013)
Italy (nine regions from North and South)	Honeybees: *Apis mellifera*	9	HS-SPME for 20 min from sample heated for 30 min at 75 °C	HS-SPME-GC-MS	Benzoic acid 1–30, β-eudesmol 2–13, α-pinene nd-13, benzyl benzoate nd-13, δ-cadinene nd-13, α-cadinol nd-10, T-cadinol nd-10, γ-cadinene nd-9, α-muurolene nd-7, germacrene D-4-ol nd-6, cubenol 1–5, *trans*-verbenol nd-5, α-bisabolol nd-5, α-terpenyl acetate nd-5, vanillin nd-5, calamenene 1–4, benzyl alcohol nd-4, γ-muurolene nd-4, γ-eudesmol nd-4, δ-cadinol 1–3, tricosane 1–3, menthol nd-3, thymol nd-3, *allo*-aromadendrene nd-3, caryophyllene oxide nd-3, cedrol nd-3, benzyl cinnamate nd-3	Pellati et al. (2013)
Mexico (Yucatán)	Honeybees: *Apis mellifera*	1	Distillation for 1 h using a Likens–Nickerson apparatus	GC and GC-MS	(Values in mg/kg) α-pinene 2, limonene 1, *trans*-verbenol 1, pinocarvone 1, β-pinene 0.3, spathulenol 0.3, α-campholenal 0.2, *trans*-pinocarveol 0.3, myrtenal 0.3, myrtenol 0.3, verbenone 0.6, caryophyllene oxide 0.6, δ-cadinene 0.5, camphene 0.1, α-copaene 0.1, β-bourbonene 0.1, β-caryophyllene 0.1, α-humulene 0.1, *allo*-aromadendrene 0.1, bicyclogermacrene 0.1, viridiflorol 0.1, α-humulene epoxide II 0.1	Pino et al. (2006)

Mexico (Yucatán)	Stingless bees: *Melipona beecheii*	1	Distillation for 1 h using a Likens–Nickerson apparatus	GC and GC-MS	(Values in mg/kg) α-pinene 82, β-caryophyllene 55, spathulenol 45, caryophyllene oxide 44, β-bourbonene 43, β-pinene 32, α-copaene 22, *trans*-verbenol 19, viridiflorol 11, bicyclogermacrene 7, *trans*-pinocarveol 6, myrtenal 6, verbenone 6, α-humulene 6, α-humulene epoxide II 6, δ-cadinene 5, myrtenol 5, β-cubebene 5, pinocarvone 4, α-cubebene 4, camphene 4, limonene 4, α-campholenal 4, *allo*-aromadendrene 3	Pino et al. (2006)
Mexico (Veracruz)	Stingless bees: *Melipona beecheii*	1	HS-SPME of 15 min from sample heated for 30 min at 45 °C	HS-SPME-GC-MS	β-Fenchene 15, styrene 9, benzaldehyde 8, (*cis*)-ocimenone 6, α-pinene 4, *m*-cymene 4, limonene 3, heptanal 3, *trans*-isocarveol 3, (4-*trans*)-decenal 3, verbenone 3	Torres-González et al. (2016)
Mexico (Guanajuato)		1	HS-SPME	GC-MS	Nonanal 19, 1-octen-3-ol 12, β-pinene 12, neryl alcohol 10, α-pinene 8, 6-methyl-3,5-heptadiene-2-one 7, sylvestrene 3, *p*-cymen-9-ol 3, *trans*-carveol 3	Rivero-Cruz et al. (2020)
Mongolia (Ulan Bator)		1	Steam distillation for 4 h followed by extraction with diethyl ether/pentane 1:3	GC-MS	Benzyl benzoate 9, calamenene 3	Bankova et al. (1994)
Mongolia (Inner Mongolia) ###		1	Hydrodistillation (H), simultaneous distillation-extraction (SDE) and dynamic headspace sampling (DHS)	GC-MS	H: 3-methyl-2-buten-1-ol 27, phenylethyl alcohol 17, 1,2,3 4,4a,5 6,7-octahydro-α, α,4a, 8-tetra-methyl-2-naphthalenemethanol 15, 2-methoxy-4-vinylphenol 10, α-bisabolol 4. SDE: α-bisabolol 20, 2-methyl-3-buten-2-ol 11, 3-methyl-2-buten-1-ol 8, azulene 5. DHS: heptadecane 7, phenanthrene 4, *ar*-tumerone 4, 1-(1, 5-dimethyl-4-hexenyl)-4-methyl-benzene 3, octadecane 3, 1-(1,5-dimethyl-4-hexenyl)-4-methyl-benzene 3, hexadecane 3	Fu et al. (2009)

(continued)

Table 6.1 (continued)

Propolis or geopropolis origin/ hive location	Hives' bee type[a]	[b]S	Volatile isolation/ type of extract	Volatile analysis	Propolis main volatile compounds (%, ≥ 3%)[c]	References
Morocco (21 locations and three commercial)	Honeybees: *Apis mellifera*	21	Hydrodistillation for 3 h	GC and GC-MS	Cedrol t-40, *n*-tricosane t-34, *ar*-curcumene t-32, β-eudesmol t-30, n-heneicosane t-28, α-eudesmol t-21, hexadecanoic acid t-20, *epi*-cedrol nd-18, *n*-pentacosane t-17, γ-eudesmol t-14, *n*-nonadecane t-14, *n*-decanal t-13, *n*-nonanal t-11, β-bisabolene nd-11, *n*-eicosane t-10, *n*-heptacosane t-10, *n*-docosane t-9, spathulenol t-8, terpinen-4-ol t-7, nonanoic acid t-7, bornyl acetate t-6, 2-*trans*-undecenal nd-6, geranyl acetone t-6, *n*-tetracosane t-6, α-pinene t-5, *n*-dodecanal t-5, bulnesol nd-5, *trans*-nerolidol t-4, *trans*-anethole t-3	El-Guendouz et al. (2019)
Morocco (Beni Mellal-Khénifra)	Honeybees: *Apis mellifera*	1	Hydrodistillation for 3 h	GC and GC-MS	*Euphorbia resinifera* propolis ##: α-pinene 16, β-pinene 7, spathulenol 6, *n*-decanal 4, sabinene 3, bornyl acetate 3	Boutoub et al. (2024)
Morocco (Souss-MassTiznit)	Honeybees: *Apis mellifera*	1	Hydrodistillation for 3 h	GC and GC-MS	*Euphorbia officinarum* propolis ##: α-pinene 15, *n*-decanal 8, β-pinene 7, *n*-nonanal 4, bornyl acetate 3, spathulenol 3	Boutoub et al. (2024)
Nigeria (Ondo and Osun)	Honeybees: *Apis mellifera*	2	Hydrodistillation for 3 h	GC-MS	*trans*-α-Bergamotene 9–10, α-copaene 5–6, cadina-1,4-diene 4–5, citral 1–5, triacontane nd-5, sulcatone 3–4, melonal 2–4, 1,5,9,9-tetramethyl-1,4,7-cycloundecatriene 2–4, α-curcumene 2–3, γ-muurolene 1–3, benzaldehyde nd-3, 2-methyl-5-propionylfuran nd-3, *cis*-α-bergamotene nd-3, 3,5,9-trimethyl-deca-2,4,8-trien-1-ol nd-3	Balogun et al. (2023)

New Zealand		72	HS-SPME for 30 min from sample heated at 75 °C	HS-SPME-GC-MS	# Prenal acetate nd-20, benzyl alcohol nd-16, benzenecarboxylic acid nd-14, 3-methyl-3-butenyl benzoate nd-13, isobutyric acid nd-11, prenyl benzoate nd-11, δ-cadinene nd-7, (*Z*)-α-bergamotene nd-6, α-cedrene nd-5, valencene nd-5, phenylethyl alcohol nd-4, benzyl acetate nd-4, (*E*)-α-bergamotene nd-4, isopentyl benzoate nd-4, α-muurolene nd-4, germacrene D nd-4, guaiol nd-4, *allo*-aromadendrene nd-4, α-methylbutyric acid nd-3, cevadic acid nd-3, α-curcumene nd-3, α-copaen-11-ol nd-3, eucalyptol nd-3	Mountford-McAuley et al. (2024)
Poland	Honeybees: *Apis mellifera carnica*	1	Steam distillation of the ethanolic extract	GC-MS	Guaiol, farnesol, dehydroeudesmol, nerolidol	Maciejewicz et al. (1983)
Poland (Lower Silesia, West Pomerania, Podkarpacie)		18	Distillation-extraction for 2 h using dichloromethane	GC-MS	Benzoic acid 1–42, benzyl benzoate 1–29, β-eudesmol 1–19, α-eudesmol nd-18, γ-eudesmol 1–17, benzyl alcohol 1–9, trans-benzyl cinnamate 1–9, caryophyllene oxide 1–6, salicyl benzoate 1–5, prenyl benzoate nd-4, bulnezol nd-4	Okińczyc et al. (2018)
Portugal (Algarve)		70	Hydrodistillation for 3 h	GC and GC-MS	Thymol t-79, viridiflorol 3–38, tricosane 3–22, nonadecane 3–18, heneicosane 2–14, labd-7-en-15-ol t-15, 15-nor-labdan-8-ol 1–11, pentacosane 1–11, ambroxide t-6, heptadecane t-6, guaiol t-5, palmitic acid t-5, ledol t-4, ethyl linoleate t-4, 8-epi-13-*nor*-ambreinolide t-4, pinocarvone t-3, β-eudesmol t-3, 13-*nor*-ambreinolide t-3, eicosane t-3, lab-8(17)-en-15-ol t-3, labd-8-en-15-ol t-3, eicosanol t-3	Miguel et al. (2013)

(continued)

Table 6.1 (continued)

Propolis or geopropolis origin/hive location	Hives' bee type[a]	[b]S	Volatile isolation/type of extract	Volatile analysis	Propolis main volatile compounds (%, ≥ 3%)[c]	References
Portugal (mainland), Azores archipelago and Madeira Island (thirty-six propolis samples collected from six different geographical locations)		36	Hydrodistillation for 3 h	GC and GC-MS	Thymol nd-64, α-bisabolol nd-45, β-eudesmol 1–26, α-eudesmol nd-25, α-pinene nd-19, γ-eudesmol 3–18, valerianol nd-14, viridiflorol nd-10, cedrol nd-10, α-bisabolol oxide B nd-10, benzyl salicylate nd-10, T-cadinol 1–9, *trans*-nerolidol nd-9, α-cadinol nd-9, nonadecane nd-9, δ-cadinene 0.3–8, hexadecanol nd-8, manool nd-8, tetracosane nd-8, *trans*-totarol nd-7, β-caryophyllene oxide nd-6, tricosane nd-6, heptadecane 0.2–5, 2,6,6-trimethyl cyclohexanone nd-5, ledol nd-5, α-muurolol nd-5, labd-7-en-15-ol nd-5, α-copaene nd-4, γ-cadinene nd-4, β-oplopenone nd-4, 1,2-dehydroviridiflorol nd-4, benzyl benzoate nd-4, phenylethyl octanoate nd-4, α-terpineol nd-3, globulol nd-3, decanoic acid nd-3, β-bourbonene nd-3, *allo*-aromadendrene nd-3, guaiol nd-3, δ-cadinol nd-3, 1,2-dehydroglobulol nd-3, hexadecanoic acid nd-3, pentacosane nd-3	Falcão et al. (2016)
South Africa (Gauteng, Northern Cape and Western Cape Provinces)		39	HS	GCxGC–ToF-MS	#### Acetic acid 0.05–62, α-pinene 1–47, β-pinene 2–22, dehydrosabinene 0.05–18, *E*-β-ocimene 0.05–18, benzaldehyde 0.2–18, styrene 0.05–14, octanal 0.05–13, 2,7-dimethyl-3-octen-5-yne 0.05–12, limonene 0.05–12, decanal 0.05–12, α-thujene 0.05–11, 1,8-cineole 0.1–11, furfural 0.05–11, 2-methoxy benzyl alcohol 0.05–11, octanoic acid methyl ester 0.05–11, hexanal 0.05–9, 2,7-dimethyl-3-octen-5-yne 0.05–9, sabinene 0.05–8, 2-butenal 0.05–5, prenal 0.05–5, heptanal 0.05–5, camphene 0.2–4, 3-carene 0.05–4, hexanoic acid methyl ester 0.05–4, 2-heptanone 0.05–3	Kamatou et al. (2019)

Tunisia (Zouarine Kef, Zelligua Kef, Beni Khalled Nabeul, Bizerte)		4	Hydrodistillation for 3 h	GC-MS	Manool 1–29, dibutyl phthalate nd-26, α-cedrol 0.2–22, totarol 0.4–12, manoyl oxide 0.2–11, β-eudesmol nd-12, α-eudesmol nd-8, thunbergol nd-6, β-elemene nd-6, tricosane 1–5, isobutyl phthalate nd-4, heptacosane nd-4, eicosane 0.1–3, camphene nd-3, γ-eudesmol nd-3, pentacosane nd-3, tetracosane nd-3	Ayari et al. (2020)
Turkey (Anatolia)		1	HS-SPME of propolis ethanolic extracts for 15 min at 50 °C	HS-SPME-GC-MS	Cedrol 4–16, α-bisabolol nd-14, phenyl ethyl alcohol 1–8, decanal nd-7, ethyl benzoate 0.1–7, benzyl alcohol 0.4–7, δ-cadinene 0.2–6, nonanal nd-5, *trans*-pinocarveol 1–4, α-eudesmol 0.3–4, verbenone nd-4, *trans*-verbenol 1–3, calamenene 1–3, β-eudesmol 1–3, γ-cadinene 0.3–3	Hames-Kocabas et al. (2013)
Turkey (39 regions)		39	HS-SPME	GC-MS	α-Pinene nd-77, limonene nd-40, benzaldehyde nd-40, α-terpineol nd-27, styrene nd-27, 2-methyl butanoic acid nd-26, bornyl acetate nd-25, *ar*-curcumene nd-25, benzyl alcohol nd-24, 1,8-cineole nd-21, β-pinene nd-20, acetic acid nd-19, isovaleric acid nd-18, phenyl ethyl alcohol nd-18, 3-methyl-3-buten-1-ol nd-17, 2-methyl propanoic acid nd-14, sabinene nd-13, 3-methyl-2-butenol nd-13, 2-methyl-2-butenoic acid nd-13, δ-3-carene nd-9, linalool nd-8, longifolene (= junipene) nd-8, *trans*-pinocarveol nd-8, carvacrol nd-8, camphene nd-7, β-caryophyllene nd-7, borneol nd-7, *p*-cymene nd-5, 2-phenylethyl acetate nd-5, 1,3,7-cyclooctatetraene nd-4, γ-terpinene nd-4, pinocarvone nd-4, pulegone nd-4, α-humulene nd-4, nonanal nd-3, 7α-(H)-silphiperfol-5-ene nd-3, terpinen-4-ol nd-3, carvacrol methyl ether nd-3, *p*-vinylanisol nd-3, α-guaiacol nd-3, β-Eudesmol nd-3	Sarıkahya et al. (2021)

(continued)

Table 6.1 (continued)

Propolis or geopropolis origin/hive location	Hives' bee type[a]	[b]S	Volatile isolation/type of extract	Volatile analysis	Propolis main volatile compounds (%, ≥ 3%)[c]	References
Turkey (Kocaeli, Aydın, Ordu, Ankara, Kilis, Van, Karaman, Rize)		8	SPME 30 min at 40 °C	SPME-GC-MS	Acetic acid 2–32, β-pinene nd-27, α-pinene nd-25, bornyl acetate 1–17, benzoic acid 0.3–17, *cis-p*-menth-2-en-1-ol nd-18, 2-acetylfuran nd-17, camphene nd-14, limonene 1–11, isopentyl acetate nd-10, verbenone nd-9, styrene 1–8, benzaldehyde 1–7, nonanal 1–6, butanoic acid 1–5, α-campholenal 1–5, hexane 1–5, decanal 1–4, pinocarvone 1–4, *m*-cymene 0.2–4, δ-cadinene 0.2–4, caryophyllene 0.1–4, verbenene nd-4, *endo*-fenchyl acetate nd-4, undecanal 1–3, myrcene nd-3, 3-carene nd-3, longifolene nd-3, *trans*-α-bergamotene nd-3, α-curcumene nd-3	Baştürk and Yavaş (2024)
Turkey (Afyonkarahisar, Bitlis, Hatay, Balıkesir, Çorum, Giresun, Antalya, Adıyaman)		8	SPME 30 min at 40 °C	SPME-GC-MS	Benzaldehyde 1–33, hexane nd-31, benzoic acid nd-27, α-pinene 2–20, acetic acid nd-16, *m*-cymene 1–14, β-pinene nd-12, carvone nd-7, 2-acetylfuran nd-6, isopentyl acetate nd-6, *cis-p*-menth-2-en-1-ol nd-6, 3-carene nd-5, 1,8-cineole nd-5, nonanal 1–4, pinocarvone nd-4, longifolene nd-4, (*E*)-caryophyllene nd-4, α-amorphene nd-4, butanoic acid 1–3, boric acid nd-3, *trans*-α-bergamotene nd-3, δ-cadinene nd-3	Yavaş and Baştürk (2024)
United States (Montana)		1	Hydrodistillation	GC and GC-MS	Nerolidol 14, cabreuva oxide VI 8, α-bisabolol 7, benzyl benzoate 6, β-eudesmol 4, α-eudesmol 3, T-cadinol 3, 2-methyl-3-buten-2-ol 3	Schepetkin et al. (2022)
Uruguay (three samples)		3	HS (static) for 10 min at 50 °C	HS-GC-MS	α-Pinene 23–53, β-pinene 24–27, limonene 2–16, *cis*-β-ocimene nd-7, 1,3-dichloro-benzene nd-6, camphene 4–5, styrene 0.1–5, 3-methyl-3-buten-1-ol 0.1–4, β-myrcene nd-4	Kaškoniene et al. (2014)
Venezuela (Falcón)	Honeybees: *Apis mellifera*	1	Hydrodistillation for 4 h	GC and GC-MS	Germacrene D 27, β-caryophyllene 10, β-elemene 8, α-cadinol 8, bicyclogermacrene 7, δ-cadinene 5, germacrene A 4, α-humulene 3, γ-muurolene 3, T-muurolol 3	Rios et al. (2014)

Venezuela (Aragua, Bolívar, Carabobo, Miranda)		7	SPME for 30 min at 60 °C	HS-SPME-GC-MS	*trans*-Nerolidol nd-26, α-cubebene nd-17, *Z*-β-farnesene nd-16, β-selinene nd-15, γ-gurjunene nd-14, γ-cadinene nd-14, β-himachalene nd-13, isocaryophyllene nd-12, aromadendrene nd-12, heptadecane nd-12, α-caryophyllene nd-11, α-selinene nd-11, T-muurolene nd-9, L-calamenene nd-6, β-caryophyllene nd-9, γ-muurolene nd-9, eucalyptol nd-6, δ-cadinene nd-6, bornyl acetate nd-5	Mohtar et al. (2018)
Venezuela (Aragua, Bolívar, Carabobo, Miranda)			DHS with charcoal-filtered air and Porapak Q column	HS-SPME-GC-MS	3-Methyldecane nd-17, acetophenone nd-16, 3-methylundecane 4–12, β-caryophyllene 3–12, 2-methyltridecane nd-12, α-curcumene nd-12, limonene 7–11, 3-methyl dodecane nd-10, 2,6,10-trimethyltridecane nd-10, 3-methylnonane nd-9, isocaryophyllene nd-8, α-caryophyllene nd-7, γ-cadinene nd-7, β-himachalene nd-6	Mohtar et al. (2018)
Wales (Cardiff)		1	Trapped on Tenax GC, desorbed and then analyzed	GC and GC-MS	6-Methyl-5-hepten-2-one 16, benzyl alcohol 14, isopentyl acetate 10, benzaldehyde 9, 3-methylbut-3-enol 6, 3-methylbut-2-enol 5, 2-methylbut-2-enoic acid 5, 2-methylbutyric acid 5, 2-phenylethyl alcohol 4, hexanal 3	Greenaway et al. (1989)

[a]Hives' bee type, if mentioned; [b]S: number of samples evaluated; [c]When compounds were quantified and unless otherwise stated; nd: not detected; *CPGC* chiral phase gas chromatography, *DHS* dynamic headspace sampling, *DTD–GC-MS* automated direct thermal desorption–gas chromatography-mass spectrometry, *GCxGC–ToF-MS* two-dimensional gas chromatography, *GC-O-MS* gas chromatography–olfactometry–mass, *HS* headspace, *MAE* microwave-assisted extraction, *NMR* nuclear magnetic resonance, *SDE* simultaneous distillation-extraction with Likens–Nickerson apparatus, *SHS-GC-MS* static headspace gas chromatography coupled to mass spectrometry, *SPME* solid-phase microextraction; #: only part of the data presented due to problems in table presentation; ##: determined by pollen analysis; ###: paper in Chinese (data from abstract only); ####: over 50 compounds listed >0.3%; for the complete list, see reference

region, as this is dependent on the flora preferences of the bees (Bankova et al. 1999; Pino et al. 2006; Bankova and Popova 2007; Ioshida et al. 2010; Huang et al. 2014; Isidorov et al. 2016; Torres-González et al. 2016; Bankova et al. 2018; Isidorov et al. 2022).

Isidorov et al. (2022) showed that although propolis volatiles were species-specific, there was a pairwise similarity in the propolis of the stingless bees *Scaptotrigona postica* and *Tetragonisca fiebrigi* versus that of *Tetragona clavipes* and *Melipona quadrifasciata quadrifasciata*, indicating that the respective species had plant preference resemblance, as shown in Table 6.1.

6.2.3 Natural Flora

Natural local flora determines the chemical composition of propolis, including that of their volatile fraction. In addition to exudates from trees, bees may also use secretions from aromatic plants (lavender, rosemary, sage and thyme) or material from other sources, such as asphalt or tar (Borčić et al. 1996; Alqarni et al. 2015). However, current knowledge on the plant sources of propolis is based more on observation studies of bee behaviour or comparative studies of the phenolic composition of plants and propolis than on the corresponding volatile composition.

Based mostly on phenolic composition, the buds of black poplar (*Populus nigra*), trembling aspen (*Populus tremula*) and silver birch (*Betula pendula*) were considered propolis sources, although other species are well-known to contribute to propolis composition such as *Baccharis* spp. (Brazilian green propolis), *Dalbergia* spp. (Brazilian red propolis), *Clusia* spp. (Clusia propolis), *Macaranga tanarius* (Pacific propolis), *Juniperus* spp. and *Cistus* spp. (Mediterranean propolis) (references in Bankova et al. 2018; and in Isidorov et al. 2024).

Indeed, although several of the studies on propolis volatiles report the natural local flora (Corsi 1981; Petri et al. 1988; Greenaway et al. 1989, 1991; Bankova et al. 1994, 1998a, b; Pino et al. 2006; de Albuquerque et al. 2008; Agüero et al. 2011; Haile et al. 2012; Miguel et al. 2013; Torres-González et al. 2016; El-Guendouz et al. (2019); Sartori et al. 2021; Ding et al. 2021; Nalbantsoy et al. 2022; Boutoub et al. 2024; Mountford-McAuley et al. 2024), as shown in Table 6.1, only some simultaneously compared the volatile profiles of propolis with those of the local flora (Maróstica Junior et al. 2008; Falcão et al. 2016; Okińczyc et al. 2018; Sartori et al. 2021; Mountford-McAuley et al. 2024).

Maróstica Junior et al. (2008) showed the chemical similarities between the volatile compounds found in Brazilian green propolis and those from its botanical origin, *Baccharis dracunculifolia* (Asteraceae/Compositae).

Using static headspace gas chromatography coupled to mass spectrometry (SHS-GC-MS), Sartori et al. (2021) compared the volatile profiles of organic and non-organic Brazilian brown propolis and resins of local *Araucaria angustifolia*, *Pinus elliott* and *Pinus taeda*. Principal component analysis (PCA) showed a resemblance between organic propolis and *A. angustifolia* volatile profiles, while the matching was less clear with the resins of *P. elliott* and *P. taeda*.

According to Falcão et al. (2016), the chemical composition of *Populus x canadensis* (Salicaceae) and *Cistus ladanifer* (Cistaceae) essential oils showed some correlation with that of certain Portuguese propolis samples, as shown in Fig. 6.1.

To test whether New Zealand propolis is poplar-type, Mountford-McAuley et al. (2024) compared the volatiles from propolis samples, as shown in Table 6.1, from different regions and the volatiles from resins from local poplar hybrids and native plant species. Although poplar resin–specific volatile compounds were present in all propolis samples, almost all samples (except for one) also contained native resin–specific volatile compounds.

6.2.4 Detection of Acaricides and Adulterants

Acaricides are used to control bee parasites and can be detected in many hive products, such as propolis and beeswax. Similarly, the use of adulterants such as benzoic acid and eugenol has been reported in propolis, with the objective of enhancing the antibacterial activity (Bonvehí et al. 1994; Woisky and Salatino 1998). Some studies on propolis volatiles reported the presence of acaricides such as Vanquard BT 2–4, Coumaran and Apiguard, among others (Bankova et al. 1998b; Bankova et al. 1999; Miguel et al. 2013; Wojtacka 2022). The long-term application of thymol and thymol based acaricides, such as Apiguard or Apiguard Multidose bee-hive gels, or Thymovar bee-hive strips may be responsible, for instance, for the high thymol content seen in some cases (Miguel et al. 2013).

6.2.5 Isolation of Volatiles

The propolis volatile fraction comprises a mixture of volatile compounds that can be obtained because they either vapourize naturally or after the use of specific techniques.

Diverse types of extraction procedures have been used, mainly with polar solvents, to obtain bioactive compounds from propolis (Bankova et al. 2021; Miłek et al. 2024). In many studies, propolis was extracted with ethanol or hydroethanolic mixtures. Within the so-called environmentally friendly solvents, natural deep eutectic solvents (NADES) are used nowadays as alternative solvents in propolis extraction (Funaria et al. 2019; Tzani et al. 2023). Although volatile compounds can be isolated with ethanol, other polar solvents and/or NADES, these are not extraction solvents and procedures specific to the isolation of volatile compounds, and, for this reason, these studies were not included in Table 6.1. According to Woisky and Salatino (1998), propolis extraction with aqueous ethanol (70%) is better than with

Fig. 6.1 (**a**) Isolation of propolis volatiles by hydrodistillation, using a Clevenger apparatus. (**b**, **c**) Variability of propolis colour and texture of some Portuguese propolis. (**d**) *Populus x canadensis* (Salicaceae) leaf buds used for comparative purposes in the study of Portuguese propolis by Falcão et al. (2016)

absolute ethanol as it provides wax-free tinctures containing higher amounts of phenolic substances. Some studies in which less polar solvents were used for propolis extraction were also included for comparison (Table 6.1).

Depending on the technique used for the isolation of propolis volatiles, as shown in Table 6.1, some authors also evaluated the composition of propolis essential oil [obtained by hydro, steam or dry distillation from any plant part or by expression of citrus species (Council of Europe 2010)], as shown in Fig. 6.1. Although they do not provide essential oils, other conventional volatile extraction techniques used included solvent extraction combined with distillation, simultaneous distillation-extraction (SDE), microwave-assisted extraction and less invasive techniques, such as static solid-phase microextraction (SPME) or dynamic headspace sampling (HS) and automated direct thermal desorption–gas chromatography-mass spectrometry (DTD–GC-MS), as shown in Table 6.1.

All these techniques provide different and complementary information on propolis volatiles, as they are based on diverse principles. Whereas some methodologies are humid-heat-based extraction procedures, such as hydrodistillation or distillation-extraction, others are solvent-free extractions, which sample the vapour phase of propolis, such as headspace techniques. Bracho et al. (1996), Fu et al. (2009), Pellati et al. (2013), Pellati (2014), Jercović et al. (2016), Mohtar et al. (2018) and Svečnjak et al. (2020) compared different methodologies of propolis extraction, namely distillation versus headspace, or even different types of SPME fibre coatings, obtaining different proportions of volatile compounds (Table 6.1).

6.2.6 *Analysis of Volatiles*

6.2.6.1 Volatile Composition

After isolation, and independently of the separation procedure, propolis volatiles are then analyzed to identify and quantify the components, as shown in Table 6.1. Gas chromatography (GC) and gas chromatography coupled mass spectrometry (GC-MS) together, or just GC-MS alone, were the techniques most frequently used for this purpose, as shown in Table 6.1. Although less often used, two-dimensional nuclear magnetic resonance (NMR) experiments (Kusumoto et al. 2001), gas chromatography time of flight mass spectrometry (GCxGC-ToF-MS) coupled to mass spectrometry (Kamatou et al. 2019) and Fourier transform mid-infrared spectroscopy (FT-MIR) (Svečnjak et al. 2020) showed high sensitivity to this complex matrix. One of the factors that impairs direct comparison between data is the quantification procedure. In some cases, no quantification was performed, and in others, quantification was performed by GC or GC-MS, and whereas sometimes the data is reported in percentages, in others, it is reported in absolute amounts (Table 6.1). Several studies involving a larger number of samples used diverse chemometric approaches to ascertain the clustering patterns of the propolis samples evaluated (Miguel et al. 2013; Falcão et al. 2016; Kamatou et al. 2019; El-Guendouz et al. 2019; Mountford-McAuley et al. 2024), as shown in Table 6.1.

One study (Simionatto et al. 2012) also evaluated the variability in enantiomeric composition of monoterpenes α-pinene, β-pinene and limonene in the volatile oils from three samples of Brazilian propolis. (−)-β-pinene predominated in all samples, whereas the enantiomeric excess of α-pinene and limonene was dependent on the harvesting location.

6.2.6.2 Sensory Evaluation

Odour, taste, colour, size, form, consistency and texture are overall attributes of flavour and appearance that are seldom evaluated in propolis. Sensory evaluation techniques are diverse and include tests of diverse types, such as sensory descriptive, sensory discriminative, sensory hedonic, temporal, as well as instrumental sensory devices and immersive techniques, before data treatment (Marques et al. 2022).

Yang et al. (2010) combined gas chromatography–olfactometry–mass spectrometry (GC-O-MS) and sniffing tests on 23 samples of Chinese propolis. In the sniffing studies, propolis odour was mainly described as fruity, pungent, acidic, grassy, herbaceous, stale and smelly, and it was confirmed that components with high concentrations do not always play a major role in odour contribution.

Twenty-eight aroma-active components were determined by dynamic headspace sampling (DHS)–coupled GC-O-MS of propolis samples from different regions of China (Cheng et al. 2013). The odour description at GC-sniffing port included acidic, sour, fruity, faint floral, rose, honey, spice, wood, balsamic, rubber and pungent, among others. In complement to the chemical characterization, as shown in Table 6.1, and olfactometry evaluation, Cheng et al. (2013) also used an electronic nose (e-nose) as an instrumental sensory device to differentiate the bee propolis samples from different botanical origins.

Using comparative aroma extract dilution analysis (cAEDA) on the volatiles obtained by solvent-assisted flavour evaporation (SAFE) from the New Jersey (United States) propolis, Tomaszewski et al. (2019) identified 48 propolis odorants, with 40 being detected in the 3-year period of the assay. Sensory analyses were run by free-choice profiling and quantitative olfactory profile analysis with experienced panellists over 3 years, with notes of clove, cinnamon and honey, among others, being consistently reported over the test period. According to Tomaszewski et al. (2019), when mixed at their natural concentrations, a mixture of 22 odorants can accurately mimic the characteristic of propolis.

Zuluaga-Dominguez et al. (2023) used the e-nose to generate a pattern of the representative volatile compounds from 53 propolis samples of *A. mellifera* and four stingless bee species (*Frieseomielitta* sp., *Melipona eburnea*, *Melipona* sp. and *Trigona* sp.) collected in Colombia. Some of the e-nose sensors were able to differentiate *A. mellifera* propolis from those of *Melipona* sp., *M. eburnea* and *Trigona* sp. but not from those of *Frieseomielitta* sp.

6.3 Main Volatile Compounds in Propolis

Because different isolation techniques and quantification methods have been used, it is not advisable to directly compare the qualitative and quantitative composition of propolis volatiles, as shown in Table 6.1. In the review of Miguel and Figueiredo (2017), comparisons were made based on the relative frequency of occurrence of a compound in different propolis volatile extracts. The authors pointed out that a compound's high frequency of occurrence in propolis does not always indicate that it is a naturally occurring component that bees have collected. As previously indicated, thymol may be present in propolis because of some types of varroa treatment, which employ an acaricide with thymol as its main constituent (Miguel et al. 2013).

The high increase in the number of studies in the last years, along with the fact that some of these studies embrace wide geographic areas, justifies the amplitude of the ranges of variation seen in several compounds, such as going from not detected to >60%, as shown in Table 6.1. These variations reflect not only the bees' preferences on the available flora in each region but also, in some cases, the analysis of raw propolis versus commercially available propolis. For this reason, herein, only a general overview will be made based on the continent of origin of the propolis (Table 6.1). This is only one possible way to categorize propolis volatiles, as some countries are considered transcontinental, such as Cyprus (Europe/Asia).

Each of the following groups include a wide range of geographic locations along with its own bee species and regional available flora: Africa (Algeria, Canary Islands, Ethiopia, Morocco, Nigeria, South Africa and Tunisia), America (Argentina, Brazil, Colombia, Cuba, Mexico, the United States, Uruguay and Venezuela), Asia (China, Cyprus, India, Mongolia and Turkey), Europe (Albania, Bulgaria, Croatia, England, Estonia, France, Greece, Hungary, Italy, Poland, Portugal and Wales) and Oceania (New Zealand).

6.3.1 Africa

Since the review of Miguel and Figueiredo (2017), the number of countries where propolis volatiles has been studied increased from three (Algeria, Canary Islands and Ethiopia) to seven, with the inclusion of samples from Morocco, Nigeria, South Africa and Tunisia, as shown in Table 6.1. From the former six propolis samples studied, a total of 78 propolis samples were now analyzed, 21 of which were from Morocco (El-Guendouz et al. 2019) and 39 from South Africa (Kamatou et al. 2019). These studies with a larger number of samples, from a broad range of proveniences, showed large variations in main components, reflecting the available local flora and bee preferences. Both monoterpenes, such as α-pinene; sesquiterpenes, such as cedrol; or diterpenes, like manool, occurred with a high range of variation. Acetic acid, often associated with the presence of acetic acid bacteria, was also detected in some studies in high levels, as shown in Table 6.1.

El-Guendouz et al. (2019) reported that the dominant flora in the Moroccan areas, where the propolis samples were obtained, had some similarities with that of Southern European countries, with species of *Ceratonia*, *Cistus*, *Eucalyptus*, *Euphorbia*, *Juniperus*, *Lavandula*, *Olea*, *Pinus*, *Pistacia*, *Quercus* and *Rosmarinus*, among others.

Boutoub et al. (2024) used palynological data to assess the contribution of different plant species in three Moroccan propolis samples. Whereas one sample had predominantly pollen grains of *Euphorbia resinifera* (59%), the other two were dominated by *E. officinarum* pollen grains (45–47%). *Genista hirsuta* (9%) in the first sample and *Hypericum elodes* (14%) and *Smilax aspera* (21%) in the other two samples were also among the dominant pollen grains found.

6.3.2 America

In total, 154 propolis samples have been collected and analysed from North American (Canada, Cuba, Mexico and the United States) and South American (Argentina, Brazil, Colombia, Uruguay and Venezuela) origins, as shown in Table 6.1.

As reported in Miguel and Figueiredo (2017), Brazilian propolis is by far the most studied, with a total of 115 samples, ranging from geopropolis to propolis of green, red and brown types, and from either honeybees or stingless bees, or both bee types, as shown in Table 6.1.

Overall, β-caryophyllene, nerolidol, cadinene and spathulenol are sesquiterpenes commonly found in American propolis volatiles, along with the α-pinene, β-pinene and limonene, as shown in Table 6.1.

Agüero et al. (2011) did not provide quantitative data but showed 17 common constituents between the dichloromethane extracts of *Larrea nitida* (Zygophyllaceae) and the corresponding propolis. *o*-Cymene and limonene were the most abundant monoterpenes present in *L. nitida*, both of which were detected in the propolis volatiles.

Maróstica Junior et al. (2008) reported α-pinene, 1-phenyl-ethanone, linalool, β-caryophyllene, δ-cadinene, nerolidol, spathulenol and globulol to be present in concentrations >1% in both *Baccharis dracunculifolia* and propolis volatiles, indicating the similarities in their volatile profiles.

In addition to *B. dracunculifolia* (Bankova et al. 1998a, b; de Albuquerque et al. 2008; Maróstica Junior et al. 2008) being reported as part of Brazil's local flora where propolis was collected, other authors reported other common local species, such as *Larrea nitida* for Argentina (Agüero et al. 2011); *Pinus elliott*, *Pinus taeda* and *Araucaria angustifolia* in Brazil (Sartori et al. 2021) and *Eupatorium adoratum*, *Viguiera dentata* and *Cedrela mexicana* in Mexico (Pino et al. 2006).

6.3.3 Asia

Overall, 99 Asian propolis volatile samples were evaluated, from China, Cyprus, India, Mongolia and Turkey, as shown in Table 6.1. Acetic acid, benzaldehyde and α-pinene are among the common compounds in several propolis Asian samples. Fatty acids and some alkanes are present in relatively high amounts, which may indicate the presence of waxes, as shown in Table 6.1.

Among the reported local flora, Bankova et al. (1994) mentioned *Populus suaveolens* in Mongolia, and Ding et al. (2021) reported variable *Apis* ssp. and flora according to the Chinese province where propolis was collected: *Ginkgo biloba*, *Populus alba*, *P. canadensis* and *Salix babylonica* in Shandong province; *Fraxinus mandshurica*, *Populus*, *Quercus mongolica*, *Tilia tuan* and *Betula* in Heilongjiang province; *Paulownia fortunei*, *Populus tomentosa*, *P. canadensis* and *P. alba* in Henan province and *F. mandshurica*, *Phellodendron amurense*, *Juglans mandshurica*, *Q. mongolica*, *Ulmus pumila* and *T. tuan* in Heilongjiang province.

6.3.4 Europe

Europe includes the largest number of countries, 13, from where propolis samples have been studied, and they belong to different geographic areas: Albania, Bulgaria, Croatia, England, Estonia, France, Germany, Greece, Hungary, Italy, Poland, Portugal and Wales.

As previously reported (Bankova et al. 2014; Miguel and Figueiredo 2017), the oxygen-containing sesquiterpene β-eudesmol is a common volatile compound among the 190 samples of European propolis studied, as shown in Table 6.1.

Other common sesquiterpenes included α-bisabolol, α-cadinol, δ-cadinene, γ-cadinene, γ-eudesmol, guaiol, ledol, T-cadinol and viridiflorol. The monoterpenes eucalyptol (1,8-cineole), α-pinene and thymol were the commonly reported volatile constituents. As previously mentioned, high thymol levels may be a consequence of the long-term use of a thymol-rich acaricide by beekeepers, as reported by Miguel et al. (2013).

Labdane diterpenes were abundant in volatiles from some propolis samples collected in Portugal, mainly in *C. ladanifer*-rich regions (Miguel et al. 2013; Falcão et al. 2016). Hydrocarbons such as heptadecane, nonadecane, heneicosane, tricosane and pentacosane were also commonly found in these types of propolis, depending on the extraction procedure. Aromatic compounds, such as benzyl benzoate and benzyl alcohol, which have frequently been identified in several European propolis volatiles, were not detected or were present in trace amounts in *P. canadensis* leaf buds and *C. ladanifer* branch essential oils (Falcão et al. 2016).

Poplar species, mainly *Populus nigra* (synonym of *Populus alba*), *P. italica* (synonym of *Populus nigra* f. *italica*), *Populus* x *canadensis* (former *Populus* × *euramericana*) and *P. tremula*, are frequently mentioned as natural local flora at propolis collection sites (Petri et al. 1988; Greenaway et al. 1989; Greenaway et al. 1991; Bankova et al. 1994; Falcão et al. 2016; Okińczyc et al. 2018), notwithstanding the presence of other species such as alder (*Alnus*), birch (*Betula*), cypress (*Cupressus*), fir (*Abies*), gum rockrose (*Cistus*), hornbeam (*Carpinus*), junipers (*Juniperus*), linden (*Tilia*), locust (*Robinia*), maple (*Acer*), oak (*Quercus*) and pines (*Pinus*), common also in the Mediterranean area (Petri et al. 1988; Miguel et al. 2013; Falcão et al. 2016). Based on propolis volatiles' variability, Svečnjak et al. (2020) divided Mediterranean propolis into two groups: non-poplar-type, dominated by the monoterpene α-pinene, and poplar-type, with cadinane-type sesquiterpenes.

6.3.5 *Oceania*

Only the study of Mountford-McAuley et al. (2024) was found on New Zealand propolis samples, as shown in Table 6.1. Prenal acetate and (*E*)-α-bergamotene were the two most common compounds identified in the 73 samples from all regions and seasons, but α-pinene and β-pinene were also present in most samples.

The authors also compared the resin and propolis samples and suggest that, in addition to poplar (*Populus* × *euramericana*), other native plants contribute to New Zealand's propolis volatiles, such as kanuka (*Kunzea ericoides*), kauri (*Agathis australis*), kawakawa (*Piper excelsum*), kōhūhū (*Pittosporum tenuifolium*), lemonwood (*Pittosporum eugenioides*), manuka (*Leptospermum scoparium*), ngaio (*Myoporum laetum*), spring rewarewa (*Knightia excelsa*) and tōtara (*Podocarpus totara*).

6.4 Conclusion

Despite the increase in the number of studies on propolis volatiles in the past years, there is still much to learn and understand on this topic. To have a better understanding of the relationship between propolis and local natural flora, more comparative research on their volatiles is needed, with some standardization of sampling and analytical approaches. The extent to which local flora volatiles can be utilized as propolis markers and in the characterization of this special bee product will be determined by additional research in this field.

Acknowledgements The invaluable help of Dr. M. Lurdes Saramago (Biblioteca da Faculdade de Ciências da Universidade de Lisboa) on providing several references is greatly acknowledged. This study was partially funded by Fundação para a Ciência e a Tecnologia (FCT) and PDR2020 support through national funds, under CE3C UIDB/00329/2020 https://doi.org/10.54499/UIDB/00329/2020, CHANGE https://doi.org/10.54499/LA/P/0121/2020 and MED https://doi.org/10.54499/UIDB/05183/2020.

References

Agüero MB, Svetaz L, Sánchez M, Luna L, Lima B, López ML, Zacchino S, Palermo J, Wunderlin D, Feresin GE, Tapia A (2011) Argentinean Andean propolis associated with the medicinal plant Larrea nitida Cav. (Zygophyllaceae). HPLC–MS and GC–MS characterization and antifungal activity. Food Chem Toxicol 49:1970–1978

Alqarni AS, Rushdi AI, Owayss AA, Raweh HS, El-Mubarak AH, Simoneit BRT (2015) Organic tracers from asphalt in propolis produced by urban honey bees, Apis mellifera Linn. PLoS One 10:e0128311

Atungulu G, Miura M, Atungulu E, Satou Y, Suzuki K (2007) Activity of gaseous phase steam distilled propolis extracts on peroxidation and hydrolysis of rice lipids. J Food Eng 80:850–858

Ayari J, Karoui IJ, Abderrabba M (2020) A comparative study between different Tunisian propolis essential oils and their antioxidant activites. Iran J Chem Chem Eng 39:217–231

Balogun OS, Oluduro AO, Liu Z (2023) Chemical composition and in vitro biological studies of volatile oils from Nigerian bee propolis. J Apic Res 62:355–363

Bankova V, Popova M (2007) Propolis of stingless bees: a promising source of biologically active compounds. Pharmacogn Rev 1:88–92

Bankova V, Christov R, Popov S, Pureb O, Bocari G (1994) Volatile constituents of propolis. Z Naturforsch 49c:6–10

Bankova V, Christov R, Kujumgiev A, Marcuccic MC, Popov S (1995) Chemical composition and antibacterial activity of Brazilian propolis. Z Naturforsch 50c:167–172

Bankova V, Boudourova-Krasteva G, Popov S, Sforcin J, Funari SC (1998a) Seasonal variations in essential oil from Brazilian propolis. J Essent Oil Res 10:693–696

Bankova VS, Christov RS, Delgado Tejera A (1998b) Lignans and other constituents of propolis from the Canary islands. Phytochemistry 49:1411–1415

Bankova V, Christov R, Popov S, Marcucci MC, Tsvetkovac I, Kujumgiev A (1999) Antibacterial activity of essential oils from Brazilian propolis. Fitoterapia 70:190–193

Bankova V, de Castro SL, Marcucci MC (2000) Propolis: recent advances in chemistry and plant origin. Apidologie 31:3–15

Bankova V, Popova M, Trusheva B (2014) Propolis volatile compounds: chemical diversity and biological activity: a review. Chem Cent J 8:28

Bankova V, Popova M, Trusheva B (2018) The phytochemistry of the honeybee. Phytochemistry 155:1–11

Bankova V, Trusheva BJ, Popova M (2021) Propolis extraction methods: a review. Apic Res 60:734–743

Baptista Pereira D, de Souza AN, Oliveira Silva E, de Menezes Epifanio NM, de Almeida S, Chaves D (2024) Extraction kinetics of Brazilian green propolis and chemical characterization of its volatiles. Chem Biodivers 21:e202400610

Baştürk A, Yavaş B (2024) Antioxidant activity, total phenolic content, and volatile component profiles of propolis from eight Turkish regions. Chem Biodivers 21:e202402269

Bertotto C, Bilck AP, Yamashita F, Anjos O, Siddique MAB, Harrison SM, Brunton NP, Carpes ST (2022) Development of a biodegradable plastic film extruded with the addition of a Brazilian propolis by-product. LWT – Food Sci Technol 157:113124

Bittencourt MLF, Ribeiro PR, Franco RLP, Hilhorst HWM, de Castro RD, Fernandez LG (2015) Metabolite profiling, antioxidant and antibacterial activities of Brazilian propolis: use of correlation and multivariate analyses to identify potential bioactive compounds. Food Res Int 76:449–457

Bonvehí JS, Coll FC, Jordà RE (1994) The composition active components of bacteriostatic activity of propolis in dietetics. JAOCS 71:529–532

Borčić I, Radonić A, Grzunov K (1996) Comparison of the volatile constituents of propolis gathered in different regions of Croatia. Flavour Fragr J 11:311–313

Boulechfar S, Zellagui A, Asan-Ozusaglam M, Bensouici C, Erenler R, Yildiz İ, Tacer S, Boural H, Demirtas I (2022) Chemical composition, antioxidant, and antimicrobial activities of two essential oils from Algerian propolis. Z Naturforsch 77:105–112

Boutoub O, El-Guendouz S, Matos I, El Ghadraoui L, Costa MC, Carlier JD, Faleiro ML, Figueiredo AC, Estevinho LM, Miguel MG (2024) Chemical characterization and biological properties assessment of *Euphorbia resinifera* and *Euphorbia officinarum* Moroccan propolis. Antibiotics 13:230

Bracho JC (2000) Constituyentes volátiles del propóleo, una realidade acerca de su rica composición química. Bol Soc Quim Peru 66:198–209

Bracho JC, Rosado A, Pino JA (1996) Comparison of isolation methods for propolis volatiles. J Essent Oil Res 8:665–668

Cheng H, Qin ZH, Guo XF, Hu XS, Wu JH (2013) Geographical origin identification of propolis using GC–MS and electronic nose combined with principal component analysis. Food Res Int 51:813–822

Clair G, Peyron L (1981) Contribuition à l'étude de l'huile essentille de propolis. Riv ital EPPOS 63:168–170

Corsi M (1981) Les huilles essentielles de la propolis. In: Meletinov C (ed) XXVIIIth international congress of apiculture. Apimondia, Acapulco, pp 407–411

Council of Europe (2010) European Pharmacopoeia, 7th edn. Council of Europe, Strasbourg, p 241

Cruz JN, da Silva AG, da Costa WA, Gurgel ESC, Campos WEO, de Silva RC, Oliveira MEC, da Silva Souza Filho AP, Pereira DS, Silva SG, de Aguiar Andrade EH, de Oliveira MS (2020) Volatile compounds, chemical composition and biological activities of *Apis mellifera* bee propolis. In: Oliveira M, de Costa WA, da Silva SG (eds) Bioactive compounds, new perspectives and applications. IntechOpen, London, p 250

Dallabona ID, de Lima GG, Cestaro BI, Tasso IS, Paiva TS, Laureanti EJG, Jorge LMM, da Silva BJG, Helm CV, Mathias AL, Jorge RMM (2020) Development of alginate beads with encapsulated jabuticaba peel and propolis extracts to achieve a new natural colorant antioxidante additive. Int J Biol Macromol 163:1421–1432

de Albuquerque IL, Alves LA, Lemos TLG, Dorneles CA, de Morais MO (2008) Constituents of the essential oil of Brazilian green propolis from Brazil. J Essent Oil Res 20:414–415

de Lima VHM, Almeida KCR, Alves CCF, Rodrigues ML, Crotti AEM, de Souza JM, Ribeiro AB, Squarisi IS, Tavares DC, Martins CHG, Miranda MLD (2019) Biological properties of volatile oil from Brazilian Brown propolis. Braz J Pharmacogn 29:807–810

de Oliveira MS, Cruz JN, Ferreira OO, Pereira DS, Pereira NS, Oliveira MEC, Venturieri GC, Guilhon GMSP, Souza Filho APS, Andrade EHA (2021) Chemical composition of volatile compounds in *Apis mellifera* propolis from the northeast region of Pará State, Brazil. Molecules 26:3462

de Souza ECA, da Silva EJG, Cordeiro HKC, Lage Filho NM, da Silva FMA, dos Reis DLS, Porto C, Pilau EJ, da Costa LAMA, de Souza ADL, Menezese C, Flach A (2018) Chemical compositions and antioxidant and antimicrobial activities of propolis produced by the *Frieseomelitta longipes* and *Apis mellifera* bees. Quim Nova 41:485–491

Ding Q, Sheikh AR, Gu X, Li J, Xia K, Sun N, Wu RA, Luo L, Zhang Y, Ma H (2021) Chinese propolis: ultrasound-assisted enhanced ethanolic extraction, volatile components analysis, antioxidant and antibacterial activity comparison. Food Sci Nutr 9:313–330

El-Guendouz S, Lyoussi B, Miguel MG, Figueiredo AC (2019) Characterization of volatiles from Moroccan propolis samples. J Essent Oil Res 31:27–33

El-Guendouz S, Lyoussi B, Miguel MG (2019a) Insight on propolis from Mediterranean countries: chemical composition, biological activities and application fields. Chem Biodivers 16:e1900094

Falcão SI, Freire C, Figueiredo AC, Villas-Boas M (2016) The volatile composition of Portuguese propolis towards its origin discrimination. Rec Nat Prod 10:176–188

Fernandes FH, Guterres ZR, Violante IMP, Lopes TFS, Garcez WS, Garcez FR (2015) Evaluation of mutagenic and antimicrobial properties of brown propolis essential oil from the Brazilian Cerrado biome. Toxicol Rep 2:1482–1488

Fernandes-Silva CC, Lima CA, Negri G, Salatini MLF, Salatino A, Mayworm MAS (2015) Composition of the volatile fraction of a sample of Brazilian green propolic and its phytotoxic activity. J Sci Food Agric 95:3091–3095

Fu Y-X, Xu Y-J, Chen B, Li Y, Luo L-P (2009) Analysis of volatile components from inner Mongolia propolis by gas chromatography-mass spectrometry. Chin J Anal Chem 37:745–748

Funaria CS, Sutton AT, Carneiro RL, Fraige K, Cavalheiro AJ, da Silva Bolzani V, Hilder EF, Arrua RD (2019) Natural deep eutectic solvents and aqueous solutions as an alternative extraction media for propolis. Food Res Int 125:108559

González M, García ME, Slanis A, Bonini A, Fiedler S, Fariña L, Dellacassa E, Condurso C, Lorenzo D, Russo M, Tereschuk ML (2019) Phytochemical findings evidencing botanical origin of new propolis type from North-West Argentina. Chem Biodivers 16:e1800442

Greenaway W, Scaysbrook T, Whatley FR (1989) Headspace volatiles from propolis. Flavour Fragr J 4:173–175

Greenaway W, May J, Scaysbrook T, Whatley FR (1991) Identification by gas chromatography-mass spectrometry of 150 compounds in propolis. Z Naturforsch 46c:111–121

Haile K, Kebede T, Dekebo A (2012) A comparative study of volatile components of propolis (bee glue) collected from Haramaya University and Assela Beekeeping Centers, Ethiopia. Bull Chem Soc Ethiop 26:353–360

Hames-Kocabas EE, Demirci B, Uzel A, Demirci F (2013) Volatile composition of Anatolian propolis by headspace-solid-phase microextraction (HS-SPME), antimicrobial activity against food contaminants and antioxidant activity. J Med Plant Res 7:2140–2149

Huang S, Zhang C-P, Wang K, Li GQ, Hu F-L (2014) Recent advances in the chemical composition of propolis. Molecules 19:19610–19632

Ikeda NY, Ambrosio CMS, Miano AC, Rosalen PL, Gloria EM, Alencar SM (2021) Essential oils extracted from organic propolis residues: an exploratory analysis of their antibacterial and antioxidant properties and volatile profile. Molecules 26:4694

Ioshida MDM, Young MCM, Lago JHG (2010) Chemical composition and antifungal activity of essential oil from Brazilian propolis. J Essent Oil Bear Plants 13:633–637

Isidorov VA, Bakier S, Pirożnikow E, Zambrzycka M, Swiecicka I (2016) Selective behaviour of honeybees in acquiring European propolis plant precursors. J Chem Ecol 42:475–485

Isidorov VA, Maslowiecka J, Szoka L, Pellizzer N, Miranda D, Olchowik-Grabarek E, Zambrzycka M, Swiecicka I (2022) Chemical composition and biological activity of Argentinian propolis of four species of stingless bees. Molecules 27:7686

Isidorov VA, Dallagnol AM, Zalewski A (2024) Chemical composition of volatile and extractive components of Canary (Tenerife) propolis. Molecules 29:1863

Jercović I, Marijanović Z, Kuś PM, Tuberoso CIG (2016) Comprehensive study of Mediterranean (Croatian) propolis peculiarity: headspace, volatiles, anti-Varroa-treatment residue, phenolics, antioxidant properties. Chem Biodivers 13:210–2018

Kamatou G, Sandasi M, Tankea S, van Vuuren S, Viljoen A (2019) Headspace analysis and characterisation of South African propolis volatile compounds using GCxGC–ToF–MS. Braz J Pharmacogn 29:351–357

Kaškoniene V, Kaškonas P, Maruška A, Kubiliene L (2014) Chemometric analysis of volatiles of propolis from different regions using static GC-MS. Cent Eur J Chem 12:736–746

Kusumoto T, Miyamoto T, Higuchi R, Doi S, Sugimoto H, Yamada H (2001) Isolation and structures of two new compounds from the essential oil of Brazilian propolis. Chem Pharm Bull 49:1207–1209

Li Y-J, Xuan H-Z, Shou Q-Y, Zhan Z-G, Lu X, Hu F-L (2012) Therapeutic effects of propolis essential oil on anxiety of restraint-stressed mice. Hum Exp Toxicol 31:157–165

Maciejewicz W, Scheller S, Daniewski M (1983) Gas chromatography-mass spectrometry investigation of propolis, analysis of sesquiterpenes. Acta Pol Pharm 40:251–253

Maróstica Junior MR, Daufsch A, Moraes CS, Queiroga CL, Pastore GM, Park YK (2008) Comparison of volatile and polyphenolic compounds in Brazilian green propolis and its botanical origin *Baccharis dracunculifolia*. Ciênc Tecnol Aliment 28:178–181

Marques C, Correia E, Dinis L-T, Vilela A (2022) An Overview of sensory characterization techniques: From classical descriptive analysis to the emergence of novel profiling methods. Food Secur 11:255

Márquez DGP, Maraschin M, Oliveira ER, Ovalle LVC, Díaz-Moreno (2023) Metabolomic analysis and antioxidant potential of tropical propolis nonpolar extracts from Colombia. J Food Process Preserv 2023:9489176

Mayworm MAS, dos Santos AB, Melo-de-Pinna GFA (2017) Monoterpenes from the essential oil from Brazilian propolis affect seedling cellular elongation. Braz J Bot 40:609–615

Melliou E, Stratis E, Chinou I (2007) Volatile constituents of propolis from various regions of Greece – antimicrobial activity. Food Chem 103:375–380

Mendonça LS, Frota VM, Pinto BJF, Moita ECS, da Hora JP, Costa MF, Fernandes JAB, Zocolo GJ, Gomes GA, do Vale JPC, Bandeira PN, dos Santos HS, Rodrigues THS (2021) Seasonality in the volatile oil composition of green propolis from the Caatinga biome. Braz J Pharmacogn 31:497–501

Miguel MG, Figueiredo AC (2017) Chapter 6: Propolis and geopropolis volatiles. In: Alvarez-Suarez JM (ed) Bee products – chemical and biological properties. Springer, pp 113–136

Miguel MG, Nunes S, Cruz C, Duarte J, Antunes MD, Cavaco AM, Mendes MD, Lima AS, Pedro LG, Barroso JG, Figueiredo AC (2013) Propolis volatiles characterization from acaricide-treated and -untreated beehives maintained at Algarve (Portugal). Nat Prod Res 27:743–749

Miłek M, Bonikowski R, Dżugan M (2024) The effect of extraction conditions on the chemical profile of obtained raw poplar propolis extract. Chem Pap 78:6709–6720

Miljanović A, Bhat RAH, Tandel RS, Pavić D, Grbin D, Dent M, Marijanović Z, Jerković I, Pedisić S, Maguire I, Bielen A (2022) Bioactive compounds in fluid propolis preparations inhibit different life stages of pathogenic oomycetes *Aphanomyces astaci* and *Saprolegnia parasitica*. Aquaculture 552:737982

Mohtar LG, Rodríguez SA, Nazareno MA (2018) Comparative analysis of volatile compound profiles of propolis from different provenances. J Sci Food Agric 98:3409–3415

Mountford-McAuley R, Robertson A, Taylor M, McCormick AC (2024) Characterisation of New Zealand propolis from different regions based on its volatile organic compounds. Molecules 29:3143

Naik DG, Vaidya HS, Namjoshi TP (2013) Essential oil of Indian propolis: chemical composition and repellency against the honeybee Apis florea. Chem Biodivers 10:649–657

Nalbantsoy A, Sarıkahya NB, Özverel CS, Barlas AB, Kırcı D, Akgün IH, Yalçın T, Düven G, Kışla D, Demirci B, Topcu G, Goren AC (2022) Chemical composition and biological activities of Cypriot propolis. J Apic Res 61:233–245

Negri G, Salatino MLF, Salatino A (2003a) "Green propolis": unreported constituents and a novel compound from chloroform extracts. J Apic Res 42:1–3

Negri G, Salatino MLF, Salatino A (2003b) Unusual chemical composition of a sample of Brazilian propolis as assessed by analysis of a chloroform extract. J Apic Res 42:53–56

Nunes CA, Guerreiro M (2012) Characterization of Brazilian green propolis throughout the seasons by headspace GC/MS and ESI-MS. J Sci Food Agric 93:433–438

Okińczyc P, Szumny A, Szperlik J, Kulma A, Franiczek R, Żbikowska B, Krzyżanowska B, Sroka Z (2018) Profile of polyphenolic and essential oil composition of Polish propolis, black poplar and aspens buds. Molecules 23:1262

Olegário LS, Andrade JKS, Andrade GRS, Denadai M, Cavalcanti RL, Pereira da Silva MAA, Narain N (2019) Chemical characterization of four Brazilian brown propolis: An insight in tracking of its geographical location of production and quality control. Food Res Int 123:481–502

Oliveira AP, França HS, Kuster RM, Teixeira LA, Rocha LM (2010) Chemical composition and antibacterial activity of Brazilian propolis essential oil. J Venom Anim Toxins Incl Trop Dis 16:121–130

Patricio EFLRA, Cruz-López L, Maile R, Tentschert J, Jones GR, Morgan ED (2002) The propolis of stingless bees: terpenes from the tibia of three *Frieseomelitta* species. J Insect Physiol 48:249–254

Pellati F (2014) Chapter 2: Innovative methods for the extraction and chromatographic analysis of honey bee products. In: Jayprakasha GK, Patil BS, Pellati F (eds) Instrumental methods for the analysis and identification of bioactive molecules, ACS symposium series. American Chemical Society, Washington, pp 33–49

Pellati F, Prencipe FP, Benvenuti S (2013) Headspace solid-phase microextraction-gas chromatography–mass spectrometry characterization of propolis volatile compounds. J Pharm Biomed Anal 84:103–111

Petri G, Lemberkovics E, Foldvari M (1988) Examination of differences between propolis (bee glue) produced from different floral environments. In: Lawrence BM, Mookherjee BD, Williams BJ (eds) Flavors and fragrances: a world perspective. Elsevier, Amsterdam, pp 439–446

Pino JA, Marbot R, Delgado A, Zumarraga C, Sauri E (2006) Volatile constituents of propolis from honey bees and stingless bees from Yucatán. J Essent Oil Res 18:53–56

Portal AS, Schiquet S, Amaral BP, Krepsky LM, Curbani L, Rebelo RA, Rau M, Althoff LM, Guedes A, de Cordova CMM (2023) Composition, antibiofilm, and antibacterial potential of volatile oils from geopropolis of different stingless bees' species. Chem Biodivers 20:e202300592

Ribeiro VP, Arruda C, Mejía JAA, Candido ACBB, dos Santos RA, Magalhães LG, Bastos JK (2021a) Brazilian southeast Brown propolis: gas chromatography method development for its volatile oil analysis, its antimicrobial and leishmanicidal activities evaluation. Phytochem Anal 32:404–411

Ribeiro VP, Símaro GV, Mejia JAA, Arruda C, Bastos JK (2021b) Anti-inflammatory and antinociceptive activities of the hydroalcoholic extract and the volatile fraction of southeastern Brazilian brown propolis. Braz J Pharmacogn 31:59–66

Ribeiro VP, Arruda C, Aldana-Mejia JA, Bastos JK, Tripathi SK, Khan SI, Khan IA, Ali Z (2021c) Phytochemical, antiplasmodial, cytotoxic and antimicrobial evaluation of a southeast Brazilian brown propolis produced by *Apis mellifera* bees. Chem Biodivers 18:e2100288

Ribeiro VP, Ccana-Ccapatinta GV, Aldana-Mejía JA, Berretta AA, Moraes LAB, Bastos JK (2022) Chemical characterization of Brazilian propolis using automated direct thermal desorption-gas chromatography-mass spectrometry. J Sci Food Agric 102:4345–4354

Rios N, Yánez C, Rojas L, Mora F, Usubillaga A, Vit P (2014) Chemical composition of essential oil of *Apis mellifera* propolis from Falcón State, Venezuela. Emir J Food Agric 26:639–642

Rivero-Cruz JF, Granados-Pineda J, Pedraza-Chaverri J, Pérez-Rojas JM, Kumar-Passari A, Diaz-Ruiz G, Rivero-Cruz BE (2020) Phytochemical constituents, antioxidant, cytotoxicity, and antimicrobial activities of the ethanolic extract of Mexican Brown propolis. Antioxidants 9:70

Salatino A, Teixeira ÉW, Negri G, Message D (2005) Origin and chemical variation of Brazilian propolis. Evid Based Complement Alternat Med 2:33–38

Salatino A, Fernandes-Silva CC, Righi AA, Salatino MLF (2011) Propolis research and the chemistry of plant products. Nat Prod Rep 28:925–936

Sarıkahya NB, Gören AC, Okkalı GS, Çöven FO, Orman B, Kırcı D, Yücel B, Kışla D, Demirci B, Altun M, Önem AN, Nalbantsoy A (2021) Chemical composition and biological activities of propolis samples from different geographical regions of Turkey. Phytochem Lett 44:129–136

Sartori AGO, Spada FP, Ribeiro VP, Rosalen PL, Ikegaki M, Bastos JK, Alencar SM (2021) An insight into the botanical origins of propolis from permanent preservation and reforestation areas of Southern Brazil. Sci Rep 11:22043

Schepetkin IA, Özek G, Özek T, Kirpotina LN, Kokorina PI, Khlebnikov AI, Quinn MT (2022) Neutrophil immunomodulatory activity of nerolidol, a major component of essential oils from *Populus balsamifera* buds and propolis. Plan Theory 11:3399

Segueni N, Khadraoui F, Moussaoui F, Zellagui A, Gherraf N, Lahouelc M, Rhouati S (2010) Volatile constituents of Algerian propolis. Ann Biol Res 1:103–107

Sena-Lopes Â, Bezerra FSB, das Neves RN, de Pinho RB, de Oliveira Silva MT, Savegnago L, Collares T, Seixas F, Begnini K, Henriques JAP, Ely MR, Rufatto LC, Moura S, Barcellos T, Padilha F, Dellagostin O, Borsuk S (2018) Chemical composition, immunostimulatory, cytotoxic and antiparasitic activities of the essential oil from Brazilian red propolis. PLoS One 13:e0191797

Siheri W, Alenezi S, Tusiimire J, Watson DG (2017) Chapter 7: The chemical and biological properties of propolis. In: Alvarez-Suarez JM (ed) Bee products – chemical and biological properties. Springer, pp 137–178

Simionatto E, Facco JT, Morel AF, Giacomelli SR, Linares CEB (2012) Chiral analysis of monoterpenes in volatile oils from propolis. J Chil Chem Soc 57:1240–1243

Son NT, Ribeiro VP, Teixeira MVS, Furtado NAJC, Bastos JK (2023) Northeastern Caatinga green propolis: volatile oil and antimicrobial activity. Chem Nat Compd 59:971–973

Souza ECA, Silva BP, Porto C, Pilau EJ, Menezes C, Flach A (2022) Molecular network-guided chemical profile and mass spectrometry, volatile compounds, and antimicrobial activity of *Scaptotrigona depilis* propolis. Rapid Commun Mass Spectrom 36:e9348

Svečnjak L, Marijanović Z, Okińczyc P, Kuś PM, Jerković I (2020) Mediterranean propolis from the Adriatic sea islands as a source of natural antioxidants: comprehensive chemical biodiversity determined by GC-MS, FTIR-ATR, UHPLC-DAD-QqTOF-MS, DPPH and FRAP assay. Antioxidants 9:337

Tomaszewski M, Dein M, Novy A, Hartman TG, Steinhaus M, Luckett CR, Munafo JP Jr (2019) Quantitation and seasonal variation of key odorants in propolis. J Agric Food Chem 67:1495–1503

Toreti VC, Sato HH, Pastore GM, Park YK (2013) Recent progress of propolis for its biological and chemical compositions and its botanical origin. Evid Based Complement Alternat Med 2013:697390

Torres RNS, Lopes JAD, Neto JMM, Citó AMGL (2008) Constituintes voláteis de própolis Piauiense. Quim Nova 31:479–485

Torres-González A, López-Rivera P, Duarte-Lisci G, López-Ramírez Á, Correa-Benítez A, Rivero-Cruz JF (2016) Analysis of volatile components from *Melipona beecheii* geopropolis from Southeast Mexico by headspace solid-phase microextraction. Nat Prod Res 30:237–240

Trusheva B, Ivanova D, Popova M, Bankova V (2017) Insights into the essential oil compositions of Brazilian red and Taiwanese green propolis. Nat Prod Commun 12:197–200

Tzani A, Pitterou I, Divani F, Tsiaka T, Sotiroudis G, Zoumpoulakis P, Detsi A (2023) Green extraction of Greek propolis using Natural Deep Eutectic Solvents (NADES) and incorporation of the NADES-extracts in cosmetic formulation. Sustain Chem 4:8–25

Valcanaia CP, Masote JBB, Sommer HF, Schiquet S, Padilha B, Krepsky L, Paganelli CJ, Borges PP, Danielli LJ, Apel MA, Soares KD, Althoff S, Alberton MD, Botelho TKR, Guedes A, de Cordova CMM (2022) Antimicrobial activity of volatile oils from Brazilian stingless bees Melipona quadrifasciata quadrifasciata and Tetragonisca angustula propolis. Chem Biodivers 19:e202200369

Woisky RG, Salatino A (1998) Analysis of propolis: some parameters and procedures for chemical quality control. J Apic Res 37:99–105

Wojtacka J (2022) Propolis contra pharmacological interventions in bees. Molecules 27:4914

Xavier JA, Valentim IB, Camatari FOS, de Almeida AMM, Goulart HF, Ferro JNS, Barreto EO, Cavalcanti BC, Bottoli CBG, Goulart MOF (2017) Polyphenol profile by UHPLC-MS/MS, anti-glycation, antioxidant and cytotoxic activities of several samples of propolis from the northeastern semi-arid region of Brazil. Pharm Biol 55:1884–1893

Yang C, Luo L, Zhang H, Yang X, Lva Y, Song H (2010) Common aroma-active components of propolis from 23 regions of China. J Sci Food Agric 90:1268–1282

Yavaş B, Baştürk A (2024) Identification of volatile compounds and evolution of certain phytochemical properties of Turkish propolis. Food Sci Nutr 00:1–15

Zuluaga-Dominguez CM, Fuenmayor CA, Quicazán MC (2023) Bioactive attributes and analysis of electronic nose feature signals of Colombian stingless bees propolis. Chem Biodivers 20:e202200952

Chapter 7
The Chemical and Biological Properties of Propolis

Weam Siheri, Sameah Alenezi, Jonans Tusiimire, and David G. Watson

Contents

Abstract Propolis is a complex sticky material collected by a variety of bees including the honeybee. The honeybee uses it to coat the surfaces in the hive and block any holes in order to suppress any microbes in the hive. Propolis is a complex and its composition depends on the plants in the vicinity of the hive. In temperate regions it is mainly collected from poplar tree buds but in other regions, particularly the tropics, a variety of plants may be utilized. It has a range of biological properties including antimicrobial, anti-protozoal, antifungal, antiviral, anti-inflammatory,

---

W. Siheri · D. G. Watson (✉)
Strathclyde Institute of Pharmacy and Biomedical Sciences, University of Strathclyde, Glasgow, UK
e-mail: d.g.watson@strath.ac.uk

S. Alenezi
Drug and Food Control, Ministry of Health, Shuwaikh Industrial, Kuwait City, Kuwait

J. Tusiimire
Department of Pharmacy, Mbara University of Science and Technology, Mbara, Uganda
e-mail: jonanstusiimire@must.ac.ug

J. M. Alvarez-Suarez (ed.), *Bee Products – Chemical and Biological Properties*,
https://doi.org/10.1007/978-3-031-89049-9_7

wound healing, anticancer, and antidiabetic properties. In addition, it has effects on cardiovascular health and also has an impact on bee health. Research on propolis is very active and there is a global interest in its potential for treating disease. Recently there has been a move to conduct double-blinded clinical trials using propolis in a variety of treatments. This chapter focuses largely on propolis studies conducted since the last edition of the book in 2017.

Keywords Propolis · Chemical properties · Biological properties · Antimicrobial · Clinical trials

7.1 General Overview of Propolis

Bees produce several different products which have health benefits. There is no doubt that honey has a highly significant status in medical treatments, while other apian materials, such as wax, royal jelly and propolis, have fewer medical applications, despite the fact that propolis has been used by people since ancient times (Kuropatnicki et al. 2013; Burdock 1998; Ghisalberti 1979). Over the past few decades, there has been a notable surge in interest in studying the potential therapeutic benefits of propolis. Recent studies have further explored the therapeutic potential of propolis providing new insights into its biological properties and pharmacological applications (Belmehdi et al. 2023; Hossain et al. 2022; Bhargava et al. 2021; Sforcin 2016; Banskota et al. 2001; Burdock 1998). The term propolis comes from two Greek words, pro (which means for or in defense of) and polis (which means the city); thus, propolis means in defense of the city or beehive (Ghisalberti 1979). Propolis is a sticky resinous substance, produced by honeybees as they process various resinous substances gathered from the buds and the bark of trees, combined with their own secretions. This mixture includes diverse natural secondary metabolites, beeswax, and numerous enzymes. It is also known as "bee glue" as bees use it to cover surfaces, seal holes, and close gaps in their hives (Hossain et al. 2022; Magnavacca et al. 2022; Anjum et al. 2019; Sforcin 2007, 2016). This maintains a healthy environment in the hive and protects them from microbes including fungi and molds (Bragança Castagnino et al. 2023; Almuhayawi 2020; Wagh 2013). It is also considered to be a potent chemical weapon against bacteria, viruses, and other pathogenic microorganisms that may invade the bee colony. Also, bees use propolis as an embalming substance, to mummify invaders such as other insects, that have been killed and are too heavy to remove from the colony (Hossain et al. 2022; Kuropatnicki et al. 2013; Wagh 2013). The literature on propolis is vast and even a lengthy review cannot cover everything but can only give an indication of the areas of interest in propolis research. Numerous reviews have examined the chemical and biological properties of propolis over the years and the most recent ones are listed here: (Belmehdi et al. 2023; Ożarowski and Karpiński 2023; Chavda et al. 2023; Hossain et al. 2022; Bhargava et al. 2021; Zulhendri et al. 2021; Santos et al. 2020; Almuhayawi 2020) Moreover, the most recent reviews cover a range of topics, beyond inflammatory and antimicrobial diseases (Ożarowski et al. 2022; Zulhendri et al. 2021; Almuhayawi 2020; Przybyłek and Karpiński 2019) including specific health issues such as renal disease (Salami et al. 2024; Anvarifard et al. 2022), cardiovascular disorders (Chavda et al. 2024; Silva et al. 2021), respiratory diseases (Ożarowski and Karpiński 2023; Magnavacca et al. 2022), and dental and oral diseases (Etebarian et al. 2024; Zulhendri et al. 2021), among others (Chavda et al. 2023; Zullkiflee et al. 2022; Zulhendri et al. 2021).

7.2 Propolis in History

The ancient Greeks, Romans, and Egyptians were the first to use propolis, with applications in wound healing and as a disinfection substance. Propolis has a long-standing history of medicinal use, which is said to date back as far as the use of other honeybee products, with evidence of its application since at least 300 BC. In Egyptian history, propolis was one of the main ingredients used in an embalming recipe for mummification, in which it serves as a preservative agent (El-Sakhawy et al. 2023; Hossain et al. 2022; Rojczyk et al. 2020; Ghisalberti 1979). Traditional Georgian medicine, also utilized propolis for treating ailments and newborns. During World War II, it was used for wound healing and the Union of Soviet Socialist Republics acknowledged its medicinal value in 1969. Greek and Roman writers, including Aristotle and Hippocrates, praised propolis for its healing benefits (Hossain et al. 2022). Many other ancient civilizations, such as Chinese, Indian, Jewish, and Arabian, all believed in the power of propolis to treat medical conditions like sores, ulcers, and some skin lesions, so it was used both internally and externally (Rojczyk et al. 2020). Despite such early use, propolis is often still considered a "folk medicine" and remains an unofficial drug in the field of pharmacy (Salatino 2022; Bhargava et al. 2021). However, over the last two decades, its use has begun to gain scientific backing. It is considered to be a promising natural source for the discovery of new pharmaceutical products to treat several types of diseases. Thus, it has been subjected to intensive studies investigating its antioxidant, antimicrobial, anti-inflammatory, immune-stimulating, and anticancer properties (Wieczorek et al. 2022). Nevertheless, propolis is still not officially classified as conventional medicine in healthcare because of a lack of standardization of its composition due to the variability of its chemical components and thus its biological activity, which varies greatly depending on the geographic origin or it could be according to other factors such as plant source, seasons harvesting, climate changes, and honeybee species at the site of collection (De Almeida-Junior et al. 2023; Hossain et al. 2022; Dezmirean et al. 2020). Additionally, there is currently insufficient data from clinical trials to confirm its therapeutic efficacy, resulting in only a few propolis products having received FDA approval (Boyacioglu et al. 2022; Hossain et al. 2022). However, a recent article indicated that standardized propolis extracts have shown success in clinical trials, including applications in dentistry and as an adjunct treatment for COVID-19 (Salatino 2022). The bioactivity of propolis mainly depends on its chemical composition. Its diverse biological properties are attributed to its complex chemical composition (Wieczorek et al. 2022). In general, the term biological activity describes the pharmacological activity of a substance in a living organism. However, when the therapeutic product is a complex mixture, the biological activity could be broadly based due to the presence of multiple active ingredients leading to various therapeutic indications. This is the case with propolis, which contains many active components, leading to numerous pharmacological activities (Belmehdi et al. 2023; Wieczorek et al. 2022; Anjum et al. 2019). Propolis has been demonstrated to be safe and non-toxic for human use. However, some cases of allergic reactions such as contact dermatitis have been reported by beekeepers. There is some variability in the toxic and safe dosages of propolis reported by different studies, probably due to a lack of standardized extraction methods (Oršolić 2022; Damodaran 2021; Mendonça et al. 2013; Burdock 1998). The effectiveness of propolis preparations is dependent on the method of preparation, including the solvents used during the extraction process (Wieczorek et al. 2022). The following sections will give an overview of some of the biological activity studies that have been conducted on propolis samples from different parts of the world.

7.3 The Antioxidant Properties of Propolis

Metabolic processes within the human body consist of multiple complex reactions that generate natural free radicals. These reactive molecules can also be produced due to factors such as inflammation and stress. Antioxidant processes that are essential for optimal biological function typically remove these free radicals. However, if these natural defense mechanisms are disrupted, free radicals can accumulate, which may contribute to various diseases that threaten human health. Antioxidants, which protect the body from these harmful effects, are categorized into natural and synthetic types. Natural antioxidants exist within the human body (endogenous) and in plants (exogenous). Endogenous antioxidants are further classified into enzymatic and non-enzymatic types. Enzymatic antioxidants include superoxide dismutase, catalase, and glutathione peroxidase. While non-enzymatic antioxidants include uric acid, glutathione, and melatonin. Exogenous antioxidants, compounds for defense against the harmful effects of the reactive oxygen species (ROS), can be further classified into lipid soluble (e.g., vitamin E) and water-soluble (e.g., vitamin C and glutathione) (Chaudhary et al. 2023; Malik et al. 2020; Di Meo and Venditti 2020). A free radical in the body is simply an atom or molecule containing one or more unpaired electrons in its outer orbital, such as the oxygen atom shown in Fig. 7.1.

The unpaired electron allows the ROS to participate in many reactions with other free radicals. In low to moderate concentrations, ROS play vital roles in the biological processes of the human body, including stimulating pathways in response to changes in the extracellular environment (cellular signaling), mitogenic response,

Fig. 7.1 Schematic showing the orbitals of the oxygen atom

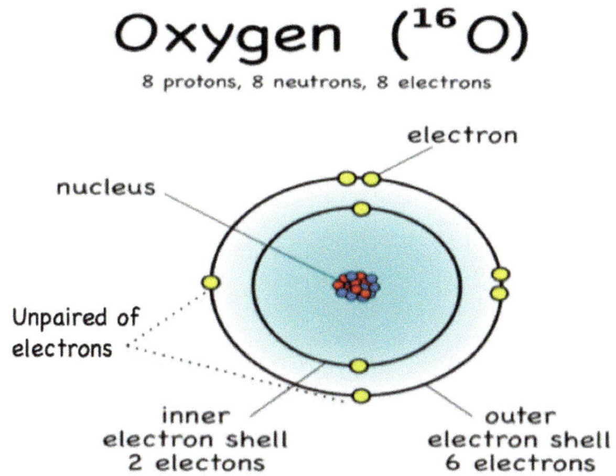

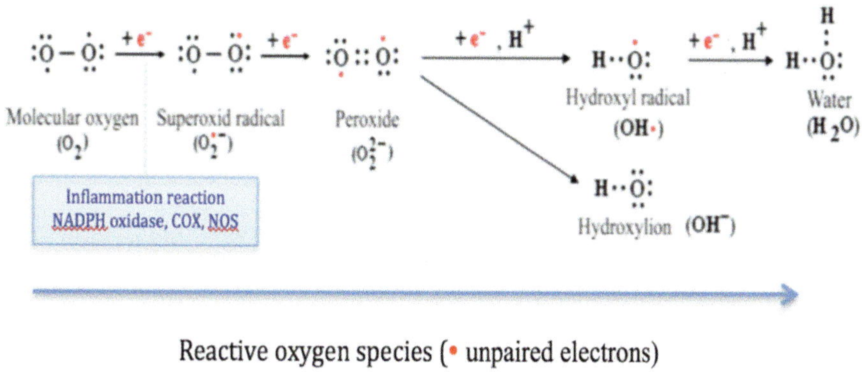

Fig. 7.2 Diagram illustrating the formation of Reactive Oxygen Species (ROS). When the oxygen molecule accepts an electron, it becomes a superoxide radical which, upon further electronation, produces peroxide. The latter can undergo further reactions with electrons and protons to produce potent hydroxyl radicals. Antioxidants act by donating protons (H+) to free radicals, leading to the formation of water, as shown in Fig. 7.3

and immune response for defense against infections in the intracellular environment (Martemucci et al. 2022; Di Meo and Venditti 2020; Halliwell and Gutteridge 2015; Valko et al. 2007).

As can be seen in Fig. 7.2, oxygen molecules can accept energy in the form of electrons as an outcome of an inflammation process, leading to the production of oxygen-centered free radicals also known as ROS.

The generation of ROS is regulated by the action of the enzyme RO synthase and over production results from both the mitochondrial electron transport chain and excessive production of nicotinamide adenine dinucleotide NAD, which generates NADH (Martemucci et al. 2022; Valko et al. 2007). Once formed, these ROS are highly reactive and produce a chain of deleterious reactions resulting in damage to

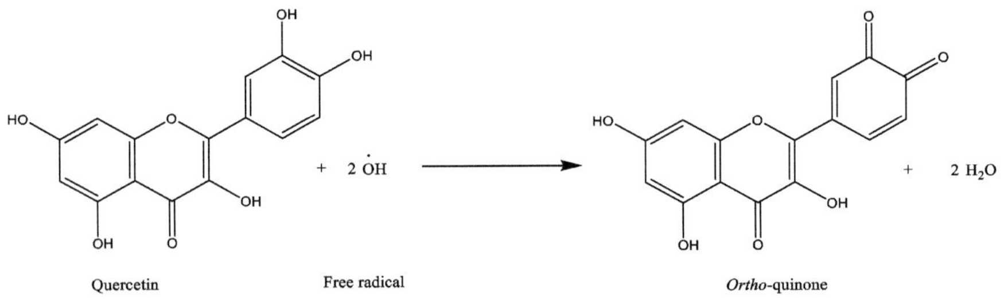

Fig. 7.3 Illustration of the antioxidant effect of quercetin, a flavonoid, on hydroxyl radicals. Quercetin reduces the free radical to water while it is oxidized to an ortho-quinone

cell structures (lipids, membranes, proteins and DNA) and modulation of many biological processes including inflammation and immune response. Oxidative stress occurs when there is an imbalance between the production of free radicals and physiologically active antioxidant metabolites in the body. The excessive production of ROS may be responsible for causing a large number of diseases, such as cancer, cardiovascular disease and inflammatory disorders such as rheumatoid arthritis. In addition, ROS can induce mutations or cause direct damage in DNA which leads to cell transformation and the possibility of developing a variety of malignant conditions. Also, active free radicals are the main factor involved in cellular aging and are responsible for the development of many CNS related medical conditions such as Parkinson's and Alzheimer's diseases. Antioxidant agents can serve as a defensive factor against free radicals in neuronal cells (Shoker et al. 2023; Martemucci et al. 2022; Malik et al. 2020; Di Meo and Venditti 2020; Valko et al. 2007).

Many scientific papers have been published on the antioxidant effects of propolis (Nazari-Bonab et al. 2023; Irigoiti et al. 2021; Martinello and Mutinelli 2021; Bittencourt et al. 2015). A recent systematic review and meta-analysis conducted by Nazari-Bonab et al. (2023) found that propolis supplementation significantly enhances antioxidant status by increasing levels of glutathione, glutathione peroxidase, and total antioxidant capacity. While propolis did not significantly affect superoxide dismutase levels, it effectively reduced malondialdehyde, an oxidative stress marker, particularly at higher doses and shorter durations, suggesting its potential benefit in managing oxidative stress-related conditions (Nazari-Bonab et al. 2023). The relationship between antioxidant activity and the chemical composition of propolis from different origins has been investigated by several authors (Pobiega et al. 2023; Kurek-Górecka et al. 2022; Asem et al. 2020). These studies demonstrated that the antioxidant activity of propolis is related to phenolic and flavonoid content, with variations in potency observed based on geographical origin and bee species, reflecting differences in phytochemical composition. These findings establish a strong correlation between the chemical composition of propolis and its antioxidant properties. Additionally, it has been reported that the essential oil constituents of and the presence of major compounds like α-pinene and β-pinene possess antioxidant properties (Ikeda et al. 2021). Several studies have further highlighted the antioxidant effects of propolis in health conditions associated with oxidative stress. A recent review examined the therapeutic properties of propolis, particularly its biologically active compounds like flavonoids and phenolic acids, which exhibit potent antioxidant and anti-inflammatory activities. These properties contribute to the management of chronic diseases such as cardiovascular disorders, rheumatoid arthritis, as well as other conditions associated with oxidative stress (Zullkiflee et al. 2022). Furthermore, Nattagh-Eshtivani et al. (2022), reviewed the impact of propolis on rheumatoid arthritis emphasizing its potential to reduce inflammation and oxidative stress through antioxidant and anti-inflammatory actions. Propolis inhibits the nuclear factor kappa B inflammatory pathway and decreases reactive oxygen species, which may improve symptoms, though further research is needed to confirm these effects (Nattagh-Eshtivani et al. 2022) the literature has reviewed the anticancer properties of propolis, emphasizing its role in reducing oxidative damage and preventing cancer progression by targeting crucial signaling pathways and triggering apoptosis through intrinsic and extrinsic mechanisms. These effects are mainly attributed to active compounds found in propolis, such as caffeic acid phenethyl ester (CAPE), artepillin C, and chrysin (Elumalai et al. 2022). Another study examined the cardiovascular benefits of propolis, highlighting its antioxidant, anti-inflammatory, anti-atherosclerotic, anti-hypertensive, and anti-hemostatic properties, particularly in Brazilian green and red varieties, suggesting its potential as a promising therapeutic agent for cardiovascular health (Silva et al. 2021).

Propolis acts as an effective natural antioxidant agent and could be particularly beneficial in treating diseases linked to oxidative stress, especially in elderly people who often face complex health issues, associated with aging. A recent study demonstrated that 12 weeks of propolis supplementation significantly reduced body fat mass increased adiponectin levels, and decreased oxidative stress in elderly females, highlighting its potential health benefits in this population (Kanazashi et al. 2023). In addition, propolis could play a key role in the management and prevention of various disease conditions in which ROS have a causative effect, such as some inflammatory disorders, cancer, cardiovascular, and immune diseases.

Moreover, a clinical study by Diniz et al. (2020) investigated the antioxidant efficacy and safety of a standardized propolis extract in healthy individuals. The study found that propolis significantly reduced oxidative stress biomarkers, such as lipid peroxidation and DNA oxidation, while enhancing oxidant enzyme activities, including dismutase (SOD) and glutathione levels. These findings suggest that propolis may offer therapeutic advantages in the management of oxidative stress-related conditions (Diniz et al. 2020).

7.4 The Antibacterial Properties of Propolis

Antimicrobial activity is recognized as an important property of propolis, particularly activity against bacteria. Several studies have been performed to evaluate this property against a large group of Gram-positive and Gram-negative bacteria; both aerobic and anaerobic types. Przybyłek and Karpiński (2019) reviewed the latest scientific reports in the field of antibacterial activity of propolis from different regions of the world, propolis samples. More than 600 bacterial strains have been tested. It was confirmed propolis is a significant antimicrobial bee product and acts both against Gram-positive and Gram-negative, as well as aerobic and anaerobic bacteria. The activity of propolis depends on its chemical composition and depends on its origin. Tables 7.1 and 7.2 show the antibacterial activity of propolis on Gram-positive and Gram-negative bacteria. \The activity is generally higher against Gram-positive bacteria.

Table 7.1 The activity of different propolis' extracts against Gram-positive bacteria

Bacteria tested	Solvent	Average (min.–max.) MIC Value _µg/mL	References
Staphylococcus aureus	DCM[a]	364 (16–950)	Bittencourt et al. (2015), Popova et al. (2017)
	DMSO[a]	930 (50–950)	Silva et al. (2012), Moncla et al. (2012)
	EEP[a]	457 (8–3100)	Bittencourt et al. (2015), Moncla et al. (2012), da Cunha et al. (2013), Mavri et al. (2012), Chen et al. (2018a, b)
	Hexan[a]	258 (16–500)	Bittencourt et al. (2015)
	MeEP[a]	266 (63–1000)	Nina et al. (2015)
	WEP[a]	883 (565–1200)	Al-Ani et al. (2018)
Staphylococcus epidermidis	DCM[a]	900	Popova et al. (2017)
	EEP[a]	345 (8–1135)	Uzel et al. (2005), Popova et al. (2009), Campos et al. (2007), Wojtyczka et al. (2013)
Streptococcus mutans	EEP[a]	511 (4–4025)	Uzel et al. (2005), Popova et al. (2009), Kim et al. (2011), Barrientos et al. (2013), Martins et al. (2018)
Streptococcus "viridans"	EEP[a]	682 (150–1370)	Uzel et al. (2005)
Streptococcus pyogenes	EEP[a]	534 (80–1556)	Bridi et al. (2015), Mavri et al. (2012), Al-Ani et al. (2018), Righi et al. (2011)
	WEP[a]	1078 (600–1556)	Al-Ani et al. (2018)
Streptococcus pneumoniae	EEP[a]	153 (80–300)	Al-Ani et al. (2018)
	WEP[a]	1003 (600–1556)	Al-Ani et al. (2018)
Streptococcus oralis	EEP[a]	167 (100–300)	Al-Ani et al. (2018)
Streptococcus agalactiae	WEP[a]	1070 (940–1200)	Al-Ani et al. (2018)
	EEP[a]	333 (100–600)	Al-Ani et al. (2018)
	WEP[a]	2150 (600–3693)	Al-Ani et al. (2018)
Streptococcus sobrinus	EEP[a]	5 (2–8)	Uzel et al. (2005)

(continued)

Table 7.1 (continued)

Bacteria tested	Solvent	Average (min.–max.) MIC Value _μg/mL	References
Enterococcus spp.	DMSO[a]	1600	Moncla et al. (2012)
	EEP[a]	544 (2–1600)	Da Cunha et al. (2013), Mavri et al. (2012), Moncla et al. (2012), Uzel et al. (2005), Campos et al. (2007), Schmidt et al. (2014), Dantas Silva et al. (2017), Al-Ani et al. (2018), Righi et al. (2011)
	SCEP[a]	698 (63–1000)	Dantas Silva et al. (2017)
	WEP[a]	250	Al-Ani et al. (2018)
Micrococcus luteus	DCM[a]	35 (8–63)	Bittencourt et al. (2015)
	EEP[a]	117 (4–400)	(2, 12), (20) Bittencourt et al. (2015), Uzel et al. (2005), Schmidt et al. (2014)
	Hexan[a]	254 (8–901)	Bittencourt et al. (2015)
Bacillus subtilis	DCM[a]	39 (16–62.5)	Bittencourt et al. (2015)
	EEP[a]	180 (21–300)	Bittencourt et al. (2015), Al-Ani et al. (2018), Righi et al. (2011)
	Hexan[a]	266 (31–500)	Bittencourt et al. (2015)
	WEP[a]	250	Al-Ani et al. (2018)
Clostridium difficile	EEP[a]	1840	Wang et al. (2016)

Adapted from Przybyłek and Karpiński (2019)
[a]*DCM* Dichloromethane, *DMSO* Dimethyl sulfoxide, *EEP* Ethanol extract of propolis, *Hexan* Hexane, *MeEP* Methanol, *WEP* Water, *SCEP* Supercritical fluid

Table 7.2 The activity of different propolis' extracts against Gram-negative bacteria

Bacteria tested	Solvent	Average (min.–max.) MIC Value _g/mL	References
Escherichia coli	DMC	1340	Popova et al. (2017)
	DSMO	3648 (3190–4940)	Silva et al. (2012)
	EEP	784 (16–5000)	Mavri et al. (2012), Wang et al. (2016), Moncla et al. (2012), Popova et al. (2013), da Cunha et al. (2013), Neto et al. (2017), Al-Ani et al. (2018), Chen et al. (2018a, b), Righi et al. (2011)
	MeEP	303 (31–1000)	Nina et al. (2015)
	WEP	2500	Al-Ani et al. (2018)
Salmonella spp.	EEP	2962 (32–14,700)	Al-Ani et al. (2018), Righi et al. (2011)
	MeEP	265 (62–1000)	Nina et al. (2015)
	WEP	2500	Al-Ani et al. (2018)

(continued)

Table 7.2 (continued)

Bacteria tested	Solvent	Average (min.–max.) MIC Value _g/ mL	References
Klebsiella spp.	DCM	1030	Popova et al. (2017)
	EEP	1006 (32–3330)	Mavri et al. (2012), Wang et al. (2016), Popova et al. (2009), Campos et al. (2007), Dantas Silva et al. (2017), Al-Ani et al. (2018), Righi et al. (2011)
	WEP	2067 (1200–2500)	Al-Ani et al. (2018)
Yersinia enterocolitica	EEP	1633 (1200–2500)	Al-Ani et al. (2018)
	MeEP	171 (63–500)	Nina et al. (2015)
Proteus mirabilis	EEP	1947 (512–3080)	Campos et al. (2007), Righi et al. (2011)
	MeEP	618 (250–1000)	Nina et al. (2015)
Shigella flexneri	EEP	1133 (300–2500)	Al-Ani et al. (2018)
	WEP	2500	Al-Ani et al. (2018)
Enterobacter cloacae	DMC	1150	Popova et al. (2017)
	EEP	1926 (300–5000)	Campos et al. (2007), Al-Ani et al. (2018)
	WEP	2500	Al-Ani et al. (2018)
Enterobacter aerogenes	EEP	34 (8–64)	Uzel et al. (2005)
Pseudomonas aeruginosa	DCM	1100	Popova et al. (2017)
	DSMO	2310 (1560–2810)	Silva et al. (2012)
	EEP	1252 (32–7910)	Uzel et al. (2005), Popova et al. (2009), Campos et al. (2007), da Cunha et al. (2013), Al-Ani et al. (2018), Righi et al. (2011), De Marco et al. (2017)
	MeEP	180 (63–500)	Nina et al. (2015), Al-Ani et al. (2018)
	WEP	2500	Al-Ani et al. (2018)
Acinetobacter baumannii	EEP	5000	Al-Ani et al. (2018)
Haemophilus influenzae	EEP	1433 (600–2500)	Al-Ani et al. (2018))
	WEP	2500	Al-Ani et al. (2018)
Campylobacter jejuni	EEP	256 (170–340)	Mavri et al. (2012), Campana et al. (2009)
Bacteroides fragilis	EEP	2460 (1840–3700)	Wang et al. (2016)
Burkholderia cepacia	EEP	2467 (1200–5000)	Al-Ani et al. (2018)
	WEP	2500	Al-Ani et al. (2018)

Adapted from Przybyłek et al. (2019)
N.B; *DCM* Dichloromethane, *DMSO* Dimethyl sulfoxide, *EEP* Ethanol extract of propolis, *Hexan* Hexane, *MeEP* Methanol, *WEP* Water, *SCEP* Supercritical fluid

Reviewing some recent studies: Extracts from different types Brazilian propolis were tested against three Gram-positive strains and four Gram-negative strains of bacteria. The minimum inhibitory concentration (MIC) values were in the range 0.06–0.2 mg/mL against Gram-negative bacteria and in the range 0.001–0.2 mg/mL against Gram-positive bacteria. The propolis extracts that exhibited the highest antimicrobial activities were from stingless bees hebora from the Distrito Federal (DF) and mandac¸aia from Santa Catarina (Gomes et al. 2024).

Evidence indicates that most types of periodontal disease are due to chronic bacterial infections, primarily caused by the overgrowth of a small group of species in dental plaque, mainly anaerobic bacteria, including *Porphyromonas gingivalis*, *Bacteroides forsythus*, and *Treponema denticola* (Campana et al. 2009). López-Valverde et al. (2021) conducted a comprehensive systematic review and meta-analysis, providing evidence of the effectiveness of propolis in treating periodontal disease. Both randomized trials involving propolis for various periodontal conditions and non-randomized clinical studies were included. A total of 224 studies were identified, with eight meeting the criteria for inclusion in the meta-analysis. Of these, three studies were selected for quantitative synthesis. The findings concluded that propolis is safe and can enhance periodontal disease treatment, significantly reducing probing pocket depth compared to placebo treatment.

Another review covered studies that investigated the antimicrobial activity of propolis extract against major periodontopathogens. The review included in vitro studies that assessed the effects of propolis against at least one periodontal pathogen. The database search yielded 1473 papers. After the selection of the best articles ultimately the review included 13 articles. *P. gingivalis* was present in all studies. The review indicated that propolis extract exhibited significant antimicrobial activity against major oral bacteria and could be a viable therapeutic option for managing periodontitis, potentially reducing the need for antibiotic therapy and lowering the risk of bacterial resistance (Melo et al. 2024).

In an advanced formulation, propolis was incorporated in polyvinylidene fluoride-co-trifluoroethylene in various material forms (2D films and 3D structures) to improve surface contact area. The findings demonstrated that adding propolis significantly altered the surface texture and mechanical properties of the materials. Particularly, 3D structures exhibited a more uniform distribution of propolis and enhanced mechanical stability compared to 2D films. Antibacterial testing on *Escherichia coli* and *Staphylococcus aureus* showed a marked reduction in bacteria, particularly when higher concentrations of propolis were applied and mechanical stimulation was used at specific frequencies, taking advantage of the surface charge changes from the piezoelectric response. In terms of antibacterial activity, the bacterial log10 reduction increased with higher propolis concentrations, with the effect being more pronounced against *S. aureus* than *E. coli*. There seemed to be a synergistic effect between the piezoelectric properties of the material and the antibacterial action of propolis, particularly against *S. aureus*. The formulation has the potential for use in wound dressings, implant coatings, and tissue engineering (Brito-Pereira et al. 2024).

A novel zinc oxide-propolis nanocomposite (ZnO P NCs) was created in a range of dimensions and was tested against both Gram-negative and Gram-positive bacteria, and compared to 28 different antibiotics and zinc oxide nanoparticles (ZnO NPs) and nanopropolis on their own. Among the bacteria tested, *Pseudomonas aeruginosa* exhibited the highest sensitivity to the biosynthesized ZnO P NCs, outperforming ZnO NPs and nanopropolis. *E. coli* was resistant to nanopropolis but showed susceptibility to ZnO NPs (31 mm) and ZnO P NCs (34 mm). SEM imaging revealed that ZnO NPs and ZnO P NCs disrupt bacterial cell membranes. Additionally, the ZnO-propolis composite demonstrated improved antimicrobial activity against a variety of bacteria, including *E. coli*, *S. typhi*, *P. aeruginosa*, *B. cereus*, and *S. aureus*. In some cases, it outperforms commercial antibiotics (Salama et al. 2024).

7.5 The Antiviral Properties of Propolis

Studies over the past three decades have provided further important information on the antiviral properties of propolis particularly after the pandemic, there has been rising interest in propolis and its compounds, with research placing more emphasis on their potential as antiviral agents (Silveira et al. 2021).

Starzyk et al. 1977; Fokt et al. 2010; Bogdanov and Bankova 2012, as summarized in Table 7.3 have reported that propolis has significant antiviral activity and interferes with the replication of some different viruses that cause human diseases, including *Herpes simplex*, *genitalis* and *zoster*, influenza, and smallpox (De Castro 2001; Bogdanov 2012; Silva-Carvalho et al. 2015).

The limited availability of effective antivirals for highly pathogenic viral strains has appeared as a major public health issue. Therefore, the development of new antiviral drugs has become critically important. Several studies have been carried out in vitro and in vivo and clinical studies and have demonstrated that propolis has broad-spectrum effects on several viral strains responsible for significant human diseases having modulatory actions on cytokine production and immune cell activation, influencing both innate and adaptive immune responses (Hegazi et al. 2017; Esposito et al. 2021). However, in many cases, the mechanisms of action behind these effects remain unclear, and the interpretation of their importance is hindered by conflicts between clinical and preclinical findings.

Toutou et al. (2024) focused on identifying the chemical compounds in Algerian raw propolis and assessing the protective potential of its inclusion complex. The chemical composition was analyzed using gas chromatography–mass spectrometry (GC/MS), Additionally, molecular docking studies were performed against the main proteases of SARS-CoV-2, and ADMET profiling was conducted. The results identified α-pinene (25.29%–11.04%) and limonene (22.46%–12.88%) as the primary components in both raw propolis with limonene and α-pinene exhibiting significant activity against SARS-CoV-2 main proteases 7N5Z, 6 LU7, and 7E6L, showing docking scores of −5.71, −5.61, and − 5.10 kcal/mol, respectively. This

Table 7.3 Antiviral activity of the different propolis constituents from different geographical origins (1992–2016)

Propolis type/plant source/ origin Type extract/isolated compounds	Species/cells/viruses	Antiviral effect
European type propolis Caffeic acid, p-coumaric acid, benzoic acid, galangin, pinocembrin, and chrysin	RC-37 cells, HSV-1 strain KOS	High anti-HSV-1 activity for both extracts when cells were treated prior to viral infection
European propolis PEE and PWE	RC-37 cells, HSV-2	High antiherpetic activity for both extracts when viruses were pretreated prior to infection
Brazilian Brown propolis/ HPE	HSV-2 strain propagated in Vero cells, female BALB/c mice	Effective against HSV-2 infection
Characteristic of Brazilian propolis Isopentyl ferulate isolate from and PEE	Influenza viruses	Suppression of influenza virus A/Hong Kong reproduction in vitro
Green Brazilian propolis/*B. dracunculifolia B. erioclada*, *Myrceugenia euosma*/Brazil, PEE	Influenza A virus	Reduction of body weight loss of infected mice and virus yields in the bronchoalveolar lavage fluids of lungs
European propolis/Populus Nigeria/ France PEE	RC-37 cells, HSV-1 strain H29S, acyclovir resistant mutant HSV1-R strain H29R, HSV-2, adenovirus type 2, poliovirus type 2, and VSV	Reduction of titer of HSV, VSV, and adenovirus, which was less susceptible; virucidal action on the enveloped viruses HSV and VSV
Geopropolis from the stingless bee *Scaptotrigona postica*/Brazil/ Hydromethanolic extract	African green monkey kidney cells (ATCC CCL-81); herpes simplex virus strain	Inhibition of HSV replication and entry into cells
Characteristic of Brazilian red and green propolis/ synthesized/ Homoisoflavonoids, especially 3-benzyl-4-chromones	BGM (Buffalo GreenMonkey) cells, coxsackie viruses B3, B4, and A9 and echovirus 30	Good antiviral activity against the coxsackie viruses s B3, B4, and A9 andechovirus 30
European propolis/*P. trichocarpa* and *P. tremuloides*/Canada/PEE	HSV-1 and HSV-2 virus replicated in MDBK cells	Impairing the ability of the virus to adsorb or to penetrate the host cells
Green propolis/*Baccharis dracunculifolia* Characteristic of Brazilian green propolis/Brazil/Water extracts, 3,4-dicaffeoylquinic acid (Isolated from Brazilian propolis)	Female BALB/c mice, Influenza A virus strain A/WSN/33 (H1N1)	Extension of the lifetime of mice. 3,4-dicaffeoylquinic acid which increases mRNA levels of tumor necrosis factor-related apoptosis-inducing and decreases H1N1 hemagglutinin mRNA

(continued)

Table 7.3 (continued)

Propolis type/plant source/ origin Type extract/isolated compounds	Species/cells/viruses	Antiviral effect
Characteristics of Brazilian green pro polis Brazil/melliferone, moronic acid, anwuweizonic acid, and betulonic acid (isolated from (isolated from Brazilian propolis)	Jurkat, uninfected human T-cell lines, and MT2 (HTLV-1 infected human T cells) cells	Moronic acid inhibiting anti-HIV replication
Mediterranean propolis/ Populus spp., Eucalyptus spp., and Castanea sativa/ Israel/PWE Characteristic of European propolis	Jurkat, uninfected human T-cell lines, and MT2 (HTLV-1 infected human T cells) cells	Inhibition of the activation of NF-KB-dependent promoter by Tax and prevention of NF-KB Tax binding to IKB and its degradation
Nanometer propolis Flavone/ Nanometer propolis Flavone provided by Binzhou Animal Science and Veterinary Medicine Academy of Shandong Province	Kidney cells (PK-15) Porcine parvovirus (PPV) Britain White guinea pigs	Inhibition of PPV infecting porcine kidney- (PK-) 15 cells. Restraining of PPV copy in lung, gonad, and blood, decrease of the impact of PPV on weight of guinea pigs and increase of hemagglutination inhibition of PPV in serum as well as improving the contents of IL-2, IL-6, and -IFN
European propolis/Populus nigra Brazil Green propolis/ Baccharis USA and China Brazil/PEE	Peripheral blood mononuclear cells obtained from blood of healthy donors, microglial cells isolated from human fetal brain tissue, HIV-1AT, HIV-1SF162	Inhibition of HIV-1 variants expression
Propolis samples Turkey (Hatay region) Hatay, Turkey/PEE	HSV-1 and HSV-2 in Hep-G2 cell cultures	Quite effective against the replication of HSV-1 and HSV-2

Adapted from Silvia-Carvalho et al. (2015)

research finding underscores the potential of propolis constituents as promising drug candidates for combating the SARS-CoV-2 (Toutou et al. 2024).

Cui et al. (2024) conducted a comprehensive in vitro study on Porcine Reproductive and Respiratory Syndrome (PRRS), a significant viral challenge in swine farming. Because of the limitations of current veterinary drugs and vaccines, this research focuses on caffeic acid phenethyl ester (CAPE), a major active component of propolis, which has been shown to effectively inhibit the proliferation of the VR-2332 strain of PRRSV in vitro. The study further examined three additional PRRSV strains stored in the same laboratory. CAPE was found to reduce the viral titers of VR-2332, GSWW-18, and GSWW-15 strains in a dose-dependent manner, with a negative correlation observed between CAPE concentrations and viral replication. Different doses of CAPE were tested in animal models to evaluate its effectiveness in treating PRRSV infections, with promising results, especially in preventive applications. As a key component of propolis, CAPE shows considerable promise as a clinical candidate, particularly for pre-exposure prophylactic treatments. This positions CAPE as a potential cornerstone in developing advanced strategies to manage future PRRSV outbreaks (Cui et al. 2024).

Despite differences in their chemical composition, influenced by Brazilian biodiversity, brown, green, and red propolis all showed antimicrobial activity (Salatino 2022). Negri et al. (2024) assessed the chemical composition and antiviral activity of an ethanolic extract of red propolis from Alagoas, Brazil, against Herpes Simplex Virus (HSV-1). Brazilian red propolis (BRP), produced by *Apis mellifera* bees, is derived from the exudates of *Dalbergia ecastaphyllum*, rich in flavonoids, and *Symphonia globulifera* and *Clusia* species, which contain abundant polyprenylated benzophenones. Previous studies have identified over 200 compounds in Brazilian red propolis, including flavonoids and polyprenylated benzophenones. HPLC-PDA-ESI-MS/MS analysis revealed the presence of flavanones, isoflavones, chalcones, pterocarpans, and polyprenylated benzophenones. Antiviral activity was evaluated through DNA quantification using quantitative real-time PCR and negative staining electron microscopy. Pretreatment, post-treatment, and virucidal assays with ethanolic extracts of red propolis at concentrations of 8, 12, 24, 48, or 96 μg/mL showed inhibition of viral binding, cell entry, and HSV-1 replication. Electron microscopy imaging revealed that HSV-1 treated with red propolis displayed disruption of the viral membrane, compared to HSV-1 treated only with phosphate-buffered saline, suggesting that the ethanolic extract directly affects the viral envelope, leading to lipid membrane degradation and/or blocking of surface proteins.

Mendonša et al. (2024), studied an aqueous extract of Brazilian propolis obtained from a *Scaptotrigona aff. postica* colony, the aim was to identify, isolate, and characterize antiviral compounds from these extracts, focusing on emerging viruses such as Zika, Chikungunya, and Mayaro viruses. The antiviral activity of both crude and purified extracts was assessed by measuring the reduction of infectious foci in Vero cell cultures. Crude propolis produced a significant reduction in viral activity, with a 64-fold decrease for Zika virus and a 128-fold reduction for Mayaro virus when a 10% v/v propolis solution was applied. Chemical characterization using high-pressure liquid chromatography (HPLC) identified a peak with antiviral activity, which was further isolated through HPLC purification and mass spectrometry. This compound showed activity against all tested viruses. The purified fraction resulted in a 16-fold reduction in Zika virus, 32-fold in Mayaro virus, and a 512-fold reduction in Chikungunya virus. Additionally, antiviral activity was found to be concentration-dependent, with stronger effects observed when propolis was added 2 h after viral infection (Mendonša et al. 2024).

A study examined the antiviral activity of six Bulgarian propolis extracts, collected from different districts across Bulgaria. The research focused on a variety of structurally distinct viruses: human coronavirus strain OC-43 (HCoV OC-43) and human respiratory syncytial virus type 2 (HRSV-2) (enveloped RNA viruses), human herpes simplex virus type 1 (HSV-1) (enveloped DNA virus), human rhinovirus type 14 (HRV-14) (non-enveloped RNA virus), and human adenovirus type 5 (HadV-5) (non-enveloped DNA virus). The research assessed the impact of the propolis extracts on the viral replication cycle using the cytopathic effect (CPE) inhibition test and evaluated virucidal activity, effects on viral adsorption to host cells, and protective effects on healthy cells. Most propolis samples demonstrated the strongest activity against HCoV OC-43 in replication. The effects on the viability of extracellular virions varied across the different viruses, with the most significant impact observed on enveloped viruses. Approximately all extracts significantly inhibited herpes virus adsorption, and to a lesser extent, coronavirus adsorption to host cells. Some extracts also showed protective effects on healthy cells when applied prior to viral infection (Vilhelmova-Ilieva et al. 2023).

The potential of a standardized hydroalcoholic extract of poplar-type propolis (sHEP) in inhibiting SARS-CoV-2 infection in vitro was assessed. The study evaluated its antiviral efficacy through viral transcription analysis using RT-PCR, infectivity assays via viral titration, and cytopathic effects (CPE) observed through optical and electron microscopy. In vitro experiments revealed that sHEP effectively inhibited SARS-CoV-2 infection. The antiviral activity of the three main components of sHEP— caffeic acid phenethyl ester (CAPE), galangin, and pinocembrin— was also assessed, both individually and in combination. It was established that standardized propolis extracts and their key components can inhibit SARS-CoV-2 infection at multiple stages (Sberna et al. 2022).

The antiviral activity of aqueous and ethanol extracts of propolis and constituents, such as flavonoids caffeic acid, p-coumaric acid, benzoic acid, galangin, pinocembrin, and chrysin, was tested against herpes simplex virus type 1 (HSV-1). Both propolis extracts demonstrated high levels of antiviral activity against HSV-1 in viral suspension tests; plaque formation was significantly reduced by >98%. Galangin and chrysin were proposed to greatly contribute to the observed activity (Schnitzler et al. 2010).

The activity of Brazilian propolis against HSV-1 infection was studied after its oral administration to infected mice three times daily and on days 0–6 after treatment. The results revealed a significant effect on the development of herpetic skin lesions (Shimizu et al. 2011).

The antiviral activities of four propolis samples from Austria, Egypt, France, and Germany were investigated against avian reovirus (ARV) and infectious bursal disease virus (IBDV). The results indicated that all propolis samples reduced the viral infectivity to a different degree and that the Egyptian propolis showed the highest antiviral activity against ARV and IBDV (El Hady and Hegazi 2002). The activity of 13 ethanol extracts of Brazilian green propolis against viruses was investigated. The extracts displayed antiviral activity against influenza virus in vitro and in vivo. The effect was attributed to the flavonoid and phenolic acid constituents (Shimizu et al. 2008).

The aqueous and ethanolic extracts of propolis were evaluated against HSV-1 and HSV-2. The anti-herpetic effect was analyzed in cell culture and both propolis extracts exhibited high levels of antiviral activity against HSV-2. Infectivity was significantly reduced by 49% and direct concentration- and time-dependent antiherpetic activity was demonstrated (Nolkemper et al. 2010).

A hydromethanolic extract of geopropolis (HMG) was evaluated using viral DNA quantification experiments and electron microscopy. The study showed a reduction of viral DNA from herpes virus by about 98% under all conditions and concentrations of HMG tested (Coelho et al. 2015).

The antiviral activity of ethanol extract of propolis (EEP) of Turkish propolis on the replication of both HSV-1 and HSV-2 was investigated. HSV-1 and HSV-2 were suppressed in the presence of 25, 50, and 100 μg/mL of propolis extract when infection of a Hep-G2 cell line was examined. Synergistic effects of propolis with acyclovir were identified against these viruses. The results showed a significant decrease in the number of viral copies (Yildirim et al. 2016).

7.6 The Anti-protozoal Properties of Propolis

The effectiveness of propolis against protozoal infections has been recently reviewed (Ebiloma et al. 2020; Asfaram et al. 2021). Protozoal pathogens which infect humans include *Trypanosoma brucei* (sleeping sickness), *Leishmania mexicana* and *donovani* (Leishmaniasis), *Falciparum malaria* (malaria), *Trichomonas vaginalis* (trichomoniasis), and *Giardia lamblia* (giardiasis). There is now abundant evidence that honeybees are commonly infected with a protozoan pathogen *Lotmaria passim*. This pathogen infects the GI tract of the bee and is transmitted via feces, so coating the hive surface with surfaces with propolis may help to reduce transmission of the pathogen. There is evidence that *L. passim* along with *Crithidia mellificae* may be harmful to bees and compromise their health (Suzuki and Sakamoto 2024; MacInnis et al. 2023; Gómez-Moracho 2020). Although bees are not known to ingest honey, honey of temperate origin does contain a lot of flavonoids which originate from poplar propolis (Martos et al. 1997; Šarić et al. 2020). Given the prevalence of protozoan parasites in bee populations it is perhaps not surprising that propolis exhibits marked anti-protozoal activity. Recent papers covering a range of anti-protozoal activities are summarized in Table 7.4.

7.7 The Antifungal Properties of Propolis

Studies of the antifungal properties of propolis have been recently reviewed (Ożarowski et al. 2022).

The antifungal activity of an ethanol extract of Spanish propolis (SEEP) was evaluated in the planktonic cells of 12 clinical isolates of *Candida glabrata*. The minimum inhibitory concentration (MIC) of propolis was determined. SEEP had antifungal activity against C. glabrata isolates, with a minimum inhibitory concentration (MIC50) of 0.2% v/v and minimum fungicidal concentration (MFC50) of 0.4% v/v, even in azole-resistant strains. An inhibitory effect on biofilm formation was also observed (Fernandez-Calderon et al. 2021).

Table 7.4 Summary of the anti-protozoal studies of propolis against different protozoa

Parasite tested	Preparation	In vivo preparation/ dose	Main results	References
T. brucei	Methanolic extract Nigerian red propolis	Rat 600, 400, 200 mg/kg 5 day	Rats treated with propolis (600 and 400 mg/kg) had a significant reduction in parasitemia, higher packed cell volumes, hemoglobin concentrations and weight gain than the DMSO control	Nweze et al. (2017)
T. cruzi	Ethanolic extracts of various types of Brazilian propolis	–	Inhibited the growth of *T. cruzi* epimastigotes.	Dantas Silva et al. (2017)
T. vaginalis	Essential oil from Brazilian red propolis	–	Exposure to propolis inhibited the growth of T. vaginalis. A concentration of 500 µg/mL can kill 100% of the T. vaginalis trophozoites	Sena-Lopes et al. (2018)
T. vaginalis	Ethanolic and hydroethanolic extracts from Brazilian brown propolis	–	Propolis-fractions showed activity against T. vaginalis and significant antioxidant capacity	De Oliveira Dembogurski et al. (2018)
T. gondii	Extracts from Egyptian propolis	Ethanolic extract Rat 28 days	Treatment with propolis showed that significantly increased the specific antibody in rats	Hegazi et al. (2017)
L. braziliensis	Hydroalcoholic extract of Brazilian red propolis. Human-derived peripheral blood mononuclear cells	–	Propolis pretreatment displays immunomodulatory effects on cells from both healthy donors. Enhancement of IL-4 and IL-17 and reduction in IL-10. Propolis controls the parasite by reducing inflammation	dos Santos Thomazelli et al. (2017)
L. braziliensis	Extracts of Brazilian green propolis. Human macrophages and bone marrow-derived murine monocytes	–	Leishmanicidal effect against both stages of *L. braziliensis* with decrease superoxide and nitric oxide levels in macrophages activated by *L. braziliensis*	Rebouças-Silva et al. (2017)
T.brucei T. congolense	Extracts and compounds isolated from Nigerian red propolis		Active at 1.66 and 4.66 µg/ml against *T.brucei* and *T. congolense*.	Alenezi et al. (2023).

(continued)

Table 7.4 (continued)

Parasite tested	Preparation	In vivo preparation/ dose	Main results	References
L.mexicana T.brucei T.congolense C.fasiculata	Ethanolic extracts of 35 samples of European propolis		Varying activity with the highest activity associated with high content of the butryrate ester of pinobanksin and methyl ethers of galangin and pinobanksin	Alotaibi et al. (2019)
Acanthamoeba	Ethanolic extracts of Turkish propolis		Killed *Acanthamoeba* at concentrations of 3–6 mg/ ml	Koloren et al. (2023)
T.brucei T. congolense C.fasiculata	Ethanolic extracts and compound from propolis from Papua New Guinea		The extract and the triterpenoids isolated from it displayed moderate to high activity	Alenezi et al. (2022)
Toxoplasma gondii	Ethanol extract of *Trigona* propolis from Indonesia	Mice infected with *T.gondii*	The propolis promoted an increased T-cell response.	Novryantoro et al. (2024)
T.brucei P.falciparum	Ethanol extracts from Libyan propolis samples and isolated compounds		The propolis and the 14 compounds isolated from it displayed moderate to high activity with the cardol bilabolol having the highest activity	Siheri et al. (2019)
T.brucei T. congolense	Extracts from Zambian and Tanzanian propolis and isolated compounds.		The propolis and isolated compounds were active. The highest activity was for Tanzanian propolis which had an IC50 of 1.2 μg/ml	Alenezi et al. (2020)
L.mexicana T.brucei	Twelve compounds isolated from temperate European propolis were tested		The most active compounds were naringenin 4′,7-dimethyl ether, 4-methoxy kaempferol and 4′7 dimethoxy kaempferol	Alotaibi et al. (2021)
Leishmania amazonis	Extracts of Brazilian red propolis		The propolis extracts exhibited anti-leishmanial in the range 23-36 μg/ml	Dutra et al. (2023)
Leishmania chagasi Leishmania amazonis	Brazilian propolis from Bahia		The hexane extract was the most active with an IC50 of 1.3 μg/ml	Cavalcante et al. (2021)
Giardia lamblia	Egyptian propolis In combination with olibanum extract	Mice infected with *Giardia*	The propolis oibanum combination significantly cleared the *Giardia* from the GI tract and improved the jejunal mucosa	Al-Ghandour et al. (2020)

Green and red Brazilian propolis was tested against Candida isolates from denture stomatitis. Minimal fungicidal concentrations were determined, and biofilm formation interference was evaluated for resistant isolates inhibition was observed in all *Candida* isolates incubated with green and red propolis extracts. At low concentrations red propolis extracts presented significant antifungal activity. Three isolates, which were highly resistant to fluconazole and itraconazole, were susceptible to low doses of red propolis extracts. These same three specimens had their biofilm formation inhibited by red propolis extract (Sokolonski et al. 2021).

The effect of green propolis extract on the adhesion and biofilm formation of range of *Candida* species in dentistry materials was studied. Green propolis extract displayed antifungal activity and inhibited both adhesion and biofilm formation down to a concentration of 2.5 μg/mL (Bezerra et al. 2021).

Paracoccidioidomycosis (PCM) is a systemic mycosis with severe clinical consequences. The direct in vitro and ex vivo effect of Brazilian red propolis (BRP) on *Paracoccidioides brasiliensis* was investigated. The in vitro experiments showed remarkable direct antifungal activity of BRP, mainly with the highest concentration employed. The results strongly suggest that BRP has four protective mechanisms in experimental paracoccidioidomycosis: activating neutrophils, exerting a direct antifungal effect, preventing fungal dissemination, and controlling excessive inflammation (Santos et al. 2021).

The activity of a Mexican propolis extract against clinical isolates of Candida species was investigated. Three Candida species were examined the inhibition halos of the propolis ranged from 7.6 to 21.43 mm, while that of the MFC and FC50 ranged from 0.312 to 1.25 and 0.014 to 0.244 mg/mL, respectively propolis extract was found to inhibit germ tube formation (IC50 ranging from 0.030 to 1.29 mg/mL) (Rivera-Yañez et al. 2022).

7.8 Studies of the Protective Effect of Propolis on Bees

The role of propolis in protecting beehives from infection has become of increasing interest. Poplar propolis has been found to be active against various bee pathogens and pests, including *Varroa* mite (Simone-Finstrom and Spivak 2010; Simone-Finstrom et al. 2017). It has been observed that bee colonies exposed to *Ascophaera apis* (chalkbrood fungus) increased their foraging for poplar propolis and that increased propolis levels in the hive reduced the intensity of infection (Simone-Finstrom and Spivak 2012).

The benefit of propolis to Varroa-infested bee colonies was tested in laboratory bioassays to investigate the effects of propolis on the fitness of infested workers. The longevity and energetic stress of adult bees kept in experimental cages and artificially infested with the mite were thus monitored over time. At the same time, in vitro experiments were performed to study the contact effects of crude propolis on Varroa mites. The results clearly demonstrate the positive effects of raw propolis on the lifespan of Varroa-infested adult bees. A narcoleptic effect (19–22%) of raw propolis on phoretic mites after 5 h was also observed. Propolis may be beneficial in this type of infestation (Pusceddu et al. 2018). The effectiveness of propolis in protecting against *V. destructor* and its associated viruses was investigated. Propolis intake of 10 field colonies was manipulated by either reducing or adding freshly collected propolis. Mite infestations, titers of deformed wing virus (DWV) and sacbrood virus (SBV), resin intake, as well as colony strength were recorded. The effect of raw propolis volatiles on mite survival in laboratory assays was also investigated. There was no significant effect of adding or removing propolis on mite survival and infestation levels. However, in relation to *V. destructor*, DWV titers increased significantly less in colonies with added propolis than in propolis depleted colonies, whereas SBV titers were similar. Colonies with added propolis were also significantly stronger than propolis-depleted colonies. These findings indicate that propolis may interfere with the dynamics of V. *destructor*-transmitted viruses, thereby further emphasizing the importance of propolis for honeybee health (Drescher et al. 2017). Propolis from Algerian honey bee subspecies were tested against Varroa mites. The most effective propolis extracts at 10% w/v killed 100% of the Varroa mites within 3–4 h in a Petri dish assay. In addition, spraying A bees infested with Varroa mites with a 10% w/v propolis in a mini-hive setup produced a high mite mortality of 85–87% without any mortality in the bees (Ayad et al. 2024).

Bees fed on a protein diet containing propolis were found to have an increased immune response when infected with *E. coli* in comparison with bees fed with the same diet without supplementation. The immune response was judged according to the level of expression of four antimicrobial peptides by the bees and the propolis produced changes of several folds in the peptides (Turcatto et al. 2018).

Honeybee colonies with propolis-rich and propolis-poor environments were compared. It was found that the honeybee microbiota from propolis-rich colonies were more similar to each other in taxonomic composition. Several bacterial groups were also found to have different relative abundances with respect to the amount of propolis in the colonies. The results suggested that propolis might support the regulation of colony microbiota by maintaining a stable microbial community. The stability of the microbiome might contribute to the ability of colonies to ward off infective microbes (Saelao et al. 2020).

The effect of propolis on the mouth part microbiome of honeybees was investigated. The results revealed that the mouthparts of worker bees in colonies with a propolis envelope exhibited a significantly lower bacterial diversity and significantly higher bacterial abundance compared to the mouthparts of bees in colonies without a propolis envelope. The propolis envelope appeared to reduce pathogenic or opportunistic microbes and promote the proliferation of putatively beneficial microbes on the honey bee mouthparts (Dalenberg et al. 2020) It was found that bees apply propolis to brood cells in beehives where it reduces the reproduction of Varroa parasites. In addition, consumption of propolis reduced the mortality of bees infested with Varroa mite (Pusceddu et al. 2021). Giant honeybees infected with *Nosema ceranae* were fed propolis which promoted an increase in hemolymph trehalose and protein levels in infected bees and improved their survival (Naree et al. 2021). Different surface textures for the encouragement of propolis deposition were studied. It was found that that rough wood boxes were most effective. However, the use of rough wood boxes did not improve colony survival overall. The viral loads for multiple viruses tended to decrease as propolis deposition increased and propolis deposition was correlated with a seasonal decrease and/or stabilization in the expression of multiple immune and bacterial genes, suggesting that propolis-rich environments contribute to hive homeostasis (Shanahan et al. 2024). The effect of propolis and various essential oils on bee lifespan was tested. Brazilian propolis extract lengthened worker life span, while Louisiana propolis extract, lemongrass oil, spearmint oil, and thyme oil exerted variable and dose-dependent effects on life span. Vitellogenin (*vg*) gene expression was reduced by Brazilian propolis extract at high doses. The results suggest that, in addition to propolis's structural and immunomodulatory, roles in the colony, it may also exert beneficial health effects when ingested (Martin Ewert et al. 2023).

Propolis was investigated for its ability to reduce *N. ceranae* infection loads in the dwarf honeybee, *Apis florea*. Experimentally infected caged bees were fed propolis extract. Propolis-treated bees had significantly lower infection loads, and higher survival in comparison to untreated bees. In addition, propolis-treated bees had significantly higher hemolymph trehalose levels and hypopharyngeal gland protein content similar to levels in uninfected bees (Suwannapong et al. 2018). The efficacy of oral administration of an ethanolic extract of propolis (EEP) against *N. ceranae* in infected worker bees was tested. Propolis extracts increased the lifespan of N. ceranae-infected bees and significantly reduced spore load. The research supports the hypothesis that propolis represents effective and control of N. ceranae. However, worker bees do not seem to use it to self-medicate when infected with this pathogen (Mura et al. 2020). Among a long list of parasites and pathogens that threaten the European honey bee (Apismellifera), European Foulbrood (EFB) has become an urgent apiary disease it is a bacterial disease of larval honey bees caused by the anaerobic bacterium, *Melissococcus plutonius*. The efficacy of propolis, honeycomb, and brood comb ethanol extracts against EFB was explored. The minimum inhibitory concentrations for each bee-derived product on *M. plutonius*, as well as two model bacterial species, *Staphylococcus saprophyticus* (gram-positive) and Escherichia coli (gram-negative) was determined. It was found that propolis extract was most effective at inhibiting the growth of gram-positive bacteria, and that *M. plutonius* was also susceptible to honey comb and brood comb extracts (Murray et al. 2023).

7.9 The Wound Healing Properties of Propolis

Propolis, along with other honeybee products, has been widely used as an external treatment for wounds and burns. Propolis is believed to possess antimicrobial, antiseptic, anti-inflammatory, antioxidant and immunomodulatory effects (Manginstar et al. 2024; Hossain et al. 2022; Zullkiflee et al. 2022; Šuran et al. 2021; Almuhayawi 2020; Castaldo and Capasso 2002). These properties probably enhance cell proliferation in the skin and activate remodeling of the skin tissue. The wound healing process is a natural reaction of the body involving a sequence of several biochemical factors and multiple cellular reactions which can be divided into four stages, namely, hemostasis (blood clotting), inflammation, proliferation of new skin tissue, and remodeling of mature tissue (Chouhan et al. 2019; Gurtner et al. 2008). Hemostasis is an initial natural reaction that occurs immediately in the first few minutes of injury. During this stage, platelets in the blood accumulate at the site of injury where they release chemical signals. These signals cause fibrin, a blood protein, to form a mesh that works as a glue to bind the platelets, leading to the formation of thrombi (blood clots), which seal the injury and control bleeding. Immediately after formation of the blood clot the inflammation process starts and immune cells such as macrophages and inflammatory cells such as neutrophils secrete large amounts of pro-inflammatory cytokines and reactive oxygen species (ROS), respectively. The generated ROS arrive in the wound lesion and act as a defense system against pathogens. However, any excess amount of superoxides may damage the surrounding normal tissue, such as occurs in diabetes. The proliferative phase then occurs 2 or 3 days after the injury and involves the movement of epithelial cells until they meet each other from either migration side. This occurs via stimulation of fibroblasts in the dermis layer along with keratinocytes in the epidermis. New blood vessels start to form to provide nutrition and oxygen to repair the injured tissue. Then, proliferative fibroblasts begin to secrete matrix proteins and collagen to build up the extracellular matrix that acts as a connective tissue to form initial dermal granulation tissue. After that, the fibroblasts differentiate into myofibroblasts which cause contraction of the wound size due to increases in collagen synthesis relative to the proliferative process. Lastly, the remodeling and maturation phase occurs, which is characterized by scar formation as a result of replacement of type III collagen by type I collagen (El-Sakhawy et al. 2023; Chouhan et al. 2019; Gurtner et al. 2008). The wound healing process is illustrated in Fig. 7.4.

Recent literature reviews have supported the therapeutic efficacy of propolis in various types of wounds, including gastric ulcers, oral ulcers, diabetic foot ulcers, surgical wounds, infected wounds, and burns (Manginstar et al. 2024; Yang et al. 2022; Da Rosa et al. 2022; Stojko et al. 2021; Rojczyk et al. 2020) Manginstar et al. recently reviewed the therapeutic benefits of propolis in burn management, highlighting its antimicrobial, antioxidant, and anti-inflammatory properties that promote wound healing through enhanced tissue regeneration, collagen synthesis, and angiogenesis. The study emphasizes propolis' potential as a complementary treatment for burns, advocating for further research to optimize formulations and clinical applications (Manginstar et al. 2024). Yang et al. highlighted the biological effects of propolis in wound healing, emphasizing its anti-inflammatory and antibacterial properties. Its effectiveness in treating different types of wounds, such as ulcer, burn, and its synergistic benefits in promoting wound healing when it combined with other drug compounds (Yang et al. 2022). Moreover, a systematic review and meta-analysis of 43 studies demonstrated that propolis effectively promotes skin regeneration and healing of cutaneous wounds such as ulcers, showing greater efficacy than traditional treatments, with increasing interest in its therapeutic

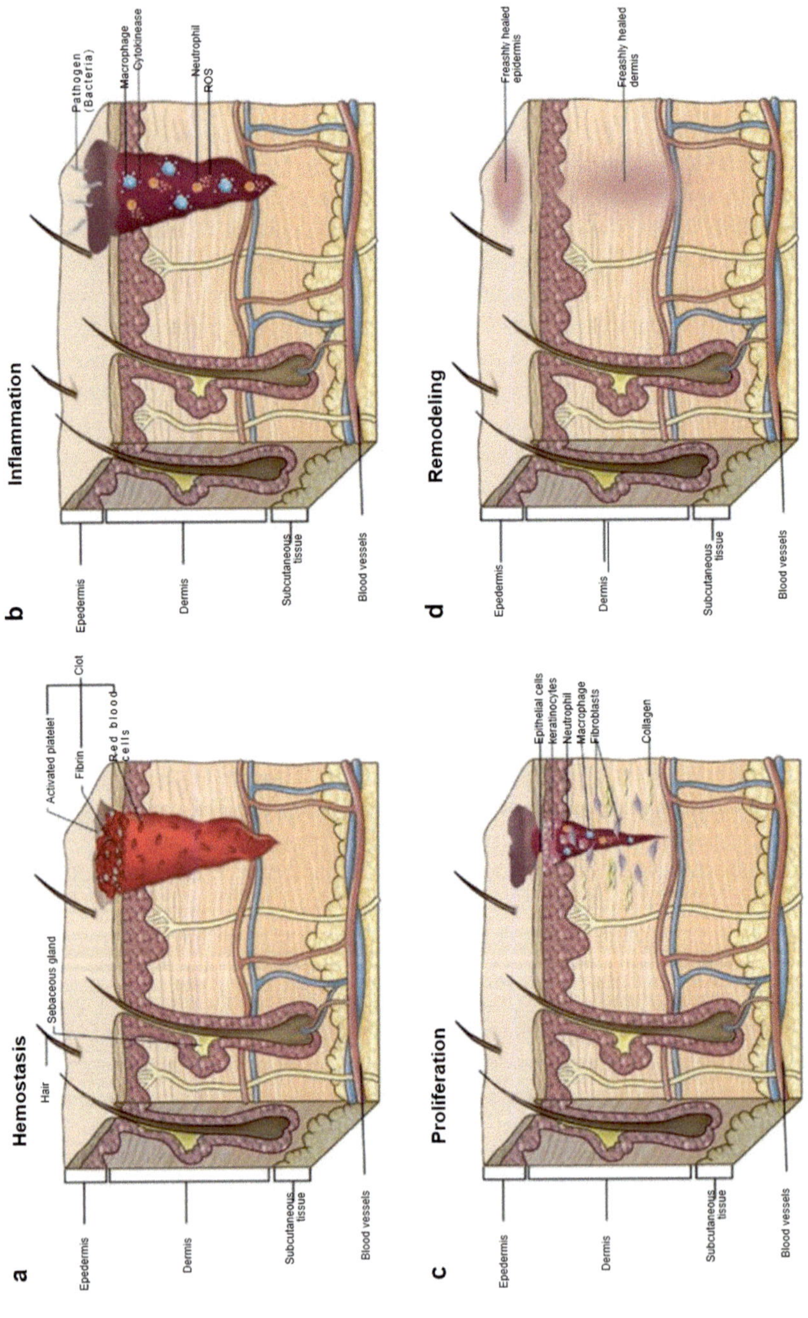

Fig. 7.4 Illustration of the process of wound healing

potential and its combination with other substances for enhanced healing outcomes (Machado Velho et al. 2023). In addition, propolis may accelerate the healing process in some medical conditions where there is a delay in wound healing, for example in patients of advanced age, or diabetic and immune-suppressed patients (El-Sakhawy et al. 2023; Da Rosa et al. 2022; Oryan et al. 2018; McLennan et al. 2008).

According to several papers, propolis may aid wound healing by activation of all the stages involved (El-Sakhawy et al. 2023; Yang et al. 2022; Martinotti and Ranzato 2015; Moura et al. 2011). Among these, there was a study that investigated the anti-inflammatory activity and wound healing potential of an alcohol extract of green Brazilian propolis in a mouse wound healing model. The findings suggest that propolis may modulate both the early wound healing processes, particularly hemostasis and inflammation including cell influx, as well as later phases involving collagen deposition (Moura et al. 2011). This was further supported by a study of Mexican propolis particularly with a Chihuahua ethanolic extract of propolis (Ch EEP), which demonstrated that topical application enhanced wound healing by promoting contraction, reducing healing time, and increasing collagen formation, highlighting its effectiveness throughout various stages of the wound healing process (Balderas-Cordero et al. 2023). Several studies carried out in vivo have reported that propolis activity in the management of burns and wound healing apparently occurs via regulation of the immune response, which involves accelerating wound closure and reducing inflammation through the modulation of cytokine production during the inflammatory phase (Manginstar et al. 2024; Ibrahim et al. 2022; Corrêa et al. 2017). Another review, based on in vivo and clinical studies, showed the therapeutic benefits of propolis in wound healing, particularly its ability to reduce the immune response, and stimulate the proliferation phase by promoting keratinocyte production (Oryan et al. 2018). Propolis was also reported to be useful during the remodeling and maturation phase in which it stimulated the repair of granulation cells in burnt skin tissue (Olczyk et al. 2013). Additionally, studies have shown that propolis accelerates wound healing by its ability to stimulate the production of transforming growth factor-β (TGF-β). This factor supports key phases of the wound repair process such as hemostasis and controlled inflammation, leading to a more efficient healing cascade. It also promotes extracellular matrix (ECM) remodeling and re-epithelization, both of which are essential for tissue repair (Martinello and Mutinelli 2021).

A study of propolis in an induced diabetic rat model found that propolis significantly accelerated wound healing by enhancing the re-epithelialization process. Additionally, this study found that the combination of propolis and honey reduced inflammation and wound contraction, contributing to improved healing outcomes. Specifically, the treatment led to increased granulation tissue formation and fibroblast activity while decreasing persistent inflammation, which is commonly associated with diabetic wounds (Behzadi et al. 2022).

A previous study on wounds in a diabetic rodent model treated with a topical cream consisting of a mixture of natural products, namely propolis, honey, royal jelly, and olive oil, showed a significant beneficial effect on increasing the rate of wound healing due to antimicrobial, anti-inflammatory, and antioxidant activities (Rashidi et al. 2016).

A recent review showed the link between the rich chemical composition of propolis, including flavonoids and phenolic compounds, and its biological activities like wound healing (Belmehdi et al. 2023). Another study investigated propolis collected from various Brazilian locations, revealing that its phenolic-rich composition, including flavonoids, phenylpropanoids, and tannins, contributes to its free radical scavenging and wound healing activities. A 10% geo-propolis extract significantly improved healing and re-epithelization in rats, particularly in samples with higher phenolic content (De Sousa-Fontoura et al. 2020). A study conducted by Olczyk et al. found that flavonoids have the ability to reduce lipid peroxidation and inhibit cellular necrosis (Olczyk et al. 2013). Flavonoid compounds in propolis, like pinocembrin and galangin, enhance its antioxidant capacity, reducing cell necrosis and lipid peroxidation, which promotes a favorable environment for wound healing (Martinotti and Ranzato 2015). Earlier, research by Rossi et al. (2002) established a link between the presence of caffeic acid in propolis samples and its anti-inflammatory effects in vivo (Rossi et al. 2002). A More recent review, further supported these findings demonstrating the anti-inflammatory effects of propolis in wound healing are attributed to flavonoids along with phenolic compounds such as caffeic and ferulic acid which further inhibit local inflammatory responses and promote wound repair (Yang et al. 2022).

The therapeutic effects of propolis in healing various types of wounds, including diabetic, venous, and surgical wounds, as well as burn injuries, are primarily attributed to its rich composition of phenolic compounds, which provide antimicrobial, anti-inflammatory, analgesic, and angiogenesis-promoting properties. Despite its low cost compared to other treatments, most studies supporting propolis-healing capabilities are based on in vitro and in vivo research. However, a review of clinical trials conducted between 1990 and 2021 on the use of propolis for skin wounds, including diabetic, venous, and surgical ulcers, highlights its healing potential, with strong evidence supporting its antimicrobial, anti-inflammatory, and angiogenesis-promoting properties (Da Rosa et al. 2022). Also, in clinical trials on patients with similar minor burns, it was found that a propolis cream improved skin tissue healing and decreased wound inflammation more effectively compared to a topical silver sulfadiazine treatment (Gregory et al. 2002). Previously, a clinical case reported on a 13-month-old infant with chronic eosinophilic oral ulcers, who had not responded to a 4-month treatment regimen, showed rapid healing after using a lanolin-based propolis ointment. After 10 days of treatment, the ulcer's size significantly reduced, and after 3 weeks, the ulcer completely healed without recurrence (Kiderman et al. 2001).

7.10 Propolis in the Treatment of Diabetes and Cardiovascular Disease

The role of propolis in the treatment of obesity and diabetes has been recently reviewed (Kitamura 2019; da Cunha et al. 2023; Tsuda and Kumazawa 2021).

Diabetic and non-diabetic rats were fed 100 or 200 mg/kg of propolis or bee pollen for a period of 16 weeks. In vivo biochemical changes associated with diabetes were induced by drinking a solution containing 10% of D-glucose. In rats, the administration of D-glucose induced hyperglycemia, increased plasma insulin levels, and an increase in insulin resistance index, and this was accompanied by dyslipidemia, elevation of hepatic enzyme levels, and a change in both serum renal biomarkers and plasmatic calcium. The co-administration of propolis and bee pollen extracts alone or in combination restored these biochemical parameters and attenuated the deleterious effects of D-glucose on liver and kidney functions. The effects were greater in diabetic rats (Laaroussi et al. 2020).

The effects of Mexican propolis on streptozotocin-induced diabetes mellitus in a murine model were studied. Mice were fed propolis at 0.3 g/kg/per day for 15 days. The results showed that an ethanolic extract of propolis significantly inhibited increases in blood glucose and the loss of body weight in diabetic mice. The extract increased plasma insulin levels in STZ-diabetic mice, whereas, in untreated diabetic mice, there was no detection of insulin. The extract increased the activities of the enzyme's superoxide dismutase, catalase, and glutathione peroxidase in diabetic mice (Rivera-Yañez et al. 2018).

The effect of Taiwanese green propolis ethanol extract (TGPE) on type 2 diabetes mellitus (T2DM) was investigated in a rat model. It was found that TGPE delayed the development and progression of T2DM and reduced the severity of β-cell failure. TGPE promoted higher levels of oxidant cytokines, leptin, and adiponectin and promoted the hepatic genes PPAR-*a* and CYP7A1, which are related to lipid catabolism and removal (Chen et al. 2018a, b).

The effect of an ethanolic extract of propolis (EEP) was tested on a mouse T2DM along with the established antidiabetic drug metformin (MET). Biochemical indexes and cytokines were measured, and the differences in metabolites in the serum were compared by ^{1}H-NMR. In addition, the diversity of intestinal flora in feces was studied. The results showed that following treatment with EEP or MET the weight loss in the diabetic mice was reduced, and the fasting blood glucose, insulin secretion level, insulin resistance index, C peptide level, and oral glucose tolerance level decreased and the insulin sensitivity index increased. Compared with untreated mice the concentrations of pro-inflammatory cytokines interleukin-1 beta (IL-1β), interleukin-6 (IL-6), and tumor necrosis factor-alpha (TNF- α) decreased significantly in EEP and MET groups, whereas the concentrations of anti-inflammatory cytokine interleukin-10 (IL-10) increased significantly. The metabolomics results revealed that EEP and MET regulated carbohydrate metabolism and restored amino acid and lipid metabolism. Treatment with EEP or MET increased the levels of beneficial bacteria in the microbiome and decreased the levels of harmful bacteria. The overall therapeutic effect of EEP was better than that of MET (Guan et al. 2023).

A study evaluated the effect of bacterial cellulose associated with Brazilian red propolis in diabetic mice as a wound healing model. Tests were performed on male diabetic mice. The animals were submitted to a surgical procedure and a single lesion was produced in the dorsal region, which was treated with the propolis/cellulose. Macroscopic assessments were performed at 2, 7, and 14 postoperative days, and biopsies were collected on days 0, 7, and 14 for histological analysis and determination of myeloperoxidase enzyme activity (MPO) and cytokines (TNF-α, IL-1β, and TGF-β). Macroscopic analysis showed treatment with propolis increased healing capabilities with an increase in leukocyte recruitment. The levels of TNF-α and TGF-β were significantly higher in wounds stimulated with red propolis. Wound healing was accelerated in the propolis-treated mice (Picolotto et al. 2019).

The effect of pinocembrin, a major component of Mexican propolis, on diabetic nephropathy was evaluated. Two groups of diabetic rats were assessed, preventive (before renal damage) and corrective (once renal damage was established). In the preventive scheme, pinocembrin treatment prevented the death of the rats, improved lipid profile, and reduced urinary markers of kidney damage. However, in the corrective scheme, pinocembrin only improved the lipid profile without showing improvement in any other parameters and pinocembrin may even exacerbate the damage (Granados-Pineda et al. 2018).

Wistar rats were given propolis orally at 50 and 100 mg/kg, respectively, for 8 weeks or no propolis in controls. Diabetes was induced in rats via injection of nicotinamide and streptozotocin (STZ). Compared to the control group, propolis greatly decreased fasting blood glucose levels, improved the glycemic status of diabetic rats, and increased the number of sperm cells and the percent of morphologically normal and viable sperm in male rats, improving their fertility. Propolis also restored the pancreatic islets, protected the testis from oxidative stress, and increased levels of reproductive hormones in the blood, especially testosterone (Ashour 2024).

The influence of propolis on the gut microbiota and intestinal mucosa in rats with diabetes was investigated. Sprague-Dawleyrats were randomly assigned to the control group, model group, and three propolis groups supplemented with 80, 160, and 240 mg/kg propolis. A high-fat diet combined with an ST abdominal injection was used to induce diabetes in the rats. After propolis treatment, compared to the model group, fasting blood glucose, and glycated hemoglobin levels declined, while the glucose tolerance and the insulin sensitivity index (ISI) increased. The levels of tight junction proteins in the ileum increased in the propolis groups. The contents of the fecal acetic acid, propionic acid, and butyrate were increased in the propolis groups. High-throughput DNA sequencing revealed that the composition of the gut microbiota of rats in the propolis supplement group was significantly improved. Thus, propolis exerted several effects in ameliorating diabetes (Xue et al. 2019).

A propolis extract was used to treat wounds made in diabetic mice. The speed and strength of wound healing were evaluated. Model induced by STZ, and the speed of closure and the wound tensile strength were evaluated by the tensiometric method, which showed that propolis extract had similar activity, improving the speed of healing and increasing the wound tensile strength needed to open the wound again. A histological analysis of the wound staining enzymes involved in wound healing indicated that these were improved by propolis application (Canales-Alvarez et al. 2024).

The effects of propolis and quercetin in treating diabetic peripheral neuropathy in rats were studied. Blood glucose and malondialdehyde (MDA) levels increased in diabetic rats and IL-6 and antioxidant status decreased. Ultrastructural findings showed degeneration of the axon and myelin sheath. The apoptotic index (AI %), TNF-α, and IL-1β immunopositivity increased significantly in the DM group. Morphological structures similar to those of the non-diabetic controls were observed in groups treated with propolis or quercetin and propolis in combination, while blood glucose and MDA levels, AI (%), TNF-α, and IL-1β immunopositivity decreased. A combination of propolis and quercetin may provide optimal protection against diabetic neuropathy (Türedi et al. 2024).

The effect of an ethanolic extract of propolis (EEP) on fecal biomarkers of T2DM was screened by using high-resolution mass spectrometry metabonomic method. The results showed that EEP improved the body weight (BW) of T2DM mice, lowered blood sugar levels, and significantly improved blood biochemical indicators related to T2DM, Liver pathology showed that EEP reversed liver damage caused by T2DM. Metabolomics data identified 27 potential biomarkers in fecal samples. EEP effectively regulated the dysfunction in the metabolic pathways of glycerophospholipids, sphingolipids, riboflavins, and sterol lipids caused by T2DM (Wang et al. 2023).

The anti-inflammatory effects of an ethanolic extract of Polish propolis (EEP) were studied using HGV-1 cells treated with lipopolysaccharide (LPS) as a model. It was found that EEP, at a concentration of 25 μg/mL, decreased the pro-inflammatory cytokine IL-6 LPS-induced HGF-1 and increased the anti-inflammatory cytokine IL10. At the same concentration, EEP increases the level of anti-inflammatory cytokine IL-10 in LPS + IFN-α-induced HGF-1. EEP reduced levels of the adhesion molecules E-selectin and endothelin 1 IFN-α -induced HGF-1. EEP and CAPE have beneficial cardiovascular effects reducing inflammation (Kurek-Górecka et al. 2024).

Cardiotoxicity as well as nephrotoxicity was induced in rats by intraperitoneal injection of doxorubicin (10 mg/kg) and protected or treated by (200 mg/kg/ day) of Egyptian propolis extract administrated by gastric intubation for 4 weeks before or after doxorubicin administration. Cardiac biomarkers such as brain natriuretic peptide (BNP), troponin (T), lactate dehydrogenase (LDH), creatinine kinase (CK), and aspartate amino transferase (AST) in addition to creatinine and urea measurements, all ameliorated in the propolis; protected or treated groups. Cardiac and renal oxidative stress was indicated by increased malondialdehyde (MDA) and decreased antioxidant enzymes such as catalase and superoxide dismutase and decreased glutathione in the animals treated with doxorubicin alone. The measured lipid profile was also ameliorated with increased HDL in the propolis-treated group. Thus, propolis exerts a cardioprotective effect (Ali et al. 2020).

The effect of Propolis on 5-fluorourcil (5-FU) -induced cardiotoxicity in rats was studied. 5-FU caused in significant cardiotoxicity shown by an increase in MDA levels, cyclooxygenase-2 (COX-2) and tumor necrosis factor-α (TNF-α) expression, cardiac enzyme levels, and histopathological degenerations. 5-FU treatment also decreased body weight, total antioxidant capacity (TAC), catalase (CAT) levels, blood cell counts, and hemoglobin levels. In addition, 5-FU disrupted ECG parameters. Treating with propolis reduced oxidative stress, cardiac enzymes, histopathological degeneration, and COX-2 expression in cardiac tissue alleviated ECG disturbances and increased the number of blood cells and TAC levels. Moreover, 5-FU-induced bodyweight loss was improved after treatment with propolis (Barary et al. 2022).

A metabolomics approach and Positron Emission Tomography (PET) imaging were used to assess the metabolic response to diet supplementation with green propolis on rats with Parkinsonism induced by 6-hydroxydopamine. The metabolomics approach revealed four cardioprotective metabolites (2-hydroxybutyric acid, 3-hydroxybutyric acid, monoacylglycerol, and alanine) that were significantly modified between animal groups. Propolis-induced changes in the level of these cardiac metabolites suggest beneficial effects of dietary intervention. PET imaging detected higher glucose metabolism in the 17 areas of the left ventricle of all rats treated with propolis compared to rats that did not consume propolis (Goncalves et al. 2022).

The effects of an extract of propolis (EP) on inflammatory cytokines, sympathetic nerve activity, and anti-endoplasmic reticulum stress were studied. A rate model of myocardial infarction (MI) was compared against controls. Inflammatory markers were significantly elevated following MI as compared with control animals. IL-1, TNF-α, CRP, and several markers of sympathetic nervous activity were also elevated. The changes produced by MI were attenuated by EP (Wang et al. 2022).

An extract of Brazilian green propolis was tested on mouse aortic endothelial cells treated with LPS. It was found that the increased levels of TNF-α, IL-6, and adhesion molecules induced by LPS were downregulated in response to the propolis extract (Xu et al. 2020).

Male Wistar rats were divided into control, stress exposure groups. The stress groups were exposed to various mild stressors daily for 5 weeks. An extract of Turkish propolis (100 mg/kg/day) was administered orally to one of the stress groups for 5 weeks. The depression-like behaviors were assessed with the forced swimming test (FST). Thoracic endothelial functions and expressions of endothelial nitric oxide synthase (eNOS), TNFα, IL-1β), heme oxygenase-1 (HO-1), and superoxide dismutase (SOD) levels were assessed. Compared to the control group, the stress group exhibited a significant decrease in endothelium-dependent relaxations, and in eNOS, SOD, and HO-1 expressions and a significant increase in the thoracic expressions of TNFα and IL-1β. Propolis ameliorated depression-like behaviors, vascular endothelial dysfunctions, and alterations of stress marker expression (Celebi et al. 2023).

The cardioprotective effect of an extract of Brazilian red propolis (BRPHE) against isoproterenol (ISO)-induced myocardial infarction (MI) in rats was investigated. Rats were pretreated with BRPHE (for 30 days, with concurrent administration of ISO (85 mg/kg) during the last 2 days. Rats with MI showed an increase in cardiac marker enzymes and various other indicators of MI. Pretreatment of rats with BRPHE significantly inhibited these effects. Histopathological findings obtained for the myocardium further confirmed the biochemical findings (Neto et al. 2022).

The efficacy of propolis against doxorubicin (DOX) -induced toxicity in female rats was assessed. Rats were treated with 100 mg/kg of propolis via oral administration for 28 days before administration of DOX on day 24 and were compared to a control group that did not receive propolis. Cardiac, renal and hepatic injury markers, apoptosis, and pro-inflammatory cytokines were assessed in serum along with GSH and MDA. In addition, histopathological and immunohistochemical examination of cytochrome-c and connexin43 on lysed myocardium, liver, kidney, and lung tissues was carried out. DOX caused marked deteriorations in the markers. However, pretreatment with propolis significantly ameliorated these alterations (Mohamed et al. 2021).

The efficacy of a propolis extract against diabetic cardiomyopathy (DCM) was assessed. Propolis improved blood glucose regulation, cardiac function, and reduced myocardial oxidative stress in rats with streptozotocin-induced diabetes. Propolis was to ameliorate DCM, improving blood glucose regulation, improving heart function, and reducing oxidative stress in heart tissue (Shi et al. 2023).

7.11 The Anticancer Properties of Propolis

Cancer, which encompasses a range of diseases marked by unchecked cell growth and spread, remains a global public health problem. With approximately 20 million new cases and 9.7 million deaths in 2022 alone, cancer is the second leading cause of death worldwide, and the leading or second leading cause of premature deaths in 112 countries (more than 50% of the 195 total) (American Cancer Society 2024).

The anticancer properties of propolis and its phenolic compounds, including the molecular targets and signaling pathways involved, have been recently reviewed (Altabbal et al. 2023; Oršolić and Jazvinšćak Jembrek 2022; Forma and Bryś 2021). Generally, propolis displays potent anticancer effects due to the presence of various phytocomponents which contribute to pro-apoptotic, cytotoxic, anti-proliferative (cell cycle arrest), anti-metastatic, anti-invasive, anti-angiogenic and anti-genotoxic or anti-mutagenic properties along with antioxidant, immunomodulatory, and anti-inflammatory functions (Chiu et al. 2020). Thus, there is an overlap between anticancer effects and the immunomodulatory effects of propolis, which to some extent seems to exert its anticancer activity via immuno-modulation.

These effects have been demonstrated using in vitro (cell lines) and in vivo (animal) models, as well as clinical trials in various types of cancers, including head and neck, lung, liver, brain, pancreas, kidney, prostate, skin, breast, oral, esophagus, gastric, colon, and bladder cancers. Propolis and its components are able to achieve their anticancer effects by modulating various signaling molecules such as cyclooxygenase (COX), lipoxygenase (LOX), inducible NOS, phosphatidylinositol 3-kinase (PI3K)/protein kinase B (Akt), NF-κB, TSG (p53 and p21), mitogen-activated protein kinases (MAPKs), metalloproteinases (MMPs), caspases, Bax/Bcl2, TNF-related apoptosis-inducing ligand (TRAIL), and nuclear factor erythroid 2-related factor 2/heme oxygenase (Chiu et al. 2020).

Some of the main phytocomponents in propolis which have been demonstrated to possess potential anticancer effects, and their reported mechanisms of action, are shown in Table 7.5.

Table 7.5 Main anticancer compounds isolated from propolis and their reported mechanisms of action

Name	Molecular formula (MW)	Chemical name (IUPAC)	Mechanism
Chrysin	C15H10O4 (254.24) 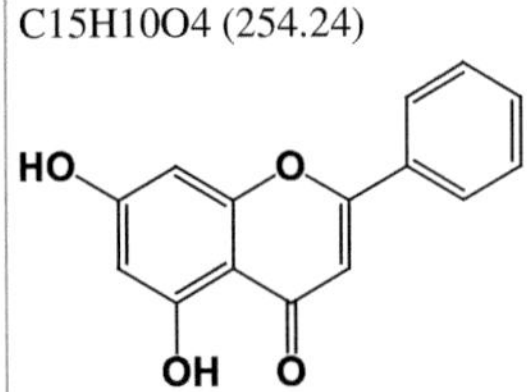	5,7-dihydroxy-2-phenylchromen-4-one	Induction of apoptosis (Celinska-Janowicz et al. 2019)
Galangin	C15H10O5 (270.24) 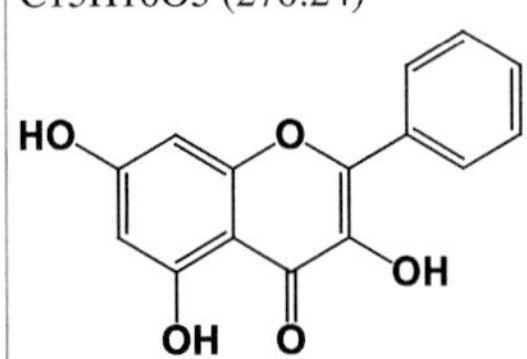	3,5,7-trihydroxy-2-phenylchromen-4-one	Induction of apoptosis, autophagy, inhibition of tumor cell proliferation (Wang and Tang 2017)
Genistein	C15H10O5 (270.24) 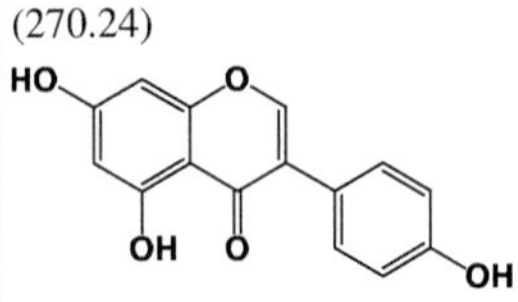	5,7-dihydroxy-3-(4-hydroxyphenyl)chromen-4-one	Inhibition of cell cycle (Zabaiou et al. 2017)
Nymphaeol A/Propolin C	C25H28O6 (424.5) 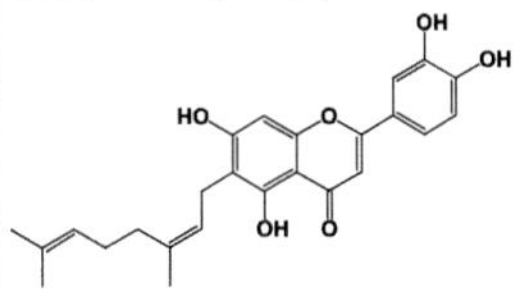	(2S)-2-(3,4-dihydroxyphenyl)-6-[(2E)-3,7-dimethylocta-2,6-dienyl]-5,7-dihydroxy-2,3-dihydrochromen-4-one	Anti-angiogenic, anti-proliferative, anti-metastatic (Pai et al. 2018)
Nymphaeol C	C30H36O6 (492.6) 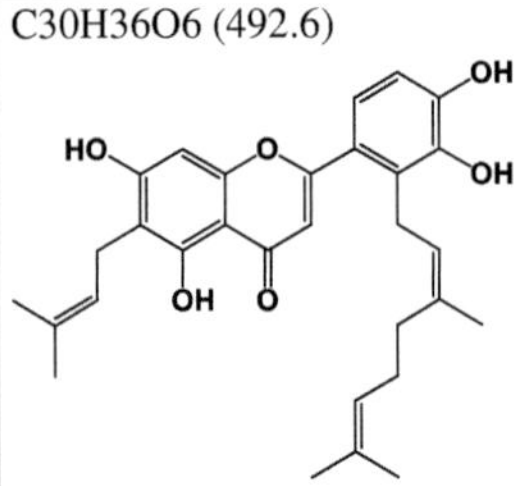	(2S)-2-[2-[(2E)-3,7-dimethylocta-2,6-dienyl]-3,4-dihydroxyphenyl]-5,7-dihydroxy-6-(3-methylbut-2-enyl)-2,3-dihydrochromen-4-one	Anti-angiogenic, anti-proliferative (Nguyen et al. 2017)

(continued)

Table 7.5 (continued)

Name	Molecular formula (MW)	Chemical name (IUPAC)	Mechanism
Vestitol	C16H16O4 (272.29) 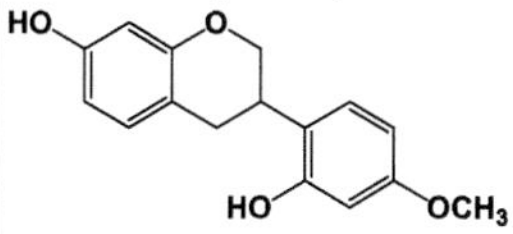	3-(2-hydroxy-4-methoxyphenyl)-3,4-dihydro-2H-chromen-7-ol	Anti-proliferative, cytotoxic, autophagy inhibition (Nani et al. 2018)
Artepillin C	C19H24O3 (300.4) 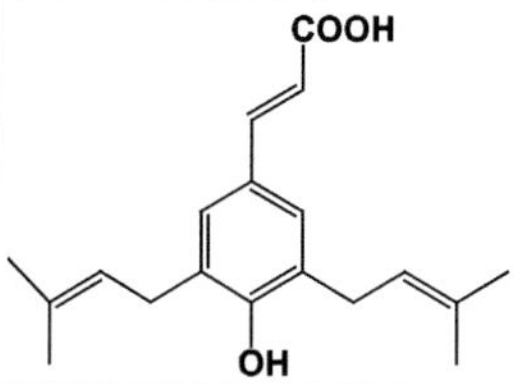	(E)-3-[4-hydroxy-3,5-bis(3-methylbut-2-enyl)phenyl] prop-2-enoic acid	Anti-proliferative, cytotoxic, autophagy inhibition (Endo et al. 2018)
Baccharin	C29H38O11 (562.6) 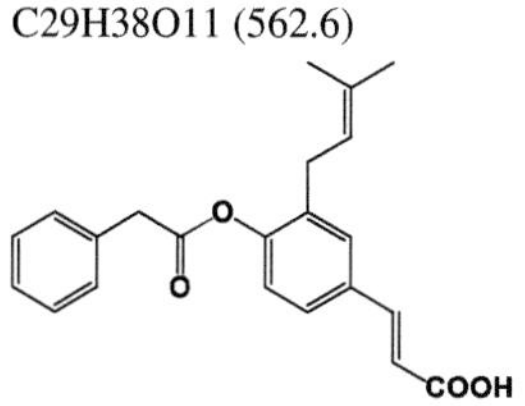	(2E)-3-{3-(3-Methyl-2-buten-1-yl)-4-[(3-phenylpropanoyl) oxy]phenyl}acrylic acid	Autophagy inhibition (do Amaral et al. 2022)
Caffeic acid	C9H8O4 (180.16) 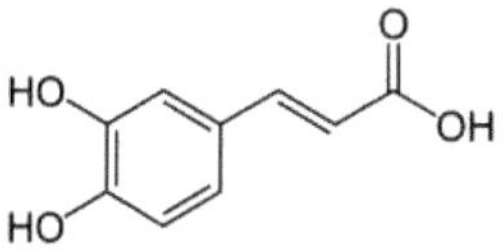	(E)-3-(3,4-dihydroxyphenyl) prop-2-enoic acid	Cell cycle arrest, apoptosis, anti-angiogenesis (Kabala-Dzik et al. 2017)
Caffeic acid phenyl ester	C17H16O4 (284.31) 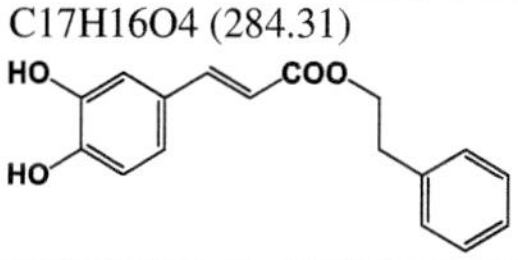	2-phenylethyl (E)-3-(3,4-dihydroxyphenyl)prop-2-enoate	Anti-proliferative Apoptosis Anti-metastatic (Kabala-Dzik et al. 2017)
Ferulic acid	C10H10O4 (194.18) 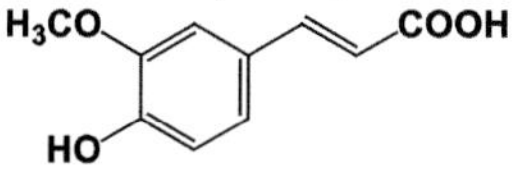	(E)-3-(4-hydroxy-3-methoxyphenyl)prop-2-enoic acid	Apoptosis (Celinska-Janowicz et al. 2019)
p-coumaric acid	C9H8O3 (164.16) 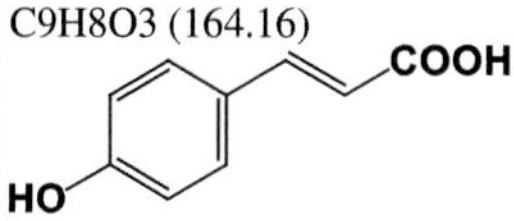	(E)-3-(4-hydroxyphenyl) prop-2-enoic acid	Cytotoxic, apoptosis (Celinska-Janowicz et al. 2019)

(continued)

Table 7.5 (continued)

Name	Molecular formula (MW)	Chemical name (IUPAC)	Mechanism
Frondoside A	C60H95NaO29S (1335.4) 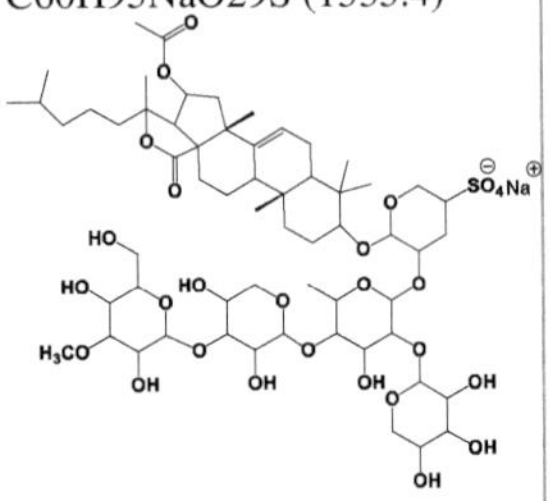	sodium;[6-[[4-acetyloxy-2,6,13,17,17-pentamethyl-6-(4-methylpentyl)-8-oxo-7-oxapentacyclo[10.8.0.02,9.05,9.013,18]icos-1(20)-en-16-yl]oxy]-5-[5-[4-[3,5-dihydroxy-6-(hydroxymethyl)-4-methoxyoxan-2-yl]oxy-3,5-dihydroxyoxan-2-yl]oxy-4-hydroxy-6-methyl-3-(3,4,5-trihydroxyoxan-2-yl)oxyoxan-2-yl]oxy-4-hydroxyoxan-3-yl] sulfate	Anti-angiogenic Anti-proliferative (Nguyen et al. 2017)
Nemorosone	C33H42O4 (502.7) 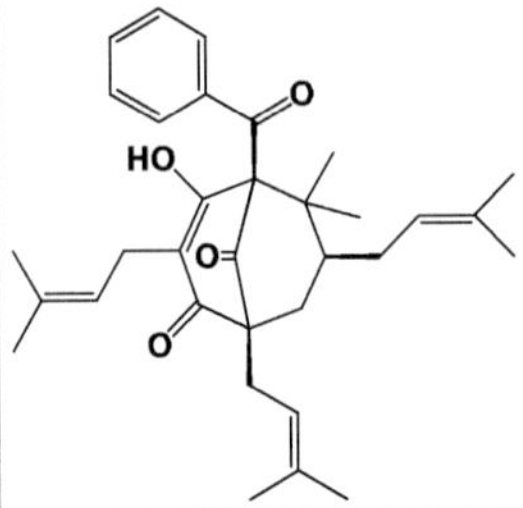	(1R,5R,7S)-1-benzoyl-4-hydroxy-8,8-dimethyl-3,5,7-tris(3-methylbut-2-enyl)bicyclo[3.3.1]non-3-ene-2,9-dione	Anti-proliferative Anti-metastatic (Frión-Herrera et al. 2019)

Note: The structures indicated are based on ChemSpider and PubChem databases

Effect on Colon and Colorectal Cancer

A study by Falcão et al. (2023) to evaluate the in vitro cytotoxic activity of the most representative propolis types, such as green, red and brown propolis, in combination with a chemotherapeutic agent (5-Fluorouracil) or CNS drug (Fluphenazine) on HT-29 colon cancer cell lines found that there was enhanced in vitro cytotoxic activity which was synergic at a concentration of 100 μg/mL.

In a related study, an animal model of early stage colorectal cancer (CRC) induced with 1,2-dimethylhydrazine (DMH) and dextran sulfate sodium (DSS) for 1 month showed that propolis potentially prevented cancer progression by increasing CD3+ and CD4+ TILs and reducing FOXP3 lymphocytes in the tumor microenvironment of early stage CRC. The study suggested a potential role for propolis as a food supplement to decrease or prevent CRC progression (Shen et al. 2023).

In another study to compare the inhibitory effect of ethanolic extract of propolis (EEP) and vitamin E on dimethylhydrazine (DMH)-induced colon lesions in rats, the EEP-exposed rats exhibited a significantly lower oxidative stress than the DMH group ($P < 0.05$). Its antioxidant effectiveness was more than that of vitamin E. Furthermore, the extract significantly reduced the level of proliferating cell nuclear antigen (PCNA), carcinoembryonic antigen (CEA), and platelet-derived growth factor (PDGF), implying that the EEP can efficiently prevent DMH-induced colonic lesions (Salehi et al. 2022).

Frión-Herrera et al. (2020a) showed that nemorosone, the main component in Cuban brown propolis, interferes in the crosstalk between colorectal cancer HT-29 cells and tumor-associated macrophages, primarily the M2 phenotype, which are involved in the progression and metastasis of this cancer. In another study, the same authors demonstrated that nemorosone has anti-proliferative and anti-metastatic effects on these cells Frión-Herrera et al. (2020b).

A methanol extract of Netherlands propolis had anti-proliferative activity toward murine colon 26-L5 carcinoma with an $_{EC50}$ value of 3.5 mg/ml. Fractionation of the extract resulted in the isolation of four flavonoids, seven cinnamic acid derivatives, and glycerol derivatives. The isolated compounds were tested for their anti-proliferative activity. The three most active had $_{EC50}$ values of 0.288, 1.76, and 0.114 μM toward colon 26-L5 carcinoma (Banskota et al. 2002).

Effect on Breast Cancer

Breast cancer is the leading cause of cancer-related deaths among women worldwide. A double-blind, randomized clinical trial among 60 patients in which the effect of propolis (250 mg, twice daily) was evaluated against a placebo found that the nutritional status and quality of life of patients on breast cancer chemotherapy were significantly better after 3 months of the intervention (Davoodi et al. 2022). Specifically, there was a significant increase in energy intake of the patients who received propolis compared to the placebo group. There were also significant improvements in the patients' quality of life in the propolis group, particularly with regard to emotional functioning, global QoL functioning, and financial difficulties compared to the control group. As a result, propolis was suggested an adequate and safe therapeutic option to improve the nutritional status and quality of life in patients with breast cancer who are on chemotherapy.

In another study to evaluate the potential of propolis as a source of epi-drugs (drugs with the ability to revert acquired epigenetic changes associated with chemoresistance of triple-negative breast cancer), three phenolic acids (caffeic, dihydrocinnamic, and p-coumaric) commonly detected in propolis were evaluated using in vitro models (Assumpção et al. 2020). The treatment with p-coumaric acid significantly reduced the cell viability of four triple-negative breast cancer cell lines (BT-20, BT-549, MDA-MB-231, and MDA-MB-436). Computational predictions by molecular docking indicated that it could interact with the MTAse domain of the human DNMT1 and directly compete with its intrinsic inhibitor S-Adenosyl-l-homocysteine (SAH). Moreover, the study also demonstrated that EEP partly demethylated the promoter region of RASSF1A in BT-549 cells and altered the RASSF1 protein expression levels. The study revealed that specific compounds in the EEP could revert the epigenetic silencing of the tumor suppressor RASSF1A, making it a promising source for epi-drugs discovery (Assumpção et al. 2020).

A study on antitumor effects of Algerian propolis in combination with doxorubicin in breast cancer showed that the combination was synergistic toward MDA-MB-231 cells while imparting significant protection in normal cells (MRC-5) against doxorubicin-mediated damage (Rouibah et al. 2021). A ten-fold decrease in the IC50 of doxorubicin was observed in the presence of propolis, implying the possibility of overcoming breast cancer resistance to treatment while solving the toxicity problems of chemotherapy in breast cancer.

The anti-proliferative effects of a water-soluble fraction of propolis were demonstrated in human breast cancer MCF-7 and human keratinocyte HaCaT cell lines, with IC50 values of 10.8 ± 0.06 and > 50 mg/ml respectively, further highlighting the potential of propolis in human breast cancer (Amalia et al. 2020).

Ethanolic extracts of propolis from the stingless bee, *Tetrigona apicalis*, were found to effectively induce apoptosis in MCF7 cells with an IC50 of 32.70 µg/mL (Mohamed et al. 2022).

A study to evaluate the effect of Withaferin-A and propolis on benz(a)pyrene-induced breast cancer in female Wistar rats found that the plasma levels of carcinoembryonic antigen (CEA)—a non-specific serum biomarker which is elevated in various types of cancers (Megahalathan and Muninathan 2022)—significantly decreased in rats which received withaferin-A and propolis combination rather than the individual compound/extract, indicating a synergistic antitumor role in mammary carcinogenesis.

Milek et al. (2022) demonstrated than ethanolic extracts of poplar-type propolis from South-Eastern Poland inhibited the migration of breast cancer cells at a low concentration (0.04 µg/mL) and inhibited the proliferation of MCF-7 cells.

A study to evaluate the effects of ethanolic extract of propolis and Withaferin-A on breast tumor initiation by benzo(a)pyrene in Wistar rats, and its effects on ATPase enzyme levels revealed that Na+/ K + -ATPase, Mg2 + -ATPase, and calcium ATPase enzyme activities were decreased in the erythrocyte membrane and tissues of breast cancer-bearing animals compared with control groups. The levels of enzyme activities were near normal in propolis-treated animals. The beneficial effect of propolis was primarily exerted during the initiation and post-initiation stages of breast carcinogenesis. Overall, the data indicated that propolis has modulating effects when administered intra-peritoneally on breast cancer-bearing animals (Thazhathuputhenpurayil and Natarajan 2022).

An ethanolic extract of Brazilian red propolis was found to significantly reduce the viability of MCF-7 breast cancer cells through the induction of mitochondrial dysfunction, caspase-3 activity, and DNA fragmentation, but did not affect these factors in a control cell line. In addition, the extract was found to promote apoptosis via endoplasmic reticulum stress (Kamiya et al. 2012).

Polyphenolic compounds isolated from propolis and a water-soluble extract of Croatian propolis were investigated for their effects on the growth and metastatic potential of mammary carcinoma in mice. Metastases in the lung were generated by intravenous injection of tumor cells. Oral dosing of the compounds and extract (50 mg/kg) significantly decreased the number of tumor nodules in the lung (Oršolić et al. 2004).

The effect of propolis in combination with paclitaxel against induced experimental breast cancer was investigated in female Sprague Dawley rats. Administration of paclitaxel and propolis effectively suppressed breast cancer, which was revealed by a decrease in the extent of lipid peroxidation with an increase in the activities of superoxide dismutase, catalase and glutathione peroxidase, and the levels of glutathione, Vitamin C and Vitamin E when compared to animals treated with either paclitaxel or propolis alone. Thus, it was proposed that co-administration could reduce the side effects of paclitaxel (Padmavathi et al. 2006).

Effect on Prostate Cancer

Prostate cancer is the second most frequent cancer, closely following lung cancer, and is the fifth leading cause of cancer deaths worldwide. Among men, prostate cancer is the most commonly diagnosed and the leading or second leading cause of cancer-related deaths worldwide (American Cancer Society 2024).

Fu et al. (2022) found that combination treatment of docetaxel with caffeic acid phenethyl ester (CAPE) effectively suppressed the proliferation, survival and tumor growth of docetaxel-resistant prostate cancer cells PC/DX25 and DU/DX50 CRPC cell lines from PC-3 and DU-145 human PCa cells, respectively in both in vitro and nude mice models. This effect was achieved via inhibition of Bcl-2 and c-Myc as well as induction of metabolism interference.

Egyptian propolis was demonstrated to possess anticancer activity through in vitro cytotoxicity in human prostate cancer (DU-145), breast adenocarcinoma (MCF-7), and colorectal adenocarcinoma (Caco-2) cell lines (Ibrahim and El-Banna 2021).

PC-3 prostate cancer cells were incubated with dimethyl sulfoxide and water extracts of Turkish propolis, and the treatments were found to significantly reduce cell viability to 24.5% and 17.7%, respectively. Statistically significant discriminatory peaks in the proteomic profiles were observed between control PC-3 cells and those treated with the dimethyl sulfoxide extract of propolis. Surface-enhanced laser desorption ionization time of flight mass spectrometry was used to examine changes in the proteome as a result of the propolis treatments and it was found that the treatment promoted changes in the proteome, suggesting that this might be a mechanism for the cytotoxic action (Barlak et al. 2011).

The effects of ethanolic extracts of two samples of Brazilian propolis and the bud resin of *B. dracunculifolia* on the proliferation of human prostate cancer cells were studied. The Brazilian propolis extracts showed significant inhibitory effects on the proliferation of human prostate cancer cells. Inhibition was achieved through regulation of protein expression of cyclins D1 and B1, and cyclin-dependent kinase, as well as expression of the p21 protein (Paredes-Guzman et al. 2007).

Effect on Head and Neck Cancer

A study evaluated the effect of Brazilian propolis on head and neck squamous cell carcinoma (HNSCC) cell lines (UM-SCC-17B and UM-SCC-74A), human keratinocytes (HK), and primary human dermal microvascular endothelial cells (HDMEC). The cells were treated with 0.5, 5.0, or 50 μg/mL green, brown, or red Brazilian propolis or vehicle control for 24, 36, and 72 h. Propolis significantly reduced cell viability of HNSCC and HDMEC cells, but not HK. Notably, red propolis caused a significant reduction in the percentage of cancer stem cells, reduced the number of orospheres, and downregulated the expression of stem cell markers.

The study suggested that propolis (particularly red propolis) might be beneficial for patients with head and neck cancer (Nör et al. 2021).

The prenylated flavonoid propolin G, isolated from Taiwanese propolis, was able to efficiently induce apoptosis in brain cancer cell lines, glioma and glioblastoma, with $_{IC50}$ values of 5 and 7.5 µg/ml, respectively, against the two cell lines. The results suggested that apoptosis might have occurred through a mitochondrial- and caspase-dependent pathway. Propolin G and Taiwanese and Brazilian propolis extracts were also found to protect against oxidative stress in rat primary cortical neurons (Huang et al. 2007).

The activity of extracts of three Mexican propolis samples against human colonic, human lung, and cervical cancer cells was examined. The extracts exhibited strong anti-proliferative activities and analysis of DNA isolated from treated cells showed the presence of intra-nucleosomal DNA cleavage (Hernandez et al. 2007).

An extract of propolis was found to inhibit the proliferation of U937 cells in a dose-dependent manner by inducing apoptosis and blocking cell cycle progression in the G2/M phase. Western blot analysis showed that propolis increased the expression of p21 and p27 proteins, and decreased the levels of cyclin B1, cyclin A, Cdk2 and Cdc2, causing cell cycle arrest. The results suggested that propolis-induced apoptosis was related to the selective activation of caspase-3 and induction of Bcl-2/Bax regulation (Motomura et al. 2008).

Effect on Pancreatic Cancer

The potential of Chinese propolis as a potential alternative treatment for pancreatic cancer was demonstrated in the human pancreatic cancer cell line, Panc-1. The study revealed that the CP induced apoptosis in these cells in a time- and dose-dependent manner with an IC50 value of approximately 50 µg/mL. A 48-h treatment with 50 µg/mL CP resulted in 34.25 ± 3.81% apoptotic cells, as compared to 9.13 ± 1.76% in the control group. Additionally, the Panc-1 cells tended to be arrested at G2/M phase after CP treatment, explaining the anti-proliferation effect of CP. The study also demonstrated suppression of Panc-1 cell migration by regulating epithelial-mesenchymal transition (EMT). Interestingly, the Hippo pathway was activated in Panc-1 cells after CP treatment, serving as a mechanism for the anti-pancreatic cancer effect of CP (Tao et al. 2021).

A study by Kuo et al. (2023) on a triple treatment combining propolis, thermal cycling-hyperthermia (TC-HT), and low-intensity ultrasound (US) found that cell viability of a human cancer cell line PANC-1 decreased to a level of 80% less than the control, without affecting the normal pancreatic cells. There was also excessive accumulation of reactive oxygen species (ROS) after the triple treatment, leading to the amplification of apoptotic pathways through the MAPK family and mitochondrial dysfunction.

A methanol extract of red Brazilian propolis was found to kill 100% of nutrient-starved human pancreatic cancer cells at a concentration of 10 µg/mL. Forty-three compounds were isolated from the extract with the most potent compound producing 100% kill at a concentration of 12.5 µM. The mechanism of cell death was believed to be via necrosis. The targeting of nutrient-deprived cells presents a method of targeting cells that are resistant to nutrient starvation (Li et al. 2008). Newly identified triterpene acids exhibited the most potent preferential cytotoxicity ($_{PC50}$ 4.3 μM) in a concentration- and time-dependent manner and induced apoptosis-like morphological changes of PANC-1 cells within 24 h of treatment (Li et al. 2010).

The anticancer effect of ethanol extracts of Chinese and Brazilian propolis on four human colon carcinoma cell lines were studied. The extracts of both Chinese and Brazilian propolis caused a marked dose-dependent growth inhibition of the cells, with $_{IC50}$ values in the range 4–41 μg/ml (Ishihara et al. 2009).

Effect on Melanomas

Melanoma is the least common but deadliest type of skin cancer, accounting for about 1% of all cancer cases globally.

A study to investigate the tumoral behavior of melanoma under treatment with the compounds baccarin and p-coumaric acid, extracted from Brazilian green propolis, in mice inoculated with B16f10 cells for 26 days showed a significant modulation in the number of inflammatory cells recruited to the tumor region and blood in the groups treated with the compounds. In addition, a significant reduction in the amount of blood vessels and mitosis in the neoplastic area was noticed. The study thus confirmed that baccarin and coumaric acid as promising anticarcinogenic compounds for the development of new antitumor agents (Gastaldello et al. 2021).

(Caetano et al. 2023) reported that the ethanol extract of Portuguese propolis from Gerês and its n-Butanol fraction were effective in decreasing cell biomass of the BRAF-mutated A375 melanoma cell line in a dose-dependent manner.

The effect of propolis on the production of IL-2, IF-y, IL-4, and IL-10 cytokines by T-helper cells in melanoma-bearing mice submitted to immobilization stress was studied. Spleen cells were assessed for cytokine expression. Stress induced a higher rate of tumor growth, while propolis-treated mice, stressed or not, showed melanoma development similar to the control. Cytokine production was inhibited in melanoma-bearing mice and propolis administration to melanoma-bearing mice submitted to stress promoted IL-2 and IFN-y production, indicating the activation of antitumor cell-mediated immunity. Propolis also stimulated IL-10 expression and production (Missima et al. 2010).

Effect on Renal Cancer

Renal cell carcinoma is proclaimed as the most lethal cancer of the urological system due to late diagnosis and treatment resistance. A Portuguese propolis sample, its fractions (n-hexane, ethyl acetate, n-butanol, and water), and three subfractions (P1-P3) were tested for their toxicity on A498, 786-O, and Caki-2 renal cell carcinoma cell lines and the non-neoplastic HK2 kidney cells. The ethyl acetate fraction showed the strongest toxicity against A498 (IC50 = 0.162 μg mL^{-1}) and 786-O (IC50 = 0.271 μg mL^{-1}) cells. With similar toxicity against 786-O, P1 (IC50 = 3.8 μg mL^{-1}) and P3 (IC50 = 3.1 μg mL^{-1}) exhibited greater effect when combined (IC50 = 2.5 μg mL^{-1}). The results demonstrated the potential of propolis and its constituents as promising coadjuvants in renal cell carcinoma treatment (Freitas et al. 2022).

Extracts from two Portuguese propolis samples were tested for their anticancer properties on human renal cell carcinoma cells, with $_{IC50}$ values of 56.5 and 56.1 μg/ml. The extracts were much less toxic against normal human renal cells (Valente et al. 2011).

Effect of Propolis on Tongue Cancer

A study of Polish propolis using ethanol, ethanol-hexane, hexane, and hexane-ethanol extracts from three different regions found that the ethanol and hexane-ethanol extractions were the most effective methods. These extracts were demonstrated to have selectivity toward tongue cancer cells in comparison to normal gingival fibroblasts.at the lowest extract concentration (100 μg/mL). The results indicated the potential of Polish propolis as a natural product with cancer-selective toxicity and anti-inflammatory effect (Wezgowiec et al. 2020).

Effect on Gastric Cancer

A study on the anti-proliferative effect of ethanolic extracts of "special Chinese propolis" from Northeast China demonstrated significant inhibition of human gastric cancer SGC-7901 cell growth, cell apoptosis, and cell cycle arrest in S phase, with increased production of reactive oxygen species (ROS) and reduced mitochondrial membrane potential (MMP) (Jiang et al. 2020). However, there were weak activities against human pancreatic cancer cell PANC1, human lung cancer cell A549, human colon cancer cell HCT116, human liver cancer cell HepG2, human bladder cancer cell T24, and human breast cancer cell MDA-MB-231. The selective inhibition of the proliferation of human gastric cancer SGC-7901 cells was postulated to arise from the induction of both death receptor-induced apoptosis and mitochondria-mediated apoptosis, and cell cycle arrest in S phase (Jiang et al. 2020).

Effect on Hepatocellular Carcinoma

Liu et al. (2021) elucidated the bioactive components and antitumor mechanism of a poplar propolis extract obtained from North China in human hepatocellular carcinoma HepG2 cells in vitro. Flavonoids and esters were the main active compounds, and the extract exerted strong anti-proliferative effects, and induced HepG2 apoptosis through a mitochondrial-dependent intrinsic pathway with elevated levels of cleaved PARP, cleaved caspase 3, and Bax and decreased the expressions of Bcl-2 and procaspase 9. The apoptosis was postulated to have been triggered through activation of the p38 MAPK and inactivation of p-ERK. Additionally, the propolis extract suppressed the Hippo pathway, leading to the inactivation of YAP/TAZ and TEAD1 and the inhibition of PI3K/AKT signaling molecules.

Effect on Lung Cancer

Stable propolis nanoparticles synthesized using the hot homogenization method exhibited remarkable anticancer properties against the A549 cell line for lung cancer, with IC50 of 0.01 mg/ml after 72 h-treatment. This suggested that besides antioxidant and antibacterial properties, these nanoparticies could be used as a drug or drug carrier for the treatment of lung cancer (Shahab-Navaei and Asoodeh 2023).

Effects on Other Cancers

Mangiferonic acid, an active compound isolated from the ethanolic extract of propolis of Homotrigona fimbriata, a stingless bee species, showed a cytotoxicity effect in three cancer cell lines MCF-7, HeLa, and Caco-2. The experimental IC50 values were 96.76 μM in MCF-7, >110.04 μM in HeLa, and >110.04 μM in Caco-2, respectively. This highlighted the potential for this propolis, and its components are a source of potent anticancer compounds (Arung et al. 2021).

A study by Alanazi et al. (2021) on Saudi propolis isolated fisetinidol and ferulic acid as the two main compounds. The propolis extracts, fractions, and isolated compounds exhibited significant cytotoxicity against U937 cells, a cellular model for histiocytic lymphoma, with IC50 > 60 μg/ml which was attributed to ferulic acid.

Apoptotic, Anti-angiogenic, Anti-proliferative, and Anti-metastatic Effects

An ethanolic extract of green Brazilian propolis (EEBP) significantly reduced the number of newly formed vessels in an in vivo angiogenesis assay. EEBP also showed anti-angiogenic effects in an in vitro tube formation assay. The major constituent of EEBP, artepillin C, was found to significantly reduce the number of newly formed vessels in an in vivo angiogenesis assay. Thus, artepillin C, at least in part, is responsible for the anti-angiogenic activity of EEBP in vivo and might prove effective in counteracting tumor angiogenesis (Ahn et al. 2007).

A study in which the effects of Apis mellifera propolis on apoptosis and invasiveness in isogenic HNSCC cell lines derived from primary (HN30 and HN4) and metastatic sites (HN31 and HN12) was investigated revealed that an ethyl acetate extract of propolis promoted apoptosis and exerted anti-invasion potential by inhibiting MMP-2 and MMP-9 activity in HNSCC cell lines. These inhibitory effects were thought to be mediated by two flavonoids, galangin, and apigenin, which were isolated from the extracts (Niyomtham et al. 2021).

Tumor necrosis factor-related apoptosis-inducing ligand (TRAIL) is an important endogenous anticancer cytokine that induces apoptosis selectively in tumor cells. However, some cancer cells are resistant to TRAIL-mediated apoptosis. Phenolic and polyphenolic compounds sensitize TRAIL-resistant cancer cells and augment the apoptotic activity of TRAIL. The cytotoxic and apoptotic effects of an ethanolic extract of Brazilian green propolis (EEP) was examined against LNCaP prostate cancer cells. The extract sensitized TRAIL-resistant prostate cancer cells to TRAIL-mediated apoptosis, and when prostate cancer cells were co-treated with 100 ng/ml TRAIL and 50 μg/ml EEP, the percentage of apoptotic cells was increased to 65.8%—several folds greater than the level induced by TRAIL alone (Szliszka et al. 2011a, b).

An ethanolic extract of stingless bee propolis from Thailand was tested for anti-proliferative activity against five cancer cell lines and two normal cell lines. The cell viability (%) and $_{IC50}$ values were calculated. The hexane extract provided the highest in vitro activity against the five cancer cell lines and the lowest cytotoxicity against the two normal cell lines. Fractionation of the extract produced a fraction with high anti-proliferative activity with $_{IC50}$ values in the range 4.09–14.7 μg/ml against cancer cells but with limited activity against normal cells (Umthong et al. 2011).

Extracts of *Apis mellifera* propolis collected in Thailand were assayed for cytotoxic activity against five human cancer cell lines and a control cell line. Crude hexane and dichloromethane extract of propolis displayed anti-prolifer-ative/cytotoxic activities with $_{IC50}$ values across the five cancer cell lines ranging from 41.3 to 52.4 μg/ml and from 43.8 to 53.5 μg/ml, respectively. Two main bioactive components, a cardanol and a cardol, were isolated and IC50 values across the five cancer cell lines ranging from 10.8 to 29.3 μg/ml were obtained for the cardanol and <3.13 to 5.97 μg/ml for the cardol. Both the compounds induced cytotoxicity and cell death without DNA fragmentation in the cancer cells, but only an anti-proliferation response in the control Hs27 cells (Teerasripreecha et al. 2012).

7.12 Clinical Trials on Propolis

The number of properly controlled clinical trials of propolis treatments has increased greatly over the past 7 years. Most of the trials involve small numbers of subjects but they are generally well designed. There are four areas where propolis has been applied as a clinical treatment: Cardiovascular disease/metabolic syndrome; viral respiratory disease, oxidative stress/inflammation and dental disease. There is some overlap between these areas.

There have been some recent reviews of clinical trials assessing the efficacy of propolis on treating metabolic parameters (Ghelflati et al. 2021; Hallajzadeh et al. 2021).

Patients with metabolic syndrome were randomized into two groups of 33 patients. One group was treated with placebo and the other group was treated with 250 mg of Iranian propolis twice daily for 12 weeks. The treatment produced a significant reduction in waist circumference but no effect on a range of other anthropometric measures. There was significant improvement in a number of subjective measures such as physical functioning and general health (Sajjadi et al. 2023).

In another study, 44 patients with non-alcoholic liver disease (NAFLD) were divided into two groups—23 patients were treated with 510 mg of poplar propolis per day and 21 were treated with placebo over 8 weeks. In addition, all patients were placed on a calorie-restricted diet. A number of biochemical and anthropometric factors were monitored. Overall, the propolis-treated group showed a greater improvement in glucose homeostasis and liver fibrosis than the control group (Nikbaf-Shandiz et al. 2022).

Ninety-two patients with NAFLD were divided into four groups. All the groups were put on a recommended diet. Group 1 received α-lipoic aid, group 2 received myoinositol (MI), group 3500 mg of Iranian propolis, and group 4 was only treated with diet. Over 8 weeks, there were slight improvements in glycemic indices in the group treated with MI, the greatest improvements in triglyceride levels were in the untreated group. There was no indication of any effect of the propolis supplementation in this study (Tutunchi et al. 2023).

The effects of propolis supplementation on glucose homeostasis, lipid profile, liver function, anthropometric indices, and meta-inflammation in patients with non-alcoholic fatty liver disease were studied in a trial with 44 patients. Patients were randomly allocated to placebo ($n = 21$) or propolis ($n = 23$). No significant changes in anthropometric measurements and lipid profile were found between the groups at the end of the intervention. Fasting blood sugar, serum insulin level and the desire to eat sweet foods and the NAFLD fibrosis score decreased significantly.

in the propolis group compared to the placebo group (Nikbaf-Shandiz et al. 2022).

The effect of supplementation with standardized poplar propolis extract powder (PPEP) on insulin homeostasis in non-diabetic insulin-resistant volunteers with obesity was studied. Two groups of nine non-diabetic insulin-resistant volunteers with obesity, aged 49 ± 7 years were treated with poplar propolis or placebo for 3 months. Propolis supplementation improved insulin sensitivity by significantly decreasing the percentage of insulin-resistant subjects. The study suggested that poplar propolis has a preventive effect on the physiopathological mechanisms of type II diabetes (Sani et al. 2023).

Clinical trials of propolis in the treatment of viral respiratory disease have been recently reviewed (Ożarowski and Karpiński 2023).

In a trial to test the efficacy of propolis in the recovery from Covid 19,124 patients with moderate to severe Covid 19 were recruited to a trial involving treatment with standardized green propolis. The patients were divided into three groups: untreated, treated with 400 mg of propolis per day, and treated with 800 mg of propolis per day. The trial was open-labeled and a placebo was not used. There was a significant reduction in hospital stay with the median length of time in hospital being: 12, 7, and 6 days for untreated, patients treated with 400 mg, and 800 mg of propolis, respectively. The patients treated with 800 mg of propolis had a lower incidence of acute kidney injury than the untreated group. The propolis treatment was well tolerated (Silveira et al. 2021).

A double-blind trial was carried out in 122 healthy adults with mild upper respiratory tract infection. 58 subjects received an oral spray containing poplar propolis and 64 received placebo. The spray was administered 2–3 times a day. After 3 days 83% of subjects had remission of symptoms whereas in the placebo group 72% of subjects had at least one remaining symptom (Esposito et al. 2021).

Fifty-two patients with newly diagnosed asthma were divided into two groups of 26 and the groups were treated either with placebo or 75 mg of Iranian brown propolis three times a day. The study ran for 1 month. The propolis-treated group had significantly improved asthma control test (ACT) scores after the treatment and reduced eosinophilia. No improvement in the ACT score was observed for the placebo group and the level of eosinophilia was slightly worse (Mirsadraee et al. 2021).

Forty-six patients with chronic obstructive airway disease were recruited to a study involving treatment with a combination of N-acetylcysteine (NAC) and propolis. Nineteen patients were treated with placebo, 13 with a combination of 80 mg of propolis with 600 mg of NAC daily and 14 with 160 mg of propolis and 1200 mg of NAC daily. Doses were given once daily for 3 months. There was a significant reduction in the incidence of acute COPD in both the treated groups. In the placebo group, the incidence of acute COPD was 52.6%; in the group treated with 80 mg of propolis and 600 mg NAC, the incidence was 15.4%; in the group treated with 160 mg propolis, and 1200 mg NAC the incidence was 7.1% (Buha et al. 2022).

Clinical trials assessing the effect of propolis on oxidative stress have been recently reviewed. Some recent examples are described below (Malekahmadi et al. 2023).

Two groups of 27 military cadets were given either 900 mg of poplar propolis daily or placebo for 4 weeks in a blind trial. The subjects underwent a 12-min running test and a sprint test to evaluate anaerobic performance. There were no significant effects on anaerobic performance but there was a significant reduction plasma levels of the inflammatory marker IL-6 and some significant effects markers of oxidative stress (Soleimani et al. 2021).

Hemodialysis patients were given 400 mg of Brazilian green propolis daily or placebo for 2 months. There were 21 patients in the group treated with propolis and 20 in the placebo group. Several makers of inflammation were monitored. In the propolis group there was a significant reduction in TNF-α and a marginally significant reduction in macrophage inflammatory protein 1-β. The results suggested that propolis supplementation might reduce inflammation in patients requiring chronic kidney dialysis (Chermut et al. 2023).

In a related study, 19 patients being treated with peritoneal dialysis for chronic kidney disease were divided into two groups. Ten patients were treated with 400 mg of Brazilian green propolis and nine were treated with placebo for 2 months. The propolis treatment produced a tendency for an increase in the expression of the anti-inflammatory gene Nrf2 in the group treated with propolis, there was no effect on the pro-inflammatory gene NF-κB. There was a reduction in TNFα in the propolis-treated group but no effect on IL-6, C-reactive protein, or the oxidative stress maker MDA (Baptista et al. 2023).

Forty-four patients with CKD were randomly assigned to receive propolis capsules at a dose of 250 mg daily or placebo for 3 months. At the end of the intervention improvement in some dimensions of health-related quality of life were reported by the group treated with propolis in comparison with the placebo group. Systolic and diastolic BP, serum creatinine, 24-h urine volume and protein, fasting blood sugar, hemoglobin A1C, insulin resistance (HOMA-IR), quantitative insulin sensitivity check index, and pro-oxidant antioxidant balance did not differ significantly between the two groups (Anvarifard et al. 2023).

Sixty patients being treated with radiation for head and neck cancer were randomized into two groups. Twenty-eight treated themselves with an oral spray containing 0.8% w/v of Brazilian organic propolis in suspension and 30 received a placebo spray. The application of treatment was continued for the duration of the radiotherapy. The propolis treatment significantly reduced levels of TNF-α and IL-1β and the level of oral candidiasis in comparison with the placebo group (Fernandes et al. 2022).

Eighty women with moderate rheumatoid arthritis were divided into two groups of 40. One group was given 508 mg of Brazilian propolis daily as tablets for 24 weeks and the other group received placebo. Effects on the disease were monitored by measuring erythrocyte sedimentation rate, C-reactive protein, ultrasonographic evaluation of synovitis, cytokine levels, and quality of life changes. There were no observed effects of the propolis treatment on the disease although the fact that the patients were already taking antiarthritic medication might have masked potential effects (Matsumoto et al. 2021).

A double-blind study evaluated the effect of Brazilian green propolis (500 mg/day for 3 months) on the level of oxidative stress of patients with HIV. Two groups of 20 patients were compared with one group being given propolis and the other placebo. Plasma MDA, carbonylation, total oxide nitric, total antioxidant capacity (TAP), superoxide dismutase, catalase, and NF-KB and NRF2 gene expression were measured. A decreased MDA concentration was seen in the propolis group as well as a slight but non-significant increase in TAP. Propolis may have some benefits in reducing oxidative stress in HIV patients (Tasca et al. 2023).

The effects of Brazilian green propolis on body fat mass and levels of adiponectin and reactive oxygen species in community-dwelling elderly females were studied altogether, 78 females aged 66–84 years were randomly assigned to the propolis (n = 39) or placebo (n = 39) groups. The propolis group was given three capsules containing 227 mg of propolis twice a day for 12 weeks. No changes were observed in absolute or relative fat mass in the placebo group but this showed a significant decline in the propolis group. The level of serum adiponectin significantly increased in the propolis group. The level of oxidative stress was lowered significantly in the propolis group and the level of superoxide dismutase was significantly increased in the propolis group. The results suggest that propolis supplementation may decrease body fat mass and oxidative stress in elderly females (Kanazashi et al. 2023).

A clinical trial was carried out on 80 participants, who were catheterized, who were assigned to propolis treatment or placebo groups. A cream containing Iranian propolis was applied to the angiocatheter insertion sites repeatedly every 12 h. The patients were observed for signs of phlebitis. The onset of phlebitis in the control group was significantly earlier than in the treatment group (Pakdaman et al. 2024).

The efficacy of propolis in treating periodontal disease has been recently reviewed (López-Valverde et al. 2021). Examples of recent clinical trials are discussed below.

A study evaluated the antimicrobial efficacy of a 2.5% Brazilian Red Propolis (BRP) dental varnish to prevent caries in children. Seventy-five children with high caries risk, aged between 36 and 71 months and with no caries were treated. The varnish containing BRP was effective in reducing the level of the cariogenic bacteria *Streptococcus mutans* for up to 180 days after application of the varnish. Varnishes containing chlorhexidine and fluoride were also effective (Rodrigues Neto et al. 2021).

A randomized clinical trial was carried out in 60 high caries risk children who were divided randomly into two groups. One group received propolis chewing gum and the other propolis mouthwash. The results showed that propolis had a significant effect on reducing plaque scores and colony counts in both vehicles (El-Allaky et al. 2020). Forty-five subjects with gingivitis were randomly assigned into one of the three different groups: Group A Propolis mouthwash; Group B Chlorhexidine mouthwash; and Group C Placebo mouthwash. Supragingival plaque and gingival inflammation were assessed by full mouth Plaque index (PI) and gingival index (GI) at baseline and after 21 days. The study was divided into three phases, each phase lasted for 21 days separated by a washout period of 15 days in between them. Groups A, B, and C were treated with 0.2% Propolis, chlorhexidine, and placebo mouthwash, respectively, in phase I. The propolis treatment produced a significant improvement in plaque index in comparison with the other two treatments (Gunjal and Pateel 2024).

References

Ahn M-R, Kunimasa K, Ohta T, Kumazawa S, Kamihira M, Kaji K, Uto Y, Hori H, Nagasawa H, Nakayama T (2007) Suppression of tumor-induced angiogenesis by Brazilian propolis: major component artepillin C inhibits in vitro tube formation and endothelial cell proliferation. Cancer Lett 252:235–243

Alanazi S, Alenzi N, Alenazi F, Tabassum H, Watson D (2021) Chemical characterization of Saudi propolis and its antiparasitic and anticancer properties. Sci Rep 11:5390

Al-Ani I, Zimmermann S, Reichling J, Wink M (2018) Antimicrobial activities of European propolis collected from various geographic origins alone and in combination with antibiotics. Medicines 5:2

Alenezi SS, Natto MJ, Igoli JO, Gray AI, Fearnley J, Fearnley H, De Koning HP, Watson DG (2020) Novel flavanones with anti-trypanosomal activity isolated from Zambian and Tanzanian propolis samples. Int J Parasitol Drugs Drug Resist 14:201–207

Alenezi SS, Alenezi ND, Ebiloma GU, Natto MJ, Ungogo MA, Igoli JO, Ferro VA, Gray AI, Fearnley J, De Koning HP, Watson DG (2022) The antiprotozoal activity of Papua New Guinea propolis and its triterpenes. Molecules 27:1622

Alenezi SS, Alenezi ND, Ebiloma GU, Natto MJ, Ungogo MA, Igoli JO, Ferro VA, Gray AI, Fearnley J, Koning HP, Watson DG (2023) The activity of red Nigerian propolis and some of its components against Trypanosoma brucei and Trypanosoma congolense. Molecules 28:622

Al-Ghandour AM, Ahmed HK, Salem AE, Tealeb AS, Mohamed RM, Yousef AM (2020) Efficacy of olibanum and propolis medicinal extracts versus metronidazole in Giardia lamblia experimentally infected mice. Microbes Infect Dis 1:209–220

Ali SY, Abdel-Moneim A, Abdel-Reheim ES (2020) Protective effect of propolis on doxorubicin induced cardio-and nephrotoxicity. Ann Cardiol Vasc Med 3:1028

Almuhayawi MS (2020) Propolis as a novel antibacterial agent. Saudi J Biol Sci 27:3079–3086

Alotaibi A, Ebiloma GU, Williams R, Alenezi S, Donachie AM, Guillaume S, Igoli JO, Fearnley J, De Koning HP, Watson DG (2019) European propolis is highly active against trypanosomatids including Crithidia fasciculata. Sci Rep 9:11364

Alotaibi A, Ebiloma GU, Williams R, Alfayez IA, Natto MJ, Alenezi S, Siheri W, AlQarni M, Igoli JO, Fearnley J, De Koning HP (2021) Activity of Compounds from Temperate Propolis against Trypanosoma brucei and Leishmania mexicana. Molecules 26:3912

Altabbal S, Athamnah K, Rahma A, Wali AF, Eid AH, Iratni R, Al Dhaheri Y (2023) Propolis: a detailed insight of its anticancer molecular mechanisms. Pharmaceuticals (Basel) 16:450

Amalia E, Diantini A, Subarnas A (2020) Water-soluble propolis and bee pollen of Trigona spp. from South Sulawesi Indonesia induce apoptosis in the human breast cancer MCF-7 cell line. Oncol Lett 20:274

American Cancer Society (2024) Global cancer facts & figures, 5th edn. American Cancer Society, Atlanta

Anjum SI, Ullah A, Khan KA, Attaullah M, Khan H, Ali H, Bashir MA, Tahir M, Ansari MJ, Ghramh HA (2019) Composition and functional properties of propolis (bee glue): a review. Saudi J Biol Sci 26:1695–1703

Anvarifard P, Anbari M, Ostadrahimi A, Ardalan M, Ghoreishi Z (2022) A comprehensive insight into the molecular and cellular mechanisms of the effects of Propolis on preserving renal function: a systematic review. Nutr Metab 19:6

Anvarifard P, Ostadrahimi A, Ardalan M, Anbari M, Ghoreishi Z (2023) The effects of propolis on pro-oxidant–antioxidant balance, glycemic control, and quality of life in chronic kidney disease: a randomized, double-blind, placebo-controlled trial. Sci Rep 13:9884

Arung ET, Ramadhan R, Khairunnisa B, Amen Y, Matsumoto M, Nagata M, Kusuma IW, Paramita S, Sukemi Y, Tandirogang N, Takemoto N, Syafrizal KYU, Shimizu K (2021) Cytotoxicity effect of honey, bee pollen, and propolis from seven stingless bees in some cancer cell lines. Saudi J Biol Sci 28:7182–7189

Asem N, Abdul Gapar NA, Abd Hapit NH, Omar EA (2020) Correlation between total phenolic and flavonoid contents with antioxidant activity of Malaysian stingless bee propolis extract. J Apic Res 59:437–442

Asfaram S, Fakhar M, Keighobadi M, Akhtari J (2021) Promising anti-protozoan activities of propolis (bee glue) as natural product: a review. Acta Parasitol 66:1–12

Ashour AM (2024) Propolis attenuates diabetes-induced testicular injury by protecting against DNA damage and suppressing cellular stress. Front Pharmacol 15:1416238

Assumpção JHM, Takeda AAS, Sforcin JM, Rainho CA (2020) Effects of propolis and phenolic acids on triple-negative breast cancer cell lines: potential involvement of epigenetic mechanisms. Molecules 25:1289

Ayad AS, Benchaabane S, Daas T, Smagghe G, Loucif-Ayad W (2024) Assessment of efficacy of algerian propolis against the parasitic mite varroa destructor and safety for honey bees by spray treatment. Insects 15:75

Balderas-Cordero D, Canales-Alvarez O, Sánchez-Sánchez R, Cabrera-Wrooman A, Canales-Martinez MM, Rodriguez-Monroy MA (2023) Anti-inflammatory and histological analysis of skin wound healing through topical application of mexican propolis. Int J Mol Sci 24:11831

Banskota AH, Nagaoka T, Sumioka LY, Tezuka Y, Awale S, Midorikawa K, Matsushige K, Kadota S (2002) Antiproliferative activity of the Netherlands propolis and its active principles in cancer cell lines. J Ethnopharmacol 80(1):67–73. https://doi.org/10.1016/S0378-8741(02)00022-3

Banskota AH, Tezuka Y, Kadota S (2001) Recent progress in pharmacological research of propolis. Phytother Res 15:561–571

Baptista BG, Fanton S, Ribeiro M, Cardozo LF, Regis B, Alvarenga L, Ribeiro-Alves M, Berretta AA, Shiels PG, Mafra D (2023) The effect of Brazilian Green Propolis extract on inflammation in patients with chronic kidney disease on peritoneal dialysis: a randomised double-blind controlled clinical trial. Phytomedicine 114:154731

Barary M, Hosseinzadeh R, Kazemi S, Liang JJ, Mansoori R, Sio TT, Hosseini M, Moghadamnia AA (2022) The effect of propolis on 5-fluorouracil-induced cardiac toxicity in rats. Sci Rep 12:8661

Barlak Y, Değer O, Çolak M, Karataylı SC, Bozdayı AM, Yücesan F (2011) Effect of Turkish propolis extracts on proteome of prostate cancer cell line. Proteome Sci 9:1

Barrientos L, Herrera CL, Montenegro G, Ortega X, Veloz J, Alvear M, Cuevas A, Saavedra N, Salazar LA (2013) Chemical and botanical characterization of Chilean propolis and biological activity on cariogenic bacteria Streptococcus mutans and Streptococcus sobrinus. Braz J Microbiol 44:577–585

Behzadi A, Delshad S, Moghadaszadeh M (2022) Amplification of wound healing by propolis and honey ointment in healthy and diabetic rat models; Histopathological and morphometric findings. Arch Razi Inst 77:1673

Belmehdi O, El Menyiy N, Bouyahya A, El Baaboua A, El Omari N, Gallo M, Montesano D, Naviglio D, Zengin G, Skali Senhaji N (2023) Recent advances in the chemical composition and biological activities of propolis. Food Rev Intl 39:6078–6128

Bezerra CR, Assunção Borges KR, Alves RD, Teles AM, Pimentel Rodrigues IV, da Silva MA, Nascimento MD, Bezerra GF (2021) Highly efficient antibiofilm and antifungal activity of green propolis against Candida species in dentistry materials. PLoS One 15:e0228828

Bhargava P, Mahanta D, Kaul A, Ishida Y, Terao K, Wadhwa R, Kaul SC (2021) Experimental evidence for therapeutic potentials of propolis. Nutrients 13:2528

Bittencourt ML, Ribeiro PR, Franco RL, Hilhorst HW, de Castro RD, Fernandez LG (2015) Metabolite profiling, antioxidant and antibacterial activities of Brazilian propolis: use of correlation and multivariate analyses to identify potential bioactive compounds. Food Res Int 76:449–457

Bogdanov S (2012) Propolis: composition, health, medicine: a review. Bee Product Science. www.bee-hexagon.net

Bogdanov S, Bankova V (2012) The propolis book. Bee hexagon

Boyacioglu D, Samanci AET, Samanci T (2022) Future prospects of propolis, bee pollen, royal jelly, and bee venom. In: *Bee products and their applications in the food and pharmaceutical industries*. Elsevier

Bragança Castagnino GL, Meana A, Cutuli De Simón MT, Batista Pinto LF (2023) Propolis and its effects on bee diseases and pests: a systematic review. J Apic Res 62:171–184

Brito-Pereira R, Moreira J, Tubio CR, Fernandes MM, Lanceros-Mendez S (2024) Enhancing antibacterial efficacy through propolis-infused electroactive polymers: a novel approach to combat antibiotic resistance. Chem Eng J 496:154223

Buha I, Mirić M, Agić A, Simić M, Stjepanović M, Milenković B, Nagorni-Obradović L, Škodrić-Trifunović V, Ilić B, Popević S, Dimic-Janjic S (2022) A randomized, double-blind, placebo-controlled study evaluating the efficacy of propolis and N-acetylcysteine in exacerbations of chronic obstructive pulmonary disease. Eur Rev Med Pharmacol Sci 26:4809–4815

Burdock GA (1998) Review of the biological properties and toxicity of bee propolis (propolis). Food Chem Toxicol 36:347–363

Caetano AR, Oliveira RD, Celeiro SP, Freitas AS, Cardoso SM, Gonçalves MST, Baltazar F, Almeida-Aguiar C (2023) Phenolic compounds contribution to portuguese propolis anti-melanoma activity. Molecules 28:3107. https://doi.org/10.3390/molecules28073107

Campana R, Patrone V, Franzini ITM, Diamantini G, Vittoria E, Baffone W (2009) Antimicrobial activity of two propolis samples against human Campylobacter jejuni. J Med Food 12:1050–1056

Campos JF, Santos UP, Rocha PD, Damião MJ, Balestieri JB, Cardoso CA, Paredes-Guzman JF, Daugsch A, Park YK, Rhim JS (2007) Antiproliferation of human prostate cancer cells by ethanolic extracts of Brazilian propolis and its botanical origin. Int J Oncol 31:601–606

Canales-Alvarez O, Canales-Martinez MM, Dominguez-Verano P, Balderas-Cordero D, Madrigal-Bujaidar E, Álvarez-González I, Rodriguez-Monroy MA (2024) Effect of Mexican propolis on wound healing in a murine model of diabetes mellitus. Int J Mol Sci 25:2201

Castaldo S, Capasso F (2002) Propolis, an old remedy used in modern medicine. Fitoterapia 73:S1–S6

Cavalcante GM, Camara CA, Silva EM, Santos MS, Leite AB, Queiroz AC, Evelyn Da Silva A, Araújo MV, Alexandre-Moreira MS, Silva TM (2021) Leismanicidal activity of propolis collected in the semiarid region of Brazil. Front Pharmacol 12:702032

Celebi G, Gocmez SS, Ozer C, Duruksu G, Yazır Y, Utkan T (2023) Propolis prevents vascular endothelial dysfunction by attenuating inflammation and oxidative damage in the chronic unpredictable stress model of depression in rats. J Pharm Pharmacol 75:1418–1429

Celinska-Janowicz K, Zareba I, Lazarek U, Teul J, Tomczyk M, Palka J, Miltyk W (2019) Constituents of propolis: chrysin, caffeic acid, p-coumaric acid, and ferulic acid induce PRODH/POX-dependent apoptosis in human tongue squamous Cell Carcinoma Cell (CAL-27). Front Pharmacol 9:336

Chaudhary P, Janmeda P, Docea AO, Yeskaliyeva B, Abdull Razis AF, Modu B, Calina D, Sharifi-Rad J (2023) Oxidative stress, free radicals and antioxidants: Potential crosstalk in the pathophysiology of human diseases. Front Chem 11:1158198

Chavda VP, Chaudhari AZ, Teli D, Balar P, Vora L (2023) Propolis and their active constituents for chronic diseases. Biomedicines 11:259

Chavda VP, Vuppu S, Balar PC, Mishra T, Bezbaruah R, Teli D, Sharma N, Alom S (2024) Propolis in the management of cardiovascular disease. Int J Biol Macromol 131219:131219

Chen LH, Chien YW, Chang ML, Hou CC, Chan CH, Tang HW, Huang HY (2018a) Taiwanese green propolis ethanol extract delays the progression of type 2 diabetes mellitus in rats treated with streptozotocin/high-fat diet. Nutrients 10:503

Chen YW, Ye SR, Ting C, Yu YH (2018b) Antibacterial activity of propolins from Taiwanese green propolis. J Food Drug Anal 26:761–768

Chermut TR, Fonseca L, Figueiredo N, de Oliveira LV, Borges NA, Cardozo LF, Leite PE, Alvarenga L, Regis B, Delgado A, Berretta AA (2023) Effects of propolis on inflammation markers in patients undergoing hemodialysis: a randomized, double-blind controlled clinical trial. Complement Ther Clin Pract 51:101732

Chiu HF, Han YC, Shen YC, Golovinskaia O, Venkatakrishnan K, Wang CK (2020) Chemopreventive and chemotherapeutic effect of propolis and its constituents: a mini-review. J Cancer Prev 25:70–78

Chouhan D, Dey N, Bhardwaj N, Mandal BB (2019) Emerging and innovative approaches for wound healing and skin regeneration: current status and advances. Biomaterials 216:119267

Coelho GR, Mendonça RZ, Vilar KS, Figueiredo CA, Badari JC, Taniwaki N, Namiyama Oliveira MI, Curti SP, Evelyn Silva P (2015) Antiviral action of hydromethanolic extract of geopropolis from Scaptotrigona postica against antiherpes simplex virus (HSV-1). Evid Based Complement Alternat Med 2015:296086

Corrêa FRS, Schanuel FS, Moura-Nunes N, Monte-Alto-Costa A, Daleprane JB (2017) Brazilian red propolis improves cutaneous wound healing suppressing inflammation-associated transcription factor NFκB. Biomed Pharmacother 86:162–171

Cui Z, Zhang J, Wang J, Liu J, Sun P, Li J, Li G, Sun Y, Ying J, Li K, Zhao Z (2024) Caffeic acid phenethyl ester: an effective antiviral agent against porcine reproductive and Respiratory Syndrome Virus. Antivir Res 225:105868

Da Cunha MG, Franchin M, Galvão L, de Ruiz A, de Carvalho JE, Ikegaki M, de Alencar SM, Koo H, Rosalen PL (2013) Antimicrobial and antiproliferative activities of stingless bee Melipona scutellaris geopropolis. BMC Complement Altern Med 13:1–9

Da Cunha GA, Carlstrom PF, Franchin M, Alencar SM, Ikegaki M, Rosalen PL (2023) A systematic review of the potential effects of propolis extracts on experimentally-induced diabetes. Planta Med 89:236–244

Da Rosa C, Bueno IL, Quaresma ACM, Longato GB (2022) Healing potential of propolis in skin wounds evidenced by clinical studies. Pharmaceuticals 15:1143

Dalenberg H, Maes P, Mott B, Anderson KE, Spivak M (2020) Propolis envelope promotes beneficial bacteria in the honey bee (Apis mellifera) mouthpart microbiome. Insects 11:453

Damodaran T (2021) Propolis. Nutraceuticals. Elsevier

Dantas Silva RP, Machado BA, Barreto GD, Costa SS, Andrade LN, Amaral RG, Carvalho AA, Padilha FF, Barbosa JD, Umsza-Guez MA (2017) Antioxidant, antimicrobial, antiparasitic, and cytotoxic properties of various Brazilian propolis extracts. PLoS One 12:e0172585

Davoodi SH, Yousefinejad V, Ghaderi B, Akbari ME, Darvishi S, Mehrabi Y, Darvishi N (2022) Oral propolis, nutritional status and quality of life with chemotherapy for breast cancer: a randomized, double-blind clinical trial. Nutr Cancer 74:2029–2037

De Almeida-Junior S, Ferraz MVF, De Oliveira AR, Maniglia FP, Bastos JK, Furtado RA (2023) Advances in the phytochemical screening and biological potential of propolis. Fundam Clin Pharmacol 37:886–899

De Castro S (2001) Propolis: biological and pharmacological activities. Therapeutic uses of this bee-product. Annu Rev Biomed Sci 3:49–83

De Marco S, Piccioni M, Pagiotti R, Pietrella D (2017) Antibiofilm and antioxidant activity of propolis and bud poplar resins versus Pseudomonas aeruginosa. Evid Based Complement Alternat Med 2017:5163575

De Oliveira Dembogurski DS, Trentin DS, Boaretto AG, Rigo GV, da Silva RC, Tasca T, Macedo AJ, Carollo CA, Silva DB (2018) Brown propolis-metabolomic innovative approach to determine compounds capable of killing Staphylococcus aureus biofilm and Trichomonas vaginalis. Food Res Int 111:661–673

De Sousa-Fontoura DM, Olinda RG, Viana GA, De Fm Costa KM, Batista JS, Serrano RM, Silva OM, Camara CA, Silva TM (2020) Wound healing activity and chemical composition of geopropolis from Melipona subnitida. Rev Bras 30:367–373

Dezmirean DS, Paşca C, Moise AR, Bobiş O (2020) Plant sources responsible for the chemical composition and main bioactive properties of poplar-type propolis. Plan Theory 10:22

Di Meo S, Venditti P (2020) Evolution of the knowledge of free radicals and other oxidants. Oxidative Med Cell Longev 2020:9829176

Diniz DP, Lorencini DA, Berretta AA, Cintra MA, Lia EN, Jordão AA Jr, Coelho EB (2020) Antioxidant effect of standardized extract of propolis (EPP-AF®) in healthy volunteers: a “Before and After” Clinical Study. Evid Based Complement Alternat Med 2020:7538232

do Amaral L, Caldas GR, Dos Santos NA, Parreira RL, Bastos JK, Dos Santos AC (2022) Baccharin from Brazilian green propolis induces neurotrophic signaling pathways in PC12 cells: potential for axonal and synaptic regeneration. Naunyn Schmiedeberg's Arch Pharmacol 395:659–672

dos Santos Thomazelli AP, Tomiotto-Pellissier F, da Silva SS, Panis C, Orsini TM, Cataneo AH, Miranda-Sapla MM, Custódio LA, Tatakihara VL, Bordignon J, Silveira GF (2017) Brazilian propolis promotes immunomodulation on human cells from American Tegumentar Leishmaniasis patients and healthy donors infected with L. braziliensis. Cell Immunol 11:22–27

Drescher N, Klein AM, Neumann P, Yañez O, Leonhardt SD (2017) Inside honeybee hives: impact of natural propolis on the ectoparasitic mite Varroa destructor and viruses. Insects 8:15

Dutra RP, de Sousa Jr MM, Mignoni MS, de Oliveira KG, Pereira EB, Figueredo AS, da Costa AA, Dias TG, Vasconcelos CC, Silva LA, Reis AS (2023) Brazilian Amazon red propolis: leishmanicidal activity and chemical composition of a new variety of red propolis. Metabolites 13:1027

Ebiloma GU, Ichoron N, Siheri W, Watson DG, Igoli JO, De Koning HP (2020) The strong antikinetoplastid properties of bee propolis: composition and identification of the active agents and their biochemical targets. Molecules 25:5155

El Hady FKA, Hegazi AG (2002) Egyptian propolis: 2. Chemical composition, antiviral and antimicrobial activities of East Nile Delta propolis. Z Naturf C 57(3–4):386–394

El-Allaky HS, Wahba NA, Talaat DM, Zakaria AS (2020) Antimicrobial effect of propolis administered through two different vehicles in high caries risk children: a randomized clinical trial. J Clin Pediatr Dent 44:289–295

El-Sakhawy M, Salama A, Tohamy H-a S (2023) Applications of propolis-based materials in wound healing. Arch Dermatol Res 316:61

Elumalai P, Muninathan N, Megalatha ST, Suresh A, Kumar KS, Jhansi N, Kalaivani K, Krishnamoorthy G (2022) An insight into anticancer effect of propolis and its constituents: a review of molecular mechanisms. Evid Based Complement Alternat Med 2022:5901191

Endo S, Hoshi M, Matsunaga T, Inoue T, Ichihara K, Ikari A (2018) Autophagy inhibition enhances anticancer efficacy of artepillin C, a cinnamic acid derivative in Brazilian green propolis. Biochem Biophys Res Commun 497:437–443

Esposito C, Garzarella EU, Bocchino B, D'Avino M, Caruso G, Buonomo AR, Sacchi R, Galeotti F, Tenore GC, Zaccaria V, Daglia M (2021) A standardized polyphenol mixture extracted from poplar-type propolis for remission of symptoms of uncomplicated upper respiratory tract infection (URTI): a monocentric, randomized, double-blind, placebo-controlled clinical trial. Phytomedicine 80:153368

Etebarian A, Alhouei B, Mohammadi-Nasrabadi F, Esfarjani F (2024) Propolis as a functional food and promising agent for oral health and microbiota balance: a review study, vol 12. Food Sci Nutr, p 5329

Falcão SI, Duarte D, Diallo M, Santos J, Ribeiro E, Vale N, Vilas-Boas M (2023) Improvement of the in vitro cytotoxic effect on HT-29 colon cancer cells by combining 5-fluorouacil and fluphenazine with green, red or brown propolis. Molecules 28:3393

Fernandes PM, Rosalen PL, Fernandes DT, Dias-Neto E, Alencar SM, Bueno-Silva B, Alves FD, Lopes MA (2022) Brazilian organic propolis for prevention and treatment of radiation-related oral acute toxicities in head and neck cancer patients: a double-blind randomized clinical trial. Front Pharmacol 13:973255

Fernandez-Calderon MC, Hernandez-Gonzalez L, Gomez-Navia C, Blanco-Blanco MT, Sanchez-Silos R, Lucio L, Perez-Giraldo C (2021) Antifungal and anti-biofilm activity of a new Spanish extract of propolis against Candida glabrata. BMC Complement Med Ther 21(1). https://doi.org/10.1186/s12906-021-03323-0

Fokt H, Pereira A, Ferreira A, Cunha A, Aguiar C (2010) How do bees prevent hive infections? The antimicrobial properties of propolis. In: Current research, technology and education. Topics in applied microbiology and microbial biotechnology, vol 1. World Scientific, Singapore, pp 481–493

Forma E, Bryś M (2021) Anticancer activity of propolis and its compounds. Nutrients 213:2594

Freitas AS, Costa M, Pontes O, Seidel V, Proenca F, Cardoso SM, Oliveira R, Baltazar F, Almeida-Aguilar C (2022) Selective cytotoxicity of portuguese propolis ethyl acetate fraction towards renal cancer cells. Molecules 27(13):4001. https://doi.org/10.3390/molecules27134001

Frión-Herrera Y, Gabbia D, Cuesta-Rubio O, De Martin S, Carrara M (2019) Nemorosone inhibits the proliferation and migration of hepatocellular carcinoma cells. Life Sci 235:116817

Frión-Herrera Y, Gabbia D, Scaffidi M, Zagni L, Cuesta-Rubio O, De Martin S, Carrara M (2020a) Cuban brown propolis interferes in the crosstalk between colorectal cancer cells and M2 macrophages. Nutrients 12:2040

Frión-Herrera Y, Gabbia D, Scaffidi M, Zagni L, Cuesta-Rubio O, De Martin S, Carrara M (2020b) The cuban propolis component nemorosone inhibits proliferation and metastatic properties of human colorectal cancer cells. Int J Mol Sci 21:1827

Fu YK, Wang BJ, Tseng JC, Huang SH, Lin CY, Kuo YY, Hour TC, Chuu CP (2022) Combination treatment of docetaxel with caffeic acid phenethyl ester suppresses the survival and the proliferation of docetaxel-resistant prostate cancer cells via induction of apoptosis and metabolism interference. J Biomed Sci 29:16

Gastaldello GH, Cazeloto ACV, Ferreira JC, Rodrigues DM, Bastos JK, Campo VL, Zoccal KF, Tefé-Silva C (2021) Green propolis compounds (Baccarin and p-coumaric acid) show beneficial effects in mice for melanoma induced by B16f10. Medicines (Basel) 8:20

Ghelflati A, Dehnavi Z, Yazdi AG, Khorasanchi Z, Raeisi-Dehkordi H, Ranjbar G (2021) The effects of propolis supplementation on metabolic parameters: a systematic review and meta-analysis of randomized controlled clinical trials. Avicenna J Phytomed 11:551

Ghisalberti E (1979) Propolis: a review. Bee World 60:59–84

Gomes KO, Messias da Silva LC, Dos Santos RD, Prado BA, da Silva Montes P, Silva Rodrigues LF, de Araújo MO, Bilac CA, Freire DO, Gris EF, Rodrigues da Silva IC (2024) Chemical characterization and antibacterial activities of Brazilian propolis extracts from Apis mellifera bees and stingless bees (Meliponini). PLoS One 19:e0307289

Gómez-Moracho T, Buendía-Abad M, Benito M, García-Palencia P, Barrios L, Bartolomé C, Maside X, Meana A, Jiménez-Antón MD, Olías-Molero AI, Alunda JM (2020) Experimental evidence of harmful effects of Crithidia mellificae and Lotmaria passim on honey bees. Int J Parasitol 50:1117–1124. https://doi.org/10.1016/j.ijpara.2020.06.009

Goncalves VC, Silva da Fonsêca V, de Paula Faria D, Izidoro MA, Berretta AA, de Almeida AC, Affonso Fonseca FL, Scorza FA, Scorza CA (2022) Propolis induces cardiac metabolism changes in 6-hydroxydopamine animal model: a dietary intervention as a potential cardioprotective approach in Parkinson's disease. Front Pharmacol 13:1013703

Granados-Pineda J, Uribe-Uribe N, García-López P, Ramos-Godinez MD, Rivero-Cruz JF, Pérez-Rojas JM (2018) Effect of pinocembrin isolated from Mexican brown propolis on diabetic nephropathy. Molecules 23:852

Gregory SR, Piccolo N, Piccolo MT, Piccolo MS, Heggers JP (2002) Comparison of propolis skin cream to silver sulfadiazine: a naturopathic alternative to antibiotics in treatment of minor burns. J Altern Complement Med 8:77–83

Guan R, Ma N, Liu G, Wu Q, Su S, Wang J, Geng Y (2023) Ethanol extract of propolis regulates type 2 diabetes in mice via metabolism and gut microbiota. J Ethnopharmacol 310:116385

Gunjal S, Pateel DG (2024) Comparative effectiveness of Propolis with chlorhexidine mouthwash on gingivitis–a randomized controlled clinical study. BMC Complementary Med Ther 24:154

Gurtner GC, Werner S, Barrandon Y, Longaker MT (2008) Wound repair and regeneration. Nature 453(7193):314–321

Hallajzadeh J, Milajerdi A, Amirani E, Attari VE, Maghsoudi H, Mirhashemi SM (2021) Effects of propolis supplementation on glycemic status, lipid profiles, inflammation and oxidative stress, liver enzymes, and body weight: a systematic review and meta-analysis of randomized controlled clinical trials. J Diabetes Metab Disord 20:831–843

Halliwell B, Gutteridge JM (2015) Free radicals in biology and medicine. Oxford University Press, New York

Hegazi AG, Al Guthami FM, Al Gethami A, Barakat AM (2017) Egyptian propolis 12: influence of propolis on cytokines of toxoplasma gondii infected rats. Int J Curr Microbiol Appl Sci 6:202–211

Hernandez J, Goycoolea FM, Quintero J, Acosta A, Castañeda M, Dominguez Z, Robles R, Vazquez-Moreno L, Velazquez EF, Astiazaran H (2007) Sonoran propolis: chemical composition and antiproliferative activity on cancer cell lines. Planta Med 73:1469–1474

Hossain R, Quispe C, Khan RA, Saikat ASM, Ray P, Ongalbek D, Yeskaliyeva B, Jain D, Smeriglio A, Trombetta D (2022) Propolis: an update on its chemistry and pharmacological applications. Chin Med 17:100

Huang W-J, Huang C-H, Wu C-L, Lin J-K, Chen Y-W, Lin C-L, Chuang S-E, Huang C-Y, Chen C-N (2007) Propolin G, a prenylflavanone, isolated from Taiwanese propolis, induces caspase-dependent apoptosis in brain cancer cells. J Agric Food Chem 55:7366–7376

Ibrahim RS, El-Banna AA (2021) Network pharmacology-based analysis for unraveling potential cancer-related molecular targets of Egyptian propolis phytoconstituents accompanied with molecular docking and in vitro studies. RSC Adv 11:11610–11626

Ibrahim H-IM, Thangavelu M, Khalifa A (2022) Honey-propolis-engineered collagen peptides as promising wound-healing matrix in mouse model. Molecules 27:7090

Ikeda NY, Ambrosio CM, Miano AC, Rosalen PL, Gloria EM, Alencar SM (2021) Essential oils extracted from organic propolis residues: an exploratory analysis of their antibacterial and antioxidant properties and volatile profile. Molecules 26:4694

Irigoiti Y, Navarro A, Yamul D, Libonatti C, Tabera A, Basualdo M (2021) The use of propolis as a functional food ingredient: a review. Trends Food Sci Technol 115:297–306

Ishihara M, Naoi K, Hashita M, Itoh Y, Suzui M (2009) Growth inhibitory activity of ethanol extracts of Chinese and Brazilian propolis in four human colon carcinoma cell lines. Oncol Rep 22:349

Jiang XS, Xie HQ, Li CG, You MM, Zheng YF, Li GQ, Chen X, Zhang CP, Hu FL (2020) Chinese propolis inhibits the proliferation of human gastric cancer cells by inducing apoptosis and cell cycle arrest. Evid Based Complement Alternat Med 2020:2743058

Kabala-Dzik A, Rzepecka-Stojko A, Kubina R, Jastrzebska-Stojko Z, Stojko R, Wojtyczka RD, Stojko J (2017) Comparison of two components of propolis: Caffeic Acid (CA) and Caffeic Acid Phenethyl Ester (CAPE) induce apoptosis and cell cycle arrest of breast cancer cells MDA-MB-231. Molecules 22:1554

Kamiya T, Nishihara H, Hara H, Adachi T (2012) Ethanol extract of Brazilian red propolis induces apoptosis in human breast cancer MCF-7 cells through endoplasmic reticulum stress. J Agric Food Chem 60:11065–11070

Kanazashi M, Iida T, Nakanishi R, Tanaka M, Ikeda H, Takamiya N, Maeshige N, Kondo H, Nishigami T, Harada T, Fujino H (2023) Brazilian propolis intake decreases body fat mass and oxidative stress in community-dwelling elderly females: a randomized placebo-controlled trial. Nutrients 15:364

Kiderman A, Torten R, Furst A, Reinus K (2001) Bi-lateral eosinophilic ulcers in an infant treated with propolis. J Dermatol Treat 12:29–31

Kitamura H (2019) Effects of propolis extract and propolis-derived compounds on obesity and diabetes: knowledge from cellular and animal models. Molecules 24:4394

Koloren Z, Ertürk Ö, Şekeroğlu ZA, Karaman Ü (2023) Amoebicidal and cytotoxic activity of propolis collected from different regions in Turkey on acanthamoeba castellanii trophozoites. Middle Black Sea J Health Sci 9:312–324

Kuo YY, Chen WT, Lin GB, Lu CH, Chao CY (2023) Study on the effect of a triple cancer treatment of propolis, thermal cycling-hyperthermia, and low-intensity ultrasound on PANC-1 cells. Aging (Albany NY) 15:7496–7512

Kurek-Górecka A, Keskin Ş, Bobis O, Felitti R, Górecki M, Otręba M, Stojko J, Olczyk P, Kolayli S, Rzepecka-Stojko A (2022) Comparison of the antioxidant activity of propolis samples from different geographical regions. Plan Theory 11:1203

Kurek-Górecka A, Kłósek M, Pietsz G, Balwierz R, Olczyk P, Czuba ZP (2024) Ethanolic extract of propolis and CAPE as cardioprotective agents against LPS and IFN-α stressed cardiovascular injury. Nutrients 16:627

Kuropatnicki AK, Szliszka E, Krol W (2013) Historical aspects of propolis research in modern times. Evid Based Complement Alternat Med 2013:964149

Laaroussi H, Bakour M, Ousaaid D, Aboulghazi A, Ferreira-Santos P, Genisheva Z, Teixeira JA, Lyoussi B (2020) Effect of antioxidant-rich propolis and bee pollen extracts against D-glucose induced type 2 diabetes in rats. Food Res Int 138:109802

Li F, Awale S, Tezuka Y, Kadota S (2008) Cytotoxic constituents from Brazilian red propolis and their structure–activity relationship. Bioorg Med Chem 16:5434–5440

Li F, Awale S, Tezuka Y, Esumi H, Kadota S (2010) Study on the constituents of Mexican propolis and their cytotoxic activity against PANC-1 human pancreatic cancer cells. J Nat Prod 73:623–627

Liu H, Li J, Yuan W, Hao S, Wang M, Wang F, Xuan H (2021) Bioactive components and mechanisms of poplar propolis in inhibiting proliferation of human hepatocellular carcinoma HepG2 cells. Biomed Pharmacother 144:112364

López-Valverde N, Pardal-Peláez B, López-Valverde A, Flores-Fraile J, Herrero-Hernández S, Macedo-de-Sousa B, Herrero-Payo J, Ramírez JM (2021) Effectiveness of propolis in the treatment of periodontal disease: updated systematic review with meta-analysis. Antioxidants 10:269

Machado Velho JC, França TA, Malagutti-Ferreira MJ, Albuquerque ER, Lívero FADR, Soares MR, Soares AEE, Ribeiro-Paes JT (2023) Use of propolis for skin wound healing: systematic review and meta-analysis. Arch Dermatol Res 315:943–955

MacInnis CI, Luong LT, Pernal SF (2023) A tale of two parasites: responses of honeybees infected with Nosema ceranae and Lotmaria passim. Sci Rep 13:22515

Magnavacca A, Sangiovanni E, Racagni G, Dell'agli M (2022) The antiviral and immunomodulatory activities of propolis: an update and future perspectives for respiratory diseases. Med Res Rev 42:897–945

Malekahmadi M, Pahlavani N, Heshmati J, Clayton ZS, Beigmohammadi MT, Navashenaq JG, Alavi-Naeini A (2023) Effect of propolis supplementation on oxidative stress markers: a systematic review of randomized controlled trials. J Herb Med 26:100679

Malik MS, Azam M, Basra MAR (2020) Impact of natural antioxidants on biological systems. Life Sci 4:139–162

Manginstar CO, Tallei TE, Niode NJ, Salaki CL, Hessel SS (2024) Therapeutic potential of propolis in alleviating inflammatory response and promoting wound healing in skin burn. Phytother Res 38:856–879

Martemucci G, Costagliola C, Mariano M, D'andrea L, Napolitano P, D'alessandro AG (2022) Free radical properties, source and targets, antioxidant consumption and health. Oxygen 2:48–78

Martin Ewert A, Simone-Finstrom M, Read Q, Husseneder C, Ricigliano V (2023) Effects of ingested essential oils and propolis extracts on honeybee (Hymenoptera: Apidae) health and gut microbiota. J Insect Sci 23:15

Martinello M, Mutinelli F (2021) Antioxidant activity in bee products: a review. Antioxidants 10:71

Martinotti S, Ranzato E (2015) Propolis: a new frontier for wound healing? Burns Trauma 3:1–7

Martins ML, de França Leite KL, Pacheco-Filho EF, de Miranda Pereira AF, Romanos MT, Maia LC, Fonseca-Gonçalves A, Padilha WW, Cavalcanti YW (2018) Efficacy of red propolis hydro-alcoholic extract in controlling Streptococcus mutans biofilm build-up and dental enamel demineralization. Arch Oral Biol 93:56–65

Martos I, Cossentini M, Ferreres F, Tomás-Barberán FA (1997) Flavonoid composition of Tunisian honeys and propolis. J Agric Food Chem 45:2824–2829

Matsumoto Y, Takahashi K, Sugioka Y, Inui K, Okano T, Mandai K, Yamada Y, Shintani A, Koike T (2021) Double-blinded randomized controlled trial to reveal the effects of Brazilian propolis intake on rheumatoid arthritis disease activity index; BeeDAI. PLoS One 16:e0252357

Mavri A, Abramovič H, Polak T, Bertoncelj J, Jamnik P, Smole Možina S, Jeršek B (2012) Chemical properties and antioxidant and antimicrobial activities of Slovenian propolis. Chem Biodivers 9:1545–1558

McLennan SV, Bonner J, Milne S, Lo L, Charlton A, Kurup S, Jia J, Yue DK, Twigg SM (2008) The anti-inflammatory agent Propolis improves wound healing in a rodent model of experimental diabetes. Wound Repair Regen 16:706–713

Meghalatha TS, Muninathan N (2022) Antitumor activity of withaferin-A and propolis in benz (a) pyrene-induced breast cancer. Bioinformation 18:841–844

Melo J, Soares DM, Santos S (2024) In-vitro antibacterial activity of propolis extract against periodontopathogenic bacteria: a systematic review. Rev Port Estomatol Med Dent Cir Maxilofac 65:3–14

Mendonça IC, Medeiros ML, Penteado RA, Parolia A, Porto IC (2013) An overview of the toxic effects and allergic reactions caused by propolis. Drugs 4:7

Mendonša R, Nascimento R, Fernandes A, Silva P Jr (2024) Antiviral action of aqueous extracts of propolis from Scaptotrigona aff. postica (Hymenoptera; Apidae) against Zica, Chikungunya, and Mayaro virus. Sci Rep 14:15289

Miłek M, Ciszkowicz E, Tomczyk M, Sidor E, Zaguła G, Lecka-Szlachta K, Pasternakiewicz A, Dżugan M (2022) The Study of Chemical Profile and Antioxidant Properties of Poplar-Type Polish Propolis Considering Local Flora Diversity in Relation to Antibacterial and Anticancer Activities in Human Breast Cancer Cells. Molecules 27:725

Mirsadraee M, Azmoon B, Ghaffari S, Abdolsamadi A, Khazdair MR (2021) Effect of Propolis on moderate persistent asthma: a phase two randomized, double blind, controlled clinical trial. Avicenna J Phytomed 11:22

Missima F, Pagliarone A, Orsatti C, Araújo J, Sforcin J (2010) The Effect of propolis on Th1/Th2 cytokine expression and production by melanoma-bearing mice submitted to stress. Phytother Res 24:1501–1507

Mohamed EK, Osman AA, Moghazy AM, Abdul-Rahman AAS (2021) Propolis protective effects against doxorubicin-induced multi-organ toxicity via suppression of oxidative stress, inflammation, apoptosis, and histopathological alterations in female albino rats. Biointerface Res Appl Chem 12:1762–1777

Mohamed WAS, Ismail NZ, Muhamad M, Omar EA, Abdul Samad N, Ooi JP, Mohamad S (2022) Q-TOF LC-MS compounds evaluation of propolis extract derived from Malaysian stingless bees, Tetrigona apicalis, and their bioactivities in breast cancer cell, MCF7. Saudi J Biol Sci 29:103403

Moncla BJ, Guevara PW, Wallace JA, Marcucci MC, Nor JE, Bretz WA (2012) The inhibitory activity of typified propolis against. Enterococcus Species. Zeitschrift für Naturforschung C 67:249–256

Motomura M, Kwon KM, Suh S-J, Lee Y-C, Kim Y-K, Lee I-S, Kim M-S, Kwon DY, Suzuki I, Kim C-H (2008) Propolis induces cell cycle arrest and apoptosis in human leukemic U937 cells through Bcl-2/Bax regulation. Environ Toxicol Pharmacol 26:61–67

Moura SAL, Negri G, Salatino A, Lima LDC, Dourado LPA, Mendes JB, Andrade SP, Ferreira MAND, Cara DC (2011) Aqueous extract of Brazilian green propolis: primary components, evaluation of inflammation and wound healing by using subcutaneous implanted sponges. Evid Based Complement Alternat Med 2011:748283

Mura A, Pusceddu M, Theodorou P, Angioni A, Floris I, Paxton RJ, Satta A (2020) Propolis consumption reduces Nosema ceranae infection of European honey bees (Apis mellifera). Insects 11:124

Murray SK, Kurkul CM, Mularo AJ, Hale VL, Adams RM, Johnson RM (2023) Antibacterial effects of propolis and brood comb extracts on the causative agent of European Foulbrood (Melissococcus plutonius) in honey bees (Apis mellifera). J Apic Res 62:813–822

Nani BD, Franchin M, Lazarini JG, Freires IA, da Cunha MG, Bueno-Silva B, de Alencar SM, Murata RM, Rosalen P.L (2018) Isoflavonoids from Brazilian red propolis down-regulate the expression of cancer-related target proteins: a pharmacogenomic analysis. Phytother Res 32:750–754

Naree S, Ponkit R, Chotiaroonrat E, Mayack CL, Suwannapong G (2021) Propolis extract and chitosan improve health of Nosema ceranae infected giant honey bees, Apis dorsata Fabricius, 1793. PathoGenetics 10:785

Nattagh-Eshtivani E, Pahlavani N, Ranjbar G, Gholizadeh Navashenaq J, Salehi-Sahlabadi A, Mahmudiono T, Nader Shalaby M, Jokar M, Nematy M, Barghchi H (2022) Does propolis have any effect on rheumatoid arthritis? A review study. Food Sci Nutr 10:1003–1020

Nazari-Bonab H, Jamilian P, Radkhah N, Zarezadeh M, Ebrahimi-Mameghani M (2023) The effect of propolis supplementation in improving antioxidant status: a systematic review and meta-analysis of controlled clinical trials. Phytother Res 37:3712–3723

Negri G, Duarte-Almeida JM, Figueiredo CA, Toledo-Piza AR, Tonelli FC, Barbosa TF, Mendonça R (2024) Antiviral activity of red propolis against herpes simplex virus-1. Braz J Pharm Sci 60:e23746

Neto MR, Tintino SR, da Silva AR, do Socorro Costa M, Boligon AA, Matias EF, de Queiroz Balbino V, Menezes IR, Coutinho HD (2017) Seasonal variation of Brazilian red propolis: Antibacterial activity, synergistic effect and phytochemical screening. Food Chem Toxicol 107:572–580

Neto JC, Paulino ET, Rodrigues AK, da Silva JC, Bernardino AC, dos Santos Oliveira JM, do Nascimento TG, de Souza Oliveira W, Santos JC, Smaniotto S, Ribeiro ÊA (2022) Cardioprotective effect of hydroalcoholic extract of Brazilian red propolis against isoproterenol-induced myocardial infarction in rats. Phytomed Plus 2:100190

Nguyen HX, Nguyen MTT, Nguyen NT, Awale S (2017) Chemical constituents of propolis from Vietnamese Trigona minor and Their antiausterity activity against the PANC-1 human pancreatic cancer cell line. J Nat Prod 80:2345–2352

Nikbaf-Shandiz M, Tutunchi H, Khoshbaten M, Bonab HN, Ebrahimi-Mameghani M (2022) Propolis supplementation in obese patients with non-alcoholic fatty liver disease: effects on glucose homeostasis, lipid profile, liver function, anthropometric indices and meta-inflammation. Food Funct 13:11568–11578

Nina N, Quispe C, Jiménez-Aspee F, Theoduloz C, Feresín GE, Lima B, Leiva E, Schmeda-Hirschmann G (2015) Antibacterial activity, antioxidant effect and chemical composition of propolis from the Región del Maule, Central Chile. Molecules 20:18144–18167

Niyomtham N, Koontongkaew S, Yingyongnarongkul BE, Utispan K (2021) Apis mellifera propolis enhances apoptosis and invasion inhibition in head and neck cancer cells. PeerJ 9:e12139

Nolkemper S, Reichling J, Sensch KH, Schnitzler P (2010) Mechanism of herpes simplex virus type 2 suppression by propolis extracts. Phytomedicine 17(2):132–138

Nör F, Nör C, Bento LW, Zhang Z, Bretz WA, Nör JE (2021) Propolis reduces the stemness of head and neck squamous cell carcinoma. Arch Oral Biol 125:105087

Novryantoro V, Septiara A, Tirtasari K, Suwanti LT, Edila R, Mufasirin M, Kuncorojakti S, Puspitasari H (2024) Immunomodulatory effect of propolis Trigona spp. in mice infected with Toxoplasma gondii. Open Vet J 14:2280

Nweze NE, Okoro HO, Al Robaian M, Omar RM, Tor-Anyiin TA, Watson DG, Igoli JO (2017) Effects of Nigerian red propolis in rats infected with Trypanosoma brucei brucei. Comp Clin Pathol 26:1129–1133

Olczyk P, Komosinska-Vassev K, Winsz-Szczotka K, Stojko J, Klimek K, Kozma EM (2013) Propolis induces chondroitin/dermatan sulphate and hyaluronic acid accumulation in the skin of burned wound. Evid Based Complement Alternat Med 2013:290675

Oršolić N (2022) Allergic inflammation: effect of propolis and its flavonoids. Molecules 27:6694

Oršolić N, Jazvinšćak Jembrek M (2022) Molecular and cellular mechanisms of propolis and its polyphenolic compounds against cancer. Int J Mol Sci 23:10479

Oršolić N, Knežević AH, Šver L, Terzić S, Bašić I (2004) Immunomodulatory and antimetastatic action of propolis and related polyphenolic compounds. J Ethnopharmacol 94(2-3):307–315. https://doi.org/10.1016/j.jep.2004.06.006

Oryan A, Alemzadeh E, Moshiri A (2018) Potential role of propolis in wound healing: biological properties and therapeutic activities. Biomed Pharmacother 98:469–483

Ożarowski M, Karpiński TM (2023) The effects of propolis on viral respiratory diseases. Molecules 28:359

Ożarowski M, Karpiński TM, Alam R, Łochyńska M (2022) Antifungal properties of chemically defined propolis from various geographical regions. Microorganisms 10:364

Padmavathi R, Senthilnathan P, Sakthisekaran D (2006) Therapeutic effect of propolis and paclitaxel on hepatic phase I and II enzymes and marker enzymes in dimethylbenz (a) anthracene-induced breast cancer in female rats. Comp Biochem Physiol Part C Toxicol Pharmacol 143:349–354

Pai JT, Lee YC, Chen SY, Leu YL, Weng MS (2018) Propolin C inhibited migration and invasion via suppression of EGFR-mediated epithelial-to-mesenchymal transition in human lung cancer cells. Evid Based Complement Altern Med 2018:7202548

Pakdaman H, Khodadadizadeh A, Bazmandegan G, Hasani H, Kazemi M (2024) The effect of propolis cream on phlebitis and catheter-related bloodstream infections: a double-blinded randomised clinical trial. J Herb Med 43:100829

Paredes-Guzman JF, Daugsch A, Park YK, Rhim JS (2007) Antiproliferation of human prostate cancer cells by ethanolic extracts of Brazilian propolis and its botanical origin. Int J Oncol 31:601–606

Picolotto A, Pergher D, Pereira GP, Machado KG, da Silva Barud H, Roesch-Ely M, Gonzalez MH, Tasso L, Figueiredo JG, Moura S (2019) Bacterial cellulose membrane associated with red propolis as phytomodulator: improved healing effects in experimental models of diabetes mellitus. Biomed Pharmacother 112:108640

Pobiega K, Kot AM, Przybył JL, Synowiec A, Gniewosz M (2023) Comparison of the chemical composition and antioxidant properties of propolis from urban apiaries. Molecules 28:6744

Popova MP, Chinou IB, Marekov IN, Bankova VS (2009) Terpenes with antimicrobial activity from Cretan propolis. Phytochemistry 70:1262–1271

Popova M, Giannopoulou E, Skalicka-Woźniak K, Graikou K, Widelski J, Bankova V, Kalofonos H, Sivolapenko G, Gaweł-Bęben K, Antosiewicz B, Chinou I (2017) Characterization and biological evaluation of propolis from Poland. Molecules 22:1159

Przybyłek I, Karpiński TM (2019) Antibacterial properties of propolis. Molecules 24:2047

Pusceddu M, Floris I, Mura A, Theodorou P, Cirotto G, Piluzza G, Bullitta S, Angioni A, Satta A (2018) The effects of raw propolis on Varroa-infested honey bee (Apis mellifera) workers. Parasitol Res 117:3527–3535

Pusceddu M, Annoscia D, Floris I, Frizzera D, Zanni V, Angioni A, Satta A, Nazzi F (2021) Honeybees use propolis as a natural pesticide against their major ectoparasite. Proc R Soc B 288:20212101
Rashidi MK, Mirazi N, Hosseini A (2016) Effect of topical mixture of honey, royal jelly and olive oil-propolis extract on skin wound healing in diabetic rats. Wound Med 12:6–9
Rebouças-Silva J, Celes FS, Lima JB, Barud HS, de Oliveira CI, Berretta AA, Borges VM (2017) Parasite killing of Leishmania (V) braziliensis by standardized propolis extracts. Evid Based Complement Alternat Med 2017:6067172
Righi AA, Alves TR, Negri G, Marques LM, Breyer H, Salatino A (2011) Brazilian red propolis: unreported substances, antioxidant and antimicrobial activities. J Sci Food Agric 91:2363–2370
Rivera-Yañez N, Rodriguez-Canales M, Nieto-Yañez O, Jimenez-Estrada M, Ibarra-Barajas M, Canales-Martinez MM, Rodriguez-Monroy MA (2018) Hypoglycaemic and antioxidant effects of propolis of Chihuahua in a model of experimental diabetes. Evid Based Complement Alternat Med 2018:4360356
Rivera-Yañez CR, Ruiz-Hurtado PA, Reyes-Reali J, Mendoza-Ramos MI, Vargas-Díaz ME, Hernández-Sánchez KM, Pozo-Molina G, Méndez-Catalá CF, García-Romo GS, Pedroza-González A, Méndez-Cruz AR (2022) Antifungal activity of Mexican propolis on clinical isolates of Candida species. Molecules 27:5651
Rodrigues Neto EM, Valadas LA, Lobo PL, Fonseca SG, Fechine FV, Lotif MA, Bandeira MA, Mendonça JF, de Mendonça KM, Fonteles MM (2021) Antimicrobial efficacy of propolis-containing varnish in children: a randomized and double-blind clinical trial. Evid Based Complement Alternat Med 5547081:1
Rojczyk E, Klama-Baryła A, Łabuś W, Wilemska-Kucharzewska K, Kucharzewski M (2020) Historical and modern research on propolis and its application in wound healing and other fields of medicine and contributions by Polish studies. J Ethnopharmacol 262:113159
Rossi A, Longo R, Russo A, Borrelli F, Sautebin L (2002) The role of the phenethyl ester of caffeic acid (CAPE) in the inhibition of rat lung cyclooxygenase activity by propolis. Fitoterapia 73:S30–S37
Rouibah H, Kebsa W, Lahouel M, Zihlif M, Ahram M, Aburmaileh B, Mansour Al Shhab MA, Al-Ameer HJ, Mustafa E (2021) Algerian propolis: between protection of normal cells and potentialisation of the anticancer effects of doxorubicin against breast cancer cells via P-glycoprotein inhibition and cell cycle arrest in the S phase. J Physiol Pharmacol 72(2). https://doi.org/10.26402/jpp.2021.2.09
Saelao P, Borba RS, Ricigliano V, Spivak M, Simone-Finstrom M (2020) Honeybee microbiome is stabilized in the presence of propolis. Biol Lett 16:2020000
Sajjadi SS, Bagherniya M, Soleimani D, Siavash M, Askari G (2023) Effect of propolis on mood, quality of life, and metabolic profiles in subjects with metabolic syndrome: a randomized clinical trial. Sci Rep 13:4452
Salama SA, Essam D, Tagyan AI, Farghali AA, Khalil EM, Abdelaleim YF, Hozzein WN, Mubarak M, Nasr FA, Eweis AA, Al-Zharani M (2024) Novel composite of nano zinc oxide and nano propolis as antibiotic for antibiotic-resistant bacteria: a promising approach. Sci Rep 14:20894
Salami F, Mohebbati R, Hosseinian S, Shahraki S, Hossienzadeh H, Rad AK (2024) Propolis and its therapeutic effects on renal diseases: a review. Iran J Basic Med Sci 27:383
Salatino A (2022) Perspectives for uses of propolis in therapy against infectious diseases. Molecules 27:4594
Salehi A, Hosseini SM, Kazemi S (2022) Antioxidant and anticarcinogenic potentials of propolis for dimethylhydrazine-induced colorectal cancjer in wistar rats. Biomed Res Int 2022:8497562
Sani L, Cardinault N, Astier J, Darmon P, Landrier JF (2023) Poplar Propolis improves insulin homeostasis in non-diabetic insulin-resistant volunteers with obesity: a crossover randomized controlled trial. Antioxidants 12:1481
Santos LM, Fonseca MS, Sokolonski AR, Deegan KR, Araújo RP, Umsza-Guez MA, Barbosa JD, Portela RD, Machado BA (2020) Propolis: types, composition, biological activities, and veterinary product patent prospecting. J Sci Food Agric 100:1369–1382

Santos LA, Rosalen PL, Dias NA, Grisolia JC, Gomes BJ, Blosfeld-Lopes L, Ikegaki M, de Alencar SM, Burger E (2021) Brazilian red propolis shows antifungal and immunomodulatory activities against Paracoccidioides brasiliensis. J Ethnopharmacol 277:114181

Šarić G, Vahčić N, Bursać Kovačević D, Putnik P (2020) The changes of flavonoids in honey during storage. PRO 8:943

Sberna G, Biagi M, Marafini G, Nardacci R, Biava M, Colavita F, Piselli P, Miraldi E, D'Offizi G, Capobianchi MR, Amendola A (2022) In vitro evaluation of antiviral efficacy of a standardized hydroalcoholic extract of poplar type propolis against SARS-CoV-2. Front Microbiol 13:799546

Schmidt EM, Stock D, Chada FJ, Finger D, Christine Helena Frankland Sawaya A, Eberlin MN, Felsner ML, Quináia SP, Monteiro MC, Torres YR (2014) A comparison between characterization and biological properties of Brazilian fresh and aged propolis. Biomed Res Int 2014:257617

Schnitzler P, Neuner A, Nolkemper S, Zundel C, Nowack H, Sensch KH, Reichling J (2010) Antiviral activity and mode of action of propolis extracts and selected compounds abstract. Phytother Res 24(S1). https://doi.org/10.1002/ptr.v24.1s; https://doi.org/10.1002/ptr.2868

Sena-Lopes A, Bezerra FS, das Neves RN, de Pinho RB, Silva MT, Savegnago L, Collares T, Seixas F, Begnini K, Henriques JA, Ely MR (2018) Chemical composition, immunostimulatory, cytotoxic and antiparasitic activities of the essential oil from Brazilian red propolis. PLoS One 13:e0191797

Sforcin J (200 7) Propolis and the immune system: a review. J Ethnopharmacol 113:1–14

Sforcin JM (2016) Biological properties and therapeutic applications of propolis. Phytother Res 30:894–905

Shahab-Navaei F, Asoodeh A (2023) Synthesis of optimized propolis solid lipid nanoparticles with desirable antimicrobial, antioxidant, and anti-cancer properties. Sci Rep 13:18290

Shanahan M, Simone-Finstrom M, Tokarz P, Rinkevich F, Read QD, Spivak M (2024) Thinking inside the box: restoring the propolis envelope facilitates honey bee social immunity. PLoS One 19:e0291744

Shen MH, Liu CY, Chang KW, Lai CL, Chang SC, Huang CJ (2023) Propolis has an anticancer effect on early stage colorectal cancer by affecting epithelial differentiation and gut immunity in the tumor microenvironment. Nutrients 15:4494

Shi P, Du T, Meng F, Xie S, Han S, Yang J, Zhang Y, Sun Y, Gong J, Yao H (2023) Ethanol extract of propolis alleviates diabetic cardiomyopathy via JAK2/STAT3 signaling pathway. J Funct Foods 107:105688

Shimizu T, Hino A, Tsutsumi A, Park YK, Watanabe W, Kurokawa M (2008) Anti-influenza virus activity of propolis in vitro and its efficacy against influenza infection in mice. Antivir Chem Chemother 19:7–13

Shimizu T, Takeshita Y, Takamori Y, Kai H, Sawamura R, Yoshida H, Watanabe W, Tsutsumi A, Park YK, Yasukawa K (2011) Efficacy of Brazilian propolis against herpes simplex virus type 1 infection in mice and their modes of antiherpetic efficacies. Evid Based Complement Alternat Med 2011:976196

Shoker RM, Al-Shammery WH, Al-Aidy S (2023) A review article: free radical and replacement synthetic antioxidant by natural antioxidant. J Res Appl Sci Biotechnol 2:206–211

Siheri W, Ebiloma GU, Igoli JO, Gray AI, Biddau M, Akrachalanont P, Alenezi S, Alwashih MA, Edrada-Ebel R, Muller S, Lawrence CE (2019) Isolation of a novel flavanonol and an alkylresorcinol with highly potent anti-trypanosomal activity from Libyan propolis. Molecules 24:1041

Silva JC, Rodrigues S, Feás X, Estevinho LM (2012) Antimicrobial activity, phenolic profile and role in the inflammation of propolis. Food Chem Toxicol 50:1790–1795

Silva H, Francisco R, Saraiva A, Francisco S, Carrascosa C, Raposo A (2021) The cardiovascular therapeutic potential of propolis A comprehensive review. Biology 10:27

Silveira MA, De Jong D, Berretta AA, dos Santos Galvão EB, Ribeiro JC, Cerqueira-Silva T, Amorim TC, da Conceicao LF, Gomes MM, Teixeira MB, de Souza SP (2021) Efficacy of Brazilian green propolis (EPP-AF®) as an adjunct treatment for hospitalized COVID-19 patients: a randomized, controlled clinical trial. Biomed Pharmacother 138:111526

Silvia-Carvalho R, Baltazar F, Almeida-Aguiar C (2015) Propolis: A complex natural product with a plethora of biological activities that can be explored for drug development. Evid Based Complement Alternat Med 20151-29.https://doi.org/10.1155/2015/206439

Simone-Finstrom M, Spivak M (2010) Propolis and bee health: the natural his Simone-Finstrom MD, Spivak M (2012) Increased resin collection after parasite challenge: a case of self-medication in honey bees? PLoS One 7:e34601

Simone-Finstrom M, Borba RS, Wilson M, Spivak M (2017) Propolis counteracts some threats to honeybee health. Insects 8:46

Sokolonski AR, Fonseca MS, Machado BA, Deegan KR, Araújo RP, Umsza-Guez MA, Meyer R, Portela RW (2021) Activity of antifungal drugs and Brazilian red and green propolis extracted with different methodologies against oral isolates of Candida spp. BMC Complementary Med Ther 21:1–4

Soleimani D, Miryan M, Hadi V, Gholizadeh Navashenaq J, Moludi J, Sayedi SM, Bagherniya M, Askari G, Nachvak SM, Sadeghi E, Rashidi AA (2021) Effect of propolis supplementation on athletic performance, body composition, inflammation, and oxidative stress following intense exercise: a triple-blind randomized clinical trial. Food Sci Nutr. 9:3631–3640

Starzyk J, Scheller S, Szaflarski J, Moskwa M, Stojko A (1977) Biological properties and clinical application of propolis. II. Studies on the antiprotozoan activity of ethanol extract of propolis. Arzneimittelforschung 27:1198

Stojko M, Wolny D, Włodarczyk J (2021) Nonwoven releasing propolis as a potential new wound healing method—a review. Molecules 26:5701

Su Z, Lin J, Prewett M, Goldstein N, Fisher P (1994) Apoptosis mediates the selective toxicity of caffeic acid phenethyl ester (CAPE) toward oncogene-transformed rat embryo fibroblast cells. Anticancer Res 15:1841–1848

Šuran J, Cepanec I, Mašek T, Radić B, Radić S, Tlak Gajger I, Vlainić J (2021) Propolis extract and its bioactive compounds—from traditional to modern extraction technologies. Molecules 26:2930

Suwannapong G, Maksong S, Phainchajoen M, Benbow ME, Mayack C (2018) Survival and health improvement of Nosema infected Apis florea (Hymenoptera: Apidae) bees after treatment with propolis extract. J Asia Pac Entomol 21:437–444

Suzuki A, Sakamoto Y (2024) Molecular detection of Lotmaria passim in intestine of Japanese honeybees (Apis cerana japonica). Entomol Sci 27:e12571

Szliszka E, Czuba ZP, Bronikowska J, Mertas A, Paradysz A, Krol W (2011a) Ethanolic extract of propolis augments TRAIL-induced apoptotic death in prostate cancer cells. Evid Based Complement Alternat Med 2011:1–11

Szliszka E, Zydowicz G, Janoszka B, Dobosz C, Kowalczyk-Ziomek G, Krol W (2011b) Ethanolic extract of Brazilian green propolis sensitizes prostate cancer cells to TRAIL-induced apoptosis. Int J Oncol 38:941–953

Tao L, Chen X, Zheng Y, Wu Y, Jiang X, You M, Li S, Hu F (2021) Chinese propolis suppressed pancreatic cancer Panc-1 cells proliferation and migration via Hippo-YAP pathway. Molecules 26:2803

Tasca KI, Conte FL, Correa CR, Santiago KB, de Oliveira CE, Manfio VM, Garcia JL, Berretta AA, Sartori AA, da Silva Honorio M, do Rosário Souza L (2023) Propolis consumption by asymptomatic HIV-individuals: Better redox state? A prospective, randomized, double-blind, placebo-controlled trial. Biomed Pharmacother 162:114626

Teerasripreecha D, Phuwapraisirisan P, Puthong S, Kimura K, Okuyama M, Mori H, Kimura A, Chanchao C (2012) In vitro antiproliferative/cytotoxic activity on cancer cell lines of a cardanol and a cardol enriched from Thai Apis mellifera propolis. BMC Complement Altern Med 12(1):1

Thazhathuputhenpurayil SM, Natarajan M (2022) Effect of propolis on membrane bound enzymes linked with breast cancer. Bioinformation 18:1192–1195

Toutou Z, Fatmi S, Chibani N, Pokajewicz K, Skiba M, Wieczorek PP, Iguerouada M (2024) Exploring the therapeutic potential of algerian propolis: GC/MS profiling, protective inclusion complex, and in silico evaluation against SARS-CoV-2 main proteases. Pept Sci 2024:e24381

Tsuda T, Kumazawa S (2021) Propolis: chemical constituents, plant origin, and possible role in the prevention and treatment of obesity and diabetes. J Agric Food Chem 69:15484–15494

Turcatto AP, Lourenço AP, De Jong D (2018) Propolis consumption ramps up the immune response in honey bees infected with bacteria. Apidologie 2049:287–296

Türedi S, Çelik H, Dağlı ŞN, Taşkın S, Şeker U, Deniz M (2024) An examination of the effects of propolis and quercetin in a rat model of streptozotocin-induced diabetic peripheral neuropathy. Curr Issues Mol Biol 46:1955–1974

Tutunchi H, Arefhosseini S, Ebrahimi-Mameghani M (2023) Clinical effectiveness of α-lipoic acid, myo-inositol and propolis supplementation on metabolic profiles and liver function in obese patients with NAFLD: a randomized controlled clinical trial. Clin Nutr ESPEN 54:412–420

Umthong S, Phuwapraisirisan P, Puthong S, Chanchao C (2011) In vitro antiproliferative activity of partially purified Trigona laeviceps propolis from Thailand on human cancer cell lines. BMC Complement Altern Med 11:1

Uzel A, Önçağ Ö, Çoğulu D, Gençay Ö (2005) Chemical compositions and antimicrobial activities of four different Anatolian propolis samples. Microbiol Res 160:189–195. https://doi.org/10.1016/j.micres.2005.01.002

Valente MJ, Baltazar AF, Henrique R, Estevinho L, Carvalho M (2011) Biological activities of Portuguese propolis: protection against free radical-induced erythrocyte damage and inhibition of human renal cancer cell growth in vitro. Food Chem Toxicol 49:86–92

Valko M, Leibfritz D, Moncol J, Cronin MT, Mazur M, Telser J (2007) Free radicals and antioxidants in normal physiological functions and human disease. Int J Biochem Cell Biol 39:44–84

Vilhelmova-Ilieva NM, Nikolova IN, Nikolova NY, Petrova ZD, Trepechova MS, Holechek DI, Todorova MM, Topuzova MG, Ivanov IG, Tumbarski YD (2023) Antiviral potential of specially selected Bulgarian propolis extracts: in vitro activity against structurally different viruses. Life 13:1611

Wagh VD (2013) Propolis: a wonder bees product and its pharmacological potentials. Adv Pharmacol Pharm Sci 2013:308249

Wang H, Tang C (2017) Galangin suppresses human laryngeal carcinoma via modulation of caspase-3 and AKT signaling pathways. Oncol Rep 38:703–714

Wang X, Sankarapandian K, Cheng Y, Woo SO, Kwon HW, Perumalsamy H, Ahn YJ (2016) Relationship between total phenolic contents and biological properties of propolis from 20 different regions in South Korea. BMC Complement Altern Med 16:1–2

Wang Q, Chen R, Tang Q, Chen JY, Wang Y, Sui DJ, Liu AD (2022) Flavonoid extract from propolis provides cardioprotection following myocardial infarction by activating PPAR-γ. Evid Based Complement Alternat Med 2022:1333545

Wang C, Zhao H, Xu K, Du Y, Liu J, Wang J, Jiang Y (2023) Fecal metabolomics reveals the positive effect of ethanol extract of propolis on T2DM mice. Food Sci Human Wellness 12:161–172

Wezgowiec J, Wieczynska A, Wieckiewicz W, Kulbacka J, Saczko J, Pachura N, Wieckiewicz M, Gancarz R, Wilk KA (2020) Polish propolis-chemical composition and biological effects in tongue cancer cells and macrophages. Molecules 25:2426

Wieczorek PP, Hudz N, Yezerska O, Horčinová-Sedláčková V, Shanaida M, Korytniuk O, Jasicka-Misiak I (2022) Chemical variability and pharmacological potential of propolis as a source for the development of new pharmaceutical products. Molecules 27:1600

Wojtyczka RD, Kępa M, Idzik D, Kubina R, Kabała-Dzik A, Dziedzic A, Wąsik TJ (2013) In vitro antimicrobial activity of ethanolic extract of Polish propolis against biofilm forming Staphylococcus epidermidis strains. Evid Based Complement Alternat Med 2013:590703

Xu X, Yang B, Wang D, Zhu Y, Miao X, Yang W (2020) The chemical composition of brazilian green propolis and its protective effects on mouse aortic endothelial cells against inflammatory injury. Molecules 25:4612

Xue M, Liu Y, Xu H, Zhou Z, Ma Y, Sun T, Liu M, Zhang H, Liang H (2019) Propolis modulates the gut microbiota and improves the intestinal mucosal barrier function in diabetic rats. Biomed Pharmacother 118:109393

Yang J, Pi A, Yan L, Li J, Nan S, Zhang J, Hao Y (2022) Research progress on therapeutic effect and mechanism of propolis on wound healing. Evid Based Complement Alternat Med 2022:5798941

Yildirim A, Gulbol Duran G, Duran N, Jenedi K, Sezgin Bolgul B, Miraloglu M, Muz M (2016) Antiviral activity of Hatay Propolis against replication of herpes simplex virus Type 1 and Type 2. Med Sci Mon 22422-22430. https://doi.org/10.12659/MSM.897282

Zabaiou N, Fouache A, Trousson A, Baron S, Zellagui A, Lahouel M, Lobaccaro JA (2017) Biological properties of propolis extracts: something new from an ancient product. Chem Phys Lipids 207:214–222

Zulhendri F, Chandrasekaran K, Kowacz M, Ravalia M, Kripal K, Fearnley J, Perera CO (2021) Antiviral, antibacterial, antifungal, and antiparasitic properties of propolis: a review. Food Secur 10:1360

Zullkiflee N, Taha H, Usman A (2022) Propolis: Its role and efficacy in human health and diseases. Molecules 27:6120

Part III
Royal Jelly

Chapter 8
Chemical Composition of Royal Jelly

Xiaofeng Xue, Liming Wu, and Kai Wang

Contents

Abstract The chemical composition of royal jelly (RJ) is highly complex, consisting of water (50–70%), proteins (9–18%), carbohydrates (11–23%), lipids (3–8%), and trace amounts of vitamins, minerals, and other compounds including several nucleotides as free bases and phosphates such as adenosine 5′-monophosphate (AMP), adenosine 5′-diphosphate (ADP), and adenosine 5′-triphosphate (ATP). Carbohydrates are primarily simple sugars like glucose and fructose. Key proteins include major royal jelly proteins (MRJPs), which support queen bee development and longevity. Lipids, particularly 10-hydroxy-2-decenoic acid (10-HDA), are unique to royal jelly and exhibit antimicrobial and anti-inflammatory properties. Royal jelly also contains B-complex vitamins, vitamin C, and minerals such as calcium, potassium, and zinc. Bioactive components like peptides and phenolic compounds contribute to its antioxidant, immunomodulatory, and neuroprotective effects. These properties make royal jelly a valuable substance in nutritional and medical research, with potential applications in health and disease management. Its composition, however, can vary based on environmental factors and bee species.

Keywords Royal jelly · Chemical composition · Proteins · Lipids · Bioactive components

X. Xue (✉) · L. Wu · K. Wang
State Key Laboratory of Resource Insect, Institute of Apicultural Research, Chinese Academy of Agricultural Science, Beijing, China

J. M. Alvarez-Suarez (ed.), *Bee Products – Chemical and Biological Properties*,
https://doi.org/10.1007/978-3-031-89049-9_8

8.1 Sugar Composition

Sugars consist mostly of fructose and glucose, with a chemical composition similar to that of honey. The oldest reports reveal that the average global composition of the glycid fraction in royal jelly (RJ) ranged from 7.5% to 15% (Rembold 1965; Rembols and Lackner 1978). Earlier gas chromatography has been used to identify simple sugars such as glucose, fructose, and saccharose in royal jelly, along with other neutral constituents (Lercker 1986). Generally, the total content of fructose and glucose accounts for 90% of the total sugars contained in royal jelly. The sucrose content varies considerably from one sample to another. Other sugars seen in much smaller quantities are maltose, trehalose, melibiose, ribose, and erlose (Lercker et al. 1992). Based on HPLC analysis of RJ, fructose and glucose content varies between 4.06% and 6.4%, and 5.07% and 8.2%, respectively, with a relatively high content of sucrose (0.8% and 3.6%). Maltose and trehalose were present in smaller amounts (0.3% for both) (Popescu et al. 2009).

In authentic French royal jelly samples, the minimal, maximal, and average values for the three main sugars were reported to be 2.3–7.6% for fructose, average 4.9%; 2.9–8.1% for glucose, average 5.5%; and not detected to 1.8% for sucrose, average 0.5%. So, the total sugar content was between 7.9% and 17.9% with an average of 13.1% (Wytrychowski et al. 2013).

The other most abundant sugars found in French royal jelly are maltose (from not detected level to 1.0%, average 0.3%), galactose (from not detected level to 0.3%, average 0.2%), and trehalose (not detected to 0.5%, average 0.1%) (Wytrychowski et al. 2013). Another earlier study on French royal jelly reported similar findings for fructose (2.3–7.3%, mean 5.0%), glucose (2.9–8.0%, mean 5.5%), sucrose (not detected to 1.7%, mean 0.4%) and total sugar content (8.1–16.8%, mean 12.4%) (Daniele and Casabianca 2012). It has been indicated that the content of sucrose and erlose in French RJ is less than 1.7 and 0.3%, respectively, but they reach 3.9 and 2.0% in some commercial samples (Daniele and Casabianca 2012). Another study reported similar findings for French royal jelly, with sucrose and erlose contents of less than 1.8 and 0.4%, respectively, and the values rise up to 3.9 and 2.0% in some commercial samples and 7.7 and 1.7% in RJs produced by sugarcane feeding (Marine Wytrychowski et al. 2013). Italian royal jelly showed minor variations compared to French RJ, i.e. fructose (2.3–6.9%, mean 4.6%), glucose (3.7–8.2%, mean 5.8%), sucrose (<0.1–2.1%, mean 1.0%) and total sugar content (6.9–16.0%, mean 11.4%) (Sesta 2006). Recent evidence from Bulgarian pure royal jelly showed that fructose (4.24–5.35% mean 4.88%) was higher than glucose (2.70–4.15% mean 3.46%), and sucrose was lesser (0.59–2.05% mean 1.53%) (Balkanska et al. 2014; Balkanska 2018). In Anatolian royal jelly samples, glucose and fructose were the main sugars, whereas, maltose, trehalose, and melibiose were less than 1.0% in all samples (Kolayli et al. 2016). In Croatian royal jelly, fructose (2.82–5.82% mean 4.29%) was slightly lesser than glucose (2.56–6.36% mean 4.63%), and sucrose level was (0.70–4.46% mean 2.14%) (Flanjak et al. 2019).

8.2 Protein Content

Proteins account for >50% of RJ dry weight and almost 80% of soluble RJ proteins are major royal jelly proteins (MRJPs), as they are considered to play specific physiological roles in queen honeybee development (Schmitzova et al. 1998; Nozaki et al. 2012, Ramanathana et al. 2018). Studies have shown that the MRJP family, of which there are nine members (MRJP1, MRJP2, MRJP3, MRJP4, MRJP5, MRJP6, MRJP7, MRJP8 and MRJP9), accounts for most of the soluble proteins in RJ (31%) (Albert and Klaudiny 2004; Drapeau et al. 2006; Schonleben et al. 2007; Tamura et al. 2009; Ramadan and Al-Ghamdi 2012; Buttstedt et al. 2013). MRJP1 to 8 have been characterized by the cloning and sequencing of their respective cDNAs (Albert and Klaudiny 2004). French RJ had a minimal and maximal protein content of 11.4 and 16.9%, respectively, with a mean protein content of 13.9% (Wytrychowski et al. 2013).

MRJP1 is a weak acidic glycoprotein (pI 4.9–6.3, 55–57 kDa as monomer) which forms a 350 kDa hexamer after gel-filtration chromatography (Kimura et al. 1995). Another study reported that MRJPs are 280 kDa proteins that can be separated as 55 kDa bands by both reducing and non-reducing SDS-PAGE, and also separated into several spots by 2-DE (pH 4.2–6.5) (Tamura et al. 2009). The two different oligomerization processes tend to be pH dependent with the pentamer formed at pH 6 and the 7 and hexamer at pH 8 (Cruz et al. 2011). These studies also indicated that the MRJP1 oligomer is comprised of 55 kDa protein subunits bound by non-covalent bonds (Santos et al. 2005; Tamura et al. 2009). The binding pattern of the MRJP1 oligomer based on Superose 12 HPLC and SDS–PAGE results indicated that the oligomer is a pentamer of MRJP1 monomers (Tamura et al. 2009; Xu and Gao 2013). MRJP1 was thought to be made up of an oligomer complex linked by noncovalent bonds under natural conditions (Tamura et al. 2009). MRJP1 contains a relatively high content of essential amino acids (48%), as it is considered a potential ingredient in functional foods (Judova et al. 2004). The secondary structure of MRJP1, based on circular dichroism (CD) measurements, consists of 9.6% α-helices, 38.3% β-sheets, and 20% β-turns (Cruz et al. 2011).

Soluble RJ proteins were completely separated into five peaks at approximately 640, 280, 100, 72, and 4.5 kDa by size-exclusion HPLC on a Superose 12 column (Tamura et al. 2009). Among these, the 280 and 72 kDa corresponded to the major components of the RJ proteins, but the peak intensity of the 280 kDa tends to differ across the original RJ samples (Tamura et al. 2009). Another study showed the elimination of water-soluble proteins in RJ from queen larva analyzed by SDS-PAGE into six bands at 85, 79, 68, 60, 56 and 49 kDa, representing glucose oxidase, MRJP5, MRJP3, MRJP4, MRJP1, and MRJP2, respectively (Xu and Gao 2013). The study also demonstrated that the main components of queen larvae proteins were mainly MRJP1 and MRJP2, but they were quite distinct from the MRJPs of RJ in composition and structure (Xu and Gao 2013).

The open reading frame (ORF) of the AmMRJP1 gene from the western honeybee is a fragment of 1299 bp, which encodes a protein of 422 amino acid residues with a predicted molecular weight of 48 kDa for a deglycosylated protein (Tamura et al. 2009). Mature MRJP1 cDNA of the Chinese honeybee (*Apis cerana cerana* MRJP1, or AccMRJP1) expressed in *Pichia pastoris* showed that recombinant AccMRJP1 was identical in molecular weight to the glycosylated AmMRJP1 from the Western honeybee (Apis mellifera) (Shen et al. 2010). The molecular weight of the recombinant AccMRJP1 decreased from 57 to 48 kDa after deglycosylation, indicating that AccMRJP1 is in the glycosylated form. MRJP2, MRJP3, MRJP4, and MRJP5 are glycoproteins of 49, 60–70, 60, and 80 kDa, respectively (Li et al. 2007; Schmitzova et al. 1998). MRJP2 to 5 exist mainly in the basic pI range (pI 6.3–8.3) (Li et al. 2007; Sano et al. 2004; Santos et al. 2005; Schonleben et al. 2007). Tamura et al. (2009) reported that the RJ protein of the 72 kDa peak could be separated into several spots that were focused at 51 kDa and ranged from pH 6.2 to 7.9; they were identified as MRJP2. The wide pI range of the MRJP proteins is generally due to post-translational modification by glycosylation and/or phosphorylation (Tamura et al. 2009). Studies indicate that methylation is the most important post-translational modification of MRJPs, which triggers polymorphism of MRJP1 to 5 in the RJ proteome (Zhang et al. 2012). Only one study on RJ of Africanized honeybees was able to identify MRJP8 (Santos et al. 2005). Both MRJP8 and 9 are rare in RJ but were detected in honeybee venom (Peiren et al. 2005, 2008).

The physiological functions of RJ proteins have been widely reported (Mureşan et al. 2022; Collazo et al. 2021). They are known to have proliferation activity in cell lines derived from humans, rats, and insects. Studies have shown that RJ proteins could stimulate the proliferation of mouse osteoblast-like MC3T3-E1 cells (Narita et al. 2006), U-937 human myeloid cell lines (Watanabe et al. 1998), and human monocytes (Kimura et al. 2003). Recombinant AccMRJP1 from the Chinese honey bees significantly stimulated Tn-5B-4 cell growth, similar to Western honey bee (*Apis mellifera*) RJ and fetal bovine serum (FBS), and affected cell shape and adhesion to the substrate (Shen et al. 2010). Natural MRJP1 from the Western honey bee (*Apis mellifera*) stimulated the proliferation of rat hepatocytes (Kamakura et al. 2001), promote hepatocyte DNA synthesis, and increase albumin production in the absence of FBS (Kamakura et al. 2001; Kamakura 2002). It has been reported that crude RJ protein could inhibit the bisphenol A-induced proliferation of human breast cancer cell lines (Nakaya et al. 2007). Okamoto et al. (2003) identified that MRJP3 could suppress IL-4 production, showing potent antiallergic properties in a mouse model of immediate hypersensitivity. In their in vitro studies, MRJP3 not only suppressed the production of IL-4, but also that of IL-2 and IFN-g by T cells, concomitant with inhibition of proliferation. MRJP3-mediated suppression of IL-4 production was also observed when lymph node cells from OVA/alum-immunized mice were stimulated with OVA plus antigen-presenting cells (Okamoto et al. 2003). In spite of its antigenicity, as MRJP3 itself is an extraneous foreign protein, intraperitoneal administration of MRJP3 inhibited serum anti-OVA IgE and IgG1 production in immunized mice. Designing MRJP3-derived antiallergic peptides by identifying the associated polypeptide regions would be of great clinical significance (Okamoto et al. 2003). A recent study showed that MRJP1 could hinder vascular smooth muscle cell proliferation by reducing the supply of energy and genetic material, revealing that the function and mechanism correlated with the anti-hypertensive activity (Fan et al. 2016). MRJP1 and MRJP2 can stimulate mouse macrophages to release TNF-α (Simuth et al. 2004). It was found that honey can stimulate TNF-α secretion from murine macrophages, whereas deproteinized honey had no effect on the release of TNF-α, indicating that the immunostimulatory effect of honey is based on its RJ protein content (Majtan et al. 2006). MRJP1 and MRJP3 could induce oxidation of low-density lipoprotein (LDL) and oxidised LDL and inhibit macrophage proliferation (Sato et al. 2021). MRJP2 and its isoform X1 exhibited potent anticancer effect in HepG2 cells via induction of caspase-dependent apoptosis and controlling the expression of both Bcl-2 and p53 (Abu-Serie et al. 2019). MRJPs could reduce caspase-3 activity and oxidative stress-induced cell apoptosis with increased cell viability. MRJPs (1 to 7) could exhibit 1,1-diphenyl-2-picrylhydrazyl radical-scavenging activity and prevent oxidative DNA damage (Park et al. 2020). MRJPs do have antibacterial effects. Glycosylated MRJP1 had antibacterial effect against vancomycin-resistant Enterococci and methicillin-resistant Staphylococcus aureus (Brudzynski et al. 2015). MRJP1 could significantly inhibit the growth of *Paenibacillus alvei*, *Brevibacillus laterosporus*, and *Melissococcus plutonius* (Vezeteu et al. 2017). N-glycosylated MRJP2 isolated from RJ inhibits the growth of the Paenibacillus larvae (Feng et al. 2021). Jelleins, the peptides derived from MRJP1, exhibited an antimicrobial effect (Fratini et al. 2016). Analog of jelleine-1 enriched in arginine and leucine exhibited antimicrobial activity against both gram-negative and gram-positive bacteria, especially against multi-drug resistant Pseudomonas aeruginosa (Zhou et al. 2021). After 1 h ultracentrifugation, SRJPs are present in the supernatant of the upper, middle, and lower layers. The upper and middle layers contained MRJP2 (52 kDa) and MRJP3 (60–70 kDa), while the lower layer contained MRJP1 (290 kDa). SRJPs separated by ultracentrifugation produced 20-fold higher levels of SRJPs compared to conventional methods involving dialysis and centrifugation, where dialysis is a time-consuming process (Nozaki et al. 2012).

8.3 Content of Free Amino Acids, Lipids, Vitamins, and Bioactive Substances

RJ is rich in amino acids, containing at least 17, including 8 essential ones plus 5 non-identified related compounds. They are valine (1.6%), glycine (3.0%), isoleucine (1.6%), leucine (3.0%), proline (3.9%), threonine (2.0%), serine (2.9%), methionine (3.7%), phenylalanine (0.5%), aspartic acid (2.8%), glutamic acid (8.3%), tyrosine (4.9%), lysine (2.9%), arginine (3.3%), and tryptophan (3.4%) (Bărnuţiu et al. 2011). Rapid ultra-performance liquid chromatography (UPLC) allowed the separation and quantification of 26 amino acids in royal jelly. The results showed that the average amounts of FAA and TAA in fresh royal jelly were 9.21 and 111.27 mg/g, respectively, in which the major FAAs were Pro, Gln, Lys, Glu, and the most abundant TAAs were Asp, Glu, Lys, and Leu. However, during storage, the total content of Met and free Gln decreased significantly and continuously, indicating that they can be used as a parameter to predict the quality of royal jelly (Wu et al. 2009).

Other compounds identified in RJ include several nucleotides as free bases (adenosine, uridine, guanosine, iridin, and cytidine) and phosphates such as adenosine 5′-mono-phosphate (AMP), adenosine 5′-diphosphate (ADP), and adenosine 5′-triphosphate (ATP) (Xue et al. 2009a, b; Zhou et al. 2012). Adenosine is a naturally occurring purine nucleoside, a component of ribonucleic acid (RNA), consisting of adenine attached to a ribofuranose via a b-N9-glycosidic bond. In vivo, adenosine is produced by the metabolic degradation of intracellular adenosine triphosphate (ATP), which is the primary cellular energy source for transport systems and enzymatic activities of proteins. ATP is converted to adenosine diphosphate (ADP) and then adenosine monophosphate (AMP) by successive dephosphorylation (Kim and Lee 2011). A UPLC method showed that ATP in RJ sequentially degrades to adenosine 5′-diphosphate (ADP), adenosine 5′-monophosphate (AMP), inosine monophosphate (IMP), inosine (HxR), and hypo- xanthine (Hx) (Zhou et al. 2009). A study has shown that significantly higher levels of ATP, ADP, and AMP were seen in fresh RJ samples, and IMP, uridine, guanosine, and thymidine were identified in commercial RJ samples (Wu et al. 2015).

RJ also contains vitamins B1, B2, B6, B5, B8, B9, C (in trace amounts), and PP (Table 8.1) (Bărnuţiu et al. 2011).

The volatile organic compound (VOC) profiles of RJ were rich in acid, ester, and aldehyde compounds, but their contents were different, as exemplified by the data for acetic acid, benzoic acid methyl ester, hexanoic acid, and octanoic acid. Esters and aldehydes were the most abundant VOCs among the 40 detected, accounting for 25 and 17.5%, respectively, and contribute the most to the RJ flavor. Other classes of compounds, such as ketones (15%), acids (10%) and alcohols (10%), were also detected (Zhao et al. 2016).

Table 8.1 Principal vitamins identified in royal jelly

Vitamins in royal jelly	mg/100 g
Vitamin A	1.10
Vitamin E	5.00
Vitamin B1	2.06
Vitamin D	0.2
Vitamin B2	2.77
Vitamin B6	11.90
Vitamin B12	0.15
Vitamin B5 (Panthotenic acid)	52.80
Niacin (PP)	42.42
Vitamin C (Ascorbic acid)	2.00
Vitamin B9 (Folic acid)	0.40

Several studies have indicated that peptides have antioxidative activity (Shahidi and Zhong 2011). Antioxidative peptides were isolated from RJ, such as Ala-Leu, Phe-Lys, Phe-Arg, Ile-Arg, Lys-Phe, Lys-Leu, Lys-Tyr, Arg-Tyr, Tyr-Asp, Tyr-Tyr, Leu-Asp-Arg, and Lys-Asn-Tyr-Pro, with strong hydroxyl radical scavenging activity. Three dipeptides (Lys-Tyr, Arg-Tyr, and Tyr-Tyr) containing Tyr residues at the C-terminal showed strong hydroxyl radical and hydrogen peroxide scavenging activity (Guo et al. 2009).

8.4 Lipid Content

A unique and interesting feature of RJ is its lipid and fatty acid content (Li et al. 2013). The lipid composition of RJ reported to date is 80–85% fatty acids, 4–10% phenols, 5–6% waxes, 3–4% steroids, and 0.4–0.8% phospholipids (Terada et al. 2011). RJ fatty acids are free short-chain fatty acids containing 8–12 carbons and are usually either hydroxyl or dicarboxylic fatty acids, different from fatty acids in animals and plants which are mainly in the tri, di, and monoglyceride forms with minor amounts of free fatty acids (Melliou and Chinou 2005). Lercker et al. (1981) originally discovered that fatty acids in RJ mainly consist of 10-hydroxy-2-decenoic (10-HDA) and 10-hydroxydecanoic acids (10-HDDA), constituting between them at least 60–80% of the total amount of organic acids found in RJ. The 10-HDA content tends to vary between 2-day RJ and 4-D RJ with a declining trend (Wang et al. 2016). The fatty acid fraction consists of 32% trans-10-hydroxy-2-decenoic acid, 24% gluconic acid, 22% 10-hydroxydecanoic acid, 5% dicarboxylic acids, and several other acids (Terada et al. 2011; Echigo et al. 1986). The major fatty acid in royal jelly is trans-10-hy-droxy-2-decenoic acid, and to date, no other natural product containing this has been reported, even in other bee products (Melliou and Chinou 2005). In Greek royal jelly, the most abundant were 10-hydroxy-2-decenoic acid, 10-hydroxydecanoic acid, and sebacic acid (Melliou and Chinou 2005). Seven different fatty acid derivatives were reported in Greek royal jelly, i.e., 10-acetoxydecanoic acid, trans-10-acetoxydec-2- enoic acid, 11-oxododecanoic acid, (11S)-hydroxydodecanoic acid, (10R, 11R)- dihydroxydodecanoic acid, 3,11-dihydroxydodecanoic acid, and (11S), 12-dihydroxydodecanoic acid (Melliou and Chinou 2005). In Saudi royal jelly, the 10-HDA content ranged from 1.68% to 6.36% (Bazeyad et al. 2022). But, in Anatolian royal jelly, the 10-HDA content ranged from 1.0% to 3.9% (Kolayli et al. 2016). In Egyptian royal jelly, the 10-HAD content ranged from 2.58% to 3.37% (Arfa et al. 2021). In Iran, the lipid content was 12% in raw RJ, but it was only 11% in commercial samples (Kazemi et al. 2019). The lipid content in RJ from Africanized honeybees fed a supplemented diet in Brazil was 3.03% to 4.21% (Sereia et al. 2013). In the past, 10-HDA content was adopted as a quality marker for RJ and used as a means to evaluate bee products containing RJ (Bloodworth et al. 1995). Beekeepers and manufacturers also utilized 10-HDA content as a freshness indicator for royal jellies, i.e., royal jellies with a 10-HDA content greater than 1.8% were considered as fresh and authentic samples

(Antinelli et al. 2003). During a storage period of 12 months, 10-HDA loss rates were 0.1 and 0.2% at −18 °C and 4 °C, respectively, for samples from France and Thailand. At room temperature storage, corresponding losses were 0.4 and 0.6% for French and Thai samples, respectively (Antinelli et al. 2003).

Fatty acids in RJ are known to pose many medicinal properties, adding commercial and medicinal value to the product (Botezan et al. 2023). Earlier studies have shown that fatty acids in RJ show anticancer properties (Townsend et al. 1961). Fatty acids in RJ also exhibited several immunomodulatory activities. Earlier studies have shown that 10-HDA stimulated transforming growth factor-β (a potent immunosuppressive cytokine) production in fibroblast cell lines (Koya-Miyata et al. 2004; Wahl et al. 2004). Dendritic cells (DCs) play an important role in the initiation and regulation of the immune response against pathogens, tumors, and other antigens. At high concentrations, 10-HDA and 3,10-dihydroxy-decanoic acid (3,10-DDA) could inhibit the proliferation of allogeneic T-cells in rat DC–T-cell cocultures, indicating their immunomodulatory activity (Vucevic et al. 2007). Lipophilic fractions of the RJs induced a strong reduction in human neuroblastoma (SH-SY5Y) cell line growth, in a dose- and time-dependent manner (Gismondi et al. 2017). I 3,10-DDA could stimulate the maturation of human monocyte-derived dendritic cells in vitro by up-regulating the expression of CD40, CD54, CD86 and CD1a, and enhance the allostimulatory and Th1 polarizing capability of these cells, which could be beneficial for anti-tumor and anti-viral immune responses (Tanja Dzopalic et al. 2011). Fatty acids such as 3-hydroxydodecanedioicacid, 11-oxododecanoicacid, and (11S)-hydroxydodecanoic acid from RJ exhibited strong broad-spectrum antimicrobial activity against pathogenic bacteria and fungi (Melliou and Chinou 2005). 3-Hydroxydodecanedioic acid, trans-10-acetoxydec-2-enoic acid, and (11S)-hydroxydodecanoic acid exhibited activity against *S. aureus*, *S. epidermidis*, *S. mutans*, and *S. viridans* with MIC values ranging from 0.17 to 0.36 mg/mL. Sebacic acid (MIC values of 0.15 to 0.20 mg/mL) and 11-oxododecanoic acid (MIC values of 0.26–0.67 mg/mL) showed higher antifungal activity against different fungi (Melliou and Chinou 2005). Transient receptor potential ankyrin 1 (TRPA1) and vanilloid 1 (TRPV1), belonging to the TRP family, are Ca2+ permeable nonselective cation channels (Dhaka et al. 2006; Caterina et al. 1997). TRPA1 and TRPV1 are expressed in dorsal root ganglion and nodose ganglion neurons innervating the stomach and in nerve fibers in the gastric wall (Patterson et al. 2003; Ward et al. 2003). Ingestion of TRPA1- and TRPV1-active food components can be effective against obesity via energy expenditure enhancement (Yoshioka et al. 1998). Terada et al. (2011) searched for novel agonists of TRPA1 and TRPV1 in RJ and found that hexane extract activates TRPA1. They found that most of the fatty acids, such as 10-HDA and 10-HDDA which make up approximately 10% of RJ, had a higher affinity for TRPA1 than TRPV1 with EC50 values for the latter being 2–3 times higher and maximal responses less than half compared to those for TRPA1 (Terada et al. 2011). Other fatty acids such as 1,8-octanedicarboxylic acid and 1,6-hexanedicarboxylic acid showed relatively strong effects on both receptors, but they only constitute a small percentage of RJ. The study indicated that 10-HDA and 10-HDDA can activate TRPA1 because they contain a 10-carbon chain that renders them moderately hydrophobic (Terada et al. 2011).

References

Abu-Serie MM, Habashy NH (2019) Two purified proteins from royal jelly with in vitro dual anti-hepatic damage potency: major royal jelly protein 2 and its novel isoform X1. Int J Biol Macromol 128:782–795

Albert S, Klaudiny J (2004) The MRJP/YELLOW protein family of Apis mellifera: identification of new members in the EST library. J Insect Physiol 50(1):51–59

Antinelli FJ, Zegganea S, Davicoa R, Rognonea C, Faucona JP, Lizzani L (2003) Evaluation of (E)-10-hydroxydec-2-enoic acid as a freshness parameter for royal jelly. Food Chem 80:85–89

Arfa A, Reyad YM, Nikeety ME (2021) Quality parameters of royal jelly in national and international standards: specifications, differences and suggestions. Ann R S C 25(3):7977–7997

Balkanska R (2018) Sugar composition of royal jelly adulterated with drone brood and bee honey. Int J Curr Microbiol App Sci 7(2):1876–1881

Balkanska R, Karadjova I, Ignatova M (2014) Comparative analyses of chemical composition of royal jelly and drone brood. Bulgarian Chem Comm 46(2):412–416

Bărnuţiu LI, Al L, Mărghitaş DDS, Mihai CM, Bobiş O (2011) Chemical composition and antimicrobial activity of royal jelly–review. J Anim Sci Biotechnol 44(2):67–72

Bazeyad AY, Al-Ghamdi AA, Alattal YZ (2022) Physicochemical characteristics of local royal jelly produced in Al-Baha region, Saudi Arabia. World J Adv Res Rev 14(01):284–292

Bloodworth BC, Harn CS, Hock CT (1995) Liquid chromatographic determination of trans-10-hydroxy-2-decenoic acid content of commercial products containing royal jelly. J AOAC Int 78:1019–1023

Botezan S, Baci GM, Bagameri L, Pasca C, Dezmirean DS (2023) Current status of the bioactive properties of royal jelly: a comprehensive review with a focus on its anticancer, anti- inflammatory, and antioxidant effects. Molecules 28:1510

Brudzynski K, Sjaarda C, Lannigan R (2015) MRJP1-containing glycoproteins isolated from honey, a novel antibacterial drug candidate with broad spectrum activity against multi-drug resistant clinical isolates. Front Microbiol 6:1–9

Buttstedt A, Moritz RF, Erler S (2013) Origin and function of the major royal jelly proteins of the honeybee (Apis mellifera) as members of the yellow gene family. Biol Rev Camb Philos Soc 89:255–269. https://doi.org/10.1111/brv.12052

Caterina MJ, Schumacher MA, Tominaga M, Rosen TA, Levine JD, Julius D (1997) The capsaicin receptor: a heat-activated ion channel in the pain pathway. Nature 389:816–824

Collazo N, Carpena M, Nuñez-Estevez B, Otero P, Simal-Gandara J, Prieto MA (2021) Health promoting properties of bee royal jelly: food of the queens. Nutrients 13:543

Cruz GCN, Garcia L, Silva AJ, Barbosa JARG, Ricart CAO, Freitas SM, Sousa MV (2011) Calcium effect and pH-dependence on self-association and structural stability of the Apis mellifera major royal jelly protein. Apidologie 42:252–269

Daniele G, Casabianca H (2012) Sugar composition of French royal jelly for comparison with commercial and artificial sugar samples. Food Chem 134:1025–1029

Dhaka A, Viswanath V, Patapoutian A (2006) TRP ion channels and temperature sensation. Annu Rev Neurosci 29:135–161

Drapeau MD, Albert S, Kucharski R, Prusko C, Maleszka R (2006) Evolution of the yellow/major royal jelly protein family and the emergence of social behavior in honey bees. Genome Res 16:1385–1394

Dzopalic T, Vucevic D, Tomic S, Djokic J, Chinou I, Colic M (2011) 3,10-Dihydroxy- decanoic acid, isolated from royal jelly, stimulates Th1 polarising capability of human monocyte-derived dendritic cells. Food Chem 126:1211–1217

Echigo T, Takenaka T, Yatsunami K (1986) Comparative studies on chemical composition of honey, royal jelly and pollen loads. Bull Fac Agric Tamagawa Univ 26:1–12

Fan P, Han B, Feng M, Fang Y, Zhang L, Hu H, Hao Y, Qi Y, Zhang X, Li J (2016) Functional and proteomic investigations reveal major royal jelly protein 1 associated with antihypertension activity in mouse vascular smooth muscle cells. Sci Rep 6:30230. https://doi.org/10.1038/srep30230

Feng M, Fang Y, Ma C, Duan X, Zhang Y, Han B, Hu H, Meng L, Wang F, Li J (2021) Mechanistic insight into royal protein inhibiting the gram-positive bacteria. Biomol Ther 11(1):1–17

Flanjak I, Primorac I, Vukadin I, Kovačić M, Puškadija Z, Rajs BB (2019) Physicochemical characteristics of Croatian royal jelly. Croat J Food Sci Technol 11(2):266–271

Fratini F, Cilia G, Mancini S, Felicioli A (2016) A royal jelly: an ancient remedy with remarkable antibacterial properties. Microbiol Res 192:130–141

Gismondi A, Trionfera E, Canuti L, Di Marco G, Canini A (2017) Royal jelly lipophilic fraction induces antiproliferative effects on SH-SY5Y human neuroblastoma cells. Oncol Rep 38:1833–1844

Guo H, Kozuma Y, Yonekura M (2009) Structures and properties of antioxidative peptides derived from royal jelly protein. Food Chem 113:238–245

Judova J, Sutka R, Klaudiny J, Liskova D, Ow DW, Simuth J (2004) Transformation of tobacco plants with cDNA encoding honeybee royal jelly MRJP1. Biol Plant 48:185–191

Kamakura M (2002) Signal transduction mechanism leading to enhanced proliferation of pri- mary cultured adult rat hepatocytes treated with royal jelly 57-kDa protein. J Biochem 132:911–912

Kamakura M, Suenobu N, Fukushima M (2001) Fifty-seven kDa protein in royal jelly enhances proliferation of primary cultured rat hepatocytes and increases albumin production in the absence of serum. Biochem Biophys Res Commun 282:865–874

Kazemi V, Eskafi M, Saeedi M, Manayi A, Hadjiakhoondi A (2019) Physicochemical properties of royal jelly and comparison of commercial with raw specimens. Jundishapur J Nat Pharm Prod 14(4):e64920

Kim J, Lee J (2011) Observation and quantification of self-associated adenosine extracted from royal jelly products purchased in USA by HPLC. Food Chem 126:347–352

Kimura Y, Washino N, Yonekura M (1995) N-linked sugar chains of 350-kDa royal jelly glycoprotein. Biosci Biotechnol Biochem 59:507–509

Kimura M, Kimura Y, Tsumura K, Okihara K, Sugimoto H, Yamad H, Yonekura M (2003) 350-kDa royal jelly glycoprotein (apisin), which stimulates proliferation of human monocytes, bears the β1-3 galactosylated N-glycan: analysis of the N-glycosylation site. Biosci Biotechnol Biochem 67:2055–2058

Kolayli S, Sahin H, Can Z, Yildiz O, Malkoc M, Asadov A (2016) A member of complementary medicinal food: anatolian royal jellies, their chemical compositions, and antioxidant properties. J Evid-Based Comp Alt Med 21(4):NP43–NP48

Koya-Miyata S, Okamoto I, Ushio S, Iwaki K, Ikeda M, Kurimoto M (2004) Identification of a collagen production-promoting factor from an extract of royal jelly and its possible mechanism. Biosci Biotechnol Biochem 68:767–773

Lercker G (1986) Carbohydrate determination of royal jelly by high resolution gas chromatography (HRGC). Food Chem 19:255–264

Lercker G, Capella P, Conte LS, Ruini F, Giordani G (1981) Components of royal jelly: I identification of the organic acid. Lipids 16(12):912–919

Lercker G, Caboni MF, Vecchi MA, Sabatini AG, Nanetti A (1992) Characterizaton of the main constituents of royal jelly. APIC 410:27–37

Li J, Wang T, Zhang Z, Pan Y (2007) Proteomic analysis of royal jelly from three strains of western honeybees (Apis mellifera). J Agric Food Chem 55:8411–8422

Li X, Huang C, Xue Y (2013) Contribution of lipids in honeybee (Apis mellifera) royal jelly to health. J Med Food 16:96–102

Majtan J, Kovacova E, Blikova K, Simuth J (2006) The immunostimulatory effect of the recombinant apalbumin 1-major honeybee royal jelly protein-on TNFα release. Int Immunopharmacol 6(2):269–278

Melliou E, Chinou I (2005) Chemistry and bioactivity of royal jelly from Greece. J Agric Food Chem 53(23):8987–8992

Mureşan CI, Dezmirean DS, Marc BD, Suharoschi R, Pop OL, Buttstedt A (2022) Biological properties and activities of major royal jelly proteins and their derived peptides. J Funct Foods 98:105286

Nakaya M, Onda H, Sasaki K, Yukiyoshi A, Tachibana H, Yamada K (2007) Effect of royal jelly on bisphenol A-induced proliferation of human breast cancer cells. Biosci Biotechnol Biochem 71:253–255

Narita Y, Nomura J, Ohta S, Inoh Y, Suzuki KM, Araki Y, Okada S, Matsumoto I, Isohama Y, Abe K, Miyata T, Mishima S (2006) Royal jelly stimulates bone formation: physiologic and nutrigenomic studies with mice and cell lines. Biosci Biotechnol Biochem 70:2508–2514

Nozaki R, Tamura S, Ito A, Moriyama T, Yamaguchi K, Kono T (2012) A rapid method to isolate soluble royal jelly proteins. Food Chem 134:2332–2337

Okamoto I, Taniguchi Y, Kunikata T, Kohno K, Iwaki K, Ikeda M, Kurimoto M (2003) Major royal jelly protein 3 modulates immune responses in vitro and in vivo. Life Sci 73:2029–2045

Park MJ, Kim BY, Deng Y, Park HG, Choi YS, Lee KS, Jin BR (2020) Antioxidant capacity of major royal jelly proteins of honeybee (Apis mellifera) royal jelly. J Asia-Pac Entomol 23(2):445–448

Patterson LM, Zheng H, Ward SM, Berthoud HR (2003) Vanilloid receptor (VR1) expression in vagal afferent neurons innervating the gastrointestinal tract. Cell Tissue Res 311:277–287

Peiren N, Vanrobaeys F, De Graaf DC, Devreese B, Van Beeumen J, Jacobs FJ (2005) The protein composition of honeybee venom reconsidered by a proteomic approach. Biochim Biophys Acta 1752:1–5

Peiren N, De Graaf DC, Vanrobaeys F, Danneels EL, Devreese B, Van Beeumen J, Jacobs FJ (2008) Proteomic analysis of the honey bee worker venom gland focusing on the mechanisms of protection against tissue damage. Toxicon 52:72–83

Popescu O, Mărghitas LA, Bobis O, Stanciu O, Bonta V, Moise A, Dezmirean D (2009) Sugar profile and total proteins content of fresh royal jelly. Bull UASVM Anim Sci Biotechnol 66:265–269

Ramadan MF, Al-Ghamdi A (2012) Bioactive compounds and health-promoting properties of royal jelly: a review. J Funct Foods 4:39–52

Ramanathana ANKG, Nair AJ, Sugunan VS (2018) A review on royal jelly proteins and peptides. J Funct Foods 44:255–264

Rembold H (1965) Biologically active substances in royal jelly. Vitam Horm 23:359–382

Rembold H, Lackner B (1978) Vergleichende Analyse von Weiselfuttersäften. Mitteilungen der Deutschen Gesellschaft für allgemeine und angewandte Entomologie 1(2/3/4):299–301

Sano O, Kunikata T, Kohno K, Iwaki K, Ikeda M, Kurimoto M (2004) Characterization of royal jelly proteins in both Africanized and European honeybees (Apis mellifera) by two-dimensional gel electrophoresis. J Agric Food Chem 52(1):15–20

Santos KS, Dos Santos LD, Mendes MA, De Souza BM, Malaspina O, Palma MS (2005) Profiling the proteome complement of the secretion from hypopharyngeal gland of Africanized nurse-honeybees (Apis mellifera L.). Insect Biochem Mol Biol 35:85–91

Sato A, Unuma H, Ebina K (2021) Royal jelly proteins inhibit macrophage proliferation: interactions with native- and oxidized-low density lipoprotein. Protein J 40(5):699–708

Schmitzova J, Klaudiny J, Albert S, Schroder W, Schreckengost W, Hanes J, Judova J, Simuth J (1998) A family of major royal jelly proteins of the honeybee Apis mellifera L. Cell Mol Life Sci 54(9):1020–1030

Schonleben S, Sickmann A, Mueller MJ, Reinders J (2007) Proteome analysis of Apis mellifera royal jelly. Anal Bioanal Chem 389(4):1087–1093

Sereia MJ, de Toledo VAA (2013) Quality of royal jelly produced by Africanized honeybees fed a supplemented diet. Food Sci Tech 33(2):304–309

Sesta G (2006) Determination of sugars in royal jelly by HPLC. Apidologie 37:84–90

Shahidi F, Zhong Y (2011) Revisiting the polar paradox theory: a critical overview. J Agric Food Chem 59:3499–3504

Shen L, Zhang W, Jin F, Zhang L, Chen Z, Liu L, Parnell LD, Lai C, Li D (2010) Expression of recombinant AccMRJP1 protein from royal jelly of Chinese honeybee in Pichia pastoris and its proliferation activity in an insect cell line. J Agric Food Chem 58:9190–9197

Simuth J, Bilikova K, Kovacova E, Kuzmova Z, Schroder W (2004) Immunochemical approach to detection of adulteration in honey: physiologically active royal jelly protein stimulating TNF-α is a regular component of honey. J Agric Food Chem 52:2154–2158

Tamura S, Amano S, Kono T, Kondoh J, Yamaguchi K, Kobayashi S, Ayabe A, Moriyama T (2009) Molecular characteristics and physiological functions of major royal jelly protein 1 oligomer. Proteomics 9:5534–5543

Terada Y, Narukawa M, Watanabe T (2011) Specific hydroxy fatty acids in royal jelly activate TRPA1. J Agric Food Chem 59:2627–2635

Townsend GF, Brown WH, Felauer EE, Hazlett B (1961) Studies on the in vitro antitumor activity of fatty acids. IV. The esters of acids closely related to 10-hydroxy-2-decenoic acids from royal jelly against transplantable mouse leukemia. Can J Biochem Physiol 39:1765–1770

Vezeteu TV, Bobis O, Moritz RFA, Buttstedt A (2017) Food to some, poison to others—honeybee royal jelly and its growth inhibiting effect on European Foulbrood bacteria. Microbiol Open 6(1):1–7

Vucevic D, Melliou E, Vasilijic S, Gasic S, Ivanovski P, Chinou I, Colic M (2007) Fatty acids isolated from royal jelly modulate dendritic cell-mediated immune response in vitro. Int Immunopharmacol 7:1211–1220

Wahl SM, Swisher J, McCartney-Francis N, Chen W (2004) TGF-{beta}: the perpetrator of immune suppression by regulatory T cells and suicidal T cells. J Leukoc Biol 76:15–24

Wang Y, Ma L, Zhang W, Cui X, Wang H, Xu B (2016) Comparison of the nutrient composition of royal jelly and worker jelly of honey bees (Apis mellifera). Apidologie 47:48–56

Ward SM, Bayguinov J, Won KJ, Grundy D, Berthoud HR (2003) Distribution of the vanilloid receptor (VR1) in the gastrointestinal tract. J Comp Neurol 465:121–135

Watanabe K, Shinmoto H, Kobori M, Tsushida T, Shinohara K, Kanaeda J, Yonekura M (1998) Stimulation of cell growth in the U-937 human myeloid cell line by honey royal jelly protein. Cytotechnology 26:23–27

Wu L, Zhou J, Xue X, Li Y, Zhao J (2009) Fast determination of 26 amino acids and their content changes in royal jelly during storage using ultra-performance liquid chromatography. J Food Comp Anal 22:242–249

Wu L, Chen L, Selvaraj JN, Wei Y, Wang Y, Li Y, Zhao J, Xue X (2015) Identification of the distribution of adenosine phosphates, nucleosides and nucleobases in royal jelly. Food Chem 173:1111–1118

Wytrychowski M, Chenavas S, Daniele G, Casabianca H, Batteau M, Guibert S, Brion B (2013) Physicochemical characterisation of French royal jelly: comparison with commercial royal jellies and royal jellies produced through artificial bee-feeding. J Food Comp Anal 29:126–133

Xu X, Gao Y (2013) Isolation and characterization of proteins and lipids from honeybee (Apis mel- lifera L.) queen larvae and royal jelly. Food Res Int 54:330–337

Xue XF, Wang F, Zhou JH, Chen F, Li Y, Zhao J (2009a) Online cleanup of accelerated solvent extractions for determination of adenosine 5′-triphosphate (ATP), adenosine 5′-diphosphate (ADP), and adenosine 5′-monophosphate (AMP) in royal jelly using high performance liquid chromatography. J Agric Food Chem 57:4500–4505

Xue X, Zhou J, Wu L, Fu L, Zhao J (2009b) HPLC determination of adenosine in royal jelly. Food Chem 115:715–719

Yoshioka M, St-Pierre S, Suzuki M, Tremblay A (1998) Effects of red pepper added to high-fat and high-carbohydrate meals on energy metabolism and substrate utilization in Japanese women. Br J Nutr 80:503–510

Zhang L, Fang Y, Li R, Feng M, Han B, Zhou T, Li J (2012) Towards posttranslational modification proteome of royal jelly. J Proteome 75:5327–5341

Zhao Y, Li Z, Tian W, Fang X, Su S, Peng W (2016) Differential volatile organic compounds in royal jelly associated with different nectar plants. J Integr Agric 15:1157–1165

Zhou J, Xue X, Chen F, Zhang J, Li Y, Wu L, Chen L, Zhao J (2009) Simultaneous determination of seven fluoroquinolones in royal jelly by ultrasonic-assisted extraction and liquid chromatography with fluorescence detection. J Sep Sci 32(7):955–964

Zhou L, Xue X, Zhou J, Li Y, Zhao J, Wu L (2012) Fast determination of adenosine 5′-triphosphate (ATP) and its catabolites in royal jelly using ultra performance liquid chromatography. J Agric Food Chem 60:8994–8999

Zhou J, Zhang L, He Y, Liu K, Zhang F, Zhang H, Lu Y, Yang C, Wang Z, Fareed MS, Liang X, Yan W, Wang K (2021) An optimized analog of antimicrobial peptide Jelleine-1 shows enhanced antimicrobial activity against multidrug resistant P. aeruginosa and negligible toxicity in vitro and in vivo. Euro J Med Chem 219:113433

Chapter 9
Volatile Compounds of Royal Jelly

Maria G. Miguel and Soukaïna El-Guendouz

Contents

Abstract Up until 2016, it was challenging to find references on the royal jelly (RJ) volatiles. Identifying these compounds in RJ has gained a lot of attention since then. This chapter provides an update on the chemical composition of volatiles in RJ since 2016. China, including Taiwan, Egypt, Italy, Malaysia, Morocco, Poland, Portugal, Spain, and the United States of America were among the many RJ samples' geographical origins. The RJ samples were sourced from *Apis mellifera* L., *A. mellifera carnica* Poll., and from hybrids of *A. mellifera* (*A. mellifera ligustica* and *A. mellifera carnica*) colonies or from commercial source. The primary techniques for separating RJ volatiles included extraction with organic solvents, hydrodistillation, and headspace solid-phase microextraction (HS-SPME), employing various extraction times and temperatures, with the divinyl benzene (DVB)/carboxy (CAR)/polydimethylsiloxane (PDMS) fiber being the most commonly used. The separation and identification of the volatiles were generally performed by using gas chromatography coupled with mass spectrometry (GC-MS). Carboxylic esters, carboxylic acids, esters/lactones, aldehydes/ketones, alcohols, hydrocarbons, terpenes, ethers among other compounds were detected in RJ samples. The most often men-

M. G. Miguel (✉)
MED—Mediterranean Institute for Agriculture, Environment and Development, CHANGE—Global Change and Sustainability Institute, Faculdade de Ciências e Tecnologia, Universidade do Algarve, Faro, Portugal
e-mail: mgmiguel@ualg.pt

S. El-Guendouz
Laboratory of Physiology-Pharmacology-Environmental Health, Faculty of Sciences Dhar El Mehraz, University Sidi Mohamed Ben Abdallah, Fez, Morocco

J. M. Alvarez-Suarez (ed.), *Bee Products – Chemical and Biological Properties*,
https://doi.org/10.1007/978-3-031-89049-9_9

tioned volatiles found in different RJ samples were: hexanoic, octanoic, nonanoic, 10-hydroxydecanoic, (*E*)-10-hydroxydec-2-enoic acids; methylbenzoate, methylsalicylate, benzaldehyde, furfural, 8-nonen-2-one, 2-nonanone, 2-heptanone, benzeneacetaldehyde, nonanal, and phenol.

Keywords Royal jelly · Volatile compounds · Gas chromatography-mass spectrometry (GC-MS) · Extraction techniques · Carboxylic acids

9.1 Introduction

Royal jelly (RJ) is a secretion rich in protein and lipids produced by the hypopharyngeal and mandibular glands, respectively, of *Apis mellifera* L. young worker honeybees (called nurses) (Fujita et al. 2013). RJ is used by nurses to feed young larvae of workers and drones for the first 3 days, and is the exclusive food of the queen bee throughout its life. After this 3-day period, worker larvae are fed a mixture of RJ, pollen, and honey (Boselli et al. 2003; Ferioli et al. 2014; Wang et al. 2016). Workers and queens are females developed from fertilized eggs, whereas unfertilized eggs become males, called drones (Wang et al. 2016). Techniques for producing royal jelly have a long history. For example, Sherlock Holmes began removing RJ from a naturally occurring queen cell using a syringe. Later on, the method consisted of the queen's removal from the colony, and the RJ was obtained directly for the new queen cells that the worker honey bees rapidly built after detecting the queen's absence. This procedure was carried out despite potential drawbacks, including a decrease in honey production and colony strength (Hu et al. 2019). This disadvantage was overcome by introducing a queen excluder that permits to maintain the production of RJ and honey with a queen (Hu et al. 2019). In order to boost RJ production, new methods have been developed. These methods include genetic selection of high RJ-producing bees and development and implementation of production techniques (Ahmad et al. 2020). For example, Hu et al. (2019) describe the method for producing RJ by artificial larvae grafting. This method occurs outside the honeycomb in artificial queen cells. This procedure is a time-consuming and labor-intensive step. For this reason, a new method that does not require larval grafting has been developed for the production of RJ, which is equally efficient. For a better understanding of the procedure, it is advised to read the work of Hu et al. (2019). RJ is a viscous and creamy product. It is relatively acidic (pH 3.1–3.9), even though there are authors that expanded this range between 3.4 and 4.5, with a high buffering capacity ranging between 4 and 7, has a density of 1.1 g/mL, is partially soluble in water, ranges in color from white to yellow, and has a slightly pungent odor and taste (Ramadan and Al-Ghamdi 2012; Nabas et al. 2014; Mureşan et al. 2016; Ramanathan et al. 2018; Kunugi and Ali 2019).

9.2 General Chemical Composition and Biological Properties of Royal Jelly

RJ is mainly composed of water (60–70%), proteins (9–18%), carbohydrates (7–18%), lipids (3–8%), minerals (0.8–3%), vitamins, phenols and amino acids (Balkanska et al. 2014). The fluctuations of these metabolites within these ranges are due to several factors such as the heterogeneous nature of RJ, different locations where RJ is produced, and different times of production (Collazo et al. 2021).

Major RJ proteins (MRJPs) are water-soluble, constituting more than 80% of RJ proteins. The MRJP family has nine members, MRJP1-MRJP9, with molecular masses of 49–87 kDa (Bărnuţiu et al. 2011; Ramadan and Al-Ghamdi 2012 and references therein). In some MRJPs, there are repetitive pentapeptide regions concentrated with nitrogen-rich amino acids, and others with repetitive tripeptide regions concentrated with methionine (MRJP5) (Drapeau et al. 2006). RJ also contains other proteins such as royalisin, which belongs to the insect defensine family presenting antimicrobial activity against a large spectrum of Gram-positive and Gram-negative bacteria and fungi. It is found in a smaller amount than MRJPs (Collazo et al. 2021; Oršolić and Jembrek 2024). Peptides are also constituents of RJ which possess antioxidant activity, mainly due to their capacity to scavenge hydroxyl free radicals, particularly those containing tyrosyl residues at their C-termini (Guo et al. 2009). Examples of peptides include jelleins (jelleins I-IV) which derive from the C-terminus of MRJP1. Jelleins I-III have antimicrobial activity, whereas jellein IV does not present such activity (Collazo et al. 2021). Free amino acids are also found in RJ, with proline, glycine, lysine and glutamic acid being the most abundant (Liming et al. 2009); although Boselli et al. (2003) also reported alanine, aspartic acid, homoserine, isoleucine, hydroxyproline, ornitine, phenylalanine, serine, threonine, tyrosine and valine as being present in considerable amounts. Fructose and glucose are the main carbohydrates of RJ (over 90% of the total sugars), although fructose is the most predominant. Sucrose (in high variable concentrations) and oligosccharides trehalose, maltose, gentiobiose, isomaltose, raffinose, erlose, and melezitose (in very small concentrations) are also carbohydrates present in RJ (Sabatini et al. 2009).

The lipid fraction is predominantly composed of organic acids (80–90%), almost all in the free form (mono and dihydroxy acids and dicarboxylic acids with 8 and 10 carbon atoms) with characteristic arrangements. Phenols (4–10%), waxes (5–6%), steroids (3–4%), and phospholipids (0.4–0.8%) are also constituents of the lipid fraction. The fatty acid fraction predominantly consists of monocarboxylic acids *trans*-10-hydroxy-2-decenoic acid (HDEA or 10-HDA) (32%), gluconic acid (24%) and 10-hydroxydecanoic acid (10-HDAA) (22%). Other monocarboxylic acids have also been reported in RJ (7- and 8-hydroxyoctanoic, 3-hydroxydecanoic, 9-hydroxydeca-noic, 9-hydroxy-2-decenoic, 10-hydroxydecanoic, and 3,10-dihydroxydecanoic acids). Dicarboxylic acids are also present in RJ such as 1,10-decanedioic acid or sebacic acid, 2-octene-1,8-dioic, and 2-decene-1,10-dioic acids). Mono- and di-esters of 10-hydroxy-2-*trans*-decenoic acid in which the hydroxyl group is esterified by another fatty acid were also reported in RJ, along with one hydroxyl-2-*trans*-decenoic acid 10-phosphate (Noda et al. 2005; Sabatini et al. 2009; Terada et al. 2011; Ramadan and Al-Ghamdi 2012; Ferioli et al. 2014). Fatty acids in RJ have a spicy, astringent, and gritty taste, that is, an overall unpleasant feeling will remain in the throat for a long time and might occasionally cause vomiting. He et al. (2020) used gelatin and arabic gum as well materials to create RJ microcapsules in order to overcome these drawbacks, and this technique resulted in a satisfactory encapsulation yield of 85.71% and encapsulation efficiency of 92.34%. The outcomes of the gastrointestinal digestion simulation demonstrated that the microcapsules were well tolerated in the gastric environment, that the astringency, bitterness, and irritant odors of the royal jelly residue had decreased, and that the sensory index of the sieve residue had been improved by microencapsulation as measured by an electronic tongue (E-tongue) and nose (He et al. 2020). In the lipid fraction of RJ, Lercker et al. (1982) also detected sterols in the unsaponifiable extracts, such as 24-methylenecholesterol, the most abundant, along with sitosterol and Δ5-avenasterol in much lower amounts. More recently, other authors (Ferioli et al. 2014; Isidorov et al. 2012) also reported the presence of such phytosterols in Italian and Polish RJ samples. Isidorov et al. (2012) reported the presence of 25-hydroxy-24-methylcholesterol in Polish RJ samples. Campesterol was also detected in Polish RJ by Isidorov et al. (2016).

Only water-soluble vitamins have been reported in RJ. Pantothenic acid (B5) is the predominant vitamin in RJ, although others may also be present but in lower amounts (e.g., thiamine (B1), riboflavin (B2), niacin (B3), pyridoxal (B6), biotin (B7), folic acid, and ascorbic acid (vitamin C) (only in trace amounts)) (Ciulu et al. 2013; Bogdanov et al. 2016).

RJ is the most valuable bee product, not only as functional food but also as being used in medicine as a medicinal product and cosmetics. Its consumption is constantly growing. It can be eaten on its own or as a useful ingredient in other food products (Ahmad et al. 2020; Oršolić and Jembrek 2024). Diverse biological properties have been described for RJ based on in vitro and in vivo assays: antimicrobial, antioxidant, immunomodulatory, aging delay, diabetes, obesity, hypertension, osteoporosis, and wound-healing activities (Ahmad et al. 2020; Oršolić and Jembrek 2024), although there is limited scientific exploration focused on RJ's effects on wound repair (Martinotti et al. 2023). Generally, the mechanisms involved in wound repair, due to the presence of some constituents, include antimicrobial and anti-inflammatory activities (sebacic acid, royalisin, 10-HDA, and 10-HDAA), stimulation of regenerative activities (MRJP1 or royalactin, royalisin, 10-HAD), stimulation of proliferation and migration of fibroblasts and mesenchymal stem cells (MRJPs, namely MRJP2, MRJP3, and MRJP7) (Oršolić and Jembrek 2024; Zheng et al. 2024).

Recently, the antioxidant activity of RJ has led to its utilization along with other bee products in meat products, for example, to improve the stability of pork patties over storage (Park et al. 2024) or to obtain nitrate/nitrite-reduced dry fermented sausages (Hospital et al. 2024). In the former instance, the authors (Park et al. 2024) found that propolis and RJ-added patties were more stable during storage than ascorbic acid, but because of their distinct taste and aroma, the patties' flavor declined. In contrast, Hospital et al. (2024) found that adding 2% of the combination of chestnut honey, propolis, and RJ to dry fermented sausages along with 75 mg/kg of nitrate and nitrate did not affect the organoleptic properties of the sausages. However, it did improve their oxidation stability similar to a standard product containing the maximum amounts of these additives allowed in the European Union (150 mg/kg each). The amounts of propolis and RJ utilized can explain the discrepancies in organoleptic qualities reported by the authors. Compared to Park et al. (2024), Hospital et al. (2024) utilized much more honey and much less propolis and RJ.

9.3 Volatiles in Royal Jelly

The volatiles in RJ were not thoroughly investigated until 2016, which is why it was difficult to find references regarding this topic (Boch et al. 1979; Lercker et al. 1981; Drijfhout et al. 2005; Nazzi et al. 2009; Isidorov et al. 2012; Zhao et al. 2016). Since then, there has been an increase in interest in discovering RJ's volatile chemicals as discussed below. Table 9.1 depicts the volatile compounds detected in RJ samples and reported by several authors. Royal jelly from colonies consisting of *Apis melliefera ligustica* and *A. mellifera carnica*, located in north eastern Italy, was extracted with diethyl ether and derivatized by silylation (Nazzi et al. 2009). After this procedure, the authors identified 26 compounds by chromatography coupled with mass spectrometry (GC-MS) (Table 9.1). The compounds were predominantly carboxylic (CAR) acids (23 aliphatic carboxylic acids and 1 aromatic carboxylic acid). The remaining two compounds identified were hydroquinone and methyl-4-hydroxybenzoate. Two other compounds were not identified. Only octanoic acid was quantified and according to the GC-MS chromatograms of the silylated diethyl ether extracts of RJ, this carbpxylic acid appears to be the most abundant volatile compound (113–252 μ/g). Seventeen compounds found by Nazzi et al. (2009) had already been reported by other authors (Boch et al. 1979; Lercker et al. 1981; Drijfhout et al. 2005). The new compounds found in RJ by Nazzi et al. (2009) were 2-methylbutanoic acid, 3-methylbutanoic acid, pentanoic acid, 2-hexenoic acid, 2-hydroxy-4-methylpentanoic acid, 2-hydroxyhexanoic acid, phenylacetic acid, hydroquinone and methyl-4-hydroxybenzoate (Table 9.1). None of the compounds reported above were detected 3 years later by Isidorov et al. (2012) in volatile fractions of RJ from two colonies of *A. mellifera carnica* situated in two distinct regions of Poland (north eastern and south western), obtained after extraction by headspace solid-phase microextraction (HS-SPME) and analyzed by GC-MS. According to the authors, the volatile fraction predominantly consisted of carbonyl compounds (37.2–43.3%), particularly ketones, with 2-heptanone alone accounting for 20%. The other carbonyl compounds identified were, in descending order of quantity: acetone >2- nonanone > benzaldehyde >2-butanone >2-pentanone, hexanal (all of them in significantly lower percentages than 2-heptanone). The second major

Table 9.1 Volatiles found in royal jelly (RJ) samples from various origins that were isolated and identified using diverse methods

Group of compounds	Compounds	Boch et al. (1979) (1)	Drijfhout et al. (2005) (2)	Nazzi et al. (2009) (3)	Isidorov et al. (2012) (4)	Zhao et al. (2016) (5)	Hamid et al. (2018) (6)	Qi et al. (2020) (7)	El-Guendouz et al. (2020) (8)	Qi et al. (2021) (9)	Chen et al. (2022) (10)	Chen et al. (2023) (11)	Montaser et al. (2023) (12)
Carboxylic acids	Acetic acid				x	x	x						
	Propiolic acid												x (Mmarjoram)
	Butanoic acid				x	x							
	Pentanoic acid			x									
	Hexanoic acid	x		x		x			x				
	Octanoic acid	x		x	x	x			x			x	x (Citrus)
	Nonanoic acid	x		x					x	x			
	Decanoic acid	x							x				
	(*E*)-Dec-2-enoic acid	x							x				
	Sebacic acid	x	x	x									
	Dodecanedioic acid		x										
	2-Decenedioic acid		x	x									
	2-Dodecenedioic acid		x										
	(*E*)-10-Hydroxydec-2-enoic acid	x	x	x			x						
	10-Hydroxydodecenoic acid		x										
	12-Hydroxy-2-dodecenoic acid		x										
	10-Hydroxydecanoic acid	x	x	x			x						
	Lauric acid	x							x				x (Clover)
	Myristic acid	x							x				x (Clover)
	Palmitic acid	x		x					x				

	Stearic acid	x											
	2-Hexenoic			x									
	2-Methylpropanoic acid										x		
	2-Methylbutanoic acid			x									
	3-Methylbutanoic acid			x									
	2-Hydroxy-4-methylpentanoic acid			x									
	2-Hydroxyhexanoic acid			x									
	Benzoic acid				x								
	Phenylacetic acid			x									
	2-Octenoic acid			x						x		x	
	3-Octenoic acid											x	
	3-Hydroxyoctanoic acid		x	x									
	7-Hydroxyoctanoic acid			x									
	8-Hydroxyoctanoic acid		x	***x***									x (Clover)
Carboxylic acids	2-Hydroxydecanoic acid						x						
	3-Hydroxydecanoic acid		x	x									
	9-Hydroxydecanoic acid		x	x									
	9-Hydroxy-2-decenoic acid			x									
	10-Hydroxydodecanoic acid		x										

(continued)

Table 9.1 (continued)

Group of compounds	Compounds	Boch et al. (1979) (1)	Drijfhout et al. (2005) (2)	Nazzi et al. (2009) (3)	Isidorov et al. (2012) (4)	Zhao et al. (2016) (5)	Hamid et al. (2018) (6)	Qi et al. (2020) (7)	El-Guendouz et al. (2020) (8)	Qi et al. (2021) (9)	Chen et al. (2022) (10)	Chen et al. (2023) (11)	Montaser et al. (2023) (12)
	11-Hydroxydodecanoic acid		x	*x*									
	12-Hydroxydodecanoic acid		x										
	12-Hydroxytetradecanoic acid		x										
	13-Hydroxytetradecanoic acid		x										
	15-Hydroxyhexadecanoic acid		x										
	11-Hydroxydodecenoic acid		x										
	13-Hydroxytetradecenoic acis		x										
	3,10-Dihydroxydecanoic acid			x									
	Linolenic acid								x				
	Oleic acid												
	Pentadecanoic acid												x (Clover)
	trans-2-Nonenoic acid								x				
	2,3-Dihydro-3,5-dihydroxy-6-methyl-1--piperazinecarboxylic acid												x (Marjoram)
	Ammonium oxalate												x (Citrus)
Esters/Lactones	Methyl acetate				x								

	Ethyl acetate				x	x					x		
	Methylbenzoate				**x**	**x**		**x**		**x**		**x**	
	Methyl-4-hydroxybenzoate			*x*									
	Methylsalicylate				**x**	**x**				**x**		**x**	
	1-Propen-2-ol, formate					x							
Esters/Lactones	Butanoic acid, ethyl ester					x							
	Acetic acid, butyl ester					x					x		
	Methyl butyrate					x							
	Ethyl octanoate								x		x		
	Methyl oleate								x				
	Methyl linolenate												x (Citrus)
	Methyl phthalate												x (Clover)
	Ethyl phthalate												x (Clover)
	Diethyl phthalate												x (Clover)
	Dibutyl sebacate								x				
	Ethyl linolenate								x				
	Methyl anthranilate									x			
	Methyl *N*-methyl anthranilate												x (Citrus)
	Ethyl oxanoate										x		
	Methyl caprylate									x			
	δ- Dodecalactone								x				

(continued)

Table 9.1 (continued)

Group of compounds	Compounds	Boch et al. (1979) (1)	Drijfhout et al. (2005) (2)	Nazzi et al. (2009) (3)	Isidorov et al. (2012) (4)	Zhao et al. (2016) (5)	Hamid et al. (2018) (6)	Qi et al. (2020) (7)	El-Guendouz et al. (2020) (8)	Qi et al. (2021) (9)	Chen et al. (2022) (10)	Chen et al. (2023) (11)	Montaser et al. (2023) (12)
	γ-Octalactone					x				x		x	
	γ-Caprolactone					x				x			
	δ-Octalactone					x				x		x	
	δ-Decalactone									x		x	
Aldehydes/Ketones	Acetone				x								
	Hydroxyacetone												x (Marjoram) x (Citrus)
	2-Butanone				x								
	3-Methylbutanal				x						x		
	2-Methylbutanal				x	x					x		
	2-Pentanone				x	x					x		
	Acetoin										x		x (Citrus)
	2-Methyl-2-butenal				x								
	Hexanal				x	x					x		
	2-Heptanone				**x**			**x**	**x**	**x**	**x**	**x**	
	2-Nonanone				**x**	**x**			**x**	**x**	**x**	**x**	x (Marjoram)
	8-Nonen-2-one					**x**		**x**		**x**		**x**	
	2-Pentanone, 4,4-dimethyl												x (Citrus)
	Benzaldehyde				**x**	**x**		**x**	**x**	**x**	**x**	**x**	
	4-Octanone					x							
Aldehydes/Ketones	Octanal					x				x	x		
	Furfural					**x**			**x**	**x**	**x**		**x** (Marjoram)

	2-Furanocarboxaldehyde, 5-methyl					x	x						
	Acetophenone					x				x			
	Ethanone, 1-(1H-pyrrol-2-yl) or acetylpyrrole					X		x		x			
	2-Furancarboxaldehyde, 5-(hydroxymethyl)-					x							
	2-Cyclopenten-i-one, 2-hydroxy						x						
	(*E,E*)-2,4-Heptadienal							x	x	x			
	Benzeneacetaldehyde							**x**	**x**	**x**		**x**	
	(*E*)-Cinnamaldehyde							x					
	5-Ethylcyclopent-1-enecarboxaldehyde							x					
	2-Nonen-4-one							x		x		x	
	Cycloheptanone							x					
	p-Benzoquinone							x		x			
	n-Nonanal								**x**	**x**	**x**	**x**	
	n-Decanal								x			x	
	(*E,E*)-2,4-Decadienal												x (Citrus)
	2-Undecanone								x			x	
	n-Hexadecanal								x	x			
	n-Octadecanal								x	x			
	(*E*)-2-Pentenal									x			
	Heptanal									x	x		
	(*E*)-2-Heptenal									x			
	(*E*)-2-Octenal									x			
	(*E,Z*)-2,6-Nonadienal									x			

(continued)

Table 9.1 (continued)

Group of compounds	Compounds	Boch et al. (1979) (1)	Drijfhout et al. (2005) (2)	Nazzi et al. (2009) (3)	Isidorov et al. (2012) (4)	Zhao et al. (2016) (5)	Hamid et al. (2018) (6)	Qi et al. (2020) (7)	El-Guendouz et al. (2020) (8)	Qi et al. (2021) (9)	Chen et al. (2022) (10)	Chen et al. (2023) (11)	Montaser et al. (2023) (12)
	2-Phenylpropenal									x			
	3-Nonen-2-one									x			
	6,10,14-Trimethyl-2-pentadecanona									x			
	6-Methyl-5-hepten-2-one										x		
	(*E*)-2-Hexenal										x		
	2-Dodecanone											x	
Alcohols	Hydroquinone			x									
	Ethanol				x	x					x		x (Marjoram)
	Phenol				**x**	**x**		**x**		**x**			
	2-Methoxyphenol (*o*-guaiacol)				x			x		x			
	2-Methoxy-*p*-cresol				x								
	2-Furanomethanol					x							
	Benzyl alcohol					x		x		x			
	Phenylethyl alcohol					x		x		x			
	4-Ethylguaiacol							x					
	2-Nonanol							x		x			
	1-Hexanol							x		x			
	1-Heptanol									x			
	(Z)-3-Nonen-1-ol									x			
	1-Octanol							x					
	1-Dodecanol								x				x (Clover)

	1-Tetradecanol												x (Clover)
	2-Hexadecanol												x (Clover)
	1-Hexadecanol									x			x (Clover)
	1-Octadecanol								x				
	2-Cresol							x		x			
	2-Methoxy-5-methylphenol									x			
	2-Methyl-1-butanol										x		
	2-Propanol										x		
	1,3-Butylene glycol												x (Marjoram)
	2,3-Butylene glycol												x (Marjoram) x (Citrus)
	Myristic alcohol												x (Clover)
	2-Hexyl-1-octanol												x (Citrus)
	(R)-(−)-14-Methyl-8-hexadecyn-1-ol												x (Citrus)
Hydrocarbons	Toluene				x	x				x			
	1-Pentadecene				x								
	Isobutane												x (Marjoram)
	Ethylbenzene					x							
	Benzene,1,2,3,5-tetramethyl-					x							
Hydrocarbons	Styrene					x				x			
	Naphthalene					x				x			

(continued)

Table 9.1 (continued)

Group of compounds	Compounds	Boch et al. (1979) (1)	Drijfhout et al. (2005) (2)	Nazzi et al. (2009) (3)	Isidorov et al. (2012) (4)	Zhao et al. (2016) (5)	Hamid et al. (2018) (6)	Qi et al. (2020) (7)	El-Guendouz et al. (2020) (8)	Qi et al. (2021) (9)	Chen et al. (2022) (10)	Chen et al. (2023) (11)	Montaser et al. (2023) (12)
	Methylnaphthalene									x			
	1,3-Di tert-butylbenzene							x		x			
	Hexadecane							x	x	x			
	Heptadecane							x	x	x			
	Octadecane							x		x			
	Nonadecane							x		x			
	Octane								x				
	Nonane								x				
	trans-1,3-Nonadiene								x				
	2-Methyldecane								x				
	Pentadecane								x	x			
	Heneicosane								x	x			
	Tricosane								x				
	Pentacosane								x				
	Heptacosane								x				
	Nonacosane								x				
	1,3-Dimethylbenzene									x			
	1-Nonene									x			
	Eicosane									x			
	10-Heneicosene									x			
Terpenes	α-Terpinyl acetate							x					
	(−)-Bornyl acetate							x					
	Allocedrol							x					

	Epicubenol							x					
	Guaiol							x					
	γ-Eudesmol							x					
	Guaiazulene							x					
	Cubenol							x					
	Fenchone								x				
	Linalool								x				
	Cuminaldehyde								x				
	Carvacrol								X				
	Carvacrol acetate								x				
	ar-Curcumene								x	x			
	β-Bisabolene								x				
	Cedrol								x				
Terpenes	T-Cadinol								x				
	α-Cadinol								x				
	Zingiberene								x				
	6-Isocedrol									x			
	α-Ionone									x			
	(*E*)- β-Ionone									x			
	trans- β-ionon-5,6-epoxide												x (Marjoram)
	1,4-p-Menthadien-7-al											x	

(continued)

Table 9.1 (continued)

Group of compounds	Compounds	Boch et al. (1979) (1)	Drijfhout et al. (2005) (2)	Nazzi et al. (2009) (3)	Isidorov et al. (2012) (4)	Zhao et al. (2016) (5)	Hamid et al. (2018) (6)	Qi et al. (2020) (7)	El-Guendouz et al. (2020) (8)	Qi et al. (2021) (9)	Chen et al. (2022) (10)	Chen et al. (2023) (11)	Montaser et al. (2023) (12)
	trans- β-Ocimene											x	
Ethers	Anethole					x			x				
	cis-2-(2-Pentenylfuran)									x			
	Oxirane												x (Marjoram)
	3,4,5,6,7,8-Hexahydro-2H-chromene												x (Citrus)
Others	4H-Pyran-4-one, 2,3-dihydro-3,5-dihydroxy-6-methyl-					x	x						
	Indol							x					
	Benzoxazole					x							
	Methane, tetranitro												x (Marjoram) x (Citrus)
	Formamidine acetate												x (Marjoram)
	Dimethylamine-D1												x (Marjoram)
	d-Alanine												x (Marjoram)
	Ammonium carbamate												x (Clover)
	Cyclopropane,1,1-dibromo-2-chloro-2--fluoro												x (Clover)
	Vinyl fluoride												x (Clover)
	Butylated hydroxytoluene								x				

(1) The distillate was collected in a cold trap and then extracted with diethyl ether. Identification made by gas chromatography (GC) and gas chromatography—mass spectrometry (GC-MS) chromatography. Royal jelly was purchased from a commercial supplier in Taiwan

(2) Extraction with dichloromethane followed by dichloromethane:methanol and then column chromatography. The fractions obtained were analyzed by GC and GC-MS after methylation with methanol-sulfuric acid followed by silylation of the hydroxyl groups with bis(trimethylsilyl)trifluoroacetamide

(3) Extraction with diethyl ether which was further reduced to dryness under a gentle stream of nitrogen, stopping immediately the solvent was completely evaporated to avoid loss of volatile acids. Derivatization of the ethyl acetate extract with *N,O*-bis(trimethylsilyl)trifluoroacetamide, at 50 °C for 1 h. Analysis by GC-MS. Royal jelly obtained from hybrids of *Apis mellifera* (*A. mellifera ligustica* and *A. mellifera carnica* from colonies in Udine (northeastern Italy))

(4) Extraction using headspace solid-phase microextraction (HS-SPME). Samples were transferred into a headspace vial and immersed into a water bath (40 °C). The SPME fiber was exposed to the RJ sample headspace. The volatiles collected on the fiber were desorbed by introducing SPME fiber [divinyl benzene (DVB)/carboxy (CAR)/polydimethylsiloxane (PDMS)] into the inject port of the GC-MS equipment. Royal jelly obtained from *Apis mellifera carnica* Poll. colonies from northeastern Poland

(5) Extraction using HS-SPME. Samples were transferred into a headspace bottle and immersed into a water bath (50 °C) for 30 min. The SPME fiber was exposed to the RJ sample headspace. The volatiles collected on the fiber were desorbed by introducing SPME fiber into the inject port of the GC-MS equipment. Royal jelly obtained from *Apis mellifera* L. apiaries from China distributed in 10 different flowering seasons

(6) Methanolic extraction followed by GC-MS analysis. Royal jelly obtained from *Apis mellifera* L. colonies from north and south regions of Malaysia

(7) Extraction using HS-SPME. Samples were transferred into a headspace vial and immersed into a water bath (70 °C) for 50 min. The SPME fiber (DVB /carboxy/PDMS) was exposed to the RJ sample headspace. The volatiles collected on the fiber were desorbed by introducing SPME fiber into the inject port of the GC-MS equipment. Royal jelly obtained from honeybees of China, Italy and American Italian bees from the United States of America

(8) Extraction by hydrodistillation for 1 h using a Clevenger-type apparatus. Analysis by GC and GC-MS. Royal jelly samples were from Morocco, Spain and Portugal

(9) Extraction using HS-SPME. Samples were transferred into a headspace vial and immersed into a water bath (70 °C) for 50 min. The SPME fiber (DVB /Carboxen/ PDMS) was exposed to the RJ sample headspace. The volatiles collected on the fiber were desorbed by introducing SPME fiber into the inject port of the GC-MS equipment. Royal jelly samples were obtained from Italy and China, and from Italian bee colonies but maintained in China

(10) Extraction using HS and the volatiles analyzed by GC-ion mobility spectrometry (IMP). Samples were transferred into a plastic tube and maintained at 50 °C for 15 min. HS samples were automatically injected. Royal jelly obtained from *Apis mellifera* L. colonies of China

(11) Extraction using HS-SPME. Samples were transferred into a headspace vial and kept at 60 °C, for 5 min. The SPME fiber (DVB/CAR/PDMS) was exposed to the RJ sample headspace. The volatiles collected on the fiber were desorbed by introducing SPME fiber into the inject port of the GC-MS equipment. Royal jelly samples were obtained from China

(12) Extraction and analysis by using HS-GC-MS. Royal jelly samples from Egypt collected during the flowering period of citrus, marjoram and Egyptian clover

component was the alkene 1-pentadecene (17–21%). Phenols were the third major group of compounds (7–12%), in descending order of quantity: phenol > methyl salicylate >2-methoxyphenol >2-methoxy-*p*-cresol. In the same work, the authors (Isidorov et al. 2012) evaluated the effect of storage temperature (−18 °C, 4 °C, and room temperature) on the volatile profile of RJ. They verified that samples kept for 10 months at room temperature showed the greatest quantitative modifications, particularly the phenol content which decreased twofold (phenol from 12% to 7% and methyl salicylate from 7% to 3%), as well as the volatile C2 and C4 acid content (acetic acid from not detected to 7%, and butyric acid from not detected to 5%) and C5 aldehyde content (2-methylbutanal, 3-methylbutanal and 2-methyl-2-butenal were absent or in trace amounts in the fresh material while after 10 months their percentages were about 1%). The authors suggested that the increase in C2 and C4 carboxylic acids in RJ kept at room temperature may have been due to rancidity (Isidorov et al. 2012).

In RJ samples purchased from China, Chen et al. (2023) reported 14 more than 24 volatiles previously identified by Isidorov et al. (2012) in the volatile fractions of RJ from Poland. The total of 37 types of volatile compounds included 6 acids, 11 ketones, 7 esters, 6 aldehydes, 2 alcohols, 1 phenol, and 4 other compounds. Octanoic acid had the highest percentage in the volatile fraction. All RJ samples also presented 3-octenoic acid and 2-octenoic acid, 2-heptanone, 8-nonene-2-one, 2-nonen-4-one, 2-dodecanone, 2-nonanone, methyl benzoate, benzaldehyde, among other components. The amounts of these compounds changed over time when stored at 4 °C and 25 °C for 7, 14, 21, and 30 days. For example, the level of octanoic acid increased by 1.5 times (44% at 4 °C) and 1.8 times (48% at 25 °C), respectively, in comparison with that of the control group (RJ stored at −20 °C (27%). The content of 2-nonanone also increased from 17%, in the control group, to 32% (at 25 °C in 7 days). Compared to 4 °C, the content of methyl benzoate was much higher at 25 °C. On the other side, 3-octenoic acid and 2-octenoic acid showed a declined content as the storage temperature and time increased, the content of 2-heptanone decreased from 20%, in the control group, to 10%, at 4 °C in 30 days, and further decreased to 8% (at 25 °C in 30 days). Similarly, there was a progressive decline in the contents of 8-nonene-2-one, 2-nonen-4-one, 2-dodecanone, and benzaldehyde (Chen et al. 2023). Beyond the identification of the volatiles compounds by HS-SPME-GC/MS, these authors also evaluated the taste and smell of RJ samples through the electronic tongue (E-tongue) system and a Gemini E-nose, respectively. They found that the taste was significantly changed in RJ stored at 25 °C, suggesting that this system may be useful to detect slight differences in quality and taste characteristics among the RJ samples. E-nose system only could distinguish the overall change of smell in RJ samples, though limited in identifying the specific volatile components, being therefore necessary to use other more accurate methods, such as GC-MS to identify and quantify the volatiles (Chen et al. 2023).

The effect of storage temperature on the volatiles' profile had already been reported by Chen et al. (2022) also for Chinese RJ samples. The RJ samples stored at 28 °C for 30 days had lower amounts of ethyl acetate, heptanal, 2-pentanone, 2-nonanone, ethyl octanoate, 2-nonanone, ethyl hexanoate, 3-methyl-1-butanol,

2-heptanone, 2-propanol, butyl acetate, *n*-nonanal, ethanol, 6-methyl-5-heptene-2--one, and benzaldehyde than those stored at −20 °C (the amounts are unknown, they are not reported by the authors) (Chen et al. 2022). On the other hand, 3-methylbutanol, 2-methylbutanal, furfural, 2-butanone, and 2-methylpropanoic acid were higher in RJ at 28 °C for 30 days than the frozen samples. The authors (Chen et al. 2022) reported that a significant amount of acid, ester, aldehyde, and ketone compounds were present in RJ held at −20 °C and that these compounds gradually broke down over the 28 °C storage. Additionally, they showed that while storage at 28 °C caused some compounds to decrease and some to produce flavor compounds, these differences were somewhat unlike from those described by Isidorov et al. (2012). For instance, these authors found that *d*-ethyl acetate increased twofold in RJ samples after 10 months of storage at room temperature, although Chen et al. (2022) found that this compound decreased over several days at 28 °C. These authors partially attributed the extraction and detection methods to the disparity in results. These authors have used headspace-gas chromatography coupled to ion mobility spectrometry (HS-GC-IMS), whereas Isidorov et al. (2012) have used HS-SPME followed by GC-MS analysis of RJ samples.

Zhao et al. (2016) used GC-MS to analyze the volatile fraction of 30 RJ samples collected from 30 apiaries during flowering seasons of different nectar plants (ten types of RJ: *Schisandra chinensis*, *Medicago sativa*, *Vitex negundo*, *Citrullus lanatus*, *Sophora alopecuroides*, *Brassica campestris*, *Robinia pseudoacacia*, *Helianthus annus*, *Litchi chinensis*, *Tilia tuan*) in China after extraction by HS-SPME. A total of 89 compounds were identified by the authors, distributed among the following compound classes: esters and aldehydes (25 and 17.5%, respectively), ketones (15%), acids (10%), and alcohols (10%). Although 89 compounds were identified, the authors prudently decided that only 40 of these could be constitutive volatile components of RJ samples (Table 9.1) because they were present in relatively high amounts and in at least 2 types of RJ. Zhao et al. (2016) found that 2-nonanone and acetic acid were common to all RJ samples, whereas octanoic acid, benzaldehyde, and phenol were present in 9 RJ types, toluene and octanoic acid were present in 8 RJ types, and 2-pentanone and (benzoic acid, methyl ester) were present in 7 RJ types. On the other hand, there were 14 compounds that were only present in 2–4 RJ types, thus making them unsuitable for classification purposes: (1-propen-2-ol, formate), ethanol, (butanoic acid, ethyl ester), (acetic acid, butyl ester), hexanal, ethylbenzene, octanal, (benzene, 1,2,3,5-tetramethyl-), (2-furancarboxaldehyde, 5-methyl-), benzoxazole, 2-furanmethanol, methyl salicylate, (4H–pyran-4-one,2,3-dihydro-3,5-dihydroxy-6-methyl-), and 2-furan-carboxaldehyde, 5-(hydroxymethyl)-). (4H–Pyran-4-one,2,3-dihydro-3,5-dihydroxy-6-methyl-) was also found by Hamid et al. (2018) in two RJ samples from Malaysia (Penang of multifloral type and Johor of monofloral type, collected from Acacia trees) (Table 9.1). Both RJ samples were picked up in similar harvesting seasons and produced by *A. mellifera*. Ethanol and furfural, which are not constitutive compounds of RJ, were present in some samples possibly due to the fermentation of carbohydrates and heating during the extraction process, respectively (Zhao et al. 2016). The results obtained by Zhao et al. (2016) allowed the classification of all samples into three main groups;

nevertheless, there was one group that only contained RJ from the flowering season of *Tilia tuan*, which was considerably different from the samples belonging to the other two groups. This sample mainly consisted of alcohols, whereas acids, esters, aldehydes, and ketones predominated in diverse proportions in the other samples, allowing them to be grouped in the remaining classes.

In short, there are differences in the chemical composition of the volatile fractions of RJ samples studied by several authors (Nazzi et al. 2009; Isidorov et al. 2012; Zhao et al. 2016). Several factors were pointed out by Zhao et al. (2016) as being responsible for these differences: honeybee species, harvesting time, region, storage method, and processing technology. In fact, when comparing the results obtained by these three groups it is possible to verify differences, even within the same group when working with different RJs of varying origins. Nevertheless, when only comparing the results of Zhao et al. (2016) and Isidorov et al. (2012), who extracted the volatile fraction of RJ samples by the same procedure (HS-SPME) but not knowing the type of fiber used by Zhao et al. (2016), a common compound is found in all samples: 2- nonanone. Subsequent investigation revealed that what first appeared to be a general characteristic of the volatile component of RJ, unaffected by the nectar plant, area, honeybee species, storage technique, and processing technology, could not be so straightforward. Later on, other research teams studying the volatiles of RJ of different harvesting locations (Morocco and Spain) also reported traces of 2-nonanone, but not in the Portuguese RJ (El-Guendouz et al. 2020); in marjoram RJ from Egypt but not in citrus RJ and clover RJ of Egypt (Montaser et al. 2023); and in tea RJ from China, but not in pagoda RJ from China (Qi et al. 2021). Such results confirm that the presence of this ketone is not a characteristic of RJ because the same teams using RJ samples from different origins may present or not 2-nonanone, being this volatile dependent on the location and floral origin.

Other components were also common to the works of Zhao et al. (2016) and Isidorov et al. (2012): ethanol, acetic acid, ethylacetate, 2-pentanone, 2-methylbutanal, toluene, hexanal, benzaldehyde, phenol, octanoic acid, and methylsalicylate; however, Zhao et al. (2016) found that these compounds could also be absent in at least one sample in the study. Such absences in RJ of different nectar origins reveal that the presence of these compounds may be dependent on the nectar plant. This is reinforced by the fact that octanoic acid was also identified in the study of Nazzi et al. (2009), despite the use of a different extraction method to that of Zhao et al. (2016). However, the group also verified that octanoic acid was absent in the flowering season of *Sophora alopecuroides*. Montaser et al. (2023) isolated and identified but not quantified 23 volatile compounds in marjoram RJ, 22 volatile compounds in clover RJ, and 20 volatile compounds in citrus RJ by HS-GC-MS. Through this method, they identified the volatiles hydroxyacetone, 2,3-butylene glycol, and methano, tetranitro- that were present in both citrus and clover RJ samples (Table 9.1). 2-Nonanone was only detected in clover RJ and octanoic acid only in citrus RJ (Table 9.1). The remaining volatiles were distinctive for every type of royal jelly. Qi et al. (2021) also reported differences in the volatiles of RJ from different botanical origins: tea (*Camellia sinensis* L.) and pagoda tree (*Sophora japonica* L.). Tea RJ volatiles were for the first time studied and then compared to that of pagoda RJ. Regarding alkanes, 16 alkanes were identified in tea RJ, including both arenes and aliphatic hydrocarbons and, according to the authors, 10 of them were first identified in RJ: 1,3-dimethylbenzene, 1-nonene, 1,3-di-*tert*-butylbenzene, 2-methylnaphthalene, α-curcumene, pentadecane, heptadecane, nonadecane, 10-heneicosene, and heneicosane. Nevertheless, α-curcumene, pentadecane, heptadecane, and heneicosane had already been reported in diverse RJs from Iberian Peninsula and Morocco (El-Guendouz et al. 2020) (Table 9.1). Of the 14 aldehydes, 11 were registered by Qi et al. (2021) for the first time in tea RJ, although some of them had been reported in other food and beverages (e.g., furfural, (*E*)-2-heptenal, benzaldehyde, (*E,E*)-2,4-heptadienal, benzene acetaldehyde, (*E*)-2-octenal, nonanal. Furfural, benzaldehyde, (*E,E*)-2,4-heptadienal, benzene acetaldehyde, and nonanal were also previously reported in some RJ samples from Iberian Peninsula and Morocco (El-Guendouz et al. 2020). Benzaldehyde was found by seven research teams in diverse RJ samples (Table 9.1). Figure 9.1 shows its chemical structure as well as additional chemical structures that at least four research teams reported in RJ samples. Regarding the ketones, beyond the presence of 2-heptanone, acetophenone, 8-nonen-2-one and 2-nonanone already reported in other RJ samples from different origin (Isidorov et al. 2012; Zhao et al. 2016; El-Guendouz et al. 2020), Qi et al. (2021) detected for the first time 3-nonen-2-one, α-ionone, (*E*)-β-ionone and 6,10,14-trimethyl-2-pentadecanone. 2-Phenylethanol and benzyl alcohol were already reported in other RJ samples (Zhao et al. 2016). Nevertheless, Qi et al. (2021) also found in tea RJ the following alcohols: 1-hexanol, 1-heptanol, 1-phenylethanol, 2-nonanol, (*Z*)-3-nonen-1-ol, 6-isocedrol, and 1-hexadecanol. New phenols were also reported for the first time for RJ samples,

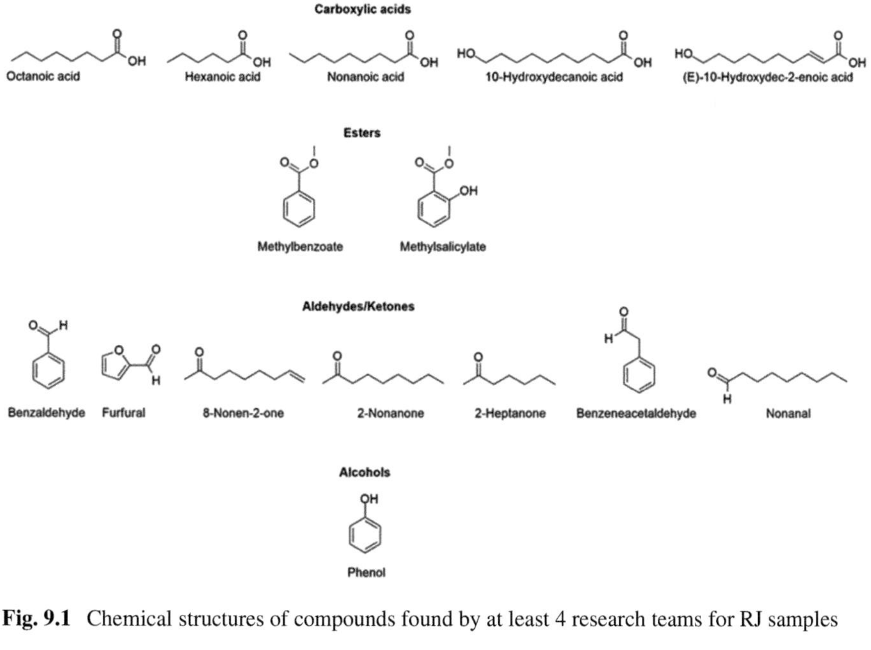

Fig. 9.1 Chemical structures of compounds found by at least 4 research teams for RJ samples

such as 2-cresol, 2-methoxy-5-methylphenol, and 2,3,5-trimethylphenol, although reported in other natural materials (Qi et al. 2021). The nonanoic acid or pelargonic acid claimed by the authors for tea RJ had also been previously reported by El-Guendouz et al. (2020) for Iberian and Moroccan RJ samples. In this case, the floral origin is unknown. Qi et al. (2021) claimed that the lactones γ-caprolactone, γ-octalactone, δ-octalactone were for the first time found in RJ samples. Nevertheless, Zhao et al. (2016) had already reported the presence of these compounds in RJ samples. However, with the respective synonym names 2,3(H)-furanone,5-ethyldihydro-, 2,3(H)-furanone,5-butyldihydro-, and 2H-pyran-2-one,tetrahydro-6-propyl-. δ-Dodecalactone had also been previously found in some samples of Spain and Morocco (El-Guendouz et al. 2020) (Table 9.1).

In the comparison between tea and pagoda RJ samples, performed by Qi et al. (2021) and based on 35 differential compounds, they noticed that 22 compounds were present in all RJ samples, whereas 4 existed exclusively in tea RJ (α-ionone, (*E*)-2-pentenal, 1-phenylethanol, and 1,3-dimethylbenzene). The two pagoda RJ samples shared 30 compounds; nevertheless, only eight were present in two pagoda RJ samples (epicubenol, allocedrol, guaiol, γ-eudesmol, β-*cis*-ocimene, β-calacorene, DL-pantolactone and (−)-bornyl acetate) (Qi et al. 2021) These data are not depicted in Table 9.1. Despite more than 20 compounds that can be found in tea and pagoda RJ samples, their amounts may profoundly vary. For example, 15 compounds are at a lower level of content in tea RJ than pagoda RJ samples, and phenolic 2-methoxyphenol, α-curcumene, 2-methoxy-5-methylphenol, and 6-isocedrol are 10 times more in pagoda RJs than in tea RJ (Qi et al. 2021). According to these results, the authors deduced that RJ volatile profiles may be attributed to the different trees that contain some specific compounds, as for α-ionone only identified in tea RJ. The presence of α-ionone in tea RJ is due to tea tree or tea pollen (Qi et al. 2021). Although this hypothesis could be considered, such must be much more studied since some compounds reported by these authors as being only detected in tea RJ, had been already reported in some RJ samples from the Iberian Peninsula or Morocco (El-Guendouz et al. 2020). Nevertheless, these authors did not perform a study on the chemical composition of the flora around the beehives where RJ samples were collected for being compared with those of RJs.

Genetic selection to increase RJ yield has been made in China from Italian *Apis mellifera ligustica* since the 1980s. Such selection has led to an RJ increase of 5–10 tenfold. Qi et al. (2020) compared the volatile fractions of RJ samples obtained from non-genetically and genetically selected bees and concluded that there are also differences in the chemical composition, analyzed by HS-SPME, beyond the RJ production. The authors also studied the chemical composition of RJ from American Italian bees. To avoid environmental and human factors on the chemical composition, the study was made with colonies of each bee strain with identical colony management, population, and brood pattern, and all were kept at the apiary of the Institute of Apicultural Research, Chinese Academy of Agricultural Sciences in Beijing. Thirty-seven volatiles were identified but only 15 of them were identified by comparing information with that of authentic standards. Those 37 entities included 5 aldehydes, 5 esters, 5 alkanes, 14 alcohols and phenols, 4 ketones, and 4 other components. The differences among the three samples were mainly in content rather than constitution (Qi et al. 2020), with the exception of guaiazulene which was only present in RJ from genetically selected bees. Guaiazulene, 2-methoxyphenol, and 2-methylphenol were the most important components that differentiated the 3 RJ samples (RJ from non-genetically and genetically selected bees, and American Italian bees). These 3 samples were arranged into two groups, in which the RJ from genetically selected bees was alone in one group. Most aldehydes except benzeneacetaldehyde occurred in significantly higher abundance in the RJ from genetically selected bees than in the remaining RJ samples. Regarding the five esters found in all samples (methyl benzoate, α-terpinyl acetate, γ-caprolactone, γ-octalactone, and bornyl acetate) there were no uniform variations. The five alkanes identified in the three RJ (1,3-di-*tert*arybutylbenzene, hexadecane, heptadecane, octadecane, and nonadecane) were in lower amounts in RJ from genetically selected bees than in the remaining RJ samples. Most of the 14 alcohols and phenols were present in higher amounts (e.g., guaiol, γ-eudesmol, cubenol, and phenylethyl alcohol) in the RJ from genetically selected bees, excepting phenol, 2-methylphenol, 2-methoxyphenol, and 2-nonanol. The content of 2-methoxyphenol was much higher (11 times) in the RJ from American Italian bees than that in RJ from genetically selected bees. The content of 2-nonen-4-one and cycloheptanone were higher in RJ from genetically selected bees, whereas 8-nonen-2-one and 2-heptanone were lower in this RJ than in the remaining samples. In the other metabolites' group (2-acetylpyrrole, p-benzoquinone, indole, and guaiazulene), indole was not detected in RJ from genetically selected bees in contrast to guaiazulene; and 2-acetylpyrrole content was higher in this sample than in the remaining ones (Qi et al. 2020).

The conclusion drawn from the findings of different research teams is that more thorough studies into RJ volatiles are required to identify potential variables, such as the type of bee used, the location and local flora, extraction methods, isolation and analysis of the volatiles, and storage conditions, that may affect the volatile composition of RJ. These results may lead to the identification of volatile markers that might be used to identify specific RJ samples that have greater profit potential.

Acknowledgments The authors acknowledged MED (https://doi.org/10.54499/UIDB/05183/2020; https://doi.org/10.54499/UIDP/05183/2020) and CHANGE (https://doi.org/10.54499/LA/P/0121/2020).

Funding This research was funded by the FCT (Fundação para a Ciência e a Tecnologia, Portugal) under the project UIDB/05183/2020.

References

Ahmad S, Campos MG, Fratini F, Altaye SZ, Li J (2020) New insights into the biological and pharmaceutical properties of royal jelly. Int J Mol Sci 21:382

Balkanska R, Karadjova I, Ignatova M (2014) Comparative analyses of chemical composition of royal jelly and drone brood. Bulg Chem Commun 46:412–416

Bărnuţiu LI, Mărghitas LA, Dezmirean DS, Mihai CM, Bobiş O (2011) Chemical composition and antimicrobial activity of royal jelly—review. J Anim Sci Biotechnol 44:67–72
Boch R, Shearer DA, Shuel RW (1979) Octanoic and other volatile acids in the mandibular glands of the honeybee and in royal jelly. J Apic Res 18:250–252
Bogdanov S (2016) Royal jelly, bee brood: composition, health, medicine: a review. Bee Product Science, from https://wwwbee-hexagon.net April 2016
Boselli E, Caboni MF, Sabatini AG, Marcazzan GL, Lercker G (2003) Determination and changes of free amino acids in royal jelly during storage. Apidologie 34:129–137
Chen D, Lu W, Ye Q, Zhang C, Ge X, Xiao C (2022) HS-GC-IMS and ATR-FT-MIR analysis reveal the differences in volatile compounds, proteins, and polyphenols of royal jelly. Adv Mater Sci Eng 2022:3223558
Chen L, Ning F, Zhao L, Ming H, Zhang J, Yu W, Yi S, Luo L (2023) Quality assessment of royal jelly based on physicochemical properties and flavor profiles using HS-SPME-GC/MS combined with electronic nose and electronic tongue analyses. Food Chem 403:134392
Ciulu M, Floris I, Nurchi VM, Panzanelli A, Pilo MI, Spano N, Sanna G (2013) HPLC determination of pantothenic acid in royal jelly. Anal Methods 5:6682–6685
Collazo N, Carpena M, Nunez-Estevez B, Otero P, Simal-Gandara J, Prieto MA (2021) Health promoting properties of bee royal jelly: food of the queens. Nutrients 13:543
Drapeau MD, Albert S, Kucharski R, Prusko C, Maleszka R (2006) Evolution of the yellow/major royal jelly protein family and the emergence of social behaviour in honey bees. Genome Res 16:1385–1394
Drijfhout FP, Kochansky J, Lin S, Calderone NW (2005) Components of honeybee royal jelly as deterrents of the parasitic varroa mite, Varroa destructor. J Chem Ecol 31:1747–1764
El-Guendouz S, Machado AM, Aazza S, Lyoussi B, Miguel MG, Mateus MC, Figueiredo AC (2020) Chemical characterization and biological properties of royal jelly samples from the Mediterranean area. Nat Prod Commun 15(2):1–13
Ferioli F, Armaforte E, Caboni MF (2014) Comparison of the lipid content, fatty acid profile and sterol composition in local Italian and commercial royal jelly samples. J Am Oil Chem Soc 91:875–884
Fujita T, Kozuka-Hata H, Ao-Kondo H, Kunieda T, Oyama M, Kubo T (2013) Proteomic analysis of the royal jelly and characterization of the functions of its derivation glands in the honeybee. J Proteome Res 12:404–411
Guo H, Kouzuma Y, Yonekura M (2009) Structure and properties of antioxidative peptides derived from royal jelly proteína. Food Chem 113:238–245
Hamid NAB, Bakar ABA, Mohamed M (2018) Phytochemical analysis and GC-MS profile of royal jelly from selected areas in Malaysia. Malays Appl Biol 47(3):101–107
He R, Ye J, Wang L, Sun P (2020) Preparation and evaluation of microcapsules encapsulating royal jelly sieve residue: flavor and release profile. Appl Sci 10:8126
Hospital XF, Hierro E, Martín-Cabrejas I, Caballero N, Jiménez B, Sánchez-Martín V, Morales P, Haza AI, Fernámdez M (2024) Bee products as na alternative for the preservation of nitrate and nitrite-reduced dry fermented sausages. Food Biosci 59:104048
Hu FL, Bíliková K, Casabianca H, Daniele G, Salmen Espindola F, Feng M, Guan C, Han B, Krištof Kraková T, Li JK et al (2019) Standard methods for *Apis mellifera* royal jelly research. J Apic Res 58:1–68
Isidorov VA, Bakier S, Grzech I (2012) Gas-chromatographic-mass spectrometric investigation of volatile and extractable compounds of crude royal jelly. J Chromatogr B Analyt Technol Biomed Life Sci 885–886:109–116
Isidorov VA, Bakier S, Stocki M (2016) GC-MS investigation of the chemical composition of honeybee drone and queen larva homogenate. J Apic Sci 60. https://doi.org/10.1515/JAS-2016-0011
Kunugi H, Ali AM (2019) Royal jelly and its components promote healthy aging and longevity: from animal models to humans. Int J Mol Sci 20:4662

Lercker G, Capella P, Conte LS, Ruini F (1981) Components of royal jelly: I. Identification of the organic acids. Lipids 16:912–919

Lercker G, Capella P, Conte LS, Ruini F, Giordani G (1982) Components of royal jelly II: the lipid fraction, hydrocarbons and sterols. J Apic Res 21:178–184

Liming W, Jinhui Z, Xiaofeng X, Yi L, Jing Z (2009) Fast determination of 26 amino acids and their content changes in royal jelly during storage using ultra-performance liquid chromatography. J Food Compos Anal 22:242–249

Martinotti S, Bonsignore G, Ranzato E (2023) Applications of beehive products for wound repair and skin care. Cosmetics 10:127

Montaser M, Sayed AM, Bishr MM, Zidan EW, Zaki MA, Hassan HM, Mohammed R, Hifnawy MS (2023) GC-MS analysis of honeybee products derived from medicinal plants. Beni-Suef Univ J Basic Appl Sci 12:63

Mureşan CI, Mărghitaş LA, Dezmirean DS, Bobiş O, Bonta V, Zacharias I, Mărgăoan R, Paşca C (2016) Quality parameters for commercialized royal jelly. Bull UASVM Anim Sci Biotechnol 73:1. https://doi.org/10.15835/buasvmcn-asb:11630

Nabas Z, Haddadin MSY, Haddadin J, Nazer IK (2014) Chemical composition of royal jelly and effects on symbiotic with two different locally isolated probiotic strains on antioxidant activities. Pol J Food Nutr Sci 64:171–180

Nazzi F, Bortolomeazzi R, della Vedova G, del Piccolo F, Annoscia D, Milani N (2009) Octanoic acid confers to royal jelly varroa - repellent properties. Naturwissenschaften 96:309–314

Noda N, Umebayashi K, Nakatani T, Miyahara K, Ishiyama K (2005) Isolation and characterization of some hydroxy fatty and phosphoric acid and esters of 10-hydroxy-2-decenoic acid from the royal jelly of honey bees (*Apis mellifera*). Lipids 40:833–838

Oršolić N, Jembrek MJ (2024) Royal jelly: biological action and health benefits. Int J Mol Sci 25:6023

Park G, Choi H, Cho J, Chang S, Lee H, Choi J (2024) Effects of bee products (propolis, royal jelly and honeybee comb extracts) on physico-chemical and storage characteristics of pork patties. J Food Nutr Res 63(1):19–28

Qi D, Ma C, Wang W, Zhang L, Hao J, Li J (2020) Gas-chromatography-mass spectrometry analysis reveals the differences in volatile components of royal jelly from different honeybee stocks. LWT Food Sci Technol 124:109143

Qi D, Ma C, Wang W, Zhang L, Hao J, Li J (2021) Gas-chromatography-mass spectrometry analysis as a tool to reveal differences between the volatile compound profile of royal jelly produced from tea and pagoda trees. Food Anal Methods 14:616–630

Ramadan MF, Al-Ghamdi A (2012) Bioactive compounds and health-promoting properties of royal jelly: a review. J Funct Foods 4:39–52

Ramanathan ANKG, Nair AJ, Sugunan VS (2018) A review on royal jelly proteins and peptides. J Funct Foods 44:255–264

Sabatini AG, Marcazzan GL, Caboni MF, Bogdanov S, de Almeida-Muradian LB (2009) Quality and standardisation of royal jelly. JAAS 1:1–6

Terada Y, Narukava M, Watanabe M (2011) Specific hydroxyl fatty acids in royal jelly activate TRPA1. J Agric Food Chem 59:2627–2635

Wang Y, Ma L, Zhang W, Cui X, Wang H, Xu B (2016) Comparison of the nutrient composition of royal jelly and worker jelly of honey bees (*Apis mellifera*). Apidologie 47:48–56

Zhao Y-z, Li Z-g, Tian W-l, Fang X-m, Su S-k, Peng W-j (2016) Differential volatile organic compounds in royal jelly associated with different nectar plants. J Integr Agric 15:1157–1165. https://doi.org/10.1016/S2095-3119(15)61274-6

Zheng Y, Pan C, Xu P, Liu K (2024) Hydrogel-mediated extracellular vesicles for enhanced wound healing: the latest progress, and their prospects for 3D bioprinting. J Nanobiotechnol 22:57

Chapter 10
Royal Jelly: Health Benefits and Uses in Medicine

Manuel Viuda-Martos, Raquel Lucas-González, José A. Pérez-Álvarez, and Juana Fernández-López

Contents

Abstract Royal jelly (RJ) is a thick and "milky" secretion from the hypopharyngeal and mandibular glands of young worker bees (Apis mellifera L.) and is used to feed the larvae. Due to the high content of bioactive compounds present in their composition, mainly peptides such as royalisin, jelleines, aspimin and royalactin, polyphenols principally phenolic acids and flavonoids, and lipids as 10-Hydroxy-2-Decenoic Acid. In addition, royal jelly exhibits lots of health-promoting properties such as antibacterial, antioxidant anti-inflammatory, vasodilative and hypotensive, hipocholesterolemiant and hepatoprotective, and antitumor among others. Therefore, royal jelly has been used extensively for improving human health in food and nutraceutical industries. This chapter will examine current literature on bioactive compounds content and health benefits of the royal jelly, with a particular focus given to their uses in medicine.

Keywords Royal jelly · Health effects · Bioactive compounds · Peptides · Polyphenolic · Lipids

M. Viuda-Martos · R. Lucas-González · J. A. Pérez-Álvarez · J. Fernández-López (✉)
IPOA Research Group. Agri-Food and Agri-Environmental Research and Innovation Institute (CIAGRO-UMH), Miguel Hernández University, Orihuela, Alicante, Spain
e-mail: j.fernandez@umh.es

J. M. Alvarez-Suarez (ed.), *Bee Products – Chemical and Biological Properties*,
https://doi.org/10.1007/978-3-031-89049-9_10

10.1 Introduction

Royal jelly (RJ) is a creamy white-yellow liquid that is secreted by the mandibular and hypopharyngeal glands of the honeybee (Apis mellifera L.) which is essential for feeding and raising broods and queens (Paganini Damiani et al. 2022). This substance is yellowish, highly acidic (pH 3.4–4.5), and partially soluble in water material with a slightly pungent odor and taste (Kazemi et al. 2019). In addition, royal jelly shows a gelatinous consistency, often not homogenous due to the presence of undissolved granules of varying sizes composed mainly of pollen grains derived from the foraging activity of honeybees (Marwaha et al. 2022). Chemical analyses reveal that this substance contains 60–70% water, 12–18% proteins (albumin, α, β, γ globulin, glycoproteins, and lipoproteins), 10–16% carbohydrates, 3–6% lipids as well as minerals and vitamins including B_1, B_2, B_3, B_5, B_6, B_7, B_9, B_{12}, E, D, A, K, and C (Moskwa et al. 2023; Ghosh and Jung 2024). In their composition, also it is possible to find a huge amount of bioactive compounds principally (*i*) peptides such as royalisin, jelleines, aspimin, and royalactin (Li et al. 2021), (*ii*) polyphenolic compounds including phenolic acids and flavonoids that are categorized into flavanones, flavonols, flavones, and isoflavones (Kumar et al. 2024), (*iii*) lipids as 10-hydroxy-2-decenoic acid (10-HDA) and 9-hydroxy-2-decenoic acid (9-HAD) (Guo et al. 2021), and (*iv*) sterols including 24-methylene cholesterol (24MET), campesterol, stigmasterol, desmosterol, β-sitosterol, and Δ5-avenasterol (Yan et al. 2022).

Nonetheless, it is important to bear in mind that the royal jelly composition may depend on several factors such as climate and environmental conditions, seasonal and regional conditions, botanical sources, honeybee breeds, age of bee larvae, and beekeeping practices among others (Khalfan Saeed Alwali Alkindi et al. 2024; Botezan et al. 2023; Ghosh and Jung 2024) Due to the high content in bioactive compounds presents in their composition, royal exhibits lots of health-promoting properties such as antibacterial, antioxidant, anti-inflammatory, vasodilative and hypotensive, hipocholesterolemiant and hepatoprotective, antidiabetic, and antitumor (Altun and Aydemir 2022; Almeer et al. 2018; Pan et al. 2019; Jovanović et al. 2023; Bagameri et al. 2023; Omer et al. 2019) among the others. Although the cause-and-effect relationship has not been established between the consumption of royal jelly and the physiological functions up to now (Agostoni et al. 2011), royal jelly has been used extensively for improving human health in the food and nutraceutical industry. However, it is also important to know the disadvantages of the use of royal jelly that include the adulteration, the fact that it is perishable, its production is seasonal, and its price is high (Collazo et al. 2021; Botezan et al. 2023).

This chapter will examine current literature on bioactive compounds content and health benefits of the Royal Jelly, with a particular focus given to their uses in medicine.

10.2 Antibacterial Activity

For many years, a vast range of different chemical and synthetic compounds have been employed as antibacterial and antifungal agents to prevent microbial spoilage. However, due to economic factors and increasing consumer concerns regarding the safety of products with synthetic chemicals, there has been a shift in focus toward natural or naturally derived compounds (Stan et al. 2021). Furthermore, the scientific community is interested in biologically active compounds obtained and isolated from natural sources for eliminating pathogenic microorganisms due to the resistance that microorganisms have built up against antibiotics (Eshboev et al. 2024). In this way, the antibacterial properties of royal jelly against Gram-positive and Gram-negative bacteria as well as several molds have been widely demonstrated by several authors. Therefore, Dinkov et al. (2016) investigated the antibacterial effect of royal jelly, obtained from beekeepers, from the region of Stara Zagora, Bulgaria, against Methicillin-resistant *Staphylococcus aureus* (MRSA). These authors reported that at least 3 decimal reductions of MRSA count were observed in Tryptone Soy broth with concentrations of 20% and 30% of royal jelly. Coutinho et al. (2018) investigated the antibacterial effect of royal jelly against periodontopathic bacteria including *Aggregatibacter actinomycetemcomitans*, *Porphyromonas gingivalis*, *Prevotella intermedia*, and *Fusobacterium nucleatum*. They reported that all the tested bacterial strains were sensitive to royal jelly. The minimum bactericidal concentration (MBC) values were identical and showed bactericidal effect in the range between 12.5 and 100 μg/ml. In other study, Altuntas et al. (2020) evaluated the antibacterial activity, at different concentrations (0.3341.67 mg/mL), of royal jelly obtained from Bursa Inegol region, in Turkey against *Listeria monocytogenes* ATCC 7644. The authors reported that the minimum inhibitory concentration $(MIC)_{90}$ was 23.85 mg/mL, which confirms the possibility of using royal jelly as an anti-biofilm agent to reduce the risk of listeriosis contamination. Bengü et al. (2020) carried out a study to analyze the antibacterial activity of Bingol royal jelly cultivated in Turkey against *Salmonella typhimurium* NRRL 441, *Escherichia coli* ATCC 25922, and *S. aureus* ATCC 6538 P. They found that Bingol royal jelly, at concentrations of 1 mg/mL, showed antimicrobial activity against *S. typhimurium*, *E. coli*, and *S. aureus* with inhibitions halos of 8.64, 9.10, and 10.73 mm, respectively. Uthaibutra et al. (2023) assessed the antibacterial activity of royal jelly from Thailand against skin pathogenic bacteria, including *Cutibacterium acnes*, methicillin-resistant *S. aureus*, *Pseudomonas aeruginosa*, *S. aureus*, *Staphylococcus epidermidis*, and *Corynebacterium* spp. and found minimum inhibitory concentration (MIC) values of 18.75, 37.50, 75.00, 37.50, 75.00, and 18.75 mg/mL, respectively while the minimum bactericidal concentration (MBC) values were 37.50, 37.50, 75.00, 75.00, 75.00, and 18.75 mg/mL, respectively.

According to Ramanathan et al. (2018), the antibacterial royal action could be attributed to the ether-soluble fraction and particularly 10-carbon molecule of fatty acid, the trans-10-hydroxy-2-decenoic acid (10-HDA). In this sense, Yang et al. (2018) reported that 10-HDA acted as potent bactericide against animal- or human-specific pathogens, including *S. aureus*, *Streptococcus alactolyticus*, *Staphylococcus intermedius* B, *Staphylococcus xylosus*, *Salmonella cholearasuis*, *Vibrio parahaemolyticus*, and *E. coli* (hemolytic). Thus, MIC of 10-HDA against Gram-positive bacteria was 23–44 μM, while the MBC was 33–66 μM. On the other hand, the MIC and MBC of 10-HDA against Gram-negative bacteria were in the ranges of 40–43 μM and 74–78 μM, respectively. Šedivá et al. (2018) analyzed the antibacterial effect of 10-HDA against different genotypes and source of *Paenibacillus larvae*. They obtained MIC values ranged between 0.60 and 2.80 μg/mL.

Others important compounds found in royal jelly with huge antibacterial properties are peptides such as jelleines (jelleine-I, jelleine-II, jelleine-III and jelleine-IV) and royalisin. Therefore, Bilikova et al. (2015) analyzed the antibacterial activity of royalisin against Gram-positive and Gram-negative bacteria and found that this peptide produced minimal inhibitory concentration values of 7.5, 9.0, 4.0 and 10.5 for *S. aureus*, *S. alactolyticus*, *S. intermedius B* and *S. xylosus*, respectively as Gram-positive bacteria whilst for Gram-negative the minimal inhibitory concentration values obtained were 10.0, 9.0 and 4.0 for *Ps. aeruginosa*, *S. cholearasuis* and *V. parahaemolyticus*, respectively. On the other hand, the antimicrobial activity of jelleines has been tested against some Gram-positive and some Gram-negative bacteria. Thus, Jia et al. (2021) reported that Jelleine-I, showed potent antimicrobial activity against bacteria strains such as *E. coli* (ATCC25922), *E. coli* ML-35 (ATCC 43837), *Ps. aeruginosa* (ATCC27853), *Klebsiella influenza* (ATCC 700603*)*, *Cronobacter sakazakii* (ATCC 29544), *S. aureus* (ATCC 29213), *Bacillus subtilis* (ATCC 23857), and *S. epidermidis* (ATCC 12228), with the MICs of 32, 32, 32, 64, 64, 128, 32, and 64 μM, respectively whilst the MBCs values were 64, 32, 256, 128, 128, 128, 64 and 64 μM, respectively. Zhou et al. (2021) informed that optimized analogs of Jelleine-I, have shown enhanced antimicrobial activity against many multi-drug-resistant strains with minimal toxicity including *Ps. aeruginosa*, extended-spectrum β-Lactamase *E. coli*, and methicillin-resistant *S. aureus*. In a similar study, Shen et al. (2023) reported that the peptide Jelleine-I exhibited antimicrobial activity against both gram-positive and gram-negative bacteria. Notably, Jelleine-I exhibited potent activity against *L. monocytogenes, Salmonella enterica* ATCC 14028, and *E. coli* CGMCC 1.8723 with MIC values of 12.5, 50, and 100 μg/mL, respectively whereas MBC values were 50, 100 and 100 μg/mL, respectively.

Antimicrobial peptides produce their antimicrobial action through one of the following action mechanisms: (*i*) Induction of Reactive Oxygen Species production; (*ii*) Binding to chromosomic DNA; (*iii*) Binding to bacterial adhesins; (*iv*) Inhibition of energy metabolism; and (*v*) altering the membrane permeability or by making canals or pores (e.g., Į-helix peptides) (Lima et al. 2024).

In reference to the antifungal activity, Moghim et al. (2021) conducted a study to analyze the antifungal activity against *Candida albicans* of royal jelly of Iranian origin. These authors reported that the MIC value and minimum fungicidal concentration value of royal jelly on *C. albicans* were, respectively, 80, and 160 mg/mL. In a similar study, Gunalp et al. (2018) found that the concentration of royal jelly, from Turkey, at 10% (0.125–1 μg/ml) and above was effective on *C. albicans*. Ulubayram and Cinar (2023) analyzed the antifungal capacity of royal jelly from Bursa (İnegöl region, Turkey) against the molds *Penicillium digitatum* and *Aspergillus flavus*. They reported that royal jelly showed antifungal activity with inhibitions halos, at 125 mg/mL, of 27.30 and 14.92 mm, for *A. flavus* and *P. digitatum*, respectively.

All of these results provide evidence for the presence of antibacterial compounds in this secretion of *Apis millifera*. These findings clearly demonstrate and confirm the effectiveness of royal jelly in inhibiting microbial activity.

10.3 Anti-inflammatory Activity

Inflammation is a natural defense response of the organism against stimuli, such as trauma, infectious agents, toxic and chemical substances, and physical and surgical operations. The inflammatory response is initiated by various chemical and/or biological factors, including pro-inflammatory enzymes and cytokines, low-molecular-weight molecules like eicosanoids, and the enzymatic breakdown of tissues (Liu 2022). This process which is the first physiological defense system in the human body, can protect against injuries caused by physical wounds, poisons, among others. This defense mechanism, known as acute inflammation, is capable of eliminating infectious microorganisms, removing irritants, and supporting normal physiological functions. However, prolonged or excessive inflammation can lead to disruptions in regular physiological processes, potentially resulting in conditions like asthma and rheumatoid arthritis (Charoenngam et al. 2020).

Royal jelly, or different bioactive compounds such as fatty acid, mainly 10-HAD, or major proteins present therein, has been proposed as an anti-inflammatory agent. The anti-inflammatory properties of royal jelly were investigated in formalin-induced rat paw edema by Arzi et al. (2015). These authors demonstrated that royal jelly had a dose-dependent relationship in inhibiting formalin-induced inflammation. Thus, royal jelly (doses 50 and 100 mg/kg) significantly inhibited formalin-induced rat paw edema as compared to the control group. Also, it was shown that there was no significant difference between groups treated with doses of 50 and 100 mg/kg and the positive control group (aspirin 300 mg/kg) in inhibiting formalin-induced swelling. Moreover, 25 mg/kg of royal jelly could not diminish paw swelling in response to formalin. Royal jelly at doses of 50 and 100 mg/kg showed comparable effects to aspirin (300 mg/kg) to inhibit paw edema up to 60%. A similar study was carried out by Gu et al. (2018) who investigated the anti-inflammatory activity of enzyme-treated royal jelly, which is an allergen-free form of royal jelly, on mice. These authors reported that mice, fed (50, 150, 300, and 500 mg/kg of enzyme-treated royal jelly) for 4 weeks and stimulated lipopolysaccharide-induced stress, significantly reduced levels of tumor necrosis factor-alpha, interleukins-1, 6, 10, 12, and interferon gamma compared to control mice. In the same line of thought, Petelin et al. (2019) examined the anti-inflammatory actions of royal jelly in asymptomatic overweight adults, considered at an early stage of developing metabolic syndrome. These authors reported that royal jelly supplementation (666 mg) decreased significantly the inflammatory marker C-reactive protein (19%), whereas significant increases were observed in the anti-inflammatory marker adiponectin (34%). More recently, Guo et al. (2022) explored the projected effects of royal jelly against Dextran-Sulfate-Sodium-Induced Colitis. They found that the administration of 1.0 g/kg, 1.5 g/kg, and 2.0 g/kg body weight royal jelly every day for 31 days reduced the expression of proinflammatory cytokine IL-6 and increased the expression of anti-inflammatory cytokine IL-10 and sIgA in C57BL/6JNifdc mice. Aslan et al. (2022a) informed that in rats fed with royal jelly, at 100 mg/kg, decreased downregulation of COX-2, Bcl-2, glycogen synthase kinase 3, and TNF-α protein expressions. Whereas increased glutathione and catalase activities, as well as, upregulated Bax, Brain-derived neurotrophic factor, caspase-3, caspase-6, and caspase-9 protein expressions in rats injured by the fluoride exposure. Nazarinia et al. (2023) pointed out that different doses (100 and 200 mg/kg) of royal jelly gave rise to a significant decrease in the level of IL-1β, TNF-α, and malondialdehyde and it restored the activities of glutathione peroxidase and superoxide dismutase enzymes in Wistar rats. In the study of Kheirdeh et al. (2023) the impact of royal jelly on inflammatory factors in the hippocampus, as well as depression and anxiety were explored. They found that royal jelly at 50 and 100 mg/kg significantly decreased IL-17, IL-23 gene expression, anxiety, and depression indices, and significantly increased IL-10 in Sprague-Dawley rats compared to the control group.

As occurs with antimicrobial activity, several components found in royal jelly including lipids mainly 10-hydroxy-2-decenoic acid (10-HAD), or proteins primarily major royal jelly proteins (MRJPs) have been shown to have anti-inflammatory activity. In reference to the anti-inflammatory activity of 10-HAD, Chen et al. (2016) evaluated the anti-inflammatory activity of the three major fatty acids present in royal: 10-HDA, 10-hydroxy decanoic acid (10-HDAA), and sebacic acid. They evaluated and compared the in vitro anti-inflammatory effects of these royal jelly fatty acids in lipopolysaccharide-stimulated RAW 264.7 macrophages. The results showed that 10-HDA, 10-HDAA, and sebacic acid had potent, dose-dependent inhibitory effects on the release of the major inflammatory mediators, nitric oxide, and interleukin-10, and only sebacic acid decreased TNF-α production. Several key inflammatory genes have also been modulated by these royal jelly fatty acids, with 10-H2DA showing distinct modulating effects as compared to the other two fatty acids. Additionally, these authors found that found that these three FAs regulated several proteins involved in mitogen-activated protein kinase and nuclear factor-κB signaling pathways. In a more recent study, Chen et al. (2018) informed that 10-HDA had potent, dose-dependent inhibitory effects on the release of major inflammatory mediators and nitric oxide. Several key inflammatory genes, including IL-1β, IL-6, MCP-1, and COX-2, have also been suppressed by 10-HAD in Lipoteichoic acid from *Staphylococcus aureus* induced RAW 264.7 macrophages. In the same sense, Yang et al. (2018) analyzed the anti-inflammatory activity of the main component of royal jelly, 10-HDA acid, on human WiDr colon cancer cells. The authors found that 10-HDA has an inhibitory effect on the production of pro-inflammatory cytokines IL-1β, IL-8, and tumor necrosis factor-α in WiDr cells.

Moreover, it induced the production of IL-1ra in a dose of 0.1–3.0 mM, as a consequence of which receptor antagonist cytokine limited the production of IL-1β. 10-HDA significantly inhibited the production of IL-8 in a dose-dependent manner at a dose of 0.5–3.0 mM and factor- kappa B in WiDr cells. You et al. (2020) pointed out that 10-HDA significantly reduced the production of pro-inflammatory mediators in LPS-treated C57BL/6J mice and microglial BV-2 cells due to 10-HDA inhibited the activation of the tumor necrosis factor-α/nuclear factor-kappa B axis and NLRP3 inflammasome-IL-1 beta pathway. Likewise, Huang et al. (2022) conducted a study to analyze the effects of 10-HDA on dextran sodium sulfate (DSS)-induced mice ulcerative colitis. They reported that the inflammatory molecules including caspase-recruitment domain, cysteinyl aspartate specific proteinase-1, IL-18, IL-1β, tumor nuclear factor-α, and cyclooxygenase-2 were significantly reduced after administration of 10-HDA at 100 or 200 mg/kg.

Other important compounds found in royal jelly with vast anti-inflammatory properties are major royal jelly proteins (MRJPs). Thus, Minegaki et al. (2022) conducted a study to analyze the anti-inflammatory effects of major royal jelly protein 3 and its derived peptides. They found that the expression of both tumor necrosis factor-α and interleukin-6 (IL-6) mRNAs in lipopolysaccharides (LPS)-stimulated THP-1 cells was reduced by the addition of major royal jelly protein-3 or its C-terminal tandem penta-peptide repeats (TPRs) sequence.

10.4 Immunomodulation Effects

Immunomodulation effects have an important effect on cancer, allergy, and inflammation. They can be activated and deactivated. In the case of royal jelly, the activating effects predominate as they have been reported by many studies. Therefore, Bouamama et al. (2021) pointed out that royal jelly improved peripheral blood mononuclear cell proliferation significantly in elderly subjects, accompanied by the increase in nitric oxide and the release of IL-2, IL-4, and IL-6 cytokines among the intracellular glutathione. Naghib et al. (2022) informed that royal jelly (1 g/day) administration during 30 days exhibited immunomodulatory properties due to a significantly reduced number of hepatitis B virus-DNA copies and led to downregulation of tumor necrosis factor-alpha, and interleukin 6, which were not substantial. Bahaaldin-beygi et al. (2022) evaluated the effects of royal jelly on the relative expression of apoptosis-associated as immune system-related molecules in patients with chronic hepatitis B infection. These authors found that the administration of royal jelly (1 g/day) increased the expression of speck-like protein, but decreased the expression of node-like receptor family pyrin domain containing1 in the patients with chronic hepatitis B infection. Relative expressions of node-like receptor family pyrin domain containing 3, S100 calcium-binding protein A4, and S100 calcium binding protein A9 were not altered following treatment with royal jelly. Alvarez et al. (2023) reported that royal jelly has demonstrated to modulate mesenchymal stem cell (MSC) differentiation and secretome, as well as decrease LPS-induced inflammation in macrophages by blocking the mitogen-activated protein kinase pathway.

Several components found in royal jelly including lipids, mainly 10-hydroxy-2-decenoic acid, or proteins primarily major royal jelly proteins have been shown to have immunomodulatory effects. In this sense, Eslami-Kaliji et al. (2021) informed that 10-HDA possesses immunomodulatory effects as it binds to pattern recognition receptors. One of these receptors is toll-like receptor4, which is a sensor for damage-associated molecular patterns. Wang et al. (2023) explored the immunomodulation effects of different components isolated from royal jelly like major royal jelly proteins on murine immune functions in cyclophosphamide-treated mice. They informed that mince which received 0.25 g/kg of major royal jelly protein, had a deep impact on the development of the spleen and liver, the quantity of peripheral blood leukocytes, immunoglobulin content, immune factor level, and the proliferation ability of spleen lymphocytes. More recently, Wu et al. (2024) tested the major royal jelly proteins on the immunological responses in mice. They showed that mice given 250 or 500 mg/kg of major royal jelly proteins improved the proliferation of splenocytes in response to mitogens. The splenocytes and mesenteric lymphocytes activated by T-cell released high levels of IL-2 but low levels of IFN-γ and IL-17A. In addition, splenocytes and mesenteric lymphocytes activated by LPS were prevented by major royal jelly proteins at the same dose as that required for producing IL-1β and IL-6, two pro-inflammatory cytokines.

10.5 Vasodilative and Hypotensive Activities

High blood pressure is one of the most important risk factors for cardiovascular diseases. Thus, around 54% of strokes and 47% of coronary heart diseases, worldwide, are attributable to elevated values of blood pressure (Fuchs and Whelton 2020). High blood pressure frequently clusters with other cardiovascular risk factors, including obesity, smoking, dyslipidemia, and type 2 diabetes (Holthuis et al. 2021). Maintaining healthy lifestyle factors (body mass index, diet, smoking, alcohol consumption, sodium excretion, and sedentary behavior) could lower systolic blood pressure BP by 3.5 mm Hg and reduce the risk of cardiovascular disease by about 30% (Ojangba et al. 2023).

Several peptides, generally composed of 2–12 amino acids with molecular weights of less than 1 kDa, have been found to reveal a practical in vivo antihypertensive effect in humans inhibiting the production of pressor active angiotensin (Ang) II or by retarding the catalytic action of angiotensin I-converting enzyme (ACE) (Kaur et al. 2021). Therefore, Liang et al. (2018) informed that royal jelly, at concentrations comprised between 0.0001 and 0.3 mg/mL caused vasorelaxation of isolated rat aortas and superior mesenteric arteries, and this effect was inhibited by atropine (10^{-5} M, 15 min) or Nitro-L-arginine methyl ester (10^{-4} M, 20 min). Additionally, the oral administration of royal jelly (100 mg/kg) augmented tail blood flow and mass in control rats 1 h after treatment without affecting velocity or systolic blood pressure. In the same sense, Pan et al. (2019) conducted a research to investigate the anti-hypertensive activity of royal jelly on hypertensive rats that received 1 g/kg via oral administration daily for 4 weeks. They reported that royal jelly significantly reduced the systolic and diastolic blood pressure as well as increased the nitric oxide levels and found that royal jelly suppresses the NE-induced intracellular Ca^{2+} releases and high K^{+}-induced extracellular Ca^{2+} influx in denuded aortic rings. The blood pressure regulating major royal jelly protein (MRJP1) to identify the ACE inhibitory peptides was examined by Tahir et al. (2020). These authors informed that the peptide "EALPHVPIFDR" showed a strong binding affinity and higher anti-hypertensive activity along with a global energy of −58.29 and docking score of 9590. The aromatic amino acids, especially tyrosine, were detected as the crucial residue to design the dietary peptides as ACE inhibitors. In a more recent study, Alcala-Escamilla et al. (2022) carried out a study to investigate the ACE inhibitory and hypotensive effects of Mucuna royal jelly, pollen royal jelly, and control royal jelly from Yucatan (Mexico). These authors found that, at 2, 3, and 4 mg/mL concentrations, a gradual increase in inhibition in ACE-1 was obtained showing the pollen royal jelly the highest inhibition values, although all samples had all the amino acids that have been associated with the anti-hypertensive peptides.

On the other hand, blood regulatory function is largely dependent on the contractibility and migrating ability of vascular smooth muscle cells (VSMCs) which are a major cell type of the arterial wall. Thus, Fan et al. (2016) reported that the expression of major royal jelly protein1 in VSMCs significantly reduced cell contraction, migration, and proliferation, suggesting a potential role in decreasing hypertension via action on VSMCs. These anti-hypertension activities were further observed in the changes in the proteome setting of mice VSMCs.

10.6 Antioxidant Activity

The inability of the reactive oxygen species scavenger to control the overproduction of oxidants leads to oxidative stress (Lushchak and Storey 2021). Imbalanced production and consumption of reactive oxygen species that produce oxidative stress are implicated in the pathophysiology of genetic and acquired human disorders including cancer, metabolic syndrome, atherosclerosis, malaria, Alzheimer's disease, rheumatoid arthritis, preeclampsia as well as neurodegenerative diseases such as Alzheimer's and Parkinson and aging processes (Kıran et al. 2023; Reddy 2023). As a result, attention has shifted toward using antioxidants, particularly natural ones, to shield tissues from damage caused by free radicals (Chaudhary et al. 2023). Thus, royal jelly has received particular attention because of studies that have reported that it is a highly efficient antioxidant and has a free radical scavenging capacity for decreasing the toxic effects of chemical agents.

In this sense, Ghanbari et al. (2018) reported that royal jelly administration at doses of 100, 200, and 400 mg/kg/body weight daily for 14 days in immature female rats produced a significant increase in the serum levels of ferric-reducing antioxidant power and a significant decrease in nitric oxide level. Similarly, Caixeta et al. (2018) mentioned that the administration of 200 mg/kg body weight of cold-stressed Wistar rats produced a reduction in reduced glutathione, Glutathione reductase, and lipoperoxidation in the liver while improving the total antioxidant capacity and superoxide dismutase activity. Asadi et al. (2019) analyzed the possible protective effects of royal jelly against varicocele-induced oxidative stress by measuring tissue biochemical and antioxidant parameters. For that, Wistar albino rats were treated for 42 consecutive days by gavage with royal jelly (200 mg/kg/day). Malondialdehyde level, catalase, glutathione peroxidase, and superoxide dismutase activities were determined. These authors found that royal jelly elicited a significant protective effect toward varicocele by decreasing the level of lipid peroxidation and increasing the activities of catalase, superoxide dismutase, and glutathione peroxidase. Likewise, Mahdivand et al. (2021) carried out a study to analyze the effects of royal jelly on oxidative stress in heat stress-exposed in male Wistar rats. The royal jelly administration (100 mg/kg/day for 48 days) decreased the malondialdehyde levels in testicular tissue, while the total antioxidant capacity level as well as the catalase activity were significantly increased compared to heat stress-exposed groups. Similarly, Chiang et al. (2021) stated that royal jelly proteins and hydrolyzed royal jelly proteins prepared with alcalase or flavorzyme exhibited significant hydroxyl radical-scavenging activity and antioxidant activity against lipid peroxidation. Nazar-Zadeh et al. (2022) conducted a study to analyze the protective role of royal jelly on testicular damage following nicotine administration due to oxidative stress and lipid peroxidation in male BALB/c mice. These authors found that the treatment with three doses (100, 150, and 200 mg/kg of body weight) of royal jelly caused a significant reduction of malondialdehyde production (83.45%, 83.64%, and 88.58% for doses of 100, 150, and 200 mg/kg of body weight, respectively), a significant reduction of nitric oxide production (37.52%, 38.35%, and 32.73% for doses of 100, 150, and 200 mg/kg of body weight, respectively), and increased the total antioxidant capacity in a concentration-dependent manner. In a similar study, Aslan et al. (2022b) investigated the protective effect of royal jelly against fluoride-induced pancreas damage. The study was conducted in 42 male Wistar albino rats who were administered 100 mg/kg royal jelly for 8 weeks. These authors found that the malondialdehyde levels were increased in all fluoride-treated rats compared to those in the control and royal jelly group, whereas the catalase and glutathione activities were reduced in all rats with fluoride-induced damage. Khoob et al. (2022) revealed that compared to the dimethylhydrazine group, the Wistar rats exposed to dimethylhydrazine + royal jelly (300 mg/kg) experienced significantly less oxidative stress and a significantly greater total antioxidant capacity level after 13 weeks. In other work, Hashemi et al. (2023) pointed out that, the royal jelly administration of 150 mg/kg/day for 14 days before induction of temporal lobe epilepsy in Wistar rats resulted in significant improvement in total antioxidant capacity and reduction of total oxidant status an malondialdehyde levels. In a comparable work, Ibrahim and Shahen (2023) investigated the effects of royal jelly on neurotoxicity caused by acrylamide in male Wistar rats. Royal jelly was given orally once daily at 150 and 300 mg/kg for 10 days. At the end of the study, treatments with royal jelly for

10 days markedly improved the levels of oxidative stress parameters. Thus, the malondialdehyde values were reduced, with respect to the group treated only with acrylamide, at 20.29 and 21.40% for the doses of 150 and 300 mg/kg, respectively whereas the superoxide dismutase and reduced glutathione activities were increased. More recently, Ayhan et al. (2024) examined the protective effect of royal jelly on toxicity and lipid peroxidation, induced by the pesticide cypermethrin, in *Allium cepa* L. root-tip cells. The authors reported that cypermethrin elevated levels of malondialdehyde, and the activity rates of the enzymes catalase and superoxide dismutase. However, when royal jelly treatment, at 250 and 500 mg/L, was applied together with cypermethrin, all negatively affected parameters due to the toxicity of cypermethrin were significantly reestablished.

In addition, the anti-oxidative effects of royal jelly have been verified in some tissues including the liver and kidney exposing them to several harmful stimuli with pro-oxidant activity. Thus, Huang et al. (2019) analyzed the protective effect of royal jelly on CCl_4-Induced liver damage in male SD rats. Thus, the rats fed with 20, 100, and 500 mL/kg of body weight/day showed CCl_4-induced liver damage, which was manifested with an increase in glutathione and superoxide dismutase activities as well as the total antioxidant capacity whereas the malondialdehyde production and lipid peroxide levels were reduced in contrast with the control group. Hamza et al. (2022) investigated the protective effect of royal against hepatotoxicity induced by molybdenum nanoparticles (MoO_3-NPs), cadmium chloride ($CdCl_2$), or their combination in male rats. These authors found that oral treatment with 85 mg/kg of body weight of royal jelly significantly ameliorated the hepatotoxicity of molybdenum nanoparticles and/or cadmium chloride increasing all hepatic vitality markers. Additionally, royal jelly was more potent and effective as an antioxidant over the oxidative damage induced by the combination of molybdenum nanoparticles and cadmium chloride. Mostafa et al. (2020) investigated the effects of royal jelly against hepato-renal damage caused by diclofenac. The adult male Wistar rats were administrated with royal jelly at doses of 150 and 300 mg/kg/day. At the end of the study administration of royal jelly resulted in a significant decrease in myeloperoxidase content and inducible nitric oxide synthase expression and increased ROS scavenging as compared to the diclofenac-control group. In a similar study, Aslan et al. (2022c) investigated the potential protective effects of royal jelly against fluoride-induced hepatotoxicity in Wistar albino rats. These authors found that oral administration of royal jelly at 100 mg/kg body weight for 8 weeks caused a significant decrease in malondialdehyde level, whereas induced a significant increase in glutathione level, and catalase activities. Jalili et al. (2022) analyzed the hepatoprotective and antioxidant activities of the royal jelly against Dichlorvos, an organophosphate insecticide, which induced acute liver damage. Wistar male rats which were orally given every day for 28 days with royal jelly doses of 50, 100, and 150 mg/kg were used for the experiment. The authors reported that royal jelly exerted a significant protective effect on liver damage as well as on oxidative stress induced by dichlorvos, resulting in an increased of total antioxidant levels (46%, 51%, and 55% for doses of 50, 100, and 150 mg/kg, respectively) with as well as a reduction in the nitric oxide compared with the control group. More recently,

Khadrawy et al. (2023) designed a study to determine the possible protective effects of royal jelly against hepatic injury induced by Gibberellic acid in male Wister albino rats. In this study, royal jelly administration at 300 mg/kg body weight during 28 days to hepatic injury-induced rats decreased the serum malondialdehyde levels and increased the superoxide dismutase, catalase, and glutathione peroxidase activities, which indicated the antioxidant properties of royal jelly.

10.7 Anti-Hypercholesterolemic Activity

The main risk factors for the development of cardiovascular disease are the high content of plasma low-density lipoprotein cholesterol (LDL-C) linked to dyslipidemia which is characterized by high levels of triacylglycerol (TG) and low levels of high-density lipoprotein cholesterol (HDL-C) (Du and Quin 2023). In addition, elevated low-density lipoprotein cholesterol levels increase the risk of atherosclerotic cardiovascular disease (Feingold and Chait 2023). Atherosclerosis-related diseases represent significant health issues among adults globally and it is associated with a poor prognosis and significantly reduce life expectancy in 60-year-old patients by 8–12 years depending on the vascular event (Agnelli et al. 2020). It is characterized by lesions called atheromas or fibro-fatty plaques that protrude into the lumen weaken the underlying media and undergo a series of complications as cardiovascular disease progresses (Prado dos Santos et al. 2021). Hypolipidemic agents comprise a varied class of medications aimed at treating hyperlipidemia. These drugs help lower overall cholesterol levels in the blood, including LDL cholesterol and triglycerides, particularly in patients at heightened risk for cardiovascular issues (Kim et al. 2022).

In the scientific literature, there are several works where the anti-hypercholesterolemic activity of royal jelly has been determined. In this sense, Lambrinoudaki et al. (2016) examine the effect of royal jelly on cardiovascular markers in clinically healthy postmenopausal women. For that, the participants received 150 mg of royal jelly daily for 3 months. The royal jelly used in this study was particularly rich in medium-chain fatty acids, which comprised 63% of the dry-weight fatty content. At the end of the study royal jelly treatment resulted in a significant increase in high-density lipoprotein cholesterol (HDL-C 60.2 mg/dL versus 64.7 mg/dL, 7.7% increase), as well as in a significant decrease in low-density lipoprotein cholesterol (LDL-C, 143.9 mg/dL versus 136.2 mg/dL 4.1% decrease) and in total cholesterol (TC) (224.4 to 216.1 3.09% decrease). Similarly, Metwally Ibrahim and Kosba (2018) conducted a study to analyze the effect, on lipid metabolism, of royal jelly administration in Sprague-Dawley male rats which were fed with 100 mg/kg body weight during 8 weeks. At the end of the study, their serum total cholesterol, triglyceride concentrations, and serum low-density lipoprotein decreased significantly (16.04%, 22.68%, and 43.72%, respectively) compared with those of the control group. Additionally, there were significant differences in serum high-density lipoprotein which was increased by 20.02% compared to the control group. More recently, Petelin et al. (2019) analyzed the effect of royal jelly (two capsules of 333 mg/capsule) on serum total lipids and cholesterol levels in 72 asymptomatic overweight adults aged 25–50. They reported that royal jelly, after 8 weeks significantly decreased serum total lipids and cholesterol levels. The author pointed out that the ingestion of two capsules of royal jelly (333 mg/capsule) per day, decreased serum total cholesterol (TC) levels by about 6.26%, the total serum lipids by about 27.14%, and LDL cholesterol by about 13.82% in the group of patients studied. Hazrati Nejad et al. (2022) reported that low-density lipoprotein, total cholesterol, and total triglycerides serum levels were reduced and high-density lipoprotein was increased in high-intensity interval training, consumption of royal jelly (1 g, once daily with a fixed diet for 8 weeks), and training plus royal jelly supplementation in overweight and obese middle-aged men. Similarly, Hamza et al. (2022) designed a study to determine the possible effect of royal jelly on the serum lipid profile in male rats. These authors found that oral treatment with 85 mg/kg of body weight of royal jelly for 30 successive days reduced serum cholesterol (0.9%), total triglycerides (5.18%), very-low-density lipoprotein (7.00%), and low-density lipoprotein (9.75%) in serum. In a recent study, Cho et al. (2024) investigated the effects of royal jelly on serum triglyceride cholesterol levels in zebrafish. These authors found that the 10% (w/w) royal jelly supplementation effectively maintained the blood lipid profile by curtailing serum triglycerides and elevating high-density lipoprotein cholesterol.

The chemical composition of royal jelly could influence the anti-hypercholesterolemic activity. Thus, the large number of proteins in RJ may decrease plasma levels of cholesterol. In this way, Thair et al. (2020) informed that several products of major royal jelly protein 1 and major royal jelly protein 3 did bind low-density lipoprotein, and that oxidized low-density lipoprotein could lead to the regression of atherosclerotic plaque by lowering plaque inflammation. In a similar study, Sato et al. (2021) reported that royalisin and several products of major royal jelly protein 1 and major royal jelly protein 3 produce an inhibition of macrophage proliferation in atherosclerotic plaque in a concentration-dependent manner.

10.8 Antitumor Activity

Nutritional factors have been suggested to play an important role in the prevention of several chronic diseases including cancers. Cancer is a leading cause of death worldwide, accounting for nearly ten million deaths in 2020, or nearly one in six deaths (WHO 2022). Natural products can offer a sustainable source of bioactive compounds with significant effectiveness in treating and overcoming numerous disorders and chronic diseases, including cancer (Hassan et al. 2022). In this sense, the role of the bioactive compound and natural products, as a source of anticancer drugs, has been marked within a collaborative, integrated, and multidisciplinary approach (Asma et al. 2022). One natural substance with anti-carcinogenic properties could be royal jelly. Thus, in the scientific literature there are several studies regarding the antitumoral action of royal jelly in different experimental cellular or animal models (Liu et al. 2022; Khoob et al. 2022; Alnomasy and Al Shehri 2022; Albalawi et al. 2022; Jovanović et al. 2023) and even in studies with humans (Miyata et al. 2020; Bahoush and Pahlevani 2020).

Focusing on studies with animals, Zhang et al. (2017) analyzed the effect on a 4T1 breast cancer mice model, of royal jelly administration (0.5 and 1.5 g/kg). These authors reported that the weight of the tumor was significantly reduced and the levels of interleukin (IL)-2 and interferon (IFN)-α, were significantly elevated in the serum. Albalawi et al. (2022) examined the effect of royal jelly against the Ehrlich solid tumors (EST) in mice; they observed that treatment of EST-suffering mice with royal jelly at doses of 200 and 400 mg/kg caused a significant reduction in tumor volume and inhibition rate through induction of apoptosis as well as a reduction on body weight and tumor markers α-fetoprotein and carcinoembryonic antigen tumor. Similarly, Khoob et al. (2022) investigated the effect of royal jelly administration on dimethylhydrazine-induced colorectal cancer in Wistar rats. For that, the animals were fed a dietary supplement of 300 mg/kg of royal jelly for 13 weeks. These authors found that the expression of proliferating cell nuclear antigen, platelet-derived growth factor, and carcinoembryonic antigen proteins significantly decreased among the animals receiving royal jelly compared to the control group. Liu et al. (2022) carried out a study to analyze the effects of royal jelly on hyperplasia of the mammary gland in rats as well as the dose-response relationship of royal jelly on the treatment of hyperplasia of the mammary gland. These authors found that royal jelly, at 0.1 and 0.8 g/kg/day, enhances the expansion of acinar and breast tissue ducts. Additionally, at 0.8 g/kg/day, the mRNA expression of hypothalamic gonadotropin-releasing hormone (GnRH) was reduced while the pituitary GnRH receptors (GnRH-R) were increased. On the other hand, at 0.1 g/kg/day, GnRH-R mRNA expression was increased.

Concentrating on experimental with cancer cell lines, Abu-Serie and Habashy (2019) reported that major royal jelly protein 2 and its novel isoform X1, two purified proteins from royal jelly, might promote caspase-dependent apoptosis and inhibit the expression of Bcl-2 and p53 in HepG2 cells. Moubarak et al. (2021) studied the cytotoxicity of royal jelly in FHs 74 Int intestinal cells and the anticancer effects in the MDA-MB-231 human breast cancer cell line. These authors found that royal jelly showed relatively non-toxic effects against MDA-MB-231 cells and FHs 74 Int small intestinal cells at concentrations below 5 μg/mL. Nevertheless, when high doses were analyzed (200 μg/mL), greater toxicity against MDA-MB-231 cells was obtained. Similarly, Ayna et al. (2021) conducted a study to assess the antiproliferative effects of royal jelly (0.3 and 0.5 mg/mL) in HT-29 colon cancer cells. They reported that increase in apoptotic index and caspase-3 activation, upregulation of levels of apoptotic genes and proteins of intrinsic and extrinsic pathway (Bax, cyt-c, p53, NFκB, caspase-3, caspase-6, caspase-8, caspase-9, caspase-10), and downregulation of antiapoptotic Bcl-2 both at transcriptional and translational levels. Alnomasy and Al Shehri (2022) planned a research to analyze the anticancer effects of royal jelly against the human hepatoma cell line HepG2. These authors found an IC_{50} value of 1.13 mg/mL for HepG2 cells and reported that royal jelly at the concentrations of ½ IC_{50} significantly increased apoptotic and necrotic cells from 0.96% to 28.3% and 9.3%, respectively, while at the concentration of IC_{50} significantly increased apoptotic and necrotic cells from 0.96% to 39.2% and 14.12%, respectively. Jovanović et al. (2023) carried out a study to analyze the effects of royal jelly, on the invasiveness of the colorectal cancer cell line HCT-116 and the expression of two pro-invasive/ epithelial-mesenchymal transition markers (Snail and vimentin). They found that royal jelly at 10 and 100 μg/mL exerted substantial anti-invasive activity against HCT-116 colorectal cell line by attenuating the Snail and vimentin invasive markers. The effects of royal jelly on human hepatocellular carcinoma-LM3 and Hep3B cell invasion, migration, proliferation, and apoptosis were investigated by Xu et al. (2023). They found that royal jelly exerts antitumor effects by affecting the glycolytic pathway and this substance might regulate the lactylation of H3K9la and H3K14la sites on H3 histone using lactate as a clue in the glycolytic pathway. More recently, Ma et al. (2024) informed that treatment with 5–320 μg/mL of royal jelly, an ethanolic extract of royal jelly (RJEE), and protein hydrolysate reduced the viability of MKN-28 gastric cancer cells, with a half-maximal inhibitory concentration of 123.22 μg/mL for ethanolic extract of royal jelly. In addition, royal jelly, ethanolic extract of royal jelly, and protein hydrolysate royal jelly increase the lactate dehydrogenase release rate and change the morphology of the cells, resulting in cell shrinkage, nucleoplasm condensation, and the formation of apoptotic bodies.

Regarding the studies of anticancer activity of royal jelly administration on humans, Miyata et al. (2020) examined the effects of oral administration of royal jelly (capsules 900 mg) in patients with high grade (Fuhrman grade 3 and 4) and high pathological stage (pT3 and 4) cancer. These authors reported that the royal jelly administration produced lower values in post- and pre-treatment ratios of the serum levels of tumor necrosis factor-α and Transforming growth factor-β and these ratios correlated with decreasing tumor size and frequency of anorexia or fatigue in patients. In another study with humans, Bahoush and Pahlevani (2020) investigated the immune response of blood cells N-chromosome following royal jelly administration (2 g) in patients with childhood ALL. They demonstrated that royal jelly can be a promising immune-enhancing supplement in patients diagnosed with acute lymphoblastic leukemia in their complete remission and maintenance therapy time.

10.9 Other Physiological Effects

As well as its antibacterial, antioxidant anti-inflammatory, vasodilative, hypotensive, hipocholesterolemiant, hepatoprotective, and antitumor activities, which have been extensively analyzed previously, royal jelly also shows other physiological effects due to its high content of bioactive compounds. In this sense, Sharif and Darsareh (2019) conducted a study to determine the effects of royal jelly on menopausal symptoms such as anxiety, irritability, sleeping problems, and hot flushes among others. These authors found that the daily consumption of oral royal jelly

(1000 mg) for 8 weeks was effective in alleviating the menopausal symptoms. Kunugi and Mohammed Ali (2019) found that the lipid components of royal jelly mostly 10-HAD 10-H2DA and Sebacic acid were capable of inhibiting the natural aging process in human cell cultures by two mechanisms: upregulation of epidermal growth factor signaling and downregulation of insulin-like growth factors. Rahnama et al. (2020) provide evidences that royal jelly may improve the preservation of rooster spermatozoa. Thus, these authors pointed out that the spermatozoa forward progressive motility, abnormality, membrane integrity, and viability were improved by the rooster feed with royal jell at 100 mg/kg/body weight group compared to the other groups after 24 and 48 h storage period at 4 °C. Guardia de Souza e Silva et al. (2020) have demonstrated that the oral treatment with royal jelly (100 mg/ml) during 2 weeks in Wistar rats submitted to streptozotocin promotes beneficial effects on cognitive functions and exhibits a neuroprotective action in experimental model of Alzheimer's disease. Sofiabadi and Samiee-Rad (2020) informed that acetic acid-induced peptic ulcers in rats may be treated significantly through the use of a daily administration of royal jelly (200 mg/kg), in comparison to 20 mg/kg of omeprazole a common anti-ulcer drug widely used, for 14 days of treatment Vajdi et al. (2023) mentioned that the daily supplementation with royal jelly, at the dosages <3 g reduces body weight and the body mass index. In another study, Kobayashy et al. (2023) informed that royal jelly can help in non-alcoholic fatty liver disease. Royal jelly enhances the excretion of saturated palmitic acid in the feces, thus decreasing its content in the blood and liver, and increasing the content of medium-chain fatty acids in the liver. In addition, improved gut dysbiosis and inflammation. Huang et al. (2024) reported that C57BL/6J mice treated with royal jelly (100 mg/kg/day) and 10-HAD (100 mg/kg/day), for 28 weeks prevent skin dysfunction at the early stage in diabetic mice by modulating the pyroptosis pathway.

10.10 Conclusions

Royal jelly is an important functional food item that possesses several health-promoting properties. Due to these properties, it has been widely used in commercial medical products, healthy foods, and cosmetics in many countries. However, to improve our knowledge of royal jelly production and composition, together with our understanding of its physiological effects on the human body, collaborative studies are needed involving the participation of researchers from different scientific areas such as chemistry, biochemistry, biotechnology, biology, physiology, nutrition, and medicine.

References

Abu-Serie MM, Habashy NH (2019) Two purified proteins from royal jelly with in vitro dual anti-hepatic damage potency: major royal jelly protein 2 and its novel isoform X1. Int J Biol Macromol 128:782–795. https://doi.org/10.1016/j.ijbiomac.2019.01.210

Agnelli G, Belch JJF, Baumgartner I, Giovas P, Hoffmann U (2020) Morbidity and mortality associated with atherosclerotic peripheral artery disease: a systematic review. Atherosclerosis 293:94–100. https://doi.org/10.1016/j.atherosclerosis.2019.09.012

Agostoni C, Bresson J, Fairweather-Tait S, Flynn A, Golly I et al (2011) EFSA Panel on dietetic products, nutrition and allergies (NDA): scientific opinion on the substantiation of health claims related to royal jelly. Eur Food Saf Author J 9(4):2083

Albalawi AE, Althobaiti NA, Alrdahe SS, Alhasani RH, Alaryani FS, BinMowyna MN (2022) Antitumor activity of royal jelly and its cellular mechanisms against Ehrlich solid tumor in mice. Biomed Res Int 2022:7233997. https://doi.org/10.1155/2022/7233997

Alcalá Escamilla KI, Moguel Ordóñez YB, Sandoval-Peraza VM, Acevedo Fernández JJ, Betancur Ancona DA (2022) Anti-hypertensive activity in vitro and in vivo on royal jelly produced by different diets. Em J Food Agric 34(1):9–15. https://doi.org/10.9755/ejfa.2022.v34.i1.2812

Almeer RS, Alarifi S, Alkahtani S, Ibrahim SR, Ali D, Moneim A (2018) The potential hepatoprotective effect of royal jelly against cadmium chloride-induced hepatotoxicity in mice is mediated by suppression of oxidative stress and upregulation of Nrf2 expression. Biomed Pharmacother 106:1490–1498. https://doi.org/10.1016/j.biopha.2018.07.089

Alnomasy SF, Al-Shehri ZS (2022) Anti-cancer and cell toxicity effects of royal jelly and its cellular mechanisms against human hepatoma cells. Pharmacogn Mag 18(79):635–640. http://www.phcog.com/text.asp?2022/18/79/635/356388

Altun SK, Aydemir ME (2022) Determination of antioxidant capacities with the phenolic and flavonoid contents of royal jelly mixtures. Bee Studies 14(1):9–15. https://doi.org/10.51458/BSTD.2022.22

Altuntas S, Cinar A, Altuntas V (2020) Modelling of *Listeria monocytogenes* growth and survival in presence of royal jelly, a promising anti-biofilm agent. J Food Nutr Res 59:7–15

Arzi A, Olapour S, Yaghooti H, Karampour NS (2015) Effect of royal jelly on formalin induced-inflammation in rat hind paw. Jundishapur J Nat Pharm Prod 10(1):e22466. https://doi.org/10.17795/jjnpp-22466

Asadi N, Kheradmand A, Gholami M, Saidi SH, Mirhadi SA (2019) Effect of royal jelly on testicular antioxidant enzymes activity, MDA level and spermatogenesis in rat experimental Varicocele model. Tissue Cell 57:70–77. https://doi.org/10.1016/j.tice.2019.02.005

Aslan A, Can MI, Beyaz S, Gok O, Parlak G, Ramazan Gundogdu R, Ozercan IH, Erman O (2022a) A new approach on the regulation of NF-κB and Bax protein signaling pathway activation by royal jelly in fluoride-induced pancreas damage in rats. Tissue Cell 79:101913. https://doi.org/10.1016/j.tice.2022.101913

Aslan A, Gok O, Beyaz S, Parlak G, Can MI, Gundogdu R, Baspinar S, Ozercan IH, Parlak AE (2022b) Royal jelly arranges apoptotic and oxidative stress pathways and reduces damage to liver tissues of rats by down-regulation of Bcl-2, GSK3 and NF-κB and up-regulation of caspase and Nrf-2 protein signalling pathways. Biomarkers 28(2):217–226. https://doi.org/10.1080/1354750X.2022.2159526

Aslan A, Gok O, Beyaz S, Can MI, Parlak G, Gundogdu R, Ozercan IH, Baspinar S (2022c) Royal jelly regulates the caspase, Bax and COX-2, TNF-α protein pathways in the fluoride exposed lung damage in rats. Tissue Cell 76:101754. https://doi.org/10.1016/j.tice.2022.101754

Asma ST, Acaroz U, Imre K, Morar A, Shah SRA, Hussain SZ, Arslan-Acaroz D, Demirbas H, Hajrulai-Musliu Z, Istanbullugil FR et al (2022) Natural products/bioactive compounds as a source of anticancer drugs. Cancers 14(24):6203. https://doi.org/10.3390/cancers14246203

Ayhan BS, Macar TK, Macar O, Yalçın E, Çavuşoğlu K, Özkan B (2024) A comprehensive analysis of royal jelly protection against cypermethrin-induced toxicity in the model organism Allium cepa L., employing spectral shift and molecular docking approaches. Pestic Biochem Physiol 203:105997. https://doi.org/10.1016/j.pestbp.2024.105997

Ayna A, Tunç A, Özbolat SN, Bengü AS, Aykutoğlu G, Canli D, Polat R, Çiftci M, Darendelioğlu E (2021) Anticancer, and antioxidant activities of royal jelly on HT-29 colon cancer cells and-melissopalynological analysis. Turk J Bot 45(8):9. https://doi.org/10.3906/bot-2109-8

Bagameri L, Botezan S, Bobis O, Bonta V, Dezmirean DS (2023) Molecular insights into royal jelly anti-inflammatory properties and related diseases. Life 13(7):1573. https://doi.org/10.3390/life13071573

Bahaaldin-beygi M, Kariminik A, Kazemi Arababadi M (2022) Royal jelly significantly alters inflammation pathways in patients with chronic hepatitis B. Indian J Exp Biol 60:875–879

Bahoush G, Pahlevani R (2020) Evaluation of the effectiveness and safety of N-Chromosome Royal Jelly on the number of peripheral white blood cells in patients with acute lymphoblastic leukemia. J Contemp Med Sci 6(6). https://doi.org/10.22317/jcms.v6i6.887

Bengü AŞ, Ayna A, Özbolat S, Tunç A, Aykutoğlu G, Çiftci M, Darendelioğlu E (2020) Content and antimicrobial activities of Bingol royal Jelly. Turk J Agric Nat Sci 7(2):480–486. https://doi.org/10.30910/turkjans.725977

Bilikova K, Huang SC, Lin IP, Simuth J, Peng CC (2015) Structure andantimicrobial activity relationship of royalisin, an antimicrobial peptide from Royal Jelly of *Apis mellifera*. Peptides 68:190–196. https://doi.org/10.1016/j.peptides.2015.03.001

Botezan S, Baci G-M, Bagameri L, Paşca C, Dezmirean DS (2023) Current status of the bioactive properties of royal jelly: a comprehensive review with a focus on its anticancer, anti-inflammatory, and antioxidant effects. Molecules 28(3):1510. https://doi.org/10.3390/molecules28031510

Bouamama S, Merzouk H, Latrech H, Charif N, Bouamama A (2021) Royal jelly alleviates the detrimental effects of aging on immune functions by enhancing the in vitro cellular proliferation, cytokines, and nitric oxide release in aged human PBMCS. J Food Biochem 45:e13619. https://doi.org/10.1111/jfbc.13619

Caixeta DC, Teixeira RR, Peixoto LG, Machado HL, Baptista NB, de Souza AV, Vilela DD, Rodrigues Franci C, Espindola FS (2018) Adaptogenic potential of royal jelly in liver of rats exposed to chronic stress. PLoS One 13(1):e0191889. https://doi.org/10.1371/journal.pone.0191889

Charoenngam N, Ponvilawan B, Rittiphairoj T, Tornsatitkul S, Wattanachayakul P, Rujirachun P, Ungprasert P (2020) Patients with asthma have a higher risk of rheumatoid arthritis: a systematic review and meta-analysis. Semin Arthritis Rheum 50(5):968–976. https://doi.org/10.1016/j.semarthrit.2020.07.015

Chaudhary P, Janmeda P, Docea AO, Yeskaliyeva B, Abdull Razis AF, Modu B, Calina D, Sharifi-Rad J (2023) Oxidative stress, free radicals and antioxidants: potential crosstalk in the pathophysiology of human diseases. Front Chem 11:1158198. https://doi.org/10.3389/fchem.2023.1158198

Chen YF, Wang K, Zhang YZ, Zheng YF, Hu FL (2016) *In vitro* anti-inflammatory effects of three fatty acids from royal jelly. Med Inf 2016:3583684. https://doi.org/10.1155/2016/3583684

Chen YF, You MM, Liu YC, Shi YZ, Wang K, Lu YY, Hu FL (2018) Potential protective effect of Trans-10-hydroxy-2-decenoic acid on the inflammation induced by lipoteichoic acid. J Funct Food 45:491–498. https://doi.org/10.1016/j.jff.2018.03.029

Chiang SH, Yang KM, Sheu SC, Chen CW (2021) The bioactive compound contents and potential protective effects of royal jelly protein hydrolysates against DNA oxidative damage and LDL oxidation. Antioxidants 10(4):580. https://doi.org/10.3390/antiox10040580

Cho KH, Nam HS, Bahuguna A, Kim JE (2024) Long-term supplementation of royal jelly (Raydel®) improves zebrafish growth, embryo production and survivability, blood lipid profile and functionality of vital organs: a 72-weeks consumption study. Pharmaceuticals 17(3):324. https://doi.org/10.3390/ph17030324

Collazo N, Carpena M, Nuñez-Estevez B, Otero P, Simal-Gandara J, Prieto MA (2021) health promoting properties of bee royal jelly: food of the queens. Nutrients 13(2):543. https://doi.org/10.3390/nu13020543

Coutinho D, Karibasappa SN, Mehta DS (2018) Royal jelly antimicrobial activity against Periodontopathic Bacteria. J Interdiscip Dent 8:17–22. https://doi.org/10.4103/jid.jid_72_17

Dinkov D, Stratev D, Balkanska R, Sergelidis D (2016) Antibacterial activity of royal jelly and rape honey against methicillin-resistant *Staphylococcus aureus* strains. J Food Health Sci 2(2):67–73

Du Z, Qin Y (2023) Dyslipidemia and cardiovascular disease: current knowledge, existing challenges, and new opportunities for management strategies. J Clin Med 12(1):363. https://doi.org/10.3390/jcm12010363

Eshboev F, Mamadalieva N, Nazarov PA, Hussain H, Katanaev V, Egamberdieva D, Azimova S (2024) Antimicrobial action mechanisms of natural compounds isolated from endophytic microorganisms. Antibiotics 13(3):271. https://doi.org/10.3390/antibiotics13030271

Eslami-Kaliji F, Sarafbidabad M, Kiani-Esfahani A, Mirahmadi-Zare SZ, Dormiani K (2021) 10-hydroxy-2-decenoic acid a bio-immunomodulator in tissue engineering; generates tolerogenic dendritic cells by blocking the toll-like receptor4. J Biomed Mater Res A 109:1575–1587. https://doi.org/10.1002/jbm.a.37152

Fan P, Bin Han B, Feng M, Fang Y, Zhang L, Hu H, Hao Y, Qi Y, Zhang X, Li J (2016) Functional and proteomic investigations reveal major royal jelly protein 1 associated with antihypertension activity in mouse vascular smooth muscle cells. Sci Rep 6:30230. https://doi.org/10.1038/srep30230

Feingold KR, Chait A (2023) Approach to patients with elevated low-density lipoprotein cholesterol levels. Best Pract Res Clin Endocrinol Metab 37(3):101658. https://doi.org/10.1016/j.beem.2022.101658

Fuchs FD, Whelton PK (2020) High blood pressure and cardiovascular disease. Hypertension 75:285–292. https://doi.org/10.1161/HYPERTENSIONAHA.119.14240Hypertension

Ghanbari E, Khazaei MR, Khazaei M, Nejati V (2018) Royal jelly promotes ovarian follicles growth and increases steroid hormones in immature rats. Int J Fertil Steril 11(4):263–269. https://doi.org/10.22074/ijfs.2018.5156

Ghosh S, Jung C (2024) Chemical composition and nutritional value of royal jelly samples obtained from honey bee (*Apis mellifera*) hives fed on oak and rapeseed pollen patties. Insects 15(3):141. https://doi.org/10.3390/insects15030141

Gu H, Song IB, Han HJ, Lee NY, Cha JY, Son YK, Kwon J (2018) Anti-inflammatory and immune-enhancing effects of enzyme-treated royal jelly. Appl Biol Chem 61:227–233. https://doi.org/10.1007/s13765-018-0349-5

Guardia de Souza e Silva T, Ferreira do Val de Paulo ME, Mamona da Silva JR, da Silva Alves A, Britto LRG, Xavier GF, Lopes Sandoval MR (2020) Oral treatment with royal jelly improves memory and presents neuroprotective effects on icv-STZ rat model of sporadic Alzheimer's disease. Heliyon 6(2):e03281. https://doi.org/10.1016/j.heliyon.2020.e03281

Gunalp R, Ulusoy M, Celebier I, Keskin N (2018) Antifungal effect of royal jelly on *Candida albicans*. J Biotechnol 280:S70–S71. https://doi.org/10.1016/j.jbiotec.2018.06.230

Guo J, Wang Z, Chen Y, Cao J, Tian W, Ma B, Dong Y (2021) Active components and biological functions of royal jelly. J Funct Foods 82:104514. https://doi.org/10.1016/j.jff.2021.104514

Guo J, Ma B, Wang Z, Chen Y, Tian W, Dong Y (2022) Royal jelly protected against dextran-sulfate-sodium-induced colitis by improving the colonic mucosal barrier and gut microbiota. Nutrients 14(10):2069. https://doi.org/10.3390/nu14102069

Hamza RZ, Al-Eisa RA, El-Shenawy NS (2022) Possible ameliorative effects of the royal jelly on hepatotoxicity and oxidative stress induced by molybdenum nanoparticles and/or cadmium chloride in male rats. Biology 11(3):450. https://doi.org/10.3390/biology11030450

Hashemi P, Moloudi MR, Vahabzadeh Z, Izadpanah E (2023) Anticonvulsant effects of royal jelly in kainic acid-induced animal model of temporal lobe epilepsy through antioxidant activity. Neurochem Res 48:2187–2195. https://doi.org/10.1007/s11064-023-03897-w

Hassan SS, Abdel-Daim MM, Behl T, Bungau S (2022) Natural products for chronic diseases: a ray of hope. Molecules 27(17):5573. https://doi.org/10.3390/molecules27175573

Hazrati Nejad H, Khajehlandi A, Mohammadi A (2022) The impact of high-intensity interval training with physical fitness course using royal jelly supplementation on lipid profile in overweight and obese middle-aged men. J Nutr Fast Health 10(2):103–109. https://doi.org/10.22038/JNFH.2022.62978.1373

Holthuis EI, Visseren FLJ, Bots ML, Peters SAE, UCC-SMART study group (2021) Risk factor clusters and cardiovascular disease in high-risk patients: the UCC-SMART Study. Glob Heart 16(1):85. https://doi.org/10.5334/gh.897

Huang Y, Shou-Ming LSM, Hong T, Shu K, Jia B, Yuan KW, Chen L, Li LX, Liu L, Liu ZY (2019) Hepatoprotective effect of royal jellies with different harvesting time on CCl_4-Induced liver damage in rats. Int J Agric Biol 22:1161–1166. https://doi.org/10.17957/IJAB/15.1182

Huang S, Tao R, Zhou J, Qian L, Wu J (2022) Trans-10-Hydroxy-2-Decenoic acid alleviates dextran sulfate sodium-induced colitis in mice via regulating the inflammasome-mediated pyroptotic pathway and enhancing colonic barrier function. Mol Nutr Food Res 66(12):e2100821. https://doi.org/10.1002/mnfr.202100821

Huang X, Xiu L, An Y, Gong Y, Li S, Chen X, Liu C, Lu J, Shan H, Chang J, Zhang M (2024) Preventive effect of royal jelly and 10-HDA on skin damage in diabetic mice through regulating keratinocyte wnt/β-catenin and pyroptosis pathway. Mol Nutr Food Res 68(19):2400098. https://doi.org/10.1002/mnfr.202400098

Ibrahim DS, Shahen EMS (2023) Effect of royal jelly on acrylamide-induced neurotoxicity in rats. J Chem Neuroanat 134:102358. https://doi.org/10.1016/j.jchemneu.2023.102358

Jalili C, Farzaei MH, Rashidi I, Mohammadnezamian A, Ghanbari A (2022) Royal jelly protects dichlorvos liver-induced injury in male Wistar rats. Res Pharm Sci 17(2):209–218. https://doi.org/10.4103/1735-5362.335178

Jia F, Wang J, Zhang L, Zhou J, He Y, Lu Y, Liu K, Yan W, Wang K (2021) Multiple action mechanism and in vivo antimicrobial efficacy of antimicrobial peptide Jelleine-I. J Pept Sci 27:e3294. https://doi.org/10.1002/psc.3294

Jovanović M, Virijević K, Arsenijević D, Pecić K, Šeklić D (2023) Bee product royal jelly suppress emt and invasiveness of HCT-116 cells. Biol Life Sci Forum 26(1):80. https://doi.org/10.3390/Foods2023-15064

Kaur A, Kehinde BA, Sharma P, Sharma D, Kaur S (2021) Recently isolated food-derived antihypertensive hydrolysates andpeptides: a review. Food Chem 346:128719

Kazemi V, Eskafi M, Saeedi M, Manayi A, Hadjiakhoondi A (2019) Physicochemical properties of royal jelly and comparison of commercial with raw specimens. Jundishapur J Nat Pharm Prod 14(4):e64920. https://doi.org/10.5812/jjnpp.64920

Khadrawy SM, Mohamed DS, Hassan RM, Abdelgawad MA, Ghoneim MM, Alshehri S, Shaban NS (2023) Royal jelly and *Chlorella vulgaris* mitigate gibberellic acid-induced cytogenotoxicity and hepatotoxicity in rats via modulation of the PPARα/AP-1 signaling pathway and suppression of oxidative stress and inflammation. Food Secur 12(6):1223. https://doi.org/10.3390/foods12061223

Khalfan Saeed Alwali Alkindi F, El-Keblawy A, Lamghari Ridouane F, Bano Mirza S (2024) Factors influencing the quality of royal jelly and its components: a review. Cogent Food Agric 10(1):2348253. https://doi.org/10.1080/23311932.2024.2348253

Kheirdeh M, Jahromi MK, Hemmatinafar M, Nemati J (2023) Additive beneficial effects of aerobic training and royal jelly on hippocampal inflammation and function in experimental autoimmune encephalomyelitis rats. Mult Scler Relat Disord 70:104527. https://doi.org/10.1016/j.msard.2023.104527

Khoob MS, Hosseini SM, Kazemi S (2022) *In vitro* and *in vivo* antioxidant and anticancer potentials of royal jelly for dimethylhydrazine-induced colorectal cancer in Wistar rats. Oxidative Med Cell Longev 2022:9506026. https://doi.org/10.1155/2022/9506026

Kim K, Ginsberg HN, Choi SH (2022) New, novel lipid-lowering agents for reducing cardiovascular risk: beyond statins. Diabetes Metab J 46:517–532. https://doi.org/10.4093/dmj.2022.0198

Kıran TR, Otlu O, Karabulut AB (2023) Oxidative stress and antioxidants in health and disease. J Lab Med 47(1):1–11. https://doi.org/10.1515/labmed-2022-0108

Kobayashi G, Okamura T, Majima S, Senmaru T, Okada H, Ushigome E, Nakanishi N, Nishimoto Y, Yamada T, Okamoto H et al (2023) Effects of royal jelly on gut dysbiosis and nafld in db/db mice. Nutrients 15(11):2580. https://doi.org/10.3390/nu15112580

Kumar R, Thakur A, Kumar S, Ahmad Hajam YA (2024) Royal jelly a promising therapeutic intervention and functional food supplement: a systematic review. Heliyon 10(17):e37138. https://doi.org/10.1016/j.heliyon.2024.e37138

Kunugi H, Mohammed Ali A (2019) Royal jelly and its components promote healthy aging and longevity: from animal models to humans. Int J Mol Sci 20:4662. https://doi.org/10.3390/ijms20194662

Lambrinoudaki I, Augoulea A, Rizos D, Politi M, Tsoltos N, Moros M, Chinou I, Graikou K, Kouskouni E, Kambani S, Panoulis K, Moutsatsou P (2016) Greek-origin royal jelly improves the lipid profile of postmenopausal women. Gynecol Endocrinol 26:1–5. https://doi.org/10.1080/09513590.2016.1188281

Li S, Tao L, Yu X, Zheng H, Wu J, Hu F (2021) Royal jelly proteins and their derived peptides: preparation, properties, and biological activities. J Agric Food Chem 69(48):14415–14427. https://doi.org/10.1021/acs.jafc.1c05942

Liang Y, Kagota S, Maruyama K, Oonishi Y, Miyauchi-Wakuda S, Ito Y, Yamada S, Shinozuka K (2018) Royal jelly increases peripheral circulation by inducing vasorelaxation through nitric oxide production under healthy conditions. Biomed Pharmacother 106:1210–1219. https://doi.org/10.1016/j.biopha.2018.07.047

Lima WG, Brito JCM, Verly RM, Med L (2024) Jelleine, a family of peptides isolated from the royal jelly of the honey bees (*Apis mellifera*), as a promising prototype for new medicines: a narrative review. Toxins 16(1):24. https://doi.org/10.3390/toxins16010024

Liu JY (2022) Editorial: Eicosanoids and cytokines: resolution of inflammation. Front Pharmacol 11(13):978331. https://doi.org/10.3389/fphar.2022.978331

Liu Y, Wu D, Wang K, Chen H, Xu H, Zong W, Zhang N, Zhao L, Lin Z, Ji T (2022) Dose-dependent effects of royal jelly on estrogen-and progesterone-induced mammary gland hyperplasia in rats. Mol Nutr Food Res 66(5):2100355. https://doi.org/10.1002/mnfr.202100355

Lushchak VI, Storey KB (2021) Oxidative stress concept updated: definitions, classifications, and regulatory pathways implicated. EXCLI J 20:956–967. https://doi.org/10.17179/excli2021-3596

Ma Y, Wang H, Hu X, Zou H, Xu T, Wang Z, Ju X, He R (2024) Inhibitory effects of royal jelly and its functional components on the proliferation of MKN-28 gastric cancer cells. J Agric Food Chem 72(38):20918–20929. https://doi.org/10.1021/acs.jafc.4c03367

Mahdivand N, Ali-Jalali A, Nejati V, Najafi G, Rahmani F (2021) Adaptogenic potential of royal jelly in reproductive system of heat stress-exposed male rats. J Therm Biol 96:102827. https://doi.org/10.1016/j.jtherbio.2020.102827

Marwaha L (2022) Royal jelly as larval food for honey bees. In: The polyandrous queen honey bee: biology and apiculture. Bentham Science Publishers, pp 67–82. https://doi.org/10.2174/9789815079128112010l

Metwally Ibrahim SEL, Kosba AA (2018) Royal jelly supplementation reduces skeletal muscle lipotoxicity and insulin resistance in aged obese rats. Pathophysiol 25(4):307–315. https://doi.org/10.1016/j.pathophys.2018.05.001

Minegaki N, Koshizuka T, Hatasa K, Kondo H, Kato H, Tannaka M, Takahashi K, Tsuji M, Inoue N (2022) The C-Terminal penta-peptide repeats of major royal jelly protein 3 ameliorate the progression of inflammation *in vivo* and *in vitro*. Biol Pharma Bull 45:583–589. https://doi.org/10.1248/bpb.b21-00922

Miyata Y, Araki K, Ohba K, Mastuo T, Nakamura Y, Yuno T et al (2020) Oral intake of royal jelly improves anti-cancer effects and suppresses adverse events of molecular targeted therapy by regulating TNF-α and TGF-β in renal cell carcinoma: a preliminary study based on a randomized double-blind clinical trial. Mol Clin Oncol 13:29. https://doi.org/10.3892/mco.2020.2099

Moghim H, Taghipour S, Kheiri S, Khabbazi H, Baradaran A (2021) Antifungal effects of Iranian propolis extract and Royal jelly against Candida albicans in-vitro. Int J Prev Med 1(12):163. https://doi.org/10.4103/ijpvm.IJPVM_420_18

Moskwa J, Naliwajko SK, Dobiecka D, Socha K (2023) Bee products and colorectal cancer—active components and mechanism of action. Nutrients 15(7):1614. https://doi.org/10.3390/nu15071614

Mostafa R, El-Marasya SA, Jaleela GAA, Bakeer RM (2020) Protective effect of royal jelly against diclofenac-induced hepato-renal damage and gastrointestinal ulcerations in rats. Heliyon 6(2):e03330. https://doi.org/10.1016/j.heliyon.2020.e03330

Moubarak MM, Chanouha N, Abou Ibrahim N, Khalife H, Gali-Muhtasib H (2021) Thymoquinone anticancer activity is enhanced when combined with royal jelly in human breast cancer. World J Clin Oncol 12(5):342–354. https://doi.org/10.5306/wjco.v12.i5.342

Naghib M, Kariminik A, Arababadi M, Rokhbakhsh-Zamin F (2022) the effects of royal jelly on the pro-inflammatory innate immunity cytokines in patients infected with hepatitis b virus and its antiviral activity. J App Biotechnol Rep 9(2):40–644. https://doi.org/10.30491/jabr.2021.295464.1421

Nazarinia D, Karimpour S, Hashemi P, Dolatshahi M (2023) Neuroprotective effects of Royal Jelly (RJ) against pentylenetetrazole (PTZ)-induced seizures in rats by targeting inflammation and oxidative stress. J Chem Neuroanat 129:102255. https://doi.org/10.1016/j.jchemneu.2023.102255

Nazar-Zadeh M, Jalili C, Fathi AN, Ghanbari A, Bakhtiari M (2022) Royal-jelly-based apitherapy can attenuate damages to male reproductive parameter following nicotine administration. Anim Mod Exp Med 5(2):133–140. https://doi.org/10.1002/ame2.12217

Ojangba T, Boamah S, Miao Y, Guo X, Fen Y, Agboyibor C, Yuan J, Dong W (2023) Comprehensive effects of lifestyle reform, adherence, and related factors on hypertension control: a review. J Clin Hypertens (Greenwich) 25(6):509–520. https://doi.org/10.1111/jch.14653

Omer K, Gelkopf MJ, Newton G (2019) Effectiveness of royal jelly supplementation in glycemic regulation: a systematic review. World J Diabetes 10(2):96–113. https://doi.org/10.4239/wjd.v10.i2.96

Paganini Damiani A, Lummertz Magenis M, Salvan Dagostin L, da Luz Beretta AC, Jacobsen Sarter R, Martins Longaretti L, de Oliveira MI, Moraes de Andrade V (2022) Royal jelly reduce DNA damage induced by alkylating agent in mice. Mutat Res/Fundamental and Molecular Mechanisms of Mutagenesis 825:111796. https://doi.org/10.1016/j.mrfmmm.2022.111796

Pan Y, Rong Y, You M, Ma Q, Chen M, Hu F (2019) Royal jelly causes hypotension and vasodilation induced by increasing nitric oxide production. Food Sci Nutr 7(4):1361–1370. https://doi.org/10.1002/fsn3.970

Petelin A, Kenig S, Kopinč R, Deželak M, Bizjak MČ, Pražnikar ZJ (2019) Effects of royal jelly administration on lipid profile, satiety, inflammation, and antioxidant capacity in asymptomatic overweight adults. Evid -Based Complement Altern Med 2019:969720. https://doi.org/10.1155/2019/4969720

Prado dos Santos V, Pozzan G, Castelli Júnior V, Caffaro RA (2021) Arteriosclerose, aterosclerose, arteriolosclerose e esclerose calcificante da média de Monckeberg: qual a diferença? J Vasc Bras 20:e20200211. https://doi.org/10.1590/1677-5449.200211

Rahnama G, Deldar H, Pirsaraei ZA, Kazemifard M (2020) Oral administration of royal jelly may improve the preservation of rooster spermatozoa. J Anim Physiol Anim Nutr 104(6):1768–1777. https://doi.org/10.1111/jpn.13415

Ramanathan ANKG, Nair AJ, Sugunan VS (2018) A review on Royal Jelly proteins and peptides. J Funct Foods 44:255–264. https://doi.org/10.1016/j.jff.2018.03.008

Reddy VP (2023) Oxidative stress in health and disease. Biomedicines 11(11):2925. https://doi.org/10.3390/biomedicines11112925

Sato A, Unuma H, Ebina K (2021) Royal jelly proteins inhibit macrophage proliferation: interactions with native- and oxidized-low density lipoprotein. Protein J 40(5):699–708. https://doi.org/10.1007/s10930-021-09998-1

Šedivá M, Laho M, Kohútová L, Mojžišová A, Majtán J, Klaudiny J (2018) 10-HDA, a major fatty acid of royal jelly, exhibits ph dependent growth-inhibitory activity against different strains of *Paenibacillus larvae*. Molecules 23(12):3236. https://doi.org/10.3390/molecules23123236

Sharif SN, Darsareh F (2019) Effect of royal jelly on menopausal symptoms: a randomized placebo-controlled clinical trial. Complement Ther Clin Prac 37:47–50. https://doi.org/10.1016/j.ctcp.2019.08.006

Shen P, Ding K, Wang L, Tian J, Huang X, Zhang M, Dang X (2023) *In vitro* and *in vivo* antimicrobial activity of antimicrobial peptide Jelleine-I against foodborne pathogen *Listeria Monocytogenes*. Int J Food Microbiol 387:110050. https://doi.org/10.1016/j.ijfoodmicro.2022.110050

Simón Álvarez S, Pamina Contreras-Kallens P, Sebastian Aguayo S, Orlando Ramírez O, Catalina Vallejos C, Jorge Ruiz J, Eva Carrasco-Gallardo E, Stefanie Troncoso-Vera S, Bernardo Morales B, Christina Schuh CMAP (2023) Royal jelly extracellular vesicles promote wound healing by modulating underlying cellular responses. Mol Ther - Nucleic Acids 31:541–552. https://doi.org/10.1016/j.omtn.2023.02.008

Sofiabadi M, Samiee-Rad F (2020) Royal jelly accelerates healing of acetate induced gastric ulcers in male rats. Gastroenterol Hepatol Bed Bench 13:18–22

Stan D, Enciu AM, Mateescu AL, Ion AC, Brezeanu AC, Stan D, Tanase C (2021) Natural compounds with antimicrobial and antiviral effect and nanocarriers used for their transportation. Front Pharmacol 6(12):723233. https://doi.org/10.3389/fphar.2021.723233

Tahir RA, Bashir A, Yousaf MN, Ahmed A, Dali Y, Khan S, Sehgal SA (2020) In Silico identification of angiotensin-converting enzyme inhibitory peptides from MRJP1. PLoS One 15(2):1–18. https://doi.org/10.1371/journal.pone.0228265

Ulubayram N, Cinar AY (2023) Microencapsulated and fresh royal jelly: monitoring 10-HDA content, antibacterial and antifungal activity at different storage periods. Braz Arch Biol Technol 66:e23220203. https://doi.org/10.1590/1678-4324-2023220203

Uthaibutra V, Kaewkod T, Prapawilai P, Pandith H, Tragoolpua Y (2023) Inhibition of skin pathogenic bacteria, antioxidant and anti-inflammatory activity of royal jelly from northern Thailand. Molecules 28(3):996. https://doi.org/10.3390/molecules28030996

Vajdi M, Musazadeh V, Khajeh M, Safaei E, Darzi M, Noshadi N, Bazyar H, Askari G (2023) The effects of royal jelly supplementation on anthropometric indices: a GRADE-assessed systematic review and dose-response meta-analysis of randomized controlled trials. Front Nutr 10:1196258. https://doi.org/10.3389/fnut.2023.1196258

Wang W, Li X, Li D, Pan F, Fang X, Peng W, Tian W (2023) Effects of major royal jelly proteins on the immune response and gut microbiota composition in cyclophosphamide-treated mice. Nutrients 15(4):974. https://doi.org/10.3390/nu15040974

WHO, World Health Organization (2022) Cancer, Key facts https://www.who.int/news-room/fact-sheets/detail/cancer. Accesed 10/03/2024

Wu H, Zhou S, Ning W, Wu X, Xu X, Liu Z, Liu W, Liu K, Shen L, Wang J (2024) Major royal-jelly proteins intake modulates immune functions and gut microbiota in mice. Food Sci Human Well 13(1):444–453. https://doi.org/10.26599/FSHW.2022.9250038

Xu H, Li L, Wang S, Wang Z, Qu L, Wang C, Xu K (2023) Royal jelly acid suppresses hepatocellular carcinoma tumorigenicity by inhibiting H3 histone lactylation at H3K9la and H3K14la sites. Phytomed 118:154940. https://doi.org/10.1016/j.phymed.2023.154940

Yan S, Wang X, Sun M, Wang W, Wu L, Xue X (2022) Investigation of the lipidomic profile of royal jelly from different botanical origins using UHPLC-IM-Q-TOF-MS and GC-MS. LWT-Food Sci Tech 169:113894. https://doi.org/10.1016/j.lwt.2022.113894

Yang YC, Chou WM, Widowati DA, Lin IP, Peng CC (2018) 10-Hydroxy-2-Decenoic acid of royal jelly exhibits bactericide and anti-inflammatory activity in human colon cancer cells. BMC Complement Altern Med 18:202. https://doi.org/10.1186/s12906-018-2267-9

You M, Miao Z, Tian J, Hu F (2020) Trans-10-hydroxy-2-decenoic acid protects against LPS-induced neuroinflammation through FOXO1-mediated activation of autophagy. Eur J Nutr 59:2875–2892. https://doi.org/10.1007/s00394-019-02128-9

Zhang S, Shao Q, Geng H, Su S (2017) The effect of royal jelly on the growth of breast cancer in mice. Oncol Lett 14:7615–7621. https://doi.org/10.3892/ol.2017.7078

Zhou J, Zhang L, He Y, Liu K, Zhang F, Zhang H, Lu Y, Yang C, Wang Z, Fareed MS, Liang X, Yan W, Wang K (2021) An optimized analog of antimicrobial peptide Jelleine-1 shows enhanced antimicrobial activity against multidrug resistant P. aeruginosa and negligible toxicity *in vitro* and *in vivo*. Eur J Med Chem 219:113433. https://doi.org/10.1016/j.ejmech.2021.113433

Part IV
Bee Pollen

Chapter 11
Chemical Composition of Bee Pollen

Otilia Bobiş, Adriana Cristina Urcan, Adriane Alexandre Machado De-Melo, and Ligia Bicudo de Almeida-Muradian

Contents

O. Bobiş (✉)
Department of Apiculture and Sericiculture, Faculty of Animal Science and Biotechnologies, University of Agricultural Sciences and Veterinary Medicine, Cluj-Napoca, Romania
e-mail: obobis@usamvcluj.ro

A. C. Urcan
Department of Microbiology and Immunology, Faculty of Animal Science and Biotechnologies, University of Agricultural Sciences and Veterinary Medicine, Cluj-Napoca, Romania

A. A. M. De-Melo
Department of Food and Experimental Nutrition, Pharmaceutical Sciences School, University of São Paulo, São Paulo, SP, Brazil

Laboratório de Análise de Alimentos da Faculdade de Ciências Farmacêuticas, Universidade de São Paulo, São Paulo, SP, Brazil

L. B. de Almeida-Muradian
Department of Food and Experimental Nutrition, Pharmaceutical Sciences School, University of São Paulo, São Paulo, SP, Brazil

J. M. Alvarez-Suarez (ed.), *Bee Products – Chemical and Biological Properties*,
https://doi.org/10.1007/978-3-031-89049-9_11

Abstract Bee pollen represents the main protein and lipid nutrient source in the hive. The chemical composition of this product depends on the botanical and geographical origin, and also the harvest time and processing conditions. The contents of carbohydrates, proteins, amino acids, lipids, fatty acids, vitamins, and minerals are presented. Bioactivity of bee pollen is given by the presence of phenolic acids, flavonoids, and carotenoids as secondary metabolites of plants, and by the presence of different probiotics, namely different *Lactobacillus* and *Bifidobacterium* species. Due to the nutritional value and biological potential of bee pollen, numerous studies have been carried out in the last decades, with the aim of characterizing samples from different regions and botanical origins, seeking to identify particularities and those samples with high levels of certain compounds. In this chapter, the results of this research are presented, with the aim of bringing together the greatest quantity of information regarding the composition of bee pollen.

Keywords Bee pollen · Chemical composition · Nutritional value · Bioactive compounds

11.1 Introduction

Bee pollen is a food produced by bees to serve as a nutrient source for the development and maintenance of the colony. The insects collect thousands of pollen grains and use salivary secretions, nectar, and/or honey to agglutinate them, making up pollen loads generally varying between 1.4 and 4 mm in size (Brazil 2001; Campos et al. 2008; Saavedra et al. 2013; Thakur and Nanda 2020). The bees transport the pollen loads adhered to their legs, specifically to the pollen basket (Pereira et al. 2013); and, once in the hive, store the pollen loads within the alveoli for subsequent consumption. Bee pollen is collected by beekeepers by setting traps at the hive entrance so that the pollen loads are retained at the moment the insects enter the hive.

Approximately 250 substances can be found in this product (Komosinska-Vassev et al. 2015; Anjum et al. 2024). Among the macronutrients, protein levels, which can reach 62 g/100 g, stand out. Carbohydrates, lipids, soluble and insoluble vitamins, minerals and fatty acids are also present, as well as secondary metabolites of plants with an antioxidant, antibacterial, antifungal, anti-inflammatory, antiallergic, hepatoprotective, and antitumor potential (Campos et al. 2008; Bogdanov 2012b; Kacániová et al. 2012; Nogueira et al. 2012; Fatrcová-Šramková et al. 2013; Pascoal et al. 2014; Liolios et al. 2018; Urcan et al. 2021).

The chemical composition varies significantly, according to the botanical and geographical origin of the pollen grains that make up the pollen loads. The composition of the grains, in their turn, suffers the influence of the soil and climate conditions, age, and nutritional state of the plant during the development of the grain. In the same plant species, there can be variations in pollen composition according to region, season of the year, and even between years (Silveira 2012). Bees visit around

200 flowers to produce one load of bee pollen, which generally contains grains from the same plant species (Bogdanov 2012a; Komosinska-Vassev et al. 2015).

It is possible to identify the botanical origin of bee pollen. Currently, the most utilized method is the morphological and structural analysis of the grains under a microscope. This method allows the identification of which species, genus, or family the grains belong to and determines the percentage of each pollen type. The frequency of taxa therefore is used to classify monofloral bee pollen, where there is a predominance of one pollen type, or heterofloral, or multifloral pollen, when there is no predominance (Barth et al. 2010).

Processing can also influence the physicochemical, microbiological, and potential biological characteristics of bee pollen. It is common that the product goes through pre-cleaning, freezing, defrosting, dehydration, aeration, final cleaning, packaging, and storing stages. As bee pollen contains a lot of water when is collected from the hive, a process of dehydration is necessary in order to avoid fermentation and spoilage (Estevinho et al. 2012). Drying is an important step in the production of commercial bee pollen and it is responsible for promoting the preservation of the product and for increasing its shelf life (Moreira et al. 2008). Currently, the great majority of beekeepers carry out this step, and the dehydration conditions vary according to production unit.

The commercialization of bee pollen is regulated in countries such as Argentina, Armenia, Bulgaria, Brazil, China, Cuba, Japan, Poland, Russia, Switzerland, Turkey, and Uraguay with physicochemical parameters generally being established (Almeida-Muradian 2009). Brazilian legislation, for example, includes moisture limits on the product in natura (30% maximum) and dehydration (4% maximum), as well as minimum protein levels of 8 g/100 g, 1.8 g/100 g of lipids, and 2 g/100 g of brute fiber, maximum levels of 4 g/100 g of ash and sugars total between 14.5 and 55 g/100 g, all on a dry base. Maximum free acidity levels of 300 mEq $kg-^{1}$ and pH of 4–6 were also established (Brazil 2001).

Due to the nutritional value and biological potential of bee pollen, a number of studies have been carried out with the aim of characterizing samples from different regions, seeking to identify particularities and those samples with elevated levels of certain compounds (Thakur and Nanda 2020; Mărgăoan et al. 2021; El Ghouizi et al. 2023). In this chapter, the results of this research are presented, with the aim of bringing together the greatest quantity of information regarding the composition of bee pollen.

11.2 Sugar Content

Carbohydrates content in bee pollen represent 13–55% of dry weight. They consist mainly of polysaccharides, represented by sporopollenin (from exine) and cellulose and pectin (from intine) (Thakur and Nanda 2020). The sugar composition consists of reducing sugars (RS) (de Arruda et al. 2017; Ares et al. 2018; Wu et al. 2019) and simple sugars (fructose, glucose, and saccharose) (Bertoncelj et al. 2018; Gardana et al. 2018; Oroian et al. 2022). Also, different studies identified some minor sugars such as maltose, trehalose, turanose, raffinose, and melezitose (Bertoncelj et al. 2018; Oroian et al. 2022). Although polysaccharides are important constituents of bee pollen, only the simple sugars contribute to their nutritional value. The content of sugars depends on the botanical origin of bee pollen, but also climatic conditions and geographic origin are important factors determining this class of compounds (Bertoncelj et al. 2024). Different amounts of simple sugars are reported in monofloral or multifloral bee pollens all over the world. Analyzing different pollens of *Brasicaceae* family, *Castanea sativa*, *Hedera helix*, *Fagopyrum aesculenum*, and multifloral, Bertoncelj et al. (2018) quantified glucose (11.94–28.49%), fructose (13.17–26.40%), traces of sucrose, and 0.16–6.03% maltose. Higher amounts were quantified by Liolios et al. (2018) in different types of pollen from Greece. In this study, the authors also demonstrated that the sample collection period influences the amount of simple sugars of the same botanical species. Monofloral bee pollen of *Sisymbrium irio* collected from April to October 2016 (in 9 collection sessions) was subjected to high-performance liquid chromatography (HPLC)-refractive index detection (RID) determination of sugars and statistically significant differences among the periods of collection were noticed. Also, not all simple sugars were detected in all periods, except 22–23 April, 15–16 May, and 4–5 June, and except fructose and glucose (Liolios et al. 2018). Commercial bee pollen from Brazil (de Arruda et al. 2017; Martins et al. 2011) exhibits lower amounts of sugars namely 2.77–21.85% glucose and 8.08–25.71% fructose.

Mărgăoan et al. (2012) quantified 4.37–16.14% glucose and 8.44–15.39% fructose in different samples of bee pollen from Romania.

Later, Oroian et al. (2022) analyzed 24 samples of bee pollen collected from the north east part of Romania, from spring to autumn, having 9 botanical species as dominant pollen (>45%), quantified fructose, glucose, sucrose, turanose, maltose, trehalose, melesitose and raffinose in different amounts depending on botanical origin of the samples. Fructose was the most abundant sugar, with values ranging between 18.46 and 21.44%, lower amounts of glucose (9.49–17.19%) and the other sugars in amounts around 1% and lower. These values were in the same range as those reported for bee pollen from Spain, Italy, and Colombia (Gardana et al. 2018).

Total carbohydrate content in bee pollen has been measured using various methodologies. Some studies calculate total carbohydrate content indirectly by subtracting the total weight of other components (e.g., protein, fat, ash, water, and alcohol) from 100%. This method inherently includes non-carbohydrate substances such as fiber and organic acids (Estevinho et al. 2012; Martins et al. 2011; Urcan et al. 2021). In contrast, other methods, like the anthrone method and the phenol-sulfuric acid method, directly assess carbohydrate content through chemical reactions (Orzáez Villanueva et al. 2002; Nielsen 2010). Anthrone method involves hydrolyzing the sample with perchloric acid ($HClO_4$), followed by spectrophotometric measurement after reacting with anthrone and sulfuric acid. The method is broadly applicable to food matrices and allows simultaneous assessment of hydrolyzed starch and soluble sugars (Orzáez Villanueva et al. 2002).

Phenol-Sulfuric acid method, this rapid colorimetric technique hydrolyzes the sample, neutralizes it, and reacts it with phenol and sulfuric acid. The resulting chromophore's absorbance is measured spectrophotometrically. Although simple, this method may require the results to be expressed relative to a specific carbohydrate due to variability in absorptive properties (Nielsen 2010).

Reducing sugars (RS) content has been measured using traditional and advanced techniques, such as the following.

Lane-Eynon Method (LEM), this titration method determines RS concentration using Fehling's solution and methylene blue as an indicator. The endpoint is observed when the RS completely reduce the copper sulfate, turning the indicator from blue to white (Orzáez Villanueva et al. 2002; González-Martín et al. 2007). Although straightforward, LEM cannot differentiate among types of RS and is prone to interference from other reducing agents.

The 3,5-dinitrosalicylic acid (DNS) method involves reacting RS with DNS, forming a reduced compound (3-amino-5-nitrosalicylic acid) that alters the sample's light absorption. The change in absorbance is measured using a spectrophotometer (Estevinho et al. 2012; Fuenmayor et al. 2014). Pre-treatment, such as hydrolysis or protein removal, is often required to avoid interference (Fuenmayor et al. 2014).

High-performance liquid chromatography (HPLC) with refractive index detection (RID) or fluorescence detection (FLD) is frequently used to quantify individual sugars such as glucose and fructose (Yang et al. 2013; Martins et al. 2011). These techniques typically employ water or water-methanol mixtures for extraction, with clarification by zinc acetate and potassium ferricyanide in some cases (Martins et al. 2011). Analytical columns with polar stationary phases (NH_2) or steric exclusion modes are used for effective separation (Fuenmayor et al. 2014; González-Martín et al. 2007).

Modern methods like near-infrared spectroscopy (NIRS) offer promising alternatives for carbohydrate analysis. For instance, results from NIRS closely align with those obtained using LEM, confirming its reliability for RS quantification (González-Martín et al. 2007). Additionally, specialized HPLC columns, such as MetaCarb Ca Plus, allow for the simultaneous determination of RS and sucrose under simple mobile phase conditions (Fuenmayor et al. 2014).

While classical techniques like the anthrone and DNS methods remain widely used for total carbohydrate content and RS analysis, modern chromatographic tools provide higher specificity and accuracy. These advanced methods reduce interference and enable precise quantification of individual carbohydrates, highlighting their utility for detailed carbohydrate profiling in bee pollen.

11.3 Protein and Amino Acid Content

Protein levels in dehydrated bee pollen can vary between 2.5 and 62 g/100 g per sample, mainly according to the botanical origin. Elevated levels of this macronutrient make some samples an alternative dietary supplement, especially for vegetarians. The Food and Agriculture Organization of the United Nations (FAO 2011), recommends that adults (>18 years) consume 0.66 g/kg/day of protein, something around 49.5 g for an individual of 75 kg. For these individuals, a 25 g portion of bee pollen (the equivalent of one soupspoon) can provide up to 31% of the recommended intake. Further, according to the FAO, it is important to consider the intake of essential amino acids, which are present in bee pollen as will be subsequently considered.

The protein content of bee pollen samples collected in different countries, with different botanical origins are presented in Table 11.1. Considering monofloral samples from different countries, bee pollen of *Brassica napus* contains protein levels between 23 g/100 g (Brazil) and 27.3 g/100 g (China); *Cistus* sp. ranging from 12.6 g/100 g (Spain) to 27.1 g/100 g (Portugal); *Helianthus annuus* contains protein ranging from 14.2 (South Africa) to 15.3 g/100 g (China) and 18.03 g/100 g (Romania); *Vicia faba* ranging from 27.9 g/100 g (China) to 29.2 g/100 g (Egypt); and *Zea mays* contains from 17.9 g/100 g (China) to 23.3 g/100 g (Egypt). Low amount of proteins was detected in Romanian bee pollen from *Zea mays* (15.58 g/100 g).

A positive correlation between protein content and *Brassica napus*, *Cestrum* and *Anadenanthera* pollen types has been observed, and a negative correlation with the *Astrocaryum*, *Baccharis*, *Gochnatia*, and *Vernonia* types (Arruda et al. 2013b; Sattler et al. 2015). However, it is not always possible to establish this type of relation, since variation in pollen composition can occur within the same species, grown under different environmental conditions. De-Melo (2015), who evaluated monofloral *Mimosa scabrella* samples collected in different regions of Brazil, observed protein content varying significantly between 11.7 and 33.9 g/100 g.

Pollen from plants that flower during spring tends to have a higher protein content than those that flower during summer or autumn (Liolios et al. 2016). Therefore, bees can produce bee pollen with a higher level of this compound during this season. The insects seem to be able to identify pollen sources with greater nutritional value (Modro et al. 2011). However, other factors, such as the quantity of resource available, the size and form of the pollen grain, the presence of competitors, and the distance between the plant source and the hive, also influence pollen harvesting by bees and in this way, the composition of the final product (Luz et al. 2010; Silveira 2012).

In terms of amino acid content, interest in this compound's profile in bee pollen goes beyond the question of nutrition, given that the level of certain amino acids could serve as an indicator of the freshness and adequacy of the drying process and storage. Further, it could also indicate species-specific profiles and quantities that could be used as a tool to determine the botanical and even geographical origin of samples (González-Paramás et al. 2006). The greater part of the amino acids found in this product are in the bound form, with only about 1/10 of the total protein coming from free amino acids (González-Paramás et al. 2006; Bogdanov 2012b). The composition of the free amino acid profile in bee pollen is influenced by the botanical origin, but also external factors, such as storage and heating conditions (Serra-Bonvehí and Escolà-Jordà 1997; Domínguez-Valhondo et al. 2011).

Table 11.1 Protein and lipid content in monofloral and heterofloral bee pollen collected from different countries

Botanical origin	Country of origin	Proteins(g/100 g)	Lipids (g/100 g)	References
Alternanthera	Brazil	8.8–10.6	5.1–6.9	De-Melo (2015)
Anadenanthera	Brazil	16–32.2	3.2–9.8	Arruda (2013), De-Melo (2015)
Arecaceae	Brazil	22.1–22.8	4.7–5.7	Arruda et al. (2013b)
Brassicaceae	Serbia	23.6	6.8	Kostic et al. (2015a)
Brassica	Brazil	14.5–23.2	8.6–14.0	Souza (2014), De-Melo (2015)
Brassica kaber	Egypt	29.0	9.7	Youssef et al. (1978)
Brassica juncea	India	18.2	13.8	Ketkar et al. (2014)
Brassica napus	China	27.3	6.6	Yang et al. (2013)
	Brazil	23.0–24.5	4.7–6.2	Souza (2014)
	Romania	20.4	5.5	Spulber et al. (2018)
Camellia japonica	China	29.0	5.3	Yang et al. (2013)
Castanea	Italy	26.6	2.1	Gabriele et al. (2015)
	Romania	20.6	1.3	Spulber et al. (2018)
Cestrum	Brazil	24.8	4.6	Arruda et al. (2013b)
Cistus	Spain	12.6–22.5	2.9–7.2	Serra-Bonvehí and Escolà-Jordà (1997), De Sá-Otero et al. (2009), Nogueira et al. (2012)
	Italy	25.9	1.9	Gabriele et al. (2015)
	Portugal	23.0–27.1	2.8–5.2	Féas et al. (2012), Nogueira et al. (2012)
Citrullus lanatus	China	20.7	3.6	Yang et al. (2013)
Cocos nucifera	Brazil	10.3–21.6	4.6–8.3	Arruda (2013), De-Melo (2015)
Crataegus monogyna	Romania	22.7	2.6	Spulber et al. (2018), Oroian et al. (2022)
Dendranthema indicum	China	14.9	4.0	Yang et al. (2013)
Encelia canescens	Peru	14.3	0.2	Saavedra et al. (2013)
Eucalyptus	Brazil	12.5–19.5	3.8–4.0	Souza (2014) and De-Melo (2015)
Fabaceae	Serbia	19.9	2.1	Kostic et al. (2015a)
Fagopyrum esculentum	China	14.3	5.2	Yang et al. (2013)
Helianthus annuus	China	15.3	1.5	Yang et al. (2013)
	South Africa	14.2	5.5	Nicolson and Human (2013)
	Romania	13.2–18.0	3.7–5.2	Spulber et al. (2018), Oroian et al. (2022)
Myrcia	Brazil	12.5–41.2	2.4–7.8	Arruda et al. (2013b), De-Melo (2015)
Nelumbo nucifera	China	17.0	5.2	Yang et al. (2013)

(continued)

Table 11.1 (continued)

Botanical origin	Country of origin	Proteins(g/100 g)	Lipids (g/100 g)	References
Onagraceae	Poland	21.5	–	Szczesna (2006b)
Papaver rhoeas	China	25.0	2.2	Yang et al. (2013)
Prosopis pallida	Peru	13.7–16.0	0.2	Saavedra et al. (2013)
Prunus sp.	Romania	24.5	5.3	Oroian et al. (2022)
Robinia pseudoacacia	Romania	23.1	6.0	Oroian et al. (2022)
Rosa rugosa	China	22.9	1.4	Yang et al. (2013),
Rubus	Italy	28.4	2.8	Gabriele et al. (2015)
Salix	Serbia	24.8	2.9	Kostic et al. (2015b)
Schisandra chinensis	China	27.4	2.4	Yang et al. (2013)
Taraxacum sp.	Romania	13.8–23.7	2.1–3.5	Spulber et al. (2018), Oroian et al. (2022)
Tilia sp.	Romania	16.3–1.2	2.3–3.3	Spulber et al. (2018), Oroian et al. (2022)
Trifolium alexandrium	Egypt	35.5	7.1	Youssef et al. (1978)
Vicia faba	China	27.9	0.7	Yang et al. (2013)
	Egypt	29.2	10.6	Youssef et al. (1978)
	China	17.9	4	Yang et al. (2013)
	Egypt	23.3	9.9	Youssef et al. (1978)
	Romania	14.1–15.6	1.7–2.2	Spulber et al. (2018), Oroian et al. (2022)
	Thailand	19.1	7.0	Chantarudee et al. (2012)
Heterofloral or ND	Brazil	8.4–40.5	0.4–14.0	Almeida-Muradian et al. (2005), Barreto et al. (2005), Martins et al. (2011), Arruda et al. (2013b), Sattler et al. (2015)
Heterofloral or ND	Spain	12.5–20.8	2.8–5.6	Nogueira et al. (2012), Fuenmayor et al. (2014)
Heterofloral	Portugal	18.8–34.2	2.3–3.3	Estevinho et al. (2012), Féas et al. (2012), Nogueira et al. (2012)
Heterofloral	Serbia	14.8–27.2	1.3–5.5	Kostic et al. (2015a)

ND not determined

Influences from botanical origin on amino acids present in bee pollen seem to be more quantitative than qualitative. With few exceptions, all essential amino acids are found (Human and Nicolson 2006; González-Paramás et al. 2006; Szczesna 2006b; Bogdanov 2012a), which represent from 12% to 43% of the total amino acid level (González-Paramás et al. 2006; Szczesna 2006b; Martins 2010; Yang et al. 2013; Komosinska-Vassev et al. 2015).

The total amino acid content in bee pollen can vary between 108.1 and 287.7 mg/g of product. The following amino acids were already quantified: aspartic acid (10.6–32.3 mg/g), alanine (6.5–14.5 mg/g), glycine (5.5–12.8 mg/g), glutamine

(5.0–14.2 mg/g), isoleucine (4.8–16.0 mg/g), phenylalanine (4.6–27.2 mg/g), arginine (4.3–25.8 mg/g), threonine (4.2–12.5 mg/g), proline (3.8–59.5 mg/g), asparagine (3.4–5.7 mg/g), glutamic acid (3.3–29.3 mg/g), histidine (2.9–44.9 mg/g), tyrosine (2.6–37.6 mg/g), serine (2.4–13.3 mg/g), leucine (2.2–23.1 mg/g), tryptophan (1.6–148.0 mg/g), lysine (1.3–37.1 mg/g), valine (1.3–15.8 mg/g), cystine (0.6–3.6 mg/g), methionine (0.2–7.2 mg/g), γ-aminobutyric acid (GABA) (0.16–0.34 mg/g), cysteine, hydroxyproline, ornithine, homoserine (González-Paramás et al. 2006; Szczesna 2006b; You et al. 2007; Martins 2010; Bogdanov 2012b; Yang et al. 2013; Taha et al. 2019).

Proline was the main amino acid found in samples produced in Spain, with an average of 21.8 mg/g per sample (Serra-Bonvehí and Escolà-Jordà 1997; González-Paramás et al. 2006; Domínguez-Valhondo et al. 2011). It was also the main amino acid found in Brazilian samples, with average concentrations of 10.7 mg/g (Martins 2010), and in samples produced in China, in which the average content was between 11.8 and 23.3 mg/g (You et al. 2007; Yang et al. 2013). In samples produced in Korea, glutamic acid (23.2 mg/g average), aspartic acid (22.5 mg/g average), and proline (21.9 mg/g average) made up 36% of total amino acids (Szczesna 2006b).

Bees directly contribute to the proline content (Louveaux et al. 1978; González-Paramás et al. 2006). The level of this amino acid tends to increase over time with storage. This is because the presence of the enzyme glutamate dehydrogenase leads to the biosynthesis of proline from glutamic acid (Serra-Bonvehí and Escolà-Jordà 1997). When, during the dehydration of bee pollen, beekeepers use high temperatures over extended periods of time, there is a reduction in the level of free amino acids, and consequently, a proportional increase in proline content (Serra-Bonvehí and Escolà-Jordà 1997).

Among essential amino acids, leucine and lysine show the greatest quantities, varying between 0.4 and 23.1 mg/g and from 0.2 to 21.1 mg/g, respectively. Valine (0.06–15.7 mg/g), isoleucine (0.02–13.3 mg/g), and methionine (ND-9.1 mg/g) occur at lower quantities (González-Paramás et al. 2006; Szczesna 2006b; You et al. 2007; Martins 2010; Bogdanov 2012b; Yang et al. 2013). Usually, tryptophan cannot be determined due to the process of hydrolysis employed during the analyses (Nicolson and Human 2013), however, quantities varying between 7 and 148 mg/g were already found in monofloral bee pollen from *Helianthus annuus* and from *Camellia japonica*, respectively (Yang et al. 2013).

In terms of the solubility of proteins present in bee pollen, at pH 7, this parameter can vary between 2.8 and 25.9 g/100 g, which could be compared with low solubility commercial soy protein products (concentrated and isolated flours) (Kostic et al. 2015a). In food technology, protein solubility should be considered, since this property is particularly important because it can affect emulsification, foaming, and gelation (Kinsella 1976). In bee pollen, there seems to be a positive correlation between protein solubility and carbohydrate levels, and a negative correlation between this parameter and lipid and ash content. In a complex matrix, such as is the case with this product, this parameter can be determined by protein composition and conformation as well as by interaction among proteins and constituents (Kostic et al. 2015a).

The Kjeldahl method is the gold standard for determining total protein content. This technique involves three primary steps: digestion, distillation, and titration. In the digestion phase, the sample is heated in sulfuric acid (H_2SO_4) with potassium sulfate (K_2SO_4) as a catalyst to convert nitrogen into ammonium sulfate. The addition of sodium hydroxide (NaOH) liberates ammonia (NH_3), which is distilled and quantified by titration (Chromý et al. 2015). Some studies employ micro-Kjeldahl setups, which allow for smaller-scale analyses (Gabriele et al. 2015; Taha 2015).

Bradford assay uses Coomassie Brilliant Blue G-250 dye, which binds to proteins and produces a measurable color change detected spectrophotometrically. The dye is dissolved in ethanol and acidified with phosphoric acid to enhance binding specificity (Bradford 1976).

Bicinchoninic acid (BCA) assay reduces copper (Cu^{2+}) to cuprous ions (Cu^{+}) in the presence of proteins. The resulting complex with BCA produces a colorimetric change that is measured spectrophotometrically. This method allows differentiation between polypeptide nitrogen and non-polypeptide nitrogen (Šarić et al. 2009).

Near-Infrared Spectroscopy (NIRS), offers a rapid, non-invasive approach for protein quantification. Using a fiber-optic probe, this method eliminates the need for extensive pre-treatment and provides results comparable to chemical methods (González-Martín et al. 2007).

Proteins can be individually analyzed using electrophoretic techniques such as sodium dodecyl sulfate-polyacrylamide gel electrophoresis (SDS-PAGE) and two-dimensional gel electrophoresis (IEF/SDS). These methods involve:

1. Extraction Buffers: Proteins are homogenized with acetone containing trichloroacetic acid (TCA) and 1,4-dithiothreitol (DTT), or SDS lysis buffers.
2. Electrophoresis: Proteins are separated based on molecular weight or isoelectric points.
3. Advanced Identification: Matrix-assisted laser desorption/ionization time-of-flight (MALDI-TOF) can identify protein profiles with high precision.

Free Amino Acids are extracted using mixtures of water and organic solvents: ethanol, methanol, or acetonitrile, commonly used for their polarity and compatibility with AAs analysis (Martins et al. 2011).

Total Amino Acids include both free and protein-bound AAs, requiring hydrolysis to release bound amino acids. Acid hydrolysis with 6N hydrochloric acid (HCl) is the method of choice, as it efficiently breaks peptide bonds.

Acid hydrolysis with 6N hydrochloric acid (HCl) is the method of choice, as it efficiently breaks peptide bonds (Pascoal et al. 2014).

To enhance detection, AAs are often derivatized:

1. Phenylisothiocyanate (PITC): Produces stable derivatives for UV detection, although successive drying steps make it time-intensive (You et al. 2007).
2. O-Phthaldialdehyde (OPA) with 2-Mercaptoethanol (MEC): Provides rapid derivatization and is compatible with fluorescence detection, although it is unsuitable for secondary amino acids like proline (Pascoal et al. 2014).
3. 2-(11H-Benzo[α]-carbazol-11-yl) ethyl chloroformate (BCEC-Cl): Enhances sensitivity for AAs like tryptophan and cystine, with high stability and derivatization efficiency (You et al. 2007).

High-Performance Liquid Chromatography (HPLC) is frequently employed with reversed-phase columns (e.g., C18) for AA separation. Detection methods include fluorescence (FLD) and UV-Vis (Pascoal et al. 2014).

AA Analyzers provide direct quantification without requiring derivatization.

Ninhydrin-Based Methods: Useful for proline quantification, using chromogenic reactions with ninhydrin (Rebelo et al. 2016).

Other methods for amino acid determination are: Electrochemical Detection (ECD) which offers high sensitivity for AAs without derivatization and BCEC-Cl Labeling improves fluorescence sensitivity, particularly for low-abundance AAs like tryptophan (You et al. 2007).

11.4 Lipids and Fatty Acids

11.4.1 Lipids

Lipid content in this food source can reach 22 g/100 g (Table 11.1). Certain steroids, carotenoids, and squalene are found in the lipid fraction, beyond a variety of fatty acids (Xu et al. 2011; Komosinska-Vassev et al. 2015). Given that the type of compound can influence the techno-functional properties of the food source, Kostic et al. (2015a) observe that bee pollen suspensions do not produce foams in assayed conditions, maybe due to the surface-active lipids present in bee pollen samples. The results of this study indicate that bee pollen can be used as a foam depressant or as a food ingredient where foaming properties are not desirable.

Lipid levels in bee pollen generally vary according to the botanical origin. However, at the time of analysis, the extraction of compounds can be influenced by the processing adopted by the beekeeper. De-Melo et al. (2016) observed that the dehydration conditions influenced the result, which corroborated the hypothesis of Domínguez-Valhondo et al. (2011) that processing can affect the structure of the grains, facilitating or not the extraction of compounds.

Studies of correlations between the botanical origins and lipid content of bee pollen suggest a positive association of this parameter with the pollen types *Astrocaryum* and *Brassica napus* and a negative correlation with the pollen types *Mimosa caesalpiniaefolia* and *Cestrum*, in samples produced in the south and southeast of Brazil (Arruda et al. 2013b; Sattler et al. 2015). These results indicate that greater or lesser percentages of these pollen types can be associated, respectively, with greater or lesser concentrations of lipids in bee pollen.

Considering monofloral samples collected in different countries, bee pollen of *Brassica napus* contains lipid content between 4.7 g/100 g (Brazil) and 6.6 g/100 g (China); of *Cistus* sp. contains values between 1.9 g/100 g (Italy) and 7.2 g/100 g (Spain); of *Helianthus annuus* ranging from 1.5 g/100 g (China) to 5.5 g/100 g (South Africa); of *Vicia faba* contains from 0.7 g/100 g (China) to 10.6 g/100 g (Egypt); and *Zea mays* contains levels varying from 4 g/100 g (China) to 9.9 g/100 g (Egypt) (Table 11.1).

The extraction of lipids from bee pollen has traditionally relied on Soxhlet extraction, employing solvents such as petroleum ether, ethyl ether, and diethyl ether (Gabriele et al. 2015; Conte et al. 2017). Soxhlet extraction is a robust and widely used method involving repeated cycles of solvent vaporization and condensation to isolate lipids. Alternative solvents, such as chloroform-methanol mixtures, have also been used, sometimes in conjunction with a hydrolysis step to enhance lipid release (Odoux et al. 2012; Rebelo et al. 2016). While effective, Soxhlet extraction can be time-consuming and may lead to the degradation of heat-sensitive unsaturated compounds (Wang et al. 2009).

The Folch method, widely recognized for its effectiveness in lipid extraction, utilizes a chloroform-methanol solvent system and is commonly applied to bee pollen. This technique is favored for its ability to extract a wide range of lipid classes, including triglycerides, phospholipids, and free fatty acids. Hsu et al. (2021) reported lipid contents in various bee pollen samples, highlighting the effectiveness of the Folch method in yielding high recovery rates of lipids. This method has been widely adopted due to its efficiency and the quality of the extracted lipids (Mărgăoan et al. 2014; Urcan et al. 2021).

Supercritical fluid extraction (SFE) using CO_2, which offers several advantages, such as preserving volatile compounds, reducing extraction time, and avoiding toxic solvents (Wang et al. 2009). CO_2, as a solvent, is non-toxic, non-explosive, and cost-effective, making it a promising alternative in lipid extraction. The use of SFE reflects a broader trend toward sustainable and efficient methodologies in food and nutrition research (Bernal et al. 2011).

11.4.2 Fatty Acids

Variable levels of fatty acids are present in the lipid fraction. Nicolson and Human (2013), analyzing samples collected in South Africa, found that half of the lipid content of monofloral bee pollen from *Helianthus annuus* consisted of long-chain fatty acids. In samples produced in Israel, the total fatty acid content in the bee pollen averaged 3.8%, ranging from a minimum of 2.3% to a maximum of 6.6% (Avni et al. 2014). In all the studies consulted, linoleic and α-linolenic acids, considered strictly essential for the human organism, were quantified. Essential fatty acids have a significant impact from the nutritional point of view, due to their being membrane components and playing an important role in metabolic processes (Perini et al. 2010). Beyond this, the adequate ingestion of these compounds helps in the prevention of cardiovascular and degenerative diseases, and certain types of cancer (Calviello et al. 2007; Carl et al. 2009). Certain fatty acids, such as linoleic, linolenic, myristic, and lauric acids, have bactericidal and antifungal properties (Yang et al. 2013).

In samples collected in Poland, China, and South Korea α-linolenic, palmitic, and linoleic acids predominate, and myristic, stearic, oleic, arachidonic, behenic, and lignoceric acids were also quantified, all at levels that varied according to the botanical origin. In this same study, the author observed unsaturated-to-saturated fatty acid ratio (UFA/SFA ratio), which ranged from 1.87 to 2.16 for the samples from Poland, from 1.21 to 2.35 for the samples from South Korea, and from 1.50 to 2.09 for the samples from China (Szczesna 2006a). This was similar to what was observed by Serra-Bonvehí and Escolà-Jordà (1997) in samples collected from Spain (1.96).

In bee pollen collected in Romania, α-linolenic, palmitic, and linoleic acids also predominated, and as well as these, caproic, caprylic, capric, lauric, myristic, stearic, oleic, elaidic, arachidonic, 11-eicosenoic, and behenic acids were also identified (Mărgăoan et al. 2014). In samples from Colombia, with a mean lipids value of 6.9 g/100 g, the most abundant fatty acids were also α-linolenic (6.8–42.0 mg/100 mg lipids), palmitic (6.3–16.6 mg/100 mg lipids) and linoleic (5.7–12.5 mg/100 mg lipids). The other fatty acids determined were: lauric (0.8–9.1 mg/100 mg lipids), myristic (0.5–2.6 mg/100 mg lipids), stearic (1.1–3.0 mg/100 mg lipids), oleic (1.3–3.9 mg/100 mg lipids), arachidic (0.6–1.5 mg/100 mg lipids), homo-γ-linolenic (0.2–5.3 mg/100 mg lipids) and tricosanoic (0.2–2.3 mg/100 mg lipids) (Fuenmayor et al. 2014).

In bee pollen produced in Brazil, Bastos et al. (2004) observed that the unsaturated compounds represented from 14% to 56% of the total fatty acids. The percentage of linoleic acid varied from 9% to 50% and that of linolenic acid from traces to 4% and the UFA/SFA ratio was <1.0 in almost all samples. In bee pollen collected in Portugal, caprylic, capric, lauric, palmitic, oleic, arachidonic, eicosenoic, linoleic, and α-linolenic acids were quantified, with the last representing between 26% and 57% of the total fatty acids. The polyunsaturated-to-saturated fatty acid ratio (PUFA/SFA ratio) ranged between 1.40 and 4.51 and the linoleic acid:α-linolenic acid proportion varied between 0.11 and 0.96 (Estevinho et al. 2012; Féas et al. 2012). Linoleic and α-linolenic acids are key compounds for cell membranes, neurotransmission, and brain function (Féas et al. 2012).

It is possible that bees seek out species whose pollen grains contain greater unsaturated than saturated fatty acid levels (Szczesna 2006a), and that this perception has a relationship with aromatic substances present in the lipid fraction which ends up related, by the bees, to pollen grain composition. The UFA/SFA ratio also shows the influence of dehydration and storing processes, during which time decomposition of unsaturated compounds can occur, explaining the values of ratios below 1 in certain samples (Bastos et al. 2004; Human and Nicolson 2006).

The level of fatty acids in the pollen grain can vary between plant species, or within the same species according to the cultivar. Environmental factors can also influence the composition of pollen grains and, consequently, bee pollen; and different extraction and analysis methods for fatty acids can reach different results. In monofloral samples of *Zea mays* produced in China, α-linolenic (52%) and palmitic (25%) acids predominated, while in samples from the same botanical origin but produced in Egypt, oleic (42%) and myristic (40%) acid predominated (Shawer et al. 1987; Yang et al. 2013). In *Helianthus annuus* monofloral bee pollen lauric (33%) and palmitic (23%) acids predominated when the samples originated from South Africa, and α-linolenic (42%) and palmitic (29%) when the samples originated from China (Nicolson and Human 2013; Yang et al. 2013). In monofloral *Vicia faba* samples, produced in China and Egypt, the same fatty acids predominated, but in different percentages in relation to the total α-linolenic (52%: China; 22%: Egypt) and palmitic (36%: China; 28%; Egypt) (Shawer et al. 1987; Yang et al. 2013).

In monofloral bee pollen from *Aloe greatheadii* var. davyana, collected in South Africa, the main fatty acids found were gadoleic (42%; 12.9 mg/g) and palmitic (14%; 4.4 mg/g). As well as these, the following saturated fatty acids were identified and quantified: myristic, palmitic, stearic, arachidic, behenic, and lignoceric; and the monounsaturated fatty acids: oleic and gadoleic; and the polysaturated fatty acids: ricinoleic, linoleic, α-linolenic, γ-linolenic, eicosadienoic, homo-g-linolenic, timnodonic, and brassic (Human and Nicolson 2006). Xu et al. (2011) analyzed monofloral bee pollen from *Nelumbo nucifera* produced in China, in which palmitic (45%) and oleic (19%) acids predominated, and other compounds were identified: myristic, pentadecanoic, palmitoleic, margaric, stearic, linoleic, α-linolenic, and behenic acids. In *Cistus ladanifer* monofloral bee pollen from Spain, the predominant fatty acids were oleic (28%) and palmitic (12%). Lauric, myristic, myristoleic, palmitoleic, margaroleic, stearic, linoleic, linolenic, arachidic, gadoleic, heneicosanoic and lignoceric acids were also identified (Domínguez-Valhondo et al. 2011).

Myristic and linolenic acids were the major acids in monofloral bee pollen of *Trifolium alexandrinum* and *Brassica kaber*, and myristic and oleic acids were the dominant acids in monofloral bee pollen of *Melilotus siculus* from Egypt (Shawer et al. 1987). In monofloral *Brassica campestres* samples, collected in China, the predominant compounds were α-linolenic and palmitic acids (Xu et al. 2009). Yang et al. (2013), who also analyzed samples produced in China, identified that in monofloral bee pollen from *Citrullus lanatus, Camellia japonica, and Schisandra chinensis*, α-linolenic and palmitic acids predominated; from *Fagopyrum esculentum, Papaver rhoeas and Rosa rugose*, α-linolenic and oleic acids predominated; from *Brassica napus*, α-linolenic and myristic acids predominated, meanwhile monofloral pollen from *Dendranthema indicum*, oleic and linoleic predominated and from *Nelumbo nucifera*, oleic and palmitic acids are predominant. Generally speaking, fatty acids present in different bee pollen samples are the same, however, the proportion of these compounds is variable.

Fatty acids are a key focus in bee pollen research, particularly omega-3 fatty acids, which are critical for human health. Their extraction mirrors that of total lipids, typically involving solvents like petroleum ether or chloroform-methanol mixtures (Conte et al. 2017; Estevinho et al. 2012). Soxhlet and Soxtec™ systems are frequently used, with Soxtec™ offering improved efficiency through an initial boiling step that accelerates solubilization and reduces solvent consumption (Odoux et al. 2012; Rebelo et al. 2016).

Before gas chromatography (GC) analysis, fatty acids are converted to methyl esters (FAMEs) via transesterification, ensuring reliable and reproducible chromatographic results (Ares et al. 2013). SFE combined with reverse-phase liquid chromatography (RPLC) and fluorescence detection (FLD) has emerged as a viable alternative, addressing limitations such as solvent toxicity and heat-induced degradation (Wang et al. 2009).

Sterols, including phytosterols and cholesterol, have been less extensively studied in bee pollen compared to other lipids (Conte et al. 2017). Extraction typically involves chloroform-methanol mixtures, followed by cold saponification to isolate sterols. Gas chromatography (GC), equipped with flame ionization detectors (FID) or mass spectrometry (MS), is the preferred analytical method for sterols, as it allows precise quantification after derivatization into trimethylsilyl derivatives (Bernal et al. 2011). Phytosterols are particularly valuable for their cholesterol-lowering properties, highlighting their dietary significance (Estevinho et al. 2012).

11.5 Vitamins and Minerals

11.5.1 Vitamins

Vitamins are a group of organic compounds that are necessary to ensure growth and maintain the health of humans. These compounds have different chemical compositions and biological functions, and are necessary for the synthesis of essential cofactors and metabolic reactions. Vitamins are essential for the maintenance of biological functions and can appear in nature as a vitamin itself or in the form of precursors such as provitamins (Ball 1998; Lopes et al. 2011).

These compounds can be classified in terms of their solubility. Lipid-soluble vitamins (e.g. vitamins A, D, E, and K) are a group of chemical substances with different structures, but they have in common solubility in organic solvents. They are stored in body fat and can be toxic when consumed in excess (hypervitaminosis). Water-soluble vitamins (e.g. vitamins B_1 or thiamine, B_2 or riboflavin, B_3 or niacin, B_5 or pantothenic acid, B_6 or pyridoxine, B_7 or biotin, B_9 or folic acid, B_{12} or cyano-cobalamin, and C or ascorbic acid) are not normally stored in significant amounts in the body, so we need a daily supply of these vitamins. These vitamins are involved in the metabolism of fats, carbohydrates, and proteins (Ball 1998). Daily requirements for these vitamins are low, which can be explained by their non-participation in body structure formation and the fact that they are not an energy source.

Bee pollen contains hydrosoluble and liposoluble vitamins: B complex, vitamin C, E and carotenes with provitamin A function (α and β-carotene) (Oliveira et al. 2009; Pereira de Melo and Almeida-Muradian 2010; Arruda et al. 2013a, b; Souza 2014; Sattler et al. 2015). According to Arruda et al. (2013a, b) and Souza (2014), bee pollen is rich in the B complex vitamins: thiamine, niacin, riboflavin, pyridoxine, pantothenic acid, folic acid, and biotin. B complex vitamins in bee pollen have also been reported by Al-Kahtani (2017) and Barbieri et al. (2020). This food is also rich in provitamin A and contains varying amounts of vitamin C and lipid-soluble vitamins. Botanical origin and processing conditions appear to be the main influences on the vitamin content in bee pollen.

The analysis of vitamins naturally present in food can be considered a challenge because they are present in low concentrations. The presence of an interfering matrix complexity further complicates the task, as well as the need for special care due to low stability. One of the main challenges of characterizing vitamin content in food is the development and establishment of simultaneous methods. Before vitamin analysis, moisture content must be determined, in order to facilitate the expression of results on the dry base. In the study by Pereira de Melo and Almeida-Muradian (2011) six methods for moisture content determination in dehydrated bee pollen samples were compared (conventional oven at 100 °C, vacuum oven at 70 °C, desiccator with sulfuric acid, drying out process with infrared light at 85 °C, lyophilization and Karl Fisher's method). Results showed that the methods which use infrared and the lyophilization processes are the most recommended.

11.5.1.1 B Complex Vitamins

Vitamins B_1 (thiamine), B_2 (riboflavin), three vitamers of B_6 (pyridoxol, pyridoxal, and pyridoxamine), and two vitamers of vitamin PP (nicotinic acid and nicotinamide) were analyzed in Brazilian bee pollen samples. For vitamin B_1 a variation from 0.59 to 1.09 mg/100 g (fresh samples) was obtained and from 0.64 to 1.01 mg/100 g for dried samples. Vitamin B_2 concentrations varied from 1.73 to 2.23 mg/100 g (fresh samples) and from 1.77 to 2.56 mg/100 g (dried samples). Vitamin PP concentrations varied from 6.43 to 15.34 mg/100 g (the fresh samples) and from 7.27 to 14.43 mg/100 g (dried samples). Vitamin B_6 content varied from 0.50 to 0.79 mg/100 g (fresh samples) and from 0.33 to 0.77 mg/100 g (dried samples). The amount of pyridoxol was very low and sometimes below the limit of quantification (0.07 mg/100 g) for either fresh or dried samples. The concentration of pyridoxal and pyridoxamine was also above that limit. Regarding the pollinic analysis, a direct proportional correlation was found between the percentages of *Myrcia* and *Vernonia* pollen types and the concentration of vitamin B_1. It was also observed that the greater the presence of the *Cestrum* pollen type, the lower the concentration of vitamin B_2. The occurrence of the *Ilex* pollen type was associated in a marginally significant way and inversely proportional to the concentration of vitamin PP. There was no association between the pollen types and vitamin B_6 (Arruda et al. 2013a, b).

Further, in southern Brazil, the B complex vitamins (vitamin B_1, B_2, and the vitamers of vitamins B_3 and B_6) in dehydrated bee pollen were studied (Souza 2014). This region has a differentiated climate, topography, and vegetation and is one of the largest Brazilian producers of bee pollen. The results were expressed on a dry basis and the following was found: vitamin B_1 (0.46–1.83 mg/100 g); vitamin B_2 (0.40–1.86 mg/100 g); for vitamin B_6, with only the pyridoxal and pyridoxamine vitamers quantified (pyridoxal = from 0.42 to 6.70 mg/100 g; pyridoxamine = from 0.26 to 0.95 mg/100 g). For 25 g of bee pollen (suggested daily intake), 15 out of 28 samples were considered a source and 2 were rich in thiamine; 19 lots were sources and 3 were rich in riboflavin, and, 2 lots were sources of and 26 were rich in pyridoxine in relation to the Daily Intake Reference (DIR) for adults, as established by Brazilian legislation (Brazil 2005).

De-Melo et al. (2016) showed B complex vitamin in heterofloral bee pollen samples from São Paulo State, Brazil, submitted to two different dehydration procedures. The results obtained for thiamine (vitamin B_1) were 0.6–1.3 mg/100 g (electric oven dehydration) and 0.5–0.9 mg/100 g (lyophilized samples); for riboflavin (vitamin B_2) the values reported were 0.4–0.5 mg/100 g (electric oven dehydration) and 0.4–0.5 mg/100 g (lyophilized samples). For niacin the results observed were 1.3–2.6 mg/100 g (electric oven dehydration) and 1.3–1.5 mg/100 g (lyophilized samples); and for niacinamide the values 3.0–3.2 mg/100 g (electric oven dehydration) and 3.7–3.8 mg/100 g (lyophilized samples) were obtained. Regarding vitamin B_6, for pyridoxamine, the values 0.4–0.6 mg/100 g (electric oven dehydration) and the mean of 0.5 mg/100 g (lyophilized samples) were obtained; for pyridoxal 1.9–3.8 mg/100 g (electric oven dehydration) and 3.4–3.6 (lyophilized samples) was reported. For pyridoxol 0.1–0.2 for both dehydration procedures was observed.

In 1980, Loper et al. reported the following vitamins in bee pollen: 0.76 mg/100 g of panthothenic acid, 7.10 mg/100 g of niacin, 0.61 mg/100 g of pyridoxine, and 188 mg/100 g of inositol. In bee pollen from France, Donadieu (1983) presented a range of 16.30–19.20 μg/g for vitamin B_2 and 0–9 μg/g (on a dry basis) for vitamin B_6 (not separating the vitamins of vitamin PP and B_6). Thiamine was analyzed in American samples from Beltsville and a maximum value of 6.5 μmol/100 g (dry weight) was obtained (Herbert et al. 1987). Sánchez (2004) found 5 μg/g of vitamin B_6 (without separation of the vitamins) in bee pollen from Mexico. Bee pollen from Romania presented a range of 6–13 mg/kg of thiamine; 6–20 mg/kg of riboflavin; 40–110 mg/kg of niacin; 5–20 mg/kg of pantothenic acid, and 2–7 mg/kg of pyridoxine (Mărgăoan et al. 2010). In Thailand, monofloral bee pollen from corn was evaluated and the authors reported the following values: 0.20 mg/100 g of thiamine; 0.50 mg/100 g for riboflavin; 7.03 mg/100 g of niacin; 0.39 mg/100 g of vitamin B_5 and 1.87 μg/100 g of vitamin B_{12}. Vitamin B_6 was not found (Chantarudee et al. 2012).

Campos et al. (2008), on behalf of the International Honey Commission, proposed a standard for bee pollen composition and established values of 0.6–1.3 mg/100 g of vitamin B_1, 0.6–2 mg/100 g of vitamin B_2, 4–11 mg/100 g of vitamin PP, and 0.2–0.7 mg/100 g of vitamin B_6.

11.5.1.2 Vitamin C

Loper et al. (1980) reported the value of 20.6 mg/100 g of vitamin C for bee pollen. The first authors who tried to analyze vitamin C in Brazilian bee pollen found no vitamin C at all (Almeida-Muradian et al. 2005) in samples from southern Brazil. Another study was carried out by Oliveira et al. (2009) which analyzed vitamin C (ascorbic acid) from fresh bee pollen samples from southeast Brazil. A range from 273.9 to 560.3 μg/g of ascorbic acid was observed. Samples collected in October (Brazil's spring) could be considered vitamin C sources. A relationship between the following botanical origins: *Anadenanthera* sp., *Arecaceae* type, and *Philodendron* sp., and vitamin C content was observed.

Other studies using southeastern Brazilian dried bee pollen (Pereira de Melo and Almeida-Muradian 2010) obtained values from 114 to 340 μg/g of vitamin C. The pollen types from *Myrtaceae* had a correlation with vitamin C content. Regarding the stability of this vitamin (Pereira de Melo and Almeida-Muradian 2010), it was observed that on average samples stored at room temperature (exposed and protected from light) presented a 50% loss in ascorbic acid concentration. According to the statistical analysis, there was a significant difference ($p < 0.05$) between these two conditions when compared to vitamin concentrations at time 0. No difference, however, was observed between the periods of analysis (6 and 12 months of storage) using either storage method at room temperature. This suggests that when bee pollen is stored at room temperature, vitamin C is rapidly lost. After this initial sharp loss, however, it becomes stable during the storage process. Samples stored in the freezer presented a lower level of vitamin C loss (22% in 6 months and 26% in 12 months) when compared to losses that occurred in storage at room temperature. Nevertheless, for initial vitamin C concentrations, there was a statistically significant difference ($p < 0.05$) after 12 months of storage in the freezer. These results indicated that the storage of dried bee pollen in the freezer until consumption was advisable.

Ascorbic acid was also evaluated in southern Brazilian bee pollen samples by Sattler et al. (2015) and an average value of 262.3 μg/g for vitamin C was obtained. Bee pollen from Poland presented the values of 12.8 and 15.21 mg% of vitamin C (Szczesna et al. 1991). Bee pollen from Romania presented a range of 70–560 mg/kg for ascorbic acid (Mărgăoan et al. 2010).

Barajas et al. (2012) studied vitamin C in dehydrated bee pollen from Colombia (samples dried at 35 and 45 °C). The first samples presented the values 40.22 mg/100 g (fresh), 31.75 mg/100 g (dried at 35 °C), and 27.35 mg/100 g (dried at 45 °C). The second sample showed the values 40.37 mg/100 g (fresh), 32.79 mg/100 g (dried at 35 °C), and 28.75 mg/100 g (dried at 45 °C). Therefore, the content of this vitamin seems to be altered by processing conditions, especially temperature.

11.5.1.3 Vitamin E

Vitamin E is one of the antioxidant vitamins, with a-tocopherol being the most studied compound. In Brazil, the first attempt to identify a-tocopherol in bee pollen was made by Almeida-Muradian et al. (2005). This vitamin, however, was not found in the commercial samples, probably due to the dehydration process. The second attempt in Brazilian bee pollen was made by Oliveira et al. (2009) and the authors used fresh bee pollen samples from southeastern Brazil. A range of 13.5–42.5 μg/g for vitamin E (a-tocopherol) was observed, and a relationship between vitamin E and the following botanical origins: *Raphanus* sp., *Eucalyptus* sp., *Macroptilium* sp., *Mimosa caesalpineafolia* was also reported. The four tocopherols (alpha, beta, gamma, and delta-tocopherol) were determined by Sattler et al. (2015), who reported the following mean values for vitamin E: 32.8 μg/g of alpha-tocopherol, 2.1 μg/g of beta-tocopherol, 4.6 μg/g of gamma-tocopherol, and 6.3 μg/g of delta-tocopherol.

De-Melo et al. (2016) studied Brazilian heterofloral bee pollen (São Paulo state) using two different dehydration processes. The results obtained for vitamin E (a-tocopherol) were 27.2–27.5 μg/g (bee pollen dehydrated in an electric oven with forced air circulation) which was lower than lyophilized samples (37.5–53.7 μg/g). The same parameters such as light, oxygen, and temperature are responsible for the degradation of tocopherols; therefore, processing conditions significantly influence vitamin E content in bee pollen.

Corn monofloral bee pollen from Thailand presented a value of 6.21 mg/100 g for vitamin E (alpha-tocopherol) (Chantarudee et al. 2012).

11.5.2 Minerals

Bee pollen contains elements that are indispensable for the development of bees and humans, which makes this food an alternative nutritional supplement. Potassium (1.4–38 g/kg), phosphorus (0.2–9.6 mg/kg), calcium (0.2–5.8 g/kg), manganese (5–430 mg/kg), zinc (5.1–340 mg/kg), copper (3–42 mg/kg), chromium (0.03–42 mg/kg), boron (8.2–14 mg/kg), molybdenum (0.1–4.6 mg/kg), and selenium (<0.01–4.5 mg/kg) were found (Youssef et al. 1978; Somerville and Nicol 2002; Bogdanov 2012a; Formicki et al. 2013; Morgano et al. 2012; Yang et al. 2013; Fuenmayor et al. 2014; Sattler et al. 2015).

The content of these nutrients varies according to the botanical origin and geographical region, and season during the year. The characteristics of the region, especially the soil type where the plant developed, are determinant for the bioelement content of bee pollen (Formicki et al. 2013; Yang et al. 2013; De-Melo 2015). Beyond this, a part of these elements can come from the nectar added by the bees at the time of pollen harvesting (Kostic et al. 2015b). Somerville and Nicol (2002) published a study comparing the mineral content of monofloral samples from a single plant species, but collected in different regions of Australia between 1991 and 1998, and the results indicated variations in bioelement concentration. On the other hand, processing and storage conditions did not influence mineral content (Youssef et al. 1978; De-Melo et al. 2016).

Researchers from the International Honey Commission, who proposed quality criteria for bee pollen, suggested parameters for the contents: potassium (4000–20,000 mg/kg), phosphorus (800–6000 mg/kg), magnesium (200–3000 mg/kg), alcium (200–3000), zinc (30–250 mg/kg), manganese (20–110 mg/kg), iron (11–170 mg/kg), and copper (2–16 mg/kg) (Campos et al. 2008).

11.5.2.1 Macro-elements

Potassium (K)

Independent of the origin of bee pollen, this mineral occurs in elevated concentrations, reaching 38 g/kg in monofloral bee pollen from *Asphodelus fistulosus* collected in Australia (Somerville and Nicol 2002).

In South American countries (Argentina, Brazil, and Colombia) K content in bee pollen varies between 1.4 and 9.9 g/kg (Morgano et al. 2012; Fuenmayor et al. 2014; De-Melo 2015; De-Melo et al. 2016). Some of the highest values were observed in monofloral samples from *Mimosa* or heterofloral samples where species from this genus were present (De-Melo 2015). In bee pollen collected in China the content of this mineral was between 2.4 and 9.5 g/kg (Szczesna 2007; Yang et al. 2013; Fuenmayor et al. 2014), with the lowest value found being in a monofloral sample from *Dendranthema indicum* and the highest value in a heterofloral product.

In samples from Saudi Arabia, K content varied from 6.2 g/kg, in monofloral bee pollen from *Helianthus annuus,* to 8.3 g/kg, and in monofloral bee pollen from *Brassica napus* (Taha 2015). In monofloral samples from *Brassica napus* collected in China and Australia the average K content was lower, at 4.7 and 5.4 g/kg, respectively (Somerville and Nicol 2002; Yang et al. 2013). In harvesting undertaken in Serbia and Poland, samples considered to be monofloral from an unidentified pollen type from the Brassicaceae family presented a K content of 3.2 and 3.6 g/kg, respectively (Szczesna 2007; Kostic et al. 2015b). In a monofloral sample from another species from the genus *Brassica* (*B. kaber*), collected in Egypt, K content was 3.8 g/kg (Youssef et al. 1978).

In Spain, a K content in bee pollen of between 2.5 and 6.4 g/kg was found (Serra-Bonvehí and Escolà-Jordà 1997; Fuenmayor et al. 2014), something close to that observed in samples produced in Poland (2.8–5.8 g/kg) (Szczesna 2007; Fuenmayor et al. 2014) and Serbia (2.5–4.2 g/kg) (Kostic et al. 2015b). In the later country, monofloral bee pollen from *Salix* had a content of 3.4 g of potassium/kg (Kostic et al. 2015b); close to that observed in bee pollen with the same botanical origin collected in Australia (3.7 g/kg) (Somerville and Nicol 2002).

The content of this bioelement in bee pollen in South Korea varied from 3.5 to 5.5 g/kg (Szczesna 2007) and similar values were identified in samples from Solvenia (5.2 g/kg), Romania (4.9 g/kg), India (4.8 g/kg), Bulgaria (4.6 g/kg), and Hungary (3.6 g/kg) (Fuenmayor et al. 2014).

Sodium (Na)

Currently, a preoccupation with the excessive intake of sodium by the population exists and is frequently related to the consumption of processed and mass-produced foods. For this reason, it is interesting that a natural food, which can be utilized as a food supplement, as is the case with bee pollen, does not contain elevated levels of Na. The results so far indicate that a 25 g portion of this food provides, at the maximum, 10% of the maximum intake of 2 g/day of sodium, recommended by the World Health Organization (WHO 2012). The K/Na ratio is also considered relevant since it is important to maintain an equilibrium of these bioelements in the organism. The high K/Na ratio found in bee pollen makes this food useful for diets with a defined electrolytic balance (Serra-Bonvehí and Escolà-Jordà 1997).

In general, the Na content in bee pollen is below 2 g/kg, but in Saudi Arabia this value reached 8.4 g/kg in monofloral bee pollen from *Brassica napus* (Taha 2015). This value was far above that observed in monofloral samples of this same pollen type collected in China (459.1 mg/kg) and Australia (30.5 mg/kg) (Somerville and Nicol 2002; Yang et al. 2013).

In samples from China, the content of this mineral varied from 125 to 2447 mg/kg (Szczesna 2007; Yang et al. 2013; Fuenmayor et al. 2014). In the product sourced from Spain, the variation was from 198 to 2375 mg/kg (Serra-Bonvehí and Escolà-Jordà 1997; Fuenmayor et al. 2014), very similar to that found in samples produced in Poland (236–2191 mg/kg) (Szczesna 2007). In bee pollen produced in Brazil Na content was between <0.004 and 1466 mg/kg (Morgano et al. 2012; De-Melo 2015; De-Melo et al. 2016), though 75% of the samples from this country had a Na content below 400 mg/kg. Szczesna (2007) found a maximum Na content in bee pollen from South Korea (1113 mg/kg) similar to that observed in Brazilian samples.

Somerville and Nicol (2002) encountered, in samples from Australia, a content for this mineral varying between 16 and 480 mg/kg, with an average of 82 mg/kg. Among the Australian samples, values far above the average were identified in monofloral bee pollen from *Banksia ericifolia* (480 mg/kg), *Eucalyptus punctata* (340 mg/kg), *Casuarina littoralis* (290 mg/kg), and *Hypochoeris radicata* (220 mg/kg).

Youssef et al. (1978) observed an average Na content of 426 mg/kg. Fuenmayor et al. (2014) found a Na content of 8.9–206 mg/kg in samples from Colombia, of 2140 mg/kg in samples from Slovenia, 219 mg/kg in samples from Hungary, 199 mg/kg in bee pollen produced in Bulgaria, 113 mg/kg in a sample collected in India, and 84 mg/kg in product from Romania. In Serbia, the content of this mineral varied from 4.9 to 54.9 mg/kg (Kostic et al. 2015b).

Phosphorus (P)

In bee pollen produced in China, Yang et al. (2013) observed phosphorus content of 2.1 g/kg in bee pollen from *Dendranthema indicum* and 9.6 g/kg in bee pollen from *Nelumbo nucifera*. In Brazil, samples had a phosphorus content between 2.2 and 8.9 g/kg (Morgano et al. 2012; Sattler et al. 2016), similar to that found in Australia (1.4–8.0 g/kg) (Somerville and Nicol 2002) or Poland (3.2–4.6 g/kg) (Matuszewska et al. 2021). On the other hand, a lower P content occurred in bee pollen from Spain (294.7–854.4 mg/kg) (Serra-Bonvehí and Escolà-Jordà 1997; Fuenmayor et al. 2014) and Saudi Arabia (234.4–468.1 mg/kg) (Taha 2015).

Bee pollen from *Echium plantagineum* contains a P content between 7.1 and 8.0 g/kg, from *Brassica napus* between 0.4 and 6.8 g/kg, from *Zea mays* between

3.7 and 4.7 g/kg, and from *Helianthus annuus* between 0.2 and 2.6 g/kg (Somerville and Nicol 2002; Yang et al. 2013; Taha 2015). These values are related to the region where they were collected, especially to the characteristics of the soil that influenced the composition of the pollen grains of each plant.

Calcium (Ca)

Calcium is a mineral of great importance for the human organism. Among its functions are constituting bones and tissues, acting in the contraction of muscular fibers, in nerve conduction, and in the liberation of hormones, beyond being an enzyme component (Forsén and Kördel 1994). A moderate lack of this bioelement is common, with the recommended intake being 1000 mg/day for adult individuals (19–50 years) (USDA 2011).

Taha (2015) identified the highest levels of Ca in bee pollen samples (2.1–5.8 g/kg). The authors evaluated products from Saudi Arabia. The bee pollen collected in Brazil had a Ca content between 0.3 and 4.7 g/kg (Morgano et al. 2012; De-Melo 2015; De-Melo et al. 2016; Sattler et al. 2016). De-Melo (2015) observed, in Brazilian products, a tendency toward greater Ca content in monofloral samples from *Cocos nucifera* or heterofloral samples with a strong participation of this pollen type, when compared to monofloral samples from *Brassica* or heterofloral samples with a significant participation of this pollen type.

Somerville and Nicol (2002) found Ca content of 0.4–3.1 g/kg in bee pollen produced in Australia. In that country, the content of this element in monofloral samples from *Eucalyptus* was from 0.8 to 3.1 g/kg, while in Brazil a value of 1.6 g/kg was observed (Somerville and Nicol 2002; De-Melo 2015). Each eucalypt species produced pollen with a characteristic composition, beyond this, the content of this biolement in the soil is determinant in the composition of the grain for all plant species.

In China, the content of this mineral varied from 0.2 to 3.1 g/kg (Szczesna 2007; Yang et al. 2013), while in Colombia it did not reach 2.5 g/kg (Fuenmayor et al. 2014), in Serbia, it varied between 0.9 and 2.0 g/kg (Kostic et al. 2015b) and in Egypt, it was between 0.7 and 2.1 g/kg (Youssef et al. 1978). In samples collected in Europe (Spain, Poland, Hungary, Slovenia, Romania, and Bulgaria) Ca content varied between 0.3 and 1.7 (Serra-Bonvehí and Escolà-Jordà 1997; Szczesna 2007; Fuenmayor et al. 2014). A lower Ca content was found in samples collected in South Korea (105–694 mg/kg) (Szczesna 2007).

Magnesium (Mg)

The worldwide preoccupation with calcium intake is well known; however, there needs to be an equilibrium in calcium and magnesium intake for the perfect functioning of the organism. According to Rosanoff et al. (2012), in the United States, almost half the population consumes less than the recommended magnesium intake.

Research into bee pollen indicated that a 25 g portion would provide from 2% to 45% of the recommended magnesium intake for an adult individual whose requirement would be 260 mg/day.

In bee pollen produced in Saudi Arabia, Mg content was between 2.4 g/kg, in monofloral samples of *Cucurbita pepo*, and 4.7 g/kg, in monofloral samples of *Phoenix dactylifera* (Taha 2015). In the product collected in Brazil, Mg content was between 0.4 and 3.6 g/kg (Morgano et al. 2012; De-Melo 2015; De-Melo et al. 2016). In this country, De-Melo (2015) observed that samples with a significant presence of the *Cocos nucifera* pollen type had a higher content of this mineral (2.1–2.3 g/kg) compared to those in which the *Brassica* type predominated (1.3–1.8 g/kg). In monofloral samples from *Brassica kaber* from Egypt, Youssef et al. (1978) found a Mg content of 1.3 g/kg, while in samples from *Brassica napus* from Australia and Saudi Arabia, the content of this mineral was 1.4 and 3.9 g/kg, respectively (Somerville and Nicol 2002; Yang et al. 2013). In monofloral bee pollen from an unidentified pollen type from Brassicaceae, the values encountered for this mineral were 832 mg/kg (in samples from Serbia) and 1.1 g/kg (in samples from Poland) (Szczesna 2007; Kostic et al. 2015b).

Szczesna (2007), Yang et al. (2013), and Fuenmayor et al. (2014) found an Mg content in samples collected in China of 0.3–2.8 g/kg. In Australia and Argentina, a similar range was observed: 0.2–2.7 g of Mg/kg and 0.4–2.2 g of mg/kg, respectively (Somerville and Nicol 2002). In bee pollen from South Korea, the content of this mineral was between 1.0 and 1.8 g/kg (Szczesna 2007). In a monofloral sample from *Artemisia* a lower Mg value (0.7 g/kg) was found, among the products evaluated from Poland on the other hand, the highest value (1.7 g/kg) was found in a sample with an undetermined botanical origin (Szczesna 2007; Fuenmayor et al. 2014). In Colombia, between 343 and 1.5 g of magnesium/kg were quantified in the samples.

Fuenmayor et al. (2014) studied a sample from each country: India (1.4 g/kg), Slovenia (1.1 g/kg), Romania (865 mg/kg), Hungary (635 mg/kg) and Bulgaria (577 mg/kg). In Spain, Mg content varied between 273 and 816 mg/kg (Serra-Bonvehí and Escolà-Jordà 1997; Fuenmayor et al. 2014) and in Serbia, it was between 503 and 964 mg/kg, with lower and higher values identified in heterofloral samples (Kostic et al. 2015b).

Sulfur (S)

Sulfur is considered macroelement present in bee-collected pollen (Somerville and Nicol 2002; Matuszewska et al. 2021). Fifty pollen samples from Australia were subjected to mineral analysis and high amounts of sulfur were determined (1100–3700 mg/kg) (Somerville and Nicol 2002). Pollen samples having *Eucalyptus* and *Echium* species as major pollen exhibit higher amounts of this element (>3000 mg/kg) in all analyzed samples. A more recent study on bee pollen from Poland (Matuszewska et al. 2021) identified and quantified a high amount of sulfur in multifloral bee pollen samples collected during May –July of 2018–2019. Values of 1800–2700 mg/kg were quantified using inductively coupled plasma-mass spectrometry and inductively coupled plasma-optical emission spectrometry (ICP-MS and ICP-OES) methods. Sulfur is an important component of different proteins, enzymes, and vitamin molecules, hence bee pollen is recommended for human consumption. These findings are very interesting, because this macroelement was not detected in other studies analyzing bee pollen from Romania, Serbia, Turkey, or Spain (Hărmănescu and Popovici 2007; Kostić et al. 2015; Altunatmaz et al. 2017; Valverde et al. 2023), thus in all bee pollen samples, sulfur-containing amino acids are present (Anjum et al. 2024).

11.5.2.2 Micro-elements

Iron (Fe)

According to the World Health Organization, iron deficiency is a widespread and common public health condition that mainly affects women and children, as much in developing and developed nations (WHO 2015). This mineral is an essential component of hemoglobin and enzymes and deficiencies lead to anemia, persistent cognitive and developmental impairment, and depressed immune function (USDA 2015). Based on the results presented in the following, considering an individual whose necessary daily iron intake would be 14 mg/day, a 25 g portion of bee pollen would provide from 2% to 220% of the recommended intake.

In Egypt, elevated content for this mineral was observed in monofloral bee pollen samples: *Vicia faba* (1290 mg/kg), *Trifolium alexandrium* (1265 mg/kg), *Brassica kaber* (820 mg/kg), and *Zea mays* (555 mg/kg) (Youssef et al. 1978). Yang et al. (2013) also evaluated Fe content in bee pollen from *Vicia faba* (158.3 mg/kg) and *Zea mays* (75.2 mg/kg), collected this time in China. Somerville and Nicol (2002) determined Fe content from *Zea mays* bee pollen (26.0 mg/kg) produced in Australia.

In bee pollen from Brazil, Fe content reached 1179 mg/kg (Morgano et al. 2012; De-Melo 2015; Sattler et al. 2016). In a monofloral sample from *Anadenanthera* collected in southeastern Brazil, the content was 466.3 mg/kg, while in a monofloral sample from this same species, this time produced in the south, it was 143 mg/kg. In monofloral samples from *Alternanthera*, collected from the same apiary in different months, Fe content was 821.0 mg/kg (harvested in autumn: the end of May and start of June), 1017.5 mg/kg (harvested during winter: end of June and start of July) and 1179 mg/kg (harvested during winter: August) (De-Melo 2015).

In Saudi Arabia, considerable Fe values were also observed (338.1–562.1 mg/kg) (Taha 2015), just as in samples collected in Australia (14.0–520.0 mg/kg) (Somerville and Nicol 2002) and South Korea (74.3–365.9 mg/kg) (Szczesna 2007). In Saudi Arabia, the greatest value was observed in a monofloral sample from *Cucurbita pepo* and the lowest value in a monofloral sample from *Phoenix dactylifera*, while in Australia the greatest and lowest values were observed in monofloral samples from *Eucalyptus punctata* and *Lavandula*, respectively.

Bee pollen produced in China contained between 59.0 and 207.8 mg/kg of Fe/kg of product (Szczesna 2007; Yang et al. 2013). In monofloral samples from

Fagopyrum esculentum, *Brassica napus*, and *Helianthus annuus* from the same country, Fe content was significantly higher (207.8, 89.4 and 88.3 mg/kg, respectively) than that found in samples from the same botanical origins produced in Australia (51, 27.5 and 40 mg/kg, respectively). In Saudi Arabia, the content in monofloral bee pollen from *Brassica napus* and *Helianthus annuus* was greater still (361.3 and 562.1 mg/kg, respectively).

In European countries, the Fe content was found between 11.1 and 169 mg/kg (Serra-Bonvehí and Escolà-Jordà 1997; Szczesna 2007; Formicki et al. 2013; Fuenmayor et al. 2014; Kostic et al. 2015b), and in Colombia it varied between 23.3 and 126.6 mg/kg (Fuenmayor et al. 2014). The iron content in bee pollen produced in Chile varied between 146.7 and 181.8 mg/kg, with one monofloral sample from *Lotus pedunculatus* having 175.4 mg/kg and a monofloral sample from *Escallonia rubra* having 181.8 mg/kg (Mejías and Montenegro 2012).

Zinc (Zn)

Bee pollen contains significant zinc levels. Based on the results presented in the following, this food source provides up to 120% of the recommended intake of 7 mg/day for an individual adult. In bee pollen collected in Australia, the average content of this mineral was 58.3 mg/kg; however, in a monofloral sample from *Echium plantagineum* the value was 340 mg/kg (Somerville and Nicol 2002). The authors analyzed eight other samples from the same species and the average content was 68 mg/kg. Though these nine bee pollen samples have been collected in different regions of Australia over different years, it was not clear why there was such a high value in only one sample.

In research carried out in Brazil, zinc levels observed varied between 5.1 and 118 mg/kg (Morgano et al. 2012; De-Melo 2015; Sattler et al. 2016). According to Morgano et al. (2012) the levels of this mineral tend to be high in samples from northeastern Brazil. Bee pollen with a significant quantity of pollen types from the genus *Brassica* tends to have a Zn content between 30 and 80 mg/kg, when produced in southern Brazil (De-Melo 2015). A positive correlation between zinc and copper content and antioxidant capacity, measured by DPPH and ORAC methods, was observed in heterofloral bee pollen collected in southeastern Brazil. The authors reported that these minerals were cofactors for copper, zinc superoxide dismutase (CueZn SOD), an important antioxidant enzyme already reported in plants; therefore, Cu and Zn might be indicators of the presence of the Cu/ZnSOD in these bee pollen samples (De-Melo et al. 2016).

In Spain, values between 18.8 and 81.1 mg/kg were observed (Serra-Bonvehí and Escolà-Jordà 1997; Fuenmayor et al. 2014), close to that quantified in samples produced in Serbia (28.8–75.9 mg/kg), Colombia (19.8–70.6 mg/kg) (Fuenmayor et al. 2014), and China (23.9–65.3 mg/kg) (Szczesna 2007; Yang et al. 2013; Fuenmayor et al. 2014). In monofloral samples from *Zea mays* collected in China, Australia, and Egypt, Zn content was, respectively, 54.6, 48, and 80 mg/kg (Youssef et al. 1978; Somerville and Nicol 2002; Yang et al. 2013).

In samples collected in Chile, Zn content was similar, being between 62.7 and 63.9 mg/kg (Mejías and Montenegro 2012). In South Korea, values between 23.7 and 60.7 mg of Zn/kg were found (Szczesna 2007). In European countries (Poland, Hungary, Slovenia, Romania, and Bulgaria) levels of this mineral between 25.6 and 53.6 mg/kg were encountered (Szczesna 2007; Fuenmayor et al. 2014), while in Saudi Arabia Zn content in bee pollen samples varied from 31.9 to 44.2 mg/kg (Taha 2015).

Copper (Cu)

Levels of copper in bee pollen are elevated. In research carried out by Somerville and Nicol (2002) on samples collected in Australia, Cu levels varied between 3 and 42 mg/kg. These authors observed little variation when comparing mono-floral products from the same botanical origin: *Hypochoeris radicata* (4–5 mg/kg), *Eucalyptus bridgesiana* (16–17 mg/kg), *Echium plantagineum* (5–7 mg/kg), *Corymbia maculata* (39–42 mg/kg), and *Brassica napus* (6–7 mg/kg). In bee pollen from *Brassica napus* collected in Saudi Arabia, Cu content was similar to that observed in Australian samples (6.3 mg/kg), however in samples collected in China, the value was greater (15.7 mg/kg) (Yang et al. 2013; Taha 2015).

In Brazil, Cu content in samples varied between 3.2 and 25.4 mg/kg (Morgano et al. 2012; De-Melo 2015; Sattler et al. 2016), similar to that observed by Szczesna (2007) and Yang et al. (2013) in samples from China (3.2–25.1 mg/kg), by Szczesna (2007) and Fuenmayor et al. (2014) in samples from Poland (5.6–23.9 mg/kg). Similar Cu levels were also observed in samples collected in Spain (4.1–15.7 mg/kg), in South Korea (5.3–14.7 mg/kg), Chile (8.8–12.5 mg/kg), Serbia (4.4–10.7 mg/kg), Egypt (6.0–10.0 mg/kg), and in Saudi Arabia (4.3–7.0 mg/kg) (Youssef et al. 1978; Serra-Bonvehí and Escolà-Jordà 1997; Szczesna 2007; Mejías and Montenegro 2012; Fuenmayor et al. 2014; Kostic et al. 2015b; Taha 2015).

Manganese (Mn)

In bee pollen from China, manganese content varied between 8.7 and 429.8 mg/kg, with the highest level being found in a sample of undetermined botanical origin (Szczesna 2007; Yang et al. 2013). In a product collected in Brazil, Mn levels were between 12 and 314 mg/kg (Morgano et al. 2012; De-Melo 2015; Sattler et al. 2016). De-Melo (2015) observed that in samples from the same apiary, located in the southeast of Brazil, collected during the winter, spring, and summer, Mn levels were high (157–215 mg/kg) when compared to the other results (21–162 mg/kg), independent of the predominant botanical origin at the time of harvesting and the season of the year. This was only seen at this apiary.

In monofloral samples collected in Egypt, the content of this mineral was 210 mg/kg (*Vicia faba* and *Trifolium alexandrium*), 205 mg/kg (*Brassica kaber*), and 189 mg/kg (*Zea mays*) (Youssef et al. 1978). Research into monofloral products from China indicated values in samples of *Camellia japonica* (357.4 mg/kg), *Rosa rugosa* (156.1 mg/kg), *Nelumbo nucifera* (99.4 mg/kg), *Zea mays* (24.1 mg/kg), in *Brassica napus* (24.2 mg/kg), and in *Helianthus annuus* (16.7 mg/kg) (Yang et al. 2013). Monofloral samples from *Zea mays* (14 mg/kg), *Brassica napus* (30–42 mg/kg), and *Helianthus annuus* (12 mg/kg) were also studied in Australia. Generally, in bee pollen collected in Australia, the values were between 5 and 110 mg/kg (Somerville and Nicol 2002).

In samples from South Korea and Serbia, similar Mn levels were found: 25.6–99.4 and 13.5–92.2 mg/kg, respectively (Szczesna 2007; Kostic et al. 2015b). The same was found in bee pollen analyzed from Poland (13.3–59.7 mg/kg) by Szczesna (2007). In Chile, Mn content varied from 52.1 to 62.9 mg/kg (Mejías and Montenegro 2012), while the lowest values occurred in samples produced in Saudi Arabia (16.6–38.6 mg/kg)(Taha 2015) and Spain (7.4–18.3 mg/kg) (Serra-Bonvehí and Escolà Jordà 1997; Fuenmayor et al. 2014).

Manganese levels in bee pollen indicated that 25 g of a sample provided up to four times more than the daily requirement of, for example, 2.3 mg/day.

Other Bioelements

Chrome and molybdenum content in bee pollen from Brazil was 2.0–42 mg/kg and from 0.2 to 4.6 mg/kg, respectively (Sattler et al. 2016); while in samples produced in China molybdenum content varied from 0.1 to 0.4 mg/kg, and chrome was not detected (Yang et al. 2013). In bee pollen produced in Chile and Serbia, a lower chrome content was observed, when compared with Brazilian samples: 0.03–0.5 and 0.2–0.5 mg/kg, respectively (Mejías and Montenegro 2012; Kostic et al. 2015b).

Altunatmaz et al. (2017) encountered boron levels varying between 0.59 to 61.3 mg/kg in bee pollen from Turkey. Adequate boron intake is necessary to promote bone and brain health (Nielsen 2008). Morgano et al. (2012) identified selenium content of <0.01–4.45 mg/kg in 154 samples across all regions of Brazil. In samples collected in China, this bioelement was not detected (Yang et al. 2013). Selenium is a component of proteins that play a critical role in reproduction, thyroid hormone metabolism, DNA synthesis, and protection from oxidative damage and infection (Sunde 2012; NIH 2016).

11.6 Bioactive Compounds

11.6.1 Phenolic Acids and Flavonoids

These compounds are present in all plant structures, including pollen grains (Simões et al. 2004; Cunha and Roque 2009a). Some compounds are widely found, such as benzoic acid derivatives, while others have a smaller-scale distribution, such as caffeic acid esters found only in plants from the Lamiaceae family. Freire et al. (2012) researched the occurrence of certain flavonoids in 25 bee pollen samples collected in northeastern Brazil, and concluded that the phenolics: isoquercetin, quercetin, and isorhamnetin were present in more than 90% of the samples from different botanical origins. The importance of identifying specific compounds is that they can be used as taxonomic markers (Simões et al. 2004, Campos & Markham 2007).

Phenolics are considered bioactive substances, in other words, not synthesized by the human organism, having a specific physiological or metabolic action, which, when regularly present in significant quantities in the diet, aid in the prevention of disease (Campos et al. 2008; Horst and Lajolo 2011; Oliveira and Bastos 2011). The action of these compounds seems to be related to an intrinsic capacity for reduction, though other mechanisms are indicated, such as the modulation of the activity of certain enzymes, interactions with receptors, and modes of signal transduction and regulation of the cellular cycles (Menezes et al. 2010; Oliveira and Bastos 2011; Bogdanov 2012b). The biological activity of phenolics should also be considered in terms of the rate of intestinal absorption, metabolization, and excretion of these substances by the organism (Oliveira and Bastos 2011).

Over the years, researchers have studied the antioxidant, antimicrobial, anti-inflammatory, anxiolytic, anti-arterosclerosis, antimutagenic, and cardioprotective action of different compounds (Angelo and Jorge 2007; Halliwell and Gutteridge 2007; Hamalainen et al. 2007; Campos et al. 2008; Yang et al. 2008; Aguirre-Hernández et al. 2010; Lopes et al. 2011; Cabrera and Montenegro 2013; Fatrcová-Šramková et al. 2013; Komosinska-Vassev et al. 2015). Aguirre-Hernández et al. (2010) attributed anxiolytic action to the flavonoids quercetin and kaempferol, which resulted in a significant reduction in the activity of male rats. Yang et al. (2008), in studies performed on rutin in different in vitro systems, concluded that it was a powerful antioxidant agent whose response is dose-dependent. Hamalainen et al. (2007) and Komosinska-Vassev et al. (2015) noted that flavonoids, phenolic acids, fatty acids, and phytosterols are determined as the main agents responsible for the anti-inflammatory capacity of bee pollen.

To determine the level of these substances in bee pollen certain methods were adopted. The colorimetric method with the Folin-Ciocalteu reagent was frequently used to indicate the levels of phenolic compounds in gallic acid equivalents (GAE), even if it was not specifically intended to quantify these compounds. Certain substances, such as ascorbic acid, amines, carbohydrates, and sulfur-containing compounds, can also react with phosphomolybdic and phosphotungstic acids of the reagent used (Huang et al. 2005). To determine the total flavonoid level, a colorimetric method based on quantifying the formation of complexes between flavonoids and the aluminum present in the reagent was frequently utilized, with the result being expressed in equivalents of quercetin (EQ) or epicatechin (EE). Currently, the colorimetric methods tend to be carried out together with other more specific methods, such as high-efficiency liquid chromatography in conjunction with mass spectrometry (HPLC-MS), which allows the individual identification and quantification of compounds. According to Cunha and Roque (2009b), the results of the mass spectrometry have a higher precision, even in complex mixtures.

The total phenolic and flavonoid content and the compounds identified and quantified in the bee pollen are presented in Table 11.2, together with the botanical origin and precedence of the samples. The total phenolic content varied from 0.5 to

213.2 mg GAE/g and the flavonoid content can represent up to 92% of the total phenolic content (Leblanc et al. 2009; Mărghitas et al. 2009; Menezes et al. 2010; Morais et al. 2011; Freire et al. 2012; Mejías and Montenegro 2012; Nogueira et al. 2012; Rebiai and Lanez 2012; Fatrcová-Šramková et al. 2013; Kaškonienė et al. 2015; Ketkar et al. 2014; Ulusoy and Kolayli 2014; De-Melo 2015; De-Melo et al. 2016).

In monofloral samples collected in Romania, total phenolic content was differentiated and decreasing in the following order: *Salix* sp. > *Taraxacum offici-nale* > *Centaurea cyanus* > *Crataegus monogyna* > *Capsella bursa pastoris* > *Matricaria chamomilla* > *Carex* sp. > *Carduus* sp. > *Helianthus annuus* > *Onobrychis viciifolia* > *Pinus* sp. > *Knautia arvensis,* while total flavonoid content diminished in the following order: *Salix* sp. > *M. chamomilla* > *C. cyanus* > *C. monogyna* > *H. annuus* > *Carduus* sp. > *C. bursa pastoris* > *O. viciifolia* > *Carex* sp. > *T. officinale* > *K. arvensis* > *Pinus* sp. (Mărghitas et al. 2009). In Chile, total phenolic content of monofloral bee pollen of *Lotus pedunculatus* was 0.5 mg GAE/g, while in the monofloral sample of *Escallonia rubra* it was 1.2 mg GAE/g (Mejías and Montenegro 2012). In bee pollen from Brazil, total phenolic and flavonoid contents were determined in a monofloral *Myrica* sample (29.2 mg GAE/g e 17.5 mg QE/g, respectively), in a monofloral *Alternanthera* sample (13.8 mg GAE/g e 1.0 mg QE/g), and in a monofloral *Cocos nucifera* sample (11.6 mg GAE/g e 3.4 mg QE/g) (De-Melo 2015) (Table 11.2).

A recent study compared bee pollen samples from Romania and Turkey (Mărgăoan et al. 2021). Following the palynological analysis, most of the samples were multifloral, with some exceptions, where Asteraceae, Brasicaceae, Fabaceae, Fagaceae, Lamiaceae, and Rosaceae were the predominant pollens. The amounts of total phenolic content (TPC) and total flavonoid content (TFC) in the bee pollen extracts (BP) were different according to botanical and geographical origin: Turkey pollen extracts having *Coronilla* sp. as predominant pollen, showed the highest TPC (41.17 mg GAE/g), followed by Cistus and Pyrus sp. (38.06 mg GAE/g), Castanea sp. (29.89 mg GAE/g) and Lamiaceae (BP4 TR) (28.20 mg GAE/g), whereas the Rosaceae BP extract showed the lowest level (16.40 mg GAE/g). Samples from Romania exhibit the highest TPC (38.53 mg GAE/g), having Rosaceae predominant pollen, followed by Brassica sp. (38.53 mg GAE/g). Lowest levels of TPC and TFC were determined from extracts having Fagaceae family (Castanea sp.) as major pollen. Samples from Portugal, Spain, Turkey, USA, Brazil, China, Egypt, New Zealand, and Greece had TPC values ranging between 0.50 and 213 mg GAE/g and TFC values ranging between 1.00 and 17.50 mg QE/g, respectively (Pascoal et al. 2014; Ulusoy and Kolayli 2014; Komosinska-Vassev et al. 2015; Araujo et al. 2017).

The variation in total phenolic and flavonoid contents of bee pollen is related to the botanical origin and the type of processing employed by the beekeeper. De-Melo et al. (2016), when studying samples collected in the southeast of Brazil, observed total phenolic content of on average 14.4 mg GAE/g of product submitted to dehydration in an electric oven with forced air circulation; and 16.1 mg GAE/g in the same product when submitted to freeze-drying. The authors attributed this result to the action of enzymes such as polyphenol oxidase and peroxidase liberated during the freezing and defrosting processes that preceded dehydration in the electric oven.

Table 11.2 Total phenolic (TP) and flavonoid contents (FC) and compounds identified and quantified (CIQ) in monofloral and heterofloral bee pollen collected from different countries

Botanical origin	Origin/ country	TP (mg GAE/g)	FC (mg QE/g)	CIQ (μg/g)	References
Eucalyptus globulus	Portugal	–	–	Quercetin-3-sophoroside, myricetin, tricetin, luteolin, 3-*O*-methylquercetin	Campos et al. (1997)
Washingtonia	USA	15.9	–	2,6-Dihydroxy-6-methylbenzaldehyde (38.0), 2-formyloxy-1- phenylethanone (20.0), 4′,5-dihydroxy-7-methoxyflavanone (58.0), anthraquinone derivative (85.0), 5-methoxy-7-methyl-1,2-naphthoquinone (70.0), 7-hydroxy-1-(2-methoxyphenyl)-9,10-anthracenedione (14.0)	LeBlanc et al. (2009)
Yucca	USA	19.5	–	Naringenin (580.0), 2-methyl-5-hydroxybenzofuran (92.0), 4′,5-dihydroxy-7-methoxyflavanone (90.0), anthraquinone derivative (62.0), 5-methoxy-7-methyl-1,2-naphthoquinone (41.0), 1,2,3,4-tetrahydro-2-(2-hydroxy-3-phenoxypropyl)-6,7-dimethoxyisoquinoline (34.0), 1-(2-methoxy phenyl)-9,10-anthracenedione (14.0)	LeBlanc et al. (2009)
Cecropia	Brazil	122.1	–	Isoquercetin, tricetin, quercetin, kaempferol	Freire et al. (2012)
Elaeis	Brazil	213.2	–	Isoquercetin, myricetin, tricetin, quercetin, luteolin, selagin, isorhamnetin, kaempferol	Freire et al. (2012)
Eupatorium	Brazil	140.1	–	Isoquercetin, quercetin, kaempferol	Freire et al. (2012)
Mimosa pudica	Brazil	77.6	–	Isoquercetin, quercetin, isorhamnetin, kaempferol	Freire et al. (2012)
Scoparia	Brazil	134.1	–	Myricetin	Freire et al. (2012)
Brassica juncea	Brazil	18.3	1.2	Kaempferol (65.4), quercetin (51.4)	Ketkar et al. (2014)
Brassica	Brazil	10.3	4.0	Rutin (292.5), kaempferol (19.2), f3-resorcylic acid (3815.0), *p*-coumaric acid (16.1), ferulic acid (189.8)	De-Melo (2015)
Mimosa verrucosa	Brazil	7.7	5.5	Rutin (97.6), naringenin (72.8), kaempferol (157.9), chlorogenic acid (42.7), vanillic acid (16.4)	De-Melo (2015)

(continued)

Table 11.2 (continued)

Botanical origin	Origin/ country	TP (mg GAE/g)	FC (mg QE/g)	CIQ (μg/g)	References
Astrocaryum	Brazil	16.9	1.2	Cinnamic acid (117.7)	De-Melo (2015)
Brassica campestris	China	–	–	Quercetin (1370.0) kaempferol (23,440.0)	Lv et al. (2015)
Zea mays	Egypt	–	–	Gallic acid, vanillic acid, synringic acid, *p*-coumaric acid, ferulic acid, caffeic acid, quercitin, rutin, catechin, epicatechin, α-Catechin, kaempferol, apigenin, 3,4-dimethoxycinnamic acid, naringenin, luteolin	Mohdaly et al. (2015)
Anadenanthera	Brazil	11.2	1.0	Catechin (6.9), vanillic acid, ampeloptin, f3-resorcylic acid (125.4), ferulic acid (10.4), quercetin (74.2), quercetin diglucoside, quercetin-3-*O*-arabinoside, quercetin-3-*O*-rutinoside, quercetin rutinoside (isomer), isorhamnetin-3-*O*-rhamnosyl-glucoside, isorhamnetin-3-*O*-rhamnosyl-glucoside (isomer), eriodictyol, rutin (36.6), 1-sinapoyl-2-feruloyl gentiobiose, naringenin hexoside, cinnamic acid (130.5), tri-*p*-coumaroyl spermidine, tri-*p*-coumaroyl spermidine (isomer), ellagic acid, naringenin (45.7), N′,N″,N‴-tris-*p*-feruloylspermidine	De-Melo (2015)
Mimosa caesalpiniaefolia	Brazil	12.7	1.3	Catechin (7.5), chlorogenic acid (45.3), caffeic acid (1.5), siringeic acid (53.6), 4-methylsulfinylbutyl glucosinolate (glucoraphanin), kaempferol-3-*O*-rhamnosyl-glucoside, isorhamnetin-3-*O*-diglucoside, isorhamnetin-3- *O*-rutinoside-7-*O*-glucoside, isorhamnetin-3-*O*-rhamnosyl-glucoside, rutin (83.9), kaempferol-3-*O*-glucoside, 1-sinapoyl-2-feruloyl gentiobiose, isorhamnetin-3-*O*-glucoside, quercetin (29.1), kaempferol-3-*O*-acetulglucoside, kaempferol, quercetin-3-methyl ether, pinobanksin-5-methylether-3-*O*--acetate, bis-methylated quercetin, isorhamnetin −3-*O*-methyl hexuronide, dihydroquercetin-3-*O*-rhamnoside	De-Melo (2015)

(continued)

Table 11.2 (continued)

Botanical origin	Origin/ country	TP (mg GAE/g)	FC (mg QE/g)	CIQ (μg/g)	References
Heterofloral	Portugal	–	–	Kaempferol-3- neohesperidoside, quercetin-3-rhamnoside	Campos et al. (1997)
Heterofloral	New Zealand	–	–	Kaempferol-3- neohesperidoside, kaempferol-3-sophoroside, 7-O-methylherbacetin-3-diglycoside, 8-O-methylherbacetin-3-diglycoside, 7-O—methylherbacetin-3-sophoroside, quercetin-3-diglycoside, highly derivatized herbacetin glycoside, isorhamnetin-3-sophoroside-diglycoside, 8-O-methylherbacetin-3-sophoroside, 8-O-methylherbacetin-3-sophoroside, 7-O-methylkaempferol-3-neohesperidoside, 8-O-methylherbacetin-3-sophoroside, 8-O-methylherbacetin-glycoside	Campos et al. (1997)
Heterofloral	Brazil	–	–	Rosmarinic acid dihexoside derivative, myricetin-3-O-rhamnosyl-glucoside, quercetin-3-O-arabinoside, isorhamnetin-3-O-diglucoside, kaempferol-3- O-rhamnosyl-glucoside, isorhamnetin-3-O-(2″,3″-dirhamnosyl)glucoside, rutin, isorhamnetin-3-O-rhamnosyl-glucoside, kaempferol-3-O-glucoside, patuletin-3-O-rhamnosylglucoside, quercitrin, chalcone, kaempferol-7-O-rhamnosyl-3-O-galloyl glucuronide, N′,N″,N‴-tris-p-coumaroylspermidine, N′,N″,N‴-tris-p-feruloylspermidine	Negri et al. (2011)
Heterofloral	Greece	–	–	Ferulic acid (149.1), o-coumaric acid (36.7), quercetin (29.6), cinnamic acid (23.5), naringenin (21.9), hesperetin (3.0), kaempferol (7.8)	Fanali et al. (2013)

Flavonoids and phenolic acids are the main phenolic compounds present in bee pollen. In samples collected in Turkey, the main compounds quantified were rutin (25.6–692.9 μg/100 g), abscisic acid (21.0–288.7 μg/100 g), syringic acid (10.6–259.5 μg/100 g), ferulic acid (36.8–230.6 μg/100 g), and p-coumaric acid (34.2–184.2 μg/100 g). The compounds: benzoic acid (46.9–1077.6 μg/100 g), epicatechin (39.2–520.0 μg/100 g), and quercetin (55.9–499.2 μg/100 g) were observed in significant quantities, however they were not present in all the samples (Ulusoy and Kolayli 2014).

Eight phenolic acids (gallic, chlorogenic, caffeic, ferulic, syringic, synapic, 3,4-dihydrobenzoic, and 2-hydroxycinnamic acid), and three flavonoids (rutin, naringenin, and quercetin) were identified in the bee pollen produced in Latvia and Spain and, with the exception of ferulic acid, they were also present in the samples collected in Lithuania and China (Kaškonienė et al. 2015). In *Zea mays* monofloral bee pollen from Egypt, 3,4-dimethoxycinnamic acid was the major phenolic component in the pollen extract (45.8 mg/mL) (Mohdaly et al. 2015).

Some researchers have put forward the hypothesis that each plant species had a profile of specific phenolic compounds (Campos et al. 2003; Ulusoy and Kolayli 2014). Campos et al. (2015) carried out exhaustive research into pollen from *Zea mays* (including corn hybrids and genetic modified samples), collected on plantations located in Portugal, Brazil, and Mexico between the years 2000 and 2012, and concluded that the flavonoid/phenolic profile does neither change over the time of sampling nor in terms of the harvesting region. Bee pollen characterization studies were undertaken, attempting to identify based on a specific profile the botanical species from which the pollen grain was harvested, which would also allow an association between the biological activity and the botanical origin of the bee pollen (Campos et al. 2003, 2008; Ulusoy and Kolayli 2014). Zhou et al. (2015) analyzed monofloral bee pollen samples produced in China and proposed, based on the presence or not and on content, the quercetin-3-*O*-β-D-glucosyl-(2 → l)-β-glucoside compounds as a marker that identifies bee pollen samples from seed watermelon, rape, camellia, corn poppy, corn, buckwheat and rose; kaempferol-3,4′-di-*O*-β-D-glucoside as a marker to evaluate the floral origins of bee pollen from sesame, broad bean, rape, motherwort and corn; and kaempferol-3-*O*-β-D-glucosyl-(2 → l)-β-D-glucoside as a marker to classify bee pollen samples from sesame, broad bean, rape, corn poppy, buckwheat, camellia, and corn.

One should also consider, however, that the pollen of some species can present different phenolic compound profiles due to the agricultural and genotypic differences, or climatic conditions (temperature, hydrological stress, and light intensity) and soil, which can influence plant growth and maintenance (Daoud et al. 2015). Monofloral bee pollen from *Mimosa scabrella* collected in different regions of Brazil was submitted to HPLC-MS analysis and the results were different profiles for each sample (De-Melo 2015). In monofloral bee pollen from *Mimosa pudica* and *Elaeis* collected monthly from the same apiary located in Brazil, between February and November, total phenolic content and the compounds identified varied significantly throughout the period (Freire et al. 2012). In another study with Brazilian samples, this time monofloral from *Eucalyptus*, a considerable variation in total phenolic content (27.1–132.4 mg GAE/g) and total flavonoids (1.8–2.5 mg EE/g) was observed throughout one year (Menezes et al. 2010).

Another variable that should be considered is the processing employed by the beekeeper, beyond storage, as previously mentioned. In monofloral bee pollen from *Cistus ladanifer* collected in Spain, quercetin-3-rutinoside content, the compound observed in greater quantities in the samples, was 7.3 mg/100 g in the product in natura and 3.3 mg/100 g in the lyophilized product (Domínguez-Valhondo et al. 2011). The influence of storage was observed by Tavdidishvili et al. (2014) in samples collected in the USA.

The extraction of phenolic compounds from bee pollen is primarily performed using solvent-based methods, where the solvent choice significantly impacts yield and composition. Commonly used solvents include:

1. Ethanol-water and methanol-water mixtures: The most widely employed solvent combinations, providing efficient extraction of both flavonoid and non-flavonoid compounds (Orzáez Villanueva et al. 2002; Rebelo et al. 2016).
2. Methanol: Used alone in certain studies for its high polarity and capacity to solubilize phenolic compounds (Pascoal et al. 2014; Taha 2015).
3. Acetone-water mixtures: Occasionally utilized to target specific phenolics, offering distinct polarity characteristics (Pereira de Melo and de Almeida-Muradian 2010).
4. Water: Applied in some experiments for simplicity and environmental considerations, although its extraction efficiency may be lower compared to alcohol-based systems (Montenegro et al. 2013).

The Folin-Ciocalteu method (FCM) is a colorimetric assay widely used for assessing total phenolic content (TPC) in bee pollen. The assay relies on the reduction of a molybdate-tungstate reagent by phenols, leading to a color change measured spectrophotometrically (UV-Vis) (Bernal et al. 2011). Although popular, FCM is not entirely specific to phenolics, as it reacts with other reducing agents.

$AlCl_3$ colorimetric assay for total flavonoid content (TFC). This method involves forming complexes between flavonoid hydroxyl groups and aluminum chloride. It is a rapid and reliable spectrophotometric technique for quantifying flavonoids (Pascoal et al. 2014).

The most used chromatographic assays are reverse-phase liquid chromatography (RPLC) and Ultra-High-Performance Liquid Chromatography (UHPLC).

RPLC uses C18 columns, which is the preferred method for separating phenolic compounds. Detection is commonly performed using UV-Vis, photodiode array (PDA), or mass spectrometry (MS) detectors (Yang et al. 2013). Compounds like rutin, quercetin, and resveratrol have been identified in bee pollen using RPLC, highlighting their potential health benefits, such as antioxidant and anti-inflammatory effects (Bernal et al. 2011).

Ultra-high-performance liquid chromatography (UHPLC), combined with PDA and MS detectors, provides enhanced resolution and sensitivity for identifying phenolic profiles, including quercitrin and isorhamnetin (Gabriele et al. 2015) (Table 11.3).

11.6.2 Carotenoids

Carotenoids such as provitamin A are considered one of the antioxidant vitamins. Almeida-Muradian et al. (2005) reported carotenoids but not β-carotene (provitamin A) in samples from southern Brazil. The authors argued that the absence of this provitamin could be due to the use of thermal processes for dehydration, which can damage vitamins because of their thermal instability. Barajas et al. (2012) specifically tested the influence of the dehydration process on carotenoid content. The authors analyzed bee pollen from Colombia dried at two different temperatures (35 and 45 °C): one sample showed the values 0.77 (fresh), 0.78 (dried at 35 °C), and 0.51 mg/g (dried at 45 °C) for carotenes; the second sample showed the values of 0.21 (fresh), 0.22 (dried at 35 °C) and 0.17 mg/g (dried at 45 °C).

Oliveira et al. (2009) analyzed beta-carotene as provitamin A of fresh bee pollen samples from southeastern Brazil (São Paulo state). The authors obtained a range of 56.3–198.9 μg/g for beta-carotene. Samples collected in April (autumn in Brazil) can be considered provitamin A sources. A relationship was observed between botanical origin and β-carotene content: *Raphanus* sp., *Macroptilium* sp., and *Mimosa caesalpineafolia*.

Sattler et al. (2015) determined α and β-carotene of dehydrated bee pollen samples from southern Brazil, and obtained the following average values: 61.3 μg/g of a-carotene and 17.7 μg/g of β-carotene. The provitamin A value varied from 3.9 to 4945.2 retinol μg/100 g.

In Thailand, a vitamin A—β-carotene value of 1530.4 μg/100 g for monofloral corn bee pollen was reported (Chantarudee et al. 2012). Bee pollen from Spain was studied by Muniategui et al. (1990), who obtained a mean of 46.7 mg of n-carotene/100 g lipids. These samples presented a mean of 6.2% for lipids. Szczesna et al. (1991) found 10.87 for provitamin A (mg%) in bee pollen from Poland. In samples from Romania, the authors observed a range of 1–200 mg/kg for ß-carotene (Mărgăoan et al. 2010).

Carotenoids have been extensively studied for their health benefits and functional roles. Solvent extraction is the primary method employed, using petroleum ether, acetone, or chloroform-methanol mixtures (Sattler et al. 2015; Montenegro et al. 2013). Quantification of total carotenoid content is generally achieved spectrophotometrically using UV-Vis detectors, such as diode array detectors (DAD) or photodiode array (PDA) systems (Oliveira et al. 2009). For individual carotenoid analysis, reverse-phase liquid chromatography (RPLC) has proven effective for separation and characterization (Conte et al. 2017).

In some cases, β-carotene has been extracted and analyzed using advanced methods like open column chromatography (OCC), which facilitates separation and purification (Montenegro et al. 2013; Pereira de Melo and Almeida-Muradian 2010). Comparative analyses have shown that OCC and liquid chromatography (LC) yield consistent results, demonstrating their reliability for quantifying carotenoids (Montenegro et al. 2013) (Table 11.3).

Table 11.3 Analytical methods for chemical composition constituents' analysis of bee pollen and bee bread

Class of compounds	Preparation method/Extraction solvent	Analytical techniques for identification and/or quantification	References
Carbohydrates	Dissolution/Methanol:Water	HPLC-RID/Acetonitrile:Water as mobile phase	Bonta et al. (2008), Liolios et al. (2018)
	Dissolution/Water:Acetonitrile		SR ISO 24382 (2024)
Lipids	Dissolution/Chloroform:Methanol (2:1 v/v)	Folch method	Mărgăoan et al. (2014), Urcan et al. (2021), Conte et al. (2017), Lau et al. (2022), Baky et al. (2023)
	Diethyl ether, Ethyl ether, Petroleum ether, n-hexane	Soxhlet extraction	De-Melo et al. (2018), Hsu et al. (2021), Oroian et al. (2022), Gabriele et al. (2021), AOAC method (2000), Denisow and Denisow-Pietrzyk (2016), Baky et al. (2023)
	Ethyl ether	Intermittent Soxhlet extraction	Sattler et al. (2015)
	Chloroform: sodium sulfate solution. (1:1)	Vanillin-Phosphoric Acid Assay	Vaudo et al. (2020), Lau et al. (2022)
Fatty acids	Hexan (Soxhlet extraction)	HPLC-MS	Yan et al. (2019), Gercek et al. (2021), Dranca et al. (2020)
	Diethyl ether (Soxhlet extraction)	GS-MS	Mayda et al. (2020)
	Chloroform/methanol (2:1 *v/v*)	Fatty acid methyl esters (FAMEs) of the total lipids/GC-2010 Plus/GC-MS/FID.	Čeksterytė et al. (2023), Villagómez et al. (2023)
	Petroleum ether (Soxhlet extraction)	FAMEs/GC-MS	Estevinho et al. (2012)
	Chloroform/methanol (1:1 *v/v*)	FAMEs/GC-FID	Jarukas et al. (2020)

Proteins	Pollen grains	Kjeldahl method Sulfuric acid (95–97%) and catalysts; neutralize with 30% NaOH, distill NH_3, trap with 4% H_3BO_3, and titrate with 0.1 M HCl.Conversion factor 5.60 or 6.25	Urcan et al. (2024), Nikkeshi et al. (2022)
	Tris/EDTA buffer (pH 8.5) Phosphate buffer (pH 6,1) NaOH 0.1 M	Bradford Assay	Türe et al. (2023), Bobiş et al. (2010), Darwish et al. (2022), Vaudo et al. (2020), Lau et al. (2022)
	Ultra-pure water	Lowry Assay	Bobiş et al. (2010), Lau et al. (2022)
	NaOH 0.1 M	BCA assays (biuret reaction)	Vanderplanck et al. (2014), Ares et al. (2022)
	Pollen grains	Dumas assay, Elementar vario MAX CN 950 °C, Conversion factor 6.25	Lau et al. (2022)
Amino acids	Amino acids EZ:faast™ kit	LC/MS/water:methanol as mobile phase GC-MS/Helium HPLC-MS	Urcan et al. (2021), Urcan et al. (2024) Martín-Gómez et al. (2022)
	Ethanol (99%), Diethyl Ether	HPLC, Automatic Amino Acid Analyzer	Nikkeshi et al. (2022)
	Protein Hydrolysis using HCl (6 M), Methanol, Derivatization	UHPLC mobile phase acetonitrile/methanol/water (45:45:10)	Stabler et al. (2018)
	Hydrolysis with HCl (6N), 0.1% phenol, 500 mM norleucine	IEX Chromatography HPLC/phosphate buffer/acetonitrile/methanol/water	Jeannerod et al. (2022), Li et al. (2024)
	Hydrolysis Derivatization: OPA and FMOC-Cl	LC-FLD/sodium acetate buffer/acetonitrile/methanol	Ares et al. (2022)
	Acetic acid (0.2 mmol)	LC-MS/MS	Szczêsna (2006), Türe et al. (2023)

(continued)

Table 11.3 (continued)

Class of compounds	Preparation method/Extraction solvent	Analytical techniques for identification and/or quantification	References
Minerals	HNO_3: HCl, 1:3 in a microwave system/ dissolution in deionized water	AAS	Darwish et al. (2022)
	HNO_3 (10%, v/v) and H_2O_2 (30%, v/v), 100 °C, 12 h/dissolution in HNO_3 (10% v/v)	ICP-AES, ICP-OES	Sattler et al. (2016), Valverde et al. (2023)
	Solvent extraction (methanol) Evaporation; Dilution (methanol)	RPLCMS/MS	Ares et al. (2016)
Total flavonoids and polyphenols	Dissolution/Methanol:Water	Spectrophotometry Folin-Ciocalteu Method $AlCl_3$ Method	Pascoal et al. (2014), Sawicki et al. (2022), Baky et al. (2023), Pełka et al. (2022), Urcan et al. (2021), Mărgăoan et al. (2021), Felek et al. (2021), Hong et al. (2015), Carpes et al. (2013), Aylanc et al. (2021), Tomás et al. (2017), Ertosun et al. (2023), Baky et al. (2023), Sonmez et al. (2023)
	Dissolution/Ethanol:Water		
	Ethanol		Gercek et al. (2021), Dranca et al. (2020), Baky et al. (2023)
	Water		Darwish et al. (2022), Keyvan et al. (2023), Urcan et al. (2023), Pełka et al. (2022)
	Dissolution/Acetone:Water		Araneda et al. (2015)
Flavonoids	Ethanol:water solution (1:1 v/v) Sonicate for 60 min.	HPLC-PAD Orthophosphoric acid (pH 2.5):Acetonitrile	Campos and Markham (2007), Urcan et al. (2018), Bakour et al. (2017)

Phenolic acids, flavonoids	Heptane for lipid removal; 50% ethanol and ethyl acetate for phenolic extraction (ultrasound)	Nano-LC-UV/0.5% formic acid:methanol/	Fanali et al. (2013)
Flavonoids, hydroxycinnamic acid amide derivatives	NS	HPLC-PAD-ESI-MS/MS/0.1% acetic acid Methanol/	Negri et al. (2018)
	Methanolic pollen extracts	HPLC-DAD/acetic acid-water and methanol.	Carpes et al. (2013)
Phenolic Compounds	Ethanol/water (80:20) water	UHPLC-MS/MS/water (0.1% formic acid) and methanol (0.1% formic acid)	Ilie et al. (2024), Urcan et al. (2023), Sonmez et al. (2023)
Phenolic compounds and phenylamide	Ethanol:Water (80:20, v/v)	LC/DAD/ESI-MS	Aylanc et al. (2021), El Ghouizi et al. (2020), Ertosun et al. (2023)
Carotenoids (general)	Petroleum ether-acetone, mixtures; chloroform and methanol mixture; acetone and petroleum ether; methanol, ethyl acetate, petroleum ether, and butylated hydroxytoluene (BHT); and ethanol, hexane, and BHT	UV-VIS Spectrophotometry; HPLC-DAD (methanol, ethyl acetate, and acetonitrile mobile phase)	Sattler et al. (2015), Montenegro et al. (2013), Conte et al. (2017), Almeida-Muradian et al. (2005), Mărgăoan et al. (2014)
β-carotene	Solvent extraction (acetone) Partition (petroleum ether) Washing (water) evaporation Dilution (petroleum ether) Evaporation and dilution (acetone)	Open Column Chromatography (OCC)	Pereira de Melo and Almeida-Muradian (2010)
Anthocyanins	Methanol–1 M HCl (95:5, v/v)	HPLC-DAD-ESI-MS/MS 0.1% trifluoracetic acid/Acetonitrile	Di Paola-Naranjo et al. (2004)

(continued)

Table 11.3 (continued)

Class of compounds	Preparation method/Extraction solvent	Analytical techniques for identification and/or quantification	References
Vitamin E (α-tocopherol; β-tocopherol; γ- tocopherol; δ-tocopherol)	Solvent extraction (acetone) Partition (petroleum ether) Washing (water) and evaporation Dilution (hexane/isopropanol)	NPLC/FLD	Sattler et al. (2015), Pereira de Melo and Almeida-Muradian (2010), De-Melo et al. (2016)
	Solvent extraction (chloroform/methanol); Dilution (hexane/isopropanol); Saponification	RPLC/FLD	Conte et al. (2017)
B complex vitamins (B_1, B_2, B_3, B_6) Bee pollen	Acid hydrolysis Enzymatic treatment	RPLC/FLD	De-Melo et al. (2016), de Arruda et al. (2013)
Vitamin C	Solvent extraction (MPA/acetic acid)	Titrimetric method	Pereira de Melo et al. (2010), Oliveira et al. (2009)
	Metaphosphoric acid solution (*w/v*)	HPLC (C18): formic acid: water (v/v), pH 2.5;	Bonta et al. (2013)
Pyrrolizidine alkaloids	70% methanol: 29.5% ultra-pure water and 0.5% formic acid.	LC-MS-MS	Dübecke et al. (2011), Xiong et al. (2009)

HPLC-RID High-Performance Liquid Chromatography with Refractive Index Detection, *HPLC-MS* High-Performance Liquid Chromatography coupled with Mass Spectrometry, *GS-MS* Gas Chromatography coupled with Mass Spectrometry, *FAMEs* fatty acid methyl esters of the total lipids, *GC-MS/FID* Gas Chromatography coupled with Mass Spectrometry (MS) and a Flame Ionization Detector, *BCA Assay* bicinchoninic acid assay, *LC-MS* Liquid Chromatography Mass Spectrometry, *UHPLC* Ultra-High-Performance Liquid Chromatography, *IEX Chromatography* IonExchange Chromatography, *LC-FLD* Liquid Chromatography with Fluorescence Detection, *LC-MS/MS* Liquid Chroatography Tandem Mass Spectrometry, *AAS* atomic absorption spectrometry, *ICP-AES* Inductively coupled argon plasma-atomic emission spectrometry, *ICP-OES* inductively coupled plasma—optical emission spectrometry, *RPCL-MS/MS* Reverse-Phase Liquid Chromatography Tandem Mass Spectrometry, *HPLC-PAD* High-Performance Liquid Chromatography with Pulsed Amperometric Detection, *HPLC-DAD* High-Performance Liquid Chromatography with Diode Array Detection, *HPLC-DAD-APCI/MS* High-Performance Liquid Chromatography with Diode Array Detection and Atmospheric Pressure Chemical Ionization Mass Spectrometry, *Nano-LC-UV* Nano Liquid Chromatography with Ultraviolet Detection, *HPLC-PAD-ESI-MS/MS* High-Performance Liquid Chromatography with Pulsed Amperometric Detection and Electrospray Ionization Tandem Mass Spectrometry, *UHPLC-MS/MS* Ultra-High-Performance Liquid Chromatography Tandem Mass Spectrometry, *OCC* Open Column Chromatography, *LC-DAD-ESI-MS* Liquid Chromatography with Diode Array Detection and Electrospray Ionization and Mass Spectrometry, *NPLC/FLD* normal-phase liquid chromatography with Fluorescence Detection, *RPLC/FLD* reverse-phase liquid chromatography with Fluorescence Detection, *CFU* Colony-Forming Units, *NS* not specified

11.6.3 Probiotics

Probiotics are different bacterial strains and yeasts with beneficial potential for the digestive system of both bees and humans (Maruščáková et al. 2020; Nawaz Ranja et al. 2021). Many types of bacteria may be classified as probiotics, but the main two groups are Lactobacillus and Bifidobacterium. Bee products are known for their bioactive properties due to their chemical composition. Bee pollen is one of the most equilibrate products of the hive in terms of composition, having a balanced percentage of sugars, proteins, including essential amino acids, lipids, including unsaturated fatty acids, vitamins, minerals, phenolic entities, and natural colorants such as carotenoids and anthocyanins (Bastos et al. 2004; De Arruda et al. 2013; Bertoncelj et al. 2018; Morgano et al. 2012; Mărgăoan et al. 2014; Mărgăoan et al. 2021).

The most important probiotics used for humans and animals are Lactobacillus and Bifidobacterium. These bacteria were isolated from the digestive tract of the honeybees (Apis mellifera), but also from bee pollen and bee bread. Lactobacillus were found the most representative in the microflora of honey (90.0%), pollen (74.6%), bee bread (83.9%), and royal jelly (93.3%) (Asama et al. 2015). Earlier, Endo and Salminen (2013) have identified in bee pollen different fructophilic lactic acid bacteria, the species being classified as *Lactobacillus kunkeei.* These bacteria were also found in honeybees and honey.

It is known that different strains of the Bacillus genus are used as probiotics (*B. subtilis, B. polyfermenticus, B. clausii, some B. cereus, B. coagulans, B. pumilus, and B. licheniformis)* (Lee et al. 2019). Recently, some of these bacteria were isolated from samples of bee pollen and bee bread from Poland (Pełka et al. 2025), and used for demonstrating their probiotic potential. These strains are safe for humans (exhibiting γ-hemolytic activity) and have the requirements for probiotics in terms of viability when are in contact with bile salts and acid conditions, hydrophobicity, and auto-aggregation, with the important human pathogenic bacteria. These isolates are resistant to penicillin but sensitive to the other tested antibiotics. Following whole-genome sequencing, two of these isolated from bee pollen and bee bread were classified as *B. subtilis* and *B. velezensis*. This study demonstrates that some of the *Bacillus* strains isolated from bee pollen or bee bread are potential probiotics.

References

Aguirre-Hernández E, González-Trujano E, Martinez AL, Moreno J, Kite G, Terrazas T, Soto-Hernández M (2010) HPLC/MS analysis and anxiolytic-like effect of quercetin and kaempferol flavonoids from *Tilia americana* var. mexicana. J Ethnopharmacol 127:91–97

Al-Kahtani SN (2017) Fatty acids and B vitamin contents in honeybee collected pollen in relation to botanical origin. Sci J King Faisal Univ 18:41–48

Almeida-Muradian LB (2009) Qualidade dos produtos apicolas e otimização quimiométrica dos métodos de análise do mel por espectroscopia no infravermelho (FT-IR ATR)

Almeida-Muradian LB, Pamplona LC, Coimbra S, Barth OM (2005) Chemical composition and botanical evaluation of dried bee pollen pellets. J Food Compos Anal 18(1):105–111. https://doi.org/10.1016/j.jfca.2003.10.008

Altunatmaz SS, Tarhan D, Aksu F, Barutçu UB, Or ME (2017) Mineral element and heavy metal (cadmium, lead and arsenic) levels of bee pollen in Turkey. Food Sci Technol Campinas 37(Suppl 1):136–141

Angelo PM, Jorge N (2007) Compostos fenólicos em alimentos – uma breve revisão. Rev Inst Adolfo Lutz 66:232–240

Anjum SI, Ullah A, Gohar F, Raza G, Khan MI, Hameed M, Ali A, Chen C-C, Tlak Gajger I (2024) Bee pollen as a food and feed supplement and a therapeutic remedy: recent trends in nanotechnology. Front Nutr 11:1371672

AOAC (2000) Official method of analysis of AOAC international, vol II, 17th edn. Association of Analytical Chemists, Washington

Araneda X, Caniullan R, Catalán C, Martínez M, Morales D, Rodríguez M (2015) Nutritional contribution of pollen from species pollinated by bees (Apis mellifera L.) in the Araucanía Region of Chile. Rev FAC UN Cuyo 47:139–144

Araujo JS, Chambo ED, Costa MAPDC, da Silva C, Peixoto SM, Lopes de Carvalho CA et al (2017) Chemical composition and biological activities of mono-and heterofloral bee pollen of different geographical origins. Int J Molec Sci 18(5):921

Ares AM, Nozal MJ, Bernal J (2013) Extraction, chemical characterization and biological activity determination of broccoli health promoting compounds. J Chromatogr A 1313:78–95

Ares AM, Ayuso I, Bernal JL, Nozal MJ, Bernal J (2016) Trace analysis of sulforaphane in bee pollen and royal jelly by liquid chromatography–tandem mass spectrometry. J Chromatogr B 1012–1013:130–136

Ares AM, Valverde S, Bernal JL, Nozal MJ, Bernal J (2018) Extraction and determination of bioactive compounds from bee pollen. J Pharm Biomed Anal 147:110–124

Ares AM, Toribio L, Tapia JA, González-Porto AV, Higes M, Martín-Hernández R, Bernal J (2022) Differentiation of bee pollen samples according to the apiary of origin and harvesting period based on their amino acid content. Food Biosci 50(May):102092. https://doi.org/10.1016/j.fbio.2022.102092

Arruda VAS (2013) Pólen apicola desidratado: composição fisico-quimica, qualidade microbiológica, compostos fenólicos e flavonoides, atividade antioxidante e origem botânica. http://www.teses.usp.br/teses/disponiveis/9/9131/tde-22062015-172352/es.php. Accessed 26 Feb 2017

Arruda VAS, Estevinho MLF, Almeida-Muradian LB (2013a) Presence and stability of B complex vitamins in bee pollen using different storage conditions. Food Chem Toxicol 51:143–148

Arruda VAS, Santos-Pereira AA, Freitas AS, Barth OM, Almeida-Muradian LB (2013b) Dried bee pollen: B complex vitamins, physicochemical and botanical composition. J Food Compos Anal 29:100–105

Asama T, Arima TH, Gomi T, Keishi T, Tani H, Kimura Y, Tatefuji T, Hashimoto K (2015) *Lactobacillus kunkeei* YB38 from honeybee products enhances IgA production in healthy adults. J Appl Microbiol 119(3):818–826

Avni D, Hendriksma HP, Dag A, Uni Z, Shafir S (2014) Nutritional aspects of honey bee-collected pollen and constraints on colony development in the eastern Mediterranean. J Insect Physiol 69:65–73

Aylanc V, Falcão SI, Ertosun S, Vilas-Boas M (2021) From the hive to the table: nutrition value, digestibility and bioavailability of the dietary phytochemicals present in the bee pollen and bee bread. Trends Food Sci Technol 109(October 2020):464–481. https://doi.org/10.1016/j.tifs.2021.01.042

Bakour M, Al-Waili NS, El Menyiy N, Imtara H, Figuira AC, Al-Waili T, Lyoussi BB, El Menyiy N, Imtara H, Figuira AC, Al-Waili T, Lyoussi BB (2017) Antioxidant activity and protective effect of bee bread (honey and pollen) in aluminum-induced anemia, elevation of inflammatory makers and hepato-renal toxicity. J Food Sci Technol 54(13):4205–4212. https://doi.org/10.1007/s13197-017-2889-9

Baky MH, Abouelela MB, Wang K, Farag MA (2023) Bee pollen and bread as a super-food: a comparative review of their metabolome composition and quality assessment in the context of best recovery conditions. Molecules 28(2):715. https://doi.org/10.3390/molecules28020715

Ball GFM (1998) Bioavailability and analysis of vitamins in foods. Chapman & Hall, London

Barajas J, Cortes-Rodriguez M, Rodríguez-Sandoval E (2012) Effect of temperature on the drying process of bee pollen from two zones of Colombia. J Food Process Eng 35:134–148

Barreto LMRC, Funari SRC, Orsi RO (2005) Composição equalidade do pólen apícola proveniente de sete estados brasileiros e do Distrito Federal. Bol Ind Anim 62:167–175

Barbieri D, Gabriele M, Summa M, Colosimo R, Leonardi D, Domenici V, Pucci L (2020) Antioxidant, nutraceutical properties and fluorescence spectral profiles of bee pollen samples from different botanical origins. Antioxidants 9:1001. https://doi.org/10.3390/antiox9101001

Barth OM, Freitas AS, Oliveira ES, Silva RA, Maester FM, Andrella RR, Cardozo GM (2010) Evaluation of the botanical origin of commercial dry bee pollen load batches using pollen analysis: a proposal for technical standardization. An Acad Bras Cienc 82:893–902

Bastos DHM, Barth OM, Rocha CI, Cunha IBS, Carvalho PO, Torres EAS (2004) Fatty acid composition and palynological analysis of bee (*Apis*) pollen loads in the states of Sao Paulo and Minas Gerais, Brazil. J Apicult Res 43:35–39

Bernal J, Mendiola JA, Ibáñez E, Cifuentes A (2011) Advanced analysis of nutraceuticals. J Pharm Biomed Anal 55:758–774

Bertoncelj J, Polak T, Pucihar T, Lilek N, Borovšak AK, Korošec M (2018) Carbohydrate composition of Slovenian bee pollens. Int J Food Sci Technol 53:1880–1888. https://doi.org/10.1111/ijfs.13773

Bertoncelj J, Lilek N, Korošec M (2024) Bee pollen carbohydrates composition and functionality. In: Bayram E, Kostic Z, Gercek CA (eds) Pollen chemistry & biotechnology. Springer, Cham. https://doi.org/10.1007/978-3-031-47563-4_3

Bobiş O, Mărghitaş L, Dezmirean D, Morar O, Bonta V, Chirila F (2010) Quality parameters and nutritional value of different commercial bee products. Bull Univ Agric Sci Vet Med Cluj–Napoca Anim Sci Biotechnol 67(1–2):91–96

Bogdanov S (2012a) The bee pollen book: chapter 1. http://www.bee-hexagon.net. Accessed 12 Nov 2014

Bogdanov S (2012b) The bee pollen book: chapter 2. http://www.bee-hexagon.net. Accessed 12 Nov 2014

Bonta V, Mărghitaş LA, Stanciu O, Laslo L, Dezmirean D, Bobiş O (2008) High-performance liquid chromatography analysis of sugars in Transylvanian honeydew honey. Bull UASVM Anim Sci Biotechnol 65:229–232

Bonta V, Mărghitaş L, Bobiş O, Mărgăoan R, Bărnuţiu L, Dezmirean DS (2013) Quantitative determination of vitamin C in honey and bee pollen using HPLC-DAD method. Bull UASVM Anim Sci Biotechnol 70(1):31–36

Bradford MM (1976) A rapid and sensitive method for the quantitation of microgram quantities of protein utilizing the principle of protein-dye binding. Anal Biochem 72:548–254

Brazil (2005) Technical regulation on the Recommended Dietary Allowance (RDA) of proteins, vitamins and minerals. http://bvsms.saude.gov.br/bvs/saudelegis/anvisa/2005/rdc0269_22_09_2005.html. Accessed 1 July 2017

Cabrera C, Montenegro G (2013) Pathogen control using a natural Chilean bee pollen extract of known botanical origin. Cienc Investig Agrar 40:223–230

Calviello G, Serini S, Piccioni E (2007) n-3 polyunsaturated fatty acids and the prevention of colorectal cancer: molecular mechanisms involved. Curr Med Chem 14:81–89

Campos M d G, Markham KR (2007) Structure information from HPLC and on-line measured absorption spectra: flavones, flavonols and phenolic acids

Campos M, Markham KR, Mitchell KA, da Cunha AP (1997) An approach to the Characterization of Bee Pollens via their Flavonoid/Phenolic Profiles. Phytochem Anal 8:181–185

Campos MGR, Webby RF, Markham KR, Mitchell KA, Da Cunha AP (2003) Age-induced diminution of free radical scavenging capacity in bee pollens and the contribution of constituent flavonoids. J Agric Food Chem 51:742–745

Campos MGR, Bogdanov S, Almeida-Muradian LB, Szczesna T, Mancebo Y, Frigerio C (2008) Pollen composition and standardisation of analytical methods. J Apicult Res Bee World 47:156–163

Campos MGR, Almaraz-Abarca N, Matos MP, Gomes NM, Arruda VA, Barth OM (2015) *Zea mays* L. pollen: an approach to its quality control. J Agric Sci Technol B 5:513–522

Campos M d G, Markham KR (2007) Structure information from HPLC and on-line measured absorption spectra: flavones, flavonols and phenolic acids

Carl JL, Richard VM, Mandeep RM, Hector OV (2009) Omega-3 polyunsaturated fatty acids and cardiovascular diseases. J Am Coll Cardiol 54:585–594

Carpes ST, de Alencar SM, Cabral ISR, Oldoni TLC, Mourão GB, Haminiuk CWI, da Luz CFP, Masson ML (2013) Polyphenols and palynological origin of bee pollen of Apis mellifera L. from Brazil. Characterization of polyphenols of bee pollen. CyTA - J Food 11(2):150–161. https://doi.org/10.1080/19476337.2012.711776

Čeksterytė V, Bliznikas S, Jaškūnė K (2023) The composition of fatty acids in bee pollen, royal jelly, buckthorn oil and their mixtures with pollen preserved for storage. Food Secur 12(17):3164. https://doi.org/10.3390/foods12173164

Chantarudee A, Phuwapraisirisan P, Kimura K, Okuyama M, Mori H, Kimura A (2012) Chemical constituents and free radical scavenging activity of corn pollen collected from *Apis mellifera* hives compared to floral corn pollen at Nan, Thailand. BMC Complement Altern Med 12:1–12

Chromý V, Vinklárková B, Šprongl L, Bittová M (2015) The Kjeldahl method as a primary reference procedure for total protein in certified reference materials used in clinical chemistry. Crit Rev Anal Chem 45:106–111

Conte G, Benelli G, Serra A, Signorini F, Bientinesi M, Nicolella C, Mele M, Canale A (2017) Lipid characterization of chestnut and willow honeybee-collected pollen: impact of freeze-drying and microwave-assisted drying. J Food Compos Anal 55:12–19. https://doi.org/10.1016/j.jfca.2016.11.001

Cunha AP, Roque OR (2009a) Compostos fenólicos: características e origem biossintética. In: Cunha AP (ed) Farmacognosia e fitoquímica. Fundação Calouste Benkian, Lisboa, pp 211–224

Cunha AP, Roque OR (2009b) Obtenção de moléculas com atividade farmacológica a partir de material vegetal e sua transformação em medicamento. In: Cunha AP (ed) Farmacognosia e fitoquímica. Fundação Calouste Benkian, Lisboa, pp 109–123

Daoud A, Malika D, Bakari S, Hfaiedh N, Mnafgui K, Kadri A (2015) Assessment of polyphenol composition, antioxidant and antimicrobial properties of various extracts of Date Palm Pollen (DPP) from two Tunisian cultivars. Arab J Chem. https://doi.org/10.1016/j.arabjc.2015.07.014

Darwish A, Abd El-Wahed A, Shehata M, El-Seedi H, Masry S, Khalifa S, Mahfouz H, El-Sohaimy S (2022) Chemical profiling and nutritional evaluation of bee pollen, bee bread, and royal jelly and their role in functional fermented dairy products. Molecules 28(1):227. https://doi.org/10.3390/molecules28010227

De Arruda VAS, Pereira AAS, de Freitas AS, Barth OM, de Almeida-Muradian LB (2013) Dried bee pollen: B complex vitamins, physicochemical and botanical composition. J Food Compos Anal 29(2):100–105. https://doi.org/10.1016/j.jfca.2012.11.004

De Arruda VAS, Vieria dos Santos A, Sampaio DF, da Silva AE, de Castro L, Peixoto A, Estevinho MLF, de Almeida-Muradian LB (2017) Microbiological quality and physicochemical characterization of Brazilian bee pollen. J Apic Res 56:231–238

De Sá-Otero MDP, Armesto-Baztan S, Díaz-Losada E (2009) Analysis of protein content in pollen loads produced in north-west Spain. Grana 48(4):290–296. https://doi.org/10.1080/00173130903149140

De-Melo AAM (2015) Perfil químico e microbiológico, cor, análise polínica e propriedades biológicas do pólen apícola desidratado. http://www.teses.usp.br/teses/disponiveis/9/9131/tde-18122015-142742/pt-br.php. Accessed 26 Feb 2017

De-Melo AAM, Estevinho MLMF, Sattler JAG, Souza BR, Silva-Freitas A, Barth OM (2016) Effect of processing conditions on characteristics of dehydrated bee-pollen and correlation between quality parameters. Food Sci Technol 65:808–815

De-Melo AAM, Estevinho LM, Moreira MM, Alex CD, Barth OM, De Almeida-muradian LB (2018) Phenolic profile by HPLC-MS , biological potential, and nutritional value of a promising food: monofloral bee pollen. (February):1–21. https://doi.org/10.1111/jfbc.12536

Denisow B, Denisow-Pietrzyk M (2016) Biological and therapeutic properties of bee pollen: a review. J Sci Food Agric. https://doi.org/10.1002/jsfa.7729

Di Paola-Naranjo RD, Sánchez-Sánchez J, González-Paramás AM, Rivas-Gonzalo JC (2004) Liquid chromatographic–mass spectrometric analysis of anthocyanin composition of dark blue bee pollen from Echium plantagineum. J Chromatogr A 1054:205-210. https://doi.org/10.1016/j.chroma.2004.05.02317

Domínguez-Valhondo D, Gil DB, Hernández MT, González-Gómez D (2011) Influence of the commercial processing and floral origin on bioactive and nutritional properties of honeybee-collected pollen. Int J Food Sci Tech 46:2204–2211

Donadieu Y (1983) Le pollen: thérapeutique naturelle, 6th edn. Maloine, Paris

Dranca F, Ursachi F, Oroian M (2020) Bee bread: physicochemical characterization and phenolic content extraction optimization. Food Secur. https://doi.org/10.3390/foods9101358

Dübecke A, Beckh G, Lüllmann C (2011) Pyrrolizidine alkaloids in honey and bee pollen. Food Addit Contam Part A 28(3):348–358. https://doi.org/10.1080/19440049.2010.541594

El Ghouizi A, El Menyiy N, Falcão SI, Vilas-Boas M, Lyoussi B (2020) Chemical composition, antioxidant activity, and diuretic effect of Moroccan fresh bee pollen in rats. Vet World 13(7):1251–1261. https://doi.org/10.14202/vetworld.2020.1251-1261

El Ghouizi A, Bakour M, Laaroussi H, Ousaaid D, El Menyiy N, Hano C, Lyoussi B (2023) Bee pollen as functional food: insights into its composition and therapeutic properties. Antioxidants (Basel) 12(3):557

Endo A, Salminen S (2013) Honeybees and beehives are rich sources for fructophilic lactic acid bacteria. Syst Appl Microbiol 36(6):444–448

Ertosun S, Aylanc V, Falcão SI, Vilas-Boas M (2023) Thermal stability and antioxidant activity of bioactive compounds in bread enriched with bee pollen and bee bread. Antioxidants (Basel) 12(9):1691. https://doi.org/10.3390/antiox12091691

Estevinho LM, Rodrigues S, Pereira AP (2012) Original article Portuguese bee pollen: palynological study, nutritional and microbiological evaluation. Int J Food Sci Technol 47:429–435. https://doi.org/10.1111/j.1365-2621.2011.02859.x

Fanali C, Dugo L, Rocco A (2013) Nano-liquid chromatography in nutraceutical analysis: determination of polyphenols in bee pollen. J Chromatogr A 1313:270–274

FAO Food and Agriculture Organization of the United Nations (2011) Dietary protein quality evaluation in human nutrition. http://www.fao.org/ag/humannutrition/35978-02317b979a686a57aa4593304ffc17f06.pdf. Accessed 12 Dec 2016

Fatrcová-Šramková K, Nôžková J, Kačániová M, Máriássyová M, Rovná K, Stričík M (2013) Antioxidant and antimicrobial properties of monofloral bee pollen. J Environ Sci Health B 48:133–138

Féas X, Vázquez-Tato MP, Estevinho L, Seijas JA, Iglesias A (2012) Organic bee pollen: botanical origin, nutritional value, bioactive compounds, antioxidant activity and microbiological quality. Molecules 17:8359–8377

Felek İ, Dündar O, Çobanoğlu DN (2021) Palynological, antioxidant and physicochemical properties of pollen loads from eastern Anatolia. Bee Stud Apic Res Inst 13(2):31–38. https://doi.org/10.51458/BSTD.2021.15

Formicki G, Gren A, Stawarz R, Zysk B, Gal A (2013) Metal content in honey, propolis, wax, and bee pollen and implications for metal pollution monitoring. Pol J Environ Stud 22:99–106

Forsén S, Kördel J (1994) Calcium in biological systems. In: Bertini I, Gray HB, Lippard SJ, Valentine JS (eds) Bioinorganic chemistry. University Science Books, Mill Valley, pp 107–166

Freire KRL, Lins ACS, Dórea MC, Santos FAR, Camara CA, Silva TMS (2012) Palynological origin, phenolic content, and antioxidant properties of honeybee-collected pollen from Bahia, Brazil. Molecules 17:1652–1664

Fuenmayor C, Zuluaga C, Díaz C, Quicazán M, Cosio M, Mannino S (2014) Evaluation of the physicochemical and functional properties of Colombian bee pollen. Rev MVZ Córdoba 19:4003–4014

Gabriele M, Parri E, Felicioli A, Sagona S, Pozzo L, Biondi C (2015) Phytochemical composition and antioxidant activity of tuscan bee pollen of different botanic origins. Ital J Food Sci 27:248–259

Gabriele M, Frassinetti S, Pucci L (2021) Antimicrobial activity and protective effect of tuscan bee pollens on oxidative and endoplasmic reticulum stress in different cell-based models. Food Secur 10(6). https://doi.org/10.3390/foods10061422

Gardana C, Del Bo C, Quicazán MC, Corrrea AR, Simonetti P (2018) Nutrients, phytochemicals and botanical origin of commercial bee pollen from different geographical areas. J Food Compos Anal 29:29–38

Gercek YC, Celik S, Bayram S (2021) Screening of plant pollen sources, polyphenolic compounds, fatty acids and antioxidant/antimicrobial activity from bee pollen. Molecules 27(1):117. https://doi.org/10.3390/molecules27010117

González-Martín I, Hernández-Hierro JM, Barros-Ferreiro N, Cordón Marcos C, García-Villanova RJ (2007) Use of NIRS technology with a remote reflectance fibre-optic probe for predicting major components in bee pollen. Talanta 72:998–1003

González-Paramás AMG, Bárez JAG, Marcos CC, García-Villanova RJ, Sánchez JS (2006) HPLC-fluorimetric method for analysis of amino acids in products of the hive (honey and bee-pollen). Food Chem 95:148–156

Halliwell B, Gutteridge J (2007) Free radicals in biology and medicine, 4th edn. Oxford University Press Inc., New York

Hamalainen M, Nieminen R, Vuorela P, Heinonen M, Moilanen E (2007) Anti-inflammatory effects of flavonoids: genistein, kaempferol, quercetin, and daidzein inhibit STAT-1 and NF-κB activations, whereas flavone, isorhamnetin, naringenin, and pelargonidin inhibit only NF-κB activation along with their inhibitory effect on iNOS expression and NO production in activated macrophages. Mediat Inflamm 2007:1–10

Hărmănescu M, Popovici D, Gergen I (2007) Mineral micronutrients composition of bee's pollen. J Agroalim Proc Technol XIII(1):175–182

Herbert JREW, Vanderslice JT, Huang M, Higgs DJ (1987) Levels of thiamine and its esters in bee collected pollen using liquid chromatography and robotics. Apidologie 18:129–136

Hong I-P, Woo S-O, Han S-M, Kim S-G, Jang H-R, Lee M, Choi Y-S, Kim H-K, Lee M-L (2015) Evaluation of total polyphenol, flavonoid and vitamin content from the crushed pollens of acorn and actinidia. J Apic 30(3):225. https://doi.org/10.17519/apiculture.2015.09.30.3.225

Horst MA, Lajolo FM (2011) Biodisponibilidade de compostos bioativos de alimentos. http://wwwfcfuspbr. Accessed 15 June 2013

Hsu P-S, Wu T-H, Huang M-Y, Wang D-Y, Wu M-C (2021) Nutritive value of 11 bee pollen samples from major floral sources in Taiwan. Food Secur 10(9):2229. https://doi.org/10.3390/foods10092229

Huang D, Ou B, Prior RL (2005) The chemistry behind antioxidant capacity assays. J Agric Food Chem 53:1841–1856

Human H, Nicolson SW (2006) Nutritional content of fresh, bee-collected and stored pollen of *Aloe greatheadii* var. davyana (Asphodelaceae). Phytochemistry 67:1486–1492

Ilie CI, Spoiala A, Geana E-I, Chircov C, Ficai A, Ditu L-M, Oprea E (2024) Bee bread: a promising source of bioactive compounds with antioxidant properties—first report on some antimicrobial features. Antioxidants (Basel) 13(3):353. https://doi.org/10.3390/antiox13030353

Jarukas L, Kuraite G, Baranauskaite J, Marksa M, Bezruk I, Ivanauskas L (2020) Optimization and validation of the GC/FID method for the quantification of fatty acids in bee products. Appl Sci 11(1):83. https://doi.org/10.3390/app11010083

Jeannerod L, Carlier A, Schatz B, Daise C, Richel A, Agnan Y, Baude M, Jacquemart AL (2022) Some bee-pollinated plants provide nutritionally incomplete pollen amino acid resources to their pollinators. PLoS One 17(8 August). https://doi.org/10.1371/journal.pone.0269992

Kacániová M, Vukovic N, Chlebo R, Haščík P, Rovná K, Cubon J (2012) The antimicrobial activity of honey, bee pollen loads and beeswax from Slovakia. Arch Biol Sci 64:927–934

Kaškonienė V, Ruočkuvienė G, Kaškonas P, Akuneca I, Maruška A (2015) Chemometric analysis of bee pollen based on volatile and phenolic compound compositions and antioxidant properties. Food Anal Method 8:1150–1163

Ketkar SS, Rathore AS, Lohidasan S, Rao L, Paradkar AR, Mahadik KR (2014) Investigation of the nutraceutical potential of monofloral Indian mustard bee pollen. J Integr Med 12:379–389

Keyvan N, Usluer MS, Kaya MM, Kahraman HA, Tutun H, Keyvan E (2023) Total phenolic content, antibacterial and antiradical properties of bee bread from Turkey. Vet J Mehmet Akif Ersoy Univ 8(2):125–129. https://doi.org/10.24880/maeuvfd.1278378

Kinsella JE (1976) Functional properties of protein foods. Crit Rev Food Sci Nutr 1:219–229

Komosinska-Vassev K, Olczyk P, Kafmierczak J, Mencner L, Olczyk K (2015) Bee pollen: chemical composition and therapeutic application. Evid Based Complement Alternat Med 2015:1–6

Kostic AZ, Barac MB, Stanojevic SP, Milojkovic-Opsenica DM, Tesic ZL, Sikoparija B, Radisic P, Prentovic M (2015a) Physicochemical composition and techno-functional properties of bee pollen collected in Serbia. Food Sci Technol 62:301–309

Kostic AZ, Pešić MB, Mosić MD, Dojčinović BP, Natić MM, Trifković JD (2015b) Mineral content of bee pollen from Serbia. Arh Hig Rada Toksikol 66:251–258

Lau P, Lesne P, Grebenok RJ, Rangel J, Behmer ST (2022, 1853) Assessing pollen nutrient content: a unifying approach for the study of bee nutritional ecology. Philos Trans R Soc Lond B Biol Sci 377. https://doi.org/10.1098/rstb.2021.0510

Leblanc BW, Davis OK, Boue S, Delucca A, Deeby T (2009) Antioxidant activity of Sonoran Desert bee pollen. Food Chem 115:1299–1305

Lee NK, Kim WS, Paik HD (2019) Bacillus strains as human probiotics: characterization, safety, microbiome, and probiotic carrier. Food Sci Biotechnol 28:1297–1305

Li L, Zhang X, Wang X, Qiu Y, Li W, Guo L, Shen Q, Dai J (2024) Quality control of pollen products in the market by quantitative analysis of total amino acids with liquid chromatography. Heliyon 10(16):e35833. https://doi.org/10.1016/j.heliyon.2024.e35833

Liolios V, Tananaki V, Dimou M, Kanelis D, Goras G, Karazafiris E (2016) Ranking pollen from bee plants according to their protein contribution to honey bees. J Apic Res 54:582–592

Liolios V, Tananaki C, Dimou M, Kanelis D, Rodopoulou MA, Thrasyvoulou A (2018) Exploring the sugar profile of unifloral bee pollen using high performance liquid chromatography. J Food Nutr Res 57:341–350

Loper GM, Standifer LN, Thompson MJ, Gilliam M (1980) Biochemistry and microbiology of bee collected almond (*Prumus dulcis*) pollen and bee bread. I. Fatty acids, vitamins and minerals. Apidologie 11:63–73

Lopes J, Stanciu OG, Campos MG, Almaraz-Abarca N, Almeida-Muradian LB, Marghitas LA (2011) Bee pollen antioxidant activity – a review: achievements and further challenges. J Pharmacol 2:25–38

Louveaux J, Maurizio M, Vorvohl G (1978) Methods of melissopalynology. Bee World 59:139–157

Luz CFP, Bacha-Junior GL, Fonseca RLS, Sousa PR (2010) Comparative pollen preferences by africanized honeybees *Apis mellifera* L. of two colonies in Pará de Minas, Minas Gerais, Brazil. An Acad Bras Ciênc 82:293–304

Lv H, Wang X, He Y, Wang H, Suo Y (2015) Identification and quantification of flavonoid aglycones in rape bee pollen from Qinghai-Tibetan Plateau by HPLC-DAD-APCI/MS. J Food Compost Anal 38:49–54

Martins MCT, Morgano MA, Vicente E, Baggio SR, Rodriguez-Amaya DB (2011) Physicochemical composition of bee pollen from eleven Brazilian states. J Apic Sci 55:107–116

Mărgăoan R, Mărghitaş LA, Dezmirean DS, Bobiş O, Mihai CM (2012) Physical-Chemical Composition of Fresh Bee Pollen from Transylvania. Bull UASVM Anim Sci Biotechnol 69:351–355

Mărgăoan R, Al Mărghitaş L, Dezmirean DS, Dulf FV, Bunea A, Socaci SA, Bobiş O (2014) Predominant and secondary pollen botanical origins influence the carotenoid and fatty acid profile in fresh honeybee-collected pollen. J Agric Food Chem 62(27):6306–6316. https://doi.org/10.1021/jf5020318

Mărgăoan R, Özkök A, Keskin Ş, Mayda N, Urcan AC, Cornea-Cipcigan M (2021) Bee collected pollen as a value-added product rich in bioactive compounds and unsaturated fatty acids: a comparative study from Turkey and Romania. LWT 149(June):111925. https://doi.org/10.1016/j.lwt.2021.111925

Mărghitas LA, Stanciu OG, Dezmirean DS, Bobis O, Popescu O, Bogdanov S (2009) *In vitro* antioxidant capacity of honeybee-collected pollen of selected floral origin harvested from Romania. Food Chem 115:878–883

Mărgăoan R, Mărghitaş LA, Dezmirean DS, Mihai CM, Bobiş O (2010) Bee Collected Pollen – General Aspects and Chemical Composition. Bull UASVM Anim Sci Biotechnol 67:254-259

Martín-Gómez B, Salahange L, Tapia JA, Martín MT, Ares AM, Bernal J (2022) Fast chromatographic determination of free amino acids in bee pollen. Food Secur 11(24):4013. https://doi.org/10.3390/foods11244013

Martins MCT (2010) Pólen apícola brasileiro: valor nutritivo e funcional, qualidade e contaminantes inorgânicos. http://www.bibliotecadigital.unicamp.br/document/?code=000477526. Accessed 26 Feb 2017

Maruščáková IC, Schusterová P, Bielik B, Toporčák J, Bilikova K, Mudroňová D (2020) Effect of application of probiotic pollen suspension on immune response and gut microbiota oh honey bees (*Apis mellifera*). Probiotics Antimicrob Proteins 12:929–936. https://doi.org/10.1007/s12602-019-09626-6

Matuszewska E, Klupczynska A, Maciołek K, Kokot Z, Matysiak J (2021) Multielemental analysis of bee pollen, propolis and royal jelly collected in west-central Poland. Molecules 26(9):2415

Mayda N, Özkök A, Ecem Bayram N, Gerçek YC, Sorkun K (2020) Bee bread and bee pollen of different plant sources: determination of phenolic content, antioxidant activity, fatty acid and element profiles. J Food Meas Charact. https://doi.org/10.1007/s11694-020-00427-y

Mejías E, Montenegro G (2012) The antioxidant activity of chilean honey and bee pollen produced in the Llaima volcano's zones. J Food Qual 35:315–322

Pereira de Melo IL, Almeida-Muradian LB (2011) Comparison of methodologies for moisture determination on dried bee pollen samples. Ciênc Tecnol Aliment 31:194–197

Menezes JDS, Maciel LF, Miranda MS, Druzian JI (2010) Compostos bioativos e potencial antioxidante do pólen apícola produzido por abelhas africanizadas (*Apis mellifera* L.). Rev Inst Adolfo Lutz 69:233–242

Modro AFH, Message D, Luz CFP, Meira-Neto JAA (2011) Flora de importância polinífera para *Apis mellifera* (L.) na região de Viçosa, MG. Rev Árvore 35:1145–1153

Mohdaly AAA, Mahmoud AA, Roby MHH, Smetanska I, Ramadan MF (2015) Phenolic extract from propolis and bee pollen: composition, antioxidant and antibacterial activities. J Food Biochem 39:538–547

Montenegro G, Pizarro R, Mejías E, Rodríguez S (2013) Biological evaluation of bee pollen from native Chilean plants. Phyton 82:7–14

Morais M, Moreira L, Feás X, Estevinho LM (2011) Honeybee-collected pollen from five Portuguese Natural Parks: palynological origin, phenolic content, antioxidant properties and antimicrobial activity. Food Chem Toxicol 49:1096–1101

Moreira L, Dias LG, Pereira JA, Estevinho L (2008) Antioxidant properties, total phenols and pollen analysis of propolis samples from Portugal. Food Chem Toxicol 46:3482–3485

Morgano MA, Martins MCT, Rabonato LC, Milani RF, Yotsuyanagi K, Rodriguez-Amaya DB (2012) A comprehensive investigation of the mineral composition of brazilian bee pollen: geographic and seasonal variations and contribution to human diet. J Braz Chem Soc 23:727–736

Muniategui S, Sancho MT, López-Hernández J, Simal-Lozano J (1990) Determination of carotenes from bee-collected pollen by high performance liquid chromatography. J Apicult Res 29:147–150

Nawaz Ranja MMA, ShafiqueB BM, Kowalczewski PL, Shehzad Q, UsmanM MMF et al (2021) Nutritional and health potential of probiotics: a review. Appl Sci 11(23):11204. https://doi.org/10.3390/app112311204

Negri G, Teixeira EW, Alves MLTMF, Moreti ACCC, Otsuk IP, Borguini RG, Salatino A (2011) Hydroxycinnamic acid amide derivatives, phenolic compounds and antioxidant activities of extracts of pollen samples from Southeast Brazil. J Agric Food Chem 59:5516–5522

Negri G, Barreto LMRC, Sper FL, de Carvalho C, Campos M (2018) Phytochemical analysis and botanical origin of Apis mellifera bee pollen from the municipality of Canavieiras, Bahia State, Brazil. Braz J Food Technol 21. https://doi.org/10.1590/1981-6723.17616

Nicolson SW, Human H (2013) Chemical composition of the 'low quality' pollen of sunflower (*Helianthus annuus*, Asteraceae). Apidologie 44:144–152

Nielsen FH (2008) Is boron nutritionally relevant? Nutr Rev 66:183–191

Nielsen SS (2010). Phenol-Sulfuric Acid Method for Total Carbohydrates. In: Nielsen, S.S. (eds) Food Analysis Laboratory Manual. Food Science Texts Series. Springer, Boston, MA. https://doi.org/10.1007/978-1-4419-1463-7_6

NIH (2016) Selenium. https://ods.od.nih.gov/factsheets/Selenium-HealthProfessional/. Accessed 12 Dec 2016

Nikkeshi A, Kuramitsu K, Yokoi T, Yamaji K (2022) Simple methods of analyzing proteins and amino acids in small pollen samples. J Apic Res 61(1):107–113. https://doi.org/10.1080/00218839.2021.1915633

Nogueira C, Iglesias A, Feás X, Estevinho LM (2012) Commercial bee pollen with different geographical origins: a comprehensive approach. Int J Mol Sci 13:11173–11187

Odoux JF, Feuille D, Aupinel P, Loublier Y, Tasei JN, Mateescu C (2012) Territorial biodiversity and consequences on physico-chemical characteristics of pollen collected by honey bee colonies. Apidologie 43:561–575

Oliveira DM, Bastos DHM (2011) Biodisponibilidade de ácidos fenólicos. Quím Nova 34:1051–1056

Oliveira KCLS, Moriya M, Azedo RAB, de Almeida-Muradian LB, Teixeira EW, Alves MLT, Moreti ACC (2009) Relationship between botanical origin and antioxidant vitamins of bee-collected pollen. Quim Nova 32:1099–1102

Oroian M, Dranca F, Ursachi F (2022) Characterization of Romanian bee pollen—an important nutritional source. Food Secur. https://doi.org/10.3390/foods11172633

Orzáez Villanueva MT, Díaz Marquina A, Bravo Serrano R, Blazquez Abellán G (2002) The importance of bee-collected pollen in the diet: a study of its composition. Int J Food Sci Nutr 53:217–224

Pascoal A, Rodrigues S, Teixeira A, Feás X, Estevinho LM (2014) Biological activities of commercial bee pollens: antimicrobial, antimutagenic, antioxidant and anti-inflammatory. Food Chem Toxicol 63:233–239

Pełka K, Bucekova M, Godocikova J, Szweda P, Majtan J (2022) Glucose oxidase as an important yet overlooked factor determining the antibacterial activity of bee pollen and bee bread. Eur Food Res Technol 248(12):2929–2939. https://doi.org/10.1007/s00217-022-04101-z

Pełka K, Hafeez AB, Worobo RW, Szweda P (2025) Probiotic potential of Bacillus Isolates from Polish Bee Pollen and Bee Bread. Probiotics & Antimicro Prot 17:364–377. https://doi.org/10.1007/s12602-023-10157-4

Pereira de Melo IL, de Almeida-Muradian LB (2010) Stability of antioxidant vitamins in bee pollen samples. Quim Nova 33:514–518

Pereira FM, Lopes MTR, Camargo RCR, Vilela SLO (2013) Produção de mel. http://www.siste--masdeproducao.cnptia.embrapa.br. Accessed 2 Mar 2014

Perini JÂL, Stevanato FB, Sargi SC, Visentainer JEL, Dalalio MMO, Matshushita MS (2010) Ácidos graxos poli-insaturados n-3 e n-6: metabolismo em mamíferos e resposta imune. Rev Nutr 23:1075–1086

Rebelo KS, Ferreira AG, Carvalho-Zilse GA (2016) Physicochemical characteristics of pollen collected by Amazonian stingless bees. Cienc Rural 46:927–932

Rebiai A, Lanez T (2012) Chemical composition and antioxidant activity of *Apis mellifera* bee pollen from northwest Algeria. J Fundam Appl Sci 4:155–163
Rosanoff A, Weaver CM, Rude RK (2012) Suboptimal magnesium status in the United States: are the health consequences underestimated? Nutr Rev 70:153–164
Saavedra KIC, Rojas CI, Delgado GEP (2013) Características polínicas y composición química del polen apícola colectado en Cayaltí (Lambayeque – Perú). Rev Chil Nutr 40:71–78
Sánchez ET (2004) Polen: producción envasado y comercialización. Diproansa, Morelos
Šarić A, Balog T, Sobočanec S, Kušić B, Šverko V, Rusak G, Likić S, Bubalo D, Pinto B, Reali D, Marotti T (2009) Antioxidant effects of flavonoid from Croatian Cystus incanus L. rich bee pollen. Food Chem Toxicol 47:547–554
Sattler JAG, de Melo ILP, Granato D, Araújo E, de Freitas AS, Barth OM, Sattler A, de Almeida-Muradian LB (2015) Impact of origin on bioactive compounds and nutritional composition of bee pollen from southern Brazil: a screening study. Food Res Int 77:82–91
Sattler JAG, De-Melo AAM, Do Nascimento KS, De Melo ILP, Mancini-Filho J, Sattler A, de Almeida-Muradian LB (2016) Essential minerals and inorganic contaminants (barium, cadmium, lithium, lead and vanadium) in dried bee pollen produced in Rio Grande do Sul State, Brazil. Food Sci Technol 36(3):505–509. https://doi.org/10.1590/1678-457X.0029
Sawicki T, Starowicz M, Kłębukowska L, Hanus P (2022) The profile of polyphenolic compounds, contents of total phenolics and flavonoids, and antioxidant and antimicrobial properties of bee products. Molecules 27(4):1301. https://doi.org/10.3390/molecules27041301
Serra-Bonvehí J, Escolà Jordà R (1997) Nutrient composition and microbiological quality of honey bee-collected pollen in Spain. J Agric Food Chem 45:725–732. https://doi.org/10.1021/jf960265q
Serra-Bonvehí J, Escolà-Jordà R (1997) Nutrient composition and microbiological quality of honeybee- collected pollen in Spain. J Agric Food Chem 45:725–732
Shawer MB, Ali SM, Abdellatif MA, El-Refai AA (1987) Biochemical studies of bee-collected pollen in Egypt 2. Fatty acids and non-saponifiables. J Apic Res 26:133–136
Silveira TA (2012) Caracterização sazonal do pólen apícola quanto à origem botânica, aspectos físico-químicos e elementos traços como bioindicadora de poluição ambiental. http://www. teses.usp.br/teses/disponiveis/11/11146/tde-20032012-090016/pt-br.php. Accessed 26 Feb 2017
Simões CMO, Guerra MP, Schenkel EP, Zannin M, Mentz LA, Bordignon SAL (2004) Farmacognosia: da planta ao medicamento, 5th edn. UFRGS, Porto Alegre
Somerville DC, Nicol HI (2002) Mineral content of honeybee-collected pollen from southern New South Wales. Aust J Exp Agric 42:1131–1136
Sonmez E, Kekecoglu M, Sahin H, Bozdeveci A, Karaoglu SA (2023) Comparing the biological properties and chemical profiling of chestnut bee pollen and bee bread collected from Anatolia. Braz J Microbiol 54:2307–2317. https://doi.org/10.1007/s42770-023-00980-w
Souza BR (2014) Quantificação das vitaminas do complexo B (B1, B2) e vitâmeros das vitaminas B3 e B6 em amostras de pólen apícola desidratado provenientes da Região Sul do Brasil. http://www.teses.usp.br/teses/disponiveis/9/9131/tde-27052015-141055/pt-br.php. Accessed 26 Feb 2017
Spulber AR, Doğaroğlu M, Băbeanu N, Popa O (2018) Physicochemical characteristics of fresh bee pollen from different botanical origins. Rom Biotechnol Lett 23(1):13357–13365
SR ISO 24382 (2024) Bee Pollen – Specifications DC
Stabler D, Power EF, Borland AM, Barnes JD, Wright GA (2018) A method for analysing small samples of floral pollen for free and protein-bound amino acids. Methods Ecol Evol 9(2):430–438. https://doi.org/10.1111/2041-210X.12867
Sunde RA (2012) Selenium. In: Ross AC, Caballero B, Cousins RJ, Tucker KL, Ziegler TR (eds) Modern nutrition in health and disease, 11th edn. Lippincott Williams & Wilkins, Philadelphia, pp 225–237
Szczêsna T (2006) Protein content and amino acids compositon of bee-collected pollen originating from Poland, South Korea and China. J Apic Sci 50(2):91–99

Szczesna T (2006a) Long-chain fatty acids composition of honeybee-collected pollen. J Apic Sci 50:81–90
Szczesna T (2006b) Protein content and amino acid composition of bee-collected pollen from selected botanical origins. J Apic Sci 50:81–90
Szczesna T (2007) Concentration of selected elements in honeybee-collected pollen. J Apic Sci 51:5–13
Szczesna T, Rybak-Chimielewska H, Bornus L (1991) Effect of storage on variation of contents vitamin C and A in pollen collected bees. Apiacta 2:1–3
Taha EKA (2015) Chemical composition and amounts of mineral elements in honeybee-collected pollen in relation to botanical origin. J Apic Sci 59:75–81
Taha EKA, Al-Kahtani S, Taha R (2019) Protein content and amino acids composition of bee-pollens from major floral sources in Al-Ahsa, eastern Saudi Arabia. Saudi J Biol Sci 26:232–237
Tavdidishvili D, Khutsidze T, Pkhakadze M, Vanidze M, Kalandia A (2014) Flavonoids in georgian bee bread and bee pollen. J Chem Chem Eng 8:676–681
Thakur M, Nanda V (2020) Composition and functionality of bee pollen: a review. Trends Food Sci Technol 98:82–106
Tomás A, Falcão SI, Russo-Almeida P, Vilas-Boas M (2017) Potentialities of beebread as a food supplement and source of nutraceuticals: botanical origin, nutritional composition and antioxidant activity. J Apic Res 56(3):219–230. https://doi.org/10.1080/00218839.2017.1294526
Türe A, Nisbet C, Koçak E (2023) Determination of protein levels and amino acid composition of bee pollen collected from different geographical regions of Turkey. J Istanbul Vet Sci 7(1):15–20. https://doi.org/10.30704/http-www-jivs-net.1171929
Ulusoy E, Kolayli S (2014) Phenolic composition and antioxidant properties of anzer bee pollen. J Food Biochem 38:73–82
Urcan AC, Criste AD, Dezmirean DS, Bobiş O, Bonta V, Dulf FV, Mărgăoan R, Cornea-Cipcigan M, Campos MG (2021) Botanical origin approach for a better understanding of chemical and nutritional composition of beebread as an important value-added food supplement. LWT 142(January):111068. https://doi.org/10.1016/j.lwt.2021.111068
Urcan AC, Criste AD, Szanto KI, Ştefan R, Zahan M, Muscă AS, Focsan M, Burtescu RF, Olah NK (2023) Antimicrobial and antiproliferative activity of green synthesized silver nanoparticles using bee bread extracts. Pharmaceutics 15(7):1797. https://doi.org/10.3390/pharmaceutics15071797
Urcan AC, Criste AD, Dezmirean DS, Bobiş O, Bonta V, Burtescu RF, Olah N-K, Cornea-Cipcigan M, Mărgăoan R (2024) Enhancing antioxidant and antimicrobial activities in bee-collected pollen through solid-state fermentation: a comparative analysis of bioactive compounds. Antioxidants (Basel) 13(3):292. https://doi.org/10.3390/antiox13030292
Urcan AC, Criste AD, Dezmirean DS, Mărgăoan R, Caeiro A, Graça Campos M. Similarity of Data from Bee Bread with the Same Taxa Collected in India and Romania. Molecules. 2018;23(10):2491. https://doi.org/10.3390/molecules23102491
USDA (2011) Dri dietary reference intakes: calcium vitamin D. https://www.nap.edu/read/13050/chapter/1. Accessed 12 Mar 2015
USDA (2015) Nutrient vitamins and minerals. https://www.ams.usda.gov/sites/default/files/media/Nutrient%20Vitamins%20TR%202015.pdf. Accessed 12 Dec 2016
Valverde S, Tapia JA, Pérez-Sanz A, González-Porto AV, Higes M, Lucena JJ, Martín-Hernández R, Bernal J (2023) Mineral composition of bee pollen and its relationship with botanical origin and harvesting period. J Food Comp Anal 119:105235
Vanderplanck M, Moerman R, Rasmont P, Lognay G, Wathelet B, Wattiez R, Michez D (2014) How does pollen chemistry impact development and feeding behaviour of polylectic bees? PLoS One 9(1):e86209. https://doi.org/10.1371/journal.pone.0086209
Vaudo AD, Tooker JF, Patch HM, Biddinger DJ, Coccia M, Crone MK, Fiely M, Francis JS, Hines HM, Hodges M, Jackson SW, Michez D, Mu J, Russo L, Safari M, Treanore ED, Vanderplanck M, Yip E, Leonard AS, Grozinger CM (2020) Pollen protein: lipid macronutri-

ent ratios may guide broad patterns of bee species floral preferences. Insects 11(2):132. https://doi.org/10.3390/insects11020132
Villagómez GN, Brachvogel RC, Kárpáti Z, Leonhardt SD, Schmitt T, Ruedenauer FA (2023) A common protocol for reliable comparison of pollen fatty acid profiles: highlighting pitfalls and proposing a methodology for ecological research. Front Ecol Evol 11(May):1–12. https://doi.org/10.3389/fevo.2023.1141832
Wang X, Wang H, Liu Y, You J, Suo Y (2009) Extraction of pollen lipids by SFE-CO_2 and determination of free fatty acids by HPLC. Eur J Lipid Sci Technol 111:155–163
WHO World Health Organization (2015) Micronutrient deficiencies: iron deficiency anaemia. http://www.who.int/nutrition/topics/ida/en/. Accessed 12 Dec 2016
WHO (2012) Guideline: Sodium intake for adults and children. Geneva, World Health Organization (WHO), https://iris.who.int/bitstream/handle/10665/77985/9789241504836_eng.pdf
Wu W, Wang K, Qiao J, Dong J, Li Z, Zhang H (2019) Improving nutrient release of wall-disrupted bee pollen with a combination of ultrasonication and high shear technique. J Sci Food Agric 99:564–575. https://doi.org/10.1002/jsfa.9216
Xiong A, Yang L, He Y, Zhang F, Wang J, Han H, Wang C, Bligh SWA, Wang Z (2009) Identification of metabolites of adonifoline, a hepatotoxic pyrrolizidine alkaloid, by liquid chromatography/tandem and high-resolution mass spectrometry. Rapid Commun Mass Spectrom 23:3907–3916. https://doi.org/10.1002/rcm.4329
Xu X, Sun L, Dong J, Zhang H (2009) Breaking the cells of rape bee pollen and consecutive extraction of functional oil with supercritical carbon dioxide. Innov Food Sci Emerg Technol 10:42–46
Xu X, Dong J, Mu X, Sun L (2011) Supercritical CO_2 extraction of oil, carotenoids, squalene and sterols from lotus (*Nelumbo nucifera* Gaertn) bee pollen. Food Bioprod Process 89:47–52
Yan S, Li Q, Xue X, Wang K, Zhao L, Wu L (2019) Analysis of improved nutritional composition of bee pollen (*Brassica campestris* L.) after different fermentation treatments. Int J Food Sci Technol 54(6):2169–2181. https://doi.org/10.1111/ijfs.14124
Yang J, Guo J, Yuan J (2008) In vitro antioxidant properties of rutin. Food Sci Technol 41:1060–1066
Yang K, Wu D, Ye X, Liu D, Chen J, Sun P (2013) Characterization of chemical composition of bee pollen in China. J Agric Food Chem 61:708–718
You J, Liu L, Zhao W, Zhao X, Suo Y, Wang H, Li Y (2007) Study of a new derivatizing reagent that improves the analysis of amino acids by HPLC with fluorescence detection: application to hydrolyzed rape bee pollen. Anal Bioanal Chem 387:2705–2718
Youssef AM, Farag RS, Ewies MA, El-Shakaa SMA (1978) Chemical studies on pollen collected by honeybees in Giza region, Egypt. J Apic Res 17:110–113
Zhou J, Qi Y, Ritho J, Zhang Y, Zheng X, Wu L (2015) Flavonoid glycosides as floral origin markers to discriminate of unifloral bee pollen by LC-MS/MS. Food Cont 57:54–62

Chapter 12
Uses and Health Benefits of Bee Pollen in Medicine

Otilia Bobiş, Adriana Cristina Urcan, Adriane Alexandre Machado De-Melo, and Ligia Bicudo de Almeida-Muradian

Contents

Abstract The bioactive compounds of bee pollen have been demonstrated through the years to have antioxidant, antibacterial, and antitumor potential. Phenolic compounds are considered the main active ingredients having these biological activities, but research studies were carried out with the aim of characterizing samples from various countries, seeking to identify those with greater biological potential and the varieties that influence this potential. Recently, some researchers have focused on in vivo studies with the inclusion of this food source in the diet of humans and ani-

O. Bobiş (✉)
Department of Apiculture and Sericiculture, Faculty of Animal Science and Biotechnologies, University of Agricultural Sciences and Veterinary Medicine, Cluj-Napoca, Romania
e-mail: obobis@usamvcluj.ro

A. C. Urcan
Department of Microbiology and Immunology, Faculty of Animal Science and Biotechnologies, University of Agricultural Sciences and Veterinary Medicine, Cluj-Napoca, Romania

A. A. M. De-Melo · L. B. de Almeida-Muradian
Department of Food and Experimental Nutrition, Pharmaceutical Sciences School, University of São Paulo, São Paulo, SP, Brazil

J. M. Alvarez-Suarez (ed.), *Bee Products – Chemical and Biological Properties*,
https://doi.org/10.1007/978-3-031-89049-9_12

mals, and in the formulation of medications with a bee pollen base. Latest studies with results of in vitro and in vivo studies of the antibacterial, antioxidant and antitumor activity of bee pollen are presented, as well as factors that influence the biological potential of this product. Recently, the biological activity of bee pollen has been complemented with studies demonstrating hepatoprotective and antidiabetic activity, as well as immunomodulatory, anti-inflammatory, and antiradiation activity, which are reviewed in this chapter.

Keywords Bee pollen · Health benefits · Antioxidant activity · Antibacterial properties · Immunomodulatory effects

12.1 Introduction

Bee pollen is the result of the agglomeration of pollen harvested by *Apis mellifera* L. bees, which use their saliva secretions or nectar to agglutinate the grains (Campos et al. 2008). There is evidence that bee pollen has been intentionally consumed by people for centuries. In Ancient Egypt, it was called the "life-giving dust." Hippocrates and Pythagoras recommended the consumption to their patients because they believed in its therapeutic potential (Campos et al. 2008). In North America, there are records of intentional consumption between 1400 BC and 200 AD by Indigenous peoples (Reinhard et al. 1991; Linskens and Jorde 1997). In recent years, consumption was driven by demand from the population for natural products, complementary to diet or with therapeutic effects (Barreto et al. 2005).

The health benefits associated with bee pollen are related to the presence of compounds with the biological potential to inhibit the growth of microorganisms and aid in the prevention of oxidative stress, identified as one of the causes of the development of chronic degenerative diseases such as cancer, cardiovascular diseases, and neuronal degeneration (Uttara et al. 2009; Rodríguez-Pólit et al. 2023). Commonly, this food is consumed as a rejuvenator or tonic, to increase physical resistance, and to help in the treatment of respiratory infirmities, hormonal disturbances, and in the prevention of cancer, though current research is insufficient to validate all of these indications (Ulbricht et al. 2009). The limitation of using bee pollen as food and/or supplement is basically the allergic reactions of some individuals. In this respect, scientists are working to overcome this limitation, and one of these is using nanotechnology, a modern technique in which different carriers of the medicine (in our case the bee product or its extract) are used to target a specific problem in the human body. The results are very promising, with the delivery of the product being safe in many trial studies (Spulber et al. 2018; Al-Yousef et al. 2020; Keskin 2022; Hanafy et al. 2023).

Research was carried out with the aim of characterizing samples from various countries, seeking to identify those with greater biological potential and the varieties that influence this potential. Recently, some researchers have focused on in vivo studies with the inclusion of this food source in the diet of animals and in the formulation of medications with a bee pollen base. In this chapter, the results of in vitro and in vivo studies of the antibacterial, antioxidant, and antitumor activity of bee pollen are presented, as well as factors that influence the biological potential of this product.

12.2 Antioxidant Activity

A considerable number of studies were carried out to evaluate the benefits associated with the inclusion of natural antioxidants in the diet. Daily intake of these compounds could help prevent oxidative stress, and, consequently, diseases and cellular aging (Lopes et al. 2011). Another possible benefit of natural antioxidants is an increase in the oxidative stability in alimentary formulations, and their use in medicines and cosmetics.

Bee pollen contains compounds of significant antioxidant capacity in variable quantities. The potential of this product is greater than that found in honey (Morais et al. 2011), and similar to that found in red fruits, that are recognized as foods with an elevated antioxidant capacity (Sousa et al. 2011; Huang et al. 2012).

Phenolic compounds, especially phenolic acids and flavonoids, seem to be the main agents of the antioxidant capacity of bee pollen (LeBlanc et al. 2009; Mărghitas et al. 2009; Negri et al. 2011; Freire et al. 2012). These substances are capable of donating hydrogen atoms to the reactive species, therefore making them stable (Rodríguez-Pólit et al. 2023). Other antioxidants, such as β-carotene (provitamin A) and vitamins E and C, are responsible for part of the antioxidant action of bee pollen (Oliveira et al. 2009; Melo and Almeida-Muradian 2010).

The type and content of each antioxidant is associated with its botanical origin (Campos et al. 2008; De-Melo 2015). Variation even occurs between monofloral bee pollen from the same botanical species (De-Melo 2015), indicating that agricultural or genotypic differences, climatic conditions, such as temperature, hydrological stress, and light intensity, as well as soil conditions, can affect the composition of the pollen grain by the plant (Daoud et al. 2015).

Processing conditions and storage should be considered. De-Melo et al. (2016) analyzed bee pollen from the state of São Paulo, Brazil, after dehydration under different conditions, and concluded that antioxidant capacity was greater in lyophilized samples than in those dehydrated in an electric oven with forced air circulation. Domínguez-Valhondo et al. (2011) also observed a negative impact due to dehydration process on the antioxidant capacity of monofloral bee pollen from *Cistus ladani-fer* collected in Chile.

The determination of the antioxidant capacity involves two steps: the extraction of the compounds and the activity test. The conditions of compound extraction, especially the solvent and its dilution, influenced the results. De-Melo (2015) tested the efficiency of the three methods in the extraction of 64 bee pollen samples from Brazil: in 70% of them, an extraction based on 50% methanol and 70% acetone was more efficient, while in 27% of the samples it was with an extraction based on 70%

ethanol in which the highest values for antioxidant capacity were observed. Results can be related to the characteristics of the compounds, and, consequently, with the affinity between these and the solvents. LeBlanc et al. (2009) compared eight solvents of different polarities and concluded that methanol and dimethylformamide were the most efficient with samples from North Tuscon, USA. Pérez-Pérez et al. (2012) found better results in ethanolic extracts than in methanolic and aqueous extracts, in bee pollen from Mérida, Venezuela.

The antioxidant capacity of a compound depends on physico-chemical characteristics, such as polarity, solubility, capacity to form hydrogen bridges, and potential oxidation-reduction (Campos et al. 2008; Oliveira and Bastos 2011).

Various methods exist for determining, in vitro, the antioxidant capacity of a sample. Examples of methods adopted in studies with bee pollen are: 1,1- diphenyl-2- picrylhydrazyl (DPPH) radical-scavenging activity assay; inhibition of lipid peroxidation using thiobarbituric acid reactive substances (TBARS); ferric reducing/antioxidant power (FRAP) assay; 2,2-azinobis-(3-ethylbenzothiazoline-6-sulphonic acid) (ABTS) assay; cupric reducing power (CUPRAC) assay; β-carotene bleaching (BCB) assay; and oxygen-radical absorbance capacity (ORAC). Generally, the studies are carried out with more than one trial for a better understanding of the antioxidant potential (Castelo-Branco and Torres 2011), expecting that different methods arrive at different conclusions. The results are frequently presented as EC_{50} (*effective concentration* 50), i.e. the sample concentration (mg/mL of extract) necessary to reduce 50% of the radical; or in equivalents of some compound, commonly Trolox (TE), per g of bee pollen.

The antioxidant capacity of bee pollen collected in Brazil varies from 0.35 to 13.4 mg/mL (extraction with 70% ethanol; analysis by DPPH method) (Carpes et al. 2009; De-Melo 2015) and from 133 to 576 μmol TE/g (70% ethanol; ORAC method) (De-Melo 2015). Similar values were found in ethanolic extracts (95% ethanol) from samples produced in Italy. Analyses using DPPH and ORAC methods indicate an antioxidant capacity of, respectively, 215 μg/mL and 544 μmol TE/g (*Castanea* sp. bee pollen), 224 μg/mL and 540 μmol TE/g (*Cistus* sp. bee pollen) and 641 μg/mL and 519 μmol TE/g (*Rubus* sp. bee pollen) (Domenici et al. 2015).

Another study on different types of Italian bee pollen of unifloral origin (predominant pollen ranging from 69% to 100%), namely *Brassica* type, *Cistus* sp., *Castanea sativa, Crataegus monogyna, Daucus and Coriandrum* sp., *Hedera helix, Helianthus annuus, Hedisarium coronarium, Lamium, Magnolia, Papaver rhoeas, Prunus* sp., *Rubus* sp., and *Trifolium* sp., was made and high values for the antioxidant activity was obtained: 48.8–224.6 μmol TE/g DW (ABTS method), 11.9–134.7 μmol TE/g DW (DPPH method), and 105.0–916.1 μmol TE/g DW (ORAC method). As in previous studies, the botanical origin of the pollen was determinant in the antioxidant activity of the extracts (the lowest activity was observed in *Magnolia* and the highest activities in *Lamium*, *Castanea,* and *Crataegus monogyna* pollens (Rocchetti et al. 2019).

In Portugal, samples collected in five Portuguese nature reserves and commercially acquired, had similar EC_{50} values using the DPPH method (2.0–6.7 mg/mL) and the BCB method (3.1–6.5 mg/mL) (Morais et al. 2011; Féas et al. 2012; Pascoal et al. 2014). In Spain, the antioxidant capacity of the samples was from 3.3 to 5.1 mg/ mL (DPPH trial) and from 1.1 to 2.2 mg/mL (TBARS trial). In all the experiments, methanolic extracts were analyzed.

In Brazil, the study undertaken by De-Melo (2015) suggested that, the greater the participation of the *Eucalyptus* genus in bee pollen, the greater the antioxidant capacity. Freire et al. (2012) found the greatest antioxidant potential in samples with 50–70% of the *Eucalyptus* genus. Menezes et al. (2010) identified the greatest antioxidant capacity values in samples with a predominance of this genus, beyond a positive correlation between this pollen type and total flavonoid and phenolic content.

In Northwest Algeria, the greatest antioxidant capacity was encountered in monofloral bee pollen from wild carrot (scientific name not given), followed by bee pollen from rosemary (scientific name not given) and by bee pollen from *Eucalyptus* (Rebiai and Lanez 2012). In monofloral bee pollen from Sonoran Desert (Tucson, USA), LeBlanc et al. (2009) reported that the antioxidant capacity was greater in samples from *Yucca* bee pollen than in samples from *Washingtonia* bee pollen. In Chile, the antioxidant capacity of monofloral bee pollen from *Escallonia rubra* was greater than in the monofloral product from *Lotus pedunculatus*. The authors found that the geographical origin of the samples had an impact on the biological potential and that, in general, the antioxidant capacity of monofloral bee pollen from these species was much greater than that of the honey with the same origin (Mejías and Montenegro 2012).

Leja et al. (2007) analyzed methanolic extracts (80% methanol) of monofloral samples from bee pollen from the Krakow area in Poland, and classified them into three groups: high (*Lupinus polyphyllus*, *Phacelia tanacetifolia*, *Trifolium* sp., *Sinapis alba, Robinia pseudoacacia,* and *Aesculus hippocastanum* bee pollen), medium (*Zea mays*, *Chamerion angustifolium,* and *Pyrus communis* bee pollen) and low antioxidant capacity (*Lamium purpureum*, *Taraxacum officinale,* and *Malus domestica* bee pollen).

In Egypt, Khider et al. (2013) encountered EC_{50} of 0.6 μg/mL in monofloral bee pollen from *Z. mays*. In Nan, Thailand, Chantarudee et al. (2012) found a value of 7.4 μg/mL in monofloral samples of this pollen type, which signified an antioxidant potential almost 29 times higher in bee pollen than in the pollen grains of the plant (212 μg/mL). It was not clear how the bees combine the pollen with other substances and the alterations that occur, which forms a product with greater biological potential.

Mărghitas et al. (2009), using the DPPH method, observed that in the Transylvania area of Romania, the antioxidant capacity of monofloral samples increased in the following order: *Pinus* sp. < *Knautia arvensis* < *Taraxacum officinale* < *Helianthus annuus* < *Onobrychis viciifolia* < *Crataegus monogyna* < *Capsella bursa pastoris* < *Matricaria chamomilla* < *Carduus* sp. < *Carex* sp. < *Centaurea cyanus* < *Salix* sp. Variations in this sequence were observed with ABTS and FRAP methods, and correlations between the DPPH and ABTS methods suggested that the same compounds act in both systems. Methanolic extracts were analyzed.

Mărgăoan et al. (2021) in a study of Romanian and Turkish bee pollen samples of different botanical origins proved the importance of pallinology in the determination of bioactive properties of bee pollen analysis. DPPH and TEAC methods were used and generally, the values for DPPH were smaller in Romanian samples (0.81 mmol Trolox/g) and higher in Turkish samples (2.93 mmol Trolox/g), while the higher TEAC values were also determined in Turkish samples (9.64 mmol Trolox/g). The samples exhibiting the highest antioxidant activity in the DPPH assay were mostly multifloral, predominating in Fabaceae and Fagaceae pollen, whereas for the TEAC assay, the highest inhibitory activities were observed for the samples having as predominant pollen Asteraceae, Brassicaceae, Fabaceae, and Fagaceae families.

Fatrcová-Šramková et al. (2013) tested, using the DPPH method, the antioxidant capacity of bee pollen collected in the Nitra region of Slovakia. In the ethanolic extracts (90% ethanol) the values indicated that the sample from *Brassica napus subsp. napus* had greater biological potential than the product with a predominance of the *Papaver somniferum* pollen type, which, in its turn, had greater potential than that observed in bee pollen from *Helianthus annuus*. The authors also used two other methods to determine the antioxidant capacity of the samples, based on the reduction of Mo (VI) to Mo (V), and another in which the protective effect of antioxidants against oxidative DNA damage was measured, with both showing the activity of the samples.

While Mărghitas et al. (2009), in Romania, observed values of 0.45 mmol TE/g in monofloral bee pollen from *Helianthus annuus*, Zhang et al. (2015) encountered 0.04 mmol TE/g in samples from this botanical origin collected in China. Beyond the possible influence of geographical origin, or of distinct varieties of plants, the different solvents utilized in the extraction could have a relationship to these results. These variables could also have influenced the different values of antioxidant capacity found in monofloral bee pollen from *Phoenix dactylifera*. In Tunisia, EC50 reached 46.6 μg/mL (Daoud et al. 2015), while in Egypt the value was 0.8 μg/mL (Khider et al. 2013).

In bee pollen produced in Rize, Turkey, EC_{50} methanolic extracts varied from 0.7 to 8.2 mg/mL (DPPH assay), from 33.1 to 91.8 μmol TE/g bee pollen (CUPRAC assay), and from 11.8 to 105.1 μmol TE/g (FRAP assay); and, of the nine phenolics identified (*p*-OH benzoic acid, vanillic acid, *trans*-cinnamic acid, syringic acid, *p*-coumaric acid, ferulic acid, rutin, *cis*, *trans*-abscisic acid, and caffeic acid), only *p*-coumaric acid had some correlation with the biological activity, which was negative (Ulusoy and Kolayli 2014).

Different commercial samples of bee pollen from Turkey were analyzed by Dulger Altiner et al. (2021) with respect to their physico-chemical characteristics and antioxidant capacity of extractable, hydrolyzable, and bio-accessible fractions using three different methods: CUPRAC, ABTS, and DPPH. In vitro bioaccessibilities of the analyzed commercial samples were different due to the diversity in the chemical composition of the bee pollen samples (botanical origin not provided) and the sensibility of the methods with respect to different bioactive compounds present in the extracts.

In samples collected in Brazil, of the 18 phenolic compounds identified (gallic acid, protocatechic acid, vanillic acid, siringeic acid, β-resorcylic acid, feru-lic acid, synapic acid, cinnamic acid, caffeic acid, *p*-coumaric acid, chlorogenic acid, catechin, epicatechin, naringenin, naringin, rutin, kaempferol, and quercetin), only quercetin had a positive correlation with the antioxidant capacity, found using the ORAC and DPPH methods (De-Melo 2015).

Su et al. (2021), investigating the antioxidant and anti-tyrosinase activity of Chinese Camellia, rape, rose and lotus bee pollen extracts and also fractions obtained using different solvents (petroleum ether, ethyl acetate, n-butanol, and water), demonstrate the importance of solvent extraction and also the botanical origin of the pollen in its bioactive properties. In this case, ethyl acetate fraction and camellia pollen had the highest antioxidant and anti-tyrosinase activity.

12.3 Antibacterial and Antitumor Activity

12.3.1 Antibacterial Activity

Antibacterials can be defined as compounds that, at low concentrations, are capable of selectively inhibiting the development of bacteria or deactivating them (Cunha and Roque 2009). These compounds can be utilized in the treatment or prevention of infections (Levin et al. 2014). In commercial production, they are added to improve growth rate and/or alimentary conversion of animals, and function as growth stimulants in this context.

The antibacterial capacity of a food or isolated compound can be determined in vitro by bioautographic tests, diffusion tests, and dilution tests. The bioautographic tests involved the use of thin-layer chromatography for the separation of the compounds, which are then placed in contact with, or immersed in wells previously inoculated with the test microorganism. The diffusion tests are based on the diffusion of the extract in a solid culture medium previously inoculated. In the dilution test, the addition of aliquots of test solution in the culture medium previously inoculated with the microorganism is carried out, and after the period of incubation, the growth is compared with a negative control. The effectiveness of the compound mixture or isolated compound is frequently presented as minimum inhibitory concentration (MIC), i.e. the lowest extract concentration capable of ceasing or slowing the growth of the test microorganism.

Bee pollen contains compounds that inhibit the growth of Gram-positive and Gram-negative bacteria. The activity varies according to the potential of compounds present in each product (Morais et al. 2011; Nogueira et al. 2012; Cabrera and Montenegro 2013; Fatrcová-Šramková et al. 2013). Substances responsible for the antibacterial capacity of this food act by deactivating enzymes, promoting alterations in the permeability of membranes and loss of cellular material, or further, by deactivating or destroying genetic material of bacterial cells (Pełka et al. 2022).

Phenolic compounds are the main antibacterial agents of bee pollen, but are not the only ones. Spermidine derivatives have considerable antibacterial potential and have been identified in bee pollen (Bassard et al. 2010; Mihajlovic et al. 2015; De-Melo 2015). Some fatty acids also possess antimicrobial action (Barbosa et al. 2006), and glucose oxidase, an enzyme produced by bees and added during the formation of the grain, can also be responsible for part of the antibacterial effect of this food (Denisow and Denisow-Pietrzyk 2016). In Brazilian bee pollen, there was a positive correlation between the content of total phenolic compounds and the inhibition of Gram-negative and Gram-positive bacterial growth (De-Melo 2015). As far as can be seen, the action is not related to the totality, but rather, to specific phenolic compounds (Campos et al. 2008; Morais et al. 2011; Pascoal et al. 2014; De-Melo 2015). Graikou et al. (2011) tested the action of isolated flavonoids from bee pollen extracts through liquid chromatography. The flavonoids exercised a strong activity against Gram-positive bacteria, which led the authors to associate a significant part of this action with the presence of these compounds.

Processing of bee pollen has an impact on its antibacterial activity. In a study with heterofloral samples from Brazil, it was observed that the dehydration conditions influenced the antibacterial activity, perhaps because the processing also influenced the phenolic content. The MIC values for *Staphylococcus aureus*, *Streptococcus pyogenes*, *Escherichia coli,* and *Klebsiella* were lower in the lyophilized samples than those observed for samples treated in an electric oven (De-Melo et al. 2016).

The antibacterial capacity of bee pollen varies according to the microorganism. Generally, Gram-positive bacteria are more sensitive to antibacterials than Gram-negatives, with some studies highlighting *S. aureus* as the microorganism most sensitive to bee pollen (Graikou et al. 2011; Khider et al. 2013; Pascoal et al. 2014; De-Melo 2015; Borycka et al. 2016). The greater resistance of Gram-negative bacteria seems related to the external membrane that covers the cell, whose chemical structure can restrict the passage of certain substances (Morais et al. 2011).

The solvent/dilution method utilized in the extraction of compounds also influences the results. With methanolic extracts (99.9% methanol) a greater inhibitory effect on *E. coli* was observed, while with ethanolic extracts (70% ethanol), the greatest effect was seen against a fungus (Kacániová et al. 2012). These variations can be related to compounds extracted by each solvent/dilution, a fact which was also observed by Graikou et al. (2011) in previous studies. Fatrcová-Šramková et al. (2013) evaluated ethanolic and methanolic extracts from bee pollen and concluded that those with an ethanol base showed greater activity.

Pascoal et al. (2014) tested, using the microdilution method on microplates, eight methanolic bee pollen extracts from Portugal and Spain on the inhibition of *Staphylococcus aureus*, *Pseudomonas aeruginosa*, and *Escherichia coli*. The *S. aureus* strains were the most vulnerable, with MIC values between 1.81 and 4.28 mg/ mL. For the other microorganisms, the authors observed between 3.71–6.96 mg/mL (*P. aeruginosa*) and 4.08–9.42 mg/mL (*E. coli*). In samples collected in five Portuguese nature reserves, Morais et al. (2011) identified that the methanolic extracts were more effective against Gram-positive bacteria (*S. aureus* and *Bacillus cereus*) than against Gram-negative bacteria (*Salmonella typhi* and *E. coli*). The method used was microdilution on microplates.

Monofloral samples collected in Montesinho Nature Reserve in Portugal were extracted in *n*-hexane and tested using the agar well diffusion method. The lipophilic extract of *Cistus ladanifer* bee pollen was the most effective at inhibiting bacterial growth, followed by the *Castanea sativa* bee pollen extract and then by lipophilic extract of *Rubus* sp. bee pollen. The extracts were essentially composed of fatty acids, sterols, long-chain aliphatic alcohols as well as smaller amounts of alkanes and alkenes. Fatty acids, such as linoleic acid, are notably antimicrobial compounds (Barbosa et al. 2006).

In Egypt, monofloral bee pollen was extracted with methanol. Extracts were analyzed using agar plate diffusion and the MIC was determined by dilution. Different results were obtained depending on the botanical origin of the samples of *Trifolium alexandrinum* bee pollen (*E. coli*, *S. aureus,* and *Listeria monocytogenes*: 320 μg/mL; *Salmonella enteritidis*: 640 μg/mL; *P. aeruginosa*: 1280 μg/mL) and *Zea mays* bee pollen (*S. enteritidis*, *E. coli*, *S. aureus* and *L. monocytogenes*: 320 μg/mL; *P. aeruginosa*: 640 μg/mL) (Khider et al. 2013). Mohdaly et al. (2015) also evaluated the methanol extracts of the monofloral bee pollen from *Z. mays* produced in Egypt, however they encountered greater values for MIC: 0.30 mg/mL (*L. monocytogenes*), 0.78 mg/mL (*S. aureus*), 1.25 mg/mL (*E. coli*) and 1.35 mg/mL (*Salmonella enterica*).

Fatrcová-Šramková et al. (2013) analyzed different extracts of fresh (in natura) monofloral samples from Slovakia using the agar well diffusion method. After 24 h of incubation, the sensitivity of the microorganisms to the ethanolic extract (ethanol 70%) from the bee pollen from *Papaver somniferum* followed the sequence: *S. aureus* (*Sa*) > *E. coli* (*Ec*) > *L. monocytogenes* (*Lm*) > *Salmonella enterica* (*Se*) > *P. aeruginosa* (*Pa*); to the ethanolic extract of *Brassica napus subs*p. *napus*: *Lm* > *Pa* and *Se* > *Sa* and *Ec*; and to the extract of *Helianthus annuus*: *Se* > *Sa* > *Lm* and *Ec* > *Pa.*

De-Melo (2015) evaluated methanol extracts from monofloral and heterofloral bee pollen extracts produced in Brazil by the microplate and microdilution methods. With the Gram-positive strains, the author observed the following MIC values: MIC: 1.1–6.5 mg/mL (*Streptococcus pyogenes*) and 1.6–7.2 mg/mL (*S. aureus*). In relation to the Gram-negative strains, the MIC values were the following: 1.9–8.4 mg/ mL (*E. coli*) and 2.4–9.4 mg/mL (*Klebsiella* sp.). Generally, a monofloral sample from *Mimosa scabrella* produced in south Brazil had greater antibacterial activity against all the microorganisms tested, and the heterofloral sample with predominance of pollen from *Mimosa caesalpiniaefolia* and *Myrcia*, produced in the southeast region, showed less antibacterial action.

In another study with Brazilian samples, all collected in south Brazil, there was no inhibitory or bactericidal activity of the ethanolic extracts (ethanol 70%) against the bacteria tested (Carpes et al. 2009). In Turkey, Basim et al. (2006) observed that methanolic extracts (bee pollen from Turkey) inhibited the growth of 13 different species of agricultural bacterial pathogens, with *Agrobacterium tumefaciens* being the most sensitive and *Pseudomonas syringae* the most resistant. These results suggest that the bee pollen extract can be an alternative to protect seeds and in this way avoid the transmission of these pathogens.Methanolic extracts from bee pollen at concentrations between 0.02% and 2.5% were investigated on different bacteria (*B. cereus, B. subtilis, E. coli, S. typhimurium, S. aureus, Y. enterocolitica, Enterococcus faecalis, L. monocitogenes*), yeasts (*Saccharomices cerevisiae, Candida rugosa*) and two molds (*Aspergillus niger* and *Rhyzopus oryzae*). In the used concentrations, pollen extracts had no antimicrobial effect on mentioned pathogenic and food-related microorganisms (Erkmen and Özcan 2008).

Seven regions of Turkey were used for pollen collection in a study from 2021 (Özkök et al. 2021), where the bioactive compounds and antimicrobial activity against different bacterial pathogens were determined. The minimum inhibitory concentration (in mg/mL) of the pollen extracts ranged between 16 to higher than 128 mg/mL. All tested Gram-positive bacteria (*S. aureus*, MRSA, *E. faecalis*) had MIC value of 16 mg/mL, while tested Gram-negative bacteria (*K. pneumonia* and *P. aeruginosa*) exhibits MIC values >128 mg/mL.

In Tunisia, different extracts of monofloral bee pollen of *Phoenix dactylifera* were effective in inhibiting the growth of *Bacillus subtilis, B. cereus, S. aureus, Enterococcus faecalis, Micrococcus luteus, L. monocytogenes, Salmonella enteric serotype* Enteritidis, *Salmonella enteric serotype* Typhimurium, *E. coli,* and *Klebsiella pneumoniae* by agar well diffusion (Daoud et al. 2015).

Samples collected in Morocco showed potential activity against the growth of *S. aureus, B. cereus, P. aeruginosa*, and *E. coli* strains, using the agar well diffusion method. These strains were isolated from human samples and were resistant to antibiotics. There was no extraction of compounds and the samples were only diluted (Abouda et al. 2011). Cabrera and Montenegro (2013) tested, by agar diffusion method, aqueous extracts from bee pollen produced in Chile. The MIC for *S. aureus* and *S. pyogenes* was 21 mg/mL, and the MIC for *P. aeruginosa* and *E. coli* was 41 mg/mL and 82 mg/mL, respectively.

In Greece, Graikou et al. (2011) evaluated methanolic extracts from bee pollen by agar dilution technique. The MIC varied from 0.74 mg/mL (*S. aureus*) to >10 mg/ mL (*E. coli*). These extracts were rich in sugars, fatty acids, fatty acid esters, and phenolic acids (*p*-coumaric acid, ferulic acid, their glycerol esters and glycerol ester of caffeic acid). The flavonoids isolated from extracts were more effective against Gram-positive bacteria (*S. aureus* and *S. epidermidis*) than against Gram-negative bacteria (*E. coli*, *Enterobacter cloacae*, *Klebsiella pneumoniae,* and *Pseudomonas aeruginosa*).

Kačániová et al. (2014) determined the antibacterial activity of ethanolic extracts (unknown dilution) of bee pollen from Slovakia, by using agar disc diffusion method, against the *Clostridium* genus. Extracts with pollen concentrations of 25% and 50% were the best to inhibit the growth of the five species (*C. butyricum, C. histolyti-cumm C. intestinale*, *C. perfringens,* and *C. ramosum*).

The antibacterial action of bee pollen has been tested in *in vivo*. Kročko et al. (2012) evaluated the antibacterial action of bee pollen in the gastrointestinal tract of broiler chickens. The product was included in the food of the birds, seeking to improve feed conversion, performance, growth and health of broiler chickens through mechanisms associated with gastrointestinal tract and bacteria colonization. There was a desirable reduction in the bacterial count for the Enterobacteriaceae family in the gastrointestinal tract.

Pollens of *Quercus ilex*, *Centaurium erythraeae*, *Punica granatum*, *Citrus aurantium*, *Ruta graveolens*, and *Coriandrum sativum*, collected from the city of Fez, Morocco (Bakour et al. 2021) were investigated against different Gram-positive and Gram-negative bacteria (*Acinetobacter baumannii*, *Enterobacter cloacae*, *Escherichia coli*, *Klebsiella pneumonae*, *Pseudomonas aeruginosa*, *Staphylococcus aureus*). The best activities were recorded for *Punica granatum* and *Quercus ilex* pollen extracts, which exhibit antibacterial activity against all bacteria (MIC(MBC) 0.31(0.31)–2.5(>2.5) and 0.31(0.31)–2.5(2.5) respectively).

An interesting study (Fernandes et al. 2024) investigated the relationship between hive health, bee-associated microbial communities, and the antimicrobial activity of the collected pollen. Three healthy hives and three hives experiencing signs of stress including absent queens, low bee population, and signs of chalkbrood were used for pollen collection. Ethanol (80%) was used as solvent for pollen extraction (from healthy and stressed beehives), and tested on eight yeast species (*Candida albicans* SC5314, *Candida metapsilosis* M1879, *Candida parapsilosis* ATCC22018, *Candida tropicalis* M230640, *Cryptococcus deuterogattii* R265, *Cryptococcus neoformans* H99, *Kluyveromyces marxianus* M1896, and *Meyerozyma guillierondii* AMMRL1830), eight mould species (*Alternaria alternata* ACC02, *Ascosphaera apis* TCI01, *Aspergillus flavus* ATCC204304, *Aspergillus fumigatus* ATCC204305, *Aspergillus lentulus* DTO341-G7, *Fusarium oxysporum* ACC06, *Penicillium* sp MCC01, and *Mucor circinelloides* 80–14–240–5007), and two bacteria species (*Staphylococcus aureus* ATCC29213 and *Paenibacillus larvae* EMAI02). Also in this study, pollen extracts exhibit strongly species-specific antifungal activity only against *Cryptococcus deuterogattii*, *Cryptococcus neoformans*, *Alternaria alternata*, *Ascosphaera apis*, *Aspergillus flavus,* and *Aspergillus fumigates.* On average, pollen from healthy hives was more active compared to those obtained from stressed hives against every susceptible species.

In the search for medications with a natural compound base, Olczyk et al. (2016) tested preparations based on bee pollen extracts and pharmaceutical products in the treatment of burns inflicted on swine. The use of preparations with a bee pollen base not only reduced the time for scarring and affected positively the general state of the animals, but it also prevented infection caused by microorganisms of recently formed tissue.

12.3.2 Antitumor Activity

Phenolic compounds from plant-based materials are responsible for their cytotoxic activities. Moreover, the types and concentrations of these natural compounds are determinants in the inhibition of growth and proliferation (invasion) of cancer cells in animals and human bodies (Aylanc et al. 2023). Bee pollen has shown greater or lesser antitumor potential against some types of cancers. This potential can be related to the antioxidant capacity of the compounds that act by removing or deactivating reactive species. There may also be a relation between the induction of apoptosis of carcinogenic cells and stimulation of the secretion of tumor necrosis factor-alpha (TNF-), or further still, with the strengthening of the immunological system (Denisow and Denisow-Pietrzyk 2016). Beyond this, this product seems to be able to reduce the toxic effects associated with the use of pharmaceuticals utilized in the treatment of cancer (Furusawa et al. 1995; Pinto et al. 2010).

The antitumor potential of bee pollen is related to the presence of certain compounds. Quercetin, a flavonoid of great prevalence in this matrix, is capable of blocking the expression of specific genes in tumor cells. Rutin and chyrisin can also act in the prevention of cancer. Kaempferol, another fairly common flavonoid, caused a reversible growth inhibition of PC-3 cancer cells. There is evidence suggesting that phytosterols, as well as β-carotene, can reduce the risk of some prostate carcinoma (Bogdanov 2012).

Apigenin, a flavonoid of the flavone class with promising anticancer activity (Shukla and Gupta 2010), was identified by De-Melo (2015) in monofloral bee pollen from *Myrcia* from the State of Rio Grande do Sul, Brazil. This compound is commonly found in fruits and vegetables and has an antitumor action. The author also reported the presence of 4-methylsulfonyl-butyl glycosinolate (glucoraphanine) in monofloral samples of *Alternanthera* (São Paulo State, Brazil), *Mimosa caesalpiniaefolia* (Sergipe State, Brazil) and *Mimosa scabrella* (Santa Catarina State, Brazil).

Aliyazicioglu et al. (2005), in research with heterofloral samples collected in Turkey, observed that bee pollen extracts (dimethyl sulfoxide extracts at concentrations of 50, 25, and 12.5 mg/mL) suppress respiratory burst within cancer cell lines (K-562).

The effects of the administration of the water phase of extracts from bee pollen were studied in syngeneic mice with Lewis lung carcinoma implanted intraperitoneally. The extracts were not cytotoxic in cell cultures at concentrations of up to 2.5 mg/ mL, and significantly prolonged the life expectancy of mice carrying the tumor without any apparent side effects at 0.5 g/kg. When administration occurred in conjunction with standard cytotoxic antitumor drugs, there was a significant increase in the survival rate of the animals, with some managing to survive in a tumor-free state until the end of the experiment. The antitumor action of the extracts seems related to the immune stimulation of the macrophages (Furusawa et al. 1995).Wu and Lou (2007) tested in vitro the activity of steroid fraction of a chloroform extract from bee pollen of *Brassica campestris* (China) against nine human cancer cell lines: androgen-independent prostate cancer PC-3, estrogen-responsive breast adenocarcinoma MCF-7, androgen-sensitive prostate cancer LNCaP, human cervix carcinoma HeLa, gastric adenocarcinoma BCG-823, hepatocarcinoma BEL-7402, squamous carcinoma KB, lung carcinoma A549 and ovarian carcinoma HO8910. The PC-3 cells were more sensitive than the others, which led the authors to recommend that the bee pollen extract from *B. campestris* should be further studied as a possible treatment for prostate cancer.

In other studies undertaken with monofloral bee pollen from *Cistus* (*C. ladaniferus*) (origin unknown), the administration of aqueous extracts (10 mg/100 g of body weight, administered orally) on rats submitted to ovariectomy resulted in a smaller decrease in mineral content, mineral density, and polar strength strain index in the femoral-metaphyseal tissues (Yamaguchi et al. 2007a); and the administration of these extracts (10 or 20 mg/100 g of body weight, administered orally) for streptozotocin-diabetic rats had a preventive effect on bone loss, beyond reducing other negative effects associated with the use of streptozotocin, such as the increase in calcium levels and a reduction of inorganic phosphorus levels in the serum (Yamaguchi et al. 2007b).

Murakami et al. (2008) utilized ethanolic extracts (95% ethanol) from bee pollen with a predominance of the genus *Cistus* (origin unknown) to investigate the efficacy and safety of 12-week intake of honeybee-collected pollen lump extract (HPLE)-supplemented food in 44 patients with benign prostatic hyperplasia. Especially in patients who received a larger dose (320 mg/day), there was improvement in some symptoms associated with benign prostatic hyperplasia.The ethanolic extract (95% ethanol) from monofloral bee pollen from *Cistus* (*C. ladanifer* and *Cistus albidus*), produced in Spain, at a concentration of 300 μg/ mL, showed suppressive effects against VEGF-induced angiogenesis in a study with human umbilical vein endothelial cells. Vascular endothelial growth factor (VEGF) is a key regulator of pathogenic angiogenesis in diseases such as cancer (Izuta et al. 2009).

Aqueous extracts of bee pollen from Beni-Suef, Egypt, were tested as antimutagenic agents against cisplatin-induced chromosomal abnormalities in bone marrow cells of mice (*Mus musculus*). The animals received, for one week, injections of cisplatin [*cis*-diammine-dichloro-platinum (II)] (CDDP), a cytotoxic antineo-plastic drug, that causes oxidative stress and a series of adverse effects. Subsequently, they received daily doses of pollen extract (140 mg/kg, orally) for 14 days. There was a significant reduction in the incidence of chromosomal damage induced by CDDP in bone marrow cells and an increase in the frequency of the mitotic indices of bone marrow cells, showing a chemoprotective effect of bee pollen against CDDP-induced genotoxicity (Abdella et al. 2009). In research carried out by Tohamy et al. (2014), also with samples of bee pollen from Egypt, similar results were found. Wang et al. (2013) extracted, fractionated, and purified polysaccharides of monofloral bee pollen from *Rosa rugosa* in Beijing, China. In this study, which was carried out in vitro, the authors found anti-proliferative activity in human colon cancer HT-29 and HCT116 cells. The extracts also showed a concentration-dependent proliferation-inhibitory effect.

In a study carried out on bee pollen from Spain and Portugal, Pascoal et al. (2014) tested the anti-genotoxic effects of methanolic extracts on *Saccharomyces cerevisiae* exposed to ethyl methanesulfonate, which produces random mutations in genetic material by nucleotide substitution. All samples showed antigenotoxic activity at concentrations of up to 0.75 mg/mL, although some were more efficient at reducing the number of gene conversion colonies and only two samples, one from Portugal and the other from Spain, significantly reverted the mutations.

Yang et al. (2014) induced, using testosterone injections, a benign prostatic hyperplasia in Sprague-Dawley rats to evaluate the effects of the administration of *B. campestris* bee pollen supercritical CO2 fluid extract (SFE-CO2) at concentrations of 21.3 or 88.7 mg/kg. Treated animals showed a significant reduction in prostatic index, as well as the DHT, 5α-reductase and COX-2 expression levels, showing that the extracts inhibited the development of the benign prostatic hyperplasia.

Ethanolic (70% ethanol) and aqueous extracts from bee pollen produced in Egypt show greater anticancer activity against hepatocellular carcinoma (HerpG2) cells and less activity against breast adenocarcinoma (MCF-7) cells. The authors also analyzed propolis extracts and in these, the activity was greater (Shady et al. 2016).

In another experiment with bee pollen from *B. campestris* produced in China, Chen et al. (2016) identified, in the blood of male ICR strain mice feed on the samples, two microRNAs (miR-166a and miR-159); it was further observed that some miRNAs present in the plants could have antitumor effects. Chin et al. (2016) concluded that the abundance of miR-159 in human serum was inversely correlated with breast cancer incidence and progression in patients.

Strong antiproliferative effects were observed when human bronchogenic carcinoma cells were treated with hydrolysates from bee pollen (Saisavoey et al. 2021). Commercial bee pollen from Thailand was subjected to hydrolyzation using different enzymes. The obtained fractions were subjected to antioxidant activity determination and further purification by RP-HPLC. Also, amino acid sequence was determined using electrospray (ESI)-Q-TOF-MS/MS technique. The study concluded that bee pollen protein hydrolysates may give selectively antiproliferative effects for ChaGo-K1 lung cancer cells.

Six monofloral Taiwanese bee pollens purity (>98%) were investigated by Nguyen et al. (2022) for their anticancer activity against three human cancer lines (MCF-7, BT-20 and Hs578T). Using MTT assay, pollen from *Bidensis pilosa* exhibit the highest anticancer activity, with cell viability of 43.5% (MCF-7), and 0% for BT-20 and Hs 578 T, being a promising anticancer agent for medicinal application.

Different mono and multifloral pollen samples from Morocco were characterized and anticancer activity against human-cancer-derived cell lines (stomach gastric adenocarcinoma AGS, epithelial colorectal adenocarcinoma CaCo2, cervical carcinoma HeLa, non-small-cell lung cancer NCl-H460 and breast adenocarcinoma MCF-7) was determined. *Coriandrum* and *Daucus* sp., together with *Olea europea* monofloral bee pollen show cytotoxic effects exclusively against HeLa and MCF-7 cell lines (GI_{50} 495 μg/mL and GI_{50} 734 μg/mL respectively). This was associated with the presence of some quercetin and kaempherol glycosides in the chemical composition (Aylanc et al. 2023).

Due to the fact that colorectal cancer is one of the most spread cancer types in the world, numerous in vitro studies have been made on this type of cancer. It is known and documented that phenolics are good anticancer agents, and bee pollen is one natural product with high amounts of phenolics. A review of the studies made on this type of cancer (Moskwa et al. 2023) demonstrates that bee pollen, as well as other bee products, have a potential role in the treatment, and also in the chemoprevention of this type of cancer.

12.4 Hepatoprotective and Antidiabetic Activity

12.4.1 Hepatoprotective Activity

In 1997, Bevzo and Grygor'eva observed that by introducing bee pollen into mice diet, their livers normalized the activity of glutathione system enzymes, preventing damage that was expected from the exposure of the animals to radiation. Eraslan et al. (2008) tested the effects of daily administration of a water-solubilized extract of monofloral bee pollen from *Brassica napus* (Kayseri, Turkey), at doses of 50 or 100 mg/kg, to female Wistar albino rats, hours after receiving 225 mg/kg of carbaryl. At the end of the experiment, the inclusion of bee pollen in the diet of Wistar rats seems to alleviate the toxic effects of oxidative stress caused by a concomitantly administered pesticide. The results of the study released by Saríc et al. (2009), in which bee pollen from Croatia was included in the diet of rats, suggested that this product is a source of compounds with health protection potential.

Chestnut bee pollen aqueous extracts from Zonguldak, Turkey were investigated for its hepatoprotective properties (Yıldız et al. 2013). Sprague-Dawley rats were used in the experiment, being divided into seven groups (G1-G7): three groups (G1-G3) of control (receiving i.p. saline solution, olive oil, and ethanol). The experimental groups (4 groups) receive i.p. CCl_4 in olive oil, G4 positive control, G5 was fed with silibinin in ethanol and the last two groups received two doses of pollen. Different liver enzyme activities (AST, ALT, SOD, MDA) and proteins from blood samples of the treated animals were measured by different biochemical methods. Also, the detection of the apoptotic cells from liver samples was made using TUNEL Analysis (Terminal deoxynucleotidyl transferase dUTP nick end labeling analysis), to detect the fragmented DNA associated with the apoptotic cells. The percentage of TUNEL-positive apoptotic cells was calculated, representing the apoptotic index. AST and ALT enzymes were high in the groups of animals treated with CCl_4 due to the toxic effect of the substance. In the case of silibinin and pollen-treated groups, the activities of the enzymes were reduced. Malondialdehyde (MDA), a lipid peroxidation product, was measured in plasma, erythrocyte, and liver tissues of treated animals to determine the damage caused by oxidative stress induced by CCI_4 treatment application. No significant change was detected in the MDA levels of the control groups, but the level was significantly increased in the positive control group. A significant decrease was noticed in the MDA levels of the animals, exposed to silibinin and also the pollen groups. No significant difference was detected in SOD activities of plasma, liver, and erythrocyte among control groups, but SOD activity was decreased in the CCI_4-exposed groups. The activity slightly increased in the groups treated with silibinin and pollen. Hepatocyte apoptosis was analyzed using TUNEL assay. A number of apoptotic nuclei were much higher in the positive control group (G4) compared to control groups; however, the number of apoptotic nuclei of silibinin- and pollen-administered groups was much reduced compared to the CCI_4 treated group.

Concluding, chestnut bee pollen could be used as a suitable alternative to silibinin in the treatment of hepatocellular pathologies (Yıldız et al. 2013).

Tohamy et al. (2014) investigated the antioxidant activity of water extracts from Egyptian bee pollen on cisplatin-induced hepatic, renal, testicular, and genotoxicity in male albino mice (*Mus musculus*). The inclusion of extracts into the diet, at doses of 140 mg/kg for 14 days, resulted in a significant reduction in the level of lipid peroxidation and the elevation in catalase activity and glutathione concentration. This effect was most pronounced in the testes. Saral et al. (2016) observed that the inclusion of bee pollen in the diet of rats effectively reduced liver damage caused by the injection of carbon tetrachloride, and that this effect seems related to the antioxidant potential of this food.

Starting from the fact that red pine pollen (*Pinus brutia* Ten.) is used against liver disease in Anatolian folk medicine; Gök et al. (2021) initiated a preclinical study on the hepatoprotective effect of this pollen extract on mice, also analyzing the phenolics of the matrix. Swiss albino mice were used in the study and concentrations of 100, 200, and 300 mg/kg (b.w.) pollen extract were fed to the animals, compared to silymarin-treated group as a reference. Different biochemical parameters from the hepatic homogenate were determined at the end of the study. The obtained data allow the authors to conclude that the red pine pollen extract displays dose-dependent hepatoprotection against CCl4-induced liver damage, this type of pollen being recommended as a promising hepatoprotective agent.

12.4.2 Antidiabetic Activity

Type 1 diabetes (T1DM) is an autoimmune disorder where insulin-producing pancreatic β-cells are not functional as a result of increase in the blood glucose level (Katsarou et al. 2017). Type 2 diabetes (T2DM) is a form of diabetes mellitus characterized by chronic high glucose in the blood stream, insulin resistance, and also a lack of this hormone which regulates blood sugar levels (Setyawan et al. 2024).

As mentioned before, insufficient pancreatic β-cell are involved in both types of diabetes (Yang et al. 2021). MIN6 cells (mice pancreatic β-cell line) were treated with 0.01, 0.1 and 1.0 mg/mL polysaccharide solution (RBPP-P) obtained from *Rosa rugosa* bee pollen. Cell viability MTT assay, insulin-secretion assay and qRT-PCR assay were prepared for analysis of alloxan induced T1DM mice. The experiment demonstrate that RBPP-P modulates β-cell production and function. Oral administration of RBPP-P exhibit potential hypoglycemic activity, improving diabetes symptoms, protecting the pancreas, the producer of β-cells and glucose homeostasis. Concluding, RBPP-P could be used as a natural functional supplement for the prevention and treatment of T1DM (Yang et al. 2021).

Zhang et al. (2022) conducted a study on the bioactive components of *Fagopyrum aesculentum* Moech bee pollen extract on high-fat-diet and streptozocin induced diabetes mellitus type 2. Different techniques were used to identify, quantify, and separate a number of flavonoids and phenolic acids from bee pollen extract (FBPE), known for their bioactive properties, anti-diabetes being one of them. After 8 weeks, blood samples were collected from the experimental groups (diabetic group, positive Metformin (MET)-treated group, and low-dose and high-dose pollen extract). Serum concentrations of insulin, total cholesterol (TC), triglyceride (TG), low-density lipoprotein-cholesterol (LDL-C), high-density lipoprotein-cholesterol (HDL-C), cereal third transaminase (ALT), aspartate transaminase (AST), alkaline phosphatase (ALP), albumin, and the levels of hepatic glycogen, TC, TG, total protein (TP), superoxide dismutase (SOD), glutathione peroxidase (GSH-Px) and malondialdehyde (MDA) were determined. The homeostasis model assessment of insulin resistance (HOMAIR), insulin sensitivity index (ISI), and homeostasis model assessment-β (HOMA-β) were calculated according to Gao et al. (2015) and Yuan et al. (2019). In FBPE-treated mice's serum biochemical analysis, TG, LDL-C were decreased, HDL-C level was increased, demonstrating that bee pollen extract could prevent lipid accumulation in blood serum. ALT and AST activity was decreased in FBPE treated mice as well as in the MET group. Results revealed that both high-dose FBPE and MET could protect liver cells and liver injury in T2DM mice. Also, high dose of FBPE increased the SOD activity after 8 weeks and GSH-Px activity was improved after low-dose FBPE treatment.

12.5 Other Health Benefits

Analgesic Effects

Bee pollen's analgesic effects are primarily attributed to its ability to inhibit cyclooxygenase (COX) enzymes, particularly the pro-inflammatory COX-2, as demonstrated by (Oszczędłowski 2023) who reported significant suppression of COX-2 activity by bee pollen extracts in inflammatory conditions. This inhibition leads to a reduction in the production of prostaglandins, which mediates pain and inflammation and allows bee pollen to effectively reduce pain associated with inflammatory conditions.

Furthermore, Oliveira Lopes et al. (2019) demonstrated that extracts from pollen collected by stingless bees exhibit both anti-inflammatory and antinociceptive activities. Their findings suggest that the anti-inflammatory mechanism involves COX-2 inhibition and a consequent reduction in prostaglandin synthesis. This aligns with the broader understanding that many natural products with anti-inflammatory properties also possess analgesic effects, making bee pollen a potential candidate for pain management therapies. In addition to its effects on COX enzymes, bee pollen is rich in antioxidants, which play an important role in reducing oxidative stress and inflammation Tohamy et al. (2014) highlighted the antioxidant capacity of bee pollen, reporting its ability to scavenge free radicals and inhibit lipid peroxidation. This antioxidant activity not only contributes to overall health but also supports the analgesic effects by mitigating oxidative damage that can exacerbate pain and inflammation. Moreover, the presence of flavonoids such as quercetin and kaempferol in bee pollen has been linked to its analgesic properties (Oliveira Lopes et al. 2019). These compounds have been shown to possess anti-inflammatory effects and may modulate pain pathways in the body (Oszczędłowski 2023).

Immunomodulatory Effect
Bee-collected pollen, often referred to as bee pollen, has been recognized for its potential immunomodulatory effects on human health. This natural product is rich in nutrients, including proteins, lipids, vitamins, and minerals, which contribute to its health-promoting properties (Khalifa et al. 2021). The immunomodulatory effects of bee pollen can be attributed to its diverse biochemical composition, which includes significant amounts of essential amino acids, vitamins, minerals, polyphenols, and flavonoids known for their antioxidant and anti-inflammatory properties (Nascimento and Luz 2018), which may contribute to its potential as an immunomodulatory substance (Tomaszewska et al. 2020). Studies indicate that the polysaccharides found in bee pollen can stimulate immune responses, thereby enhancing the overall health of various organisms, including humans and animals (Wang et al. 2013; Yang et al. 2021). A key mechanism through which bee pollen exerts its immunomodulatory effects is by promoting the production of immunoglobulins, especially IgA, with research demonstrating that specific *Lactobacillus* strains derived from bee pollen can substantially enhance IgA synthesis, thereby supporting mucosal immunity (Komosinska-Vassev et al. 2015). This immunoglobulin is vital for the first line of defense against pathogens entering through mucosal surfaces. Additionally, the antioxidant properties of bee pollen, attributed to its high content of flavonoids and phenolic compounds, further bolster its immunomodulatory effects by reducing oxidative stress, which is known to impair immune function (Asama et al. 2015; Klarić et al. 2018).

Furthermore, the impact of bee pollen on the immune system is not limited to enhancing antibody production, it has been observed that bee pollen supplementation can improve cellular immunity by increasing lymphocyte proliferation and enhancing the activity of various immune cells (Kurek-Górecka et al. 2018; Attia et al. 2019). In animal models, the administration of bee pollen has resulted in improved splenic and hepatic histology, indicating a healthier immune response and better overall physiological state (Attia et al. 2019). This is particularly relevant in the context of stressors such as infections or nutritional deficiencies, where the immune system's ability to respond effectively is crucial.

Its anti-inflammatory and antimicrobial properties are well-documented, making it a valuable component in addressing a wide range of health conditions (Di Pasquale et al. 2013; MacInnis 2023) Bee pollen has been utilized in wound healing due to its ability to facilitate granulation and tissue regeneration, further underscoring its multifaceted health benefits (Hesami and Haghighian 2019). The presence of bioactive compounds such as caffeic acid phenethyl ester in bee pollen also contributes to its anti-inflammatory effects, which can help mitigate chronic inflammatory conditions

(MacInnis 2023). The synergistic effects of bee pollen with other bee products, such as propolis, have also been explored. Both substances share similar bioactive components, which can enhance their individual immunomodulatory effects when used in combination (Kurek-Górecka et al. 2018; Bhat 2023).

Moreover, the immunomodulatory effects of bee pollen may vary depending on its botanical origin and the specific compounds present, the study of Soares de Arruda et al. (2020) highlighted that different types of bee pollen exhibit varying levels of phenolic content and antioxidant properties, which can influence their effectiveness in modulating immune responses. This suggests that the health benefits of bee pollen may be optimized by selecting specific types based on their immunomodulatory profiles.

Antiradiation Effect

Bee pollen may exert its antiradiation effects through its antioxidant properties, antioxidants play a crucial role in neutralizing free radicals generated by radiation exposure, which can lead to cellular damage and increased risk of cancer (Darwish et al. 2022).

According to Denisow and Denisow-Pietrzyk (2016) bee pollen exhibits various therapeutic properties, including antioxidant and anti-inflammatory effects, which are crucial for cellular protection and recovery following radiation exposure. This suggests that regular consumption of bee pollen may bolster the body's resilience against the harmful effects of radiation. Inflammation is a common response to cellular damage, and chronic inflammation can exacerbate the harmful effects of radiation, and by reducing inflammation, bee pollen may help mitigate some of the adverse effects associated with radiation exposure (Alkhatib et al. 2017; Filannino et al. 2021). Denisow and Denisow-Pietrzyk (2016) indicated that the anti-inflammatory compounds in bee pollen can inhibit the production of pro-inflammatory cytokines, thereby promoting a healthier cellular environment which may be beneficial in the context of radiation exposure.

Research into the potential antiradiation effects of bee pollen is still in its early stages, and while preliminary findings are promising, further studies are needed to fully elucidate the mechanisms and efficacy of bee pollen in this context.

References

Abdella EM, Tohamy A, Ahmad RR (2009) Antimutagenic activity of Egyptian propolis and bee pollen water extracts against cisplatin-induced chromosomal abnormalities in bone marrow cells of mice. Iran J Cancer Prev 2:175–181

Abouda Z, Zerdani I, Kalalou I, Faid M, Ahami MT (2011) The antibacterial activity of Moroccan bee bread and bee-pollen (fresh and dried) against pathogenic bacteria. Res J Microbiol 6:376–384

Aliyazicioglu Y, Deger O, Ovali E, Barlak Y, Hosver I, Tekelioglu Y (2005) Effects of Turkish pollen and propolis extracts on respiratory burst for K-562 cell lines. Int Immunopharmacol 5:1652–1657. https://doi.org/10.1016/j.intimp.2005.04.005

Alkhatib A, Tsang C, Tiss A, Bahorun T, Arefanian H, Barake R, Khadir A, Tuomilehto J (2017) Functional foods and lifestyle approaches for diabetes prevention and management. Nutrients 9(12):1–18. https://doi.org/10.3390/nu9121310

Al-Yousef HM, Amina M, Alqahtani AS, Alqahtani MS, Malik A, Hatshan MR et al (2020) Pollen bee aqueous extract-based synthesis of silver nanoparticles and evaluation of their anti-cancer and anti-bacterial activities. PRO 8:524. https://doi.org/10.3390/pr8050524

Asama T, Arima T-H, Gomi T, Keishi T, Tani H, Kimura Y, Tatefuji T, Hashimoto K (2015) Lactobacillus kunkeei YB38 from honeybee products enhances IgA production in healthy adults. J Appl Microbiol 119(3):818–826. https://doi.org/10.1111/jam.12889

Attia YA, Bovera F, Abd-Elhamid AE-H E, Calabrò S, Mandour M, Al-Harthi MA, Hassan SS (2019) Evaluation of the carryover effect of antibiotic, bee pollen and propolis on growth performance, carcass traits and splenic and hepatic histology of growing rabbits. J Anim Physiol Anim Nutr (Berl) 103(3):947–958. https://doi.org/10.1111/jpn.13068

Aylanc V, Larbi S, Calhelha R, Barros L, Rezouga F, Rodríguez-Flores MS et al (2023) Evaluation of antioxidant and anticancer activity of mono- and polyfloral moroccan bee pollen by characterizing phenolic and volatile compounds. Molecules 28(2):835. https://doi.org/10.3390/molecules28020835

Bakour M, Laaroussi H, Ousaaid D, Oumokhtar B, Lyoussi B (2021) Antioxidant and antibacterial effects of pollen extracts on human multidrug-resistant pathogenic bacteria. J Food Qual 2021:11. https://doi.org/10.1155/2021/5560182

Barbosa SITR, Silvestre AJD, Simões MMQ, Estevinho MLMF (2006) Composition and antibacterial activity of the lipophilic fraction of honeybee pollen from native species of Montesinho Natural Park. Int J Agr Res 1:471–479. https://doi.org/10.3923/ijar.2006.471.479

Barreto LMRC, Funari SRC, Orsi RO (2005) Composição e qualidade do pólen apícola proveniente de sete Estados brasileiros e do Distrito Federal. Bol Ind Anim 62:167–175

Basim E, Basim H, Özcan M (2006) Antibacterial activities of Turkish pollen and propolis extracts against plant bacterial pathogens. J Food Eng 77:992–996. https://doi.org/10.1016/j.jfoodeng.2005.08.027

Bassard JE, Ullmann P, Bernier F, Werck-Reichhart D (2010) Phenolamides: bridging polyamines to the phenolic metabolism. Phytochemistry 71:1808–1824. https://doi.org/10.1016/j.phytochem.2010.08.003

Bevzo VV, Grygor'eva NP (1997) Effect of bee pollen extract on glutathione system activity in mice liver under X-ray irradiation. Ukr Biokhim Zh 69:115–117

Bhat C (2023) Immobilization of bee pollen extract on polyethylene terephthalate (PET) fabric for wound dressing. Biosci Biotechnol Res Asia 20(3):875–881. https://doi.org/10.13005/bbra/3139

Bogdanov S (2012) The bee pollen book: chapter 2. http://www.bee-hexagon.net/. Accessed 12 Nov 2014

Borycka K, Grabek-Lejko D, Kasprzyk I (2016) Antioxidant and antibacterial properties of commercial bee pollen products. J Apicul Res 54:491–502. https://doi.org/10.1080/00218839.2016.1185309

Cabrera C, Montenegro G (2013) Pathogen control using a natural Chilean bee pollen extract of known botanical origin. Cien Investig Agrar 40:223–230

Campos MGR, Bogdanov S, Almeida-Muradian LB, Szczesna T, Mancebo Y, Frigerio C (2008) Pollen composition and standardisation of analytical methods. J Apicult Res Bee World 47:156–163. https://doi.org/10.3896/IBRA.1.47.2.12

Carpes ST, Mourão GB, Alencar SM, Masson ML (2009) Chemical composition and free radical scavenging activity of *Apis mellifera* bee pollen from Southern Brazil. Braz J Food Technol 12:220–229

Castelo-Branco VN, Torres AG (2011) Capacidade antioxidante total de óleos vegetais comestíveis: determinantes químicos e sua relação com a qualidade dos óleos. Rev Nutr 24:173–187. https://doi.org/10.1590/S1415-52732011000100017

Chantarudee A, Phuwapraisirisan P, Kimura K, Okuyama M, Mori H, Kimura A (2012) Chemical constituents and free radical scavenging activity of corn pollen collected from *Apis mellifera* hives compared to floral corn pollen at Nan, Thailand. BMC Complement Altern Med 12:1–12. https://doi.org/10.1186/1472-6882-12-45

Chen X, Dai G, Ren Z, Tong Y, Yang F, Zhu Y (2016) Identification of dietetically absorbed rapeseed (*Brassica campestris* L.) bee pollen micrornas in serum of mice. Biomed Res Int 2016:1–5. https://doi.org/10.1155/2016/5413849

Chin AR, Fong MY, Somlo G, Wu J, Swiderski P, Wu X (2016) Cross-kingdom inhibition of breast cancer growth by plant miR159. Cell Res 26:1–12. https://doi.org/10.1038/cr.2016.13

Cunha AP, Roque OR (2009) Antibióticos de origem natural, compostos semi-sintéticos e sintéticos com actividade antibacteriana e antibióticos citotóxicos. In: Cunha AP (ed) Farmacognosia e fitoquímica. Lisboa, Fundação Calouste Benkian, pp 608–642

Daoud A, Malika D, Bakari S, Hfaiedh N, Mnafgui K, Kadri A (2015) Assessment of polyphenol composition, antioxidant and antimicrobial properties of various extracts of Date Palm Pollen (DPP) from two Tunisian cultivars. Arab J Chem 12(8):3075–3086. https://doi.org/10.1016/j.arabjc.2015.07.014

Darwish A, Abd El-Wahed A, Shehata M, El-Seedi H, Masry S, Khalifa S, Mahfouz H, El-Sohaimy S (2022) Chemical profiling and nutritional evaluation of bee pollen, bee bread, and royal jelly and their role in functional fermented dairy products. Molecules 28(1):227. https://doi.org/10.3390/molecules28010227

De-Melo AAM (2015) Perfil químico e microbiológico, cor, análise polínica e propriedades biológicas do pólen apícola desidratado. http://www.teses.usp.br/teses/disponiveis/9/9131/tde-18122015-142742/pt-br.php. Accessed 26 Feb 2017

De-Melo AAM, Estevinho MLMF, Sattler JAG, Souza BR, Silva-Freitas A, Barth OM (2016) Effect of processing conditions on characteristics of dehydrated bee-pollen and correlation between quality parameters. Food Sci Technol 65:808–815. https://doi.org/10.1016/j.lwt.2015.09.014

Denisow B, Denisow-Pietrzyk M (2016) Biological and therapeutic properties of bee pollen: a review. J Sci Food Agric 96:4303–4309. https://doi.org/10.1002/jsfa.7729

Di Pasquale G, Salignon M, Le Conte Y, Belzunces LP, Decourtye A, Kretzschmar A, Suchail S, Brunet J-L, Alaux C (2013) Influence of pollen nutrition on honey bee health: do pollen quality and diversity matter? PLoS One 8(8):e72016. https://doi.org/10.1371/journal.pone.0072016

Domenici V, Gabriele M, Parri E, Felicioli A, Sagona S, Pozzo L, Biondi C (2015) Phytochemical composition and antioxidant activity of Tuscan bee pollen of different botanic origins. Ital J Food Sci 27:248–259. https://doi.org/10.14674/1120-1770/ijfs.v191

Domínguez-Valhondo D, Gil DB, Hernández MT, González-Gómez D (2011) Influence of the commercial processing and floral origin on bioactive and nutritional properties of honeybee-collected pollen. Int J Food Sci Tech 46:2204–2211. https://doi.org/10.1111/j.1365-2621.2011.02738.x

Dulger Altiner D, Sandikci Altunatmaz S, Sabuncu M, Aksu F, Sahan Y (2021) In-vitro bioaccesibility of antioxidant properties of bee pollen in Turkey. Food Sci Technol 41:133–141. https://doi.org/10.1590/fst.10220

Eraslan G, Kanbur M, Silici S, Liman B, Altinordulu S, Karabacak M (2008) Evaluation of protective effect of bee pollen against propoxur toxicity in rat. Ecotoxol Environ Saf 72:931–937. https://doi.org/10.1016/j.ecoenv.2008.06.008

Erkmen O, Özcan MM (2008) Antimicrobial effects of Turkish propolis, pollen, and laurel on spoilage and pathogenic food-related microorganisms. J Med Food 11:587–592. https://doi.org/10.1089/jmf.2007.0038

Fatrcová-Šramková K, Nôžková J, Kačániová M, Máriássyová M, Rovná K, Stričík M (2013) Antioxidant and antimicrobial properties of monofloral bee pollen. J Environ Sci Health B 48:133–138. https://doi.org/10.1080/03601234.2013.727664

Féas X, Vázquez-Tato MP, Estevinho L, Seijas JA, Iglesias A (2012) Organic bee pollen: botanical origin, nutritional value, bioactive compounds, antioxidant activity and microbiological quality. Molecules 17:8359–8377. https://doi.org/10.3390/molecules17078359

Fernandes KE, Frost EA, Krantz M, Carter DA (2024) Pollen products collected from honey bee hives experiencing minor stress have altered fungal communities and reduced antimicrobial properties. FEMS Microbiol Ecol 100:fiae091. https://doi.org/10.1093/femsec/fiae091

Filannino P, Di Cagno R, Vincentini O, Pinto D, Polo A, Maialetti F, Porrelli A, Gobbetti M (2021) Nutrients bioaccessibility and anti-inflammatory features of fermented bee pollen: a comprehensive investigation. Front Microbiol 12:622091. https://doi.org/10.3389/fmicb.2021.622091

Freire KRL, Lins ACS, Dórea MC, Santos FAR, Camara CA, Silva TMS (2012) Palynological origin, phenolic content, and antioxidant properties of honeybee-collected pollen from Bahia, Brazil. Molecules 17:1652–1664. https://doi.org/10.3390/molecules17021652

Furusawa E, Chou SC, Hirazumi A (1995) Antitumour potential of pollen extract on lewis lung carcinoma implanted intraperitoneally in syngeneic mice. Phytother Res 9:255–259

Gao YF, Zhang MN, Wu TC, Xu MY, Cai HN, Zhang ZS (2015) Effects of d-pinitol on insulin resistance through the pi3k/akt signaling pathway in type 2 diabetes mellitus rats. J Agric Food Chem 63(26):6019. https://doi.org/10.1021/acs.jafc.5b01238

Gök HN, Gül H, Gülfraz M, Asad MJ, Öztürk N, Şanal F, Orhan IE (2021) Preclinical study on the hepatoprotective effect of Pollen extract of *Pinus brutia* ten (red pine) in mice and phenolic acid analysis. Turkish J Pharmaceut Sci 18(3):319–325. https://doi.org/10.4274/tjps.galenos.2020.47154

Graikou K, Kapeta S, Aligiannis N, Sotiroudis G, Chondrogianni N, Gonos E (2011) Chemical analysis of Greek pollen—antioxidant, antimicrobial and proteasome activation properties. Chem Cent J 5:1–9. https://doi.org/10.1186/1752-153X-5-33

Hanafy NA, Salim EI, Mahfouz ME, Eltonouby EA, Hamed IH (2023) Fabrication and characterization of bee pollen extract nanoparticles: their potential in combination therapy against human A549 lung cancer cells. Food Hydrocoll Health 3:100110. https://doi.org/10.1016/j.fhfh.2022.100110

Hesami S, Haghighian HK (2019) The therapeutic effects of bioactive compounds in honeybee products. J Qazvin Univ Med Sci 190:190–203. https://doi.org/10.32598/jqums.22.6.190

Huang WY, Zhang HC, Liu WX, Li CY (2012) Survey of antioxidant capacity and phenolic composition of blueberry, blackberry, and strawberry in Nanjing. J Zhejiang Univ 13:94–102. https://doi.org/10.1631/jzus.B1100137

Izuta H, Shimazawa M, Tsuruma K, Araki Y, Mishima S, Hara H (2009) Bee products prevent VEGF-induced angiogenesis in human umbilical vein endothelial cells. BMC Complement Altern Med 9:1–10. https://doi.org/10.1186/1472-6882-9-45

Kacániová M, Vukovic N, Chlebo R, Hascík P, Rovná K, Cubon J, Dzugan M, Pasternakiewicz A (2012) The antimicrobial activity of honey, bee pollen loads and beeswax from Slovakia. Arch Biol Sci 64:927–934. https://doi.org/10.2298/ABS1203927K

Kačániová M, Vatľák A, Vuković N, Petrová J, Brindza J, Nôžková J (2014) Antimicrobial activity of bee collected pollen against clostridia. Anim Sci Biotechnol 47:362–365. https://doi.org/10.1080/03601234.2011.589322

Katsarou A, Gudbjörnsdottir S, Rawshani A, Dabelea D, Bonifacio E, Anderson BJ et al (2017) Type 1 diabetes mellitus. Nat Rev Dis Primers 3:17016. https://doi.org/10.1038/nrdp.2017.16

Keskin M (2022) Synthesis, characterization and antidiabetic potential of bee pollen based silver nanoparticles. El-Cezeri 9:266–275. https://doi.org/10.31202/ecjse.963670

Khalifa SA, Elashal MH, Yosri N, Du M, Musharraf SG, Nahar L, Sarker SD, Guo Z, Cao W, Zhang X, El-Wahed AA, Xiao J, Omar HA, Hegazy MF, El-Seedi HR (2021) Bee pollen: current status and therapeutic potential. Nutrients 13(6):1876. https://doi.org/10.3390/nu13061876

Khider M, Elbanna K, Mahmoud A, Owayss A (2013) Egyptian honeybee pollen as antimicrobial, antioxidant agents, and dietary food supplements. Food Sci Biotechnol 22:1–9. https://doi.org/10.1007/s10068-013-0238-y

Klarić I, Pavić M, Miškulin I, Blažičević V, Dumić A, Miškulin M (2018) Influence of dietary supplementation of propolis and bee pollen on liver pathology in broiler chickens. Animals 8(4):54. https://doi.org/10.3390/ani8040054

Komosinska-Vassev K, Olczyk P, Kaźmierczak J, Mencner L, Olczyk K (2015) Bee pollen: chemical composition and therapeutic application. Evidence-based complement. Altern Med 2015:1–6. https://doi.org/10.1155/2015/297425

Kročko M, Čanigová M, Bezeková J, Lavová M, Haščík P, Ducková V (2012) Effect of nutrition with propolis and bee pollen supplements on bacteria colonization pattern in gastrointestinal tract of broiler chickens. Anim Sci Biotechnol 45:63–67

Kurek-Górecka A, Balwierz R, Marciniak D, Dzierżewicz Z, Więcławek J, Sarecka-Hujar B, Karolewicz B, Delijewski M (2018) Use of bee pollen in the prevention and treatment of many diseases. Farm Pol 74(3):158–163. https://doi.org/10.32383/farmpol/119190

LeBlanc BW, Davis OK, Boue S, Delucca A, Deeby T (2009) Antioxidant activity of Sonoran Desert bee pollen. Food Chem 115:1299–1305. https://doi.org/10.1016/j.foodchem.2009.01.055

Leja M, Mareczek A, Wyzgolik G, Klepacz-Baniak J, Czekonska K (2007) Antioxidative properties of bee pollen in selected plant species. Food Chem 100:237–240. https://doi.org/10.1016/j.food.chem.2005.09.047

Levin ASS, Kobata CHP, Litvoc MN (2014) Princípios do uso de antimicrobianos: perguntas e respostas. Rev Med 93:63–68. https://doi.org/10.11606/issn.1679-9836.v93i2p63-68

Linskens HF, Jorde W (1997) Pollen as food and medicine—a review. Econ Bot 51:78–87

Lopes J, Stanciu OG, Campos MG, Almaraz-Abarca N, Almeida-Muradian LB, Marghitas LA (2011) Bee pollen antioxidant activity—a review: achievements and further challenges. J Pharmacog 2:25–38. https://doi.org/10.9735/0976-884X.2.2.25-38

MacInnis CI (2023) A tale of two parasites: responses of honey bees infected with *Nosema ceranae* and *Lotmaria passim*. Sci Rep 13:22515. https://doi.org/10.1038/s41598-023-49189-9

Mărgăoan R, Özkök A, Keskin S, Mayda N, Urcan AC, Cornea-Cipcigan M (2021) Bee collected pollen as a value-added product rich in bioactive compounds and unsaturagted fatty acids: a comprehensive study from Turkey and Romania. LWT 149:111952. https://doi.org/10.1016/j.lwt.2021.111925

Mărghitas LA, Stanciu OG, Dezmirean DS, Bobis O, Popescu O, Bogdanov S (2009) *In vitro* antioxidant capacity of honeybee-collected pollen of selected floral origin harvested from Romania. Food Chem 115:878–883. https://doi.org/10.1016/j.foodchem.2009.01.014

Mejías E, Montenegro G (2012) The antioxidant activity of Chilean honey and bee pollen produced in the Llaima volcano's zones. J Food Qual 35:315–322. https://doi.org/10.1111/j.1745-4557.2012.00460.x

Melo ILP, Almeida-Muradian LB (2010) Stability of antioxidants vitamins in bee pollen samples. Quím Nova 33:514–518. https://doi.org/10.1590/S0100-40422010000300004

Menezes JDS, Maciel LF, Miranda MS, Druzian JI (2010) Compostos bioativos e potencial antioxidante do pólen apícola produzido por abelhas africanizadas (*Apis mellifera* L.). Rev Inst Adolfo Lutz 69:233–242

Mihajlovic L, Radosavljevic J, Burazer L, Smiljanic K, Velickovic TC (2015) Composition of polyphenol and polyamide compounds in common ragweed (*Ambrosia artemisiifolia* L.) pollen and sub-pollen particles. Phytochemistry 109:125–132. https://doi.org/10.1016/j.phytochem.2014.10.022

Mohdaly AAA, Mahmoud AA, Roby MHH, Smetanska I, Ramadan MF (2015) Phenolic extract from propolis and bee pollen: composition, antioxidant and antibacterial activities. J Food Biochem 39:538–547. https://doi.org/10.1111/jfbc.12160

Morais M, Moreira L, Feás X, Estevinho LM (2011) Honeybee-collected pollen from five Portuguese Natural Parks: palynological origin, phenolic content, antioxidant properties and antimicrobial activity. Food Chem Toxicol 49:1096–1101. https://doi.org/10.1016/j.fct.2011.01.020

Moskwa J, Naliwajko SN, Dobiecka D, Socha K (2023) Bee products and colorectal cancer—active components and mechanism of action. Nutrients 15:1614. https://doi.org/10.3390/nu15071614

Murakami M, Tsukada O, Okihara K, Hashimoto K, Yamada H, Yamaguchi H (2008) Beneficial effect of honeybee-collected pollen lump extract on benign prostatic hyperplasia (BPH)—a double-blind, placebo-controlled clinical trial. Food Sci Technol Res 14:306–310

Nascimento AM, Luz GE (2018) Bee pollen properties: uses and potential pharmacological applications—a review. J Anal & Pham Res 7(4):513–515. https://doi.org/10.15406/japlr.2018.07.00276

Negri G, Teixeira EW, Alves MLTMF, Moreti ACCC, Otsuk IP, Borguini RG (2011) Hydroxycinnamic acid amide derivatives, phenolic compounds and antioxidant activities of extracts of pollen samples from Southeast Brazil. J Agric Food Chem 59:5516–5522. https://doi.org/10.1021/jf200602k

Nguyen HC, Liu LC, Wu MC, Lin TP, Yang CY, Huang MY (2022) Chemical constituents, antioxidant and anticancer activities of bee pollen from various floral sources in Taiwan. Not Bot Hort Agrobot 50(2):12644. https://doi.org/10.15835/nbha50212644

Nogueira C, Iglesias A, Feás X, Estevinho LM (2012) Commercial bee pollen with different geographical origins: a comprehensive approach. Int J Mol Sci 13:11173–11187. https://doi.org/10.3390/ijms130911173

Olczyk P, Koprowski R, Kaźmierczak J, Mencner L, Wojtyczka R, Stojko J (2016) Bee pollen as a promising agent in the burn wounds treatment. Evid Based Complement Alternat Med 2016:1–12. https://doi.org/10.1155/2016/8473937

Oliveira DM, Bastos DHM (2011) Biodisponibilidade de ácidos fenólicos. Quím Nova 34:1051–1056. https://doi.org/10.1590/S0100-40422011000600023

Oliveira Lopes AJ, Vasconcelos CC, Nascimento Pereira FA, Moraes Silva RH, Dos Queiroz PF, Fernandes CV et al (2019) Anti-inflammatory and antinociceptive activity of pollen extract collected by stingless bee *Melipona fasciculata*. Int J Mol Sci 20(18):4512. https://doi.org/10.3390/ijms20184512

Oliveira KCLS, Moriya M, Azedo RAB, Almeida-Muradian LB, Teixeira EW, Alves MLTMF (2009) Relationship between botanical origin and antioxidants vitamins of bee-collected pollen. Quím Nova 32:1099–1102

Oszczędłowski P (2023) All that glitters is not gold: assessment of bee pollen supplementation effects on gastric mucosa. Nutrients 16(1):37. https://doi.org/10.3390/nu16010037

Özkök A, Koru Ö, Bedir O, Çetinkaya S, Gençay Çelemli Ö, Özenirler Ç (2021) Total bioactive compounds and antimicrobial capacities of bee pollen with different botanical origins Bull UASVM. Food Sci Technol 78(1):57–67. https://doi.org/10.15835/buasvmcn-fst:2020.0061

Pascoal A, Rodrigues S, Teixeira A, Féas X, Estevinho LM (2014) Biological activities of commercial bee pollens: antimicrobial, antimutagenic, antioxidant and anti-inflammatory. Food Chem Toxicol 63:233–239. https://doi.org/10.1016/j.fct.2013.11.010

Pełka K, Bucekova M, Godocikova J, Szweda P, Majtan J (2022) Glucose oxidase as an important yet overlooked factor determining the antibacterial activity of bee pollen and bee bread. Eur Food Res Technol 248:2929–2939. https://doi.org/10.1007/s00217-022-04101-z

Pérez-Pérez EM, Vit P, Rivas E, Sciortino R, Sosa A, Tejada D (2012) Antioxidant activity of four color fractions of bee pollen from Merida, Venezuela. Arch Latinoam Nutr 62:375–380

Pinto B, Caciagli F, Riccio E, Reali D, Saric A, Balog T (2010) Antiestrogenic and antigenotoxic activity of bee pollen from *Cystus incanus* and *Salix alba* as evaluated by the yeast estrogen screen and the micronucleus assay in human lymphocytes. Eur J Med Chem 45:4122–4128. https://doi.org/10.1016/j.ejmech.2010.06.001

Rebiai A, Lanez T (2012) Chemical composition and antioxidant activity of *Apis mellifera* bee pollen from northwest Algeria. J Fundam Appl Sci 4:155–163

Reinhard KJ, Hamilton DL, Hevly RH (1991) Use of pollen concentration in palaeopharmacology: coprolite evidence in medical plants. J Ethnobiol 11:117–132

Rocchetti G, Castiglioni S, Maldarizzi G, Carloni P, Lucini L (2019) UHPLC-ESI-QTOF-MS phenolic profiling and antioxidant capacity of bee pollen from different botanical origin. Int J Food Sci Technol 54:335–346. https://doi.org/10.1111/ijfs.13941

Rodríguez-Polit C, Gonzalez-Pastor R, Heredia-Moya J, Carrera-Pacheco SE, Castgillo-Solis F, Vallejo-Imbaquingo R et al (2023) Chemical properties and biological activity of bee pollen. Molecules 28(23):7768. https://doi.org/10.3390/molecules28237768

Saisavoey T, Sangtanoo P, Srimongkol P, Reamtong O, Kranchanatat A (2021) Hydrolysates from bee pollen could induced apoptosis in human bronchogenic carcinoma cells (ChaGo-K-1). J Food Sci Technol 58:752–763. https://doi.org/10.1007/s13197-020-04592-2

Saral Ö, Yildiz O, Aliyazicioğlu R, Yuluğ E, Canpolat S, Öztürk F (2016) Apitherapy products enhance the recovery of CCL4-induced hepatic damages in rats. Turk J Med Sci 46:194–202. https://doi.org/10.3906/sag-1411-35

Saríc A, Balog T, Sobocanec S, Kusic B, Sverko V, Rusak G (2009) Antioxidant effects of flavonoid from Croatian *Cystus incanus* L. rich bee pollen. Food Chem Toxicol 47:547–554. https://doi.org/10.1016/j.fct.2008.12.007

Setyawan AB, Rao USM, Sairazi NSM (2024) Biochemical evaluation of hypoglycaemic effect of stingless bee pollen in high fat diet fed-low dose STZ induced experimental Type 2 diabetes in rats. Asian J Med Biomed 8(1):98–110. https://doi.org/10.37231/ajmb.2024.8.1.713

Shady HMA, Mohamed WF, Sayed-Ahmed EF, Amer SA (2016) A comparative study on propolis and pollen extracts: chemical profile analysis, antioxidant and anticancer activity. Int J Curr Microbiol Appl Sci 5:397–414. https://doi.org/10.20546/ijcmas.2016.503.047

Shukla S, Gupta S (2010) Apigenin: a promising molecule for cancer prevention. Pharm Res 27:962–978. https://doi.org/10.1007/s11095-010-0089-7

Soares Arruda VA, dos Santos AV, Sampaio DF, da Silva Araújo E, de Peixoto AL, Estevinho LMLM et al (2020) Brazilian bee pollen: phenolic content, antioxidant properties and antimicrobial activity. J Apic Res 60(5):775–783. https://doi.org/10.1080/00218839.2020.1840854

Sousa MSB, Vieira LM, Lima A (2011) Fenólicos totais e capacidade antioxidante *in vitro* de resíduos de polpas de frutas tropicais. Braz J Food Technol 14:202–210. https://doi.org/10.4260/BJFT2011140300024

Spulber R, Chifiriuc C, Fleancu M, Popa O, Băbeanu N (2018) Antibacterial activity of magnetite nanoparticles coated with bee pollen extracts. Proc Agriculture for Life, Life for Agriculture Conference, pp 579–585. https://doi.org/10.2478/alife-2018-0091

Su J, Yang X, Lu Q, Liu R (2021) Antioxidant and anti-tyrosinase activities of bee pollen and identification of active components. J Apic Res 60(2):297–307. https://doi.org/10.1080/00218839.2020.1722356

Tohamy AA, Abdella EM, Ahmed RR, Ahmed YK (2014) Assessment of anti-mutagenic, anti-histopathologic and antioxidant capacities of egyptian bee pollen and propolis extracts. Cytotechnology 66:283–297. https://doi.org/10.1007/s10616-013-9568-0

Tomaszewska E, Knaga S, Dobrowolski P, Lamorski K, Jabłoński M, Tomczyk-Warunek A et al (2020) The effect of bee pollen on bone biomechanical strength and trabecular bone histomorphometry in tibia of young Japanese quail (*Coturnix japonica*). PLoS One 15(3):e0230240. https://doi.org/10.1371/journal.pone.0230240

Ulbricht C, Conquer J, Giese N, Khalsa KPS, Sklar J, Weissner W (2009) An evidence-based systematic review of bee pollen by the natural standard research collaboration. J Diet Suppl 6:290–312. https://doi.org/10.1080/19390210903081381

Ulusoy E, Kolayli S (2014) Phenolic composition and antioxidant properties of anzer bee pollen. J Food Biochem 38:73–82. https://doi.org/10.1111/jfbc.12027

Uttara B, Singh AV, Zamboni P, Mahajan RT (2009) Oxidative stress and neurodegenerative diseases: a review of upstream and downstream antioxidant therapeutic options. Curr Neuropharmacol 7:65–74. https://doi.org/10.2174/157015909787602823

Wang B, Diao Q, Zhang Z, Liu Y, Gao Q, Zhou Y (2013) Antitumor activity of bee pollen polysaccharides from *Rosa rugosa*. Mol Med Rep 7:1555–1558. https://doi.org/10.3892/mmr.2013.1382

Wu YD, Lou YJ (2007) A steroid fraction of chloroform extract from bee pollen of *Brassica campestris* induces apoptosis in human prostate cancer PC-3 cells. Phytother Res 21:1087–1091. https://doi.org/10.1002/ptr.2235

Yamaguchi M, Uchiyama S, Nakagawa T (2007a) Preventive effects of bee pollen *Cistus ladaniferus* extract on bone loss in ovariectomized rats *in vivo*. J Health Sci 53:571–575

Yamaguchi M, Hamamoto R, Uchiyama S, Ishiyama K, Hashimoto K (2007b) Preventive effects of bee pollen *Cistus ladaniferus* extract on bone loss in streptozotocin-diabetic rats in vivo. J Health Sci 53:190–195

Yang K, Wu D, Ye X, Liu D, Chen J, Sun P (2013) Characterization of chemical composition of bee pollen in China. J Agric Food Chem 61(3):708–718. https://doi.org/10.1021/jf304056b

Yang B, Jin L, Yang Y, Li K, Peng D (2014) Inhibitory effect of rape pollen supercritical CO2 fluid extract against testosterone-induced benign prostatic hyperplasia in rats. Exp Ther Med 8:31–37. https://doi.org/10.3892/etm.2014.1680

Yang S, Qu Y, Chen J, Chen S, Sun L, Zhou Y, Fan Y (2021) Bee pollen polysaccharide from *Rosa rugosa* thunb (Rosaceae) promotes pancreatic β-cell proliferation and insulin secretion. Front Pharmacol 12:12. https://doi.org/10.3389/fphar.2021.688073

Yıldız O, Can Z, Saral Ö, Yuluğ E, Öztürk F, Aliyazıcıoğlu R et al (2013) Hepatoprotective potential of chestnut bee pollen on carban tetrachloride-induced hepatic damages in rats. Evid Based Compl Altern Med 213:9. https://doi.org/10.1155/2013/461478

Yuan Y, Zheng YF, Zhou JH, Geng YT, Zou P, Li YQ et al (2019) Polyphenol rich extracts from brown macroalgae lessonia trabeculate attenuate hyperglycemia and modulate gut microbiota in high-fat diet and streptozotocin-induced diabetic rats. J Agric Food Chem 2019:12472–12480. https://doi.org/10.1021/acs.jafc.9b05118

Zhang H, Wang X, Wang K, Li C (2015) Antioxidant and tyrosinase inhibitory properties of aqueous ethanol extracts from monofloral bee pollen. J Apic Sci 59:119–128. https://doi.org/10.1515/jas-2015-0013

Zhang J, Cao W, Zhao H, Guo S, Wang Q, Cheng N, Bai N (2022) Protective mechanism of *Fagopyrum aesculentum* moench. Bee Pollen EtOH extract against Type II diabetes in a high-fat-diet/streptozocim-induced C57BL/6. J Mice Front Nutr 9:2022. https://doi.org/10.3389/fnut.2022.925351

Part V
Bee Bread

Chapter 13
Chemical Composition of Bee Bread

Adriana Cristina Urcan

Contents

Abstract Bee bread is a biologically active fermented product with a complex and highly bioavailable chemical composition, essential for both colony sustenance and potential human nutritional applications. This chapter provides a comprehensive analysis of the chemical composition of bee bread, covering its macronutrient and micronutrient profiles, bioactive compounds, and the biochemical transformations occurring during fermentation. The primary components—carbohydrates, proteins, and lipids—play vital roles in energy metabolism, structural development, and physiological processes. Bee bread is also a significant source of essential amino acids, vitamins (B-complex, A, E, and K), and minerals (calcium, magnesium, iron, and potassium), enhancing its nutritional value. Bioactive compounds such as flavonoids, phenolic acids, and probiotics contribute to its antioxidant, antimicrobial, and immunomodulatory properties. The fermentation process, facilitated by lactic acid

A. C. Urcan (✉)
Department of Microbiology and Immunology, Faculty of Animal Science and Biotechnologies, University of Agricultural Sciences and Veterinary Medicine, Cluj-Napoca, Romania
e-mail: adriana.urcan@usamvcluj.ro

J. M. Alvarez-Suarez (ed.), *Bee Products – Chemical and Biological Properties*,
https://doi.org/10.1007/978-3-031-89049-9_13

bacteria and enzymatic activity, improves digestibility, nutrient bioavailability, and stability. Additionally, variations in bee bread composition, influenced by botanical origin, geographical location, and hive conditions, impact its nutritional and functional properties. This chapter consolidates scientific knowledge on bee bread's chemical complexity, positioning it as a valuable candidate for functional food and nutraceutical applications.

Keywords Bee bread · Chemical composition · Bioactive compounds · Fermentation process · Nutritional value.

13.1 Introduction

Bee bread, a fermented mixture of bee pollen, nectar, and honey produced by honeybees (*Apis mellifera*), plays an essential role in the nutrition and overall health of the bee colony. This unique substance results from a complex transformation that occurs after bees gather pollen and store it within the hive, where it undergoes fermentation facilitated by the addition of nectar, honey, and glandular secretions (Ivanišová et al. 2015; Zuluaga et al. 2015; Semkiw and Skubida 2021; Ilie et al. 2024). While bee pollen serves as the initial foraged material, bee bread represents its enhanced form—both in terms of preservation and nutritional bioavailability—owing to the biochemical transformations during fermentation (Prđun et al. 2021; Urcan et al. 2021; Barta et al. 2022; Alcalá-Orozco et al. 2024). This process not only protects the bee pollen from microbial spoilage but also makes it more digestible and nutritionally potent, essential for nourishing young worker bees and larvae (Didaras et al. 2021; Khalifa et al. 2020; Todorov et al. 2024).

The chemical composition of bee bread is characterized by a complex array of macronutrients and micronutrients that support the metabolic needs of honeybee colonies and offer potential health benefits for humans. Bee bread is characterized by a robust macronutrient profile that includes carbohydrates, proteins, and lipids in a balanced proportion. Research indicates that bee bread typically contains approximately 24–35% carbohydrates, primarily in the form of simple sugars such as glucose and fructose, as well as polysaccharides (Kabakci 2024; Aksoy et al. 2024). These carbohydrates serve as a key energy source for honeybees, especially during periods of high metabolic demand or when environmental conditions limit foraging. Protein content in bee bread is estimated at around 20%, encompassing a broad spectrum of essential and nonessential amino acids (Kabakci 2024). This substantial protein content is vital for the growth and development of bee larvae and is crucial for the overall health and resilience of the hive. The fermentation process to which bee bread is subjected enhances the digestibility and bioavailability of these proteins, rendering them more accessible than those in raw pollen (Didaras et al. 2021; Aylanc et al. 2021).

Though present in smaller amounts, lipids comprise an essential 3–10% of bee bread's composition (Zuluaga et al. 2015; Ćirić et al. 2022). These lipids include essential fatty acids that contribute to the energy reserves of bees and support numerous physiological processes. The lipid content in bee bread also facilitates the absorption of fat-soluble vitamins, enhancing the product's overall nutritional profile.

In addition to macronutrients, bee bread is rich in micronutrients, including a wide range of vitamins and minerals. Noteworthy for its vitamin content, bee bread contains substantial levels of B vitamins (such as B1, B2, B6, and B12), along with vitamins A, E, and K (Mohammad et al. 2020; Suleiman et al. 2021). These vitamins play crucial roles in metabolic regulation, immune response, and general health. Vitamin K, for instance, is essential for blood coagulation, while B vitamins are indispensable for energy metabolism and neurological functions (Mitra et al. 2022; Brancaccio et al. 2022).

Minerals are another significant component of bee bread's micronutrient profile. It is known to contain key minerals, including calcium, iron, magnesium, and potassium (Ćirić et al. 2022; Bakour et al. 2023). These minerals support vital biological functions such as bone integrity, oxygen transport, and muscle function, collectively contributing to the nutritional benefits associated with bee bread and making it a highly valued dietary supplement (Habryka et al. 2020).

Beyond its fundamental macronutrient and micronutrient composition, bee bread is enriched with bioactive compounds, including flavonoids and phenolic acids with documented antioxidant properties (Tomás et al. 2017). These flavonoids, such as quercetin, kaempferol, and myricetin, have been associated with numerous health benefits, including anti-inflammatory and anticancer effects. The presence of these bioactive compounds not only enhances the nutritional profile of bee bread but also underscores its therapeutic potential (Sobral et al. 2017; Bakour et al. 2023).

The fermentation process responsible for transforming pollen into bee bread plays an instrumental role in developing its chemical complexity. During this process, lactic acid bacteria and bee-derived enzymes break down complex carbohydrates and proteins, producing a product that is not only more digestible but also enriched with various metabolites that may have additional health-promoting effects (Urcan et al. 2024; Zhang et al. 2022).

It is essential to recognize that the chemical composition of bee bread can vary widely, influenced by several factors such as the botanical origin of the pollen, geographical location, environmental conditions, and the fermentation processes involved in its production (Urcan et al. 2021; Uslinawaty 2023). Each of these elements contributes to the chemical composition and nutritional profile of bee bread, leading to significant differences in its characteristics. This variability is significant because each of these factors contributes uniquely to bee bread's nutrient profile, impacting its potential health benefits and applications in human nutrition.

One of the primary influences on bee bread composition is the botanical origin of the pollen. Different plant species provide varying amounts of proteins, carbohydrates, lipids, and micronutrients to bee bread, resulting in a diverse nutrient profile. Pollen from distinct floral sources has been shown to contain unique amino acid profiles and levels of bioactive compounds, such as flavonoids and phenolic acids, which directly impact the quality and health properties of the final product (da Silva et al. 2014; Suleiman et al. 2021). The availability of certain plants is also seasonally dependent, further influencing the nutrient composition of pollen collected by bees at different times of the year (Bleha et al. 2019).

Geographical and environmental factors also play crucial roles, soil type, climate, and local environmental conditions influence the mineral content and other nutritional properties of the plants that bees forage, which in turn affects the chemical makeup of bee bread, for example, bee bread from various geographical regions shows distinct variations in mineral content, often linked to specific flora and soil characteristics unique to those areas (da Silva et al. 2014; Bertoncelj et al. 2018). Additionally, climatic conditions impact flowering times and pollen availability, contributing further to differences in bee bread composition (Ispirli and Dertli 2021).

The fermentation process within the hive is another critical factor in determining the final composition of bee bread. The microbial communities involved, primarily lactic acid bacteria and yeasts, vary according to environmental conditions and the types of pollen being fermented (Hudz et al. 2019; Bramasta 2023). These microbial populations play essential roles in breaking down complex carbohydrates and proteins, enhancing the digestibility and bioavailability of nutrients. Variations in these microbial communities and their metabolic activity directly influence the nutrient profile, pH, and overall quality of bee bread (Sobral et al. 2017). Factors such as fermentation duration and storage conditions, including moisture, temperature, and light exposure, also affect nutrient stability and the preservation of bioactive compounds (Dranca et al. 2020; Pełka et al. 2023).

Furthermore, internal hive dynamics add another layer of variability, differences in how bees handle, store, and manage pollen within various frames or sections of the hive can result in slight compositional differences even within the same hive, for example, the clustering of tasks among different groups of bees can influence the nutritional content of bee bread stored in specific hive locations (Donkersley et al. 2017).

It is important to consider all these factors—botanical origin, geographical and environmental conditions, fermentation processes, and hive dynamics—when discussing the chemical composition of bee bread because each element plays a unique role in shaping its nutritional profile, bioactive potential, and health benefits. Recognizing the influence of these interrelated factors is crucial for accurately characterizing bee bread's composition and for leveraging its value as a functional food.

The significance of bee bread extends beyond its role in sustaining the colony. It has garnered increasing scientific interest due to its rich chemical composition and potential health benefits for humans (El Ghouizi et al. 2023; Bakour et al. 2022). The fermentation process not only preserves the nutritional integrity of the pollen but also enhances its content of bioactive compounds, thereby amplifying its antioxidant, antimicrobial, and anti-inflammatory properties (Barta et al. 2022; Rodríguez-Pólit et al. 2023; Urcan et al. 2024). Bee bread's composition, enriched with proteins, essential amino acids, carbohydrates, lipids, vitamins, minerals, and a variety of bioactive phytochemicals, makes it a highly nutritious substance (Mayda et al. 2020; Aylanc et al. 2023). These nutrients, many of which are synthesized or augmented during fermentation, serve not only the metabolic needs of bees but also present a potential source of health benefits for humans. This growing interest in bee bread stems from its unique nutrient profile and its capacity to support various biological functions, including immune modulation, antioxidative defense, and potential nutrigenomic effects (Kieliszek et al. 2018; Aylanc et al. 2021).

Bee bread's chemical composition varies depending on the plants from which bees gather pollen, as well as the geographical and environmental conditions surrounding the hive (Urcan et al. 2021; Iorizzo et al. 2024). This variability means that different types of bee bread may offer distinct therapeutic benefits. Studies have linked bee bread to potential improvements in immune function, digestive health, and even specific conditions related to inflammation and oxidative damage, making it an exciting area of research in natural health and functional nutrition (Bakour et al. 2019).

Studies have indicated that bee bread possesses a higher antioxidant capacity compared to raw pollen, particularly after gastrointestinal digestion, which suggests its effectiveness in combating oxidative damage (Bleha et al. 2019; Aylanc et al. 2021). Research has shown that bee bread can positively influence immune responses (Denisow et al. 2016). It has been linked to the modulation of inflammatory markers and the enhancement of immune cell activity, which may help in managing conditions characterized by chronic inflammation (Othman et al. 2021). The presence of phenolic compounds, such as flavonoids, in bee bread is particularly noteworthy, as these compounds are known for their anti-inflammatory and antioxidant properties (Keyvan et al. 2023). Additionally, studies have demonstrated that bee bread can ameliorate vascular inflammation and improve endothelial function, suggesting its potential role in cardiovascular health (Kacemi and Campos 2023). Furthermore, the digestive health benefits of bee bread are supported by its rich composition of dietary fibers and probiotics, which can promote gut health and enhance nutrient absorption (Mohammad et al. 2021; Aylanc et al. 2021). The presence of beneficial microorganisms, which are introduced during the fermentation process, contributes to the maintenance of a healthy gut microbiome, potentially leading to improved digestive function and overall well-being (Didaras et al. 2021). This multifaceted approach to health, encompassing immune support, anti-inflammatory effects, and digestive health, positions bee bread as a promising candidate in the realm of natural health products and functional foods.

13.2 Distinction Between Bee Pollen and Bee Bread

The distinction between bee pollen and bee bread is a topic of considerable interest within apiculture and nutritional science. Both products are derived from the activities of honey bees, yet they exhibit significant differences in composition, nutritional value, and functional properties. Bee pollen is essentially the male gametophyte of flowering plants, collected by bees and mixed with nectar and saliva, while bee bread is a fermented product created when bees store pollen in the hive, mixed with honey and enzymes, undergoing lactic acid fermentation (Mărgăoan et al. 2019).

Bee pollen is primarily composed of flower pollen grains, which contain a rich array of nutrients, including proteins, carbohydrates, lipids, vitamins, and minerals. The protein content of bee pollen can vary significantly, typically ranging from 13% to 55%, while lipids and carbohydrates can account for 1% to 20% and 2% to 60%, respectively (Hsu et al. 2021). This nutrient density makes bee pollen a highly sought-after dietary supplement, often marketed for its purported health benefits, including antioxidant properties and immune system support (Bakour et al. 2023). Furthermore, bee pollen is known for its high flavonoid content, which contributes to its antioxidant capacity and may influence its taste and pigmentation (Mărgăoan et al. 2016). In contrast, bee bread is the result of a fermentation process that occurs within the hive (Urcan et al. 2017). When bees mix pollen with honey and their saliva, the mixture is stored in the combs, where it undergoes anaerobic fermentation, primarily facilitated by lactic acid bacteria (Habryka et al. 2020).

This fermentation process alters the chemical composition of the pollen and enhances its digestibility and nutritional value. Bee bread typically has a lower pH (ranging from 3.5 to 4.3) compared to bee pollen (which ranges from 4.1 to 5.9), largely due to the production of lactic acid during fermentation (Degirmenci 2024; Suleiman et al. 2021; Urcan et al. 2024). This acidic environment can inhibit the growth of microorganisms, thereby enhancing the safety and shelf-life of bee bread as a food source for both bees and humans (Urcan et al. 2018a).

The nutritional profiles of bee pollen and bee bread also differ significantly. While bee pollen is rich in proteins and lipids, studies have shown that the protein and lipid content of bee bread may decrease slightly compared to bee pollen (Thakur and Nanda 2020; Prđun et al. 2021; Aylanc et al. 2023). However, bee bread compensates for this by containing higher levels of certain vitamins, particularly vitamin K, as well as increased fiber content and reduced sugar levels (Özkök et al. 2022). This makes bee bread a valuable food source, particularly for young bees and larvae, who rely on its nutrient composition for growth and development (Kankasemsuk 2023).

Moreover, the antioxidant capacity of bee bread has been shown to be higher than that of bee pollen after simulated gastrointestinal digestion, indicating that the fermentation process may enhance the bioavailability of certain bioactive compounds (Aylanç et al. 2021). This is particularly relevant in the context of functional foods, as the health benefits associated with bee bread may be more pronounced due to the presence of bioactive molecules that are released during fermentation (Bakour et al. 2022).

Additionally, bee bread has been found to exhibit antimicrobial properties, which can be attributed to its unique composition and the presence of various phenolic compounds (Sobral et al. 2017; Pełka et al. 2021b). The microbial dynamics within bee bread also play a crucial role in its characteristics. The fermentation process leads to a diverse community of microorganisms, including lactic acid bacteria and yeasts, which contribute to the product's stability and potential health benefits (Disayathanoowat et al. 2020; Detry et al. 2020). Studies have indicated that the microbial composition of bee bread can vary based on environmental factors and the floral sources of the pollen, which in turn affects its nutritional and functional properties (Donkersley et al. 2018; Ghosh et al. 2022). This variability underscores the importance of understanding the ecological context in which bee bread is produced.

In terms of palatability, research has shown that honey bees do not exhibit a significant preference for bee bread over fresh pollen, suggesting that both products serve essential but distinct roles in the hive's nutrition (Bridgett et al. 2014; Kieliszek et al. 2018). This finding highlights the importance of both bee pollen and bee bread in the diet of honey bees, as they provide complementary nutritional benefits (Baky et al. 2023). The therapeutic potential of bee bread is gaining recognition, with studies suggesting that it may possess anti-inflammatory, antioxidant, and antimicrobial properties, making it a candidate for functional food applications (Bakour et al. 2019; Dervişoğlu et al. 2022). The presence of various bioactive compounds, including flavonoids and phenolic acids, contributes to these health benefits, further distinguishing bee bread from bee pollen (Sawicki et al. 2022; Sobral et al. 2017; Urcan et al. 2021).

While bee pollen and bee bread share a common origin, their differences in composition, nutritional value, and functional properties are significant. Bee pollen serves as a nutrient-dense food source rich in proteins and antioxidants, while bee bread, through the process of fermentation, offers enhanced digestibility, a unique microbial profile, and potential health benefits that make it a valuable product in both apiculture and human nutrition. Understanding these distinctions not only enriches our knowledge of bee products but also opens avenues for their application in health and wellness.

In this chapter, the results of extensive research on bee bread are presented, aiming to consolidate the most comprehensive information available on its chemical composition. This includes an in-depth analysis of macronutrients, micronutrients, and bioactive compounds within bee bread, offering a thorough understanding of the elements that contribute to its nutritional and therapeutic potential.

13.3 Carbohydrates Content

Carbohydrates play a crucial role in the nutritional profile of bee bread, offering energy and supporting various health-promoting properties. Derived from nectar and transformed through hive-based fermentation, the carbohydrate content in bee bread includes both simple and complex forms, each contributing uniquely to its functionality. Simple sugars like fructose and glucose are readily available sources of energy, essential for honeybees' high metabolic demands and potentially beneficial for human consumption as well (Aksoy et al. 2024; Tomás et al. 2017; Urcan et al. 2018a, b).

In addition to these quick energy sources, bee bread's complex carbohydrates, including polysaccharides and dietary fibers, provide longer-term health benefits. Through fermentation, polysaccharides are broken down into oligosaccharides, creating prebiotic compounds that promote gut health by supporting beneficial bacterial growth (Bakour et al. 2023). This dual role of carbohydrates—rapid energy from simple sugars and digestive support from fibers and prebiotics—positions bee bread as a functional food with broad health applications for both bees and humans (Didaras et al. 2021; Kacemi and Campos 2023).

The carbohydrate content of bee bread generally ranges between 24.40% and 34.80% of its dry weight, depending on factors like floral sources and environmental conditions (Mărgăoan et al. 2014; Tomás et al. 2017; Baky et al. 2023). The primary simple sugars found in bee bread are fructose and glucose, derived from the nectar and the enzymatic breakdown of sucrose during fermentation. Fructose represents the largest portion of these sugars, at approximately 57.51%, while glucose makes up around 42.59% (Aksoy et al. 2024). The fermentation process results in a minimal sucrose content, generally around 0.12%, due to its conversion into these monosaccharides (Aksoy et al. 2024; Urcan et al. 2021). These simple sugars serve as an immediate energy source, making bee bread a highly nutritious food for both bees and humans. In addition to simple sugars, bee bread contains complex carbohydrates, mainly polysaccharides, which contribute to its carbohydrate profile and provide various health benefits. The polysaccharides in bee bread are known to promote digestive health, as they act as prebiotics that support the growth of beneficial gut bacteria (Alvarez-Suarez 2017; Bakour et al. 2022).

Glucose is one of the predominant simple sugars in bee bread, often constituting a significant portion of its carbohydrate content. Studies have reported that glucose levels in bee bread can average around 11.5 g/100 g, depending on the floral sources of the pollen used (Mohammad et al. 2020). Glucose serves as a quick source of energy for bees, particularly for foraging and brood rearing.

Fructose is another major sugar component of bee bread, typically found alongside glucose. The ratio of glucose to fructose can vary based on the nectar source, with some floral types yielding higher fructose concentrations. The presence of fructose in bee bread contributes to its sweet taste and energy content, making it an attractive food source for bees (Fallah et al. 2022; Urcan et al. 2018a, b).

Sucrose, a disaccharide composed of glucose and fructose, is also present in bee bread, although in lower concentrations compared to the monosaccharides. The enzymatic activity of bees converts sucrose into glucose and fructose during the fermentation process, which enhances the digestibility of the sugars (Özkök et al. 2022).

The fermentation process that transforms pollen into bee bread enhances the digestibility of simple sugars. The presence of enzymes from bee saliva and lactic acid bacteria facilitates the breakdown of complex carbohydrates into simpler sugars, making them more bioavailable for absorption (Othman et al. 2021).

Additionally, fermentation may yield oligosaccharides, which further enhance prebiotic effects (Bakour et al. 2022). Dietary fiber is another essential component of bee bread's carbohydrate profile, with fiber content reported to range from 8.30% to 19% (Beykaya et al. 2021; Fallah et al. 2022). This fiber supports digestive health by aiding regular bowel movements and preventing constipation. It also contributes to a sense of fullness, which may assist in weight management (Beykaya et al. 2021). Furthermore, dietary fiber plays a role in cardiovascular health by helping regulate blood cholesterol levels, an added advantage for human consumers (Fallah et al. 2022). Beyond its role as an energy source, the carbohydrate content of bee bread influences its functional properties. Carbohydrates, particularly in the form of sugars and fiber, enhance bee bread's water retention capacity, which is critical for its texture and stability during storage (Bleha et al. 2019), by acting as hydrocolloids, improving moisture absorption and thus extending shelf-life and preserving product quality (Bleha et al. 2019).

The simple sugars provide rapid energy, making bee bread an ideal food for individuals needing quick energy replenishment, such as athletes (Aksoy 2024). Meanwhile, the fiber content supports gut health and may aid in blood sugar regulation, suggesting potential benefits for individuals with diabetes (Beykaya et al. 2021; Fallah et al. 2022). Additionally, bee bread's carbohydrate and polyphenol content contribute to its antioxidant properties, which may help mitigate oxidative stress and reduce the risk of chronic diseases (Khalifa et al. 2020).

Polysaccharides in bee bread include various forms of carbohydrates such as cellulose, hemicellulose, and pectin. These polysaccharides are essential for maintaining the structural integrity of the bee bread and contribute to its functional properties. The presence of polysaccharides can enhance the texture and mouthfeel of bee bread, making it a desirable food product (Pełka et al. 2021b). Additionally, polysaccharides have been shown to possess immunomodulatory and antioxidant properties, further enhancing the health benefits of bee bread (Dervişoğlu et al. 2022).

The high fiber content in bee bread plays a vital role in promoting digestive health by supporting regular bowel movements and preventing constipation. Soluble fibers in bee bread can also contribute to cholesterol management and positively influence gut microbiota composition, both of which are critical for overall digestive health (Dranca et al. 2020).

Bee bread's polysaccharides have been linked to antioxidant activity, which helps mitigate oxidative stress and inflammation. The fermentation process enhances bee bread's antioxidant capacity, creating a potent source of bioactive compounds that may offer protection against chronic diseases (Jarušaitienė et al. 2012). Additionally, specific polysaccharides in bee bread demonstrate immunomodulatory effects, potentially enhancing immune responses in both bees and humans. These compounds may play a regulatory role in immune function, aiding in the defense against infections (Zakaria et al. 2021).

Given its rich content of fiber and polysaccharides, bee bread is increasingly recognized as a functional food with broad health applications. Incorporating bee bread into the diet may support metabolic health and reduce the risk of chronic diseases (Hudz et al. 2019). The fiber and polysaccharide composition in bee bread can vary considerably based on factors such as the botanical origin of the pollen, the bee species involved, and environmental conditions during production. For example, bee bread derived from pollen collected from specific floral sources may contain higher concentrations of certain polysaccharides (Othman et al. 2021).

Plant sources contribute varying types and quantities of fibers and polysaccharides to bee bread. Pollen from leguminous plants may enhance the fiber content, while pollen from other flowering plants may contribute unique polysaccharide profiles (Donkersley et al. 2017). Geographic location and climatic conditions also play a role; bee bread sourced from regions with diverse flora often displays a more complex fiber and polysaccharide composition compared to that from monoculture environments (Dervişoğlu et al. 2022).

The sugar content of bee bread can vary significantly based on several factors, including the botanical origin of the pollen, the species of bees involved, and environmental conditions. Different plants contribute varying types and amounts of sugars to bee bread. For example, nectar from certain flowers may yield higher concentrations of fructose, while others may provide more glucose (Dervişoğlu et al. 2022). The geographical location and climatic conditions can also influence the sugar content of bee bread. Studies have shown that bee bread collected from regions with diverse flora tends to have a richer sugar profile compared to those from monoculture environments (Disayathanoowat et al. 2020).

The transformation of bee pollen into bee bread begins when bees collect pollen and mix it with nectar and their saliva, which contains enzymes. This mixture is then packed into honeycomb cells, where it undergoes anaerobic fermentation facilitated by lactic acid bacteria and other microorganisms (Dervişoğlu et al. 2022; Bakour et al. 2022). During this fermentation process, several biochemical changes occur, such as conversion of sugars, the enzymes present in bee saliva and those produced by lactic acid bacteria break down complex carbohydrates into simpler sugars. Polysaccharides such as starch are hydrolyzed into glucose and other simple sugars, which are more readily absorbed by the organism (Prđun et al. 2021). This enzymatic activity significantly increases the availability of simple sugars in bee bread compared to raw pollen. The fermentation process leads to the production of lactic acid, which lowers the pH of the bee bread. This acidic environment is conducive to the growth of beneficial bacteria while inhibiting the growth of spoilage organisms (Özkök et al. 2022; Urcan et al. 2024). The presence of lactic acid not only contributes to the preservation of bee bread but also enhances its digestibility. As bee pollen transforms into bee bread, the carbohydrate composition undergoes significant changes, and the fermentation process results in an increase in the concentration of simple sugars such as glucose and fructose, which are essential for providing energy to bees. Studies have reported that bee bread can contain significant amounts of carbohydrates, with a notable portion being simple sugars (Aldarf and Sultaneh 2024). The breakdown of polysaccharides during fermentation leads to a decrease in the overall complex carbohydrate content of bee bread. This transformation is beneficial as it enhances the digestibility of the carbohydrates, making them more accessible for absorption by bees (Sobral et al. 2017). The transformation of carbohydrates during the conversion of bee pollen to bee bread has several nutritional implications: The fermentation process improves the digestibility of carbohydrates, allowing bees to efficiently utilize the energy provided by the sugars. This is particularly important for young bees and larvae, who rely on bee bread as a primary food source (Dervişoğlu et al. 2022). The increased concentration of simple sugars in bee bread provides a quick and readily available energy source for honey bees, supporting their high metabolic demands during foraging and brood care (Urcan et al. 2017).

While bee bread is primarily consumed by bees, its rich content of simple sugars may also offer health benefits for human consumers. The natural sugars found in bee bread can provide a quick source of energy and may have lower glycemic indices compared to refined sugars, making them a healthier alternative (Kutlu et al. 2024).

13.4 Protein and Amino Acid Content

The protein content of bee bread is substantial, typically ranging from 10% to 35% of its dry weight, depending on variables such as the botanical origin of the pollen and environmental conditions during its formation (Bleha et al. 2019; Mohammad et al. 2020; Fallah et al. 2022; Kankasemsuk 2023). This high protein level is crucial for honeybees, particularly for the growth and development of larvae. The fermentation process that bee bread undergoes not only preserves the proteins but also enhances their digestibility, thereby improving bioavailability compared to raw pollen (Bakour et al. 2019; Tawfik et al. 2023; Urcan et al. 2024).

A comparison of protein content between bee pollen and bee bread highlights important differences that impact their nutritional roles for both honeybee diets and potential human consumption. While both bee pollen and bee bread are products of honeybee activity, they differ in composition due to their respective processing methods. Bee pollen, which bees collect directly from flowers, has a protein content that typically ranges from 20% to 37% of its dry weight, depending on the floral source and environmental factors (Alaux et al. 2011). This protein is primarily derived from the pollen grains collected by bees, which are rich in essential amino acids.

In contrast, bee bread, formed through the fermentation of pollen mixed with honey and bee saliva, generally contains a lower protein concentration than raw pollen (Fallah et al. 2022; Svečnjak 2023). The fermentation process alters the protein structure and can lead to a slight reduction in total protein content. However, it enhances the digestibility of proteins, making them more bioavailable than in their raw pollen form (DeGrandi-Hoffman et al. 2012). Thus, although bee bread may have a somewhat lower protein content compared to bee pollen, the fermentation process increases its functional value by improving protein accessibility and nutritional efficacy.

Bee bread is recognized for its rich amino acid content, which can play a significant role in promoting bought bees and human health (Nicolson 2011; Yan et al. 2019; Thakur and Nanda 2020). The amino acids present in bee bread are derived from the fermentation of pollen, nectar, and bee saliva, resulting in a complex mixture that includes both essential and nonessential amino acids (Baky et al. 2023). Notably, bee bread contains a variety of amino acids, such as proline, glutamic acid, aspartic acid, arginine, leucine, and lysine, among others (Bakour et al. 2022; Aksoy et al. 2024). Proline, which is often the most abundant amino acid in bee bread, is known for its role in collagen synthesis and cellular function, making it vital for skin health and tissue repair (Urcan et al. 2024; Aksoy et al. 2024). Glutamic acid and aspartic acid, both of which are prevalent in bee bread, are important for neurotransmission and can contribute to cognitive function (Bakour et al. 2022). The presence of essential amino acids, such as lysine and leucine, is particularly beneficial as they are crucial for protein synthesis, muscle repair, and overall metabolic health (Dervişoğlu et al. 2022; Giampieri et al. 2022). The fermentation process that transforms pollen into bee bread enhances the bioavailability of these amino acids, making them more accessible for absorption in the human body. This process involves the action of specific proteolytic enzymes that break down proteins into free amino acids, thereby increasing their concentration in the final product (DeGrandi-Hoffman et al. 2012; Giampieri et al. 2022).

Bee bread from various floral sources reported the following average values for essential amino acids: Arginine (1.10 g/100 g), leucine (0.80 g/100 g), lysine (0.70 g/100 g), and valine (0.60 g/100 g) (Aylanc et al. 2023). Another study highlighted that the total amino acid content in bee bread can range from 8.0% to 28% of its total composition, depending on the floral sources of the pollen and the specific environmental conditions (Pełka et al. 2021a, b).

The fermentation process that bee bread undergoes not only enhances its digestibility but may also influence the amino acid composition. The action of specific proteolytic enzymes during fermentation can lead to the cleavage of proteins into free amino acids, thereby increasing their bioavailability (DeGrandi-Hoffman et al. 2012). This characteristic distinguishes bee bread from raw pollen, which may have a lower digestibility and amino acid bioavailability due to its rigid cell walls.

Additionally, microbial activity during fermentation can contribute to the synthesis of certain amino acids, further enriching the amino acid profile of bee bread (DeGrandi-Hoffman et al. 2012). Research has indicated that the amino acid composition of bee bread can vary depending on the floral sources of the pollen used in its production; different plant species contribute varying amounts and types of proteins, which can affect the overall nutritional profile of bee bread (Mohammad et al. 2020; Özkök et al. 2022). For example, bee bread derived from certain flowers may exhibit higher concentrations of specific amino acids, enhancing its health benefits (Mohammad et al. 2020; Othman et al. 2021). The health benefits of the amino acids found in bee bread extend beyond basic nutrition. Studies have shown that the consumption of bee bread can support immune function, improve metabolic health, and even exhibit antioxidant properties due to the presence of amino acids along with other bioactive compounds (Othman et al. 2019). The antioxidant effects of amino acids, combined with the vitamins and minerals present in bee bread, contribute to its potential role in reducing oxidative stress and inflammation in the body (Habryka et al. 2023). Furthermore, the amino acids in bee bread have been associated with various therapeutic effects, including anti-inflammatory and antimicrobial activities. This makes bee bread not only a nutritious food source but also a functional food that may aid in the prevention and management of certain health conditions (Dervişoğlu et al. 2022; Sariguzel et al. 2022).

Bee bread's amino acid profile includes both essential and nonessential amino acids, which play critical roles in various metabolic pathways for both bees and humans (Fallah et al. 2021; Aksoy et al. 2024). Essential amino acids (EAAs) such as leucine, isoleucine, valine, lysine, methionine, phenylalanine, threonine, and tryptophan are particularly abundant (Hsu et al. 2021; Aylanc et al. 2023). Leucine is often highlighted as one of the most prevalent EAAs in bee bread, contributing significantly to its nutritional value (Aylanc et al. 2021; Urcan et al. 2021). The presence of EAAs is vital for honeybee development and colony health and has implications for human nutrition, as these amino acids cannot be synthesized by the human body and must be sourced from the diet (Hou and Wu 2018; Bryś and Strachecka 2024). Studies on the essential amino acid index (EAAI) indicate that bee bread provides a high-quality protein source comparable to other dietary proteins (Mayda et al. 2020).

The fermentation process in bee bread production plays a crucial role in increasing the concentration of free amino acids, which are readily absorbable by the body. This transformation enhances the bioavailability of amino acids, making bee bread a more efficient protein source (Urcan et al. 2024). Common free amino acids in bee bread include glutamic acid, aspartic acid, proline, and alanine, each with specific metabolic functions. Glutamic acid contributes to the umami flavor profile, while proline is linked to collagen synthesis and antioxidant activity, thus contributing to the health benefits associated with bee bread consumption (da Silva et al. 2014; Erol 2024). The presence of free amino acids not only elevates the nutritional quality of bee bread but also imparts characteristic flavors and enhances its digestibility (Bakour et al. 2019).

Bee bread's protein and amino acid content underscore its value as a functional food with potential dietary benefits. The high level of essential and free amino acids supports applications in human nutrition, particularly for individuals seeking natural protein sources with enhanced digestibility. The fermentation process's effect on protein breakdown further makes bee bread suitable for those with digestive sensitivities or those aiming to optimize nutrient absorption (Aksoy et al. 2024; Urcan et al. 2021). Research highlights bee bread's potential health benefits, including its support for immune function, muscle repair, and overall metabolic health, driven by its comprehensive amino acid profile (Erol 2024). Additionally, the presence of bioactive compounds such as phenolic acids and flavonoids contributes to its antioxidant and anti-inflammatory effects, enhancing its health-promoting potential (da Silva et al. 2014).

Proline is the most abundant amino acid in bee bread, significantly contributing to its unique nutritional profile and functional benefits (Szabados and Savouré 2010), proline plays important roles in cellular structure and repair, particularly through its involvement in collagen synthesis, which supports skin, joint, and cardiovascular health in humans (Domínguez-Valhondo et al. 2011). For honeybees, proline serves as a readily accessible energy source, especially during high-energy activities like foraging and hive maintenance (Mohammad et al. 2020; Erol 2024).

The concentration of proline in bee bread also serves as a potential marker of freshness. Fresh bee bread typically contains higher levels of free proline, which can decrease over time due to natural degradation processes. Therefore, monitoring proline levels offers insight into the quality and storage stability of bee bread, with higher proline content indicating a fresher product and optimal nutrient preservation (Campos et al. 2015; Oroian et al. 2022).

The digestibility of proteins and amino acids is a critical factor in evaluating the nutritional value of bee pollen and bee bread. Although bee pollen has a higher protein content, the digestibility of its proteins can be limited due to the tough outer wall of pollen grains, which is composed of sporopollenin and cellulose (Di Pasquale et al. 2013). In contrast, the fermentation process involved in the production of bee bread breaks down these tough structures, making the proteins and amino acids more accessible for digestion (DeGrandi-Hoffman et al. 2012). From a human nutritional perspective, both bee pollen and bee bread offer unique benefits. Bee pollen is often marketed as a superfood due to its high protein and amino acid content, making it a popular dietary supplement for those seeking natural sources of protein (Alcalá-Orozco et al. 2024); meanwhile, bee bread's enhanced digestibility and bioavailability of amino acids make it an attractive option for individuals looking for functional foods that support overall health and well-being (Aylanç et al. 2021).

While bee pollen and bee bread both serve as important nutritional sources for honey bees and have potential health benefits for humans, they differ in their protein and amino acid content, with bee pollen characterized by its higher protein concentration and rich essential amino acid profile, making it a valuable food source. Conversely, bee bread, although lower in total protein, offers enhanced digestibility and bioavailability of amino acids due to the fermentation process. Understanding these differences is essential for leveraging the nutritional benefits of these bee products in both apiculture and human nutrition (Table 13.1).

13.5 Lipid and Fatty Acid Content

The lipid content of bee bread is a significant aspect of its nutritional profile, contributing to its overall energy value and physiological functions, cellular structure, and metabolic processes (Fallah et al. 2021). Various studies have reported that the lipid content in bee bread typically ranges from 3% to 15%, depending on factors such as the botanical source of the pollen, specific fermentation conditions, environmental conditions, and the specific bee species involved in its production (Nagai et al. 2004; Bobis et al. 2017; Kankasemsuk 2023; Mohammad et al. 2020).

The lipid profile of bee bread is further shaped by the floral sources of the pollen, as different plants contribute varying lipid levels to the final product (Bobis et al. 2017; Urcan et al. 2021; Aldarf and Sultaneh 2024). This botanical influence underscores the role of pollen origin in determining bee bread's nutritional composition, particularly its lipid content (Tudor et al. 2020).

A study by Mohammad et al. (2020) reported that bee bread typically contains between 6% and 13% lipids, with variations depending on the botanical source of the pollen collected by bees This variability is further supported by research indicating that different plant species contribute distinct lipid profiles to the pollen, which

Table 13.1 Comparative analysis of amino acids in bee bread

	References							
	(Aksoy et al. 2024)	(Ghosh et al. 2024)	(Hashemirad et al. 2024)	(Bayram et al. 2021)	(Kutlu et al. 2024)	(Urcan et al. 2021)	(Mohammad et al. 2020)	(Aylanc et al. 2023)
Geographic origin	Turkey	South Korea	Iran	Turkey	Turkey	Romania	Malaysia	Portugal
Amino acid								
Alanine	678.57–1627.32 mg/100 g	1.16–1.95 g/100 g	11.63 mg/100 g	505.07– 1730.49 μg/g	981.80–1086.28 mg/kg	505.07–1730.49 μg/g	1.004–1.059 g/100 g	10.90–24.26 mg/100 g
Arginine	519.25–1393.56 mg/100 g	1.10–1.85 g/100 g	4.25 mg/100 g	2475.48–5891.12 μg/g	125.25–140.31 mg/kg	4.25 mg/100 g	A 1.959–2.660 g/100 g	0.05–0.07 mg/100 g
Aspartic acid	1785.48–3884.55 mg/100 g	2.09–3.27 g/100 g	8.00 mg/100 g	2833.36–5207.37 μg/g	411.92–426.58 mg/kg	2833.36–5207.37 μg/g	1.363–1.577 g/100 g	0.22–0.31 mg/100 g
Asparagine	–	–	0.01 mg/100 g	2475.48–5891.12 μg/g	–	–	–	0.65–1.82 mg/100 g
Cystine	66.22–114.71 mg/100 g	–	–	–	–	0.04–0.06 mg/100 g	–	2.02–2.53 mg/100 g
Cysteine	–	0.14–0.30 g/100 g	0.04 mg/100 g	–	–	–	–	0.52–3.45 mg/100 g
Glutamic acid	1439.47–3247.31 mg/100 g	2.53–4.02 g/100 g	6.62 mg/100 g	374.38–2626.95 μg/g	245.75–254.30 mg/kg	6.62 mg/100 g	1.491–1.742 g/100 g	0.03–0.09 mg/100 g
Glycine	576.79–1270.73 mg/100 g	0.98–1.77 g/100 g	15.17 mg/100 g	472.35–1053.58 μg/g	175.40–176.07 mg/kg	472.35–1053.58 μg/g	1.121–1.479 g/100 g	–
Gamma-aminobutyric acid (GABA)	–	0.04–0.92 mg/100 g	0.03 mg/100 g	2703.27–5079.35 μg/g	–	–	–	–
Histidine	294.98–632.17 mg/100 g	0.52–0.83 g/100 g	3.17 mg/100 g	87.82–1263.70 μg/g	270.96–262.13 mg/kg	3.17 mg/100 g	0.778–1.151 g/100 g	0.27–0.44 mg/100 g
Isoleucine	286.18–854.51 mg/100 g	0.96–1.62 g/100 g	–	357.61–918.07 μg/g	37.88–41.29 mg/kg	0.77 mg/100 g	0.760–0.938 g/100 g	0.66–1.56 mg/100 g
Leucine	819.94–2251.99 mg/100 g	1.62–2.69 g/100 g	3.62 mg/100 g	59.44–2120.40 μg/g	225.76–232.02 mg/kg	3.62 mg/100 g	1.606–1.834 g/100 g	0.34–2.03 mg/100 g
Lysine	716.94–1625.74 mg/100 g	1.51–2.75 g/100 g	7.29 mg/100 g	544.18–852.02 μg/g	290.51–310.25 mg/kg	7.29 mg/100 g	0.617–0.752 g/100 g	0.39–0.68 mg/100 g
Methionine	213.85–528.42 mg/100 g	0.24–0.35 g/100 g	1.39 mg/100 g	98.79–384.79 μg/g	17.49–18.92 mg/kg	1.39 mg/100 g	0.365–0.525 g/100 g	19.04–38.28 mg/100 g
Ornithine	1.47–8.04 mg/100 g	–	0.19 mg/100 g	–	–	–	–	–
Phenylalanine	485.90–1219.80 mg/100 g	0.97–1.69 g/100 g	2.40 mg/100 g	1308.44–3345.67 μg/g	129.67–150.37 mg/kg	1308.44–3345.67 μg/g	1.843–2.796 g/100 g	4.42–6.18 mg/100 g
Hydroxyproline	–	–	0.30 mg/100 g	–	–	–	0.259–0.581 g/100 g	–
Proline	2193.09–3453.15 mg/100 g	1.93–2.78 g/100 g	19.57 mg/100 g	4939.23–22212.82 μg/g	19992.67–21449.88 mg/kg	4939.23–22212.82 μg/g	1.478–1.711 g/100 g	0.18–0.57 mg/100 g

(continued)

Table 13.1 (continued)

	References							
	(Aksoy et al. 2024)	(Ghosh et al. 2024)	(Hashemirad et al. 2024)	(Bayram et al. 2021)	(Kutlu et al. 2024)	(Urcan et al. 2021)	(Mohammad et al. 2020)	(Aylanc et al. 2023)
Serine	726.90–1427.24 mg/100 g	1.09–1.94 g/100 g	8.33 mg/100 g	625.51–2625.31 μg/g	105.60–127.75 mg/kg	105.60–127.75 μg/g	1.783–2.182 g/100 g	2.53–5.35 mg/100 g
Threonine	519.68–1150.40 mg/100 g	0.92–1.72 g/100 g	3.23 mg/100 g	62.08–503.59 μg/g	11.29–14.81 mg/kg	11.29–14.81 mg/100 g	1.519–2.001 g/100 g	0.07–0.11 mg/100 g
Tyrosine	208.89–723.19 mg/100 g	0.71–1.14 g/100 g	1.44 mg/100 g	656.02–1628.17 μg/g	209.28–211.53 mg/kg	209.28–211.53 μg/g	1.100–1.634 g/100 g	1.68–3.25 mg/100 g
Tryptophan	–	–	–	377.12–2829.06 μg/g	5.76–6.81 mg/kg	377.12–2829.06 μg/g	–	–
Valine	414.83–1104.71 mg/100 g	1.07–1.86 g/100 g	–	–	93.58–118.18 mg/kg	93.58–118.18 mg/100 g	1.004–1.117 g/100 g	6.70–17.36 mg/100 g
Taurine	49.67–54.21 mg/100 g	–	–	83.70–180.41 μg/g	–	–	–	–

in turn affects the lipid content of the resulting bee bread. Mărgăoan et al. (2014) demonstrated that the lipid content in pollen can vary widely among plant species, with some species like *Taraxacum officinale* exhibiting lipid levels as high as 18.9%, while others, such as *Eucalyptus marginata*, show much lower levels at around 0.8%. This suggests that the source of pollen directly impacts the lipid content of bee bread. Further emphasizing the role of botanical origin, a comparative study of bee bread from different regions in Ukraine found that the lipid content varied significantly, correlating with the predominant floral sources in those areas (Bleha et al. 2019). The study highlighted that bee bread derived from specific plants not only differed in lipid content but also in the types of fatty acids present, which are crucial for various biological functions. In another study focusing on bee bread from different regions in Indonesia, it was reported that the lipid content ranged from 3.5% to 10.2%, depending on the botanical origin of the pollen (Erwan et al. 2021). This finding underscores the importance of local flora in determining the nutritional composition of bee bread, including its lipid content. The study also noted that climatic conditions and seasonal variations could further influence these values, indicating a complex interplay between environmental factors and botanical sources. Moreover, the fermentation process that bee bread undergoes can also affect its lipid profile. The transformation of pollen into bee bread involves microbial activity that can alter the lipid composition, potentially enhancing the bioavailability of certain fatty acids (Miłek et al. 2023a). This fermentation process may lead to a reduction in some lipid fractions while increasing others, depending on the initial lipid content of the pollen used (Hudz et al. 2019).

Climatic factors such as temperature, humidity, and seasonal variations can also impact the lipid content of bee bread. Research indicates that these environmental conditions can alter the chemical composition of the pollen collected by bees, which in turn affects the lipid content of the resulting bee bread (Donkersley et al. 2017).

Although relatively low in lipids, bee bread's nutritional profile is balanced by higher proportions of proteins and carbohydrates, making it a well-rounded food source for bees and a potentially valuable addition to human diets. In terms of energy content, bee bread is reported to have a caloric value of approximately 412 Kcal per 100 g, which is higher than many other bee products, such as honey and royal jelly (Fallah et al. 2022; Oroian et al. 2022). This high-energy density is attributed to its balanced composition of carbohydrates, proteins, and lipids, making it an excellent nutritional supplement. The lipid content specifically contributes to this energy density, providing a concentrated source of calories that can support the metabolic needs of both bees and humans (Oroian et al. 2022; Othman et al. 2021; Urcan et al. 2021).

The lipid composition of bee bread is not only quantitatively significant but also qualitatively diverse. Research has identified a variety of fatty acids present in bee bread, including both saturated and unsaturated fatty acids, which are crucial for various physiological functions in both bees and humans (Isidorov et al. 2009; Kaplan et al. 2016; Kieliszek et al. 2018; Urcan et al. 2021). The presence of essential fatty acids like linoleic acid and oleic acid has been documented, which play vital roles in cellular structure and function (Bakour et al. 2019; Kalpana et al.

2019). The fatty acid profile can vary significantly based on the floral sources from which the pollen is collected, indicating that the nutritional value of bee bread can be tailored through careful management of bee foraging practices (Aldarf and Sultaneh 2024; Mohammad et al. 2020; Urcan et al. 2021).

The primary omega-3 fatty acid found in bee bread is alpha-linolenic acid (ALA), which is essential for various biological functions. Studies have shown that bee bread can contain significant amounts of ALA, contributing to its nutritional value as a source of essential fatty acids (Čeksterytė et al. 2016; Trinkl et al. 2020). The ratio of omega-3 to omega-6 fatty acids in bee bread can be beneficial for cognitive functions in honey bees, highlighting the importance of these fatty acids in their diet (Trinkl et al. 2020). Furthermore, a study by Čeksterytė et al. (2016) reported that bee bread derived from specific floral sources, such as clover, exhibited a favorable omega-3 fatty acid profile, with ratios reaching as high as 8.42 in certain samples.

Linoleic acid (LA) is the primary omega-6 fatty acid present in bee bread (Urcan et al. 2021). The presence of linoleic acid in bee bread has been linked to improved cognitive performance in honey bees, emphasizing its importance in their diet (Trinkl et al. 2020). The omega-6 fatty acid content can vary significantly based on the botanical origin of the pollen, with some studies indicating that bee bread can contain substantial amounts of LA, which is essential for the bees' overall health (Arien et al. 2018) (Table 13.2).

Oleic acid, an omega-9 fatty acid, is also found in bee bread; while not classified as essential (since the body can synthesize it), oleic acid plays a vital role in maintaining heart health and reducing inflammation. The presence of oleic acid in bee bread contributes to its overall lipid profile and enhances its nutritional benefits (Kaplan et al. 2016). The levels of oleic acid can vary depending on the floral sources of the pollen, similar to the other fatty acids.

Beyond the omega-3, omega-6, and omega-9 fatty acids, bee bread contains a diverse array of other fatty acids. Research has identified up to 37 different fatty acids in bee bread, including palmitic acid (saturated), stearic acid (saturated), and various unsaturated fatty acids (Bakour et al. 2019; Isidorov et al. 2009; Markiewicz-Zukowska et al. 2013; Oroian et al. 2022; Urcan et al. 2021). These fatty acids contribute to the overall lipid profile and nutritional value of bee bread, enhancing its role as a functional food (Albar 2022).

The predominant fatty acids include palmitic acid, oleic acid, linoleic acid, and stearic acid, with oleic acid being particularly abundant (Dranca et al. 2020). The presence of polyunsaturated fatty acids (PUFAs), such as linoleic acid, is especially noteworthy due to their recognized health benefits, including anti-inflammatory properties and roles in cardiovascular health (Zakaria et al. 2022). Unsaturated fatty acids, particularly omega-3 and omega-6 fatty acids, are essential for human health as they cannot be synthesized by the body and must be obtained through diet (Bárbara et al. 2015). These fatty acids are crucial for maintaining cell membrane integrity, supporting brain function, and regulating inflammatory responses (Zakaria et al. 2022).

Table 13.2 Distribution of fatty acids in bee bread based on geographic and botanical origins

	References							
	(Urcan et al. 2021)	(Hashemirad et al. 2024)	(Bakour et al. 2019)	(Dranca et al. 2020)	(Mayda et al. 2020)	(Kaplan et al. 2016)	(Ciric et al. 2019)	(Aylanc et al. 2023)
Geographic origin/ botanic origin	Romania/*Brassicaceae—Brassica* sp., *Poaceae, Fabaceae—Trifolium, Tiliaceae—Tilia* sp., *Fabaceae, Rosaceae, Plantaginaceae—Plantago, Fabaceae—Lotus, Asteraceae—Taraxacum officinale, Asteraceae, Lamiaceae, Asteraceae—Centaurea montana*	Iran/*Asteraceae Fabaceae Brassicaceae Rosaceae Lamiaceae*	Morocco/ *Bupleurum spinosum, Anethum graveolens, Calendula officinalis, Anacyclus, Quercus ilex, Eucalyptus, Punica granatum, and Acacia*	Romania/Not determined	Turkey/*Asteraceae, Fabaceae, Plantaginaceae and Rosaceae*	Turkey/*Citrus* spp. *Asteraceae, Fabaceae, Brassicaceae*	Serbia/not determined	Portugal/*Plantago* sp., *Crepis capillaris (Asteraceae family), Cytisus striatus (Fabaceae family) and Castanea sativa Castanea sativa and Rubus* sp. *Jasione montana, Cytisus striatus and Castanea sativa*
Fatty acids								
Caprylic acid (8:0)	–	–	0.22 ± 0.0%	0.19%	0.91%	0.09 ± 0.02%	–	–
Capric acid (10:0)	0.01%–0.14%	–	0.18 ± 0.01%	0.08%	0.94%	0.05 ± 0.01%	–	–
Undecanoic acid (C11:0)	–	–	0.89 ± 0.01%	0 1.13%	–	0.09 ± 0.00%	–	–
Lauric acid (12:0)	0.01%–0.35%	–	3.2 ± 0.2%	0 0.30%	5.22%	0.60 ± 0.02%	–	0.06 g/100 g
Tridecanoic acid (C13:0)			0.39 ± 0.02%	0.11%	–	0.07 ± 0.00%	–	–
Myristic acid (14:0)	0.05%–0.26%	13.43%	0.37 ± 0.04%	0.25%	1.78%	0.65 ± 0.02%	0.63 ± 0.13%	–
Pentadecanoic acid (C15:0)	–	–	0.161 ± 0.002%	0.68%	–	0.09 ± 0.06%	0.28 ± 0.04%	0.06 g/100 g
Palmitic acid (16:0)	10.48%–46.66%	20.31%	10.6 ± 0.4%	0.24%	23.32%	17.83 ± 8.39%	23.00 ± 2.80%	1.01–1.14 g/100 g

(continued)

Table 13.2 (continued)

	References							
	(Urcan et al. 2021)	(Hashemirad et al. 2024)	(Bakour et al. 2019)	(Dranca et al. 2020)	(Mayda et al. 2020)	(Kaplan et al. 2016)	(Ciric et al. 2019)	(Aylanc et al. 2023)
C16:1	–	–	1.09 ± 0.05	21.10%	–	0.12 ± 0.01%	0.30 ± 2.80%	–
Heptadecanoic acid (C17:0)	–	–	0.197 ± 0.004%	0.57%	–	0.49 ± 0.01%	–	0.11–0.16 g/100 g
Stearic acid (18:0)	0.14%–2.43%	2.38%	3.5 ± 0.1%	0.11%	3.21%	6.56 ± 0.05%	5.54 ± 0.25%	–
Oleic acid (18:1, ω-9)	0.30% - 14.95%	10.89%	5.7 ± 0.3%	3.63%	–	0.07 ± 0.03%	17.74 ± 0.03%	0.43–0.55 g/100 g
Elaidic acid (18:1, 9 t, ω-9)	0.02%–0.54%	–	–	10.28%	8.76%	–	–	–
Linoleic acid (18:2, ω-6)	1.29%–36.99%	24.04%	14.2 ± 0.1%	30.68%	6.02%	–	–	0.96 g/100 g
Alpha-linolenic acid (18:3, ω-3)	18.11%–82.77%	24.40%	25 ± 1%	–	–	–	21.20 ± 2.79%	–
Arachic acid (20:0)	0.04%–2.28%	–	4.4 ± 0.3%	1.21%	–	3.35 ± 0.01%	–	–
Heneicosanoic acid (C21:0)	–	–	0.403 ± 0.005	11.86%	–	0.45 ± 0.02%	1.99 ± 0.04%	–
Gadoleic acid (C20:1)	–	–	0.277 ± 0.004%	0.36	–	8.51 ± 0.12%	–	–
Eicosadienoic acid (C20:2)	–	–	0.316 ± 0.002%	0.27	–	–	–	–
Dihomo-γ-linolenic acid (C20:3n6)	–	–	0.49 ± 0.03%	0.68%	–	–	0.39 ± 0.01%	–
Arachidonic acid (C20:4n6)	–	–	23.2 ± 0.5%	0.03	2.65	–	–	–
Eicosatrienoic acid (omega-3) (C20:3n3)	–	–	1.01 ± 0.04%	0.33	–	0.52 ± 0.05%	–	–
Behenic acid (22:0)	0.08%–1.23%	–	0.01%–1.16%	0.01	1.33	0.09 ± 0.00%	–	0.32 g/100 g
Lignoceric acid (C24:0)	–	–	1.56 ± 0.02	0.46%	0.6	0.13 ± 0.01%	3.97 ± 0.48%	0.35–0.47 g/100 g
Nervonic acid (C24:1)	–	–	0.107 ± 0.003	0.04%	–	0.34 ± 0.01%	0.15 ± 0.01%	–

This diversity underscores the importance of the botanical origin of the pollen, as different plants contribute distinct fatty acid profiles to the bee bread. For example, bee bread derived from rape or willow pollen has been shown to have particularly high levels of specific fatty acids, including both omega-3 and omega-6 (Kaplan et al. 2016).

13.6 Vitamins and Minerals

13.6.1 Vitamins

Bee bread is an exceptionally nutrient-dense food, containing a rich array of vitamins that contribute to its value as a functional food (Hashemirad et al. 2024). These vitamins play critical roles in cellular function, energy metabolism, immune support, and antioxidative defense, making bee bread an ideal supplement for both humans and bees (Bakour et al. 2023). The vitamin composition of bee bread is influenced by pollen sources, fermentation processes within the hive, and environmental conditions, leading to variations in its nutritional profile (Ćirić et al. 2022).

Primarily, B-complex vitamins and vitamin C are abundant in bee bread. These vitamins play crucial roles in energy metabolism, red blood cell formation, and overall cellular function (Baky et al. 2023).

The vitamin content of bee bread includes several essential vitamins, notably those from the B-complex group, as well as vitamins A, C, E, and K. Research indicates that bee bread is a significant source of B vitamins, including B1 (thiamine), B2 (riboflavin), B3 (niacin), B5 (pantothenic acid), B6 (pyridoxine), B7 (biotin), B9 (folic acid), and B12 (cobalamin) (Nagai et al. 2004; Suleiman et al. 2021). The concentrations of these vitamins can vary based on factors such as the floral source of the pollen, geographical location, and environmental conditions during production (Othman et al. 2019; Uslinawaty 2023). Vitamin B1 (Thiamine): This vitamin is essential for carbohydrate metabolism and nerve function. In a study made by Suleiman et al. 2021). Malaysian bee bread was found to contain significant levels of thiamine, although specific quantitative values were not provided. Other studies have indicated that bee bread can contain thiamine concentrations ranging from 0.1 to 0.5 mg per 100 g (Dervişoğlu et al. 2022).

Vitamin B2 (Riboflavin): Riboflavin is vital for energy production and cellular function. Research indicates that bee bread contains riboflavin levels between 0.2 to 0.6 mg per 100 g (Dervişoğlu et al. 2022). This vitamin's presence contributes to the overall nutritional value of bee bread, supporting metabolic processes.

Vitamin B3 (Niacin): Niacin is involved in DNA repair and the production of steroid hormones in the adrenal gland. Studies have shown that bee bread can provide niacin in amounts ranging from 1.0 to 3.0 mg per 100 g, depending on the pollen sources used (Dranca et al. 2020). This vitamin is crucial for maintaining healthy skin and nerves. The natural fermentation process of bee bread enhances the bioavailability of these vitamins, making them more easily absorbed and utilized by both bees and humans (Barta et al. 2022).

The concentration of vitamin C in Malaysian bee bread samples has been reported to range from 0.1087 to 0.1152 mg per gram (Erol 2024); this vitamin is crucial for collagen synthesis, immune function, and as an antioxidant. Vitamin E, primarily in the form of tocopherols, has been detected in Moroccan bee bread samples, with concentrations varying between 10.5 and 0.40 mg per 100 g (Erol 2024). Vitamin E is known for its antioxidant properties and its role in protecting cell membranes from oxidative damage.

Vitamin K is another important vitamin found in bee bread, which plays a vital role in blood coagulation and bone health, studies have indicated that bee bread contains higher levels of vitamin K compared to raw pollen, although specific quantitative data on its concentration in bee bread is less frequently reported (Bakour et al. 2021; Çelik et al. 2022). The presence of vitamin K in bee bread underscores its potential as a functional food ingredient. The B vitamins present in bee bread are particularly noteworthy due to their involvement in various metabolic processes, B1 (thiamine) is essential for energy metabolism, while B2 (riboflavin) is crucial for the metabolism of fats, drugs, and steroids (Nagai et al. 2005). B5 (pantothenic acid) is involved in the synthesis of coenzyme A, which is important for fatty acid metabolism. The concentrations of these B vitamins in bee bread have been reported to be significant, contributing to its overall nutritional value (Bakour et al. 2023). In addition to the vitamins mentioned, bee bread also contains other bioactive compounds, including carotenoids and flavonoids, which may enhance the bioavailability of vitamins and contribute to their antioxidant effects (Mărgăoan et al. 2019; Miłek et al. 2023a). The fermentation process that bee bread undergoes not only increases the digestibility of its components but may also enhance the stability and bioavailability of vitamins, making them more accessible for absorption in the human body (Khalifa et al. 2020; Zakaria et al. 2024. The vitamin content of bee bread can be influenced by the botanical origin of the pollen, as different plants have varying vitamin profiles, and the pollen collected from certain flowers may yield higher concentrations of specific vitamins, thereby affecting the final composition of the bee bread (Uslinawaty 2023). Research has shown that the vitamin content of bee bread is generally higher than that of raw pollen, which may be attributed to the fermentation process that enhances the nutritional profile of the product (Bakour et al. 2019; Baky et al. 2023; Keskin 2021; Poyraz et al. 2023). This characteristic makes bee bread a valuable dietary supplement, particularly for individuals seeking to increase their intake of essential vitamins.

Vitamin B1 (Thiamine): This vitamin is essential for carbohydrate metabolism and nerve function. In a study made by Suleiman et al. 2021), Malaysian bee bread was found to contain significant levels of thiamine, although specific quantitative values were not provided. Other studies have indicated that bee bread can contain thiamine concentrations ranging from 0.1 to 0.5 mg per 100 g (Dervişoğlu et al. 2022).

Vitamin B2 (Riboflavin): Riboflavin is vital for energy production and cellular function. Research indicates that bee bread contains riboflavin levels between 0.2 to 0.6 mg per 100 g (Dervişoğlu et al. 2022). This vitamin's presence contributes to the overall nutritional value of bee bread, supporting metabolic processes.

Vitamin B3 (Niacin): Niacin is involved in DNA repair and the production of steroid hormones in the adrenal gland. Studies have shown that bee bread can provide niacin in amounts ranging from 1.0 to 3.0 mg per 100 g, depending on the pollen sources used (Dranca et al. 2020). This vitamin is crucial for maintaining healthy skin and nerves.

The natural fermentation process of bee bread enhances the bioavailability of these vitamins, making them more easily absorbed and utilized by both bees and humans (Barta et al. 2022).

13.6.2 Minerals

Bee bread, a nutrient-rich product created by the fermentation of pollen collected by honeybees, contains a variety of essential minerals that contribute to its nutritional profile (Khalifa et al. 2020). The mineral content in bee bread can vary widely, depending on factors such as the botanical origin of the pollen and environmental conditions, and play crucial roles in cellular functions, enzyme activities, and overall metabolic health, making them an important aspect of its composition (Bakour et al. 2019). However, studies have consistently shown that bee bread contains significant quantities of macro-minerals, including potassium, calcium, and magnesium, as well as trace elements like iron, zinc, manganese, and selenium, all of which are essential for human health (Kieliszek et al. 2018; Mohammad et al. 2021).

The mineral content of bee bread is diverse, comprising essential elements such as potassium (K), calcium (Ca), magnesium (Mg), sodium (Na), iron (Fe), zinc (Zn), copper (Cu), manganese (Mn), and selenium (Se) (da Silva et al. 2014; Suleiman et al. 2021; Fallah et al. 2022). The concentrations of these minerals can vary significantly based on several factors, including the floral sources of the pollen, geographical location, and environmental conditions under which the bee bread is produced (Çelik et al. 2022; Uslinawaty 2023). Suleiman et al. (2021) indicated that bee bread can contain up to 1.5% potassium, which is particularly notable given its importance in maintaining fluid balance and supporting cellular functions calcium is another critical mineral found in bee bread, with reported concentrations ranging from 0.1% to 0.5% This mineral plays a vital role in bone health, muscle function, and nerve transmission. Magnesium is also present in significant amounts, with studies showing levels around 0.2–0.4% (Habryka et al. 2020). Magnesium is essential for numerous biochemical reactions in the body, including energy production and protein synthesis. Iron content in bee bread has been documented to vary, with concentrations typically ranging from 0.01% to 0.1% (Fallah et al. 2022). Iron is crucial for hemoglobin formation and oxygen transport in the blood. The presence of zinc, which is vital for immune function and wound healing, has been reported in concentrations of approximately 0.01–0.05% (Suleiman et al. 2021). Copper and manganese are also found in smaller amounts, often less than 0.01%, but they are essential for various enzymatic processes in the body (da Silva et al. 2014). The mineral profile of bee bread is influenced by the botanical origin of the pollen used in its production, for example, pollen collected from certain plants may have higher concentrations of specific minerals, leading to variations in the final mineral content of the bee bread (da Silva et al. 2014; Özkök et al. 2022). Additionally, environmental factors such as soil composition and climate can affect the mineral uptake by plants, thereby influencing the mineral content of the pollen and, subsequently, the bee bread (Uslinawaty 2023). Research has shown that bee bread typically contains higher mineral concentrations compared to raw pollen. This is attributed to the fermentation process, which enhances the bioavailability of minerals, making them more accessible for absorption in the human body (Suleiman et al. 2021; Fallah et al. 2022). The fermentation process also contributes to the overall digestibility of bee bread, allowing for better utilization of its nutritional components. In terms of specific mineral concentrations, a study conducted on Malaysian bee bread revealed the following average values: Potassium (1.5%), calcium (0.3%), magnesium (0.2%), iron (0.05%), and zinc (0.02%) (Suleiman et al. 2021). These values highlight the potential of bee bread as a mineral-rich food source. Furthermore, the presence of trace minerals such as selenium, although in smaller amounts, adds to the overall nutritional profile of bee bread (Suleiman et al. 2021; Fallah et al. 2022). The mineral content of bee bread can also be affected by processing and storage conditions. For instance, improper storage may lead to mineral degradation or loss, emphasizing the importance of proper handling to maintain the nutritional quality of bee bread (Jarušaitienė et al. 2012; Uslinawaty 2023). Additionally, the method of fermentation can influence the mineral profile, as different fermentation conditions may result in varying degrees of mineral solubility and bioavailability (Araneda Durán et al. 2014) (Table 13.3).

The bioavailability of these macro- and microelements in bee bread is enhanced by the natural fermentation process it undergoes within the hive. Fermentation improves the digestibility and absorption of minerals, allowing bees and humans to efficiently utilize these nutrients (Zhang et al. 2022).

13.7 Bioactive Compounds

13.7.1 Phenolic Acids and Flavonoids

Bee bread, a natural product created through the fermentation of pollen by honey bees, has garnered significant attention in recent years for its rich nutritional and bioactive profile (Ilie et al. 2024). Among its various health-promoting compounds, flavonoids stand out for their antioxidant properties and potential therapeutic benefits. These bioactive molecules, derived from plant sources, vary widely in type and concentration depending on factors such as pollen origin and environmental conditions (Beykaya et al. 2021). Bee bread is rich in a diverse array of flavonoids and phenolic acids, many of which exist in glycosylated forms—compounds where a sugar molecule is bonded to a non-sugar moiety, enhancing the stability and bioavailability of these bioactive components. This glycosylation plays a crucial role in the antioxidant and therapeutic potential of bee bread, which is attributed to the complex interaction of these compounds (Urcan et al. 2018a, b).

Table 13.3 Chemical composition of bee bread: ranges and variability based on botanical origin

Class of compounds	Geographical origin/Botanical origin	Concentration range	References
Moisture	Turkey/*Castanea sativa, Trifolium* spp., *Gossypium hirsutum, Citrus* spp., *Helianthus annuus*	11.41 ± 0.01–15.89 ± 0.01 (%)	Kaplan et al. (2016)
	Malaysia/*Mimosa pudica, Sphagneticola trilobata, Bidens ilosa, Cassia sp, Areca catechu, Peltophorum pterocarpum, Phaleria capitata, Cassia siamea, Citrus aurantifolia,* and *Ageratum conyzoides*	12.51 ± 0.07 (%)	Mohammad et al. (2020)
	Turkey/Asteraceae, Fabaceae, Plantaginaceae, Rosaceae families	18.4 ± 0.2 (%)	Mayda et al. (2020)
	Iran/Not determined	11.81 ± 0.90 (g/100 g)	Fallah et al. (2021)
	Ukraine/*Papaver* sp. *Trifolia* sp., *Helianthus annuus, Fagopyrum esculentum*	8.81–11.65 (% m/m)	Bleha et al. (2019)
	Romania/*Salicaceae (Salix* sp.*), Fabaceae (Onobrychis viciifolia), Brassicaceae, Asteraceae, Boraginaceae (Symphytum* sp.*), Rosaceae (Prunus* sp.*), Apiaceae, Fagaceae (Quercus* sp.*), Plantaginaceae, Lamiaceae, Polygonaceae, Cornaceae (Cornus* sp.*), Betulaceae (Carpinus betulus), Liliaceae*	9.62 ± 0.21–14.15 ± 0.06 (%)	Urcan et al. (2021)
	Egypt/Not determined	25.92 ± 0.14 (g/100 g)	Darwish et al. (2022)
	Portugal/*Salix* sp., *Castanea sativa, Lavandula* sp., *Erica* sp., *Echium* sp., *Rubus* sp., *Plantago* sp., *Prunus* sp., *Cistus ladanifer*, *Crataegus monogyna*, *Cytisus striatus*, *Raphanus raphanistrum*, *Sesamoides* sp., *Anthemis arvensis*, *Quercus* sp., *Hedera helix Eucalyptus* sp., etc.	2.5 ± 0.20–19.0 ± 0.1 (%)	Tomás et al. (2017)

(continued)

Table 13.3 (continued)

Class of compounds	Geographical origin/Botanical origin	Concentration range	References
Ash	Turkey/*Castanea sativa, Trifolium* spp., *Gossypium hirsutum, Citrus* spp. *and Helianthus annuus*	1.93 ± 0.02–2.62 ± 0.01 (%)	Kaplan et al. (2016)
	Morocco/*Bupleurum spinosum, Anethum graveolens, Calendula officinalis, Anacyclus, Quercus ilex, Eucalyptus, Punica granatum,* and *Acacia*	3.32 ± 0.01 (g/100 g)	Bakour et al. (2019)
	Romania/Not determined	3.42 (%)	Dranca et al. (2020)
	Malaysia/*Mimosa pudica, Sphagneticola trilobata, Bidens ilosa, Cassia sp, Areca catechu, Peltophorum pterocarpum, Phaleria capitata, Cassia siamea, Citrus aurantifolia,* and *Ageratum conyzoides*	2.46 ± 0.02 (%)	Mohammad et al. (2020)
	Turkey/Asteraceae, Fabaceae, Plantaginaceae, and Rosaceae families	2.55 ± 0.04 (%)	Mayda et al. (2020)
	Iran/*Not determined*	2.11 ± 0.15 (g/100 g)	Fallah et al. (2021)
	Romania/*Salicaceae (Salix* sp.*), Fabaceae (Onobrychis viciifolia), Brassicaceae, Asteraceae, Boraginaceae (Symphytum* sp.*), Rosaceae (Prunus* sp.*), Apiaceae, Fagaceae (Quercus* sp.*), Plantaginaceae, Lamiaceae, Polygonaceae, Cornaceae (Cornus* sp.*), Betulaceae (Carpinus betulus), Liliaceae*	2.65 ± 0.22–4.02 ± 0.10 (%)	Urcan et al. (2021)
	Ukraine/*Papaver* sp. *Trifolia* sp., *Helianthus annuus, Fagopyrum esculentum*	3.17–5.15 (% m/m)	Bleha et al. (2019)
	Serbia/*Not determined*	3.05 (%)	Anđelković et al. (2012)
	Egypt/Not determined	3.02 ± 0.07 (g/100 g)	Darwish et al. (2022)
	Portugal/Salix sp., *Castanea sativa, Lavandula* sp., *Erica* sp., *Echium* sp., *Rubus* sp., *Plantago* sp., *Prunus* sp., *Cistus ladanifer, Crataegus monogyna, Cytisus striatus, Raphanus raphanistrum, Sesamoides* sp., *Anthemis arvensis, Quercus* sp., *Hedera helix Eucalyptus* sp., etc.	2.1 ± 0.01–7.7 ± 0.3 (%)	Tomás et al. (2017)

Total carbohydrates	Morocco/*Apiaceae (Bupleurum spinosum; Anethum graveolens), Asteraceae 224 (Calendula officinalis; Anacyclus), Fagaceae (Quercus ilex), Myrtaceae (Eucalyptus), Punicaceae (Punica granatum), Mimosaceae (Acacia)*; 5% other types of pollen grains	74.82 ± 0.04 (g/100 g)	Bakour et al. (2019)
	Romania/Not determined	72.83 (%)	Dranca et al. (2020)
	Malaysia/*Mimosa pudica, Sphagneticola trilobata, Bidens ilosa, Cassia sp, Areca catechu, Peltophorum pterocarpum, Phaleria capitata, Cassia siamea, Citrus aurantifolia,* and *Ageratum conyzoides*	57.06 ± 2.09–58.89 ± 0.28 (%)	Mohammad et al. (2020)
	Romania/Brassicaceae—*Brassica* sp., *Poaceae, Fabaceae—Trifolium, Tiliaceae—Tilia* sp., *Fabaceae, Rosaceae, Plantaginaceae—Plantago, Fabaceae—Lotus, Asteraceae—Taraxacum officinale, Asteraceae, Lamiaceae, Asteraceae—Centaurea montana*	62.06 ± 0.35–70.71 ± 0.19 (%)	Urcan et al. (2021)
	Iran/Not determined	55.60 ± 3.30 (g/100 g)	Fallah et al. (2021)
	Egypt/Not determined	26.78 ± 1.80 (%)	Darwish et al. (2022)
	Not determined	24–35 (%)	Campos et al. (2010), Khalifa et al. (2020)
	Portugal/*Salix* sp., *Castanea sativa, Lavandula* sp., *Erica* sp., *Echium* sp., *Rubus* sp., *Plantago* sp., *Prunus* sp., *Cistus ladanifer*, *Crataegus monogyna, Cytisus striatus, Raphanus raphanistrum*, *Sesamoides* sp., *Anthemis arvensis*, *Quercus* sp., *Hedera helix, Eucalyptus* sp., etc.	58 ± 0.00–78 ± 0.00 (%)	Tomás et al. (2017)
	Turkey/Not determined	24.40–34.80 (%)	Aksoy et al. (2024)
	Iran/*Not determined*	51.60 ± 4.30 (g/100 g)	Hashemirad et al. (2024)

(continued)

Table 13.3 (continued)

Class of compounds	Geographical origin/Botanical origin	Concentration range	References
Fructose	Romania/*Salicaceae (Salix* sp.*), Fabaceae (Onobrychis viciifolia), Brassicaceae, Asteraceae, Boraginaceae (Symphytum* sp.*), Rosaceae (Prunus* sp.*), Apiaceae, Fagaceae (Quercus* sp.*), Plantaginaceae, Lamiaceae, Polygonaceae, Cornaceae (Cornus* sp.*), Betulaceae (Carpinus betulus), Liliaceae*	15.46 ± 0.09–20.98 ± 0.12 (%)	Urcan et al. (2021)
	Romania/Brassicaceae—*Brassica* sp.*, Poaceae, Fabaceae—Trifolium, Tiliaceae—Tilia* sp.*, Fabaceae, Rosaceae, Plantaginaceae—Plantago, Fabaceae—Lotus, Asteraceae—Taraxacum officinale, Asteraceae, Lamiaceae, Asteraceae—Centaurea montana*	19.32 ± 0.01–13.97 ± 0.05 (%)	Urcan et al. (2018b)
	Romania/Not determined	19.73 (%)	Dranca et al. (2020)
	Malaysia/*Mimosa pudica, Sphagneticola trilobata, Bidens ilosa, Cassia sp, Areca catechu, Peltophorum pterocarpum, Phaleria capitata, Cassia siamea, Citrus aurantifolia, and Ageratum conyzoides*	1.48 ± 0.14–0.39 ± 0.01 (g/100 g)	Mohammad et al. (2020)
	Portugal/*Salix* sp., *Castanea sativa, Lavandula* sp., *Erica* sp., *Echium* sp., *Rubus* sp., *Plantago* sp., *Prunus* sp., *Cistus ladanifer*, *Crataegus monogyna, Cytisus striatus, Raphanus raphanistrum, Sesamoides* sp., *Anthemis arvensis*, *Quercus* sp., *Hedera helix, Eucalyptus* sp., etc.	94 ± 2.00–223 ± 0.01 (mg/g)	Tomás et al. (2017)
	Turkey/Not determined	16.35–11.07 (g/100 g)	Aksoy et al. (2024)
	Morocco/Not determined	11.8 (g/100 g)	Bakour et al. (2017)

Glucose	Romania/*Salicaceae (Salix* sp.*), Fabaceae (Onobrychis viciifolia), Brassicaceae, Asteraceae, Boraginaceae (Symphytum* sp.*), Rosaceae (Prunus* sp.*), Apiaceae, Fagaceae (Quercus* sp.*), Plantaginaceae, Lamiaceae, Polygonaceae, Cornaceae (Cornus* sp.*), Betulaceae (Carpinus betulus),* and *Liliaceae*	7.26 ± 0.12–12.95 ± 0.08 (%)	Urcan et al. (2021)
	Romania/Brassicaceae—*Brassica* sp., *Poaceae, Fabaceae—Trifolium, Tiliaceae—Tilia* sp., *Fabaceae, Rosaceae, Plantaginaceae—Plantago, Fabaceae—Lotus, Asteraceae—Taraxacum officinale, Asteraceae, Lamiaceae, Asteraceae—Centaurea montana*	6.40 ± 0.01–13.19 ± 0.01 (%)	Urcan et al. (2018b)
	Romania/Not determined	8.82 (%)	Dranca et al. (2020)
	Malaysia/*Mimosa pudica, Sphagneticola trilobata, Bidens ilosa, Cassia sp, Areca catechu, Peltophorum pterocarpum, Phaleria capitata, Cassia siamea, Citrus aurantifolia,* and *Ageratum conyzoides*	12.397 ± 0.98–10.270 ± 0.14 (g/100 g)	Mohammad et al. (2020)
	Portugal/*Salix* sp., *Castanea sativa*, *Lavandula* sp., *Erica* sp., *Echium* sp., *Rubus* sp., *Plantago* sp., *Prunus* sp., *Cistus ladanifer*, *Crataegus monogyna*, *Cytisus striatus*, *Raphanus raphanistrum*, *Sesamoides* sp., *Anthemis arvensis*, *Quercus* sp., *Hedera helix Eucalyptus* sp., etc	67 ± 4.00–173 ± 6.00 (mg/g)	Tomás et al. (2017)
	Turkey/Not determined	6.34–9.55 (g/100 g)	Aksoy et al. (2024)
	Morocco/Not determined	5.7 (g/100 g)	Bakour et al. (2017)

(continued)

Table 13.3 (continued)

Class of compounds	Geographical origin/Botanical origin	Concentration range	References
Proteins	Morocco/Apiaceae (*Bupleurum spinosum*; *Anethum graveolens*), Asteraceae 224 (*Calendula officinalis*; *Anacyclus*), Fagaceae (*Quercus ilex*), Myrtaceae (*Eucalyptus*), Punicaceae (*Punica granatum*), Mimosaceae (*Acacia*); 5% other types of pollen grains	19.96 ± 0.08 (g/100 g)	Bakour et al. (2019)
	Turkey/*Castanea sativa, Trifolium* spp., *Gossypium hirsutum, Citrus* spp., and *Helianthus annuus*	14.82 ± 0.01–24.26 ± 0.01 (%)	Kaplan et al. (2016)
	Romania/Salicaceae (*Salix* sp.), Fabaceae (*Onobrychis viciifolia*), Brassicaceae, Asteraceae, Boraginaceae (*Symphytum* sp.), Rosaceae (*Prunus* sp.), Apiaceae, Fagaceae (*Quercus* sp.), Plantaginaceae, Lamiaceae, Polygonaceae, Cornaceae (*Cornus* sp.), Betulaceae (*Carpinus betulus*), Liliaceae	18.41 ± 0.11–23.11 ± 0.08 (%)	Urcan et al. (2021)
	Romania/Not determined	18.60 (%)	Dranca et al. (2020)
	Egypt/Not determined	22.57 ± 0.07 (g/100 g)	Darwish et al. (2022)
	Serbia/Not determined	29.93 (%)	Anđelković et al. (2012)
	Malaysia/*Mimosa pudica, Sphagneticola trilobata, Bidens ilosa, Cassia sp, Areca catechu, Peltophorum pterocarpum, Phaleria capitata, Cassia siamea, Citrus aurantifolia, and Ageratum conyzoides*	21.70 ± 0.08–23.33 ± 0.48 (%)	Mohammad et al. (2020)
	Portugal/*Salix* sp., *Castanea sativa*, *Lavandula* sp., *Erica* sp., *Echium* sp., *Rubus* sp., *Plantago* sp., *Prunus* sp., *Cistus ladanifer*, *Crataegus monogyna*, *Cytisus striatus*, *Raphanus raphanistrum*, *Sesamoides* sp., *Anthemis arvensis*, *Quercus* sp., *Hedera helix, Eucalyptus* sp., etc	14.1 ± 0.5–21.8 ± 0.2 (%)	Tomás et al. (2017)
	Iran/Not determined	18.50 ± 1.50 (g/100 g)	Hashemirad et al. (2024)
	Thailand/Not determined	1.48–7.03 (g/100 g)	Kankasemsuk (2023)
	Indonesia/Not determined	23.1 (g/100 g)	Erwan et al. (2021)
	Iran/Not determined	8.0–28 (%)	Fallah et al. (2021)
	Colombia/Not determined	19.1–27.3 (%)	Zuluaga et al. (2015)

Lipids	Morocco/Apiaceae (*Bupleurum spinosum*; *Anethum graveolens*), Asteraceae (*Calendula officinalis*; *Anacyclus*), Fagaceae (*Quercus ilex*), Myrtaceae (*Eucalyptus*), Punicaceae (*Punica granatum*), Mimosaceae (*Acacia*); 5% other types of pollen grains	1.90 ± 0.086 (g/100 g)	Bakour et al. (2019)
	Turkey/*Castanea sativa, Trifolium* spp., *Gossypium hirsutum, Citrus* spp., and *Helianthus annuus*	5.93–11.55 (g/100 g)	Kaplan et al. (2016)
	Romania/*Salicaceae (Salix* sp.*), Fabaceae (Onobrychis viciifolia), Brassicaceae, Asteraceae, Boraginaceae (Symphytum* sp.*), Rosaceae (Prunus* sp.*), Apiaceae, Fagaceae (Quercus* sp.*), Plantaginaceae, Lamiaceae, Polygonaceae, Cornaceae (Cornus* sp.*), Betulaceae (Carpinus betulus), Liliaceae*	4.89 ± 0.12–14.26 ± 0.10 (%)	Urcan et al. (2021)
	Romania/Not determined	5.15 (%)	Dranca et al. (2020)
	Malaysia/*Mimosa pudica, Sphagneticola trilobata, Bidens ilosa, Cassia sp, Areca catechu, Peltophorum pterocarpum, Phaleria capitata, Cassia siamea, Citrus aurantifolia,* and *Ageratum conyzoides*	58.89 ± 0.28–5.95 ± 0.10 (%)	Mohammad et al. (2020)
	Serbia/Not determined	4.92 (%)	Anđelković et al. (2012)
	Portugal/*Salix* sp., *Castanea sativa*, *Lavandula* sp., *Erica* sp., *Echium* sp., *Rubus* sp., *Plantago* sp., *Prunus* sp., *Cistus ladanifer*, *Crataegus monogyna*, *Cytisus striatus*, *Raphanus raphanistrum*, *Sesamoides* sp., *Anthemis arvensis*, *Quercus* sp., *Hedera helix*, *Eucalyptus* sp., etc	3.6 ± 0.1–14.9 ± 0.1 (%)	Tomás et al. (2017)
	Iran/Not determined	11.40 ± 0.90 (g/100 g)	Hashemirad et al. (2024)
	Malaysia/Not determined	4.51 (%)	Othman et al. (2019)
	Indonesia/Not determined	3.4 (g/100 g)	Erwan et al. (2021)
	Colombia/Not determined	1.65–5.50 (%)	Zuluaga et al. (2015)
	Portugal/Not determined	3.6–16.8 (g/100)	Bonvehı and Escola (1997)

(continued)

Table 13.3 (continued)

Class of compounds	Geographical origin/Botanical origin	Concentration range	References
pH	Poland/Not determined	3.99 ± 0.01–3.99 ± 0.01	Miłek et al. (2023b)
	Not determined	3.8–4.2	Khalifa et al. (2020), Kieliszek et al. (2018)
	Not determined	4.3	Barene et al. (2014)
	Malaysia/Not determined	4.0–4.5	Suleiman et al. (2021)
	Belgium/Not determined	4	Detry et al. (2020)
	Swedish/Not determined	4.3	Vásquez and Olofsson (2009)
	Lithuania/Not determined	4.12–4.44	Adaškevičiūtė et al. (2019)
	Portugal/*Plantago* sp., *Crepis capillaris (Asteraceae family), Cytisus striatus (Fabaceae family)* and *Castanea sativa Castanea sativa* and *Rubus* sp. *Jasione montana, Cytisus striatus,* and *Castanea sativa*	4.0–4.7	Aylanc et al. (2023)
Lactic acid	Poland/Not determined	3.24 ± 0.46%–4.70 ± 0.43 (%)	Miłek et al. (2023b)
	Turkey/Not determined	7.08–37.86 (mg/g)	Aksoy et al. (2024)
	Not determined	3.06–3.20 (%)	Barene et al. (2014)
	Romania/Not determined	3.06–3.20 (%)	Dranca et al. (2020)
Vitamins			
Vitamin C	Malaysia (*Bidens pilosa*)	0.1087–0.1152 (mg/g)	Mohammad et al. (2020)
	Morocco (*Apiaceae, Asteraceae, Fagaceae, Myrtaceae*)	6.9–12.4 (mg/100 g)	Bakour et al. (2019)
	Poland (*Multifloral*)	11.0–17.4 (mg/100 g)	Sawicki et al. (2022)
	Malaysia/Not determined	0.1087–0.1152 (mg/g)	Erol (2024)

Vitamin E (Tocopherols)	Portugal/*Salix* sp., *Castanea sativa*, *Lavandula* sp., *Erica* sp., *Echium* sp., *Rubus* sp., *Plantago* sp., *Prunus* sp., *Cistus ladanifer*, *Crataegus monogyna*, *Cytisus striatus*, *Raphanus raphanistrum*, *Sesamoides* sp., *Anthemis arvensis*, *Quercus* sp., *Hedera helix, Eucalyptus* sp., etc	0.04–302 (mg/g)	Tomás et al., (2017
	Morocco (*Apiaceae, Asteraceae, Fagaceae, Myrtaceae*)	10.5–40 (mg/100 g)	Bakour et al. (2019)
Beta-carotene (Provitamin A)	Brazil/*Multifloral*	5–1233 (µg/g)	Sattler et al. (2016)
Vitamin B1 (Thiamine)	Turkey/Not determined	0.1–0.5 (mg/100 g)	Dervişoğlu et al. (2022)
Vitamin B2 (Riboflavin)	Turkey/Not determined	0.2–0.6 (mg/100 g)	Dervişoğlu et al. (2022)
Vitamin B3 (Niacin)	Romania/Not determined	1.0–3.0 (mg/100 g)	Dranca et al. (2020)
Minerals			
Macroelements Potassium (K)	Morocco/*Apiaceae, Asteraceae, Fagaceae, Myrtaceae*	338 ± 8 (mg/100 g)	Bakour et al. (2019)
	Lithuania/Not determined	800–3750 (mg/kg)	Adaškevičiūtė et al. (2019)
	Poland/*Multifloral*	350.8–617.8 (mg/100 g)	Habryka et al. (2020)
	Malaysia/*Asteraceae, Fabaceae, Rutaceae*	670.59 (mg/100 g)	Mohammad et al. (2020)
	Serbia/Not determined	7487 ± 381.50 (mg/kg)	Ćirić et al. (2022)
	Brazil/*Mimosaceae, Fabaceae*	5918.513–366.6 (mg/kg)	da Silva et al. (2014)
	Turkey/*Asteraceae, Fabaceae, Plantaginaceae and Rosaceae*	5429.27–8023.91 (mg/kg)	Mayda et al. (2020)
Calcium (Ca)	Morocco/*Apiaceae, Asteraceae, Fagaceae, Myrtaceae*	198 ± 4 (mg/100 g)	Bakour et al. (2019)
	Poland/*Multifloral*	21.5–50.1 (mg/100 g)	Habryka et al. (2020)
	Malaysia/*Asteraceae, Fabaceae, Rutaceae*	21 ± 3 (mg/kg)	Mohammad et al. (2020)
	Brazil/*Mimosaceae, Fabaceae*	1864.1–3424.9 (mg/kg)	da Silva et al. (2014)
	Turkey/*Asteraceae, Fabaceae, Plantaginaceae,* and *Rosaceae*	189.69–447.13 (mg/kg)	Mayda et al. (2020)

(continued)

Table 13.3 (continued)

Class of compounds	Geographical origin/Botanical origin	Concentration range	References
Magnesium (Mg)	Morocco/*Apiaceae, Asteraceae, Fagaceae, Myrtaceae*	61 ± 2 (mg/100 g)	Bakour et al. (2019)
	Poland/*Multifloral*	13.1–46.2 (mg/100 g)	Habryka et al. (2020)
	Lithuania (*Rape, Clover*)	65.4 ± 1.2 (mg/100 g)	Khalifa et al. (2020)
	Malaysia/*Asteraceae, Fabaceae, Rutaceae*	1441.8–2000.2 (mg/kg)	Mohammad et al. (2020)
	Turkey/*Asteraceae, Fabaceae, Plantaginaceae,* and *Rosaceae*	688.82–1399.43 (mg/kg)	Mayda et al. (2020)
Phosphorus (P)	Morocco/*Apiaceae, Asteraceae, Fagaceae, Myrtaceae*	251 ± 4 (mg/100 g)	Bakour et al. (2019)
	Turkey/*Asteraceae, Fabaceae, Plantaginaceae,* and *Rosaceae*	4742.31–5948.96 (mg/kg)	Mayda et al. (2020)
	Malaysia (*Asteraceae, Fabaceae,* and *Rutaceae*)	73.2 ± 3.1 (mg/kg)	Mohammad et al. (2020)
	Lithuania/Not determined	2820–4840 (mg/kg)	Adaškevičiūtė et al. (2019)
Sodium (Na)	Morocco/*Apiaceae, Asteraceae, Fagaceae,* and *Myrtaceae*	14.2 ± 0.1 (mg/100 g)	Bakour et al. (2019)
	Poland/*Multifloral*	7.5–8.5 (mg/100 g)	Habryka et al. (2020)
	Turkey/*Asteraceae, Fabaceae, Plantaginaceae,* and *Rosaceae*	184.62–367.34 (mg/kg)	Mayda et al. (2020)
Microelements Iron (Fe)	Morocco (*Apiaceae, Asteraceae, Fagaceae,* and *Myrtaceae*)	27.3 ± 0.3 (mg/100 g)	Bakour et al. (2019)
	Turkey/*Asteraceae, Fabaceae, Plantaginaceae,* and *Rosaceae*	84.06–317.02 (mg/kg)	Mayda et al. (2020)
	Serbia (*Dandelion, Deadnettle*)	121.99 ± 31.86 (mg/kg)	Anđelković et al. (2012)
	Lithuania (*Rape, Clover*)	110 ± 2 (mg/100 g)	Khalifa et al. (2020)
	Turkey/*Asteraceae, Fabaceae, Plantaginaceae,* and *Rosaceae*	84.06–317.02 (mg/kg)	Mayda et al. (2020)
	Brazil/*Mimosaceae, Fabaceae*	16.4–33.5 (mg/kg)	da Silva et al. (2014)

Zinc (Zn)	Morocco (*Apiaceae, Asteraceae, Fagaceae,* and *Myrtaceae*)	3.31 ± 0.04 (mg/100 g)	Bakour et al. (2019)
	Serbia (*Dandelion, Deadnettle*)	44.09 ± 3.86 (mg/kg)	Anđelković et al. (2012)
	Turkey/*Asteraceae, Fabaceae, Plantaginaceae,* and *Rosaceae*	39.37–73.96 (mg/kg)	Mayda et al. (2020)
	Brazil/*Mimosaceae, Fabaceae*	36.4–71.2 (mg/kg)	da Silva et al. (2014)
	Malaysia (*Asteraceae, Fabaceae, Rutaceae*)	32.5 ± 0.8 (mg/kg)	Mohammad et al. (2020)
Manganese (Mn)	Morocco (*Apiaceae, Asteraceae, Fagaceae, Myrtaceae*)	2.6 ± 0.1 (mg/100 g)	Bakour et al. (2019)
	Serbia (*Dandelion, Deadnettle*)	29.92 ± 0.60 (mg/kg)	Anđelković et al. (2012)
	Turkey/*Asteraceae, Fabaceae, Plantaginaceae,* and *Rosaceae*	17.40–109.94 (mg/kg)	Mayda et al. (2020)
	Lithuania (*Rape, Clover*)	15.3 ± 0.5 (mg/100 g)	Khalifa et al. (2020)
Copper (Cu)	Poland/*Multifloral*	0.10 (mg/100 g)	Habryka et al. (2020)
	Lithuania/Not determined	1–5 (mg/100 g)	Adaškevičiūtė et al. (2019)
Selenium (Se)	Poland/*Multifloral*	0.26 ± 0.1064 (mg/kg)	Habryka et al. (2020)
Plumb (Pb)	Lithuania/Not determined	0.096–0.433 (mg/kg)	Adaškevičiūtė et al. (2019)

Multiple studies have identified flavonoids in bee bread, including quercetin, kaempferol, rutin, naringin, and isorhamnetin. A study on Georgian bee bread samples conducted by (Habryka et al. 2020) utilized HPLC-UV-Vis analysis to detect three flavonoids: Naringin, rutin, and quercetin, though specific concentrations were not quantified, underscoring the flavonoid diversity within bee bread.

Sobral et al. (2017) explored the flavonoid profile of bee bread from Northeast Portugal, identifying six flavonoids: chrysin, naringenin, kaempferol, isorhamnetin, apigenin, and quercetin. Although exact quantitative data was not provided, the presence of these compounds suggests the bioactive potential associated with bee bread consumption. In a comparative analysis by Urcan et al. (2018a, b) flavonoids such as kaempferol and quercetin were identified in both bee pollen and bee bread. This study highlighted that these flavonoids are predominantly present in glycoside forms, which may impact their bioavailability and antioxidant efficacy, further supporting bee bread as a notable source of these beneficial phytonutrients. Additionally, Aylanc et al. (2021) assessed bioactive compounds in both bee pollen and bee bread, detecting flavonol derivatives such as quercetin, kaempferol, and isorhamnetin.

Regarding total flavonoid content (Beykaya et al. 2021) examined bee bread samples from various Anatolian regions, reporting a total flavonoid content ranging from 1.30 to 6.03 mg CE/g. This variation is likely due to differences in floral sources and environmental conditions that influence bee bread's chemical composition.

Hudz et al. (2019) developed a quantification approach for total flavonoid content in bee bread using a quercetin calibration curve, confirming flavonoids as prominent components, though detailed concentrations were not disclosed. Further supporting regional variability, a study by Beykaya et al. (2021) reported total flavonoid content in bee bread from various Ukrainian regions ranging from 13.56 to 18.24 mg QE/g, this variability underscores the impact of geographical and botanical factors on flavonoid concentrations.

Dranca et al. (2020) reported the presence of gallic acid, caffeic acid, and ferulic acid in bee bread, with total phenolic content quantified at approximately 24.5 mg GAE/g. This indicates a significant level of phenolic compounds that contribute to the antioxidant capacity of bee bread. A study by Beykaya et al. (2021) investigated the phenolic profile of Anatolian bee bread, reporting a total phenolic content ranging from 12.36 to 25.44 mg GAE/g. This analysis identified key phenolic acids, such as caffeic acid and p-coumaric acid, compounds recognized for their potential health benefits. Also, in the case of phenolic compounds, the observed variability in phenolic content was attributed to differences in floral sources and environmental conditions, both of which significantly influence the chemical composition of bee bread (Mărgăoan et al. 2021). Keyvan's research on Turkish bee bread identified a range of phenolic compounds, including flavonoids and phenolic acids, with a total phenolic content measured at 15.5 mg GAE/g (Keyvan 2023).

Quercetin, one of the primary flavonoids found in bee bread, has been reported in samples from diverse regions. For example, Sobral et al. identified quercetin in bee bread from Northeast Portugal, which supports its common occurrence and fundamental role within the bioactive matrix of bee bread (Sobral et al. 2017).

Kaempferol, another major flavonoid with known health-promoting properties, has been frequently identified in bee bread samples. Hudz et al., for instance, confirmed its presence through spectrophotometric analysis, while Othman et al. highlighted kaempferol's considerable contribution to the antioxidant capacity of bee bread (Hudz et al. 2017; Othman et al. 2019). Adding to this profile, rutin, a glycoside composed of quercetin and rutinose, plays an important role in bee bread's bioactivity. Detected by Sobral et al. (2017) in Portuguese bee bread and later confirmed by Tavdidishvili et al. (2014) in Georgian bee bread using HPLC-UV-Vis, rutin contributes substantially to the product's biochemical richness.

Caffeic acid and its derivatives, a potent antioxidant, was reported in Anatolian bee bread by Beykaya et al., who measured the total phenolic content at 12.36–25.44 mg GAE/g, underscoring its vital role in boosting the antioxidant potential of bee bread Beykaya et al. 2021, bat also in bee bread samples from Romania and India (Urcan et al. 2018a, b). Ferulic acid, known for its antioxidative and anti-inflammatory properties, was also identified by Othman et al. (2019) highlighting its contribution to reducing oxidative stress.

The range of flavonoids in bee bread is further expanded with isorhamnetin, a methylated form of quercetin, which adds complexity to its bioactive profile (Othman et al. 2019). Apigenin, another widely studied flavonoid, was reported by Hudz et al. alongside kaempferol, illustrating the presence of these compounds in tandem within bee bread (Hudz et al. 2017). Furthermore, chrysin, a flavonoid recognized for its antioxidant effects, was identified in bee bread by Hudz et al. (2017) underscoring its role within this bioactive matrix.

The diversity of bee bread's bioactives is enriched by naringin, another glycoside commonly found in citrus fruits and detected in Georgian bee bread by Tavdidishvili et al. through HPLC-UV-Vis analysis, adding further depth to its composition (Tavdidishvili et al. 2014).

Additional phenolic compounds, such as gallic acid and p-coumaric acid, further contribute to bee bread's bioactive spectrum. Beykaya et al. highlighted p-coumaric acid as a notable component in Anatolian bee bread, underscoring the chemical diversity and therapeutic potential of bee bread (Beykaya et al. 2021). The consistent presence of these flavonoids and phenolic acids across various regions and floral sources reflects bee bread's complex biochemical profile, positioning it as a functional food with promising health applications.

Quercetin, a primary flavonoid in bee bread, frequently appears as glycosides, such as quercetin-3-O-rutinoside and quercetin-3-O-hexoside, highlighting the biochemical richness of the product (Dranca et al. 2020). Similarly, kaempferol, another key flavonoid, is commonly found in glycosylated forms like kaempferol-3-O-rutinoside, contributing significantly to the antioxidant profile of bee bread (Mărgăoan et al. 2019), isorhamnetin, a methylated quercetin derivative, also exists as isorhamnetin-3-O-rutinoside, adding further complexity and enhancing the bioactivity of the flavonoid matrix (Campos et al. 2008).

Among these glycosides, rutin—a glycoside of quercetin—has been frequently identified in bee bread. Sobral et al. reported its presence in bee bread from Northeast Portugal, underscoring its nutritional relevance (Sobral et al. 2017). Naringin, another flavonoid glycoside typically found in citrus fruits, has also been identified in Georgian bee bread, contributing additional diversity to its composition (Sobral et al. 2017). Chrysin, a flavonoid recognized for its potential health benefits, occurs in glycosylated forms in bee bread as well, enriching its bioactive profile (Hudz et al. 2017).

The presence of these diverse glycosides across different flavonoids and phenolic acids underscores the nutritional and medicinal significance of bee bread. This complex glycosylation pattern not only supports the product's stability but also enhances its functional properties, positioning bee bread as a promising natural source of bioactive compounds with potential therapeutic benefits (Table 13.4).

13.7.2 Probiotics

Bee bread, a fermented product produced by honey bees through the natural fermentation of pollen mixed with nectar and bee saliva, has attracted significant scientific interest, not only for its rich nutritional content but increasingly for its probiotic potential (Vergalito et al. 2020). The fermentation process that transforms raw pollen into bee bread creates a unique biochemical environment where beneficial microorganisms thrive, producing compounds that can positively influence human health, this process, mediated by the hive's microflora and the conditions within it, not only enhances the digestibility and bioavailability of bee bread's nutrients but also introduces a complex microbiota with strains that exhibit probiotic qualities (Poyraz et al. 2023).

In particular, the microbial richness of bee bread includes diverse bacterial species with potential health benefits, highlighting its value as a natural source of probiotics (Ghosh et al. 2022). Studies have identified strains such as *Bacillus*, *Lactiplantibacillus*, and other lactic acid bacteria (LAB) that are known to positively impact gut health (Fidanza et al. 2021).

Recent studies emphasize the role of *Bacillus* species in bee bread's probiotic profile. Certain *Bacillus* strains isolated from bee bread, such as *Bacillus licheniformis* and *Bacillus subtilis*, have demonstrated notable resilience under gastrointestinal conditions, surpassing traditional probiotic strains in terms of survival through the acidic environment of the stomach and effective colonization of the gut (Toutiaee et al. 2022; Pełka et al. 2023; Hafeez 2024). These strains are additionally recognized for producing antimicrobial compounds that inhibit pathogenic bacteria, thereby contributing to gut health (Pełka et al. 2021a, b).

Beyond *Bacillus*, various lactic acid bacteria (LAB) have also been identified in bee bread, including *Lactiplantibacillus plantarum* and *Fructobacillus fructosus*, these LAB species are well-documented for their ability to enhance gut microbiota diversity and support digestive health (de Simone et al. 2022; Ben-Miled et al. 2023). The presence of these bacteria in bee bread suggests that it could serve as a natural source of probiotics, potentially offering similar health benefits to those of commercially available probiotic supplements (Dinu 2023). With a functional dose

Table 13.4 Total phenolic and flavonoid contents and compounds identified and quantified in monofloral and heterofloral bee pollen collected from different countries

Geographic origin/Botanical origin	Total phenolic content	Total flavonoid content	Compounds identified and quantified	References
Turkey/Not determined	11.70–18.35 (mg GAE/g)	2.08–2.82(mg QE/g)	Gallic acid (19.54–39.97 µg/g), Protocatechuic acid (0.07–0.17 µg/g), Catechin (0.45–0.72 µg/g), Hydroxybenzoic acid (6.78 µg/g), Vanillic acid (0.13–0.45 µg/g), Gentisic acid (0.34 µg/g), p-Coumaric acid (0.02 µg/g), Rutin (0.60 µg/g), Ferulic acid (0.24–3.25 µg/g), Naringin (0.13 µg/g), o-Coumaric acid (0.29 µg/g), Neohesperidin (0.37–0.95 µg/g), Coumarin (0.28 µg/g), Quercetin (0.13–0.36 µg/g), Hesperidin (0.04–0.60 µg/g), Alizarin (0.13 µg/g)	Aksoy et al. (2024)
Turkey/Not determined	3.56–14.48 (mg GAE/g)	1.22–8.65 (mg QE/g)	Gallic acid (34.65–347.37 mg/100 g), protocatechuic acid (166.61 mg/100 g), chlorogenic acid (3.89–36.09 mg/100 g), caffeic acid (17.86–102.09 mg/100 g), p-coumaric acid (28.70–142.44 mg/100 g), ferulic acid (2.47–107.86 mg/100 g), quercetin (381.10–3918.12 mg/100 g), kaempferol (112.94–2681.20 mg/100 g), isorhamnetin (1227.93 mg/100 g), rutin (1225.54–12613.49 mg/100 g), luteolin (21.91–3490.83 mg/100 g), myricetin (15.32–702.42 mg/100 g), salicylic acid (19.27–65.20 mg/100 g), phlorizin (78.6 mg/100 g), resveratrol (122.81 mg/100 g)	Bayram et al. (2021)
Turkey/Not determined	5.15–7.29 (mg GAE/g)	0.63–1.85 (mg QE/g)	Gallic acid (22.50–57.86 µg/kg), protocatechuic acid (63.85–225.38 µg/kg), salicylic acid (367.90–1112.33 µg/kg), caffeic acid (8675.56–19,594.88 µg/kg), chlorogenic acid (135.87–290.13 µg/kg), p-coumaric acid (38,332.49–67,739.53 µg/kg), rutin (214.19–1355.65 µg/kg), ferulic acid (1335.69–4963.34 µg/kg), sinapic acid (198.89–739.84 µg/kg), hesperidin (113.61–449.56 µg/kg), ethyl gallate (2.81–23.62 µg/kg), myricetin (117.53–441.31 µg/kg), quercetin (454.54–1398.78 µg/kg), abscisic acid (281.33–835.32 µg/kg), luteolin (658.78–1006.46 µg/kg), naringenin (1343.14–21,195.16 µg/kg), kaempferol (1076.40–11,995.74 µg/kg), isorhamnetin (1347.48–21846.17 µg/kg), jasmonic acid (9.21–30.70 µg/kg), apigenin (364.17–675.27 µg/kg)	Kutlu et al. (2024)

(continued)

Table 13.4 (continued)

Geographic origin/Botanical origin	Total phenolic content	Total flavonoid content	Compounds identified and quantified	References
Turkey/*Asteraceae, Fabaceae, Plantaginaceae, Rosaceae, Apiaceae, Brassicaceae, Lamiaceae, Centaurea* sp., *Castanea sativa, Teucrium* sp., *Cichorium* sp., *Ranunculaceae*	30.933 ± 0.513 (mg GAE/g)	1.81 ± 0.04–4.44 ± 0.12 (mg QE/g)	–	Mayda et al. (2020)
Greece/*Castanea sativa* *Cistus* spp. *Hedera helix* *Borago* spp. *Brassicaceae*	9.56 ± 0.02 (mg GAEs/g)	3.88 ± 0.12 (mg QE/g)	–	Didaras et al. (2021)
Malaysia/Not determined	17.44 ± 0.93 (mg GAE/g)	21.34 ± 0.83 (mg QE/g)	–	Suleiman et al. (2021)
Malaysia/Not determined	14.19 (mg GAE/g)	3.92 (mg QE/g)	–	Othman et al. (2019)
Turkey/*Actinidiaceae, Asteraceae, Balsaminaceae, Ericaceae, Fabaceae, Hederaceae, Salicaceae*	11.90–14.77 (mg GAE/g)	1.30–6.03 (mg CE/g)	–	Beykaya et al. (2021)
Romania/Not determined	101.2–128.5 (mg GAE/L)	759.4–1026.1 (mg QE/L)	Kaempferol (31.25 mg/L), Myricetin (3.15 mg/L), Luteolin (1.17 mg/L), Rosmarinic acid (0.23 mg/L), Caffeic acid (0.10 mg/L), p-Coumaric acid (0.11 mg/L), Quercetin (0.06 mg/L)	Dranca et al. (2020)

Poland/Not determined	0.47–32.52 (mg GAE/g)	0.07–11.77 (mg QE/g)	Gallic acid (21.3–69.2), ellagic acid (2.2–6.6), neochlorogenic acid (0.7–1.5), chlorogenic acid (0.7–13.2), protocatechuic acid (0.1–0.8), sakuranetin dimer (4.6), caffeic acid (0.8), rutin (5.1–10.0), sinapic acid (1.8–42.8), isorhamnetin 3-O-rutinoside (2.5–5.8), 3,4-di-O-caffeoylquinic acid (1.9), quercetin 3-O-glucuronide (0.3–7.9), orientin (7.1), vitexin (15.2), quercetin (2.2), epicatechin (2.6–7.2), kaempferol (6.0), protocatechuic acid-O-hexoside (0.2–3.4), pinobanksin (0.7–85.7), apigenin (3.0)	Sawicki et al. (2022)
Romania/*Asteraceae, Fabaceae, Brassicaceae, Rosaceae, Lamiaceae, Salicaceae, Amaryllidaceae, Rutaceae, Apiaceae, Onagraceae, Boraginaceae, Primulaceae, Violaceae, Betulaceae, Ericaceae, Colchicaceae, Pinaceae, Amaranthaceae, Convolvulaceae, Campanulaceae, Oleaceae, Resedaceae, Cornaceae, Fagaceae, Ranunculaceae*	7.10–18.30 (mg GAE/g)	0.45–1.86 (mg QE/g)	Cinnamic acid (8.65–64.67 mg/g), p-Coumaric acid (58.02–109.13 mg/g), ferulic acid (7.00–41.46 mg/g), caffeic acid (52.27–84.85 mg/g), chlorogenic acid (38.86–54.10 mg/g), 4-hydroxybenzoic acid (44.84–50.22 mg/g), vanillic acid (30.17–39.39 mg/g), syringic acid (30.19–30.84 mg/g), gallic acid (58.02–109.13 mg/g), catechin (34.54–36.01 mg/g), epicatechin (51.24–51.48 mg/g), quercetin (993.36–1192.04 mg/g), apigenin (30.88–322.38 mg/g), galangin (7.58–341.12 mg/g), kaempferol (762.24–1892.27 mg/g), isorhamnetin (466.10–1333.09 mg/g), chrysin (62.28–82.79 mg/g), pinocembrin (50.68–67.87 mg/g), rutin (56.51–383.49 mg/g), hesperidin (52.17 mg/g)	Ilie et al. (2024)
Morocco/*Apiaceae (Bupleurum spinosum, Anethum graveolens), Asteraceae (Calendula officinalis, Anacyclus), Fagaceae (Quercus ilex), Myrtaceae (Eucalyptus), Punicaceae (Punica granatum), Mimosaceae (Acacia)*	0.47–9.2 (mg GAE/g)	–	Kaempferol-O-hexosyl-O-rutinoside (0.57 mg/g), Quercetin-O-hexosyl-O-hexoside (0.80–0.95 mg/g), Methylherbacetrin-O-dihexoside (0.54 mg/g), Isorhamnetin-O-hexosyl-O-rutinoside (1.48 mg/g), Quercetin-O-pentosyl-hexoside (0.58 mg/g), Quercetin-3-O-rutinoside (0.53 mg/g), Methylherbacetrin-3-O-rutinoside (0.51 mg/g), Isorhamnetin-O-pentosyl-hexoside (0.55–0.93 mg/g), Kaempferol-3-O-rutinoside (0.51 mg/g), Isorhamnetin-O-rhamnoside-hexoside (0.56 mg/g), Isorhamnetin-3-O-rutinoside (0.68 mg/g)	Bakour et al. (2019)
Poland/Not determined	13.94–21.05 (mg GAE/g)	–	–	Pełka et al. (2021a)

(continued)

Table 13.4 (continued)

Geographic origin/Botanical origin	Total phenolic content	Total flavonoid content	Compounds identified and quantified	References
Romania/Brassicaceae—*Brassica* sp., *Poaceae, Fabaceae—Trifolium, Tiliaceae—Tilia* sp., *Fabaceae, Rosaceae, Plantaginaceae—Plantago, Fabaceae—Lotus, Asteraceae—Taraxacum officinale, Asteraceae, Lamiaceae, Asteraceae—Centaurea montana*	6.30 ± 0.11–12.83 ± 0.08 (mg GAE/g)	23.64 ± 0.1–36.19 ± 0.05 (mg Qe/g)	Kaempferol-3-O-glycoside (0.86–2.85), Quercetin-3-O-glycoside (0.81–2.98), Isorhamnetin-3-O-glycoside (0.91), Hydroxycinnamic acid derivatives (0.04–9.35), Herbacetin-3-O-glycoside (0.27–4.30), Myricetin (0.44), Luteolin (6.98)	Urcan et al. (2018b)
Portugal/*Plantaginaceae (Plantago* sp.*), Fabaceae (Cytisus striatus, Trifolium* sp.*), Fagaceae (Castanea sativa), Asteraceae (Crepis capillaris, Centaurea* sp.*), Campanulaceae (Jasione montana), Rosaceae (Rubus* sp.*), Brassicaceae (Raphanus raphanistrum), Boraginaceae (Echium* sp.*, Pentaglottis sempervirens), Ericaceae (Erica* sp.*), Myrtaceae (Eucalyptus* sp.*),* and *Poaceae (Zea mays)*	2.4 ± 0.1–4.3 ± 0.1 (mg GAE/g)	0.6 ± 0.1–2.7 ± 0.1 (mg QE/g)	Quercetin-O-hexosyl-O-rutinoside (0.08 mg/g), methyl herbacetin-O-dihexoside (0.04–0.43 mg/g), kaempferol-O-dihexoside (0.08 mg/g), isorhamnetin-O-pentosyl-hexoside (0.06–0.22 mg/g), kaempferol-3-O-rutinoside (0.01–0.03 mg/g), quercetin-3-O-glucoside (0.01–0.02 mg/g), luteolin (1.00 mg/g), quercetin-diglucoside (0.03–0.05 mg/g), isorhamnetin-O-deoxyhexoside (0.31 mg/g), isorhamnetin-3-O-malonyl-glucoside (0.06 mg/g), hesperetin (0.05 mg/g), kaempferol-O-malonyl-hexoside (0.07–0.09 mg/g), di-p-coumaroylspermidine (0.20–0.72 mg/g), N1, N5, N10-tri-caffeoylspermidine (0.20–4.15 mg/g), N1, N5-di-p-coumaroyl-N10-caffeoylspermidine (0.13–2.12 mg/g), N1, N5, N10-tri-p-coumaroylspermidine (0.14–3.20 mg/g), tetracoumaroyl spermine (0.31–9.11 mg/g), feruloyl dicoumaroyl spermidine (0.32–0.93 mg/g)	Aylanç et al. (2021)
Colombia/Not determined	8.9 (mg GAE/g)	3.2 (mg QE/g)	–	Zuluaga et al. (2015)
Ukraine/Not determined	12.36–25.44 (mg GAE/g)	13.56–18.24 (μg QE/g)	–	Ivanišová et al. (2015)

Lithuania/Not determined	19.63–22.16 (mg RUE/10 g)	7.88–15.67 (mg RUE/10 g)	–	Adaškevičiūtė et al. (2019)
Ukraine/Not determined	–	8.3 ± 6.24%–195.3 ± 1.35% (QEmg/L); 28.8 ± 19.33–603.3 ± 4.64 (RUEmg/L)	–	Hudz et al. (2019)
Portugal/Not determined	18.5–29.6 (mg GAE/g)	5.3–12.4 (mg QE/g)	Myricetin-3-O-rutinoside (41–322 µg/g), Quercetin-O-hexosyl-O-rutinoside (156 µg/g), Kaempferol-O-hexosyl-O-rutinoside (69 µg/g), Quercetin-O-hexosyl-O-hexoside (74–211 µg/g), Isorhamnetin-O-hexosyl-O-rutinoside (2615 µg/g), Quercetin-3-O-rutinoside (88–377 µg/g), Kaempferol-O-dihexoside (91–1167 µg/g), Methyl herbacetin-3-O-rutinoside (71–435 µg/g), Kaempferol-3-O-rutinoside (56–1627 µg/g), Quercetin-3-O-glucoside (52–248 µg/g), Quercetin-3-O-rhamnoside (168–3029 µg/g), Isorhamnetin-3-O-rutinoside (836 µg/g), Isorhamnetin-3-O-glucoside (64–199 µg/g), Kaempferol-3-O-rhamnoside (29–141 µg/g)	Sobral et al. (2017)

of probiotics often recommended around 1×10^9 CFU per serving, bee bread appears capable of meeting or exceeding this threshold, reinforcing its viability as a probiotic-rich food (Ben-Miled et al. 2023).

One of the most frequently studied bacterial species in bee bread is *Lactobacillus kunkeei*, which has been consistently isolated from various bee bread samples. This bacterium is particularly notable for its resilience in the acidic environment of the bee gut, a trait essential for driving the fermentation process that converts pollen into bee bread (Disayathanoowat et al. 2020; Ispirli and Dertli 2021). Alongside *L. kunkeei*, other lactic acid bacteria such as *L. plantarum and F. fructosus* are often detected in bee bread, especially in samples that contain both pollen and nectar. These bacterial species are recognized for their probiotic properties, including their ability to inhibit pathogenic microorganisms and promote a balanced gut microbiota, which highlights their potential health benefits when consumed (Ispirli and Dertli 2021).

In addition to *Lactobacillus* species, other beneficial bacteria such as *Bifidobacterium* and *Oenococcus* have been identified in bee bread, these bacteria are well-regarded for their probiotic capabilities, particularly in producing beneficial metabolites during fermentation, which may enhance the functional food properties of bee bread, aligning it closely with the benefits of commercial probiotic supplements (Ramos et al. 2020).

Further insights into the microbiological richness of bee bread were provided by Ghosh et al. (2022) who used high-throughput sequencing to compare bacterial communities in bee bread and bee pollen. Their findings revealed that bee bread exhibits greater species richness and microbial diversity than bee pollen, an important factor for maintaining a robust gut microbiome This microbial diversity plays a crucial role in the fermentation process, as various bacterial species interact synergistically to produce health-promoting compounds such as short-chain fatty acids and other bioactive metabolites, which are known to support digestive health (Prajapati et al. 2023).

The role of yeasts in bee bread fermentation is also noteworthy, with species such as *Starmerella* and *Zygosaccharomyces* commonly identified in fresh and aged bee bread, respectively (Detry et al. 2020). These yeasts contribute to the fermentation process and may also enhance the flavor and nutritional profile of the final product. The complex interactions between bacteria and yeasts in bee bread significantly shape its characteristics, providing additional health benefits through their combined metabolic activities (Pełka et al. 2021a, b).

The composition of bee bread's microbial community is not static; it is influenced by a variety of factors, including the bees' diet and environmental conditions. Research by Disayathanoowat et al. 2020 demonstrated that core gut bacteria commonly found in the hive environment, such as those in stored bee bread, play an integral role in the microbial profile of bee bread. This dynamic microbial composition can vary depending on the floral sources of pollen and specific fermentation conditions, creating variability that may contribute to the range of health benefits associated with bee bread. Such diversity allows for the presence of bacterial strains with complementary functions that support gut health and overall microbial balance (Lozo et al. 2015; Yun et al. 2018; Asoutis Didaras et al. 2024).

This diversity not only enhances the health benefits of bee bread but also suggests that different bacterial strains provide complementary effects on gut health and immune function (Dimov et al. 2021; Tsadila et al. 2023). The dynamic microbial community in bee bread implies that its probiotic composition may vary with source and processing, a critical consideration for its use as a dietary supplement (Mattila et al. 2012; Disayathanoowat et al. 2020).

Lozo et al. (2015) have emphasized the importance of understanding the microbial communities within bee bread, as these microorganisms are not only central to bee bread's fermentation process but also play a critical role in the health of honey bee colonies. The specific bacterial consortia present in bee bread contribute to transforming pollen into a more digestible and nutrient-dense food, benefiting both bees and potentially humans as well.

Taking into account all this, bee bread may serve as a rich source of probiotics, primarily due to the presence of beneficial bacteria such as *Bacillus* and *Lactiplantibacillus* species (Erban et al. 2017). The fermentation process enhances both its nutritional and probiotic content, establishing bee bread as a promising functional food.

References

Adaškevičiūtė V, Kaškonienė V, Kaškonas P, Barčauskaitė K, Maruška A (2019) Comparison of physicochemical properties of bee pollen with other bee products. Biomol Ther 9(12):819. https://doi.org/10.3390/biom9120819

Aksoy, A., Altunatmaz, S. S., Aksu, F., Tokatlı Demirok, N., Yazıcı, K., & Yıkmış, S. (2024). Bee bread as a functional product: phenolic compounds, amino acid, sugar, and organic acid profiles. Foods, 13(5), 795. https://doi.org/10.3390/foods13050795

Aksoy A, Altunatmaz SS, Aksu F, Tokatlı Demirok N, Yazıcı K, Yıkmış S (2024) Bee bread as a functional product: phenolic compounds, amino acid, sugar, and organic acid profiles. Food Secur 13(5):795. https://doi.org/10.3390/foods13050795

Alaux C, Le Dantec C, Parrinello H, Le Conte Y (2011) Nutrigenomics in honey bees: digital gene expression analysis of Pollen's nutritive effects on healthy and varroa-parasitized bees. BMC Genomics. https://doi.org/10.1186/1471-2164-12-496

Albar SA (2022) Dietary Omega-6/Omega-3 polyunsaturated fatty acid (PUFA) and Omega-3 are associated with general and abdominal obesity in adults: UK National Diet and Nutritional Survey. Cureus 14(10). https://doi.org/10.7759/cureus.30209

Alcalá-Orozco M, Lobo-Farfan I, Tirado DF, Mantilla-Escalante DC (2024) Enhancing the nutritional and bioactive properties of bee pollen: a comprehensive review of processing techniques. Food Secur 13(21):3437. https://doi.org/10.3390/foods13213437

Aldarf RJ, Sultaneh A (2024) The Difference in the source of bee bread (Citrus, Anise) in its chemical composition. Uttar Pradesh J Zool 45(5):52–58. https://doi.org/10.56557/upjoz/2024/v45i53930

Alvarez-Suarez JM (2017) Bee products - chemical and biological properties. Springer International Publishing, Cham

Anđelković B, Jevtić G, Mladenović M, Marković J, Petrović M, Nedić N (2012) Quality of pollen and honey bee bread collected in spring. J Hyg Eng Des 1:275–277

Araneda Durán X, Quezada Mardones I, Martínez Gutiérrez M, Morales Ulloa D (2014) Polifenoles totales en pan de abeja (Apis mellifera L.) de colmenas de la Región de La Araucanía. Idesia (Arica) 32(1):107–111. https://doi.org/10.4067/S0718-34292014000100013

Arien Y, Dag A, Shafir S (2018) Omega-6:3 ratio more than absolute lipid level in diet affects associative learning in honey bees. Front Psychol. https://doi.org/10.3389/fpsyg.2018.01001
Asoutis Didaras N, Karaiskou I, Nikolaidis M, Siaperopoulou C, Georgi I, Tsadila C, Karatasou K, Amoutzias GD, Mossialos D (2024) Contribution of microbiota to bioactivity exerted by bee bread. Pharmaceuticals (Basel) 17(6):761. https://doi.org/10.3390/ph17060761
Aylanc V, Falcão SI, Ertosun S, Vilas-Boas M (2021) From the hive to the table: nutrition value, digestibility and bioavailability of the dietary phytochemicals present in the bee pollen and bee bread. Trends Food Sci Technol 109(October 2020):464–481. https://doi.org/10.1016/j.tifs.2021.01.042
Aylanç V, Tomás A, Russo-Almeida P, Falcão SI, Vilas-Boas M (2021) Assessment of bioactive compounds under simulated gastrointestinal digestion of bee pollen and bee bread: bioaccessibility and antioxidant activity. Antioxidants (Basel). https://doi.org/10.3390/antiox10050651
Aylanc V, Falcão SI, Vilas-Boas M (2023) Bee pollen and bee bread nutritional potential: chemical composition and macronutrient digestibility under in vitro gastrointestinal system. Food Chem 413(January):135597. https://doi.org/10.1016/j.foodchem.2023.135597
Bakour M, Al-Waili NS, El Menyiy N, Imtara H, Figuira AC, Al-Waili T, Lyoussi B (2017) Antioxidant activity and protective effect of bee bread (honey and pollen) in aluminum-induced anemia, elevation of inflammatory makers and hepato-renal toxicity. J Food Sci Technol 54(13):4205–4212. https://doi.org/10.1007/s13197-017-2889-9
Bakour M, Fernandes Â, Barros L, Sokovic M, Ferreira ICFR, Badiaa (2019) Bee bread as a functional product: chemical composition and bioactive properties. LWT 109(February):276–282. https://doi.org/10.1016/j.lwt.2019.02.008
Bakour M, Hammas N, Laaroussi H, Ousaaid D, Fatemi HEL, Aboulghazi A, Soulo N, Lyoussi B (2021) Moroccan bee bread improves biochemical and histological changes of the brain, liver, and kidneys induced by titanium dioxide nanoparticles. Biomed Res Int 2021. https://doi.org/10.1155/2021/6632128
Bakour M, Laaroussi H, Ousaaid D, Es-safi I, Mechchate H, Lyoussi BBB, El Ghouizi A, Es-safi I, Mechchate H, Lyoussi BBB (2022) Bee bread as a promising source of bioactive molecules and functional properties: an up-to-date review. Antibiotics (Basel) 11(2):203. https://doi.org/10.3390/antibiotics11020203
Baky MH, Abouelela MB, Wang K, Farag MA (2023) Bee pollen and bread as a super-food: a comparative review of their metabolome composition and quality assessment in the context of best recovery conditions. Molecules 28(2):715. https://doi.org/10.3390/molecules28020715
Bárbara MS, Machado CS, Sodré S, Dias LG, Estevinho LM, Alfredo C, De Carvalho L (2015) Microbiological assessment, nutritional characterization and phenolic compounds of bee pollen from Mellipona mandacaia Smith, 1983. 1:12525–12544. https://doi.org/10.3390/molecules200712525
Barene I, Daberte I, Siksna S (2014) Investigation of bee bread and development of its dosage forms. Med Teor ir Prakt 21(1):16–22. https://doi.org/10.15591/mtp.2015.003
Barta DG, Cornea-Cipcigan M, Margaoan R, Vodnar DC, Barta G, Cornea-Cipcigan M, Mărgăoan R, Vodnar DC, Barta DG, Cornea-Cipcigan M, Margaoan R, Vodnar DC (2022) Biotechnological processes simulating the natural fermentation process of bee bread and therapeutic properties—an overview. Front Nutr 9(April). https://doi.org/10.3389/fnut.2022.871896
Bayram NE, Gercek YC, Çelik S, Mayda N, Kostić AŽ, Dramićanin AM, Özkök A (2021) Phenolic and free amino acid profiles of bee bread and bee pollen with the same botanical origin – similarities and differences. Arab J Chem 14(3):103004. https://doi.org/10.1016/j.arabjc.2021.103004
Ben-Miled H, Semmar N, Castellanos MS, Ben-Mahrez K, Benoit-Biancamano M, Réjiba S (2023) Effect of honey bee forage plants in Tunisia on diversity and antibacterial potential of lactic acid bacteria and bifidobacteria from Apis mellifera intermissa and its products. https://doi.org/10.21203/rs.3.rs-2743999/v1
Bertoncelj J, Polak T, Pucihar T, Lilek N, Borovšak AK, Korošec M (2018) Carbohydrate composition of Slovenian bee pollens. Int J Food Sci Technol. https://doi.org/10.1111/ijfs.13773

Beykaya M, Tanugur Samanci AE, Samanci T, Önder EY, Uzun EM, Tosun F, Tanuğur Samancı AE, Samancı T, Önder EY, Uzun EM, Tosun F, Tanugur Samanci AE, Samanci T, Önder EY, Uzun EM, Tosun F (2021) Investigation of nutritional and antioxidant properties of Anatolian bee bread. J Apic Sci 65(2):255–263. https://doi.org/10.2478/jas-2021-0017

Bleha R, Shevtsova TS, Kružík V, Škorpilová T, Saloň I, Erban V, Brindza J, Brovarskyi V, Sinica A (2019) Bee breads from two regions of eastern Ukraine: composition, physical properties and biological activities. Czech J Food Sci. https://doi.org/10.17221/201/2018-cjfs

Bobis O, Dezmirean DS, Pasca C (2017) Beebread from Apis mellifera and Apis dorsata . Comparative chemical composition and bioactivity. Bull UASVM Anim Sci Biotechnol 74(1):43–50. https://doi.org/10.15835/buasvmcn-asb

Bonvehi JS, Escola R (1997) Nutrient composition and microbiological quality of honeybee-collected pollen in Spain. J. Agric. Food Chem 45(3), 725–732.

Bramasta D (2023) Melissopalynology and vegetation analysis surrounding sunggau of giant honey bee Apis dorsata in Belitung Regency. Hayati J Biosci. https://doi.org/10.4308/hjb.30.6.1167-1174

Brancaccio M, Mennitti C, Cesaro A, Fimiani F, Vano M, Gargiulo B, Caiazza M, Amodio F, Coto I, D'Alicandro G, Mazzaccara C, Lombardo B, Pero R, Terracciano D, Limongelli G, Calabrò P, D'Argenio V, Frisso G, Scudiero O (2022) The biological role of vitamins in athletes' muscle, heart and microbiota. Int J Environ Res Public Health 19(3):1249. https://doi.org/10.3390/ijerph19031249

Bridgett RJ, Kirk WD, Drijfhout FP (2014) The use of a within-hive replication bioassay method to investigate the phagostimulatory effects of pollen, bee bread and pollen extracts, on free-flying honey bee colonies. Apidologie. https://doi.org/10.1007/s13592-014-0324-z

Bryś MS, Strachecka A (2024) The key role of amino acids in pollen quality and honey bee physiology—a review. Molecules 29(11):2605. https://doi.org/10.3390/molecules29112605

Campos MGR, Bogdanov S, de Almeida-Muradian LB, Szczesna T, Mancebo Y, Frigerio C, Ferreira F (2008) Pollen composition and standardisation of analytical methods. J Apic Res 47(2):154–161. https://doi.org/10.1080/00218839.2008.11101443

Campos MGR, Frigerio C, Lopes J, Bogdanov S (2010) What is the future of bee-pollen? J ApiProduct ApiMedical Sci 2(4):131–144. https://doi.org/10.3896/IBRA.4.02.4.01

Campos MGR, Bogdanov S, De Almeida-muradian LB, Szczesna T, Mancebo Y, Frigerio C, Ferreira F, Campos MGR, Bogdanov S, De Almeida-muradian LB, Szczesna T, Mancebo Y, Frigerio C, Ferreira F (2015) Pollen composition and standardisation of analytical methods Pollen composition and standardisation of analytical methods. 8839. https://doi.org/10.1080/00218839.2008.11101443

Čeksterytė V, Navakauskienė R, Treigytė G, Jansen EE, Kurtinaitienė B, Dabkevičienė G, Balžekas J, Čeksteryte V, Navakauskiene R, Treigyte G, Jansen EE, Kurtinaitiene B, Dabkevičiene G, Balžekas J (2016) Fatty acid profiles of monofloral clover beebread and pollen and proteomics of red clover (*Trifolium pratense*) pollen. Biosci Biotechnol Biochem 80(11):2100–2108. https://doi.org/10.1080/09168451.2016.1204218

Çelik S, Gerçek YC, Özkök A, Ecem Bayram N (2022) Organic acids and their derivatives: minor components of bee pollen, bee bread, royal jelly and bee venom. Eur Food Res Technol 248(12):3037–3057. https://doi.org/10.1007/s00217-022-04110-y

Ciric J, Spiric D, Baltic T, Janjic J, Petronijevic R, Simunovic S, Djordjevic V (2019) Element concentration and fatty acid composition of Serbian bee bread. IOP Conf Ser Earth Environ Sci 333(1):012050. https://doi.org/10.1088/1755-1315/333/1/012050

Ćirić J, Haneklaus N, Rajić S, Baltić T, Lazić IB, Đorđević V (2022) Chemical composition of bee bread (perga), a functional food: a review. J Trace Elem Miner 2(August):100038. https://doi.org/10.1016/j.jtemin.2022.100038

da Silva GR, da Natividade TB, Camara CA, da Silva EMS, dos Santos F d AR, Silva TMS (2014) Identification of sugar, amino acids and minerals from the pollen of Jandaíra stingless bees (Melipona subnitida). Food Nutr Sci 5(11):1015–1021. https://doi.org/10.4236/fns.2014.511112

Darwish A, Abd El-Wahed A, Shehata M, El-Seedi H, Masry S, Khalifa S, Mahfouz H, El-Sohaimy S (2022) Chemical profiling and nutritional evaluation of bee pollen, bee bread, and royal jelly and their role in functional fermented dairy products. Molecules 28(1):227. https://doi.org/10.3390/molecules28010227

de Simone N, Rocchetti MT, la Gatta B, Spano G, Drider D, Capozzi V, Russo P, Fiocco D (2022) Antimicrobial properties, functional characterisation and application of Fructobacillus fructosus and Lactiplantibacillus plantarum isolated from artisanal honey. Probiotics Antimicrob Proteins. https://doi.org/10.1007/s12602-022-09988-4

Degirmenci A (2024) The process of pollen transformation into bee bread: changes in bioactivity, bioaccessibility, and microbial dynamics. Food Funct. https://doi.org/10.1039/d3fo04466a

DeGrandi-Hoffman G, Eckholm B, Huang M-B (2012) A comparison of bee bread made by Africanized and European honey bees (Apis mellifera) and its effects on hemolymph protein titers. Apidologie. https://doi.org/10.1007/s13592-012-0154-9

Denisow B, Denisow-Pietrzyk M, Denisow-Pietrzyk M (2016) Biological and therapeutic properties of bee pollen: a review. J Sci Food Agric 96(13):4303–4309. https://doi.org/10.1002/jsfa.7729

Dervişoğlu G, Çobanoğlu DN, Yelkovan S, Karahan D, Çakır Y, Koçyığıt S (2022) Comprehensive study on BeeBread: palynological analysis, chemical composition, antioxidant and cytotoxic activities. Int J Second Metab. https://doi.org/10.21448/ijsm.1066884

Detry R, Simon-Delso N, Bruneau É, Daniel H-M, Simon-Delso N, Bruneau E, Daniel H-M, Simon-Delso N, Bruneau É, Daniel H-M (2020) Specialisation of yeast genera in different phases of bee bread maturation. Microorganisms 8(11):1789. https://doi.org/10.3390/microorganisms8111789

Di Pasquale G, Salignon M, Le Conte Y, Belzunces LP, Decourtye A, Kretzschmar A, Suchail S, Brunet J-L, Alaux C (2013) Influence of pollen nutrition on honey bee health: do pollen quality and diversity matter? PLoS One 8(8):e72016. https://doi.org/10.1371/journal.pone.0072016

Didaras NA, Kafantaris I, Dimitriou TG, Mitsagga C, Karatasou K, Giavasis I, Stagos D, Amoutzias GD, Hatjina F, Mossialos D (2021) Biological properties of bee bread collected from apiaries located across Greece. Antibiotics (Basel). https://doi.org/10.3390/antibiotics10050555

Dimov S, Zagorchev L, Iliev MN, Dekova T, Ilieva R, Kitanova M, Miteva D, Dimitrov M, Peykov S (2021) A snapshot picture of the fungal composition of bee bread in four locations in Bulgaria, differing in anthropogenic influence. J Fungi (Basel). https://doi.org/10.3390/jof7100845

Dinu L-D (2023) The modulation effect of a fermented bee pollen postbiotic on cardiovascular microbiota and therapeutic perspectives. Biomedicines. https://doi.org/10.3390/biomedicines11102712

Disayathanoowat T, Li H-Y, Supapimon N, Suwannarach N, Lumyong S, Chantawannakul P, Guo J (2020) Different dynamics of bacterial and fungal communities in hive-stored bee bread and their possible roles: a case study from two commercial honey bees in China. Microorganisms. https://doi.org/10.3390/microorganisms8020264

Domínguez-Valhondo D, Bohoyo Gil D, Hernández MT, González-Gómez D (2011) Influence of the commercial processing and floral origin on bioactive and nutritional properties of honeybee-collected pollen. Int J Food Sci Technol 46(10):2204–2211. https://doi.org/10.1111/j.1365-2621.2011.02738.x

Donkersley P, Rhodes G, Pickup RW, Jones KC, Power EF, Wright GA, Wilson K (2017) Nutritional composition of honey bee food stores vary with floral composition. Oecologia. https://doi.org/10.1007/s00442-017-3968-3

Donkersley P, Rhodes G, Pickup RW, Jones KC, Wilson K (2018) Bacterial communities associated with honeybee food stores are correlated with land use. Ecol Evol. https://doi.org/10.1002/ece3.3999

Dranca F, Ursachi F, Oroian M (2020) Bee bread: physicochemical characterization and phenolic content extraction optimization. Food Secur. https://doi.org/10.3390/foods9101358

El Ghouizi A, Bakour M, Laaroussi H, Ousaaid D, El Menyiy N, Hano C, Lyoussi B (2023) Bee pollen as functional food: insights into its composition and therapeutic properties. Antibiotics (Basel) 12(3):557. https://doi.org/10.3390/antiox12030557

El Ghouizi, A., Bakour, M., Laaroussi, H., Ousaaid, D., El Menyiy, N., Hano, C., & Lyoussi, B. (2023). Bee pollen as functional food: Insights into its composition and therapeutic properties. Antioxidants, 12(3), 557. https://doi.org/10.3390/antiox12030557

Erban T, Ledvinka O, Kamler M, Hortova B, Nesvorna M, Tyl J, Titera D, Markovic M, Hubert J (2017) Bacterial community associated with worker honeybees (Apis mellifera) affected by European foulbrood. PeerJ 5(9):e3816. https://doi.org/10.7717/peerj.3816

Erol E (2024) Quantitative analysis of bioaccessible phenolic compounds in Aegean bee bread using LC-HRMS coupled with a human digestive system model. Chem Biodivers. https://doi.org/10.1002/cbdv.202301497

Erwan E, Franti LD, Purnamasari DK, Muhsinin M, Agussalim A (2021) Preliminary study on moisture, fat, and protein contents of bee bread from Apis cerana from different regions in North Lombok Regency, Indonesia. Ternak Trop J Trop Anim Prod. https://doi.org/10.21776/ub.jtapro.2021.022.01.5

Fallah M, Najafi F, Kavoosi G (2021) Bee propolis, bee bread, and royal jelly: proximate analysis, fatty acid composition, nutritional quality, and anti-amylase activity. https://doi.org/10.21203/rs.3.rs-927023/v1

Fallah M, Najafi F, Kavoosi G (2022) Proximate analysis, nutritional quality and anti-amylase activity of bee propolis, bee bread and royal jelly. Int J Food Sci Technol. https://doi.org/10.1111/ijfs.15605

Fidanza M, Panigrahi P, Kollmann TR (2021) Lactiplantibacillus plantarum–nomad and ideal probiotic. Front Microbiol 12(October):1–13. https://doi.org/10.3389/fmicb.2021.712236

Ghosh S, Namin SM, Jung C (2022) Differential bacterial community of bee bread and bee pollen revealed by 16s rRNA high-throughput sequencing. Insects 13(10):863. https://doi.org/10.3390/insects13100863

Ghosh S, Sun S, Jang H, Jung C (2024) Amino acid dynamics in bee feed: comparative account of amino acid composition among pollen, bee-pollen and bee bread. J Apic 39(2):127–132. https://doi.org/10.17519/apiculture.2024.06.39.2.127

Giampieri F, Quiles JL, Cianciosi D, Forbes-Hernández TY, Orantes-Bermejo FJ, Alvarez-Suarez JM, Battino M (2022) Bee products: an emblematic example of underutilized sources of bioactive compounds. J Agric Food Chem 70(23):6833–6848. https://doi.org/10.1021/acs.jafc.1c05822

Habryka C, Socha R, Juszczak L (2020) The influence of honey enrichment with bee pollen or bee bread on the content of selected mineral components in multifloral honey. Potravin Slovak J Food Sci 14(October):874–880. https://doi.org/10.5219/1329

Habryka C, Socha R, Juszczak L (2023) The influence of bee bread on antioxidant properties, sensory and quality characteristics of multifloral honey. Appl Sci. https://doi.org/10.3390/app13137913

Hafeez A (2024) In silico safety assessment of bacillus isolated from polish bee pollen and bee bread as novel probiotic candidates. Int J Mol Sci. https://doi.org/10.3390/ijms25010666

Hashemirad F-S, Behfar M, Kavoosi G (2024) Proximate composition, physico-chemical, techno-functional, amino acid profile, fatty acid profile, nutritional quality, antioxidant, anti-amylase and anti-lipase properties of bee bread, royal jelly, and bee propolis. LWT 200(May):116190. https://doi.org/10.1016/j.lwt.2024.116190

Hou Y, Wu G (2018) Nutritionally essential amino acids. Adv Nutr 9(6):849–851. https://doi.org/10.1093/advances/nmy054

Hsu P-S, Wu T-H, Huang M-Y, Wang D-Y, Wu M-C (2021) Nutritive value of 11 bee pollen samples from major floral sources in Taiwan. Food Secur 10(9):2229. https://doi.org/10.3390/foods10092229

Hudz N, Ivanova R, Brindza J, Grygorieva O, Schubertová Z, Ivanišová E (2017) Approaches to the determination of antioxidant activity of extracts from bee bread and safflower leaves and flowers. Potravin Slovak J Food Sci. https://doi.org/10.5219/786

Hudz N, Yezerska O, Grygorieva O, Brindza J, Felšöciová S, Kačániová M, Wieczorek P (2019) Analytical procedure elaboration of total flavonoid content determination and antimicrobial activity of bee bread extracts. Acta Pol Pharm - Drug Res. https://doi.org/10.32383/appdr/103745

Ilie C-I, Spoiala A, Geana E-I, Chircov C, Ficai A, Ditu L-M, Oprea E (2024) Bee bread: a promising source of bioactive compounds with antioxidant properties—first report on some antimicrobial features. Antioxidants (Basel) 13(3):353. https://doi.org/10.3390/antiox13030353

Iorizzo M, Albanese G, Letizia F, Testa B, Di Criscio D, Petrarca S, Di Martino C, Ganassi S, Avino P, Pannella G, Aturki Z, Tedino C, De Cristofaro A (2024) Diversity of plant pollen sources, microbial communities, and phenolic compounds present in bee pollen and bee bread. Environ Sci Pollut Res. https://doi.org/10.1007/s11356-024-34517-x

Isidorov VA, Isidorova AG, Sczczepaniak L, Czyzewska U (2009) Gas chromatographic-mass spectrometric investigation of the chemical composition of beebread. Food Chem 115(3):1056–1063. https://doi.org/10.1016/j.foodchem.2008.12.025

Ispirli H, Dertli E (2021) Detection of fructophilic lactic acid bacteria (FLAB) in bee bread and bee pollen samples and determination of their functional roles. J Food Process Preserv 45(5):1–11. https://doi.org/10.1111/jfpp.15414

Ivanišová E, Kačániová M, Frančáková H, Petrová J, Hutková J, Brovarskyi V, Velychko S, Adamchuk L, Schubertová Z, Musilová J (2015) Bee bread - perspective source of bioactive compounds for future. Potravinarstvo 9(1):592–598. https://doi.org/10.5219/558

Jarušaitienė D, Jankauskienė J, Čeksterytė V (2012) Changes of ocular signs in children with thyroid diseases after intake of bee bread. Chem Technol. https://doi.org/10.5755/j01.ct.61.3.2660

Kabakci D (2024) Comparison of some biological activities of propolis and bee bread samples obtained from Apis mellifera Anatoliaca and its Muğla and Efe ecotypes. J Hell Vet Med Soc. https://doi.org/10.12681/jhvms.30859

Kacemi R, Campos MG (2023) Translational research on bee pollen as a source of nutrients: a scoping review from bench to real world. Nutrients 15(10):2413. https://doi.org/10.3390/nu15102413

Kalpana D, Han JH, Park WS, Lee SM, Wahab R, Lee YS (2019) Green biosynthesis of silver nanoparticles using Torreya nucifera and their antibacterial activity. Arab J Chem 12(7):1722–1732. https://doi.org/10.1016/j.arabjc.2014.08.016

Kankasemsuk S (2023) Total protein content of bee bread in Apis cerana combs and Tetragonula pegdeni storage pots of different plant sources from Chanthaburi Province, Thailand. J Curr Sci Technol. https://doi.org/10.59796/jcst.v13n3.2023.1139

Kaplan M, Karaoglu Ö, Eroglu N, Silici S (2016) Fatty acid and proximate composition of bee bread. Food Technol Biotechnol 54(4):497–504. https://doi.org/10.17113/ftb.54.04.16.4635

Keskin M (2021) Synthesis, characterization and antidiabetic potential of bee pollen based silver nanoparticles. El-Cezeri Fen ve Mühendislik Derg 9(1):266–275. https://doi.org/10.31202/ecjse.963670

Keyvan N, Usluer MS, Kaya MM, Kahraman HA, Tutun H, Keyvan E (2023) Total phenolic content, antibacterial and antiradical properties of bee bread from Turkey. Vet J Mehmet Akif Ersoy Univ 8(2):125–129. https://doi.org/10.24880/maeuvfd.1278378

Keyvan, N., Usluer, M. S., Kaya, M. M., Kahraman, H. A., Tutun, H., & Keyvan, E. (2023). Total phenolic content, antibacterial and antiradical properties of bee bread from Turkey. Veterinary Journal of Mehmet Akif Ersoy University, 8(2), 125-129. https://doi.org/10.24880/maeuvfd.1278378

Khalifa SAM, Elashal M, Kieliszek M, Ghazala NE, Farag MA, Saeed A, Xiao J, Zou X, Khatib A, Göransson U, El-Seedi HR (2020) Recent insights into chemical and pharmacological studies of bee bread. Trends Food Sci Technol 97(May 2019):300–316. https://doi.org/10.1016/j.tifs.2019.08.021

Kieliszek M, Piwowarek K, Kot AM, Błażejak S, Chlebowska-Śmigiel A, Wolska I (2018) Pollen and bee bread as new health-oriented products: a review. Trends Food Sci Technol 71(October 2017):170–180. https://doi.org/10.1016/j.tifs.2017.10.021

Kutlu N, Gerçek YC, Çelik S, Bayram S, Ecem Bayram N (2024) An optimization study for amino acid extraction from bee bread using choline chloride-acetic acid deep eutectic solvent and determination of individual phenolic profile. J Food Meas Charact 18(2):1026–1037. https://doi.org/10.1007/s11694-023-02273-0

Lozo J, Berić T, Terzić-Vidojević A, Stanković S, Fira D, Stanisavljević L (2015) Microbiota associated with pollen, bee bread, larvae and adults of solitary bee Osmia cornuta (Hymenoptera: Megachilidae). Bull Entomol Res 105(4):470–476. https://doi.org/10.1017/S0007485315000292

Mărgăoan R, Al Mărghitaş L, Dezmirean DS, Dulf FV, Bunea A, Socaci SA, Bobiş O (2014) Predominant and secondary pollen botanical origins influence the carotenoid and fatty acid profile in fresh honeybee-collected pollen. J Agric Food Chem 62(27):6306–6316. https://doi.org/10.1021/jf5020318

Mărgăoan R, Zăhan M, Mărghitaş LA, Dezmirean DS, Erler S, Bobiş O (2016) Antiproliferative activity and apoptotic effects of Filipendula ulmaria pollen against C26 mice colon tumour cells. J Apic Sci 60(1):135–144. https://doi.org/10.1515/JAS-2016-0014

Mărgăoan R, Stranţ M, Varadi A, Topal E, Yücel B, Cornea-Cipcigan M, Campos MG, Vodnar DC (2019) Bee collected pollen and bee bread: bioactive constituents and health benefits. Antioxidants (Basel) 8(12):568. https://doi.org/10.3390/antiox8120568

Mărgăoan R, Özkök A, Keskin Ş, Mayda N, Urcan AC, Cornea-Cipcigan M (2021) Bee collected pollen as a value-added product rich in bioactive compounds and unsaturated fatty acids: a comparative study from Turkey and Romania. LWT 149(June):111925. https://doi.org/10.1016/j.lwt.2021.111925

Markiewicz-Zukowska R, Naliwajko SK, Bartosiuk E, Moskwa J, Isidorov V, Soroczyńska J, Borawska MH (2013) Chemical composition and antioxidant activity of beebread, and its influence on the glioblastoma cell line (U87MG). J Apic Sci 57(2):147–157. https://doi.org/10.2478/jas-2013-0025

Mattila HR, Ríos D, Walker-Sperling VE, Roeselers G, Newton ILG (2012) Characterization of the active microbiotas associated with honey bees reveals healthier and broader communities when colonies are genetically diverse. PLoS One. https://doi.org/10.1371/journal.pone.0032962

Mayda N, Özkök A, Ecem Bayram N, Gerçek YC, Sorkun K (2020) Bee bread and bee pollen of different plant sources: determination of phenolic content, antioxidant activity, fatty acid and element profiles. J Food Meas Charact. https://doi.org/10.1007/s11694-020-00427-y

Miłek M, Mołoń M, Kielar P, Sidor E, Bocian A, Marciniak-Lukasiak K, Pasternakiewicz A, Dżugan M (2023a) The comparison of honey enriched with laboratory fermented pollen vs. natural bee bread in terms of nutritional and antioxidant properties, protein in vitro bioaccessibility, and its Genoprotective effect in yeast cells. Molecules 28(15):5851. https://doi.org/10.3390/molecules28155851

Miłek M, Mołoń M, Kula-Maximenko M, Sidor E, Zaguła G, Dżugan M (2023b) Chemical composition and bioactivity of laboratory-fermented bee pollen in comparison with natural bee bread. Biomol Ther 13(7):1025. https://doi.org/10.3390/biom13071025

Mitra S, Paul S, Roy S, Sutradhar H, Bin Emran T, Nainu F, Khandaker MU, Almalki M, Wilairatana P, Mubarak MS (2022) Exploring the immune-boosting functions of vitamins and minerals as nutritional food bioactive compounds: a comprehensive review. Molecules 27(2):555. https://doi.org/10.3390/molecules27020555

Mohammad SM, Mahmud-Ab-Rashid N-K, Zawawi N (2020) Botanical origin and nutritional values of bee bread of stingless bee (Heterotrigona itama) from Malaysia. J Food Qual 2020:1–12. https://doi.org/10.1155/2020/2845757

Mohammad SM, Mahmud-Ab-Rashid N-K, Zawawi N (2021) Probiotic properties of bacteria isolated from bee bread of stingless bee Heterotrigona itama. J Apic Res 60(1):172–187. https://doi.org/10.1080/00218839.2020.1801152

Nagai T, Nagashima T, Myoda T, Inoue R (2004) Preparation and functional properties of extracts from bee bread. Nahrung/Food 48(3):226–229. https://doi.org/10.1002/food.200300421

Nagai T, Nagashima T, Suzuki N, Inoue R (2005) Antioxidant activity and angiotensin I-converting enzyme inhibition by enzymatic hydrolysates from bee bread. Z Naturforsch C J Biosci 60(1–2):133–138. https://doi.org/10.1515/znc-2005-1-224

Nicolson SW (2011) Bee food: the chemistry and nutritional value of nectar, pollen and mixtures of the two. African Zool 46(2):197–204. https://doi.org/10.1080/15627020.2011.11407495

Oroian M, Dranca F, Ursachi F (2022) Characterization of Romanian bee pollen—an important nutritional source. Food Secur. https://doi.org/10.3390/foods11172633

Othman ZA, Wan Ghazali WS, Noordin L, Mohd Yusof NA, Mohamed M (2019) Phenolic compounds and the anti-atherogenic effect of bee bread in high-fat diet-induced obese rats. Antioxidants (Basel). https://doi.org/10.3390/antiox9010033

Othman ZA, Zakaria Z, Suleiman JB, Nna VU, Che Romli A, Wan Ghazali WS, Mohamed M, Romli AC, Wan Ghazali WS, Mohamed M (2021) Bee bread ameliorates vascular inflammation and impaired vasorelaxation in obesity-induced vascular damage rat model: the role of eNOS/NO/cGMP-signaling pathway. Int J Mol Sci 22(8):4225. https://doi.org/10.3390/ijms22084225

Özkök A, Bakhspour M, Mayda N, Denizli A, Sorkun K (2022) Microscopic examination and comparison of exine layer of bee pollen and bee bread (Perga). Istanbul J Pharm. https://doi.org/10.26650/istanbuljpharm.2022.932126

Pełka K, Otłowska O, Worobo RW, Szweda P (2021a) Bee bread exhibits higher antimicrobial potential compared to bee pollen. Antibiotics (Basel) 10(2):125. https://doi.org/10.3390/antibiotics10020125

Pełka K, Worobo RW, Walkusz J, Szweda P (2021b) Bee pollen and bee bread as a source of bacteria producing antimicrobials. Antibiotics (Basel) 10(6):713. https://doi.org/10.3390/antibiotics10060713

Pełka K, Bin HA, Worobo RW, Szweda P (2023) Probiotic potential of bacillus isolates from polish bee pollen and bee bread. Probiotics Antimicrob Proteins. https://doi.org/10.1007/s12602-023-10157-4

Poyraz F, Yalmanci D, İspirli H, Dertli E (2023) Characterization of bee bread produced with defined starter cultures mimicking the natural fermentation process. Fermentation 9(2). https://doi.org/10.3390/fermentation9020174

Prajapati N, Patel J, Singh S, Yadav VK, Joshi C, Patani A, Prajapati D, Sahoo DK, Patel A (2023) Postbiotic production: harnessing the power of microbial metabolites for health applications. Front Microbiol 14(December):1–16. https://doi.org/10.3389/fmicb.2023.1306192

Prđun S, Svečnjak L, Valentić M, Marijanović Z, Jerković I (2021) Characterization of bee pollen: physico-chemical properties, headspace composition and FTIR spectral profiles. Food Secur 10(9):2103. https://doi.org/10.3390/foods10092103

Ramos OY, Basualdo M, Libonatti C, Vega MF (2020) Current status and application of lactic acid bacteria in animal production systems with a focus on bacteria from honey bee colonies. J Appl Microbiol 128(5):1248–1260. https://doi.org/10.1111/jam.14469

Rodríguez-Pólit C, Gonzalez-Pastor R, Heredia-Moya J, Carrera-Pacheco SE, Castillo-Solis F, Vallejo-Imbaquingo R, Barba-Ostria C, Guamán LP (2023) Chemical properties and biological activity of bee pollen. Molecules 28(23):1–27. https://doi.org/10.3390/molecules28237768

Sariguzel FM, Silici S, Koç A, Sağıroğlu P, Dinç B (2022) Does dry or fresh bee bread contain clinically significant, and antimicrobial agents resistant microorganisms? Tarım Bilim Derg. https://doi.org/10.15832/ankutbd.1067297

Sattler JAG, de Melo AAM, do Nascimento KS, de Melo ILP, Mancini-Filho J, Sattler A, de Almeida-Muradian LB (2016) Essential minerals and inorganic contaminants (barium, cadmium, lithium, lead and vanadium) in dried bee pollen produced in Rio Grande do Sul State, Brazil. Food Sci Technol 36(3):505–509. https://doi.org/10.1590/1678-457X.0029

Sawicki T, Starowicz M, Kłębukowska L, Hanus P (2022) The profile of polyphenolic compounds, contents of total phenolics and flavonoids, and antioxidant and antimicrobial properties of bee products. Molecules 27(4):1301. https://doi.org/10.3390/molecules27041301

Semkiw P, Skubida P (2021) Bee bread production—a new source of income for beekeeping farms? Agriculture. https://doi.org/10.3390/agriculture11060468

Sobral F, Calhelha RC, Barros L, Dueñas M, Tomás A, Santos-Buelga C, Vilas-Boas M, Ferreira ICFRFR, Santos-Buelga C, Vilas-Boas M, Ferreira ICFRFR, Santos-Buelga C, Vilas-Boas M, Ferreira ICFRFR (2017) Flavonoid composition and antitumor activity of bee bread collected in Northeast Portugal. Molecules 22(2):1–12. https://doi.org/10.3390/molecules22020248

Suleiman JB, Mohamed M, Abu Bakar AB, Nna VU, Zakaria Z, Othman ZA, Aroyehun AB (2021) Chemical profile, antioxidant properties and antimicrobial activities of Malaysian Heterotrigona itama bee bread. Molecules. https://doi.org/10.3390/molecules26164943

Svečnjak L (2023) Metric and spectral insight into bee-pollen-to-bee-bread transformation process. Food Secur. https://doi.org/10.3390/foods12224149

Szabados L, Savouré A (2010) Proline: a multifunctional amino acid. Trends Plant Sci 15(2):89–97. https://doi.org/10.1016/j.tplants.2009.11.009

Tavdidishvili D, Khutsidze T, Pkhakadze M, Vanidze M, Kalandia A (2014) Flavonoids in Georgian bee bread and bee pollen. J Chem Chem Eng 8:676–681

Tawfik AI, Ahmed ZH, Abdel-Rahman MF, Moustafa AM (2023) Effect of some bee bread quality on protein content and antioxidant system of honeybee workers. Int J Trop Insect Sci 43(1):93–105. https://doi.org/10.1007/s42690-022-00888-2

Thakur M, Nanda V (2020) Composition and functionality of bee pollen: a review. Trends Food Sci Technol 98(November 2019):82–106. https://doi.org/10.1016/j.tifs.2020.02.001

Todorov SD, Alves MV, Bueno GCA, Alves VF, Ivanova IV (2024) Bee-associated beneficial microbes—importance for bees and for humans. Insects 15(6):430. https://doi.org/10.3390/insects15060430

Tomás A, Falcão SI, Russo-Almeida P, Vilas-Boas M (2017) Potentialities of beebread as a food supplement and source of nutraceuticals: botanical origin, nutritional composition and antioxidant activity. J Apic Res 56(3):219–230. https://doi.org/10.1080/00218839.2017.1294526

Toutiaee S, Mojgani N, Harzandi N, Moharrami M, Mokhberosafa L (2022) In vitro probiotic and safety attributes of Bacillus spp. isolated from beebread, honey samples and digestive tract of honeybees Apis mellifera. Lett Appl Microbiol. https://doi.org/10.1111/lam.13650

Trinkl M, Kaluza BF, Wallace HM, Heard TA, Keller A, Leonhardt SD (2020) Floral species richness correlates with changes in the nutritional quality of larval diets in a stingless bee. Insects. https://doi.org/10.3390/insects11020125

Tsadila C, Amoroso C, Mossialos D (2023) Microbial diversity in bee species and bee products: pseudomonads contribution to bee well-being and the biological activity exerted by honey bee products: a narrative review. Diversity 15(10):1088. https://doi.org/10.3390/d15101088

Tudor C, Bohn T, Iddir M, Dulf FV, Focşan M, Rugină DO, Pintea A (2020) Sea buckthorn oil as a valuable source of bioaccessible xanthophylls. Nutrients 12(1):1–13. https://doi.org/10.3390/nu12010076

Urcan A, Al Mărghitaş L, Dezmirean DS, Bobiş O, Bonta V, Mureşan CI, Mărgăoan R (2017) Chemical composition and biological activities of beebread – review. Bull Univ Agric Sci Vet Med Cluj-Napoca Anim Sci Biotechnol 74(1):6–14. https://doi.org/10.15835/buasvmcn-asb

Urcan AC, Criste A, Dezmirean DS, Bobiş O, Mărghitaş LA, Mărgăoan R, Hrinca A, Bobiş O, Mărghitaş LA, Mărgăoan R, Hrinca A (2018a) Antimicrobial activity of bee bread extracts against different bacterial strains. Bull Univ Agric Sci Vet Med Cluj-Napoca Anim Sci Biotechnol 75(2):85. https://doi.org/10.15835/buasvmcn-asb:2018.0004

Urcan AC, Criste AD, Dezmirean DS, Mărgăoan R, Caeiro A, Campos MG (2018b) Similarity of data from bee bread with the same taxa collected in India and Romania. Molecules 23(10). https://doi.org/10.3390/molecules23102491

Urcan AC, Criste AD, Dezmirean DS, Bobiş O, Bonta V, Dulf FV, Mărgăoan R, Cornea-Cipcigan M, Campos MG (2021) Botanical origin approach for a better understanding of chemical and nutritional composition of beebread as an important value-added food supplement. LWT 142(January):111068. https://doi.org/10.1016/j.lwt.2021.111068

Urcan AC, Criste AD, Dezmirean DS, Bobiş O, Bonta V, Burtescu RF, Olah N-K, Cornea-Cipcigan M, Mărgăoan R (2024) Enhancing antioxidant and antimicrobial activities in bee-collected pollen through solid-state fermentation: a comparative analysis of bioactive compounds. Antioxidants (Basel) 13(3):292. https://doi.org/10.3390/antiox13030292

Uslinawaty Z (2023) Characteristics of the composition of stingless bee bread with health benefits originating from the Kulisusu Subdistrict, North Buton Regency. J Aisyah J Ilmu Kesehat. https://doi.org/10.30604/jika.v8i3.2057

Vásquez A, Olofsson TC (2009) The lactic acid bacteria involved in the production of bee pollen and bee bread. J Apic Res 48(3):189–195. https://doi.org/10.3896/IBRA.1.48.3.07

Vergalito F, Testa B, Cozzolino A, Letizia F, Succi M, Lombardi SJ, Tremonte P, Pannella G, Di Marco R, Sorrentino E, Coppola R, Iorizzo M (2020) Potential application of Apilactobacillus kunkeei for human use: evaluation of probiotic and functional properties. Food Secur 9(11). https://doi.org/10.3390/foods9111535

Yan S, Li Q, Xue X, Wang K, Zhao L, Wu L (2019) Analysis of improved nutritional composition of bee pollen (Brassica campestris L.) after different fermentation treatments. Int J Food Sci Technol 54(6):2169–2181. https://doi.org/10.1111/ijfs.14124

Yun JH, Jung MJ, Kim PS, Bae JW (2018) Social status shapes the bacterial and fungal gut communities of the honey bee. Sci Rep 8(1):1–11. https://doi.org/10.1038/s41598-018-19860-7

Zakaria Z, Othman ZA, Suleiman JB, Che Jalil NA, Ghazali WSW, Nna VU, Mohamed M, Wan Ghazali WS, Nna VU, Mohamed M (2021) Hepatoprotective effect of bee bread in metabolic dysfunction-associated fatty liver disease (MAFLD) rats: impact on oxidative stress and inflammation. Antioxidants (Basel) 10(12):2031. https://doi.org/10.3390/antiox10122031

Zakaria Z, Othman ZA, Suleiman JB, Fadzli Mustaffa KM, Che Jalil NA, Wan Ghazali WS, Zulkipli NN, Mohamed M, Kamaruzaman KA (2022) Therapeutic effects of Heterotrigona itama (stingless bee) bee bread in improving hepatic lipid metabolism through the activation of the Keap1/Nrf2 signaling pathway in an obese rat model. Antioxidants (Basel). https://doi.org/10.3390/antiox11112190

Zakaria, Z., Othman, Z. A., Kamaruzaman, K. A., & Mohamed, M. (2024). Discovering the Potential of Bee Bread from Malaysian Stingless Bee as Vitamins B1, B3, and B6-Rich Source for Health and Disease. In Hydrophilic Vitamins in Health and Disease (pp. 87-103). Cham: Springer International Publishing.

Zhang H, Lu Q, Liu R (2022) Widely targeted metabolomics analysis reveals the effect of fermentation on the chemical composition of bee pollen. Food Chem 375(December 2021):131908. https://doi.org/10.1016/j.foodchem.2021.131908

Zuluaga CM, Serrato JC, Quicazan MC (2015) Chemical, nutritional and bioactive characterization of Colombian bee-bread. Chem Eng Trans 43:175–180. https://doi.org/10.3303/CET1543030

Chapter 14
Health Benefits and Uses of Bee Bread in Medicine

Adriana Cristina Urcan

Contents

Abstract Bee bread, a fermented hive product rich in bioactive compounds, has gained increasing recognition for its diverse health benefits and medicinal applications. This chapter provides a review of the therapeutic properties of bee bread, highlighting its antioxidant, antimicrobial, antitumor, hepatoprotective, antidiabetic, immunomodulatory, and anti-allergic effects. The antioxidant capacity of bee bread, primarily attributed to its high content of phenolic compounds, flavonoids, and vitamins, plays a crucial role in mitigating oxidative stress and reducing inflammation, key factors in chronic disease prevention. Its potent antimicrobial activity against Gram-positive and Gram-negative bacteria underscores its potential as a natural antimicrobial agent, particularly in the context of rising antibiotic resistance. Emerging research suggests that bee bread exhibits selective cytotoxicity against cancer cells, modulating key cellular pathways involved in tumor progression. Additionally, its hepatoprotective and antidiabetic properties, linked to its ability to regulate metabolic pathways and modulate oxidative stress, highlight its potential in managing metabolic disorders. Furthermore, its immunomodulatory effects and ability to reduce allergic responses position bee bread as a promising functional

---

A. C. Urcan (✉)
Department of Microbiology and Immunology, Faculty of Animal Science and Biotechnologies, University of Agricultural Sciences and Veterinary Medicine, Cluj-Napoca, Romania
e-mail: adriana.urcan@usamvcluj.ro

J. M. Alvarez-Suarez (ed.), *Bee Products – Chemical and Biological Properties*,
https://doi.org/10.1007/978-3-031-89049-9_14

food with therapeutic applications. This chapter compiles current scientific knowledge on bee bread, emphasizing its health benefits and its potential for therapeutic use in various medical and nutritional contexts.

Keywords Bee bread · Health benefits · Antioxidant properties · Antimicrobial activity · Immunomodulatory effects

14.1 Introduction

Bee bread, a unique product of the hive, is formed through the natural fermentation of pollen combined with nectar and bee saliva, resulting in a nutrient-dense substance with a long history of use in traditional medicine (Mărgăoan et al. 2019). In recent years, scientific investigations have illuminated the broad spectrum of health benefits and medicinal applications associated with bee bread, establishing it as a valuable supplement within modern healthcare frameworks (Kowalski and Makarewicz 2017; Gomaa 2017; Khalifa et al. 2020; Zakaria et al. 2022). Bee bread's composition is notably rich in proteins, carbohydrates, vitamins, and bioactive compounds that contribute to its therapeutic potential (Bakour et al. 2019; Hashemirad et al. 2024). Specifically, it contains considerable amounts of essential amino acids and fatty acids and a variety of enzymes, along with phenolic compounds and vitamins, including B1, B2, C, E, and K, underscoring its role as a potent nutritional supplement (Kankasemsuk 2023; Kabakci 2024).

The antioxidant capacity of bee bread is one of its most distinguished attributes. Studies have shown that bee bread exhibits robust antioxidant activity, essential for mitigating oxidative stress and inflammation within the body (Ilie et al. 2024). This property is largely attributed to its high levels of phenolic compounds and flavonoids, which effectively neutralize free radicals (Sobral et al. 2017; Dervişoğlu et al. 2022; Urcan et al. 2024). Further research indicates that the bioavailability of nutrients in bee bread exceeds that of bee pollen, suggesting a more efficient delivery of health benefits to the human body (Mărgăoan et al. 2020; Beykaya et al. 2021).

Beyond its antioxidant properties, bee bread has demonstrated significant antimicrobial activity against a variety of pathogens. Evidence suggests that bee bread inhibits the growth of both Gram-positive and Gram-negative bacteria, including clinically relevant strains such as *Staphylococcus aureus* and *Escherichia coli* (Abouda et al. 2011; Akhir et al. 2017; Didaras et al. 2020; Pełka et al. 2021; Suleiman et al. 2021; Urcan et al. 2018). This antimicrobial efficacy is particularly relevant given the increasing challenge of antibiotic resistance, positioning bee bread as a potential natural alternative for infection prevention and treatment. The presence of natural antimicrobial agents and bioactive molecules in bee bread underscores its potential as a functional food with therapeutic value (Bakour et al., 2021; Khalifa et al. 2020).

The effects of bee bread on metabolic health have also gained interest in recent studies, which indicate that bee bread may influence the regulation of appetite-related hormones, such as leptin and ghrelin, suggesting its potential for managing obesity and metabolic disorders (Doğanyiğit et al. 2020; Zakaria et al. 2021). Moreover, bee bread's favorable fatty acid profile, characterized by a high ratio of unsaturated to saturated fatty acids, supports its application as a dietary supplement beneficial for cardiovascular health and the reduction of obesity-related complications (Doğanyiğit et al. 2020).

Bee bread also exhibits notable anti-inflammatory properties, which are especially pertinent given the role of chronic inflammation in the pathogenesis of numerous diseases, including diabetes (Bakour et al. 2019; Mărgăoan et al. 2019). Polyphenols and phytosterols in bee bread have demonstrated anti-inflammatory effects that may help lower the risk of these diseases (Martiniaková et al. 2021). Additionally, studies link bee bread consumption to enhanced immune function, reinforcing its role as a valuable dietary component for supporting overall health and resilience against disease (Jarušaitienė et al. 2012; Hudz et al. 2019). This chapter reviews in vitro and in vivo studies on the antibacterial, antioxidant, and antitumor activities of bee bread, with an emphasis on the factors modulating its bioactivity. A synthesis of current research underscores bee bread's complex bioactive profile and its potential applications in therapeutic contexts.

14.2 Antioxidant Activity

The antioxidant activity of bee bread is primarily attributed to its rich composition of bioactive compounds, including phenolic acids, flavonoids, vitamins, and other phytochemicals (Bakour et al. 2017).

Various methodologies have been applied to evaluate its antioxidant capacity, including the 2,2-diphenyl-1-picrylhydrazyl (DPPH), ferric reducing antioxidant power (FRAP), Trolox equivalent antioxidant capacity (TEAC), and oxygen radical absorbance capacity (ORAC) assays, each of which provides unique insights into different mechanisms of antioxidant action, ultimately underscoring bee bread's health-promoting properties (Stanciu et al. 2016; Mărgăoan et al. 2021; Kacemi and Campos 2023).

Reactive oxygen species (ROS), while essential in cellular signaling, can lead to DNA damage when their levels surpass the body's antioxidant capacity, triggering processes that promote cancer initiation and progression, bee bread's ability to neutralize ROS through its robust antioxidant properties is well-documented and has been validated through assays such as DPPH and FRAP, which consistently demonstrate its capacity to scavenge free radicals effectively (Elsayed et al. 2021; Bakour et al. 2022).

The study of Sawicki et al. (2022) demonstrated that bee bread has a significantly higher concentration of bioactive compounds than honey, including elevated levels of vitamins C and E, which contribute to its strong scavenging ability against superoxide anion radicals. In addition, their research revealed that bee bread possesses higher reducing values in both FRAP and ORAC tests, establishing it as a superior antioxidant source compared to other bee-derived products (Sawicki et al. 2022). Similarly, Suleiman et al. (2021) reported substantial antioxidant properties in Malaysian bee bread, largely due to its diverse content of essential amino acids, phenols, and flavonoids, compounds known to counter oxidative stress and related health issues effectively.

DPPH assay is one of the most widely used methods for assessing the radical scavenging ability of antioxidants. Studies have shown that bee bread exhibits significant DPPH radical scavenging activity, indicating its potential to neutralize free radicals (Kaškonienė et al. 2018; Miłek et al. 2023; Hashemirad et al. 2024). For example, Hashemirad et al. (2024) reported an IC50 value of 0.45 mg/mL, indicating bee bread's strong antioxidant properties. Similarly, Kaškonienė et al. (2018) observed that DPPH radical scavenging activity varied widely among bee bread samples, with IC50 values ranging from 0.02 to 13 mg/mL, reflecting differences in antioxidant capacity across samples. Miłek et al. (2023) also found that bee bread exhibited significant DPPH activity, with IC50 values between 0.5 and 1.5 mg/mL, underscoring its role as a natural source of antioxidants.

Martinello and Mutinelli (2021) highlighted that bee bread's antioxidant activity surpassed that of other bee products, such as honey and propolis, with DPPH inhibition percentages exceeding 80% at concentrations of 1 mg/mL. This finding is supported by Akhir et al. (2017) who reported strong DPPH scavenging activity in stingless bee bread, with an IC50 value of 0.31 mg/mL, further emphasizing its efficacy as an antioxidant source. Additionally, Bleha et al. (2019) confirmed that bee bread samples showed antioxidant activity comparable to that of bee pollen, with DPPH inhibition around 70% at 0.5 mg/mL, supporting its potential as a functional food with health-promoting properties.

Similarly, Bakour et al. (2017) reported that bee bread exhibited substantial antioxidant activity, with DPPH inhibition values around 70% at a concentration of 0.5 mg/mL, effectively neutralizing free radicals and providing protection against oxidative damage. These findings are further supported by Aylanc et al. (2021) who observed that bee bread maintained a high antioxidant capacity of approximately 65–75% DPPH inhibition during simulated gastrointestinal digestion, indicating its stability, bioavailability, and effectiveness within the human digestive system. Sobral et al. (2017) reported a scavenging activity of Portuguese bee bread samples, ranging from 0.02 to 13 mg/mL, expressed through EC50 values.

In addition to the DPPH assay, the FRAP assay has also been employed to assess the reducing power of bee bread. This assay measures the ability of antioxidants to reduce ferric ions ($Fe^{3+)}$ to ferrous ions (Fe^{2+}), providing an indication of antioxidant capacity (Stanciu et al. 2007). The study made by Doğanyiğit et al. (2020) demonstrated that bee bread exhibited a notable FRAP value of approximately 1.7 mmol $Fe^{2+/}$g, highlighting its effectiveness in combating oxidative stress. Similarly, research by Uslinawaty (2023) revealed that the antioxidant capacity of stingless bee bread, as measured by the FRAP assay, was strongly correlated with its phenolic content, suggesting that phenolic compounds play a crucial role in its antioxidant properties.

In a related study, Othman et al. (2019) linked the antioxidant effects of bee bread specifically to the presence of phenolic compounds such as isorhamnetin, apigenin, and caffeic acid, known for their strong antioxidant activities. This underscores the importance of specific phenolic compounds in boosting bee bread's antioxidant potential. Moreover, findings indicate that the total phenolic content of bee bread is positively correlated with its antioxidant activity; however, studies such as that by Simone et al. (2022) suggest that interactions between different classes of phenolics may contribute synergistically to its overall antioxidant potential, further enhancing its efficacy.

The TEAC assay, which evaluates the ability of antioxidants to scavenge the ABTS radical, has been utilized to assess the antioxidant capacity of bee bread. In a comparative study, bee bread demonstrated high TEAC values, around 2.5 mmol Trolox equivalents (TE)/g, indicating its strong capacity for free radical neutralization (Martinello and Mutinelli 2021). This finding aligns with observations by Zucca et al. (2013) who reported that TEAC values often surpass those obtained from DPPH assays, suggesting that various assays may yield differing antioxidant capacity results. The TEAC assay's ability to measure both water-soluble and lipid-soluble antioxidants makes it particularly effective for assessing the comprehensive antioxidant potential of complex matrices such as bee bread (Mărgăoan et al. 2021).

The ORAC assay, a method for evaluating antioxidant capacity via hydrogen atom transfer, assesses the ability of antioxidants to quench peroxyl radicals, thereby providing insights into their protective effects against lipid peroxidation. Research by Anand et al. (2018) indicated that bee bread exhibits significant ORAC values, approximately 3.5 mmol Trolox equivalents (TE)/g, suggesting its strong potential to counteract oxidative damage. This high ORAC value is closely correlated with the total phenolic content, as noted by Valdenegro et al. (2021) which reinforces the essential role of phenolic compounds in enhancing the antioxidant activity of bee bread.

The antioxidant properties of bee bread are primarily attributed to its diverse array of phenolic compounds, vitamins, and other phytochemicals, which work synergistically to neutralize free radicals and protect the body from oxidative stress (Kaškonienė et al. 2018). One of the key components responsible for the antioxidant activity of bee bread is its high phenolic content. Phenolic compounds, such as flavonoids, are known for their ability to scavenge free radicals and inhibit oxidative processes (Khalifa et al. 2020). Research by Sobral et al. identified several flavonol derivatives in bee bread, including quercetin, kaempferol, and isorhamnetin, which are well-documented for their antioxidant properties (Sobral et al. 2017). Gallic acid is present in glycosylated forms, adding to the total phenolic content and health-promoting potential of the product (Suleiman et al. 2021). Similarly, ferulic acid, another potent antioxidant, occurs in glycosylated forms, providing additional stability and bioactivity (Othman et al. 2019). Unique compounds, such as methylherbacetin glycosides, also contribute to the chemical diversity, making bee bread a rich source of bioactive glycosides with varied potential health applications (Dranca et al. 2020).

These compounds not only contribute to the radical scavenging ability of bee bread but also play a role in reducing inflammation and preventing chronic diseases associated with oxidative stress (Sobral et al. 2017). In addition to flavonoids, bee bread is rich in vitamins, particularly vitamins C and E, which are potent antioxidants. According to Sawicki et al. (2022), bee bread contains higher levels of these vitamins compared to honey, which enhances its overall antioxidant capacity. The presence of vitamins C and E in bee bread underscores its potential as a functional food that can support health and well-being (Ćirić et al. 2022). The antioxidant activity of bee bread has been quantitatively assessed using various assays, including DPPH, FRAP, and ABTS. The synergistic effect of the various bioactive compounds in bee bread is also significant. Research has shown that the combined action of phenolic compounds, vitamins, and other antioxidants can lead to a greater overall antioxidant capacity than the sum of their individual effects. This synergism is particularly important in the context of dietary interventions aimed at reducing oxidative stress and promoting health (Aylanç et al. 2021; Aksoy 2024). These findings highlight the importance of bioactive compounds in bee bread, which contribute to its ability to combat oxidative stress.

14.3 Antimicrobial Activity

Bee bread, a naturally fermented product of pollen, nectar, and bee secretions, is a complex and nutrient-rich substance. Its composition includes a high concentration of proteins and amino acids, which under ordinary circumstances could provide an ideal substrate for microbial growth. However, despite its nutrient-dense profile, bee bread exhibits notable antibacterial properties, a paradox that underscores the unique interplay of its bioactive components. The antimicrobial activity of bee bread is primarily attributed to its diverse composition of bioactive compounds, including phenolics, flavonoids, and organic acids, which collectively enhance its effectiveness against a broad spectrum of bacterial strains, these compounds contribute to both its bacteriostatic and bactericidal effects, demonstrating the potential of bee bread as a natural antimicrobial agent (Didaras et al. 2020; Ilie et al. 2024).

The disc diffusion method has been frequently employed to evaluate the antibacterial activity of bee bread against both Gram-positive and Gram-negative bacteria. However, reports on specific inhibition zones, such as the claimed 15 mm zone for *Escherichia coli* (Anderson et al. 2013), remain inconsistent and lack direct experimental support. Such discrepancies highlight the necessity for further validation using standardized methodologies to ensure the reproducibility and reliability of the reported findings. For Gram-positive bacteria, bee bread extracts have shown considerable activity. Studies have reported inhibition zones for *Staphylococcus aureus* ranging between 12 mm and 18 mm when methanolic extracts were used, demonstrating its strong efficacy (Suleiman et al. 2021). Similarly, for *Bacillus subtilis*, inhibition zones have been observed in the range of 10 mm to 15 mm when ethanolic extracts were applied (Didaras et al. 2020), further emphasizing the potency of bee bread against Gram-positive pathogens.

The activity against Gram-negative bacteria is comparatively lower, reflecting the inherent resistance of these organisms due to the structural complexity of their outer membrane. For *Escherichia coli*, inhibition zones ranging from 8 mm to 13 mm have been reported (Keyvan 2023). Similarly, for *Salmonella enterica*, inhibition zones of 7 mm to 12 mm were observed for both aqueous and methanolic extracts, showcasing variability in antibacterial activity based on extraction methodology (Gabriele et al. 2021). These findings underscore the variability in inhibition zone diameters, influenced by factors such as bacterial strain, extraction solvent, and bioactive compound concentration (Rodríguez-Pólit et al. 2023). While bee bread exhibits strong antimicrobial activity, particularly against Gram-positive bacteria, further research employing consistent and standardized protocols is necessary to fully elucidate its potential as a natural antimicrobial agent.

Other studies have consistently reported that bee bread exhibits significant antibacterial activity, particularly against Gram-positive bacteria such as *Staphylococcus aureus* and *Bacillus subtilis* (Pełka et al. 2021). Urcan et al. (2018) demonstrated that bee bread extracts exert both bactericidal and bacteriostatic effects against *S. aureus*, with minimum inhibitory concentrations (MICs) ranging between 0.5 and 1.0 mg/mL. Similarly, Keyvan (2023) highlighted that bee bread exhibits notable activity against Gram-positive bacteria, reporting MIC values as low as 1.617 μg/mL against *S. aureus*. These findings underscore the potent antibacterial properties of bee bread and its selective efficacy against Gram-positive strains. MIC values reported by the same author (Keyvan 2023) for *Klebsiella pneumoniae* (1.914 μg/mL), *E. coli* (1.923 μg/mL), *Shigella* (1.813 μg/mL), *and Salmonella typhi* (1.617 μg/mL) indicate significant activity against these Gram-negative pathogens as well.

Keyvan (2023) also noted that bee bread extracts displayed a broader spectrum of activity against Gram-negative bacteria compared to Gram-positive strains, consistent with findings of Gabriele et al. (2021). The same authors reported that *Cistus* bee pollen, which shares similar bioactive properties with bee bread, inhibited the growth of Gram-negative bacteria with MIC values around 10 mg/mL, compared to approximately 5 mg/mL for Gram-positive bacteria.

Didaras et al. (2020) further emphasized the antimicrobial potential of bee bread, noting its targeted activity against Gram-positive pathogens, including *S. aureus*. Their study confirmed the ability of bee bread extracts to inhibit the growth of multiple Gram-positive bacterial strains, reinforcing its potential application as a natural antimicrobial product. For Gram-negative bacterial strains, ethanolic extracts of bee bread have shown inhibition zones of 8–14 mm and 10–16 mm, respectively, with MICs ranging from 0.5 to 2 mg/mL for *P. aeruginosa* and around 1 mg/mL for *K. pneumoniae* (Didaras et al. 2020). These effects are attributed to its rich composition of phenolics and flavonoids, which vary with extraction methods and floral origin. Bee bread's activity against these pathogens highlights its potential as a natural agent for managing resistant bacterial infections and its role in developing functional antimicrobials.

The extraction method has been identified as a critical factor influencing the antimicrobial efficacy of bee bread. Suleiman et al. (2021) found that alcoholic extracts of bee bread demonstrated significantly stronger antibacterial activity against Gram-positive bacteria compared to aqueous extracts, highlighting the importance of solvent selection in optimizing bioactive compound extraction.

In addition to its effects on Gram-positive bacteria, bee bread has demonstrated substantial antimicrobial activity against Gram-negative bacteria, including *Escherichia coli* and *Salmonella enterica* and *Pseudomonas aeruginosa* (Abouda et al. 2011; Maželiene et al. 2022). Suleiman et al. (2021) reported MIC values as low as 6.40 μg/mL against *E. coli*, indicating strong antimicrobial potential. The same study highlighted the effectiveness of bee bread extracts against *S. enterica*, further supporting its use as a natural preservative in food products.

Furthermore, Akhir et al. (2017) reported that ethanolic extracts of *Heterotrigona itama* bee bread exhibited stronger antibacterial activity compared to hexane extracts, with Gram-positive bacteria such as *Bacillus cereus* showing greater sensitivity, while Bakour et al. (2019) highlighted that the bioactive properties of bee bread, including its antimicrobial activity, are strongly influenced by its chemical composition, which varies according to the floral origin of the pollen. Their findings revealed that methanolic extracts of bee bread exhibited MIC values of 0.5 mg/mL against *S. aureus* and 1 mg/mL against *E. coli.* These results were further corroborated by inhibition zones ranging from 12 mm to 18 mm for *Staphylococcus aureus* and 8 mm to 13 mm for *S. enterica*. These data underscore the importance of both extraction methods and the botanical source in determining the antimicrobial efficacy of bee bread, further solidifying its role as a superior natural antimicrobial agent.

A comparative analysis conducted by Pełka et al. (2021) demonstrated that bee bread exhibits superior antimicrobial potential compared to bee pollen. Ethanolic extracts of bee bread showed inhibition zones of up to 15 mm for *S. aureus* and 10 mm for *E. coli*, highlighting its enhanced efficacy as a natural antimicrobial agent. Furthermore, the study reported minimum inhibitory concentrations (MICs) ranging from 2.5% to 10% (v/v) against *S. aureus* strains ATCC 25923 and ATCC 29213, emphasizing the potency of bee bread in inhibiting bacterial growth.

Rzepecka-Stojko et al. (2015) reported a MIC of 1.25 mg/mL for *C. albicans*, indicating its potential as a natural antifungal. Similarly, Suleiman et al. highlighted the significant antifungal activity of Malaysian bee bread extracts against various *Candida* strains, with ethanolic extracts showing superior efficacy compared to aqueous extracts, although specific MIC values were not provided.

Didaras et al. (2020) emphasized the broad-spectrum antifungal properties of bee bread, attributing its activity to the synergistic effects of its bioactive compounds (Hudz et al. 2019). Furthermore, Ispirli and Dertli (2021) noted that bee bread's antifungal potential is enhanced by its microbial community, particularly lactic acid bacteria and yeasts, which contribute to its effectiveness against opportunistic fungal pathogens like *Candida* (İspirli and Dertli 2021).

Moreover, Hudz et al. (2019) reported that extracts of bee bread showed moderate antifungal activity against various fungal strains, including those from the genera *Aspergillus* and *Penicillium*, with inhibition percentages reaching up to 93.60% when using 70% ethanolic extracts. BB exhibits even greater antimicrobial potential, partly due to the presence of lactic acid produced by lactic acid bacteria (LAB), such as *Lactobacillus kunkeei* and *L. plantarum*. These LAB play a crucial role during the 2-week fermentation process that transforms BP into BB (Sonmez et al. 2023) and by lowering the pH, lactic acid creates an acidic environment that inhibits the growth of many pathogens, including *Pseudomonas aeruginosa* and *Candida albicans* (Vasquez and Olofsson 2009). During this time, bacteriocins and antimicrobial metabolites are produced, and lipid hydrolysis occurs, enhancing the overall bioactivity of BB (Vásquez and Olofsson 2009; Iorizzo et al. 2022).

Lactic acid also disrupts bacterial membrane integrity, leading to intracellular acidification and metabolic dysfunction, which are key mechanisms in its antibacterial action (Vergalito et al. 2020; Poyraz et al. 2023; Aksoy et al. 2024). Furthermore, its presence supports the bioavailability of other bioactive compounds in bee bread, such as phenolics and flavonoids, further amplifying its antimicrobial properties (Salazar-González and Díaz-Moreno 2016).

The choice of solvent plays a critical role in determining the antimicrobial activity of BB and BP, methanol extracts consistently demonstrated superior efficacy due to their ability to extract a broader spectrum of bioactive compounds compared to ethanol or water (Sonmez et al. 2023). This highlights the importance of optimizing extraction techniques to maximize the antimicrobial potential of these natural products (Pełka et al. 2021).

Bee bread demonstrates significant antimicrobial activity against both Gram-positive and Gram-negative bacteria, with MIC values highlighting its potential as a therapeutic agent and natural preservative. The antimicrobial properties of bee bread are the result of a synergistic interaction between its phytochemical constituents and microbial-derived metabolites. Its bioactivity against a broad spectrum of pathogens, combined with the influence of extraction techniques on its efficacy, underscores the importance of bee bread as a functional food and potential natural antimicrobial agent. These findings provide a strong foundation for further research into its applications in food preservation and therapeutic contexts.

14.4 Antitumoral Activity

Among the diverse range of natural products derived from bees, bee bread stands out not merely for its role in hive nutrition but for the growing body of evidence supporting its therapeutic potential. This fermented product, a result of pollen fermentation facilitated by honey bees and their symbiotic microbes, offers a complex matrix of bioactive compounds, including phenolics, flavonoids, vitamins, and amino acids, which together drive its diverse health benefits (Bakour et al. 2019; Mărgăoan et al. 2019; Mohammad et al. 2021). Among its most compelling attributes are its antitumor and anticancer properties, which are underpinned by its ability to modulate oxidative stress, inflammation, and immune responses—key factors in the pathogenesis and progression of cancer (Denisow et al. 2016; Sobral et al. 2017; Elsayed et al. 2021). The bioactivity of bee bread against cancer is closely tied to its rich phenolic and flavonoid content, which collectively influence key cellular pathways involved in tumorigenesis (Uțoiu et al. 2018).

The chemical composition of bee bread (BB) is influenced by its botanical and geographical origins, as well as the conditions under which it is stored, this variation plays a critical role in determining its bioactive properties, particularly its antioxidant and antitumor potential, which are closely linked to its polyphenol and unsaturated fatty acid content (Komosinska-Vassev et al. 2015). Notably, bee bread is a rich source of α-linolenic acid, which accounts for up to 65.7% of its fatty acid composition and has been shown to affect membrane properties and signaling pathways critical for cancer progression (Campos et al. 2018; Negri et al. 2018).

The antitumor activity of bee bread can also be attributed to its fatty acid content, particularly polyunsaturated fatty acids (PUFAs) like linoleic and linolenic acids. These compounds disrupt the lipid structure of cancer cell membranes, impair nutrient uptake, and induce apoptosis. Additionally, sterols such as β-sitosterol modulate hormonal pathways and inhibit tumor growth, particularly in hormone-sensitive cancers (Elsayed et al. 2021; Khalifa et al. 2020; Mărgăoan et al. 2021).

Phytosterols, another key component of bee bread, have been shown to inhibit cholesterol absorption in the intestine and exhibit anticancer effects in various clinical trials (Ostlund et al. 2003; Bradford and Awad 2007). β-sitosterol and avenasterol have also been identified in high concentrations in bee bread samples, and research highlights the inhibitory effects of specific sterol fractions from bee bread extracts on prostate cancer cell lines, such as DU145 and LNCaP (Mărgăoan et al. 2019). These findings underscore the potential role of bee bread in modulating cancer cell growth while sparing healthy tissues (Vanderplanck et al. 2014).

Further studies have characterized sterol profiles in bee bread derived from different botanical sources, such as *Salix* spp. and *Calluna vulgaris*, with total sterol concentrations ranging from 2.5 to 9.6 mg/g of dry weight (Kaplan et al. 2016). The unique composition of these sterols enhances the therapeutic potential of bee bread, particularly in managing hormone-sensitive cancers (Furse et al. 2023).

Bee bread's cytotoxic effects have been extensively evaluated using human cancer cell lines. Mărgăoan et al. (2016) demonstrated the antiproliferative activity of methanolic extracts from bee bread against C26 colorectal cancer cell lines. The highest apoptotic index (30%) was observed with a 1 mg/mL concentration after 24–48 hours of treatment. This study also reported morphological changes in cancer cells, including cytoplasmic vacuolization and granulation, indicative of cell death.

Moreover, bee bread exerts pro-apoptotic effects on cancer cells. By modulating gene expression, it enhances the activity of pro-apoptotic markers such as Bax and suppresses anti-apoptotic proteins like Bcl-2, leading to the activation of intrinsic apoptotic pathways (Sonmez et al. 2023). This selective targeting of tumor cells while sparing healthy ones makes bee bread particularly promising in the realm of natural anticancer agents (Khider et al. 2013).

Markiewicz-Zukowska et al. (2013) tested ethanolic extracts of Polish bee bread on glioblastoma cells (U87MG) over 24, 48, and 72 hours, observing a reduction in cell viability by 49–66%, with the strongest inhibitory effects after 72 hours. Sobral et al. (2017) investigated bee bread from northeastern Portugal against several cancer cell lines, including NCI-H460 (lung cancer), HepG2 (liver cancer), HeLa (cervical cancer), and MCF-7 (breast cancer), reporting low to moderate cytotoxicity (IC50 = 68 to >400 μg/mL) while showing no toxicity to normal liver cells. These findings highlight bee bread's selective action against tumor cells while sparing healthy tissues.

The antitumor effects of bee bread have been investigated using the MTT assay on cancer cell lines, including colorectal adenocarcinoma (Caco-2), prostate adenocarcinoma (PC3), hepatocellular carcinoma (HepG-2), and normal lung cells (WI-38) was tested by Elsayed et al. (2021). Bee bread exhibited dose-dependent cytotoxicity, with IC50 values of 262 μg/mL for Caco-2, 314 μg/mL for PC3, and 386 μg/mL for HepG-2 cells. The strongest inhibition was observed in Caco-2 cells, where viability was reduced to 19.5%, 22.58%, and 51.39% at concentrations of 1000, 500, and 250 μg/mL, respectively, followed by PC3 and HepG-2.

In another study by Borawska et al. (2014), ethanolic extracts of bee bread derived from *Salix* spp. were tested on diffuse astrocytoma and glioblastoma multiforme (U87MG) cell lines. Results showed significant cytotoxicity, with cell viability reduced to 26.5% after 72 hours of exposure. When combined with the chemotherapeutic agent temozolomide (TMZ), bee bread further inhibited DNA synthesis and reduced cell proliferation, suggesting its potential to enhance conventional cancer treatments.

The selective action of bee bread against cancer cells while sparing normal cells is particularly noteworthy, five different bee bread samples were tested on human tumor cell lines, including HeLa (cervical cancer), HepG2 (liver cancer), MCF-7 (breast cancer), and NCI-H460 (lung cancer), as well as non-tumor liver cells (PLP2) by Sobral et al. (2017). Among the tested samples, BB3 inhibited all tumor cell lines without affecting normal cells, while other samples exhibited cell line-specific activity. Samples BB1 and BB2 showed efficacy against MCF-7 and BB4 was active against NCI-H460 and HeLa cells. These findings highlight the importance of sample-specific differences in bee bread's chemical composition and bioactivity.

Bee bread's diverse chemical composition, including polyphenols, fatty acids, and phytosterols, underpins its significant antitumor potential. Its ability to selectively target cancer cells, reduce oxidative stress and inhibit angiogenesis positions it as a promising natural therapeutic agent. However, future research should focus on standardizing bee bread's composition and conducting large-scale clinical trials to validate its safety and efficacy in humans. These efforts will pave the way for its integration into modern oncological practices as a complementary treatment option.

14.5 Other Health Benefits

14.5.1 Hepatoprotective Activity and Antidiabetic Activity

The hepatoprotective effects of bee bread have been increasingly recognized in recent literature, with several studies highlighting its potential to protect liver function against various forms of damage. Bee bread, a fermented mixture of pollen, honey, and bee saliva, is rich in bioactive compounds, including antioxidants, flavonoids, and phenolic acids, which contribute to its protective effects on the liver (Komosinska-Vassev et al. 2015; Khalifa et al. 2020). One notable study by Zakaria et al. (2021) investigated the hepatoprotective effects of bee bread in a rat model of metabolic dysfunction-associated fatty liver disease (MAFLD). The researchers found that administration of bee bread significantly reduced oxidative stress markers and inflammation in the liver, specifically bee bread was shown to decrease the expression of Keap1, which is involved in the regulation of antioxidant responses, thereby promoting the translocation of Nrf2 and enhancing the activity of antioxidant enzymes in the liver Zakaria et al. (2021). This suggests that bee bread may help mitigate the oxidative damage associated with fatty liver disease. In another study, Bakour et al. (2017) examined the protective effects of bee bread against aluminum-induced hepatotoxicity. The study reported that bee bread administration led to a significant reduction in liver enzyme levels (ALT, AST, and ALP) that were elevated due to aluminum exposure. Additionally, histological analysis revealed that bee bread helped restore normal liver architecture, alleviating congestion and other pathological changes induced by aluminum chloride (Bakour et al. 2017). This indicates that bee bread may possess hepatoprotective properties by counteracting the toxic effects of environmental pollutants. Dervişoğlu et al. (2022) also contributed to the understanding of bee bread's hepatoprotective potential by demonstrating its antioxidant and cytotoxic activities and showed that bee bread extracts did not exhibit toxicity in non-tumor liver cells while effectively inhibiting the growth of various cancer cell lines, suggesting a selective protective effect on healthy liver cells. This selective action is critical in the context of liver health, as it implies that bee bread can support liver function without causing additional harm. Furthermore, Othman et al. (2019) highlighted the antioxidant properties of bee bread, noting its ability to neutralize free radicals and reduce oxidative stress, which is a key factor in liver damage. The presence of various antioxidants in bee bread, including phenolic compounds, contributes to its potential as a hepatoprotective agent, by reducing oxidative stress, bee bread may help prevent liver cell damage and support overall liver health (Mărgăoan et al. 2019; Zakaria et al. 2022). In summary, the literature indicates that bee bread exhibits significant hepatoprotective effects, primarily through its antioxidant and anti-inflammatory properties. These findings suggest that bee bread could be a valuable natural supplement for promoting liver health and preventing liver-related diseases.

Several studies have demonstrated that bee bread can positively influence metabolic parameters associated with diabetes. Zucker diabetic fatty (ZDF) rats indicated that bee bread supplementation significantly reduced lipid abnormalities and improved bone morphology, suggesting its potential role in managing complications associated with type 2 diabetes mellitus (T2DM) (Martiniaková et al. 2021). The administration of bee bread at varying doses resulted in lowered levels of triglycerides and cholesterol, which are critical factors in diabetes management (Martiniaková et al. 2021). Additionally, the anti-inflammatory properties of bee bread, attributed to its high phenolic content, may help mitigate the chronic inflammation often observed in diabetic conditions (Doğanyiğit et al. 2020). The mechanism by which bee bread exerts its antidiabetic effects appears to involve the inhibition of key enzymes related to carbohydrate metabolism; specifically, bee bread has been shown to inhibit α-amylase activity in vitro, which is crucial for the digestion of carbohydrates and the subsequent absorption of glucose (Fallah et al. 2022; Bakour et al. 2022). This inhibition can lead to a slower release of glucose into the bloodstream, thereby preventing spikes in blood sugar levels after meals. Furthermore, the presence of antioxidants in bee bread may enhance insulin sensitivity, further aiding in blood glucose regulation (Doğanyiğit et al. 2020; Khalifa et al. 2020).

14.5.2 Immunomodulatory Effect and Anti-Allergy Potential

The potential anti-allergic effects of bee bread have been a subject of interest in recent literature, particularly due to its rich composition of bioactive compounds, including flavonoids, phenolic acids, and various vitamins. These components are known for their anti-inflammatory and immunomodulatory properties, which may contribute to alleviating allergic reactions (Parvez et al. 2006; Yin et al. 2024). Also, the nutritional value of bee bread, which includes essential amino acids, vitamins, and minerals, further underscores its role as a functional food that can enhance overall health and immunity (Dervişoğlu et al. 2022; Urcan et al. 2017, 2024). The combination of proteins, lipids, and antioxidants not only supports immune function but also plays a role in metabolic health by reducing inflammation and promoting cellular repair mechanisms (Jarušaitienė et al. 2012).

Bee bread has been shown to possess significant anti-inflammatory properties. Recent studies indicate that flavonoids present in bee bread, such as kaempferol and isorhamnetin, can reduce the levels of pro-inflammatory cytokines like TNF-α and NF-κβ in various models, including those induced by obesity (Othman et al. 2019, 2021). This reduction in inflammatory markers suggests that bee bread may help mitigate allergic responses, which are often exacerbated by inflammation. The anti-inflammatory effects are further supported by evidence indicating that bee bread enhances the activity of interleukin-10 (IL-10), an anti-inflammatory cytokine, thereby promoting a balanced immune response (Othman et al. 2021). The presence of high levels of phenolic compounds in bee bread is associated with scavenging reactive oxygen species (ROS), which play a role in the pathophysiology of allergic reactions. By reducing oxidative stress, bee bread may help prevent the activation of pathways that lead to allergic inflammation (Zakaria et al. 2022). This is particularly relevant as oxidative stress has been implicated in various allergic conditions, including asthma and allergic rhinitis (Han et al. 2021a).

Research has indicated that in addition to its anti-inflammatory and antioxidant properties, bee bread's immunomodulatory effects may also play a crucial role in its potential anti-allergic activity, which can influence the expression of hormones such as leptin and ghrelin, which are involved in regulating immune responses (Doğanyiğit et al. 2020). This modulation of the immune system could help in balancing the Th1/Th2 response, which is often skewed in allergic individuals, leading to a more favorable outcome in managing allergies.

The antimicrobial properties of bee bread may indirectly enhance its anti-allergic potential through supporting a healthy gut microbiome, thereby strengthening immune function and reducing the likelihood of allergic sensitization (Suleiman et al. 2021). A balanced gut microbiome is crucial for the development of a robust immune system that can appropriately respond to allergens without overreacting (Wu and Wu 2012; Han et al. 2021a, b). While the literature provides promising insights into the potential anti-allergic effects of bee bread, it is essential to note that most studies have been conducted in animal models or in vitro, therefore, further clinical research is necessary to establish the efficacy and safety of bee bread as a natural remedy for allergies in humans.

References

Abouda Z, Zerdani I, Kalalou I, Faid M, Ahami MT (2011) The antibacterial activity of Moroccan bee bread and bee-pollen (fresh and dried) against pathogenic bacteria. Res J Microbiol 6(4):376–384. https://doi.org/10.3923/jm.2011.376.384

Akhir RAM, Bakar MFA, Sanusi SB (2017) Antioxidant and antimicrobial activity of stingless bee bread and propolis extracts. AIP Conf Proc 1891. https://doi.org/10.1063/1.5005423

Aksoy A, Altunatmaz SS, Aksu F, Tokatlı Demirok N, Yazıcı K, Yıkmış S (2024) Bee bread as a functional product: phenolic compounds, amino acid, sugar, and organic acid profiles. Food Secur 13(5):795. https://doi.org/10.3390/foods13050795

Aksoy A, Altunatmaz SS, Aksu F, Tokatlı Demirok N, Yazıcı K, Yıkmış S (2024) Bee Bread as a Functional Product: Phenolic Compounds, Amino Acid, Sugar, and Organic Acid Profiles. Foods 13(5):795. https://doi.org/10.3390/foods13050795

Anand S, Pang E, Livanos G, Mantri N (2018) Characterization of physico-chemical properties and antioxidant capacities of bioactive honey produced from Australian grown Agastache rugosa and its correlation with colour and poly-phenol content. Molecules 23(1):108. https://doi.org/10.3390/molecules23010108

Anderson KE, Sheehan TH, Mott BM, Maes P, Snyder L, Schwan MR, Walton A, Jones BM, Corby-Harris V (2013) Microbial ecology of the hive and pollination landscape: bacterial associates from floral nectar, the alimentary tract and stored food of honey bees (Apis mellifera). PLoS One 8(12):e83125. https://doi.org/10.1371/journal.pone.0083125

Aylanc V, Falcão SI, Ertosun S, Vilas-Boas M (2021) From the hive to the table: nutrition value, digestibility and bioavailability of the dietary phytochemicals present in the bee pollen and bee bread. Trends Food Sci Technol 109:464–481. https://doi.org/10.1016/j.tifs.2021.01.042

Bakour M, Al-Waili NS, El Menyiy N, Imtara H, Figuira AC, Al-Waili T, Lyoussi BB, El Menyiy N, Imtara H, Figuira AC, Al-Waili T, Lyoussi BB (2017) Antioxidant activity and protective effect of bee bread (honey and pollen) in aluminum-induced anemia, elevation of inflammatory makers and hepato-renal toxicity. J Food Sci Technol 54(13):4205–4212. https://doi.org/10.1007/s13197-017-2889-9

Bakour M, Fernandes Â, Barros L, Sokovic M, Ferreira ICFR (2019) Bee bread as a functional product: chemical composition and bioactive properties. LWT 109(February):276–282. https://doi.org/10.1016/j.lwt.2019.02.008

Bakour M, Hammas N, Laaroussi H, Ousaaid D, Fatemi HEL, Aboulghazi A, Soulo N, Lyoussi B (2021) Moroccan bee bread improves biochemical and histological changes of the brain, liver, and kidneys induced by titanium dioxide nanoparticles. Biomed Res Int 2021. https://doi.org/10.1155/2021/6632128

Bakour M, Laaroussi H, Ousaaid D, Es-safi I, Mechchate H, Lyoussi BBB, El Ghouizi A, Es-safi I, Mechchate H, Lyoussi BBB (2022) Bee bread as a promising source of bioactive molecules and functional properties: an up-to-date review. Antibiotics 11(2):203. https://doi.org/10.3390/antibiotics11020203

Beykaya M, Tanugur Samanci AE, Samanci T, Önder EY, Uzun EM, Tosun F, Tanuğur Samancı AE, Samancı T, Önder EY, Uzun EM, Tosun F, Tanugur Samanci AE, Samanci T, Önder EY, Uzun EM, Tosun F (2021) Investigation of nutritional and antioxidant properties of Anatolian bee bread. J Apic Sci 65(2):255–263. https://doi.org/10.2478/jas-2021-0017

Bleha R, Shevtsova TS, Kružík V, Škorpilová T, Saloň I, Erban V, Brindza J, Brovarskyi V, Sinica A (2019) Bee breads from two regions of eastern Ukraine: composition, physical properties and biological activities. Czech J Food Sci. https://doi.org/10.17221/201/2018-cjfs

Borawska MH, Markiewicz-Żukowska R, Naliwajko SK, Moskwa J, Bartosiuk E, Socha K, Surażyński A, Kochanowicz J, Mariak Z (2014) The interaction of bee products with temozolomide in human diffuse astrocytoma, glioblastoma multiforme and astroglia cell lines. Nutr Cancer 66(7):1247–1256. https://doi.org/10.1080/01635581.2014.951735

Bradford PG, Awad AB (2007) Phytosterols as anticancer compounds. Mol Nutr Food Res 51(2):161–170. https://doi.org/10.1002/mnfr.200600164

Campos MG, Machado J, Costa ML, Lino S, Correia F, Maltez F (2018) Case report: severe hematological, muscle and liver toxicity caused by drugs and artichoke infusion interaction in an elderly polymedicated patient. Curr Drug Saf 13(1):44–50. https://doi.org/10.2174/1574886312666170912163746

Ćirić J, Haneklaus N, Rajić S, Baltić T, Lazić IB, Đorđević V (2022) Chemical composition of bee bread (perga), a functional food: a review. J Trace Elem Miner 2(August):100038. https://doi.org/10.1016/j.jtemin.2022.100038

de Simone N, Rocchetti MT, la Gatta B, Spano G, Drider D, Capozzi V, Russo P, Fiocco D (2022) Antimicrobial properties, functional characterisation and application of *Fructobacillus fructosus* and *Lactiplantibacillus plantarum* isolated from Artisanal Honey. Probiotics Antimicrob Proteins 15:1406. https://doi.org/10.1007/s12602-022-09988-4

Denisow B, Denisow-Pietrzyk M, Denisow-Pietrzyk M (2016) Biological and therapeutic properties of bee pollen: a review. J Sci Food Agric 96(13):4303–4309. https://doi.org/10.1002/jsfa.7729

Dervişoğlu G, Çobanoğlu DN, Yelkovan S, Karahan D, Çakır Y, Koçyığıt S (2022) Comprehensive study on beebread: palynological analysis, chemical composition, antioxidant and cytotoxic activities. Int J Second Metab. https://doi.org/10.21448/ijsm.1066884

Didaras NA, Karatasou K, Dimitriou TG, Amoutzias GD, Mossialos D (2020) Antimicrobial activity of bee-collected pollen and beebread: state of the art and future perspectives. Antibiotics 9(11):811. https://doi.org/10.3390/antibiotics9110811

Doğanyiğit Z, Yakan B, Soylu M, Kaymak E, Silici S (2020) The effects of feeding obese rats with bee bread on leptin and ghrelin expres. Turkish J Zool 44(2):114–125. https://doi.org/10.3906/zoo-1911-16

Dranca F, Ursachi F, Oroian M (2020) Bee bread: physicochemical characterization and phenolic content extraction optimization. Food Secur 9. https://doi.org/10.3390/foods9101358

Elsayed N, El-Din HS, Altemimi AB, Ahmed HY, Pratap-Singh A, Abedelmaksoud TG (2021) In vitro antimicrobial, antioxidant and anticancer activities of Egyptian citrus beebread. Molecules 26(9):2433. https://doi.org/10.3390/molecules26092433

Fallah M, Najafi F, Kavoosi G (2022) Proximate analysis, nutritional quality and anti-amylase activity of bee propolis, bee bread and royal jelly. Int J Food Sci Technol. https://doi.org/10.1111/ijfs.15605

Furse S, Koch H, Wright GA, Stevenson PC (2023) Sterol and lipid metabolism in bees. Metabolomics 19(9):1–19. https://doi.org/10.1007/s11306-023-02039-1

Gabriele M, Frassinetti S, Pucci L (2021) Antimicrobial activity and protective effect of tuscan bee pollens on oxidative and endoplasmic reticulum stress in different cell-based models. Food Secur 10(6). https://doi.org/10.3390/foods10061422

Gomaa EZ (2017) Antimicrobial, antioxidant and antitumor activities of silver nanoparticles synthesized by Allium cepa extract: a green approach. J Genet Eng Biotechnol 15(1):49–57. https://doi.org/10.1016/j.jgeb.2016.12.002

Han M, Lee D, Lee SH, Kim TH (2021a) Oxidative stress and antioxidant pathway in allergic rhinitis. Antioxidants 10(8):1266. https://doi.org/10.3390/antiox10081266

Han P, Gu J-Q, Li L-S, Wang X-Y, Wang H-T, Wang Y, Chang C, Sun J-L (2021b) The association between intestinal bacteria and allergic diseases—cause or consequence? Front Cell Infect Microbiol 11(April):1–22. https://doi.org/10.3389/fcimb.2021.650893

Hashemirad F-S, Behfar M, Kavoosi G (2024) Proximate composition, physico-chemical, techno-functional, amino acid profile, fatty acid profile, nutritional quality, antioxidant, anti-amylase and anti-lipase properties of bee bread, royal jelly, and bee propolis. LWT 200(May):116190. https://doi.org/10.1016/j.lwt.2024.116190

Hudz N, Yezerska O, Grygorieva O, Brindza J, Felšöciová S, Kačániová M, Wieczorek P (2019) Analytical procedure elaboration of total flavonoid content determination and antimicrobial activity of bee bread extracts. Acta Pol Pharm - Drug Res 76:439. https://doi.org/10.32383/appdr/103745

Ilie C-I, Spoiala A, Geana E-I, Chircov C, Ficai A, Ditu L-M, Oprea E (2024) Bee bread: a promising source of bioactive compounds with antioxidant properties—first report on some antimicrobial features. Antioxidants 13(3):353. https://doi.org/10.3390/antiox13030353

Iorizzo M, Ganassi S, Albanese G, Letizia F, Testa B, Tedino C, Petrarca S, Mutinelli F, Mazzeo A, De Cristofaro A (2022) Antimicrobial activity from putative probiotic lactic acid bacteria for the biological control of American and European foulbrood diseases. Vet Sci 9(5):236. https://doi.org/10.3390/vetsci9050236

Ispirli H, Dertli E (2021) Detection of fructophilic lactic acid bacteria (FLAB) in bee bread and bee pollen samples and determination of their functional roles. J Food Process Preserv 45(5):1–11. https://doi.org/10.1111/jfpp.15414

Jarušaitienė D, Jankauskienė J, Čeksterytė V (2012) Changes of ocular signs in children with thyroid diseases after intake of bee bread. Chem Technol 61. https://doi.org/10.5755/j01.ct.61.3.2660

Kabakci D (2024) Comparison of some biological activities of propolis and bee bread samples obtained from Apis Mellifera Anatoliaca and its Muğla and Efe ecotypes. J Hell Vet Med Soc 74:6451. https://doi.org/10.12681/jhvms.30859

Kacemi R, Campos MG (2023) Translational research on bee pollen as a source of nutrients: a scoping review from bench to real world. Nutrients 15(10):2413. https://doi.org/10.3390/nu15102413

Kankasemsuk S (2023) Total protein content of bee bread in Apis Cerana combs and Tetragonula Pegdeni storage pots of different plant sources from Chanthaburi Province, Thailand. J Curr Sci Technol. https://doi.org/10.59796/jcst.v13n3.2023.1139

Kaplan M, Karaoglu Ö, Eroglu N, Silici S (2016) Fatty acid and proximate composition of bee bread. Food Technol Biotechnol 54(4):497–504. https://doi.org/10.17113/ftb.54.04.16.4635

Kaškonienė V, Katilevičiūtė A, Kaškonas P, Maruška A (2018) The impact of solid-state fermentation on bee pollen phenolic compounds and radical scavenging capacity. Chem Pap 72(8):2115–2120. https://doi.org/10.1007/s11696-018-0417-7

Keyvan N (2023) Total phenolic content, antibacterial and antiradical properties of bee bread from Turkey. Vet J Mehmet Akif Ersoy Univ. https://doi.org/10.24880/maeuvfd.1278378

Khalifa SA, Elashal MH, Kieliszek M, Ghazala NE, Farag MA, Saeed A et al (2020) Recent insights into chemical and pharmacological studies of bee bread. Trends Food Sci Technol 97(May 2019):300–316. https://doi.org/10.1016/j.tifs.2019.08.021

Khider M, Elbanna K, Mahmoud A, Owayss AA (2013) Egyptian honeybee pollen as antimicrobial, antioxidant agents, and dietary food supplements. Food Sci Biotechnol 22(5):1–9. https://doi.org/10.1007/s10068-013-0238-y

Komosinska-Vassev K, Olczyk P, Kaźmierczak J, Mencner L, Olczyk K (2015) Bee pollen: chemical composition and therapeutic application. Evid Based Complement Altern Med 2015:1–6. https://doi.org/10.1155/2015/297425

Kowalski S, Makarewicz M (2017) Functional properties of honey supplemented with bee bread and propolis. Nat Prod Res 31(22):2680–2683. https://doi.org/10.1080/14786419.2017.1286481

Mărgăoan R, Zăhan M, Mărghitaş LA, Dezmirean DS, Erler S, Bobiş O (2016) Antiproliferative activity and apoptotic effects of Filipendula ulmaria pollen against C26 mice colon tumour cells. J Apic Sci 60(1):135–144. https://doi.org/10.1515/JAS-2016-0014

Mărgăoan R, Stranţ M, Varadi A, Topal E, Yücel B, Cornea-Cipcigan M, Campos MG, Vodnar DC (2019) Bee collected pollen and bee bread: bioactive constituents and health benefits. Antioxidants 8(12):568. https://doi.org/10.3390/antiox8120568

Mărgăoan R, Cornea-Cipcigan M, Topal E, Kösoğlu M (2020) Impact of fermentation processes on the bioactive profile and health-promoting properties of bee bread, mead and honey vinegar. Processes 8(9):1081. https://doi.org/10.3390/pr8091081

Mărgăoan R, Özkök A, Keskin Ş, Mayda N, Urcan AC, Cornea-Cipcigan M (2021) Bee collected pollen as a value-added product rich in bioactive compounds and unsaturated fatty acids: a comparative study from Turkey and Romania. LWT 149(June):111925. https://doi.org/10.1016/j.lwt.2021.111925

Markiewicz-Zukowska R, Naliwajko SK, Bartosiuk E, Moskwa J, Isidorov V, Soroczyńska J, Borawska MH (2013) Chemical composition and antioxidant activity of beebread, and its influence on the glioblastoma cell line (U87MG). J Apic Sci 57(2):147–157. https://doi.org/10.2478/jas-2013-0025

Martinello M, Mutinelli F (2021) Antioxidant activity in bee products: a review. Antioxidants 10. https://doi.org/10.3390/antiox10010071

Martiniaková M, Blahová J, Kováčová V, Babikova M, Mondockova V, Kalafová A, Capcarová M, Omelka R (2021) Bee bread can alleviate lipid abnormalities and impaired bone morphology in obese Zucker diabetic rats. Molecules 26:2616. https://doi.org/10.3390/molecules26092616

Maželienė Ž, Aleksandravičienė A, Pašvenskaitė M, Viliušienė I, Šakienė D, Dailidaitė E (2022) Antimicrobial activity of royal jelly, honey, and their mixture. Biologija 68(3):159–164. https://doi.org/10.6001/biologija.v68i3.4786

Miłek M, Mołoń M, Kielar P, Sidor E, Bocian A, Marciniak-Lukasiak K, Pasternakiewicz A, Dżugan M (2023) The comparison of honey enriched with laboratory fermented pollen vs. natural bee bread in terms of nutritional and antioxidant properties, protein in vitro bioaccessibility, and its genoprotective effect in yeast cells. Molecules 28(15):5851. https://doi.org/10.3390/molecules28155851

Mohammad SM, Mahmud-Ab-Rashid NK, Zawawi N (2021) Stingless bee-collected pollen (bee bread): chemical and microbiology properties and health benefits. Molecules 26. https://doi.org/10.3390/molecules26040957

Negri G, Ruv Carelli Barreto LM, Lugli Sper F, de Carvalho C, Campos MDGR (2018) Phytochemical analysis and botanical origin of Apis mellifera bee pollen from the municipality of Canavieiras, Bahia State, Brazil Análise fitoquímica e origem botânica de uma amostra de pólen apícola. J Food Technol 21:e2016176. https://doi.org/10.1590/1981-6723.17616

Ostlund RE, Racette SB, Stenson WF (2003) Inhibition of cholesterol absorption by phytosterol-replete wheat germ compared with phytosterol-depleted wheat germ. Am J Clin Nutr 77(6):1385–1389. https://doi.org/10.1093/ajcn/77.6.1385

Othman ZA, Wan Ghazali WS, Noordin L, Mohd Yusof NA, Mohamed M (2019) Phenolic compounds and the anti-atherogenic effect of bee bread in high-fat diet-induced obese rats. Antioxidants 9:33. https://doi.org/10.3390/antiox9010033

Othman ZA, Zakaria Z, Suleiman JB, Nna VU, Che Romli A, Wan Ghazali WS, Mohamed M, Romli AC, Wan Ghazali WS, Mohamed M (2021) Bee bread ameliorates vascular inflammation and impaired vasorelaxation in obesity-induced vascular damage rat model: the role of eNOS/NO/cGMP-signaling pathway. Int J Mol Sci 22(8):4225. https://doi.org/10.3390/ijms22084225

Parvez S, Malik KA, Ah Kang S, Kim H-Y (2006) Probiotics and their fermented food products are beneficial for health. J Appl Microbiol 100(6):1171–1185. https://doi.org/10.1111/j.1365-2672.2006.02963.x

Pełka K, Otłowska O, Worobo RW, Szweda P (2021) Bee bread exhibits higher antimicrobial potential compared to bee pollen. Antibiotics 10(2):125. https://doi.org/10.3390/antibiotics10020125

Poyraz F, Yalmanci D, İspirli H, Dertli E (2023) Characterization of bee bread produced with defined starter cultures mimicking the natural fermentation process. Fermentation 9(2). https://doi.org/10.3390/fermentation9020174

Rodríguez-Pólit C, Gonzalez-Pastor R, Heredia-Moya J, Carrera-Pacheco SE, Castillo-Solis F, Vallejo-Imbaquingo R, Barba-Ostria C, Guamán LP (2023) Chemical properties and biological activity of bee pollen. Molecules 28(23):1–27. https://doi.org/10.3390/molecules28237768

Rzepecka-Stojko A, Stojko J, Kurek-Górecka A, Górecki M, Kabała-Dzik A, Kubina R, Moździerz A, Buszman E (2015) Polyphenols from bee pollen: structure, absorption, metabolism and biological activity. Molecules 20(12):21732–21749. https://doi.org/10.3390/molecules201219800

Salazar-González C, Díaz-Moreno C (2016) The nutritional and bioactive aptitude of bee pollen for a solid-state fermentation process. J Apic Res 55(2):161–175. https://doi.org/10.1080/00218839.2016.1205824

Sawicki T, Starowicz M, Kłębukowska L, Hanus P (2022) The profile of polyphenolic compounds, contents of total phenolics and flavonoids, and antioxidant and antimicrobial properties of bee products. Molecules 27(4):1301. https://doi.org/10.3390/molecules27041301

Sawicki T, Starowicz M, Kłębukowska L, Hanus P (2022) The Profile of Polyphenolic Compounds, Contents of Total Phenolics and Flavonoids, and Antioxidant and Antimicrobial Properties of Bee Products. Molecules 27(4):1301. https://doi.org/10.3390/molecules27041301

Simone N de, Rocchetti MT, Gatta B la, Spano G, Drider D, Capozzi V, Russo P, Fiocco D (2022) Antimicrobial Properties, Functional Characterisation and Application of Fructobacillus Fructosus and Lactiplantibacillus Plantarum Isolated From Artisanal Honey. Probiotics Antimicrob Proteins. https://doi.org/10.1007/s12602-022-09988-4

Sobral F, Calhelha RC, Barros L, Dueñas M, Tomás A, Santos-Buelga C, Vilas-Boas M, Ferreira ICFRFR, Santos-Buelga C, Vilas-Boas M, Ferreira ICFRFR, Santos-Buelga C, Vilas-Boas M, Ferreira ICFRFR (2017) Flavonoid composition and antitumor activity of bee bread collected in Northeast Portugal. Molecules 22(2):1–12. https://doi.org/10.3390/molecules22020248

Sonmez E, Kekecoglu M, Sahin H, Bozdeveci A, Karaoglu SA (2023) Comparing the biological properties and chemical profiling of chestnut bee pollen and bee bread collected from Anatolia. Braz J Microbiol 54(3):2307–2317. https://doi.org/10.1007/s42770-023-00980-w

Stanciu O, Mărghitas LA, Dezmirean D (2007) Examination of antioxidant capacity of bee-bread extracts by different complementary assays. 64(1–2):63–64. https://doi.org/10.15835/buasvmcn-asb:64:1-2:2247

Stanciu OG, Dezmirean DS, Campos MG (2016) Bee pollen in Transylvania (Romania): palynological characterization and ORACFL values of lipophilic and hydrophilic extracts of monofloral pollen pellets. J Agric Sci Technol A 6(1):18–37. https://doi.org/10.17265/2161-6256/2016.01.003

Suleiman JB, Mohamed M, Abu Bakar AB, Nna VU, Zakaria Z, Othman ZA, Aroyehun AB (2021) Chemical profile, antioxidant properties and antimicrobial activities of Malaysian Heterotrigona Itama Bee Bread. Molecules 26:4943. https://doi.org/10.3390/molecules26164943

Urcan AC, Criste A, Dezmirean DS, Bobiş O, Mărghitaş LA, Mărgăoan R, Hrinca A, Bobiş O, Mărghitaş LA, Mărgăoan R, Hrinca A (2018) Antimicrobial activity of bee bread extracts against different bacterial strains. Bull Univ Agric Sci Vet Med Cluj-Napoca Anim Sci Biotechnol 75(2):85. https://doi.org/10.15835/buasvmcn-asb:2018.0004

Urcan AC, Criste AD, Dezmirean DS, Bobiş O, Bonta V, Burtescu RF, Olah N-K, Cornea-Cipcigan M, Mărgăoan R (2024) Enhancing antioxidant and antimicrobial activities in bee-collected pollen through solid-state fermentation: a comparative analysis of bioactive compounds. Antioxidants 13(3):292. https://doi.org/10.3390/antiox13030292

Uslinawaty Z (2023) Characteristics of the composition of stingless bee bread with health benefits originating from the Kulisusu Subdistrict, North Buton Regency. J Aisyah J Ilmu Kesehat. https://doi.org/10.30604/jika.v8i3.2057

Urcan A, Mărghitaş L Al, Dezmirean DS, Bobiş O, Bonta V, Mureşan CI, Mărgăoan R (2017) Chemical Composition and Biological Activities of Beebread – Review. Bull Univ Agric Sci Vet Med Cluj-Napoca Anim Sci Biotechnol 74(1):6–14. https://doi.org/10.15835/buasvmcn-asb

Uțoiu E, Matei F, Toma A, Diguță C, Ştefan L, Mănoiu S et al (2018) Bee collected pollen with enhanced health benefits, produced by fermentation with a Kombucha Consortium. Nutrients 10(10):1365. https://doi.org/10.3390/nu10101365

Valdenegro M, Bernales M, Knox M, Vinet R, Caballero E, Ayala-Raso A et al (2021) Characterization of fruit development, antioxidant capacity, and potential vasoprotective action of Peumo (Cryptocarya alba), a native fruit of Chile. Antioxidants 10(12):1997. https://doi.org/10.3390/antiox10121997

Vanderplanck M, Moerman R, Rasmont P, Lognay G, Wathelet B, Wattiez R, Michez D (2014) How does pollen chemistry impact development and feeding behaviour of polylectic bees? PLoS One 9(1):e86209. https://doi.org/10.1371/journal.pone.0086209

Vásquez A, Olofsson TC (2009) The lactic acid bacteria involved in the production of bee pollen and bee bread. J Apic Res 48(3):189–195. https://doi.org/10.3896/IBRA.1.48.3.07

Vergalito F, Testa B, Cozzolino A, Letizia F, Succi M, Lombardi SJ, Tremonte P, Pannella G, Di Marco R, Sorrentino E, Coppola R, Iorizzo M (2020) Potential application of Apilactobacillus kunkeei for human use: evaluation of probiotic and functional properties. Food Secur 9(11). https://doi.org/10.3390/foods9111535

Wu H-J, Wu E (2012) The role of gut microbiota in immune homeostasis and autoimmunity. Gut Microbes 3(1):4–14. https://doi.org/10.4161/gmic.19320

Yin S, Li QQ, Tao Y, Zhou E, Wang K, Chen W, Li X, Wu LM (2024) Allergen degradation of bee pollen by lactic acid bacteria fermentation and its alleviatory effects on allergic reactions in BALB/c mice. Food Sci Human Wellness 13(1):349–359. https://doi.org/10.26599/FSHW.2022.9250029

Zakaria Z, Othman ZA, Suleiman JB, Che Jalil NA, Ghazali WSW, Nna VU, Mohamed M, Wan Ghazali WS, Nna VU, Mohamed M (2021) Hepatoprotective effect of bee bread in metabolic dysfunction-associated fatty liver disease (MAFLD) rats: impact on oxidative stress and inflammation. Antioxidants 10(12):2031. https://doi.org/10.3390/antiox10122031

Zakaria Z, Othman ZA, Suleiman JB, Fadzli Mustaffa KM, Che Jalil NA, Wan Ghazali WS, Zulkipli NN, Mohamed M, Kamaruzaman KA (2022) Therapeutic effects of Heterotrigona Itama (Stingless bee) bee bread in improving hepatic lipid metabolism through the activation of the Keap1/Nrf2 signaling pathway in an obese rat model. Antioxidants 11:2190. https://doi.org/10.3390/antiox11112190

Zucca P, Rosa A, Tuberoso CIG, Piras A, Rinaldi AC, Sanjust E et al (2013) Evaluation of antioxidant potential of "Maltese mushroom" (*Cynomorium coccineum*) by means of multiple chemical and biological assays. Nutrients 5(1):149–161. https://doi.org/10.3390/nu5010149

Part VI
Beeswax

Chapter 15
Chemical Composition of Beeswax

Ge Chen, Jingjing Qian, Yang Bei, Zexiu Qi, Danila Cianciosi, Yasmany Armas Diaz, Manuela Cassotta, José M. Álvarez-Suárez, Maurizio Battino, and Francesca Giampieri

Contents

Abstract Beeswax is a complex natural substance consisting of more than 300 components. Currently, beeswax has many applications in cosmetics, medicine, food, and handicrafts because of its natural composition. The main chemical components of beeswax are presented in this chapter, such as esters, hydrocarbons, free acids, and free alcohols. As the largest component in beeswax, the main molecules, monoesters, diesters, and hydroxyesters, and the minor components, triesters and acidic esters, of esters compounds are discussed. In addition, other minor substances in beeswax are also described: aromatic and volatile compounds, a few minerals,v small amounts of phenolic compounds, flavonoids, and trace chemical elements such as Ca, Fe, Mn, Zn, and P.

G. Chen · Y. Bei · Z. Qi · D. Cianciosi · Y. Armas Diaz
Department of Clinical Sciences, Polytechnic University of Marche, Ancona, Italy

Joint Laboratory on Food Science, Nutrition, and Intelligent Processing of Foods, Polytechnic University of Marche, Italy, Universidad Europea del Atlántico Spain and Jiangsu University, China, Polytechnic University of Marche, Ancona, Italy

J. Qian
Department of Clinical Sciences, Polytechnic University of Marche, Ancona, Italy

J. M. Alvarez-Suarez (ed.), *Bee Products – Chemical and Biological Properties*,
https://doi.org/10.1007/978-3-031-89049-9_15

Keywords Beeswax · Chemical composition · Esters · Hydrocarbons · Bioactive compounds

15.1 Introduction

Beeswax is a substance secreted by four pairs of wax glands in the abdomen of honeybees (worker bees) and is used as the material for constructing beehives (Bogdanov 2009). In history, beeswax has played a crucial role in human civilization, serving as a key component in cosmetics, medicine, food, and handicrafts (Bernardini et al. 2012; Stacey 2011). However, the detailed chemical composition of beeswax has been difficult to determine until the advent of modern analytical

M. Cassotta
Research Group on Food, Nutritional Biochemistry and Health, Universidad Europea del Atlántico, Santander, Spain

Joint Laboratory on Food Science, Nutrition, and Intelligent Processing of Foods, Polytechnic University of Marche, Italy, Universidad Europea del Atlántico Spain and Jiangsu University, China, Universidad Europea del Atlántico, Santander, Spain

J. M. Álvarez-Suárez
Laboratorio de Investigación en Ingeniería en Alimentos (LabInAli), Departamento de Ingeniería en Alimentos, Colegio de Ciencias e Ingenierías, Universidad San Francisco de Quito (USFQ), Quito, Ecuador

Laboratorio de Bioexploración, Colegio de Ciencias Biológicas y Ambientales, Universidad San Francisco de Quito (USFQ), Quito, Ecuador

M. Battino
Department of Clinical Sciences, Polytechnic University of Marche, Ancona, Italy

Research Group on Food, Nutritional Biochemistry and Health, Universidad Europea del Atlántico, Santander, Spain

International Joint Research Laboratory of Intelligent Agriculture and Agri-Products Processing, Jiangsu University, Zhenjiang, China

Joint Laboratory on Food Science, Nutrition, and Intelligent Processing of Foods, Polytechnic University of Marche, Italy, Universidad Europea del Atlántico Spain and Jiangsu University, China, Polytechnic University of Marche, Ancona, Italy

F. Giampieri (✉)
Department of Clinical Sciences, Polytechnic University of Marche, Ancona, Italy

Research Group on Food, Nutritional Biochemistry and Health, Universidad Europea del Atlántico, Santander, Spain

Joint Laboratory on Food Science, Nutrition, and Intelligent Processing of Foods, Polytechnic University of Marche, Italy, Universidad Europea del Atlántico Spain and Jiangsu University, China, Polytechnic University of Marche, Ancona, Italy

International Research Center for Food Nutrition and Safety, Jiangsu University, Zhenjiang, China
e-mail: f.giampieri@univpm.it

techniques, such as chromatography, spectroscopy, and mass spectrometry, which have revealed the complex content of bioactive compounds, more than 350, that make up beeswax (Garnier et al. 2002).

Beeswax is a complex natural substance renowned for its intricate chemical composition, which underpins its various properties and applications. The major components of beeswax can be divided into four categories: esters, hydrocarbons, free acids, and free alcohols, as shown in Table 15.1 (Fratini et al. 2016).

However, due to differences in bee species, pollen sources, plant origins, extraction methods, and other variables, there are certain variations in its composition (Giampieri et al. 2022; Hepburn et al. 2014a).

15.2 Ester Compounds

Esters are the most important class of compounds found in beeswax, constituting a large part of its chemical composition, as presented in Table 15.1. These compounds play a crucial role in determining many characteristics and properties of beeswax, thus giving it value and utility (Münstedt and Bogdanov 2009).

Esters account for about 67% of the total weight of beeswax (Tulloch 1980). Beeswax contains a variety of esters formed through esterification of fatty acids with alcohols (Bogdanov 2009). These esters can be classified into several categories based on their chemical structure and properties. Among them, monoesters are the simplest form of ester compounds in beeswax, consisting of a single fatty acid molecule esterified with an alcohol molecule (Stránský and Streibl 1971). On the other hand, diesters involve the esterification of hydroxy acids with diol molecules. Additionally, beeswax may include hydroxyesters that, aside from the ester portion, also have a hydroxyl (-OH) functional group (Tulloch 1971).

15.2.1 Monoesters

Monoesters are an important class of ester compounds found in beeswax, accounting for approximately 35% of its composition (Hepburn et al. 2014a). These compounds are formed through the esterification of fatty acids with various alcohols, contributing to their chemical complexity and functional characteristics such as melting point, hardness, and compatibility with other substances (Hepburn et al. 2014b; Tewo et al. 2019).

In beeswax, monoesters predominantly consist of straight-chain fatty acids, such as palmitic acid and oleic acid, esterified with aliphatic alcohols to form the C38 monoester docosanyl palmitate and the C46 monoester triacontyl palmitate (the chemical structures are shown in Fig. 15.1a) (Hepburn et al. 2014a). Monoesters are usually separated by thin-layer chromatography (TLC) and appear as C40–C48 esters, which are palmitic acid esters of C24–C32 alcohols (Tulloch 1980).

Table 15.1 Composition of beeswax

Main component	Content (min–max %)		
Monoesters	27.5–41.1		
Diesters	7.4–15.7		
Hydroxymonoesters	9.1–23.6		
Hydroxydiesters	1.1–3.9		
Acidic esters	1.0–3.0		
Hydrocarbons	11.4–27.8		
Fatty acids	0.8–4.9		
Fatty alcohols	0.4–1.8		
Others	4.0–8.0		
Monoesters (%)			
C40	10.1–10.3	C46:1	0.05–0.17
C42	5.8–6.0	C48	8.1–8.3
C42:1	0.22–0.24	C48:1	0.31–0.33
C44	6.4–6.6	C50	0.61–0.77
C44:1	0.1–0.3	C50:1	0.59–0.81
C46	14.4–16.3		
Hydrocarbons (%)			
H17	0.134–0.138	H31:1	0.81–2.777
H19	0.129–0.225	H31:1b	1.12–1.14
H20	0.045–0.356	H31	2.262–7.327
H21	0.27–0.424	H32:1	0.076–0.08
H22	0.03–0.147	H32	0.05–0.369
H23:1	0.037–0.039	H33:2	0.051–0.053
H23	0.704–0.716	H33:1	1.756–4.206
H24	0.0632–0.188	H33	0.397–0.743
H25:1	0.066–0.068	H34	0.111–0.119
H25	1.725–5.409	H35:1	0.104–0.226
H26	0.191–0.386	H35:1b	0.06–0.064
H27:1	0.062–0.066	H35	0.0149–0.441
H27:1b	0.17–0.178	H36	0.053–0.159
H27	3.164–15.718	H37	0.06–0.094
H28	0.173–0.392	H38	0–0.018
H29:1	0.123–0.131	H39	0.027–0.071
H29	1.982–9.416	H40	0.092–0.102
H30:1	0.055–0.057	H41	0.016–0.066
H30:1b	0.0268–0.0276	H42	0.021–0.037
H30	0.13–0.315	H43	0.028–0.058
H31:2	0.021–0.023	H44	0.004–0.012
Total acids (%)			
C14	0.09–0.11	C24	4.43–4.51
C16	17.15–17.41	C26	1.8–1.92

(continued)

Table 15.1 (continued)

Main component	Content (min–max %)		
C18:2	0.08–0.1	C28	1.84–1.94
C18:1	2.54–2.6	C30	1.77–1.91
C18	0.62–0.72	C32	1.7–1.8
C20	0.09–0.11	C34	1.79–1.87
C22	0.49–0.51	C36	0.27–0.29
Total alcohols (%)			
C22-OH	0.073–0.081	C32:1-OH	0.61–0.67
C24-OH	6.39–6.65	C32-OH	9.62–9.66
C26-OH	5.26–5.4	C34:1-OH	0.2–0.24
C28-OH	5.9–6.02	C34-OH	1.49–1.55
C30-OH	10.45–11.37		
Chemical elements (mg/kg)			
Al	7.7–12.1	Mg	18–24.4
As	0.013–0.017	Mn	0.34–0.68
Au	0.6–2.6	P	45.6–49
Ca	86–230	Pb	1.4–5.6
Cd	0.009–0.031	Se	0.003–0.027
Co	0–0.008	Si	0–2.2
Fe	18.5–22.9	V	0.017–0.025
Hg	0.4–3	Y	0–0.007
K	82–94	Zn	9.8–10.8

Modified from Giampieri et al. (2022), Jiménez et al. (2006), Navarro-Hortal et al. (2019)

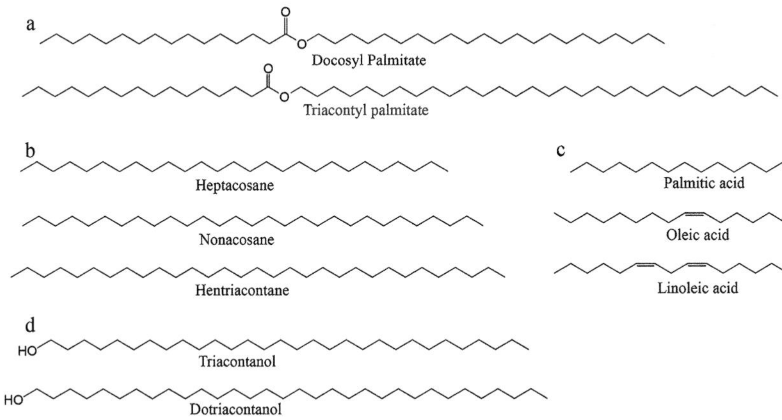

Fig. 15.1 Chemical structures of the main components of beeswax

15.2.2 Diesters

The diesters account for approximately 14% of beeswax composition, producing hydroxy acids and diols upon hydrolysis (Hepburn et al. 2014a). In beeswax, the diesters are identified as C56–C64 compounds through gas–liquid chromatography (GLC) analysis, typically involving hydroxyhexadecanoic acid, hydroxyoctadecanoic acid, and hydroxyeicosanoic acid, as well as C24–C32 diols (Tulloch 1971). Tulloch proposed the structure of diesters in beeswax as acylated hydroxy acid esters of 80% long-chain alcohols and 20% diol diesters (Tulloch 1980). These diesters generally have high molecular weights, contributing to the structural integrity and stability of beeswax. The presence of diesters affects key properties of beeswax, including its viscosity, plasticity, and resistance to environmental factors (Han et al. 2022).

15.2.3 Hydroxyesters

Hydroxyesters are of two types: hydroxy monoesters and hydroxy polyesters, accounting for approximately 12% of beeswax components (Hepburn et al. 2014a). These compounds are formed through the esterification of fatty acids with hydroxyl-containing alcohols, such as diols or polyols (Puleo 1991). Hydroxy monoesters consist of a structure where long-chain alcohols are esterified with hydroxy acids and the primary hydroxyl of a diol; they are identified as C40–C50 components. On the other hand, the structure of hydroxy polyesters is more complex, speculated to consist of hydroxy acid chains and diol structures, possibly including polyesters, such as pentaesters and hexaesters (Tulloch 1980). In beeswax, hydroxyesters play a crucial role in enhancing its moisturizing properties and adhesive characteristics, making beeswax an effective emollient and moisturizer in cosmetic formulations (Nong et al. 2023).

15.2.4 Minor Components of Ester Compounds

In addition, ester compounds also include minor components, such as triesters (3%), acidic monoesters (1%), and acidic polyesters (2%). Among them, the acidic monoglycerides contain C32–C44 components and are mainly composed of hydroxy acids derived from palmitic acid esterification.

15.3 Hydrocarbons

Hydrocarbons are organic compounds composed exclusively of carbon and hydrogen atoms. Hydrocarbons are an important component of beeswax, typically accounting for around 14% of its total composition (Table 15.1) (Tulloch 1980). In beeswax, hydrocarbons are primarily made up of saturated hydrocarbons (alkanes) and unsaturated hydrocarbons with one or two double bonds (alkenes and dienes) (Hepburn et al. 2014a). Moreover, hydrocarbons in beeswax mainly exist in the form of straight-chain alkanes, with carbon chain lengths usually ranging from C20 to C35 (Waś et al. 2014a). Among them, the components with a relatively high proportion of saturated hydrocarbons are C27, C29, and C31 (the chemical structures are shown in Fig. 15.1b) (Hepburn et al. 2014a). The content of the various components of hydrocarbons present in beeswax may vary depending on factors such as the botanical source of the nectar, environmental climate conditions, and the species of bees (European, Asian, or African), among others (Waś et al. 2014b; Tanleque-Alberto et al. 2023).

With advancements in technology, precise qualitative and quantitative analysis of hydrocarbons in beeswax have been conducted through gas chromatography-mass spectrometry and thin-layer chromatography techniques (Garnier et al. 2002; Jiménez et al. 2004; Jiménez et al. 2006). It has been discovered that in pure beeswax, the content of odd-numbered carbon chain n-alkanes is significantly higher than that of even-numbered carbon chain n-alkanes, which can be utilized to detect adulteration in beeswax (Bonvehi and Bermejo 2012; Jiménez et al. 2007).

15.4 Total Fatty Acids

Free fatty acids are organic compounds composed of a carboxylic acid group (COOH) connected to a long alkyl chain. Free fatty acids account for a relatively small proportion in beeswax, approximately 12–15% of its total composition, among which stearic acid is the most abundant component in the free fatty acid fraction of beeswax (Table 15.1) (Hepburn et al. 2014a; Bogdanov 2004).

In beeswax, the carbon chain length of fatty acids typically ranges from C14 to C36 (Giampieri et al. 2022). The structure of free fatty acids in beeswax varies depending on the specific fatty acids present, such as palmitic acid (saturated fatty acids), oleic acid, and linoleic acid (unsaturated fatty acids) (the chemical structures are shown in Fig. 15.1c), which can also serve as clues for honeybee nest recognition (Buchwald et al. 2009; Brockmann et al. 2003).

15.5 Total Alcohols

As shown in Table 15.1, beeswax contains a variety of alcohols, with the total monoalcohols being even-chain primary alcohols of chain length C22-C34. Most of the fatty monoalcohols are combined with acidic compounds, while free fatty alcohols (constituting approximately 1% or less of the total beeswax composition) are minor compounds with chain lengths of C24-C34. The main components are C30-OH and C32-OH (the chemical structures are shown in Fig. 15.1d), followed by C28-OH, C26-OH, and C24-OH (Svečnjak et al. 2019). J.J. Jiménez and others have also found four diols in hydrolyzed and acetylated beeswax, which form hydroxy monoesters and monoesters in beeswax. Furthermore, compounds related to 1,2,3-propanetriol have been discovered in beeswax (Jiménez et al. 2003).

15.6 Other Minor Substances

It has been reported that more than 50 aromatic and volatile compounds have been identified in beeswax, which play a role in the fragrance and potential biological activity of beeswax (Ferber and Nursten 1977). Furthermore, the analysis of ash content and mineral composition in beeswax revealed that yellow beeswax contains very few minerals, while white beeswax does not contain any minerals (Bernal et al. 2005). Navarro-Hortal's report mentioned that beeswax also contains small amounts of phenolic compounds, flavonoids, and trace chemical elements such as Ca, Fe, Mn, Zn, and P (Table 15.1) (Navarro-Hortal et al. 2019).

Acknowledgments Danila Cianciosi was supported by a Fondazione Umberto Veronesi fellowship.

References

Bernal JL, Jiménez JJ, del Nozal MJ et al (2005) Physico-chemical parameters for the characterization of pure beeswax and detection of adulterations. Eur J Lipid Sci Technol 107:158–166. https://doi.org/10.1002/ejlt.200401105

Bernardini F, Tuniz C, Coppa A et al (2012) Beeswax as dental filling on a Neolithic human tooth. PLoS One 7:e44904. https://doi.org/10.1371/journal.pone.0044904

Bogdanov S (2004) Quality and standards of pollen and beeswax. Apiacta 38:334–341

Bogdanov S (2009) Beeswax: production, properties composition and control. Bee Product Sci

Bonvehi JS, Bermejo FO (2012) Detection of adulterated commercial Spanish beeswax. Food Chem 132:642–648. https://doi.org/10.1016/j.foodchem.2011.10.104

Brockmann A, Groh C, Fröhlich B (2003) Wax perception in honeybees: contact is not necessary. Sci Nat 90:424–427. https://doi.org/10.1007/s00114-003-0442-3

Buchwald R, Breed MD, Bjostad L et al (2009) The role of fatty acids in the mechanical properties of beeswax. Apidologie 40:585–594. https://doi.org/10.1051/apido/2009035

Ferber CE, Nursten HE (1977) The aroma of beeswax. J Sci Food Agric 28:511–518. https://doi.org/10.1002/jsfa.2740280608
Fratini F, Cilia G, Turchi B et al (2016) Beeswax: a minireview of its antimicrobial activity and its application in medicine. Asian Pac J Trop Med 9:839–843. https://doi.org/10.1016/j.apjtm.2016.07.003
Garnier N, Cren-Olivé C, Rolando C et al (2002) Characterization of archaeological beeswax by electron ionization and electrospray ionization mass spectrometry. Anal Chem 74:4868–4877. https://doi.org/10.1021/ac025637a
Giampieri F, Quiles JL, Cianciosi D et al (2022) Bee products: an emblematic example of underutilized sources of bioactive compounds. J Agric Food Chem 70:6833–6848. https://doi.org/10.1021/acs.jafc.1c05822
Han W, Chai X, Liu Y et al (2022) Crystal network structure and stability of beeswax-based oleogels with different polyunsaturated fatty acid oils. Food Chem 381:131745. https://doi.org/10.1016/j.foodchem.2021.131745
Hepburn H, Pirk C, Duangphakdee O et al (2014a) The chemistry of beeswax. Honeybee Nests: Composition, Structure, Function, pp 319–339. https://doi.org/10.1007/978-3-642-54328-9_16
Hepburn H, Pirk C, Duangphakdee O et al (2014b) Synthesis of beeswax. Honeybee Nests: Composition, Structure, Function, pp 341–365. https://doi.org/10.1007/978-3-642-54328-9_17
Jiménez JJ, Bernal JL, Aumente S et al (2003) Quality assurance of commercial beeswax: II. Gas chromatography–electron impact ionization mass spectrometry of alcohols and acids. J Chromatogr A 1007:101–116. https://doi.org/10.1016/S0021-9673(03)00962-2
Jiménez J, Bernal J, Aumente S et al (2004) Quality assurance of commercial beeswax: part I. Gas chromatography–electron impact ionization mass spectrometry of hydrocarbons and monoesters. J Chromatogr A 1024:147–154. https://doi.org/10.1016/j.chroma.2003.10.063
Jiménez J, Bernal J, del Nozal MJ et al (2006) Sample preparation methods for beeswax characterization by gas chromatography with flame ionization detection. J Chromatogr A 1129:262–272. https://doi.org/10.1016/j.chroma.2006.06.098
Jiménez JJ, Bernal JL, del Nozal MJ et al (2007) Detection of beeswax adulterations using concentration guide-values. Eur J Lipid Sci Technol 109:682–690. https://doi.org/10.1002/ejlt.200600308
Münstedt K, Bogdanov S (2009) Bee products and their potential use in modern medicine. J ApiProd ApiMed Sci 1:57–63. https://doi.org/10.3896/IBRA.4.01.3.01
Navarro-Hortal MD, Orantes-Bermejo FJ, Sánchez-González C et al (2019) Industrial-scale decontamination procedure effects on the content of acaricides, heavy metals and antioxidant capacity of beeswax. Molecules 24:1518. https://doi.org/10.3390/molecules24081518
Nong Y, Maloh J, Natarelli N et al (2023) A review of the use of beeswax in skincare. J Cosmet Dermatol 22:2166–2173. https://doi.org/10.1111/jocd.15718
Puleo S (1991) Beeswax minor components: a new approach. Cosmetics & Toiletries 106:83–89
Stacey R (2011) The composition of some Roman medicines: evidence for Pliny's Punic wax? Anal Bioanal Chem 401:1749–1759. https://doi.org/10.1007/s00216-011-5160-7
Stránský K, Streibl M (1971) On natural waxes. XVIII. Simple alkyl esters of the wax of the honeybee (Apis mellifera L). Collect Czechoslov Chem Commun 36:2267–2280. https://doi.org/10.1135/cccc19712267
Svečnjak L, Chesson LA, Gallina A et al (2019) Standard methods for Apis mellifera beeswax research. J Apic Res 58:1–108. https://doi.org/10.1080/00218839.2019.1571556
Tanleque-Alberto F, Vargas M, Juan-Borrás M et al (2023) Thermal properties and hydrocarbon composition of beeswax from Mozambique and other geographical origins. J Apic Res 62:883–892. https://doi.org/10.1080/00218839.2021.1907086
Tewo RK, Rutto HL, Focke W et al (2019) Formulations, development and characterization techniques of investment casting patterns. Rev Chem Eng 35:335–349. https://doi.org/10.1515/revce-2017-0068
Tulloch A (1971) Beeswax: structure of the esters and their component hydroxy acids and diols. Chem Phys Lipids 6:235–265. https://doi.org/10.1016/0009-3084(71)90063-6

Tulloch AP (1980) Beeswax—composition and analysis. Bee World 61:47–62. https://doi.org/10.1080/0005772X.1980.11097776

Waś E, Szczęsna T, Rybak-Chmielewska H (2014a) Determination of beeswax hydrocarbons by gas chromatography with a mass detector (GC-MS) technique. J Apic Sci 58:145–157. https://doi.org/10.2478/jas-2014-0015

Waś E, Szczęsna T, Rybak-Chmielewska H (2014b) Hydrocarbon composition of beeswax (Apis mellifera) collected from light and dark coloured combs. J Apic Sci 58:99–106. https://doi.org/10.2478/jas-2014-0026

Chapter 16
Health Benefits and Uses of Beeswax in Medicine

Jingjing Qian, Ge Chen, Yang Bei, Zexiu Qi, Danila Cianciosi, Yasmany Armas Diaz, Manuela Cassotta, José M. Álvarez-Suárez, Maurizio Battino, and Francesca Giampieri

Contents

Abstract Beeswax (BW) is the substance that forms the structure of a honeycomb, made by bees to store honey. Compared with honey and propolis, beeswax as a by-product possesses several favorable therapeutic properties and is of great interest to the scientific community. In this chapter, the various health benefits of beeswax and its applications are systematically introduced, including mainly the protection of skin disease, wound healing, oxidative damage, antimicrobial activities, and other

J. Qian · G. Chen · Y. Bei · Z. Qi · D. Cianciosi
Department of Clinical Sciences, Polytechnic University of Marche, Ancona, Italy

Y. Armas Diaz
Department of Clinical Sciences, Polytechnic University of Marche, Ancona, Italy

Joint Laboratory on Food Science, Nutrition, and Intelligent Processing of Foods, Polytechnic University of Marche, Italy, Universidad Europea del Atlántico Spain and Jiangsu University, China, Polytechnic University of Marche, Ancona, Italy

M. Cassotta
Research Group on Food, Nutritional Biochemistry and Health, Universidad Europea del Atlántico, Santander, Spain

Joint Laboratory on Food Science, Nutrition, and Intelligent Processing of Foods, Polytechnic University of Marche, Italy, Universidad Europea del Atlántico Spain and Jiangsu University, China, Universidad Europea del Atlántico, Santander, Spain

J. M. Alvarez-Suarez (ed.), *Bee Products – Chemical and Biological Properties*,
https://doi.org/10.1007/978-3-031-89049-9_16

health applications, with the aim to make more common the use of this by-product for human health. Beeswax, due to its beneficial components and special structural properties, could even assist in preventing and treating some types of cancers or some metabolic diseases, such as nonalcoholic fatty liver disease. Therefore, this chapter aims to update the major scientific works that have highlighted the various health benefits exerted by beeswax in different types of pathological conditions.

Keywords Beeswax · Health Benefits · Wound healing · Antimicrobial activity · Oxidative protection

16.1 Introduction

The use of beeswax (BW) for medical purposes dates back to ancient times. In ancient Egypt, about 1550 BC, it was found in 32 prescriptions in papyrus and was traditionally used for embalming in mummification, retaining the permanency of wig curls, or preserving the papyrus scrolls (Bogdanov 2002). In ancient China, beeswax was recorded as a top medicinal ingredient for antiaging and treating wounds in skin, in the most famous medicine book *The Shen Nong Book of Herbs*, from the first to the second century BC (Jing 2022). Apart from the original use of beeswax, it was also commonly used for the preparation of cosmetics (Nong et al.

J. M. Álvarez-Suárez
Laboratorio de Investigación en Ingeniería en Alimentos (LabInAli), Departamento de Ingeniería en Alimentos, Colegio de Ciencias e Ingenierías, Universidad San Francisco de Quito (USFQ), Quito, Ecuador

Laboratorio de Bioexploración, Colegio de Ciencias Biológicas y Ambientales. Universidad San Francisco de Quito (USFQ), Quito, Ecuador

M. Battino
Department of Clinical Sciences, Polytechnic University of Marche, Ancona, Italy

Research Group on Food, Nutritional Biochemistry and Health, Universidad Europea del Atlántico, Santander, Spain

International Joint Research Laboratory of Intelligent Agriculture and Agri-Products Processing, Jiangsu University, Zhenjiang, China

Joint Laboratory on Food Science, Nutrition, and Intelligent Processing of Foods, Polytechnic University of Marche, Ancona, Italy

F. Giampieri (✉)
Department of Clinical Sciences, Polytechnic University of Marche, Ancona, Italy

Research Group on Food, Nutritional Biochemistry and Health, Universidad Europea del Atlántico, Santander, Spain

Joint Laboratory on Food Science, Nutrition, and Intelligent Processing of Foods, Polytechnic University of Marche, Ancona, Italy

International Research Center for Food Nutrition and Safety, Jiangsu University, Zhenjiang, China
e-mail: f.giampieri@univpm.it

2023), pharmaceutical products (Molina et al. 2015), candles (Bogdanov 2004), superhydrophobic films/coatings (Hosseini et al. 2023), and other applications in recent years.

To date, beeswax is primarily known for its versatility that has not only demonstrated plenty of applications in craftsmanship but has also exhibited promising medicinal potentials. It is worthy to mention that the US Food and Drug Administration has designated the yellow and white beeswax as "Generally Recognized as Safe" as a food additive or when used in food packaging. For example, regarding skin health, beeswax performs a natural emollient, locking in moisture and creating a protective barrier against environmental pollutants. The anti-inflammatory and antioxidant properties of beeswax make it an excellent ingredient for soothing and healing skin ailments (Bayir et al. 2019). In the pharmaceutical products of beeswax, which even have pharmacopeia standards, the representative beeswax alcohols (D-002) were used to treat osteoarthritis (OA) (Puente et al. 2014), gastroprotection/colitis (Molina et al. 2015), and nonalcoholic fatty liver disease (Illnait et al. 2013) because of their effective antioxidant, anti-inflammatory, and anti-ulcer effects. Moreover, it was used in the treatment of Alzheimer's disease, burns, wounds, abscesses, and dental problems (Bernardin et al. 2012; Gümüş and Özlü 2017; Romero et al. 2020; Zaheri-Abdevand and Badr 2023). Furthermore, beeswax was also used to mix with some drugs as coat pills or edible coating, facilitating ingestion but retarding dissolution as a time-release controller (Kaur et al. 2023). It could also be used in the conservation and preservation of food as edible coatings due to its antimicrobial activity (Farag et al. 1993; Formiga et al. 2019). Meanwhile, beeswax is widely utilized in crafting candles, as it burns cleanly and emits a subtle honey scent. Its adhesive qualities make it an essential component in wood/leather work, providing strength and durability to finished products. To summarize, beeswax stands as a gift from nature, offering a multitude of health benefits across different domains (Table 16.1).

16.2 Protection of Skin Disease

Beeswax has a rich history in skincare dating back to ancient civilizations. Ancient Egyptian, Greek, and Roman cultures utilized beeswax as a key ingredient in skincare preparations (Hepburn 1986). In ancient Egypt, beeswax was combined with other natural ingredients like plant oils to create ointments and balms for protecting and moisturizing the skin. In ancient Greek and Roman cultures, beeswax was also extensively used in creams and salves to shield the skin from environmental factors. During the medieval period in Europe, beeswax found its place in treating wounds and various skin ailments.

Atopic dermatitis (AD) and psoriasis are two common chronic inflammatory skin diseases that are usually accompanied by severe itching and trigger a rapid turnover of skin cells, leading to thickening of the outermost layer of the skin. The skin barrier dysfunction and immune dysregulation are at the basis of the pathogenesis of AD. However, the underlying complex mechanism of AD is still not clearly

Table 16.1 Application and health benefits of beeswax: evidence in vitro and in vivo

Samples	Activity	Experimental models	Dose and duration	Main outcome	Reference
Carnauba wax, beeswax, pumpkin seed oil (PSO-NLC)	Skin health activity	In vitro activity	–	Synergistic effect among the PSO-NLC with Uvinul® A Plus B	Chu et al. (2020)
Cera Flava (CF)	Skin health activity	Mouse model of atopic dermatitis (AD)	The topical application of CF for 6 weeks	CF decreased the mRNA levels of inflammatory factors, including interleukin (IL)-1β, IL-4, IL-13, IL-8, TARC, MDC, and RANTES	Park et al. (2021)
A mixture containing honey, olive oil, and beeswax	Skin health activity	Thirty-seven patients with pityriasis versicolor (PV)	Apply on the lesions three times daily for a maximum of 4 weeks	Mycological cure was obtained in 75%, 71%, and 62% of patients with PV, tinea cruris, and tinea corporis, respectively	Al-waili (2004)
A mixture containing honey, olive oil, and beeswax	Skin health activity	Twelve infants suffering from diaper dermatitis	Four times daily with gentle rubbing for a maximum of 7 days	The severity of erythema was significantly decreased	Al-waili (2005b)
Sunscreen formula: 75% almond oil (64.0 g), 9% beeswax (7.6 g), and 16% zinc oxide (13.4 g)	Skin health activity	Five volunteers using standard Solar protection factor (SPF) testing	At a dose of 2 mg cm^{-2} for 15 min	An SPF of approximately 15	Miyar et al. (2014)
Beeswax (BW) and cinnamon oil (CO) into the hydrogel matrix	Wound-healing activity	Antimicrobial assay in vitro	–	Strong resistance and inhibition effects against both bacteria (*Staphylococcus aureus and Klebsiella pneumonia*) and fungi (*Candida albicans and Aspergillus niger*) strains	Zandraa et al. (2021)
Beeswax, olive oil, and butter (BOB)	Wound-healing activity	Second-degree burn model in rats	Bandages applied on the third, seventh, and fourteenth days	Encouraged skin renewal via modulating tissue tumor growth factor (TGF)–β1 and vascular endothelial growth factor (VEGF)-α	Bayir et al. (2019)

$Ca(OH)_2$ 33%, sesame oil 50%, and beeswax 17%	Wound-healing activity	A deep second-degree burn in rats	2 g (8 g/kg) of materials twice daily for 24 days	VEGF, TGF-β1 genes, and matrix metalloproteinase enzyme (MMP)-9 and 2 genes had a significant peak of expression on day 17 and a rapid reduction of expression on day 24	Ebrahimpour et al. (2020)
A mixture of honey, beeswax, and olive oil (MHBO)	Wound-healing activity	Dogs	1% MHBO for 27 consecutive days	Inflammatory reaction and exudation were less	Moustafa and Atiba (2015)
Beeswax	Antibacterial and antioxidant activities	Microorganism and in vitro activity	–	The antioxidant activity of beeswax extracts is higher than that of honey PCL assay	Sawicki et al. (2022)
Two by-products from beeswax recycling processes (MUD1 and 2)	Antioxidant activity	Human dermal fibroblast cells	24 h at 250 and 750 μg/mL	Significantly recovering the apoptosis level, reactive oxygen species, and nitrite production	Giampieri et al. (2018a)
Two by-products from beeswax recycling processes (MUD1 and 2)	Antioxidant activity	In vitro and HepG2 cells	Cells were treated with 0.25, 0.50, and 1 mg/ml of MUD1 for 24 h	Reducing cellular viability, increasing intracellular reactive oxygen species (ROS) levels, and affecting mitochondrial functionality	Giampieri et al. (2018b)
A mud by-product obtained from the recycling of beeswax	Antioxidant and neuroprotective activities	In vitro and the *Caenorhabditis elegans* model	All extracts (0, 0.1, 1, 10, 50, and 100 μg/mL)	A delay in paralysis caused by the accumulation of beta amyloid (Aß 1–42)	Romero et al. (2020)
Beeswax alcohol (BWA) and fermented black rice bran (BRB-F)	Antioxidant activity	In vitro and zebrafish	10% mixtures of BWA + BRB-F (MIX)	MIX supplement decreased the hepatic ROS production and interleukin-6 expression	Han et al. (2023)
A natural mixture of high-molecular-weight alcohols (D-002)	Anti-inflammatory activity	Rats	Orally administered with 25, 50, 100, and 200 mg/kg in a single dose or over 6 days	Significantly decreased the leukotriene B (LTB) levels in the exudate and significantly diminished the granuloma weight	Carbajal et al. (1998)
D-002	Antioxidant activity	Human plasma	D-002 (50 mg/day) for 12 weeks	Significantly reduced plasma total hydroxyperoxides (TOHP) and malondialdehyde (MDA) and increased total antioxidant status (TAS)	López et al. (2008)

(continued)

Table 16.1 (continued)

Samples	Activity	Experimental models	Dose and duration	Main outcome	Reference
Beeswax alcohol	Anti-oxidative activity	In vitro and zebrafish and clinical study	12 weeks of BWA (100 mg/day) supplementation	Efficiently diminished serum MDA and total hydroperoxides and enhanced total antioxidant status by 25%, 27%, and 22%, respectively	Cho et al. (2024)
D-002	Antioxidant activity	Clinical study	D-002 (50 mg/day) for 12 weeks	The lag-phase of oxidation was prolonged by 1.5-fold. D-002 oral treatment decreased Thiobarbituric acid reactive substances (TBARS) and increased plasma TAS	Menéndez et al. (2001)
Honey and beeswax, ethanolic extract of propolis	Antimicrobial activity	Microorganism	Then 100 μL of honey products and/or herbs extract were added to each well	Ethanolic extracts of propolis and beeswax were more effective than a single extract showing a 1.5-fold increase of inhibition against *S. aureus ATCC255923* and *C. albicans NCTC2708*	Ghanem (2011)
Apis mellifera beeswax	Antibacterial activity	Microorganism	A final volume of 200 μL was incubated for 24 h	*Pseudomonas aeruginosa* and *Listeria monocytogenes* were inhibited by all beeswax samples	Felicioli et al. (2019)
Propolis, bee pollen loads, and beeswax samples	Antimicrobial activity	Microorganism	A final volume of 200 μL was incubated for 48 h	Most sensitive to beeswax extracts were Escherichia coli (*E. coli*), Aspergillus niger (*A. niger*),and Candida Tropicalis(*C. tropicalis*)	Kacániová et al. (2012)
A mixture containing honey, olive oil, and beeswax	Antimicrobial activity	Microorganism	12.5%, 25%, 33%, 50%, and 66% (v/v) honey, olive oil, or beeswax	Honey and honey mixture apparently could inhibit the growth of *S. aureus* or *C. albicans*	Al-waili (2005a)
A semisolid beeswax by-product (BBR) and beeswax	Antimicrobial activity	Microorganism	0.1–0.01–0.001 mg/mL for 16 h	*Klebsiella pneumoniae* and *Salmonella enterica* were inhibited at concentrations 2 0.001 mg/mL, while *Enterococcus faecalis* and methicillin-resistant *Staphylococcus aureus* from 0.01 mg/mL	Peron et al. (2023)
Beeswax for a composite fibrous material	Antibacterial, antifungal, and antitumor activities	Microorganism and in cell lines	PEO/BW = 70/30, [PEO/BW = 80/20]/NQ, [PEO/BW = 70/30]/NQ and [PEO/BW =60/40] /NQ for 24 h	Antibacterial effects against *S. aureus* and *E. coli*, antifungal effect against *C. albicans*, and selective anticancer activity against HeLa and SH-4 cell lines	Kyuchyuk et al. (2022)

D-002	Anti-lipid peroxidative activity in gastric mucosa	Rats	Single doses of D-002 (5, 25, or 100 mg/kg) or 10 days	Significantly reduced the gastric erosion index induced by ischemia-reperfusion (I-R)	Molina et al. (2001)
D-002 (beeswax alcohols)	Anti-inflammatory activity in gastrointestinal symptoms	Rats	1, 5, and 25 mg/kg at a single administration	The gastric damage at the same doses showed a significant increase in neutral glycoproteins and sulfated macromolecules	Carbajal et al. (2000)
D-002 (beeswax alcohols)	Anti-inflammatory activity in gastrointestinal symptoms	Rats	D-002 (5, 25, and 100 mg/kg) in a single oral administration	D-002 (5–100 mg/kg) significantly and dose-dependently reduced indomethacin-induced ulceration to 75%	Perez et al. (2013)
D-002 (beeswax alcohols)	Anti-inflammatory and antinociceptive activities	Mice	D-002 (25, 50, and 200 mg/kg) as single or repeat (15 days) once-daily doses	D-002 (25, 50, and 200 mg/kg) significantly decreased xylene-induced ear oedema (44.7%, 60.8%, and 76.4%, respectively) and the increase of myeloperoxidase (MPO) activity induced by xylene (38.0%, 47.0%, and 57.0%, respectively)	Ravelo et al. (2011)
D-002 (beeswax alcohols) combined with green-lipped mussel (*Perna canaliculus*) (GME) extracts	Anti-inflammatory activity in osteoarthritis (OA) symptoms	Clinical study	GME capsules (50 mg of the extract/capsule) + placebo or D-002 (50 mg) tablets for 6 weeks	GME + D-002 decreased significantly the total Western Ontario and McMaster Individual Osteoarthritis Index (WOMAC) score, pain and physical function WOMAC sub-scores, and visual analogy scale (VAS) as compared to GME + placebo	Puente et al. (2014)
D-002 (beeswax alcohols)	Anti-inflammatory activity in osteoarthritis (OA) symptoms	Clinical study	D-002 (40 or 100 mg/day) for 14 days	The frequency of patients treated with D-002, 40 and 100 mg/day, reporting acidity was lower ($P < 0.05$) than in placebo	Illnait et al. (2005)
D-002 (beeswax alcohols)	Nonalcoholic fatty liver disease	Clinical study	D-002 (100 mg/day) for 24 weeks	A significant ultrasonography-detected reduction of liver fat infiltration versus a placebo	Illnait et al. (2013)

known. Beeswax was used to treat AD, focusing on restoring the skin barrier through increased skin moisture. Chu et al. (2020) found that beeswax could be used in the formulation of novel nanostructured lipid carriers with antioxidant activity to display photoprotective effects in the skin. In addition, the mixture of beeswax, honey, and olive oil (1:1:1) played a role in pityriasis versicolor and various tineas through its antifungal effects to specifically inhibit the growth of *Staphylococcus aureus* and *Candida albicans* (Al-waili 2004, 2005a). Meanwhile, Cera Flava, which is a kind of yellow beeswax obtained from beehives extract, could significantly decrease the mRNA levels of inflammatory factors, including interleukin (IL)-1β, IL-4, IL-13, IL-8, TARC, MDC, and RANTES in the dorsal skin tissue of mice (Park et al. 2021). This observation suggested that Cera Flava influenced skin barrier function and immune regulation to alleviate AD symptoms via regulating the TLR2/MyD88/ TRAF6/ ERK pathway. Another study proved that a mixture containing honey (50%), olive oil (29%), and beeswax (21%) could significantly recover the diaper dermatitis in infants (Al-waili 2005b). Another study showed that an affordable sunscreen, composed of beeswax and two other natural ingredients, zinc oxide and almond oil, displayed a similar narrowband ultraviolet B radiation (UVB) and minimal erythema dose with photoprotective effects from UVB in five volunteers (Miyar et al. 2014). These studies showed that beeswax could be useful for treating psoriasis by its antimicrobial activity. Therefore, beeswax is commonly applied for general skin disease and skin barrier health, which could be used both independently and in combination with other skincare ingredients as a therapeutic treatment.

16.3 Wound-Healing Activity

Beeswax is thought to be particularly effective for wound healing in the treatment of burns due to the restoration and maintenance of the skin barrier. Research has indicated that beeswax holds significant potential in promoting wound healing, especially burns by fire or boiling liquid and second-degree burns (Gümüş and Özlü 2017; Ebrahimpour et al. 2020). The primary hydrocarbons, fatty acids, and esters in beeswax make it display antimicrobial, anti-inflammatory, and moisturizing properties, thereby aiding in the wound-healing process. Studies suggested that bioactive compounds found in beeswax extracts, such as beeswax alcohol and beeswax esters, could stimulate cell proliferation and collagen generation, thereby accelerating wound closure (Bayir et al. 2019; Dumitru et al. 2022).

Wound healing is a complex and dynamic process that involves the restoration of the skin anatomical continuity and function and regulates the generation and physiological activities of extracellular matrix mediators and different cells such as fibroblasts and keratinocytes. Beeswax can form a protective barrier, preventing wound contamination by bacteria and external environmental factors, thereby providing an optimal wound-healing environment. In this context, the mumio-based hydrogel dressing containing mumio, beeswax, and cinnamon oil showed a favorable wound-healing effect through the inhibition of microbial infection (Zandraa et al. 2021).

Beeswax inhibited the growth of different bacteria (*S. aureus* and *K. pneumoniae*) and fungi (*C. albicans* and *A. niger*). In addition, the ethanolic re-extract combination of beeswax, olive oil, and nine medicinal plants was suggested to increase wound-healing activity in fibroblast cells (Zbuchea et al. 2016).

Moreover, a representative nutritive product, beeswax, olive oil, and butter (BOB), was proved to increase tumor growth factor (TGF)–β1 and vascular endothelial growth factor (VEGF)-α expressions compared to a burn group in rats to treat second-degree burns (Bayir et al. 2019). BOB could promote skin regeneration and increase vesicles, bullae, and fibroblast activity to improve burn wound healing. A mixture of honey, beeswax, and olive oil (MHBO) was efficient in the treatment of second-degree burns in dogs (Moustafa and Atiba 2015). Additionally, the famous traditional Iranian medicine lime salve (L.S.), which is composed of a combination of refined calcium hydroxide powder (33%), beeswax (17%), and sesame oil (50%), could effectively accelerate the healing process of deep second-degree burn wounds by regulating important cytokine and matrix metalloproteinase enzyme (MMP) expressions (Ebrahimpour et al. 2020). Overall, beeswax possess wound-healing activity and heals burns via stimulating or inhibiting the release of some related important factors including cytokines, inflammatory factors, and the MMP-9 protein to promote reepithelialization and extracellular matrix remodeling or generating barriers to create a wound-healing environment.

16.4 Protection of Oxidative Damage

Physiological metabolic processes in the human body can generate many unstable free radicals. These highly unstable free radicals try to pair up with other molecules, atoms, or individual electrons to create a stable compound, which would further generate reactive oxygen species (ROS)–induced oxidative stress and damage. Meanwhile, oxidative damage causes lots of chronic and degenerative diseases such as inflammatory disorders, cancer, tumors, aging, rheumatoid arthritis, and cardiovascular, immune, and Alzheimer's diseases (Martinello and Mutinelli 2021). Recently, many natural bioactive substances have drawn lots of attention to protect tissues from damage by free radicals, including bee products. Navarro et al. (2019) first evaluated the antioxidant capacity of the beeswax of the hydrophilic and lipophilic extracts. Subsequently, Sawicki et al. (2022) determined the antioxidant activity of four kinds of bee products including bee pollen, bee bread, beeswax, and honey using different assays in vitro. The results showed that the antioxidant activity of beeswax extracts is higher than that of honey under photo chemiluminescence (PCL) assay. Furthermore, the by-products from beeswax recycling are rich sources of proteins, minerals, and polyphenols that have potential in biomedical properties, as well as organic and inorganic waste, protecting from oxidative stress in human dermal fibroblast cells with strong total antioxidant capacity and low levels of toxins (Giampieri et al. 2018a, b). Beeswax could play significant roles in protection against oxidative damage in vivo. Five hydrophilic extracts from a mud by-product obtained from the recycling of beeswax were proved to exert antioxidant and neuroprotective effects to decrease the β-amyloid toxicity in a *C. elegans* model of Alzheimer's disease (Romero et al. 2020, 2023). The by-product of beeswax could also act against oxidative damage to protect from related diseases. Additionally, the combinations of beeswax alcohol and fermented black rice bran synergistically exert antioxidant and anti-inflammatory activities to alleviate hepatic injury and dyslipidemia by regulating the ROS and IL-6 expression levels and decreasing the blood total cholesterol (TC) and triglyceride (TG) levels (Han et al. 2023).

ROS could further cause dysfunction of cells to induce inflammatory responses. The beeswax alcohol, D-002, was proved to exert anti-inflammatory activity by inducing a reduction of leukotriene B4 (LTB4) levels (Carbajal et al. 1998). The most abundant component of D-002, triacontanol, could inhibit both enzymatic and nonenzymatic lipid peroxidation in vitro. Triacontanol has been suggested to be an effective anti-inflammatory drug as a natural compound, which could inhibit soybean lipoxygenase-I (LOX-1) and regulate lipid peroxidation to exert anti-inflammatory action in animals (Ramanarayan et al. 2000). In addition, the beeswax alcohol D-002 (50 mg/day) played an important role in lipid peroxidation in middle-aged and older subjects by reducing significantly plasma total hydroxy-peroxides and malondialdehyde levels, increasing the total antioxidant status (López et al. 2008). Cho et al. (2024) also suggested that beeswax alcohol prevented the low-density lipoprotein (LDL) oxidation in zebrafish embryos and humans. LDL would be affected and oxidized by oxidative stress, resulting in several detrimental health effects. This further expands previous knowledge on the antioxidant effects of beeswax in humans. It was reported that D-002 has been already proved to exert antioxidant effects in many experimental and clinical studies (Menéndez et al. 2001; Molina et al. 2001; Pérez et al. 2002). All these findings clearly showed the effectiveness of beeswax against oxidative damage.

16.5 Antimicrobial Activity

Antimicrobial activity refers to the ability of a substance to inhibit the growth and proliferation of bacteria, thereby preventing infections and contamination. This activity can be exhibited by a wide range of agents, including antibiotics, antimicrobial peptides, and natural substances such as essential oils and extracts from plants and animals. Recently, there is growing interest in the antimicrobial properties of beeswax, although there are still few studies in the literature focused only on the action of beeswax (Fratini et al. 2016). Studies have demonstrated that beeswax exhibits inhibitory effects on both Gram-positive and Gram-negative bacteria (Ghanem 2011; Felicioli et al. 2019). The ethanolic extract of beeswax effectively inhibited *S. aureus ATCC255923* and *C. albicans NCTC2708* to exert antimicrobial activity. It further showed the inhibited effects with solvent-dependent that beeswax ethanolic extract in 70% concentration inhibited *E. coli, A. niger,* and *C. tropicalis* more sensitively than that of beeswax or other concentration extractions (Kačániová et al. 2012). Many studies are focused on beeswax synergy with other natural products that also exhibited favorable antimicrobial activity. For instance, beeswax and propolis in synergy (1:1, v/v) exerted strong antimicrobial activity used in medicine that effectively inhibited the growth of *S. aureus ATCC25923*, *Staphylococcus epidermidis ATCC12228*, *Bacillus subtilis ATCC27853,* and *C. albicans NCTC270* (Castaldo and Capasso 2002). In addition, the beeswax, honey, and olive oil mixture (1:1:1, v/v) is useful to inhibit the growth of *S. aureus* and *C. albicans*, which is used for patients with various tinea in clinical trials and even infants with diaper dermatitis (Al-waili 2004, 2005b). Beeswax with antimicrobial activity could also be used as an encapsulation vector protection in food or medicine to improve stability and solubility (Trevisani et al. 2017; Mumuni et al. 2022). Beeswax antimicrobial coating and dressings could extend the shelf life of vegetables and fruits (Agyemang et al. 2022; Betancur et al. 2024). It was even used in textiles or papers for enhancing antimicrobial activity for the skin health or improving substantially, respectively (Zhang and Xiao 2013; Szulc et al. 2020). Moreover, the by-products of beeswax processing, which included flavonoids, phenolic acids, carbohydrates, hydrocarbons, and minerals, inhibited *Klebsiella pneumoniae* and *Salmonella enterica* with a significant antibacterial activity, and they could be sustainable alternatives to propolis as food preservatives or nutraceuticals (Peron et al. 2023). The component flavonoids and esters of phenolic acids in beeswax might be the bioactive compounds that are responsible for the antimicrobial activity. However, beeswax composition is very complicated, so more investigation is need to understand these different effects exerted by this matrix.

16.6 Other Health Benefits and Applications

In addition to the above common activities, beeswax also shows other health benefits because of its rich and complex bioactive components. Beeswax also possessed antitumor activity against HeLa (human cervical adenocarcinoma cells) and SH-4 (human melanoma cells) cell lines (Kyuchyuk et al. 2022). The representative application in medicine, D-002, a mixture of beeswax alcohols, alleviated gastrointestinal symptoms that increased the secretion of gastric mucus; reduced the hydroxyl radical, lipid peroxidation, protein oxidation, and neutrophil infiltration; and upregulated antioxidant enzymes on the gastric mucosa as a dual cyclooxygenase (COX)/5-lipoxygenase (LOX) inhibitor (Molina et al. 2015; Perez et al. 2013). The gastric gavage of D-002 also exerted anti-inflammatory and antinociceptive effects in xylene-induced ear edema mouse and increased significantly the myeloperoxidase (MPO) activity (Ravelo et al. 2011). D-002 also has been shown to ameliorate experimental OA and reduce OA symptoms in clinical trials by its effective anti-inflammatory effects (Puente et al. 2014; Illnait et al. 2005). Additionally, Illnait et al. (2013) suggested that D-002 contributed to treating nonalcoholic fatty liver disease in patients. The multiple functional beeswax could be also used in drugs, pills, capsules as binding agent, carrier or controlling release factor. It is also used as a color carrier in food supplements and as a food additive (E901) (Fernando et al. 2007).

Acknowledgments Danila Cianciosi was supported by a Fondazione Umberto Veronesi fellowship.

References

Agyemang B, Grabulos J, Hubert O et al (2022) Properties of beeswax antifungal coatings obtained by high pressure homogenisation and their application for preserving bananas during storage. Int J Food Sci Technol 57:5349–5356. https://doi.org/10.1111/ijfs.15865

Al-waili NS (2004) An alternative treatment for pityriasis versicolor, tinea cruris, tinea corporis and tinea faciei with topical application of honey, olive oil and beeswax mixture: an open pilot study. Complement Ther Med 12(1):45–47. https://doi.org/10.1016/j.ctim.2004.01.002

Al-waili NS (2005a) Mixture of honey, beeswax and olive oil inhibits growth of *Staphylococcus aureus* and *Candida albicans*. Arch Med Res 36(1):10–13. https://doi.org/10.1016/j.arcmed.2004.10.002

Al-waili NS (2005b) Clinical and mycological benefits of topical application of honey, olive oil and beeswax in diaper dermatitis. Clin Microbiol Infect 11(2):160–163. https://doi.org/10.1111/j.1469-0691.2004.01013.x

Bayir Y, Un H, Ugan RA et al (2019) The effects of beeswax, olive oil and butter impregnated bandage on burn wound healing. Burns 45(6):1410–1417. https://doi.org/10.1016/j.burns.2018.03.004

Bernardin F, Tuniz C, Coppa A et al (2012) Beeswax as dental filling on a neolithic human tooth. PLoS One 7(9):44904. https://doi.org/10.1371/journal.pone.0044904

Betancur DAM, Perez CE, Barrera MC et al (2024) Antimicrobial activity, mechanical and thermal properties of cassava starch films incorporated with beeswax and propolis. J Food Sci Technol 61(4):782–789. https://doi.org/10.1007/s13197-023-05878-x

Bogdanov S (2002) Beeswax: history, uses and trade. Beeswax

Bogdanov S (2004) Beeswax: quality issues today. Bee World 85(3):46–50. https://doi.org/10.1080/0005772X.2004.11099623

Carbajal D, Molina V, Valde S et al (1998) Anti-inflammatory activity of D-002: an active product isolated from beeswax. Prostaglandins Leukot Essent Fatty Acids 59(4):235–238. https://doi.org/10.1016/s0952-3278(98)90135-1

Carbajal D, Molina V, Noa M et al (2000) Effect of D-002 on gastric mucus composition in ethanol-induced ulcer. Pharmacol Res 42(4):329–332. https://doi.org/10.1006/phrs.2000.0693

Castaldo S, Capasso F (2002) Propolis, an old remedy used in modern medicine. Fitoterapia 73:1–6. https://doi.org/10.1016/S0367-326X(02)00185-5

Cho KH, Baek SH, Nam HS et al (2024) Beeswax alcohol prevents low-density lipoprotein oxidation and demonstrates antioxidant activities in zebrafish embryos and human subjects: a clinical study. Curr Issues Mol Biol 46(1):409–429. https://doi.org/10.3390/cimb46010026

Chu CC, Hasan ZA, Chua SK et al (2020) Formulation and characterization of novel nanostructured lipid carriers with photoprotective properties made from carnauba wax, beeswax, pumpkin seed oil, and UV filters. J Am Oil Chem Soc 97(5):531–542. https://doi.org/10.1002/aocs.12340

Dumitru CD, Neacsu IA, Alexandru MG et al (2022) Bee-derived products: chemical composition and applications in skin tissue engineering. Pharmaceutics 14(4):750. https://doi.org/10.3390/pharmaceutics14040750

Ebrahimpour N, Mehrabn M, Iranpour M et al (2020) The efficacy of a traditional medicine preparation on second-degree burn wounds in rats. J Ethnopharmacol 252:112570. https://doi.org/10.1016/j.jep.2020.112570

Farag RS, Hassan MNA, Ali F (1993) Bees wax and its unsaponifiables as natural preservative for butter and cottonseed oils. Grasas y aceites (Sevilla) 44(3):183–189. https://doi.org/10.3989/gya.1993.v44.i3.1091

Felicioli A, Cilia G, Mamcini S et al (2019) In vitro antibacterial activity and volatile characterisation of organic Apis mellifera ligustica (Spinola, 1906) beeswax ethanol extracts. Food Biosci 29:102–109. https://doi.org/10.1016/j.fbio.2019.04.004

Fernando AH, Dekant KE, Christian LC et al (2007) Beeswax (E 901) as a glazing agent and as carrier for flavours scientific opinion of the panel on food additives, flavourings, processing aids and materials in contact with food (AFC). EFSA J 5(12):615. https://doi.org/10.2903/j.efsa.2007.615

Formiga AS, Pinsetta JJ, Pereira EM et al (2019) Use of edible coatings based on hydroxypropyl methylcellulose and beeswax in the conservation of red guava 'Pedro Sato'. Food Chem 29:144–151. https://doi.org/10.1016/j.foodchem.2019.03.142

Fratini F, Cilia G, Turchi B et al (2016) Beeswax: a minireview of its antimicrobial activity and its application in medicine. Asian Pac J Trop Med 9(9):839–843. https://doi.org/10.1016/j.apjtm.2016.07.003

Ghanem NB (2011) The antimicrobial activity of some honey bee products and some saudi folkloric plant extracts. J King Saud Univ Sci 23(2):47–62. https://doi.org/10.4197/Sci.23-2.4

Giampieri F, Gasparrini M, Forbes HTY et al (2018a) Beeswax by-products efficiently counteract the oxidative damage induced by an oxidant agent in human dermal fibroblasts. Int J Mol Sci 19(9):2842. https://doi.org/10.3390/ijms19092842

Giampieri F, Quiles JL, Orantes BFJ et al (2018b) Are by-products from beeswax recycling process a new promising source of bioactive compounds with biomedical properties? Food Chem Toxicol 112:126–133. https://doi.org/10.1016/j.fct.2017.12.041

Gümüş K, Özlü ZK (2017) The effect of a beeswax, olive oil and *Alkanna tinctoria (L.) Tausch* mixture on burn injuries: an experimental study with a control group. Complement Ther Med 34:66–73. https://doi.org/10.1016/j.ctim.2017.08.001

Han Y, Zee S, Cho KH (2023) Beeswax alcohol and fermented black rice bran synergistically ameliorated hepatic injury and dyslipidemia to exert antioxidant and anti-inflammatory activity in ethanol-supplemented zebrafish. Biomolecules 13(1):136. https://doi.org/10.3390/biom13010136

Hepburn (1986) Chapter 2: The origin of beeswax: an historical perspective. In: Honeybees and wax. Springer-Verlag Press

Hosseini SF, Mousavi Z, Mcclements DJ (2023) Beeswax: a review on the recent progress in the development of superhydrophobic films/coatings and their applications in fruits preservation. Food Chem 424:136404. https://doi.org/10.1016/j.foodchem.2023.136404

Illnait J, Terry H, Mas R et al (2005) Effects of D-002, a product isolated from beeswax, on gastric symptoms of patients with osteoarthritis treated with piroxicam: a pilot study. J Med Food 8(1):63–68. https://doi.org/10.1089/jmf.2005.8.63

Illnait J, Rodriguez I, Mendoza S et al (2013) Effects of D-002, a mixture of high molecular weight beeswax alcohols, on patients with nonalcoholic fatty liver disease. Korean J Intern Med 28(4):439–448. https://doi.org/10.3904/kjim.2013.28.4.439

Jing S (2022) The divine farmers materia medica a translation of the Shen Nong Ben Cao Jing. Shanghai Science & Technology Press, The Shen Nong Book of Herb

Kacániová M, Vukovic N, Chlebo R et al (2012) The antimicrobial activity of honey, bee pollen loads and beeswax from Slovakia. Arch Biol Sci 64(3):927–934. https://doi.org/10.2298/ABS1203927K

Kaur K, Gupta N, Mahajan M et al (2023) A novel edible coating of beeswax impregnated with karonda polyphenol rich extract maintains the chemical and bioactive potential of fresh ber fruit during storage at low temperature. Hortic Environ Biotechnol 64(6):1001–1014. https://doi.org/10.1007/s13580-023-00533-y

Kyuchyuk S, Paneva D, Karashanova D et al (2022) Core-sheath-like poly (ethylene oxide)/beeswax composite fibers prepared by single-spinneret electrospinning. Antibacterial, antifungal, and antitumor activities. Macromol Biosci 22(6):2200015. https://doi.org/10.1002/mabi.202200015

López E, Illnait J, Molina V et al (2008) Effects of D-002 (beeswax alcohols) on lipid peroxidation in middle-aged and older subjects. Lat Am J Pharm 27(5):695–703

Martinello M, Mutinelli F (2021) Antioxidant activity in bee products: a review. Antioxidants 10(1):71. https://doi.org/10.3390/antiox10010071

Menéndez R, Más R, Amor AM et al (2001) Antioxidant effects of D002 on the in vitro susceptibility of whole plasma in healthy volunteers. Arch Med Res 32(5):436–441. https://doi.org/10.1016/S0188-4409(01)00315-0

Miyar ME, Diven DG, Cosulich MT et al (2014) Serving the underserved: creating a low-cost sunscreen with natural ingredients for humanitarian medical trips to the developing world. Br J Dermatol 171(2):415–417. https://doi.org/10.1111/bjd.12993

Molina V, Valdes S, Carbajal D et al (2001) Antioxidant effect of D-002 on gastric mucosa of rats with experimentally induced injury. J Med Food 4(2):79–83. https://doi.org/10.1089/109662001300341734

Molina V, Mas R, Carbajal D (2015) D-002 (beeswax alcohols): concurrent joint health benefits and gastroprotection. Indian J Pharm Sci 77(2):127–134. https://doi.org/10.4103/0250-474x.156542

Moustafa A, Atiba A (2015) The effectiveness of a mixture of honey, beeswax and olive oil in treatment of canine deep second-degree burn. Glob Vet 14(2):244–250. https://doi.org/10.5829/idosi.gv.2015.14.02.9361

Mumuni MA, Amarachi CS, Victor AC et al (2022) A novel formulation design based on heterotemplated solid lipid microparticles to improve the solubility of anti-inflammatory piroxicam for oral administration. New J Chem 46:3961–3965. https://doi.org/10.1039/dinj05281k

Navarro HMD, Orantes BJ, Sanchez GC et al (2019) Industrial-scale decontamination procedure effects on the content of acaricides, heavy metals and antioxidant capacity of beeswax. Molecules 24(8):1518. https://doi.org/10.3390/molecules24081518

Nong Y, Maloh J, Natarelli N et al (2023) A review of the use of beeswax in skincare. J Cosmet Dermatol 22(8):2166–2173. https://doi.org/10.1111/jocd.15718

Park G, Moon BC, Choi G et al (2021) Cera Flava alleviates atopic dermatitis by activating skin barrier function via immune regulation. Int J Mol Sci 22(14):7531. https://doi.org/10.3390/ijms22147531

Pérez RM, González AM, Amor S et al (2002) D-002 on antioxidant enzymes in liver and brain of rats. Rev CENIC Cienc Biol 33:3

Perez Y, Oyarzabal A, Mas R et al (2013) Protective effect of D-002, a mixture of beeswax alcohols, against indomethacin-induced gastric ulcers and mechanism of action. J Nat Med 67(1):182–189. https://doi.org/10.1007/s11418-012-0670-y

Peron G, Carmo DSNA, Ferrarese I et al (2023) The beeswax processing by-product: a potential antibacterial ingredient for food and nutraceutical applications. Int J Food Sci Technol 58(10):5549–5556. https://doi.org/10.1111/ijfs.16520

Puente R, Illnait J, Mas R et al (2014) Effects of D-002, a mixture of beeswax alcohols, coadministered with green-lipped mussel extract, on osteoarthritis symptoms. Int J Pharm Sci Rev Res 29(2):191–202. https://doi.org/10.3904/kjim.2013.28.4.439

Ramanarayan K, Bhat A, Shripathi V et al (2000) Triacontanol inhibits both enzymatic and nonenzymatic lipid peroxidation. Phytochemistry 55(1):59–66. https://doi.org/10.1016/s0031-9422(00)00201-6

Ravelo Y, Molina V, Carbajal D et al (2011) Evaluation of anti-inflammatory and antinociceptive effects of D-002 (beeswax alcohols). J Nat Med 65(2):330–335. https://doi.org/10.1007/s11418-010-0496-4

Romero MJM, Navarro HMD, Varela LA et al (2020) Usefulness of beeswax recycling by-products in the treatment of β-amyloid toxicity in a C. elegans model of Alzheimer. Med J Nutrition Metab 13(2):163–173. https://doi.org/10.3233/MNM-200404

Romero MJM, Navarro HMD, Orantes FJ et al (2023) In vivo anti-Alzheimer and antioxidant properties of avocado (persea americana mill.) honey from southern Spain. Antioxidants 12(2):404. https://doi.org/10.3390/antiox12020404

Sawicki T, Starowicz M, Klebukowska L et al (2022) The profile of polyphenolic compounds, contents of total phenolics and flavonoids, and antioxidant and antimicrobial properties of bee products. Molecules 27(4):1301. https://doi.org/10.1007/s00217-022-04041-8

Szulc J, Machnowski W, Kowalska S et al (2020) Beeswax-modified textiles: method of preparation and assessment of antimicrobial properties. Polymers (Basel) 12(2):344. https://doi.org/10.3390/polym12020344

Trevisani M, Cecchini M, Siconolfi D et al (2017) Effects of beeswax coating on the oxidative stability of long-ripened italian salami. J Food Qual 20:171–175. https://doi.org/10.1155/2017/8089135

Zaheri-Abdevand Z, Badr P (2023) A review of the wound-healing properties of selected plant exudates. J Herb Med 41:100715. https://doi.org/10.1016/j.hermed.2023.100715

Zandraa O, Ngwabebhoh FA, Patwa R et al (2021) Development of dual crosslinked mumio-based hydrogel dressing for wound healing application: physico-chemistry and antimicrobial activity. Int J Pharm 607:120952. https://doi.org/10.1016/j.ijpharm.2021.120952

Zbuchea A, Lungu L, Popa C et al (2016) An innovative ointment made of natural ingredients with increased wound healing activity. Rom Biotechnol Lett 21(2):11427–11437

Zhang D, Xiao H (2013) Dual-functional beeswaxes on enhancing antimicrobial activity and water vapor barrier property of paper. ACS Appl Mater Interfaces 5(8):3464–3468. https://doi.org/10.1021/am400585m

Part VII
Bee Venom

Chapter 17
Health Benefits and Uses of Bee Venom in Medicine

Sok Cheon Pak, Soo Liang Ooi, and Benjamin Song Pak

Contents

Abstract The medicinal properties of melittin, apamin, phospholipase A_2, and crude bee venom have been extensively investigated in medical science to understand their mechanistic actions. This chapter provides an overview of the current knowledge regarding the mechanisms of these enzymes and peptides, as studied in vitro and in animal models. The exploration encompasses various disorders, including Alzheimer's disease, Parkinson's disease, vascular dementia, peripheral nerve injury, arthritis, multiple sclerosis, asthma, cancer, liver fibrosis, atopic dermatitis, and pain. The roles of these compounds are expected to continue expanding in the design and development of novel therapeutic drugs.

Keywords Bee venom · Health benefits · Melittin · Phospholipase A_2 · Therapeutic applications

S. C. Pak (✉) · S. L. Ooi · B. S. Pak
School of Dentistry and Medical Sciences, Charles Sturt University, Bathurst, NSW, Australia
e-mail: spak@csu.edu.au

J. M. Alvarez-Suarez (ed.), *Bee Products – Chemical and Biological Properties*,
https://doi.org/10.1007/978-3-031-89049-9_17

17.1 Introduction

The medicinal promise of honey bee venom (BV) has intrigued the medical community for over a century. Its historical recognition as a remedy dates back to times when our understanding of BV relied on anecdotal evidence and traditional wisdom. Nevertheless, the advent of advanced venomics techniques has unveiled a wealth of bioactive compounds within this venom, propelling it into a new realm of pharmacological significance. This breakthrough has paved the way for the refinement of therapeutic strategies, harnessing the potential of active BV compounds such as melittin, apamin, and phospholipase A_2 (PLA_2).

In recent years, the application of BV has expanded far beyond its conventional antinociceptive effects, extending into the domain of degenerative nervous system diseases. This expansion can be attributed to the unique properties of venom enzymes and peptides, including their inherent stability as injectable solutes, their remarkable precision in targeting specific tissues, and their capacity to facilitate cell-cell interactions to enhance biological effects.

This chapter comprehensively overviews the latest research findings on BV's diverse biological functions. Our discussion will primarily revolve around the most recent approaches in studies that emphasize the health benefits and medical applications of BV compounds. By doing so, we aim to contribute to the growing body of knowledge about BV from an applied science perspective, providing insights that will guide further research endeavors toward harnessing the potential of these valuable peptides for treating various human ailments. In this updated version, we extend our exploration of BV's therapeutic potential, covering publications not included in the previous book edition, thus ensuring our position at the forefront of this evolving field.

17.2 Bee Venom Uses in Medicine

17.2.1 Alzheimer's Disease

Recent data indicate that the incidence of dementia is projected to increase twofold in Europe and triple globally (Scheltens et al. 2021). This projection rises further to threefold when utilizing a biological definition of Alzheimer's disease (AD) instead of a clinical one. The cellular phase of AD occurs concurrently with the accumulation of amyloid β (Aβ), which triggers the propagation of tubulin associated unit (tau) phosphorylation. The susceptibility to AD is influenced by hereditary factors in the range of 60–80%, with over 40 genetic risk loci associated with AD already identified, among which the apolipoprotein E gene (or APOE) alleles exhibit the most significant correlation with the condition (Scheltens et al. 2021). While lifestyle factors may not exert a direct impact on the pathology of AD, they can nevertheless play a role in fostering favorable outcomes for individuals afflicted by the condition. Pharmacological interventions are progressing through advanced phases of clinical trials, encompassing approaches targeting Aβ, tau, and inflammatory pathways.

To establish an animal model of neuroinflammatory cognitive impairment, Gu et al. (2015) administered lipopolysaccharide (LPS) intraperitoneally to mice. Prior to the LPS treatment, two doses of BV were injected. The choice of 7–8-week-old mice was based on the occurrence of AD pathological changes within 7–10 weeks after LPS administration. Memory decline was evaluated using the Morris water maze test and passive avoidance test. In the context of the in vitro experiment, BV and LPS were co-applied to astrocytes and microglial cells. The results showcased BV's capacity to mitigate amyloidogenesis, neuroinflammation, and memory loss. The researchers substantiated these findings by observing the expressions of inflammatory response genes, including glial fibrillar acid protein (GFAP), IBA-1, COX-2, and iNOS. Notably, BV treatment led to a reduction in the levels of these proteins. Moreover, BV exhibited an ability to suppress the nuclear factor kappa B (NF-κB) signaling pathway both in mice brain tissue and cultured astrocytes and microglial cells. Consequently, BV emerges as a promising candidate for addressing conditions characterized by amyloidogenesis and neuroinflammation, such as AD.

Similarly, LPS was employed in a separate study to activate microglial cells and induce neuroinflammation (Park et al. 2020). Apamin was demonstrated to suppress LPS-induced expressions of tumor necrosis factor (TNF)-α, cluster of differentiation (CD)11b, interleukin (IL)-1β, IL-6, and COX-2. Apamin's efficacy was mainly attributed to its inhibition of LPS-induced SK2/K_{Ca}2.2 expression, as TNF-α is generated in SK2/K_{Ca}2.2 channel-activated microglia. Given that LPS binds to Toll-like receptor 4 (TLR4) on the surface of microglial cells, apamin exerted its inhibitory effect on the subcellular interaction between TLR4 and TNF-α in LPS-stimulated microglial cells. Furthermore, apamin effectively prevented the translocation of the transcription factor NF-κB p65 and activated STAT3 from the cytoplasm to the nucleus following LPS stimulation. This inhibitory action extended to the LPS-induced phosphorylation of extracellular signal-regulated kinases (ERK) and c-Jun N-terminal kinases (JNK), demonstrating apamin's potential as an anti-neuroinflammatory agent.

Several key AD features can be replicated within the triple-transgenic AD (3xTg-AD) mouse model, which expresses pathogenic variants of the amyloid precursor protein, presenilin, and tau. At 18 months, 3xTg-AD mice exhibit a comprehensive recapitulation of both histological and behavioral indications of AD. As a result, the 3xTg-AD mouse is widely acknowledged as a complete transgenic model that faithfully reproduces the pathological aspects of AD (Javonillo et al. 2021). In their study exploring the neuroprotective potential of PLA_2, Ye et al. (2016b) employed the 3xTg-AD mouse model and assessed various parameters, including Aβ levels within the hippocampus, brain glucose metabolism, and microglial activation. Given the established role of CD4 regulatory T cells (Treg) in modulating neuroinflammation-mediated disorders, the infiltration of Treg into the hippocampus was quantified using immunostaining with an anti-CD4 antibody. As evidenced by PET imaging, PLA_2 demonstrated the capacity to enhance cognitive function, accompanied by increased brain glucose metabolism. Additionally, PLA_2 administration led to a significant reduction in the levels of Aβ deposits. This neuroprotective effect of PLA_2 was further correlated with the deactivation of microglia and a decrease in the infiltration of Treg. Intriguingly, the neuroprotective impact of PLA_2 was nullified in mice depleted of Treg. The observation suggests that the augmentation of the Treg population by PLA_2 might hinder AD progression in the 3xTg-AD mouse model.

The same team of researchers (Ye et al. 2016b) conducted a follow-up study a few years later (Baek et al. 2018) to ascertain if PLA_2 treatment could stimulate Treg and improve AD pathology in the 3xTg-AD mouse model. They sought to achieve this outcome without eliciting undesirable T-cell–mediated inflammation after Aβ vaccination. The results indicate that mobilization of Treg through PLA_2 intervention, combined with co-immunization using the $Aβ_{42}$ vaccine, could lead to distinct impacts on immunosuppressive reactions. This combination appears to promote a harmonious equilibrium in AD-related homeostasis. The augmentation of Treg by PLA_2 was linked to enhancements in cognitive function, decreased Aβ deposition, reduced production of inflammatory cytokines, and heightened glucose uptake within Aβ-vaccinated 3xTg-AD mice. PLA_2-induced Treg also played a role in diminishing inflammatory infiltration in the brain and spinal cord, a response observed after Aβ immunization and the introduction of pertussis toxin as a model for neuroinflammation. These findings substantiate the theory that utilizing PLA_2-induced Treg in conjunction with an Aβ vaccine could potentially serve as a therapeutic approach to address central nervous system (CNS) inflammation in the case of AD.

The $Aβ_{23–25}$ fragment has been proven to possess a similar extent of oxidative neurotoxicity as the full-length $Aβ_{1–42}$, as shown by Pike et al. (1995). In a recent study by Nguyen and Lee (2021), the researchers examined the protective effects of melittin on neuronal cells subjected to $Aβ_{23–25}$-induced oxidative stress. Additionally, they investigated the impact of melittin on memory impairment induced by $Aβ_{23–25}$ intracerebroventricular injections in a mouse model. Immunofluorescence analysis revealed that the administration of melittin at varying concentrations ranging from 0.3 to 3 μM exhibited a dosage-dependent improvement in the apoptotic condition of HT22 mouse hippocampal cells. Melittin treatment also effectively restored the expression levels of both proapoptotic and antiapoptotic factors in response to the $Aβ_{23–25}$ stress challenge. These data were substantiated utilizing Western blot analysis, which shows the restoration of crucial apoptosis proteins and the suppression of pivotal oxidative stress markers, including protein carbonyl. The administration of $Aβ_{23–25}$ also exhibited a notable reduction in the migration of Nrf2 into the nucleus of the cells, consequently leading to the disruption in the synthesis of the 1-HO antioxidant enzyme. Conversely, the treatment with melittin demonstrated a dose-dependent promotion of Nrf2 nuclear translocation. Additional results indicated that melittin exhibited a substantial restorative effect on the phosphorylation levels of p-TrkB, p-CREB, and expression of BDNF in comparison to the impact of $Aβ_{23–25}$ treatment. In an in vivo model, melittin treatment caused a significant reduction in the escape latency and crossing time of mice in the Morris water maze test compared to the treatment with $A\beta_{23-25}$. These findings suggest a successful recovery from memory impairment. Moreover, the administration of melittin led to a reversal of reduced neurogenesis in the dentate gyrus region of the hippocampus and improved levels of reactive oxygen species (ROS), NO, and malondialdehyde (MDA) in the hippocampus and serum. Furthermore, melittin normalized the down-regulation of M1 muscarinic acetylcholine receptors, mitigated the excessive activation of acetylcholinesterase, and increased levels of acetylcholine. The same group of researchers conducted a similar study (Nguyen et al. 2022b) with comparable findings but employed BV and $A\beta_{1-42}$ instead of melittin and $A\beta_{23-25}$. Similarly, BV and melittin were observed to effectively restore impaired learning and memory functions in a neurodegenerative model.

The TgCRND8 mouse, a widely recognized model for AD, demonstrates impaired cholinergic activation in the prefrontal attention circuitry, leading to early attention deficits. This impairment is attributed to SK(Ca)-mediated inhibition, which exerts a potent negative feedback on nicotinic excitation, thereby attenuating attention-relevant signaling in the TgCRND8 mouse brain. Researchers from Canada (Proulx et al. 2020) have successfully demonstrated the efficacy of the SK channel blocker apamin in reversing the inhibitory constraint on nicotinic receptors within the prefrontal cortical attention circuit.

The typical route of administration for biological efficacy studies involving BV is transdermal injection. However, this method may have several disadvantages in terms of pharmacokinetics. Recently, researchers have proposed using a microneedles patch (MN) as an alternative delivery method for biopharmaceuticals such as BV. The MN delivery system allows for controlled drug release with stable pharmacokinetic profiles. In a study by Nguyen et al. (2022a), the administration of BV via subcutaneous injection (SC) was compared to MN administration. Both SC and MN were found to have significant neuroprotective effects in a mouse model of scopolamine-induced neurodegeneration. Notably, MN required a lower concentration of BV to elicit a more substantial and comprehensive neuroprotective effect than SC. This finding was supported by data from water maze experiments and the analysis of key neuronal regulatory proteins.

17.2.2 Parkinson's Disease

Neurological disorders, such as Parkinson's disease (PD), pose a significant and growing global health issue, contributing substantially to the overall disease burden across the world. Between 1990 and 2019, for example, the disability-adjusted life years associated with PD and AD experienced a significant rise when compared to other neurological diseases (GBD 2019 Stroke Collaborators 2021). Despite this, accurately diagnosing PD continues to be a challenging task, and recognizing the initial stages of the disease presents a considerable obstacle.

Dantas et al. (2022) showed that Africanized BV demonstrated neuroprotective effects in a rodent model of PD induced by 6-hydroxydopamine (6-OHDA). 6-OHDA causes motor behavior impairment by forming dopaminergic lesions, leading to the loss of dopaminergic cells and fibers, which can be assessed by tyrosine-hydroxylase (TH) immunohistochemistry. Injection of 6-OHDA into the striatal tissue also degenerates the synaptic terminals of the dopaminergic neurons. Then, the astrocytes become activated with subsequent hypertrophic morphology by increased production of intermediate filaments called glial fibrillar acid protein (GFAP). Via chromatographic analysis, the authors confirmed that the Africanized BV utilized in the study contains large amounts of melittin, PLA_2, and apamin. BV was administered subcutaneously at different dosages every other day for 3 weeks. BV reversed the 6-OHDA-induced decrease in spontaneous motor activity and loss of TH-positive dopaminergic neurons in the substantia nigra and dopaminergic terminals in the striatum. In addition, BV prevented astroglial reaction in the lesioned animals, evidenced by a significant decrease in the mean of GFAP compared to those treated with 6-OHDA. Hence, the peptide-rich Africanized BV could protect against neurodegeneration in a progressive mouse model of PD.

Similarly, Maurice et al. (2015) utilized 6-OHDA to induce a rat model of PD. Their objective was to investigate whether BV exhibits a symptomatic action by targeting the pathological functioning of the basal ganglia. The researchers conducted motor behavior analyses using the cylinder test, apomorphine-induced rotations, and haloperidol-induced catalepsy. The results showed that BV treatment significantly mitigated the rats' contralateral forelimb akinesia, rotations, and catalepsy. This antiparkinsonian effect was also observed with apamin, a blocker of Ca^{++}-activated K^+ channels of small (SK) conductance. Interestingly, the effect of apamin was counteracted by CyPPA, a positive modulator of SK channels that increases their apparent Ca^{++} sensitivity by acting at the calmodulin-binding domain. This observation suggests that SK channels are involved in the antiparkinsonian mechanism of BV. The study also revealed that BV enhanced the functional impact of the direct striatonigral pathway, which is known to be hypoactive in the PD state. This finding suggests that BV may positively influence the cortico-basal ganglia circuits that underlie motor PD symptoms. Thus, BV could have potential implications for the symptomatic treatment of PD.

In another PD rodent model induced by reserpine, Ahmed-Farid et al. (2021) showed that BV could be a potential adjuvant treatment for PD. When used as a potent inhibitor of the vesicular monoamine transporter 2, reserpine reduces the storage of central catecholamines. This, in turn, results in rats displaying symptoms such as muscular rigidity, tremors, and akinesia. The study investigated the effects of treating reserpine-administered male rats with BV, L-Dopa, and dopamine (DA)-nanoparticles. L-Dopa and DA-nanoparticles were orally administered for 30 days, whereas BV was administered intraperitoneally every other day over 30 days. The study provided evidence that BV and DA-loaded nanoparticles effectively restored monoamines (norepinephrine, dopamine, and serotonin), achieved a balance in neurotransmitters by increasing GABA levels and reducing glutamate levels, and efficiently halted DNA fragmentation in the hippocampus and striatum. In addition, both treatments exhibited a significant reduction in pro-inflammatory cytokines (TNF-α, IL-1β) and an increase in levels of endogenous antioxidant (PON1) and neurotrophic factor (BDNF) when compared to the group of rats injected with reserpine. It is worth noting that the ameliorative effects of BV were significantly superior to those of DA-nanoparticles. The findings from this reserpine-induced PD rat study justify using BV- and DA-nanoparticles for treating PD.

α-Synuclein (α-Syn) plays a critical role in microglia-mediated neuroinflammation, leading to the development of PD. Researchers from South Korea (Ye et al. 2016a) investigated whether PLA_2 treatment could improve PD's motor dysfunction and pathological features in human A53T α-Syn mutant transgenic (A53T Tg) mice. The motor dysfunction in A53T Tg mice was assessed using the pole test. The levels of α-Syn, microglia, and the M1/M2 phenotype in the spinal cord were evaluated through immunofluorescence. Pro-inflammatory cytokines released by activated microglia and recruited peripheral macrophages contribute to neurodegeneration in PD. The activated microglia/macrophages are typically classified into M1 or M2 phenotypes. The study revealed that the progressive degeneration observed in A53T Tg mice and the improvements observed with PLA2 treatment were associated with α-Syn accumulation in the spinal cord. At the behavioral level, these changes were demonstrated by significantly reduced T-turn and T-locomotor times, indicative of a prolonged trend representing characteristic motor dysfunction in PD mice. PLA_2 treatment effectively mitigated these abnormal behaviors and was accompanied by distinct changes in the expression of inflammatory mediators in the spinal cords. Specifically, PLA_2 treatment reduced the expression of the microglial activation marker CD11b and the M1 marker CD86 in the spinal cords of A53T Tg mice and increased the expression of the M2 marker arginase 1. These in vivo findings were further corroborated by cell culture experiments using microglial cells. In vitro PLA_2 treatment promoted M2 differentiation while inhibiting M1 differentiation. These compelling findings support using PLA_2 as a promising and innovative strategy for treating PD.

17.2.3 Vascular Dementia

The etiology of vascular cognitive impairment and dementia is diverse, and the latest diagnostic criteria categorize it into four phenotypic classifications: Subcortical ischemic vascular dementia, poststroke dementia, multi-infarct dementia, and mixed dementia. In terms of key strategies for averting vascular cognitive impairment and dementia, it is advisable to focus on controlling blood pressure, managing cholesterol levels, and regulating blood glucose during midlife (Wong and Chui 2022). Vascular dementia is commonly regarded as the second most prevalent cause of age-related dementia. It contributes to approximately 20% of all dementia causes, and its prevalence tends to align with the risk of stroke (Bir et al. 2021).

Researchers who had previously focused on amyotrophic lateral sclerosis redirected their attention to utilizing the bilateral common carotid artery occlusion

(BCCAO) animal model (Cai et al. 2017). This model was employed to explore the potential therapeutic effects of BV on vascular dementia. Within this experimental framework, the reduction in blood flow within the cerebral cortex triggers a neuroinflammatory response akin to what is observed in cases of vascular dementia. BCCAO animals were treated with BV acupuncture (BVA) at the ST36 acupoint (Zusanli), known for its anti-inflammatory properties, four times every other day. This BVA treatment improved spatial memory as determined by a Y-maze test, compared to BCCAO animals that only received saline. Beyond its positive effect on cognitive dysfunction, BV also decreased neuroinflammatory protein expression in the hippocampus. These proteins include Iba-1, TLR4, CD14, TNF-α, iNOS, and Bax. Conversely, BVA contributed to an increased expression of BDNF through the activation of ERK. As a critical neurotrophic factor, BDNF plays an essential role in supporting neuronal survival and maintenance. Therefore, BV can potentially enhance cognitive function and mitigate the neuroinflammation associated with dementia.

17.2.4 Peripheral Nerve Injury

The peripheral nerve system (PNS) consists of an extensive network of nerves that play a vital role in integrating various body parts with the CNS. When it comes to neurological conditions, peripheral nerve injury (PNI) is particularly concerning as it poses a serious threat to patients, often leading to significant and long-lasting functional and physiological disabilities. Treatment options for PNI encompass a range of approaches, such as surgical methods, pharmacological therapies, and cell-based therapies, either independently or in conjunction with biomaterials. Nevertheless, it is worth noting that the PNS has a greater regenerative capacity compared to the CNS, mainly due to their distinct functional environments (Lopes et al. 2022).

BV was studied to assess its potential impact on functional recovery and reinnervation following PNI (Er-Rouassi et al. 2023). The left buccal and marginal mandibular branches of the facial nerve were section-sutured to induce PNI in mice. Locally collected and prepared BV was analyzed and verified to contain melittin, PLA_2, and apamin. The evaluation of nerve regeneration functionality involved recording videos to assess the activity of the vibrissae and nasal deviation. In rodents, the facial nerve is responsible for innervating the muscles that control the movement of the vibrissae. Additionally, the animals underwent retrograde labeling of facial motoneurons to evaluate the reinnervation of vibrissae muscles. The results demonstrated the beneficial effect of early administration of whole BV at the injured site following a distal facial nerve section-suture. As for behavioral recovery, BV treatment proved to be more effective than a combination of PLA_2 and melittin or betamethasone. The mice treated with BV showed faster vibrissae movement and nasal deviation completely disappeared compared to mice in the other treatment groups. Further morphological analysis revealed that BV-treated mice exhibited a normal fluorogold labeling of the facial motoneurons 4 weeks after the facial nerve section-suture, indicating successful restoration. Based on these findings, BV emerged as a promising therapeutic candidate for treating PNI.

17.2.5 Arthritis

Arthritis, whether acute or chronic, impacts one or multiple joints, with clinical manifestations encompassing joint swelling, deformities, pain, and stiffness. Compelling evidence indicates that stress related to arthritis influences the CNS. When combined with inflammatory changes in the peripheral areas, this stress can trigger central sensitization, potentially resulting in the development of chronic pain and an exacerbation of emotional distress. Cytokines and chemokines play a role in joint inflammation and central sensitization processes (Harth and Nielson 2019).

Park et al. (2004) examined the antiarthritic properties of BV in a rat model involving both acute inflammatory edema and chronic arthritis. In the initial phase, they induced acute edema by injecting carrageenan into the subplantar region of the right hind paw. BV was directly administered to the plantar area of the right hind paw 30 min before the carrageenan injection. Subsequently, in the chronic adjuvant-induced arthritis (AIA) model, they injected *Mycobacterium butyricum* into the rat tail. BV was administered to the plantar surface of the right hind paw from the first day of AIA induction until day 14. To further investigate the suppressive effects of BV and melittin on the expression of inflammatory genes, the researchers utilized a murine macrophage cell line and synoviocytes derived from patients with rheumatoid arthritis for in vitro testing. Treatment with BV led to a significant reduction in edema formation in the acute arthritis model and a decrease in tissue swelling and formation of osteophytes in the chronic arthritis model. These antiarthritic actions of BV were further investigated at the molecular level. Both BV and melittin demonstrated the ability to suppress the expression of genes induced by LPS, including cyclooxygenase 2, cytosolic PLA_2, and inducible nitric oxide (NO) synthase. Additionally, they hindered the generation of PGE_2 and NO, along with the regulation of intracellular calcium levels. Notably, BV and melittin also prevented LPS-triggered transcriptional and DNA-binding activity of NF-κB. This preventive effect was accomplished through the inhibition of IkB release and the translocation of p50. Remarkably, BV and melittin exhibited direct binding to p50. As a result, the inhibition of NF-κB activation mediated by the direct interaction with p50 emerged as a crucial molecular mechanism underlying BV's capacity to counteract inflammatory responses in arthritis.

After assessing the BV collected in Turkey with an HPLC-variable wavelength detector, which contained apamin, PLA_2, and melittin at concentrations of 3.85%, 13.51%, and 70.49%, respectively, Sevin et al. (2023b) conducted an experiment in which they cultured mammalian macrophages for 24 h, both with and without LPS, in the presence and absence of BV. Subsequently, they examined the supernatants through ELISA to determine the levels of production of inflammatory cytokines. The results indicated that BV did not possess the capacity to induce the production of inflammatory cytokines independently without LPS, suggesting that BV lacks immunostimulatory properties. However, when LPS-stimulated mammalian macrophages were exposed to BV, it demonstrated the ability to suppress the production of inflammatory cytokines such as TNF, IL6, and IL12p40. Given the pivotal role that macrophages play in the development of arthritis, BV likely showcases its anti-inflammatory properties in the context of inflammatory and autoimmune diseases by inhibiting the production of cytokines in macrophages. This suppression helps to mitigate the aggressive and detrimental inflammatory responses associated with these conditions.

17.2.6 Multiple Sclerosis

Multiple sclerosis (MS) is an autoimmune disease characterized by progressive inflammation, demyelination, and neurodegeneration. Pathologically, MS is marked by perivascular infiltrates of mononuclear inflammatory cells, axonal loss, and gliosis. This process leads to focal and diffuse brain and spinal cord abnormalities. The condition primarily affects the optic nerves, brainstem, spinal cord, cerebellar, and periventricular white matter. However, significant damage to cortical and subcortical gray matter is also observed. As a result, many individuals with MS experience chronic progressive disability (Barkhof and Koeller 2020).

Experimental autoimmune encephalomyelitis (EAE) has been extensively utilized as a model for studying MS. In this model, mice are typically injected subcutaneously with myelin oligodendrocyte glycoprotein peptide in Complete Freund's Adjuvant containing killed *Mycobacterium tuberculosis* emulsion. Subsequently, two intraperitoneal injections of pertussis toxin are administered. In a study conducted by Jelodar et al. (2021), the effects of Iranian honey BV on the progression of EAE in mice were investigated. Once clinical signs appeared, the animals received intraperitoneal treatment with BV. Histopathological and immunological analyses were performed. The findings showed a noteworthy rise in IL-27 serum levels in BV-treated mice, indicating that this cytokine holds promise as an effective response modifier for autoimmune disease immunotherapy. Moreover, BV treatment led to a reduction in the number of inflammatory cells in the brain parenchyma. These results indicate that BV might be suitable for treating inflammatory diseases such as MS (Jelodar et al. 2021).

On the other hand, Lee et al. (2016) utilized myelin basic protein (MBP) to induce EAE in a rat model. In this research, they investigated the effects of BVA on EAE. BVA was administered by bilaterally injecting BV subcutaneously into the ST36 acupoint. For pretreatment in acute EAE rats, BVA was administered 20 min before immunization with the MBP peptide and subsequently injected every 3 days for a total of 21 days. As for the posttreatment approach, BVA was administered daily from day 10 to 15 after the initial immunization. The study found that pretreatment with BVA showed greater effectiveness than posttreatment with BVA in terms of mitigating neurological impairment and preventing loss of body weight. Furthermore, pretreatment with BVA was observed to suppress demyelination, glial activation, expressions of several inflammatory cytokines, chemokines, and inducible NO synthase, as well as the activation of p38 mitogen-activated protein kinase (MAPK) and NF-κB signaling pathways in the spinal cord. Additionally, the study observed an upregulation of regulatory T cells and a suppression of T-helper cell (Th) 17 responses, along with Th1 responses. Based on these findings, it can be concluded that BVA has a beneficial effect on both the induction and progression of EAE.

17.2.7 Asthma

Asthma encompasses enduring and repetitive wheezing, coughing, and breathing challenges. These symptoms are linked to reversible airflow blockage and increased sensitivity of the airways. Asthma emerges from diverse and intricate interactions between genes and the environment, resulting in a range of clinical presentations, inflammation, and morphological changes in the airways. Achieving the best possible asthma management entails utilizing specific drug treatments customized for individual patients alongside measures to prevent exposure to triggers. Consistent monitoring is essential to uphold symptom management, sustain lung function, and identify any adverse effects caused by the treatments (Papadopoulos et al. 2022).

Given that a perturbation in the equilibrium between Th1 and Th2 immune responses serves as the fundamental trigger for the inflammatory response in asthma, Choi et al. (2013) conducted an assessment of BV's impact on the generation of cytokines associated with Th1, Th2, and Th17 responses through an in vitro culture system. Subsequently, they employed a mouse model of allergic asthma induced by ovalbumin (OVA) to appraise the ramifications of BV treatment. The study revealed that BV treatment led to an augmentation in the Treg population while concurrently suppressing the production of Th1, Th2, and Th17-related cytokines. Detailed examination of the cellular composition of the bronchoalveolar lavage and histopathological analysis showcased a noteworthy reduction in peribronchial and perivascular inflammatory cell infiltration following BV treatment. BV also exhibited a mitigating effect on airway hyperresponsiveness. Importantly, the favorable outcomes linked to BV treatment for these facets of allergic asthma were nullified upon Treg depletion induced by injection of anti-C25 antibody. This implies that Treg cells were the primary therapeutic target for BV.

Having established that the mechanism through which BV influences allergic asthma involves Treg-mediated immune tolerance, the researchers proceeded with their investigation to identify the specific constituents within BV responsible for its ability to suppress allergic asthma. To this end, Park et al. (2015) employed PLA_2 to regulate OVA-induced allergic asthma in mice via Treg. Similarly, the application of PLA_2 prompted the activation of Treg and resulted in a decrease in the infiltration of inflammatory cells into the lungs of mice subjected to OVA challenges. Cytokines associated with Th2 responses in the bronchoalveolar lavage fluid exhibited reduced levels in mice treated with PLA_2. The effectiveness of Treg regulation in mitigating asthmatic symptoms was reaffirmed. Moreover, the role of CD206, a recognized receptor for PLA_2, in alleviating airway inflammation was highlighted, as the modulating effects were absent in PLA_2-treated mice with OVA-induced allergic asthma that lacked CD206.

17.2.8 Cancer

The research volume exploring the influence of BV and its constituents on various types of cancer, along with their respective mechanisms, is on the rise. While the clinical application of BV remains limited, the continuous investigation in this field undeniably brings us closer to acknowledging these compounds as potential contenders for integration into therapeutic strategies against cancer.

Having initially observed the selective cytotoxic impact of BV on specific cancer cells, Tetikoglu and Celik-Uzuner (2023) delved into the cellular mechanism underlying this phenomenon. To achieve this, they evaluated the extent of H2AX phosphorylation as an indicator of DNA damage and examined the intracellular distribution of β-actin through immunofluorescence analysis. This comparative study involved human metastatic breast cancer cells and human liver cancer cells versus normal embryonic mouse fibroblast cells. The staining levels of H2AX phosphorylation decreased in normal cells but increased in cancer cells upon BV treatment. Following BV treatment, most of the β-actin was localized within the cytoplasm of normal cells, whereas it was predominantly observed within the nucleus in cancer cells. These findings confirm that BV elicits distinct responses in normal and cancerous cells.

Obeidat et al. (2023) employed Jordanian crude BV to evaluate its cytotoxic effects on leukemia cell lines (K562). They chose BV collected during the springtime for their investigation due to its notably higher melittin content than venoms gathered in other seasons. The cytotoxicity assay revealed that both BV and melittin demonstrated the ability to hinder the growth of K562 cells at relatively low concentrations, with IC_{50} values of 3.7 ± 0.37 μg/ml and 1.84 ± 0.75 μg/ml, respectively. In terms of the mode of cellular death, melittin-treated cells displayed late apoptotic demise, accompanied by a moderate cellular arrest in the G0/G1 phase and an increase in cell numbers at the G2/M phase, indicating G2/M arrest. NF-κB plays a role in enabling tumors to evade cell death by inhibiting apoptosis signaling pathways. BV downregulated the expression of the NF-κB RELA (p65) gene and the MAPK14 gene, known for its antiapoptotic function. Both BV and melittin reduced c-MYC expression, a factor associated with drug resistance in leukemia cells. An increase in the expression of the proto-oncogene JUN was also observed following treatment with BV or melittin, which can lead to apoptosis.

Triple-negative breast cancer (TNBC) is characterized as a subtype of breast cancer that lacks expression of the estrogen receptor, progesterone receptor, and human epidermal growth factor receptor 2, rendering it resistant to endocrine therapy, molecular-targeted therapy, and chemotherapy. Turkish scientists have employed BV to assess its impact on human TNBC cells (Sevin et al. 2023a). To evaluate the apoptotic effects of BV, they conducted the annexin V assay and acridine orange staining. The results showed that BV-induced apoptotic cell death in TNBC cells at a lower dose (0.47 μg/ml). The observation was further confirmed by nuclear fragmentation, apoptotic blebbing, and cell contraction. However, BV did not trigger cell cycle arrest. These findings suggest that BV may have the potential for cancer treatment by promoting apoptosis.

Hwang et al. (2022) utilized commercially available BV to treat human lung cancer lines. They observed a concentration-dependent inhibition of cancer cell growth with BV treatment, a finding that correlated with observable morphological changes. They also conducted a Western blot analysis to investigate the expression of apoptosis marker proteins. Their results revealed that BV treatment increased proapoptotic proteins, including cleaved caspase-9, cleaved PARP-1, p53, and Bax, while simultaneously decreasing the expression of the Bcl-2, an antiapoptotic protein.

In another study, Yu et al. (2022) examined the role of BV in apoptosis from a novel perspective. Their investigation focused on BV's potential impact on autophagy in lung cancer cells, exploring the connection between autophagy and apoptosis, as both processes are recognized to collaborate in cancer cell death induction. They began by assessing microtubule-associated protein light chain 3 (LC3) levels through immunoblotting. LC3-I undergoes conversion to LC3-II through phosphatidylethanolamine, with LC3-II being a widely accepted autophagy marker. BV treatment led to an increase in LC3-II levels and the accumulation of LC3 puncta, which are indicators of autophagosome formation. The boost in LC3-II levels and autophagosome formation can result from either autophagy induction or autophagic flux blockade. To further elucidate this, the researchers employed hydroxychloroquine (HCQ), a known inhibitor of autophagosome degradation. When BV treatment was combined with HCQ, LC3-II and LC3 puncta levels further increased. The finding was corroborated by positive staining of autophagic vacuoles in BV-treated cancer cells, confirming enhanced autophagic flux. The mechanistic link to autophagy was reinforced by the observations related to mTOR, a well-known autophagy suppressor. Downstream factors of mTOR, such as p70 S6 kinase 1 (S6K1) and eIF4E-binding protein 1 (4E-BP1), were examined. BV treatment was found to reduce the phosphorylation of both S6K1 and 4E-BP1 in a dose-dependent manner, underscoring the induction of autophagy by BV via the mTOR pathway. An intriguing finding was that HCQ significantly diminished the levels of cleaved caspase-3, cleaved caspase-9, and Bax induced by BV while concurrently elevating the expression of the Bcl-2, as indicated by Western blot analysis. Additionally, BV treatment caused the depletion of ATG5, a pivotal autophagy regulator. Collectively, these data indicate that BV-induced autophagy may be an essential component of

BV-induced apoptosis in cancer cells, with the mTOR pathway playing a significant role in this underlying mechanism.

Glioblastoma multiforme (GBM) is a common malignant glioma. Treatment typically combines surgery, radiotherapy, and chemotherapy, but survival is often short. Therefore, developing new therapies for GBM is crucial. The transient receptor potential melastatin 2 (TRPM2), a Ca^{++}-permeable nonselective cation channel, plays a role in regulating cell proliferation and survival, making it a promising target for cancer treatment. Ertilav and Naziroglu (2023) explored the potential synergistic effect of melittin and cisplatin, a chemotherapy drug used in GBM treatment, on glioblastoma cell death. The combination therapy resulted in a significant increase in cytosolic and mitochondrial reactive oxygen species, along with mitochondrial membrane depolarization due to the accumulation of Ca^{++} and Zn^{++} in the cytosol. There was also an upregulation of apoptosis through increased activation of caspase-3, caspase-8, and caspase-9 with melittin and cisplatin treatment. However, the use of a TRPM2 antagonist led to a reduction in caspase activation and apoptosis levels. Melittin treatment further amplified cisplatin-induced oxidative and apoptotic actions in the cells by stimulating TRPM2.

In their recent study, Małek et al. (2022), hailing from Poland, sought to assess the anticancer properties of BV on glioblastoma cells with a focus on matrix metalloproteinases (MMPs). MMPs, specifically gelatinases like MMP-2 and MMP-9, play a crucial role in the invasion and metastasis of cancerous cells. The activities of MMP-2 and MMP-9 were determined using gelatin zymography. After 72 h of incubation with BV at a concentration of 1.0 μl/ml and higher, a notable reduction in the secretion of MMP-2 and MMP-9 was observed. It is important to note that this reduction was not replicated in mouse hippocampal cells, which were used as control cells in the study. This observation suggests that BV exerts its anticancer effects on glioblastoma cells in vitro by inhibiting the secretion of both gelatinases.

A group of Canadian researchers also examined the fundamental molecular mechanisms underlying the anticancer properties of BV and melittin in glioblastoma cells (Lebel et al. 2021). Their research, which included flow cytometry analysis, revealed that BV and melittin triggered cell death in glioma cells through both apoptosis and necrosis. Notably, a synergistic effect was observed when temozolomide, the primary chemotherapy drug for GBM, was combined with BV or melittin. The study further delved into the apoptotic pathway using immunoblotting. The results showed a significant increase in the levels of Bak, Bax, and the voltage-dependent anion channel (a proapoptotic protein). Meanwhile, there were lower expression levels of caspase-3, cleaved caspase-8, and caspase-9, confirming that BV and melittin induced the early stages of apoptosis in glioma cells through proapoptotic proteins. This study confirms that BV and melittin induce apoptosis in glioma cells via proapoptotic proteins.

Researchers in Germany (Erkoc et al. 2022) conducted an extensive in vitro screening of various melittin variants derived from both honeybees and earlier solitary bee lineages in the context of inflammation and cancer. Some of these melittin variants, along with the original melittin, were found to induce Ca^{++} release for the induction of calcium signaling apoptosis, whereas others had inhibitory effects on it in human embryonic kidney cells. Moreover, almost all variants decreased cAMP levels, except for one particular variant. A melittin variant sourced from carpenter bees demonstrated the ability to prevent LPS-induced NO release in monocyte/macrophage-like cells. Additionally, several melittin derivatives exhibited superior bioactivity on the TNBC cells compared to melittin. This perspective suggests that less toxic melittin variants might offer a more direct and effective approach to treating conditions such as breast cancer.

Ferroptosis is a distinct form of cell death, separate from apoptosis and necrosis, and closely tied to intracellular oxidative stress. The crucial factor in ferroptosis is the balance between oxidative damage and antioxidant defense. Li et al. (2022) delved into the core mechanisms behind melittin-induced cell death in lung cancer cells. Melittin was found to increase intracellular levels of ROS and MDA, along with elevated levels of Fe^{++}. In relation to the glutathione (GSH)-glutathione peroxidase 4 (GPX4) antioxidant system, melittin reduced the expression of GPX4 and depleted intracellular GSH levels. Moreover, deferoxamine and ferrostatin-1, both inhibitors of ferroptosis, mitigated lipid peroxidation and cell death and reduced ROS generation and the accumulation of MDA induced by melittin. Excessive intracellular ROS can also lead to endoplasmic reticulum (ER) stress, which subsequently triggers the expression of the apoptotic factor known as C/EBP homologous protein (CHOP). Melittin induced the activation of the ER stress-CHOP apoptotic signal. This signal's involvement in apoptosis was further confirmed by using a specific ER stress inhibitor, 4-PBA, which inhibited the expression of CHOP. These findings suggest a potential molecular mechanism for melittin-induced cell death, at least in the context of lung cancer cells, involving both ferroptosis and ER stress-medicated apoptosis.

Little is currently known about the impact of BV on pancreatic cancer. Zhao et al. (2022) assessed BV effects on cell cycle, apoptosis, and migration using the human metastatic pancreatic cancer cell line. BV treatment led to S phase arrest in cancer cells by suppressing the expression of cyclin A and D1 proteins. BV treatment resulted in higher protein expression of p21 and p53. BV also increased both early and late apoptosis, as evidenced by increased expression of cleaved caspase-3, cleaved caspase-9, cleaved PARP, and Bax, along with decreased expression of Bcl-2. Given that cell migration plays a critical role in cancer metastasis, a scratch wound healing assay was conducted, which revealed that BV treatment effectively reduced the number of migrating cells. Consequently, it can be inferred that the antitumor effect of BV on pancreatic cancer may be linked to cell cycle arrest, the induction of cell apoptosis, and the inhibition of cell migration.

A Portuguese BV was utilized in another anticancer study after undergoing chemical evaluation, with melittin, PLA_2, and apamin present at $65.6 \pm 0.3\%$, $4.0 \pm 0.2\%$, and $1.2 \pm 0.1\%$, respectively (Duarte et al. 2022). The findings revealed that BV exhibited antitumor properties against both breast and colon cancer cells employed in this study, with a more substantial impact observed in breast cancer cells. Combining BV with fluoxetine, sertraline, and thioridazine yielded superior results compared to BV combined with 5-FU and fluphenazine. Moreover, combining BV with repurposed drugs demonstrated better anticancer efficacy than when combined with doxorubicin. The combination of chemotherapeutic and repurposed drugs with BV presents a promising avenue for the treatment of colorectal and breast cancer.

The outer surface of healthy cell membranes consists of electrically neutral phosphatidylcholine (PC), primarily presenting its choline group. This choline-exposed surface acts as a positively charged shield, repelling basic protein toxins. Melittin approaches these membranes with its helix aligned longitudinally, and its hydrophobic side chains help overcome electrostatic repulsion from the choline shield. This electrostatic attraction positions melittin at the interface between the polar lipid heads and the non-polar alkyl chains, preventing it from penetrating the inner membrane layer. In cancer cells, acidic phospholipids accumulate, usually located on the inner monolayer, which now extends to the outer monolayer. This leads to heightened electrostatic attraction between melittin and the acidic phospholipids, such as phosphatidylserine (PS) and phosphatidic acid (PA), causing melittin to interact more avidly with cancer cell membranes. An international group of scientists examined the interaction between melittin and membrane-containing acidic phospholipids (Gasanoff et al. 2021). They used a pure PC membrane to represent a healthy cell model, while a PC membrane-containing PS or PA served as a model for cancerous cells. The study calculated melittin's binding affinity to PC, PS, and PA. The results suggest that melittin is more efficient in penetrating cancer cells' plasma membrane than healthy cells. Consequently, the increased permeability of melittin through the plasma membrane is believed to underlie its cytotoxic effect against cancer cells.

Hepatocellular carcinoma (HCC) is the most common form of primary liver cancer, and sorafenib (Sorf) serves as the chemotherapeutic agent for advanced HCC. Sorf is a multitargeted tyrosine kinase inhibitor with anti-angiogenic and antiproliferative properties. Despite its clinical effectiveness, it is associated with adverse side effects and the development of drug resistance. Egyptian scientists aimed to assess the potential use of a combination of BV and melittin as part of an HCC treatment regime to mitigate the undesirable side effects of Sorf and prevent chemoresistance (Mansour et al. 2021). The combined treatments of BV/Sorf and melittin/Sorf exhibited dose-dependent antiproliferative effects, indicating a synergistic response. Following the combined treatments, the expression levels of pro-apoptotic genes, including p53 and Bax, as well as apoptotic-execution genes, such as Cas3 and Cas7, showed significant upregulation. Additionally, the tumor-suppressor gene PTEN exhibited a marked increase. Conversely, the antiapoptotic marker Bcl2, the cell cycle regulating gene CCND1, the hypoxia-inducible gene HIF1a, the angiogenic-related gene vascular endothelial growth factor (VEGF), as well as markers associated with invasion, migration, metastasis (Rac1 and MMP9), and the inflammatory and immunomodulatory marker NF-κB all displayed significant downregulation when subjected to the combined treatments. Furthermore, the activity of MDA, a marker for lipid peroxidation, increased with the combined treatments, although this was not the case for the activity of antioxidant enzymes. Notably, the combined treatments induced cell cycle arrest at the G2/M phase. Overall, the anticancer effects of Sorf were significantly enhanced when combined with BV and melittin, as they targeted different cellular signaling pathways that collectively led to cell death. The findings suggest that BV and melittin may present a promising approach for enhancing HCC treatment.

Gastric cancer typically carries an unfavorable prognosis, particularly when it progresses to the metastatic stage, often resulting in patient fatality. There is limited available evidence regarding the potential of melittin to modulate the migration and invasion of human gastric cancer. A study conducted by a group of Taiwanese researchers utilized the human gastric cancer cell line known as AGS (Huang et al. 2021). In this study, melittin, at concentrations ranging from 0.05 to 0.15 μM, demonstrated the capacity to inhibit cell motility, a phenomenon evaluated through wound healing assay. This reduction in cell motility was further corroborated by observing a decrease in the expression of vimentin, a protein associated with alterations in cell shape and enhanced motility. Furthermore, after AGS cells were treated with melittin, there was a noticeable decrease in the number of adherent cells and the formation of cell colonies. It was observed through a transwell chamber assay that melittin also effectively reduced the migration and invasion of AGS cells. This anti-metastatic property of melittin was further substantiated by the reduced expression of PI3K, p-Akt, and N-cadherin, along with an increase in E-cadherin. The PI3K/AKT/mTOR signaling pathways are recognized for their involvement in tumor metastasis, with N-cadherin enhancing tumor cells' migratory and invasive capacity. At the same time, E-cadherin serves a role in suppressing migration. Additionally, melittin treatment was found to reduce both secretions and gene expression of gelatinases in AGS cells. Melittin also had the effect of decreasing the levels of bone morphogenetic protein, Smad1/5/8, Wnt-5α, and β-catenin, which are crucial components for cells in acquiring a migratory phenotype through the process of epithelial-mesenchymal transition. Consequently, it can be concluded that melittin inhibits metastasis by modulating multiple signaling pathways in gastric cancer.

N-methyl nitrosourea (MNU) was administered to female rats to induce mammary tumors (El-Beltagy et al. 2021). After confirming the presence of positive tumor markers, female rats were paired with males for mating. MNU led to notable histopathological manifestations in the breast and ovaries. However, the joint administration of *Annona muricata* (graviola) juice and BV reversed these histological alterations, restoring normalcy. This combined therapeutic approach also reduced serum MMP-1, NF-κB, and TNF-α levels, accompanied by an increase in serum caspase 3. In terms of antioxidant enzymes, graviola juice and BV contributed to elevated serum levels of catalase and superoxide dismutase. These effects are attributed to the antioxidant, anti-inflammatory, and antiproliferative components in graviola juice and BV.

Uzuner et al. (2021) examined how BV affected metastatic breast and hepatocellular cancer cell lines. The application of BV led to changes in mitochondrial membrane permeability and mitochondrial DNA copy numbers, resulting in noticeable alterations in cell morphology, a phenomenon not observed in normal cells. This finding confirms that BV treatment induces distinct patterns of epigenetic reprogramming in these cells.

Head and neck squamous cell carcinoma (HNSCC) comprises a diverse group of malignancies that originate from the squamous epithelium of the pharynx, larynx, and oral cavity. The prognosis for individuals with HNSCC is generally poor, and the limited use of cisplatin because of its adverse effects highlights the need for significant advancements in current treatment approaches. Egyptian researchers utilized four different types of HNSCC cell lines to evaluate the synergistic effects of BV in combination with cisplatin (Grawish et al. 2020). The study revealed strong synergistic effects in inhibiting cell growth, which were further substantiated by a significant increase in Bax levels and a notable decrease in the gene expressions of Bcl2 and epidermal growth factor receptor (EGFR). EGFR is a tyrosine kinase receptor family frequently activated in cancer through receptor overexpression or mutations. The combined treatment's impact was observed during the G2/M phase. These findings suggest that BV can exert its cytotoxic effects on HNSCC and could potentially enhance the anticancer effects of cisplatin while mitigating its adverse effects when combined.

Melittin, derived from an indigenous honey bee in Thailand, was studied to understand its apoptotic mechanisms in human malignant melanoma (A375) cells (Sangboonruang et al. 2020). The induction of apoptosis by melittin was demonstrated by a significant increase in DNA fragmentation, accompanied by a notable upregulation of cytochrome-c mRNA and increased caspase-9 and caspase-3 mRNA levels. Moreover, cytochrome-c translocated from the mitochondrial membrane to the cytoplasm was observed. In addition to the release of cytochrome-c, there was an increase in the protein expression of caspase-9 and activation of caspase-3. Furthermore, melittin inhibited the migration and invasion of A375 cells. These properties were further confirmed by the thickening of F-actin stress fibers and an increased accumulation of F-actin. Furthermore, the presence of melittin led to a reduction in the expression of EGFR. These findings confirm that melittin induces apoptosis in melanoma cells, causing disorganization of F-actin and disruption of EGFR signaling, thus ultimately inhibiting cell migration and invasion through the intrinsic mitochondrial pathway.

Salama et al. (2020) assessed BV's cytotoxic, apoptotic, and anti-inflammatory characteristics on cell lines associated with liver, breast, and cervical cancer. The cytotoxic impact of BV exhibited a dose- and time-dependent pattern, which correlated with elevated IL-10 and reduced TNF-α levels in liver cancer cells, low levels of both IL-10 and TNF-α, along with high interferon (IFN)-γ levels in breast cancer cells, and elevated IL-10 and IFN-γ levels, along with reduced TNF-α in cervical cancer cells. Additionally, the expression of caspase-3 was enhanced in all the cancer cell lines tested when treated with BV.

Cervical cancer, which ranks among the most prevalent malignancies affecting women, can be attributed to various factors, with human papillomavirus (HPV) infection being one of the prominent causes. Squamous cell carcinoma is more frequently encountered than adenocarcinoma in cervical cancer types. As per international guidelines, the standard approach to treating cervical cancer involves surgery for early stage tumors, concomitant chemoradiation or neoadjuvant chemotherapy, followed by radical surgery for locally advanced disease, and chemotherapy as a standalone treatment for metastatic or recurrent conditions. A study by Kim et al. (2020) explored the therapeutic mechanism of BV in cervical cancer cells. The experiment utilized two HPV-positive cervical cancer cell lines, Caski and HeLa cells, as well as one HPV-negative cervical cancer cell line, C33A, in an in vitro setting. BV demonstrated the ability to impede the proliferation of Caski and HeLa cells, concomitant with a reduction in mRNA and protein expressions of HPV E6 and HPV E7. Simultaneously, the expressions of tumor-suppressor proteins, p53, and Rb, were increased. BV treatment also induced apoptosis by upregulating Bax, cleaved caspase-3, and cleaved caspase-9 while downregulating Bcl-2 and Bcl-XL. Furthermore, BV restrained cell motility. Moreover, BV downregulated mitogenic signaling pathways by reducing the levels of proteins such as Akt and JNK, p38, and p44/42.

BV has also been used to suppress cervical tumor growth in mice by Lee et al. (2015). Caski cells were subcutaneously injected into female BALB/c nude mice for xenografts. The inhibitory effect of BV on cervical tumor growth was significant in terms of tumor weight and volume. The expressions of proapoptotic proteins such as DR3, DR6, FAS, and cleaved caspase-3 were concomitantly increased, but NF-κB activity and the expression of nucleus p50 and p65 were inhibited in tumor tissues by BV treatment. Collectively, BV may play a role in suppressing the tumorigenesis of cervical cancer.

A team of Polish researchers administered melittin to leukemia cell lines derived from both acute lymphoblastic leukemia and chronic myelogenous leukemia (Ceremuga et al. 2020). They observed the dose-dependent cytotoxic effects of melittin on the tested leukemia cells. One of the events associated with the initiation of apoptosis is the depolarization of the transmembrane potential. The treated cells exhibited a significant reduction in mitochondrial membrane potential. This apoptotic process was further verified by increases in annexin V fluorescence and 530 nm fluorescence. The rise in 530 nm fluorescence indicates an enhanced proteolytic activity of caspase 3/7 in leukemia cells. These findings clearly showed that melittin triggered apoptosis in leukemia cells.

Jeong et al. (2019) explored the signaling pathway of BV concerning epidermal growth factor (EGF)-induced epithelial-mesenchymal transition (EMT) in non-small cell lung cancer (NSCLC) cells. The research unveiled that BV effectively impeded EGF-induced F-actin reorganization and cell invasion, ultimately suppressing EGF-induced EMT, a crucial process linked to tumor metastasis in NSCLC. Notably, BV promoted the upregulation of E-cadherin and the concurrent downregulation of vimentin. Furthermore, BV exhibited inhibitory effects on EGF-induced phosphorylation of ERK, JNK, FAK, and mTOR in NSCLC cells, leading to the suppression of EMT-related transcription factors ZEB2 and Slug. In a separate study, Zhang and Chen (2017) applied melittin on NSCLC cells, concentrating on vascular endothelial growth factor (VEGF), a key player in angiogenesis development. As anticipated, melittin reduced the expression of VEGF protein and hypoxia-inducible factor (HIF)-1α, validating the primary regulation of VEGF expression by HIF-1α. Subsequently, this in vitro evidence was corroborated in a tumor xenograft mouse model, where melittin notably restrained tumor growth.

Furthermore, NSCLC cells were treated with natural killer (NK) cells pre-treated with BV (Kollipara et al. 2014). This combined approach for anticancer activity on lung cancer cells demonstrated increased expression of apoptotic proteins and death receptors (DR) proteins, along with the inactivation of NF-κB. However, the knockdown of DRs with si-RNA in NSCLC cells reversed the NF-κB inactivation. This innovative approach is noteworthy, as contemporary cancer immunotherapy, centered around NK cells, aims to overcome NK cell dysfunction by employing chemotherapeutic agents like BV to activate them. In summary, these findings underscore the potential of both BV and melittin as anti-metastatic agents for NSCLC.

Fascinating results from Soliman et al. (2019) showcase the swift impact of melittin on gastric and colorectal cancer cells. Within just 1 min of treatment, melittin triggered cell membrane damage, leading to observed granulation, blebbing, and cell swelling, followed by shrinkage, culminating in complete apoptosis within 15 min. Understanding the rapid kinetics of cytolysis represents a crucial advancement in moving closer to the realization of targeted melittin-based cancer therapy.

The clinical utility of melittin is hindered by its nonspecific cytotoxicity and hemolytic activity when administered in high treatment doses. A study conducted by Shaw et al. (2019) in Belgium sought to address this limitation by combining plasma-treated phosphate-buffered saline solution (PT-PBS) with melittin. The aim was to investigate the feasibility of reducing the melittin dose while still effectively eliminating cancer cells. Interestingly, PT-PBS was found to induce cancer cell death through pathways mediated by oxidative stress. The researchers utilized in vitro, in ovo, and in silico approaches involving both melanoma and breast cancer cells. The combined treatment exhibited clear synergism in cell death analysis, suggesting a potential reduction in the toxic dose of melittin. Furthermore, this combined approach induced lipid peroxidation in both cancer cell lines, with a notable increase in MDA production. The majority of lipid peroxidation products were localized around the cellular membrane, indicating a change in the physical properties of the membranes. Tissue sections from malignant solid tumors obtained through chicken chorioallantoic membrane assay were stained for the proliferative marker Ki-67. The combined treatment demonstrated a detrimental effect on cell proliferation by reducing the number of Ki-67-positive cells. Consistent with histological analysis of tumor specimens, tumor weight also showed a reduction in size upon combined treatment. Computer simulations suggested that plasma-induced oxidation of the phospholipid bilayer led to a decrease in the free energy barrier for melittin translocation. Consequently, the increased rate of melittin translocation through the cell membrane allowed penetration into the cell interior at lower treatment doses, supporting the concept of synergism between PT-PBS and melittin.

Lim et al. (2019) explored the molecular mechanisms underlying the impact of BV and melittin on various melanoma cells. The hindrance of melanoma cell proliferation by BV and melittin was substantiated through both MTT and colony formation assays. Furthermore, wound healing and Matrigel invasion assays confirmed the impediment of melanoma cell migration and invasion. The downregulation of melanin formation in melanoma cells stimulated by α-MSH demonstrated the anti-melanogenic properties of BV and melittin. Treatment of melanoma cells with BV and melittin for 24 h significantly increased the total number of early and late apoptotic cells compared to controls. This effect was caspase-dependent, attributed to the heightened expression of cleaved caspase-3 and caspase-9. Notably, melittin exhibited a superior ability to induce apoptosis compared to BV. Additionally, BV and melittin exhibited significant suppression of the phosphorylation of oncogenic PI3K, Akt, and mTOR, as well as MAPKs such as ERK and p38. This inhibition of signaling pathways led to a decrease in the expression of microphthalmia-associated transcription factor, a melanogenesis regulator, as well as MMP-2 and MMP-9, crucial players in melanoma metastasis. The combined inhibitory effect of melittin and temozolomide (TMZ), a chemotherapeutic agent, was demonstrated on melanoma cell growth and invasion, surpassing the effects of single-agent treatments. These data suggest that melittin holds promising therapeutic potential in overcoming chemoresistance to TMZ in melanoma patients.

Raman spectroscopy is an analytical technique that allows real-time live cell monitoring without invasive labeling. The anticancer effect of BV on breast cancer cells has been assessed using Raman spectroscopy (Jung et al. 2018). The Raman spectrum of cells treated with 3.0 μg/ml of BV for 48 h revealed a decrease in signal intensity, corresponding to Raman bands associated with DNA and protein. This decline was linked to nuclear fragmentation and protein degradation, ultimately resulting in cell death. Notably, the findings obtained through Raman spectroscopy align with those obtained through traditional biological tests.

Hesperidin (Hes) is a plant flavonoid in citrus fruits, recognized for its anti-inflammatory, antioxidant, proapoptotic, and antiproliferative properties. Piperine (Pip) is a dietary phytochemical derived from black and long pepper, displaying anti-inflammatory and anti-metastatic attributes. Tamoxifen (Tam) is widely employed in the treatment of estrogen receptor (ER)-positive breast cancer, resulting in a notable reduction in recurrence and mortality, albeit with the caveat of eventual serious side effects due to prolonged use. Khamis et al. (2018) proposed a hypothesis suggesting that the combined use of Hes, Pip, and BV could enhance the therapeutic efficacy of Tam. Two ER-positive breast cancer cell lines were utilized in the study. The combination of Tam, Hes, Pip, and BV demonstrated a robust synergistic cytotoxicity against breast cancer cells. This was further substantiated by a downregulation in the expression levels of Bcl-2, ERα (a major receptor highly expressed in ER-positive breast cancer cells), and EGFR genes. Additionally, there was a significant upregulation in the expression levels of the Bax gene. The findings suggest that natural products, including BV, could serve as safe adjuvants to Tam in the treatment of breast cancer.

Another study investigated combination therapy by integrating melittin into cisplatin, a primary treatment for ovarian cancer (Alonezi et al. 2017). This therapeutic approach holds promise for enhancing drug tolerability, diminishing cancer recurrence, and mitigating drug resistance. In this study, researchers employed metabolomics within the oncological context to assess the impact of combination therapy on metabolic alterations in ovarian cancer cells. Concerning changes in metabolites along the purine and pyrimidine pathways, a substantial increase was observed in adenine metabolite, hypoxanthine, and guanine levels. Conversely, the levels of purine metabolites such as AMP, ADP, and GTP decreased. Similarly, the levels of pyrimidine metabolites, including orotate, dihydrothymine, dihydrouracil, and uridine triphosphate, exhibited a reduction. Moreover, the number of metabolites grouped under the arginine and proline pathways decreased following the combination treatments. Depletion was also noted in cellular cysteine and GSH metabolism. With the combination therapy, there was a decrease in the levels of serine, glycine, ATP, and NAD^+. In summary, the combination treatment induced notable changes in ovarian cells, manifesting as reduced levels of metabolites in the Krebs cycle, oxidative phosphorylation, purine, and pyrimidine metabolism, and the arginine/proline pathway. Notably, the same international research team employed melittin monotherapy on ovarian cancer cells and reported profound reductions in amino acid and carbohydrate metabolism (Alonezi et al. 2016).

The transmembrane glycoprotein recognized as CD147 is extensively expressed in malignant breast tumor cells. CD147 plays a crucial role in enhancing the expression of the MMP family. CD147 also functions as the receptor for cyclophilin A (CypA), a cytosolic protein that may be released in vesicles triggered by inflammation or oxidative stress. A potential therapeutic strategy for cancer thus involves hindering carcinoma cell invasion by simultaneously inhibiting both CypA and CD147-induced MMP expression. When administered to breast cancer cells, melittin led to a reduction in MMP-9 expression and secretion induced by CypA (Wang et al. 2017). Melittin also counteracted the CD147 expression induced by CypA. Consequently, melittin diminished the invasion level of breast cancer cells by suppressing CD147 and MMP-9 by inhibiting CypA expression.

TNF-related apoptosis-inducing ligand (TRAIL) triggers apoptosis in cancer cells without harming normal cells. It is known that TRAIL binds to death receptors (DRs) such as TRAIL-R1 (DR4) and TRAIL-R2 (DR5) on the surface of cancer cells, activating apoptotic pathways. Zheng et al. (2015) explored the potential mechanism of BV on colon cancer cells. BV treatment led to increased expression of proapoptotic proteins, including Bax, cleaved caspase-3, cleaved caspase-8, and cleaved caspase-9, along with the expression of DRs such as DR3, DR4, DR5, and Fas, a cell surface receptor protein involved in apoptosis. BV treatment also inhibited NF-κB in terms of its DNA-binding activity. Consistent with NF-κB inhibition, BV treatment suppressed cytosolic phosphorylation of IkB, as well as the nucleus translocation of p50 and p65 in colon cancer cells. When colon cancer cells were transfected with DR4 siRNA, DR5 siRNA, and TRAIL siRNA, followed by BV treatment, BV-induced cell growth inhibition was partially reversed. A similar opposing effect was observed with p53 siRNA or a p53 inhibitor on the growth of colon cancer cells, with marked decreases in the expressions of DR4 and DR5 upon transfection. Combination treatment with BV and p50 siRNA transfection enhanced BV-induced cell growth inhibition. Similarly, combining BV and an NF-κB inhibitor intensified BV-induced cell growth inhibition. These findings offer mechanistic evidence that BV treatment induces apoptosis in colon cancer cells by upregulating DR4 and DR5 while concurrently suppressing NF-κB.

Cervical cancer cells harbor two viral oncogenic proteins, E6 and E7, that play a crucial role in the development of cervical cancer. Kim et al. (2015) utilized both an

HPV-negative cervical cancer cell line (C33A) and HPV-infected cervical cancer cell lines (CaSki and HeLa) to assess the underlying mechanism of BV's antiviral effect. BV treatment resulted in a more pronounced suppression of cell growth in HPV16-positive cells (CaSki) and a comparatively lower suppression in HPV18-infected cells (HeLa). Conversely, C33A cells exhibited less suppression. At a concentration of 10 μg/ml BV, mRNA expression and protein levels of HPV16 E6 and E7 significantly decreased in CaSki cells, while a less substantial suppression was observed in HeLa cells. This in vitro data aligned with findings from an in vivo study, where tumor growth in mice was restricted, accompanied by a corresponding reduction in mRNA and protein expressions of HPV16 E6 and E7 in tumors due to BV treatment. BV is demonstrated to inhibit cervical cancer cell growth through the downregulation of HPV16/18 E6/E7 proteins, with the extent of suppression depending on the specific cancer cell line.

Esophageal squamous cell carcinoma (ESCC) represents the primary histopathological type of esophageal cancer. Radiotherapy is a vital option for treating ESCC without surgery, but tumor resistance leads to recurrence and poor survival rates. Melittin was used in conjunction with radiation to explore its potential in sensitizing radiotherapy for ESCC both in vitro and in vivo, along with elucidating the underlying mechanism (Zhu et al. 2014). ESCC cells were exposed to X-ray irradiation (2, 4, and 6 Gy) from a medical linear accelerator at room temperature. Male BALB/C nude mice were subcutaneously injected with ESCC cells in the right leg to establish a tumor xenograft model. The clonogenic assay revealed that ESCC cells developed radioresistance in hypoxic conditions, but melittin reversed this effect. The combined positive effect of melittin and irradiation on apoptosis was confirmed by increased Bax and decreased Bcl-2 protein expressions. In the in vivo setting, tumor-bearing mice received a single fraction of 8 Gy irradiation after 2 days of melittin injection, and this combination treatment effectively delayed tumor growth. Moreover, researchers from Egypt investigated the effect of melittin on irradiated mice with Ehrlich ascites carcinoma (EAC) tumors (El Bakary et al. 2020). The mice underwent whole-body exposure to 1 Gy of γ-radiation. Melittin treatment effectively inhibited tumor proliferation, as evidenced by reduced mRNA expressions for angiogenic factors (TNF-α and VEGF) and decreased MMP-2 and caspase-9 activities. Additionally, improvements were observed in apoptotic regulators, such as caspase-3 activity and histopathological manifestations. Collectively, these two studies highlight the promising potential of melittin as a valuable adjuvant therapy to enhance the effects of radiotherapy for both ESCC and EAC, positioning it as a novel radiosensitizer.

Phosphatase and tensin homolog (PTEN) functions as a tumor suppressor by acting as plasma membrane lipid phosphatase. The histone deacetylases (HDACs), a superfamily of enzymes, are implicated in carcinogenesis when the equilibrium between histone acetyltransferases and HDACs is disturbed. Various cancers have shown different types of HDAC overexpression, leading to the observation that HDAC inhibitors demonstrate antitumor activities. Researchers hypothesized that melittin inhibits the proliferation of hepatocellular carcinoma cells by upregulating the PTEN gene, potentially associated with the reduction of HDAC2 expression

(Zhang et al. 2014). In liver cancer cells, melittin reinstated the expression of the tumor-suppressor gene PTEN and downregulated HDAC2 expression, confirming the regulatory role of HDAC2 in PTEN expression. This melittin-induced HDAC-mediated PTEN upregulation was further validated through the downregulation of the PI3K/Akt signaling pathways.

EGF/EGFR signaling has been recognized as a crucial factor in various tumor types. The anti-metastatic effects of BV, melittin, and apamin were investigated in EGF-stimulated breast cancer cells (Jeong et al. 2014). BV and melittin significantly inhibited EGF-induced cell migration and invasion, whereas apamin exhibited a comparatively weaker effect. Additionally, BV and melittin hindered EGF-induced cell motility by preventing F-actin reorganization. The study revealed that BV and melittin reversed the EGF-induced increase in MMP-9 protein expression and FAK phosphorylation. Moreover, melittin inhibited the nuclear translocation of p65, a subunit of NF-κB, stimulated by EGF. Furthermore, melittin restored E-cadherin expression, which EGF reduced, while it counteracted the EGF-induced vimentin expression. Lastly, melittin demonstrated a suppressive effect on EGF-induced PI3K/Akt/mTOR phosphorylation. Melittin thus emerges as a potential antitumor invasion agent for breast cancer therapy.

Palladium complexes have been assessed for their antitumor properties across various cancer types. The collaborative impact of BV with a palladium metal-based component was investigated in T-cell lymphoblastic leukemia cells (Safaeinejad et al. 2013). The cytotoxic effect of the combined treatment exhibited a dose- and time-dependent pattern. Apoptotic cell death was evidenced by observable cellular morphological alterations, including condensed nuclei, and occurred in a caspase-3-dependent manner.

In the sequence of tumor development, metastasis, and invasion, angiogenesis assumes a crucial role, wherein the vascular endothelial growth factor (VEGF) is regulated by hypoxia-inducible factor-1 (HIF-1) at the transcriptional level. The molecular mechanisms underlying the anti-angiogenesis effects of melittin were assessed in cervical cancer cells (Shin et al. 2013). Treatment with EGF (20 ng/ml) and $CoCl_2$ (cobalt chloride, 200 μg/ml) notably elevated the HIF-1α protein level, while the HIF-1β protein level was affected only by $CoCl_2$ in cervical cancer cells. EGF induces an increase in HIF-1α through activating EGFR signaling, whereas $CoCl_2$ functions as a hypoxia-mimetic agent. Melittin treatment exhibited a dose-dependent reduction in EGF-induced HIF-1α protein but did not alter the expression of $CoCl_2$-induced HIF-1α protein. This observation suggests that melittin's inhibitory effect on HIF-1α protein expression involves EGF-specific regulation, impeding the progression of angiogenesis. When cervical cancer cells were treated with melittin and cycloheximide, a protein synthesis inhibitor, the EGF-induced HIF-1α protein level was rapidly decreased. The half-life of melittin-treated EGF-induced HIF-1α was shortened by 10 min compared to cycloheximide-treated EGF-induced HIF-1α, confirming that melittin caused the inhibition of HIF-1α protein expression by reducing its half-life. The suppression of EGF-induced HIF-1α protein synthesis by melittin was found to occur via the inhibition of ERK and mTOR/p70S6K signaling pathways. Melittin also decreased VEGF mRNA and protein expressions in

EGF-stimulated cervical cancer cells, aligning with the suppressed binding of HIF-1α to the VEGF promoter region. Furthermore, melittin suppressed EGF-induced migration and angiogenesis of cervical cancer cells. The study's novelty lies in discovering that melittin can inhibit EGF-induced HIF-1α and VEGF protein expressions in cervical tumors.

17.2.9 Liver Fibrosis

Dramatically unfolding within the intricate landscape of liver ailments, liver fibrosis emerges as a consequence of hepatic damage across a spectrum of liver diseases. In a symphony of biochemical events, the damaged hepatocytes set the stage, releasing a cascade of reactive oxygen species and cytokines. This tumultuous release triggers the unleashing of TNF-α and fibroblast growth factor, serving as catalysts for the activation of hepatic stellate cells (HSCs) and subsequent collagen deposition. A sinister cycle ensues as the activated HSCs now awakened, unleash a storm of pro-inflammatory cytokines, orchestrate the elevation of adhesion molecules, and hasten the recruitment of inflammatory cells, perpetuating the relentless progression of liver fibrosis (Roehlen et al. 2020).

Cholestatic liver disease is marked by the gradual breakdown of biliary epithelial cells (BECs), leading to the development of fibrosis, cirrhosis, and eventual liver failure. The primary cellular contributors to increased collagen deposition in biliary fibrosis are activated HSCs and portal fibroblasts. Researchers (Kim et al. 2017a) sought to determine if apamin could inhibit biliary fibrosis and the proliferation of HSCs in mice subjected to 3,5-diethoxycarbonyl-1,4-dihydrocollidine (DDC) feeding. Cellular assays were also conducted on HSC-T6 cells (rat-immortalized HSCs). Apamin treatment led to reduced liver injury and decreased levels of pro-inflammatory cytokines induced by DDC feeding. Apamin also hindered collagen deposition, proliferation of BECs, and the expression of fibrogenic genes in DDC-fed mice. In HSC-T6 cells, apamin inhibited HSC activation by suppressing the Smad signaling pathway. Another team of researchers used the same murine model of DDC diet feeding to assess the therapeutic effect of PLA_2 on cholestatic liver disease (Kim et al. 2021). Biochemical tests and histological examinations revealed that PLA_2 reduced myofibroblast accumulation and suppressed the TGF-β signaling cascade. Administration of PLA_2 also inhibited hepatocyte apoptosis and suppressed caspase-3 activation. Additionally, PLA_2 reduced cytokine production and inhibited the nuclear factor kappa-β pathway. The number of regulatory T cells increased, while the number of neutrophils, macrophages, and CD8+ T cells decreased in response to PLA_2 treatment. Both apamin and PLA_2 are promising therapeutic agents for cholestatic liver disease.

Overdosing on acetaminophen (APAP), a common remedy for fever and pain can lead to liver damage with limited therapeutic options available. The effect of apamin on APAP-induced hepatotoxicity was assessed in a study involving mice (Jang et al. 2023). Through intraperitoneal injection, apamin mitigated histological abnormalities and lowered serum levels of liver enzymes in mice subjected to APAP administration. Apamin increased GSH levels, reversing oxidative stress and inhibiting caspase-3 activation, thereby reducing apoptosis. Additionally, apamin decreased both serum and hepatic levels of cytokines. These findings suggest that apamin mitigates APAP-evoked hepatotoxicity by suppressing oxidative stress, apoptosis, and inflammation.

A study by Naji et al. (2021) involved a small international team of researchers who utilized isoniazid and rifampicin, primary drugs for treating tuberculosis, to induce hepatotoxicity in rats. The objective was to evaluate the effects of melittin. The researchers employed biochemical analysis as well as hematological and histological examinations. For the experiment, male rats subjected to melittin treatment post hepatotoxicity induction exhibited noticeable increases in both body weight and liver weight. The administration of melittin also led to reductions in direct bilirubin, total bilirubin, aspartate aminotransferase, alanine aminotransferase, lactate dehydrogenase, alkaline phosphatase, and total serum protein levels. Additionally, melittin-induced minimal manifestations of hemorrhage, congestion, ballooned hepatocytes, inflammation, necrosis, fibrosis, and amyloids in liver tissues, aligning with an elevation in serum GSH level. It is evident that melittin has a protective effect against hepatotoxicity induced by antituberculosis drugs.

Liver fibrosis and renal fibrosis share similarities in the pathological processes that contribute to tissue damage and repair. Renal fibrosis, in particular, can lead to the irreversible impairment of renal function, ultimately progressing to end-stage kidney disease. In a study conducted by Gwon et al. (2021), renal fibrosis was induced in mice through unilateral ureteral obstruction (UUO), while renal fibroblast activation was stimulated with TGF-β1 in normal rat renal interstitial fibroblast cells. The application of apamin effectively suppressed the elevated expression of fibrotic-related genes, such as α-SMA, vimentin, and fibronectin. Moreover, apamin administration mitigated injury to renal tubular cells and reduced tubular atrophy. Additionally, apamin demonstrated a dampening effect on fibroblast activation, ECM synthesis, and inflammatory cytokines. These beneficial effects were achieved through the suppression of the TGF-β1/Smad2/3 and STAT3 signaling pathways. The UUO model was also employed by An et al. (2015) to investigate the mechanism of BV in ameliorating renal fibrosis. BV administration significantly reduced the number of infiltrated inflammatory cells in the kidney tissues. Following BV treatment, the expression levels of TNF-α, IL-1β, TGF-β1, fibronectin, and α-SMA decreased. In a different approach, Park et al. (2014) employed an in vitro model of renal fibrosis. TGF-β was added to the culture of rat kidney fibroblast cells to induce profibrotic gene expressions. Melittin suppressed the expressions of PAI-1, type I collagen, and fibronectin by inhibiting the TGF-βRII-Smad2/3-ERK1/2 and JNK signaling pathways. Kidney failure can also be caused by diabetic nephropathy (DN). Lotfy et al. (2023) induced DN in male adult albino rats using streptozotocin (STZ) and treated them with BV. BV administration ameliorated the harmful effects resulting from STZ, improving the biochemical, histopathological, and molecular parameters, suggesting a protective role of BV against DN. In summary, BV, melittin, and apamin show promise as potential treatments for chronic kidney disease based on their demonstrated effects in mitigating renal fibrosis and associated pathological processes.

17.2.10 Atopic Dermatitis

Atopic dermatitis, more commonly known as eczema, is a chronic, noninfectious, inflammatory skin condition characterized by persistent itching. This prolonged and recurring condition significantly diminishes the quality of life for those affected and places a financial strain on patients. The underlying mechanisms driving eczema are intricate and involve multiple factors, including genetic predisposition, impairment of the skin's protective barrier, dysregulated immune responses, and disruption in the balance of skin microorganisms. Consequently, novel treatments are being developed targeting precise molecules implicated in the pathogenesis of eczema (Sroka-Tomaszewska and Trzeciak (2021).

The study conducted by Sur et al. (2016) aimed to assess the therapeutic potential of BVA to improve eczema using a mouse model with trimellitic anhydride (TMA)-induced skin impairment. TMA, employed as an allergen in the research, is recognized as a respiratory sensitizer that triggers T-cell–dependent contact hypersensitivity in mice. Mice were initially sensitized on the skin of the dorsal flank using TMA to establish the eczema model. Subsequently, T-cell–dependent skin inflammation was induced by applying TMA topically to the dorsal surfaces of both ears after the sensitization phase. The administration of BVA involved subcutaneous injections on both sides at the BL40 acupoints located behind the knee. BVA treatment demonstrated a substantial reduction in the expression levels of both Th1 and Th2 cytokines within the ear skin and lymph nodes of mice subjected to TMA treatment. The clinical manifestations resembling eczema symptoms, encompassing the ear skin symptoms, inflammation, and lymph node weight, exhibited noteworthy improvement following BVA treatment. This treatment also hindered the proliferation and infiltration of T cells, suppressed the production of Th1 and Th2 cytokines, and restrained the synthesis of IL-4 and IgE. Intriguingly, the inhibitory impact of BVA was more prominently observed at the BL40 acupoint compared to a non-acupoint located at the base of the tail. These findings underscore the effectiveness of BV injection at specific acupoints in mitigating eczema-like skin lesions, achieved through the suppression of inflammatory and allergic responses in a TMA-induced contact hypersensitivity mouse model.

In contrast, Jung et al. (2017) utilized freeze-dried crude *Dermatophagoides farinae* extract from house dust mites as an antigen, along with 2.4-dinitrochlorobenzene (DNCB) as a sensitizer, to induce eczema in mice. Posttreatment with PLA_2, they assessed parameters such as epidermal thickness, immune cell infiltration, serum immunoglobulin levels, and cytokine concentrations. Applying PLA_2 topically to the ear led to a notable reduction in the apparent eczema symptoms, including ear thickness, serum IgE levels, inflammatory cytokines, and histological alterations. Additionally, it impeded the infiltration of mast cells into the ear tissue. At the same time, the anti-atopic effects of PLA_2 were nullified upon depletion of Treg cells, implying a Treg-dependent mechanism for the actions of PLA_2. Collectively, data disclosed that topical administration of PLA_2 reduced inflammatory responses in atopic skin, supporting the potential of PLA_2 as a candidate for eczema treatment. The same research group conducted a similar experiment (Shin et al. 2018). While they arrived at a similar conclusion, they delved further into the underlying mechanism of PLA_2. Indeed, all the anti-atopic effects of PLA_2 were abolished in CD206 mannose receptor-deficient mice. It was subsequently revealed that PLA_2 alleviates atopic skin inflammation by interacting with CD206. In summary, the effects of PLA_2 involve both Treg-dependent and CD206-dependent pathways for its anti-atopic impacts in mice.

On the other hand, Lee et al. (2020b) employed phthalic anhydride (PA) to induce eczema in mice. They conducted an investigation into the anti-inflammatory and anti-eczema effects of BV on Balb/c mice induced with PA. To initiate eczema, animals received a topical application of 5% PA on their dorsal skin and ears. After 24 h, BV was applied to the back and ear skin three times a week for a duration of 4 weeks. The topical application of BV effectively mitigated ear epidermal hyperplasia, reduced IgE levels, and diminished the infiltration of inflammatory cells in the skin tissues. Furthermore, the levels of inflammatory cytokines in the serum were markedly decreased in the mice treated with BV. BV also inhibited the expression of iNOS and COX-2, including the activation of MAPK and NF-κB induced by PA in the skin tissues. In addition, through in vitro assays, BV demonstrated the ability to suppress LPS-induced inflammatory responses in RAW 264.7 murine macrophages, as well as TNF-α/IFN-γ-dependent inflammatory responses in HaCaT human keratinocytes, by reducing MAPK and NF-κB activities. The data indicates that BV mitigates the activation of MAPK and NF-κB, which in turn leads to a decrease in the expression of iNOS and COX-2, ultimately resulting in a reduction in eczema-related skin damage.

In a study conducted by Kim et al. (2017b), an in vitro inflammatory condition of eczema was induced in keratinocytes using TNF-α and IFN-γ. The primary objective was to illustrate the anti-inflammatory effect of apamin. Apamin demonstrated its ability to alleviate the inflammatory state by suppressing Th2-related chemokines, including thymus and activation-regulated chemokine (TARC/CCL17) and macrophage-derived chemokine (MDC/CCL22). Regarding the inflammation-induced activation of signal transducer and activator of transcription (STAT), apamin not only mitigated the increased protein expressions of pSTAT1 and pSTAT3 but also impeded the nuclear translocation of STAT1. Apamin also reduced the expression of pro-inflammatory cytokines such as IL-1β and IL-6, mediated by suppressing the NF-κB signaling pathway. Apamin exhibited the potential to ameliorate eczema by improving the inflammatory condition.

In addition to intricate immune responses, the complement system also appears to affect eczema. A clinical study noted an elevation in complement components such as C3, C4, and C3a in the serum of eczema patients compared to healthy non-atopic controls (Kapp et al. 1985). When the skin is exposed to ultraviolet rays, it can generate its own supply of C3 complement, leading to the recruitment of macrophages into the dermal-epidermal junction and resulting in inflammation similar to eczema (Zhuang and Lyga 2014). Furthermore, in DNCB-induced eczema mice, an upregulation of the anaphylatoxin C5a receptor expression was observed in skin tissue. Treatment with a C5a receptor antagonist caused a reversal of eczema-related symptoms, possibly by suppressing the C5a receptor-mediated cascade action of mast cells (Dang et al. 2015). As such, Kim et al. (2019) explored the influence of BV and melittin on the regulation of the complement system. Since the complement cascades are governed by membrane-bound complement inhibitory proteins, including decay-accelerating factor (CD55), researchers studied whether THP-1 cells, a human leukemia monocytic cell line, expressed CD55 on their cell membranes when exposed to BV. The results indicated that BV treatment increased both mRNA and protein levels of CD55 in THP-1 cells. Subsequent experiments revealed that BV achieved this effect by activating ERKs in THP-1 cells. Following these findings, the authors examined the levels and activities of C3 convertase (C3C), an inflammatory peptide, and membrane attack complex (MAC) in the serum of mice with eczema-like symptoms after being treated with BV and melittin. Both MAC and C3C have the potential to exacerbate eczema lesions and worsen eczema symptoms. However, when mice with eczema-like lesions were treated with either BV or melittin, their serum exhibited decreased levels of MAC and C3C, along with an increase of CD55 levels. Consequently, it was proposed that BV and melittin alleviate eczema by deactivating the complement system, primarily through the induction of CD55.

17.2.11 Pain

Neuropathic pain (NP) arises from injuries or diseases affecting the somatosensory system. It is often characterized by alterations in sensation, including heightened sensitivity, such as hyperalgesia, allodynia, and hyperpathia. As the effective treatment of NP remains a significant unmet medical challenge, a longstanding practice has been the use of combination therapy, which involves the administration of more than two drugs. However, this approach has been adopted without substantial empirical support. Guidelines for managing NP should endeavor to provide recommendations regarding research on combination therapy, prioritizing specific combinations for analysis over others. This approach aims to advance the search for more robust evidence and enhance our understanding of effective treatment for NP (Serrano Afonso et al. 2021).

In the study conducted by Yeo et al. (2016), they utilized the chemotherapeutic agent oxaliplatin to induce a pronounced painful peripheral neuropathy in mice, leading to the depletion of intraepidermal nerve fibers (IENFs). The research aimed to explore the antinociceptive impact of repetitive BVA on oxaliplatin-induced mechanical allodynia and to determine whether this intervention caused a notable reduction in the loss of IENFs. To achieve this, BV was subcutaneously administered into ST36 on the right hind limb once a day for a total of 18 days, commencing on day 15 following oxaliplatin injection. Using immunohistochemistry, the researchers measured the quantity and density of IENFs in the glabrous skin of the hind paw footpad. Additionally, mechanical allodynia was developed by placing mice in an acrylic cylinder on an elevated metal mesh grid, where an ascending series of von Frey filaments were employed to apply incremental forces. Following a single BVA session, a temporary increase in mechanical threshold was observed 1 h later. Notably, this effect was further intensified with repeated BV treatments over time. Moreover, repetitive BVA significantly reduced the loss of IENFs caused by oxaliplatin. Of particular interest, repetitive subcutaneous administration of the α_2-adrenergic antagonist as a pretreatment completely thwarted both the anti-allodynic effects and the increase in IENFs that were observed in mice subjected to repetitive BVA. This firmly establishes that BV-induced antinociceptive effects are mediated through the activation of the descending ceruleospinal noradrenergic pathway, subsequently stimulating spinal α_2 adrenoceptors.

A separate team of researchers in Korea took a different approach to NP treatment, combining BVA and morphine (Lee et al. 2020a). They used acetone and von Frey filaments to evaluate cold and mechanical allodynia induced by intraperitoneal oxaliplatin injection, respectively. Subcutaneous BVA was administered at doses of 0.25, 1.0, and 2.5 mg/kg on ST36, while morphine was injected intraperitoneally at doses of 0.5, 2.0, and 5 mg/kg, each with varying concentrations. To uncover the spinal analgesic mechanism of simultaneously administered BVA and morphine in oxaliplatin-treated mice, they introduced several antagonists through intrathecal injection 20 min prior to the combination treatment. These antagonists included the opioid receptor antagonist naloxone, the selective α_2-adrenergic receptor antagonist idazoxan, and the 5-HT$_3$ receptor antagonist MDL-72222. Three days after oxaliplatin injection, noticeable signs of allodynia were observed. These signs were alleviated when either BVA or morphine was administered individually, and the extent of relief was dependent on the dosage. However, when BVA and morphine were combined at intermediate doses, the effect was more significant and long-lasting compared to administering either BVA or morphine individually at the highest dose. Pretreatment with intrathecal naloxone or MDL-72222 blocked this additive effect, whereas pretreatment with idazoxan did not have the same blocking effect. The observed combined anti-allodynic effect of BVA and morphine is suggested to be mediated by spinal opioidergic and 5-HT$_3$ receptors. This mechanism provides a robust and enduring analgesic action against oxaliplatin-induced NP.

The BVA treatment was also utilized to address mechanical hyperalgesia and heightened spinal neuronal activity in rats induced by paclitaxel, another chemotherapy drug (Choi et al. 2017). Administering paclitaxel intraperitoneally at a dose of 2 mg/kg on four alternative days led to notable mechanical hyperalgesia symptoms, which were assessed using a von Frey filament. BVA was subcutaneously injected at the right LI11 (Quchi) and ST36 acupoints. Additionally, melittin (0.5 mg/kg) and PLA$_2$ (0.12 mg/kg) were injected at ST36. To further investigate the mechanism of BVA, idazoxan, and prazosin (α_1-adrenergic receptor antagonist) were administered intrathecally as pretreatment. BVA treatment at ST36, but not at LI11, provided relief from mechanical hyperalgesia for at least 2 h and suppressed spinal wide dynamic range neurons' hyperexcitability triggered by brush, press, or pinch stimulation. According to behavioral assessments, melittin exhibited a more potent analgesic effect compared to BVA or PLA_2 against paclitaxel-induced mechanical hyperalgesia. Intrathecal pretreatment with idazoxan, but not prazosin, blocked the analgesic effect of BVA, confirming its mechanism through spinal α_2-adrenergic receptors.

The analgesic effect of serotonin is believed to be facilitated through the descending pain inhibitory system, involving the periaqueductal gray, nucleus raphe magnus, and spinal serotonergic receptors. Within the various subtypes of serotonin receptors (5-HT_{1-7}), it is well-established that 5-HT_1, 5-HT_2, and 5-HT_3 receptors are the most frequently associated with pain modulation at the spinal cord level. Lee et al. (2014) conducted a study to investigate whether the suppressive effect of BVA on cold allodynia in oxaliplatin-injected rats is mediated by the serotonergic inhibitory system. Cold allodynia signs were assessed by immersing the rat's tails in cold water (4 °C) and measuring tail withdrawal latency (TWL). BVA was subcutaneously administered into the GV3 (Yaoyangguan) acupoint. To determine if the serotonergic system is involved in the anti-allodynic effect of BVA, DL-p-chlorophenylalanine (PCPA) was used to deplete serotonin. The levels of 5-HT in the spinal cord, measured by ELISA, significantly increased after BVA treatment, but this effect was reversed by PCPA. PCPA pretreatment eliminated the pain-relieving effect of BVA, resulting in reduced TWL. It shows that the serotonergic system plays a significant role in mediating the analgesic effect of BVA on oxaliplatin-induced cold allodynia. A systemic injection of methysergide (a mixed 5-HT_1/5-HT_2 receptor antagonist) or MDL-72222 (a 5-HT_3 receptor antagonist) prevented the pain-relieving effect of BVA on cold allodynia. When administered intrathecally, MDL-72222 blocked the effect of BVA, whereas NAN-190 (a 5-HT_{1A} receptor antagonist) or ketanserin (a 5-HT_{2A} receptor antagonist) did not affect BVA's action. These findings strongly suggest that spinal 5-HT_3 receptors play a crucial role in the BVA-induced anti-allodynic action in oxaliplatin-injected rats.

While many studies related to NP have predominantly employed BVA in animal models, Li et al. (2015) took a different approach by utilizing both PLA_2 and BV in a mouse model of oxaliplatin-induced cold and mechanical allodynia. To assess cold and mechanical sensitivity, they conducted the acetone test and von Frey hair test, respectively. In order to deplete serotonin and noradrenaline, they administered PCPA and *N*-(2-Chloroethyl)-*N*-ethyl-2-bromobenzylamine hydrochloride (DSP4), respectively. Furthermore, both prazosin and idazoxan were administered to investigate which adrenergic receptor subtype mediated the anti-allodynic effects of PLA_2. The pronounced signs of allodynia that were evident 1 day following an oxaliplatin injection showed a significant reduction after receiving five consecutive daily administrations of PLA_2. Notably, the analgesic effect of PLA_2 was more potent than that of BV. Furthermore, it is worth noting that the analgesic effect of PLA_2 was effectively blocked by DSP4 but not by PCPA, indicating the involvement of the noradrenergic system rather than the serotonergic system. Additionally, the complete blockage of PLA_2-induced anti-allodynic action was achieved with idazoxan, whereas prazosin did not have the same effect. These results strongly suggest that PLA_2 treatment alleviates cold and mechanical allodynia induced by oxaliplatin in mice through the activation of the noradrenergic pain inhibitory pathway via α_2-adrenergic receptors.

17.3 Conclusion

The exploration of BV in modern medicine has revealed a vast and promising landscape of therapeutic possibilities. Through meticulous research, the bioactive components of BV have been identified as potent agents capable of addressing an array of medical conditions. This chapter has meticulously examined the therapeutic applications of BV, melittin, apamin, and PLA_2, spanning a diverse range of diseases and conditions, from neurodegenerative disorders like AD and PD to inflammatory diseases such as arthritis and multiple sclerosis, to systemic conditions including asthma, cancer, liver fibrosis, atopic dermatitis, and chronic pain.

The therapeutic potential of BV is underscored by its ability to modulate immune responses, reduce inflammation, and enhance neural regeneration. In the context of neurodegenerative diseases, BV compounds have shown promise in mitigating symptoms and potentially slowing disease progression. For instance, melittin and PLA_2 have demonstrated neuroprotective effects that are particularly beneficial in AD and PD, where they help in reducing amyloid plaques and protecting dopaminergic neurons, respectively. Similarly, in vascular dementia and peripheral nerve injuries, BV's bioactive peptides facilitate neuroprotection and regeneration, offering a novel approach to managing these challenging conditions.

In autoimmune and inflammatory diseases like multiple sclerosis and arthritis, the anti-inflammatory and immunomodulatory properties of BV compounds provide significant therapeutic benefits. By reducing cytokine production and inhibiting inflammatory pathways, they can alleviate symptoms and improve the quality of life for patients suffering from these chronic conditions. The same properties extend to its use in asthma and atopic dermatitis, where BV compounds help in modulating immune responses and reducing hypersensitivity reactions.

BV's anticancer potential is another area of profound interest. Studies have highlighted the ability of BV components to induce apoptosis in cancer cells, inhibit tumor growth, and enhance the effectiveness of conventional cancer therapies. This makes BV a compelling candidate for integrative cancer treatment strategies, potentially improving outcomes and reducing side effects.

Moreover, BV's role in managing liver fibrosis showcases its versatility. The antifibrotic properties of BV compounds can help in reducing liver damage and promoting tissue regeneration, providing a new avenue for treating chronic liver diseases. Pain management, one of the most traditional uses of BV, continues to benefit from the advancements in understanding BV's mechanisms of action, leading to more effective and targeted analgesic therapies.

In conclusion, the diverse and potent biological activities of honey bee venom present a significant opportunity for developing novel treatments for a wide range of medical conditions. The continuous research into BV's pharmacological properties not only reinforces its historical significance but also propels it into a promising future of medical applications. As we stand at the frontier of BV research, the insights gained from this chapter underscore the importance of continued exploration and innovation in harnessing the full therapeutic potential of bee venom. This ongoing journey promises to unlock new horizons in medicine, offering hope and improved health outcomes for countless patients worldwide (Table 17.1).

Table 17.1 Application of honey bee venom (BV) components on different disease types

Disease type	Component	References
Alzheimer's disease	BV	Gu et al. (2015), Nguyen et al. (2022b), Nguyen et al. (2022a)
	Apamin	Park et al. (2020)
	Melittin	Nguyen and Lee (2021), Nguyen et al. (2022b)
	Phospholipase A_2	Ye et al. (2016b), Baek et al. (2018)
Parkinson's disease	BV	Dantas et al. (2022), Maurice et al. (2015), Ahmed-Farid et al. (2021)
	Phospholipase A_2	Ye et al. (2016a)
Vascular dementia	BV acupuncture	Cai et al. (2017)
Peripheral nerve injury	BV	Er-Rouassi et al. (2023)
Arthritis	BV	Park et al. (2004)
	Melittin	Park et al. (2004)
Multiple sclerosis	BV	Jelodar et al. (2021)
	BV acupuncture	Lee et al. (2016)
Asthma	BV	Choi et al. (2013)
	Phospholipase A_2	Park et al. (2015)
Cancer	BV	Tetikoglu and Celik-Uzuner (2023), Obeidat et al. (2023), Sevin et al. (2023a), Hwang et al. (2022), Yu et al. (2022), Sevin et al. (2023b), Małek et al. (2022), Lebel et al. (2021), Zhao et al. (2022), Duarte et al. (2022), Mansour et al. (2021), El-Beltagy et al. (2021), Uzuner et al. (2021), Grawish et al. (2020), Salama et al. (2020), Kim et al. (2020), Lee et al. (2015), Jeong et al. (2019), Kollipara et al. (2014), Lim et al. (2019), Jung et al. (2018), Khamis et al. (2018), Zheng et al. (2015), Kim et al. (2015), Jeong et al. (2014), Safaeinejad et al. (2013)
	Melittin	Ertilav and Naziroglu (2023), Lebel et al. (2021), Erkoc et al. (2022), Li et al. (2022), Gasanoff et al. (2021), Mansour et al. (2021), Huang et al. (2021), Sangboonruang et al. (2020), Ceremuga et al. (2020), Zhang and Chen (2017), Soliman et al. (2019), Shaw et al. (2019), Lim et al. (2019), Alonezi et al. (2017), (2016), Wang et al. (2017), Zhu et al. (2014), Zhang et al. (2014), Jeong et al. (2014), Shin et al. (2013)

(continued)

Table 17.1 (continued)

Disease type	Component	References
Liver fibrosis	BV	An et al. (2015), Lotfy et al. (2023)
	Apamin	Kim et al. (2017a), Jang et al. (2023), Gwon et al. (2021)
	Melittin	Naji et al. (2021), Park et al. (2014)
	Phospholipase A_2	Kim et al. (2021)
Atopic dermatitis	BV	Lee et al. (2020b), Kim et al. (2019)
	BV acupuncture	Sur et al. (2016)
	Apamin	Kim et al. (2017b)
	Melittin	Kim et al. (2019)
	Phospholipase A_2	Jung et al. (2017), Shin et al. (2018)
Pain	BV	Li et al. (2015)
	BV acupuncture	Yeo et al. (2016), Lee et al. (2020a), Choi et al. (2017), Lee et al. (2014)
	Melittin	Choi et al. (2017)
	Phospholipase A_2	Choi et al. (2017), Li et al. (2015)

References

Ahmed-Farid OA, Taha M, Bakeer RM, Radwan OK, Hendawy HAM, Soliman AS, Yousef E (2021) Effects of bee venom and dopamine-loaded nanoparticles on reserpine-induced Parkinson's disease rat model. Sci Rep 11:21141. https://doi.org/10.1038/s41598-021-00764-y

Alonezi S, Tusiimire J, Wallace J, Dufton MJ, Parkinson JA, Young LC, Clements CJ, Park JK, Jeon JW, Ferro VA, Watson DG (2016) Metabolomic profiling of the effects of melittin on cisplatin resistant and cisplatin sensitive ovarian cancer cells using mass spectrometry and biolog microarray technology. Metabolites 6(4):35. https://doi.org/10.3390/metabo6040035

Alonezi S, Tusiimire J, Wallace J, Dufton MJ, Parkinson JA, Young LC, Clements CJ, Park JK, Jeon JW, Ferro VA, Watson DG (2017) Metabolomic profiling of the synergistic effects of melittin in combination with cisplatin on ovarian cancer cells. Metabolites 7(2):14. https://doi.org/10.3390/metabo7020014

An HJ, Kim KH, Lee WR, Kim JY, Lee SJ, Pak SC, Han SM, Park KK (2015) Anti-fibrotic effect of natural toxin bee venom on animal model of unilateral ureteral obstruction. Toxins 7(6):1917–1928. https://doi.org/10.3390/toxins7061917

Baek H, Lee CJ, Choi DB, Kim NS, Kim YS, Ye YJ, Kim YS, Kim JS, Shim I, Bae H (2018) Bee venom phospholipase A2 ameliorates Alzheimer's disease pathology in Abeta vaccination treatment without inducing neuro-inflammation in a 3xTg-AD mouse model. Sci Rep 8(1):17369. https://doi.org/10.1038/s41598-018-35030-1

Barkhof F, Koeller KK (2020) Demyelinating diseases of the CNS (brain and spine). In: Hodler J, Kubik-Huch RA, von Schulthess GK (eds) Diseases of the brain, head and neck, spine 2020-2023: diagnostic imaging, IDKD Springer series. Springer, Cham, pp 165–176. https://doi.org/10.1007/978-3-030-38490-6_13

Bir SC, Khan MW, Javalkar V, Toledo EG, Kelley RE (2021) Emerging concepts in vascular dementia: a review. J Stroke Cerebrovasc Dis 30(8):105864. https://doi.org/10.1016/j.jstrokecerebrovasdis.2021.105864

Cai M, Lee JH, Yang EJ (2017) Bee venom ameliorates cognitive dysfunction caused by neuroinflammation in an animal model of vascular dementia. Mol Neurobiol 54(8):5952–5960. https://doi.org/10.1007/s12035-016-0130-x

Ceremuga M, Stela M, Janik E, Gorniak L, Synowiec E, Sliwinski T, Sitarek P, Saluk-Bijak J, Bijak M (2020) Melittin – a natural peptide from bee venom which induces apoptosis in human leukaemia cells. Biomol Ther 10(2):247. https://doi.org/10.3390/biom10020247

Choi MS, Park S, Choi T, Lee G, Haam KK, Hong MC, Min BI, Bae H (2013) Bee venom ameliorates ovalbumin induced allergic asthma via modulating CD4+CD25+ regulatory T cells in mice. Cytokine 61(1):256–265. https://doi.org/10.1016/j.cyto.2012.10.005

Choi J, Jeon C, Lee JH, Jang JU, Quan FS, Lee K, Kim W, Kim SK (2017) Suppressive effects of bee venom acupuncture on paclitaxel-induced neuropathic pain in rats: Mediation by spinal alpha(2)-adrenergic receptor. Toxins 9(11):351. https://doi.org/10.3390/toxins9110351

Dang L, He L, Wang Y, Xiong J, Bai B, Li Y (2015) Role of the complement anaphylatoxin C5a-receptor pathway in atopic dermatitis in mice. Mol Med Rep 11(6):4183–4189. https://doi.org/10.3892/mmr.2015.3301

Dantas CG, da Paixao AO, Nunes T, Silva IJF, Dos SLB, Araujo AAS, de Albuquerque-Junior RLC, Gramacho KP, Padilha FF, da Costa LP, Severino P, Cardoso JC, Souto EB, Gomes MZ (2022) Africanized bee venom (Apis mellifera linnaeus): Neuroprotective effects in a Parkinson's disease mouse model induced by 6-hydroxydopamine. Toxics 10(10). https://doi.org/10.3390/toxics10100583

Duarte D, Falcao SI, El Mehdi I, Vilas-Boas M, Vale N (2022) Honeybee venom synergistically enhances the cytotoxic effect of cns drugs in HT-29 colon and MCF-7 breast cancer cell lines. Pharmaceutics 14(3):511. https://doi.org/10.3390/pharmaceutics14030511

El Bakary NM, Alsharkawy AZ, Shouaib ZA, Barakat EMS (2020) Role of bee venom and melittin on restraining angiogenesis and metastasis in gamma-irradiated solid ehrlich carcinoma-bearing mice. Integr Cancer Ther 19. https://doi.org/10.1177/1534735420944476

El-Beltagy A, Elsyyad HIH, Abdelaziz KK, Madany AS, Elghazaly MM (2021) Therapeutic role of Annona muricata fruit and bee venom against MNU-induced breast cancer in pregnant rats and its complications on the ovaries. Breast Cancer 13:431–445. https://doi.org/10.2147/BCTT.S306971

Erkoc P, von Reumont BM, Luddecke T, Henke M, Ulshofer T, Vilcinskas A, Furst R, Schiffmann S (2022) The pharmacological potential of novel melittin variants from the honeybee and solitary bees against inflammation and cancer. Toxins 14(12):818. https://doi.org/10.3390/toxins14120818

Er-Rouassi H, Bakour M, Touzani S, Vilas-Boas M, Falcao S, Vidal C, Lyoussi B (2023) Beneficial effect of bee venom and its major components on facial nerve injury induced in mice. Biomol Ther 13(4):680. https://doi.org/10.3390/biom13040680

Ertilav K, Naziroglu M (2023) Honey bee venom melittin increases the oxidant activity of cisplatin and kills human glioblastoma cells by stimulating the TRPM2 channel. Toxicon 222:106993. https://doi.org/10.1016/j.toxicon.2022.106993

Gasanoff E, Liu Y, Li F, Hanlon P, Garab G (2021) Bee venom melittin disintegrates the respiration of mitochondria in healthy cells and lymphoblasts, and induces the formation of non-bilayer structures in model inner mitochondrial membranes. Int J Mol Sci 22(20):11122. https://doi.org/10.3390/ijms222011122

GBD 2019 Stroke Collaborators (2021) Global, regional, and national burden of stroke and its risk factors, 1990-2019: a systematic analysis for the Global Burden of Disease Study 2019. Lancet Neurol 20(10):795–820. https://doi.org/10.1016/S1474-4422(21)00252-0

Grawish ME, Mourad MI, Esmaeil DA, Ahmed RA, Ateia IM, Hany E, Elkhier MTA (2020) Emerging therapeutic modality enhancing the efficiency of chemotherapeutic agents against head and neck squamous cell carcinoma cell lines. Cancer Treat Res Commun 25:100242. https://doi.org/10.1016/j.ctarc.2020.100242

Gu SM, Park MH, Hwang CJ, Song HS, Lee US, Han SB, Oh KW, Ham YW, Song MJ, Son DJ, Hong JT (2015) Bee venom ameliorates lipopolysaccharide-induced memory loss by preventing NF-kappaB pathway. J Neuroinflammation 12:124. https://doi.org/10.1186/s12974-015-0344-2

Gwon MG, An HJ, Gu H, Kim YA, Han SM, Park KK (2021) Apamin inhibits renal fibrosis via suppressing TGF-beta1 and STAT3 signaling in vivo and in vitro. J Mol Med 99(9):1265–1277. https://doi.org/10.1007/s00109-021-02087-x

Harth M, Nielson WR (2019) Pain and affective distress in arthritis: relationship to immunity and inflammation. Expert Rev Clin Immunol 15(5):541–552. https://doi.org/10.1080/1744666X.2019.1573675

Huang JY, Peng SF, Chueh FS, Chen PY, Huang YP, Huang WW, Chung JG (2021) Melittin suppresses epithelial-mesenchymal transition and metastasis in human gastric cancer AGS cells via regulating Wnt/BMP associated pathway. Biosci Biotechnol Biochem 85(11):2250–2262. https://doi.org/10.1093/bbb/zbab153

Hwang YN, Kwon IS, Na HH, Park JS, Kim KC (2022) Dual cytotoxic responses induced by treatment of A549 human lung cancer cells with sweet bee venom in a dose-dependent manner. J Pharmacopuncture 25(4):390–395. https://doi.org/10.3831/KPI.2022.25.4.390

Jang HJ, Leem J, Kim GM (2023) Protective effects of apamin on acetaminophen-induced hepatotoxicity in mice. Curr Issues Mol Biol 45(5):4389–4399. https://doi.org/10.3390/cimb45050279

Javonillo DI, Tran KM, Phan J, Hingco E, Kramar EA, da Cunha C, Forner S, Kawauchi S, Milinkeviciute G, Gomez-Arboledas A, Neumann J, Banh CE, Huynh M, Matheos DP, Rezaie N, Alcantara JA, Mortazavi A, Wood MA, Tenner AJ, MacGregor GR, Green KN, LaFerla FM (2021) Systematic phenotyping and characterization of the 3xTg-AD mouse model of Alzheimer's disease. Front Neurosci 15:785276. https://doi.org/10.3389/fnins.2021.785276

Jelodar S, Zare Mirakabadi A, Oryan S, Mohammadnejad L (2021) Effect of honey bee venom on experimental autoimmune encephalomyelitis (EAE) as a model for multiple sclerosis (MS). Arch Razi Inst 76(6):1727–1733. https://doi.org/10.22092/ARI.2021.126291.1342

Jeong YJ, Choi Y, Shin JM, Cho HJ, Kang JH, Park KK, Choe JY, Bae YS, Han SM, Kim CH, Chang HW, Chang YC (2014) Melittin suppresses EGF-induced cell motility and invasion by inhibiting PI3K/Akt/mTOR signaling pathway in breast cancer cells. Food Chem Toxicol 68:218–225. https://doi.org/10.1016/j.fct.2014.03.022

Jeong YJ, Park YY, Park KK, Choi YH, Kim CH, Chang YC (2019) Bee venom suppresses EGF-induced epithelial-mesenchymal transition and tumor invasion in lung cancer cells. Am J Chin Med 47(8):1869–1883. https://doi.org/10.1142/S0192415X19500952

Jung KH, Baek H, Kang M, Kim N, Lee SY, Bae H (2017) Bee venom phospholipase A2 ameliorates house dust mite extract induced atopic dermatitis like skin lesions in mice. Toxins 9(2):68. https://doi.org/10.3390/toxins9020068

Jung GB, Huh JE, Lee HJ, Kim D, Lee GJ, Park HK, Lee JD (2018) Anti-cancer effect of bee venom on human MDA-MB-231 breast cancer cells using Raman spectroscopy. Biomed Opt Express 9(11):5703–5718. https://doi.org/10.1364/BOE.9.005703

Kapp A, Wokalek H, Schopf E (1985) Involvement of complement in psoriasis and atopic dermatitis–measurement of C3a and C5a, C3, C4 and C1 inactivator. Arch Dermatol Res 277(5):359–361. https://doi.org/10.1007/BF00509233

Khamis AAA, Ali EMM, El-Moneim MAA, Abd-Alhaseeb MM, El-Magd MA, Salim EI (2018) Hesperidin, piperine and bee venom synergistically potentiate the anticancer effect of tamoxifen against breast cancer cells. Biomed Pharmacother 105:1335–1343. https://doi.org/10.1016/j.biopha.2018.06.105

Kim YW, Chaturvedi PK, Chun SN, Lee YG, Ahn WS (2015) Honeybee venom possesses anticancer and antiviral effects by differential inhibition of HPV E6 and E7 expression on cervical cancer cell line. Oncol Rep 33(4):1675–1682. https://doi.org/10.3892/or.2015.3760

Kim JY, An HJ, Kim WH, Park YY, Park KD, Park KK (2017a) Apamin suppresses biliary fibrosis and activation of hepatic stellate cells. Int J Mol Med 39(5):1188–1194. https://doi.org/10.3892/ijmm.2017.2922

Kim WH, An HJ, Kim JY, Gwon MG, Gu H, Lee SJ, Park JY, Park KD, Han SM, Kim MK, Park KK (2017b) Apamin inhibits TNF-alpha- and IFN-gamma-induced inflammatory cytokines and chemokines via suppressions of NF-kappaB signaling pathway and STAT in human keratinocytes. Pharmacol Rep 69(5):1030–1035. https://doi.org/10.1016/j.pharep.2017.04.006

Kim Y, Lee YW, Kim H, Chung DK (2019) Bee venom alleviates atopic dermatitis symptoms through the upregulation of decay-accelerating factor (DAF/CD55). Toxins 11(5):239. https://doi.org/10.3390/toxins11050239

Kim DH, Lee HW, Park HW, Lee HW, Chun KH (2020) Bee venom inhibits the proliferation and migration of cervical-cancer cells in an HPV E6/E7-dependent manner. BMB Rep 53(8):419–424. https://doi.org/10.5483/BMBRep.2020.53.8.031

Kim JY, Jang HJ, Leem J, Kim GM (2021) Protective effects of bee venom-derived phospholipase A(2) against cholestatic liver disease in mice. Biomedicines 9(8):992. https://doi.org/10.3390/biomedicines9080992

Kollipara PS, Kim JH, Won D, Lee SM, Sung HC, Chang HS, Lee KT, Lee KS, Park MH, Song MJ, Song HS, Hong JT (2014) Co-culture with NK-92MI cells enhanced the anti-cancer effect of bee venom on NSCLC cells by inactivation of NF-kappaB. Arch Pharm Res 37(3):379–389. https://doi.org/10.1007/s12272-013-0319-8

Lebel AA, Kisembo MV, Soucy MN, Hebert MPA, Morin PJ, Boudreau LH (2021) Molecular characterization of the anticancer properties associated with bee venom and its components in glioblastoma multiforme. Chem Biol Interact 347:109622. https://doi.org/10.1016/j.cbi.2021.109622

Lee JH, Li DX, Yoon H, Go D, Quan FS, Min BI, Kim SK (2014) Serotonergic mechanism of the relieving effect of bee venom acupuncture on oxaliplatin-induced neuropathic cold allodynia in rats. BMC Complement Altern Med 14:471. https://doi.org/10.1186/1472-6882-14-471

Lee HL, Park SH, Kim TM, Jung YY, Park MH, Oh SH, Yun HS, Jun HO, Yoo HS, Han SB, Lee US, Yoon JH, Song MJ, Hong JT (2015) Bee venom inhibits growth of human cervical tumors in mice. Oncotarget 6(9):7280–7292. https://doi.org/10.18632/oncotarget.3110

Lee MJ, Jang M, Choi J, Lee G, Min HJ, Chung WS, Kim JI, Jee Y, Chae Y, Kim SH, Lee SJ, Cho IH (2016) Bee venom acupuncture alleviates experimental autoimmune encephalomyelitis by upregulating regulatory t cells and suppressing Th1 and Th17 responses. Mol Neurobiol 53(3):1419–1445. https://doi.org/10.1007/s12035-014-9012-2

Lee JH, Gang J, Yang E, Kim W, Jin Y-H (2020a) Bee venom acupuncture attenuates oxaliplatin-induced neuropathic pain by modulating action potential threshold in a-fiber dorsal root ganglia neurons. Toxins 12(12):737. https://doi.org/10.3390/toxins12120737

Lee YJ, Oh MJ, Lee DH, Lee YS, Lee J, Kim DH, Choi CH, Song MJ, Song HS, Hong JT (2020b) Anti-inflammatory effect of bee venom in phthalic anhydride-induced atopic dermatitis animal model. Inflammopharmacology 28(1):253–263. https://doi.org/10.1007/s10787-019-00646-w

Li D, Lee Y, Kim W, Lee K, Bae H, Kim SK (2015) Analgesic effects of bee venom derived phospholipase A(2) in a mouse model of oxaliplatin-induced neuropathic pain. Toxins 7(7):2422–2434. https://doi.org/10.3390/toxins7072422

Li X, Zhu S, Li Z, Meng YQ, Huang SJ, Yu QY, Li B (2022) Melittin induces ferroptosis and ER stress-CHOP-mediated apoptosis in A549 cells. Free Radic Res 56(5–6):398–410. https://doi.org/10.1080/10715762.2022.2131551

Lim HN, Baek SB, Jung HJ (2019) Bee venom and its peptide component melittin suppress growth and migration of melanoma cells via inhibition of PI3K/AKT/mTOR and MAPK pathways. Molecules 24(5). https://doi.org/10.3390/molecules24050929

Lopes B, Sousa P, Alvites R, Branquinho M, Sousa AC, Mendonca C, Atayde LM, Luis AL, Varejao ASP, Mauricio AC (2022) Peripheral nerve injury treatments and advances: One health perspective. Int J Mol Sci 23(2). https://doi.org/10.3390/ijms23020918

Lotfy MM, Dowidar MF, Ali HA, Ghonimi WAM, Al-Farga A, Ahmed AI (2023) Effect of selenium nanoparticles and/or bee venom against STZ-induced diabetic cardiomyopathy and nephropathy. Meta 13:3. https://doi.org/10.3390/metabo13030400

Małek A, Kocot J, Mitrowska K, Posyniak A, Kurzepa J (2022) Bee venom effect on glioblastoma cells viability and gelatinase secretion. Front Neurosci 16:792970. https://doi.org/10.3389/fnins.2022.792970

Mansour GH, El-Magd MA, Mahfouz DH, Abdelhamid IA, Mohamed MF, Ibrahim NS, Hady AAWA, Elzayat EM (2021) Bee venom and its active component Melittin synergistically potentiate the anticancer effect of Sorafenib against HepG2 cells. Bioorg Chem 116:105329. https://doi.org/10.1016/j.bioorg.2021.105329

Maurice N, Deltheil T, Melon C, Degos B, Mourre C, Amalric M, Kerkerian-Le Goff L (2015) Bee venom alleviates motor deficits and modulates the transfer of cortical information through the basal ganglia in rat models of Parkinson's disease. PLoS One 10(11):e0142838. https://doi.org/10.1371/journal.pone.0142838

Naji KM, Al-Khatib BY, Al-Haj NS, D'Souza MR (2021) Hepatoprotective activity of melittin on isoniazid- and rifampicin-induced liver injuries in male albino rats. BMC Pharmacol Toxicol 22(1):39. https://doi.org/10.1186/s40360-021-00507-9

Nguyen CD, Lee G (2021) Neuroprotective activity of melittin – the main component of bee venom-against oxidative stress induced by Abeta(25-35) in in vitro and in vivo models. Antioxidants 10(11):1654. https://doi.org/10.3390/antiox10111654

Nguyen CD, Yoo J, An EJ, Sung CY, Jeong DH, Park SY, Kim JH, Lee G (2022a) Pharmacokinetic improvement provided by microneedle patch in delivering bee venom, a case study in combating scopolamine-induced neurodegeneration in mouse model. Drug Deliv 29(1):2855–2867. https://doi.org/10.1080/10717544.2022.2116129

Nguyen CD, Yoo J, Hwang SY, Cho SY, Kim M, Jang H, No KO, Shin JC, Kim JH, Lee G (2022b) Bee venom activates the Nrf2/HO-1 and TrkB/CREB/BDNF pathways in neuronal cell responses against oxidative stress induced by Abeta(1-42). Int J Mol Sci 23(3):1193. https://doi.org/10.3390/ijms23031193

Obeidat M, Al-Khraisat IF, Jaradat DMM, Ghanim BY, Abdallah QM, Arqoub DA, Sabbah D, Al-Sanabra OM, Arafat T, Qinna NA (2023) Mellitin peptide quantification in seasonally collected crude bee venom and its anticancer effects on myelogenous K562 human leukaemia cell line. BMC Complement Med Ther 23(1):132. https://doi.org/10.1186/s12906-023-03897-x

Papadopoulos NG, Miligkos M, Xepapadaki P (2022) A current perspective of allergic asthma: from mechanisms to management. Handb Exp Pharmacol 268:69–93. https://doi.org/10.1007/164_2021_483

Park HJ, Lee SH, Son DJ, Oh KW, Kim KH, Song HS, Kim GJ, Oh GT, Yoon DY, Hong JT (2004) Antiarthritic effect of bee venom: inhibition of inflammation mediator generation by suppression of NF-kappaB through interaction with the p50 subunit. Arthritis Rheum 50(11):3504–3515. https://doi.org/10.1002/art.20626

Park SH, Cho HJ, Jeong YJ, Shin JM, Kang JH, Park KK, Choe JY, Park YY, Bae YS, Han SM, Moon SK, Kim WJ, Choi YH, Chang YC (2014) Melittin inhibits TGF-beta-induced profibrotic gene expression through the suppression of the TGFbetaRII-Smad, ERK1/2 and JNK-mediated signaling pathway. Am J Chin Med 42(5):1139–1152. https://doi.org/10.1142/S0192415X14500712

Park S, Baek H, Jung KH, Lee G, Lee H, Kang GH, Lee G, Bae H (2015) Bee venom phospholipase A2 suppresses allergic airway inflammation in an ovalbumin-induced asthma model through the induction of regulatory T cells. Immun Inflamm Dis 3(4):386–397. https://doi.org/10.1002/iid3.76

Park J, Jang KM, Park KK (2020) Apamin suppresses LPS-induced neuroinflammatory responses by regulating SK channels and TLR4-mediated signaling pathways. Int J Mol Sci 21(12):4319. https://doi.org/10.3390/ijms21124319

Pike CJ, Walencewicz-Wasserman AJ, Kosmoski J, Cribbs DH, Glabe CG, Cotman CW (1995) Structure-activity analyses of beta-amyloid peptides: contributions of the beta 25-35 region to aggregation and neurotoxicity. J Neurochem 64(1):253–265. https://doi.org/10.1046/j.1471-4159.1995.64010253.x

Proulx E, Power SK, Oliver DK, Sargin D, McLaurin J, Lambe EK (2020) Apamin improves prefrontal nicotinic impairment in mouse model of Alzheimer's disease. Cereb Cortex 30(2):563–574. https://doi.org/10.1093/cercor/bhz107

Roehlen N, Crouchet E, Baumert TF (2020) Liver fibrosis: mechanistic concepts and therapeutic perspectives. Cells 9(4):875. https://doi.org/10.3390/cells9040875

Safaeinejad Z, Nabiuni M, Nazari Z (2013) Potentiation of a novel palladium (II) complex lethality with bee venom on the human T-cell acute lymphoblastic leukemia cell line (MOLT-4). J Venomous Anim Toxins Incl Trop Dis 19:25. https://doi.org/10.1186/1678-9199-19-25

Salama MA, Younis MA, Talaat RM (2020) Cytokine and inflammatory mediators are associated with cytotoxic, anti-inflammatory and apoptotic activity of honeybee venom. J Complement Integr Med 18(1):75–86. https://doi.org/10.1515/jcim-2019-0182

Sangboonruang S, Kitidee K, Chantawannakul P, Tragoolpua K, Tragoolpua Y (2020) Melittin from apis florea venom as a promising therapeutic agent for skin cancer treatment. Antibiotics 9(8). https://doi.org/10.3390/antibiotics9080517

Scheltens P, De Strooper B, Kivipelto M, Holstege H, Chetelat G, Teunissen CE, Cummings J, van der Flier WM (2021) Alzheimer's disease. Lancet 397(10284):1577–1590. https://doi.org/10.1016/S0140-6736(20)32205-4

Serrano Afonso A, Carnaval T, Videla Ces S (2021) Combination therapy for neuropathic pain: a review of recent evidence. J Clin Med 10(16):3533. https://doi.org/10.3390/jcm10163533

Sevin S, Deveci Ozkan A, Tutun H, Kivrak I, Turna O, Guney Eskiler G (2023a) Determination of the effects of bee venom on triple negative breast cancer cells in vitro. Chem Biodivers 20(3):e202201263. https://doi.org/10.1002/cbdv.202201263

Sevin S, Kivrak I, Tutun H, Uyar R, Ayaz F (2023b) Apis mellifera anatoliaca venom exerted anti-inflammatory activity on LPS-stimulated mammalian macrophages by reducing the production of the inflammatory cytokines. Appl Biochem Biotechnol 195(5):3194–3205. https://doi.org/10.1007/s12010-022-04284-x

Shaw P, Kumar N, Hammerschmid D, Privat-Maldonado A, Dewilde S, Bogaerts A (2019) Synergistic effects of melittin and plasma treatment: a promising approach for cancer therapy. Cancers 11(8):1109. https://doi.org/10.3390/cancers11081109

Shin JM, Jeong YJ, Cho HJ, Park KK, Chung IK, Lee IK, Kwak JY, Chang HW, Kim CH, Moon SK, Kim WJ, Choi YH, Chang YC (2013) Melittin suppresses HIF-1alpha/VEGF expression through inhibition of ERK and mTOR/p70S6K pathway in human cervical carcinoma cells. PLoS One 8(7):e69380. https://doi.org/10.1371/journal.pone.0069380

Shin D, Choi W, Bae H (2018) Bee venom phospholipase A2 alleviate house dust mite-induced atopic dermatitis-like skin lesions by the CD206 mannose receptor. Toxins 10(4):146. https://doi.org/10.3390/toxins10040146

Soliman C, Eastwood S, Truong VK, Ramsland PA, Elbourne A (2019) The membrane effects of melittin on gastric and colorectal cancer. PLoS One 14(10):e0224028. https://doi.org/10.1371/journal.pone.0224028

Sroka-Tomaszewska J, Trzeciak M (2021) Molecular mechanisms of atopic dermatitis pathogenesis. Int J Mol Sci 22(8):4130. https://doi.org/10.3390/ijms22084130

Sur B, Lee B, Yeom M, Hong JH, Kwon S, Kim ST, Lee HS, Park HJ, Lee H, Hahm DH (2016) Bee venom acupuncture alleviates trimellitic anhydride-induced atopic dermatitis-like skin lesions in mice. BMC Complement Altern Med 16:38. https://doi.org/10.1186/s12906-016-1019-y

Tetikoglu S, Celik-Uzuner S (2023) Bee venom induces the interaction between phosphorylated histone variant, H2AX, and the intracellular site of beta-actin in liver and breast cancer cells. Chem Biodivers 20(6):e202300401. https://doi.org/10.1002/cbdv.202300401

Uzuner SC, Birinci E, Tetikoglu S, Birinci C, Kolayli S (2021) Distinct epigenetic reprogramming, mitochondrial patterns, cellular morphology, and cytotoxicity after bee venom treatment. Recent Pat Anticancer Drug Discov 16(3):377–392. https://doi.org/10.2174/1574892816666210422125058

Wang J, Li F, Tan J, Peng X, Sun L, Wang P, Jia S, Yu Q, Huo H, Zhao H (2017) Melittin inhibits the invasion of MCF-7 cells by downregulating CD147 and MMP-9 expression. Oncol Lett 13(2):599–604. https://doi.org/10.3892/ol.2016.5516

Wong EC, Chui HC (2022) Vascular cognitive impairment and dementia. Continuum (Minneap Minn) 28(3):750–780. https://doi.org/10.1212/con.0000000000001124

Ye M, Chung HS, Lee C, Song JH, Shim I, Kim YS, Bae H (2016a) Bee venom phospholipase A2 ameliorates motor dysfunction and modulates microglia activation in Parkinson's disease alpha-synuclein transgenic mice. Exp Mol Med 48(7):e244. https://doi.org/10.1038/emm.2016.49

Ye M, Chung HS, Lee C, Yoon MS, Yu AR, Kim JS, Hwang DS, Shim I, Bae H (2016b) Neuroprotective effects of bee venom phospholipase A2 in the 3xTg AD mouse model of Alzheimer's disease. J Neuroinflammation 13:10. https://doi.org/10.1186/s12974-016-0476-z

Yeo JH, Yoon SY, Kwon SK, Kim SJ, Lee JH, Beitz AJ, Roh DH (2016) Repetitive acupuncture point treatment with diluted bee venom relieves mechanical allodynia and restores intraepidermal nerve fiber loss in oxaliplatin-induced neuropathic mice. J Pain 17(3):298–309. https://doi.org/10.1016/j.jpain.2015.10.018

Yu JE, Kim Y, Hong DE, Lee DW, Chang JY, Yoo SS, Kim MJ, Son DJ, Yun J, Han SB, Hong JT (2022) Bee venom triggers autophagy-induced apoptosis in human lung cancer cells via the mTOR signaling pathway. J Oncol 2022:8916464. https://doi.org/10.1155/2022/8916464

Zhang SF, Chen Z (2017) Melittin exerts an antitumor effect on non-small cell lung cancer cells. Mol Med Rep 16(3):3581–3586. https://doi.org/10.3892/mmr.2017.6970

Zhang H, Zhao B, Huang C, Meng XM, Bian EB, Li J (2014) Melittin restores PTEN expression by down-regulating HDAC2 in human hepatocelluar carcinoma HepG2 cells. PLoS One 9(5):e95520. https://doi.org/10.1371/journal.pone.0095520

Zhao J, Hu W, Zhang Z, Zhou Z, Duan J, Dong Z, Liu H, Yan C (2022) Bee venom protects against pancreatic cancer via inducing cell cycle arrest and apoptosis with suppression of cell migration. J Gastrointest Oncol 13(2):847–858. https://doi.org/10.21037/jgo-22-222

Zheng J, Lee HL, Ham YW, Song HS, Song MJ, Hong JT (2015) Anti-cancer effect of bee venom on colon cancer cell growth by activation of death receptors and inhibition of nuclear factor kappa B. Oncotarget 6(42):44437–44451. https://doi.org/10.18632/oncotarget.6295

Zhu H, Yang X, Liu J, Ge Y, Qin Q, Lu J, Zhan L, Liu Z, Zhang H, Chen X, Zhang C, Xu L, Cheng H, Sun X (2014) Melittin radiosensitizes esophageal squamous cell carcinoma with induction of apoptosis in vitro and in vivo. Tumour Biol 35(9):8699–8705. https://doi.org/10.1007/s13277-014-2146-z

Zhuang Y, Lyga J (2014) Inflammaging in skin and other tissues – the roles of complement system and macrophage. Inflamm Allergy Drug Targets 13(3):153–161. https://doi.org/10.2174/1871528113666140522112003